Further Pure Mathematics for Advanced Level

In the same series

Pure Mathematics for Advanced Level, J. A. H. Shepperd and C. J. Shepperd
(This book is available either as a complete volume or in two parts.)

Statistics for Advanced Level Mathematics, I. Gwyn Evans
Applied Mathematics for Advanced Level, W. E. Williams and A. Waltham

Further Pure Mathematics for Advanced Level

J. A. H. Shepperd

Department of Mathematics, University of Manchester
Institute of Science and Technology

C. J. Shepperd

Head of Mathematics, Hinchingbrooke School, Huntingdon

HODDER AND STOUGHTON
LONDON SYDNEY AUCKLAND TORONTO

British Library Cataloguing in Publication Data

Shepperd, J. A. H.
Further pure mathematics for advanced level.
1. Mathematics—Examinations, questions, etc.
I. Title II. Shepperd, C. J.
510'.76 QA43

ISBN 0 340 35244 2

First printed 1985

Typeset in Times New Roman (Monophoto) by Macmillan India Ltd., Bangalore

Printed in Great Britain for
Hodder and Stoughton Educational,
a division of Hodder and Stoughton Ltd.,
Mill Road, Dunton Green, Sevenoaks, Kent TN13 2YD,
by Page Bros (Norwich) Ltd.

Contents

Preface

Further Pure Mathematics for Advanced Level builds on the authors' previous book *Pure Mathematics for Advanced Level*, Hodder and Stoughton, 1983. Material in *Pure Mathematics for Advanced Level* is frequently referred to in this book and is given in the form *Pure Mathematics* §0.0. Together, the two books provide material for a complete Advanced Level G.C.E. Course in Pure Mathematics. They also provide the Pure Mathematics half of a course in the double A-level subject Mathematics and Further Mathematics. The material covers most of the Pure Mathematics contained in the syllabuses of all the G.C.E. Boards, in particular, those of the London Board and the Joint Matriculation Board are completely covered.

The same pattern as was used in *Pure Mathematics for Advanced Level* has again been used. The chapters begin with revision material and practice exercises, where this is appropriate. New work is illustrated by many worked examples and the exercises in the sections are carefully graded to lead the student towards an in depth understanding. It is intended that all these exercises should be completed. Further Mathematics needs a deeper understanding of mathematical ideas and of the concept of proof. Rigorous treatment has been given in all proofs. There are some cases where a result is required but the proof is beyond the range of this book. In these cases, the result is quoted as a theorem and it is clearly indicated that the proof is not given. In an introductory chapter, the concept of proof is studied and it is hoped that this will be useful for the student to refer to while working on the later chapters. The nature of some of the material in algebra has meant that some chapters are long. It is felt that this form of presentation will be useful since some of the long proofs are broken down into small digestible steps.

Miscellaneous exercises, at the end of the chapters, contain more practice material, which is not graded in difficulty and which may contain applications of work from previous chapters. Many exercises are from the past papers of the Associated Examining Board (AEB), the University of Cambridge Local Examinations Syndicate (C), the Joint Matriculation Board (JMB), the University of London University Entrance and School Examinations Council (L), the Oxford and Cambridge Schools Examination Board School's Mathematics Project (SMP). We gratefully acknowledge these Boards' permission to reproduce their questions. We give our own brief answers for which the Boards bear no responsibility.

Once again we express our grateful thanks for the support which our wives have given us, without it the work would never have been finished.

Notation

$=$	is equal to
$\approx$	is approximately equal to
$\neq$	is not equal to
$\cong$	is isomorphic to
$<$	is less than
$\leqslant$	is less than or equal to
$>$	is greater than
$\geqslant$	is greater than or equal to
$\nless$	is not less than
$\ngtr$	is not greater than
$\sqrt{\ }$	the positive square root
$\Rightarrow$	implies that
$\Leftarrow$	is implied by
$\Leftrightarrow$	implies and is implied by
$\{a, b, c, \ldots\}$	the set with elements $a, b, c, \ldots$
$\in$	is an element of
$\notin$	is not an element of
:	such that
$\{x: a < x < b\}$	the set of x such that a is less than x and also x is less than b
$n(A)$	the number of elements in the set A
$\mathscr{E}$	the universal set
ϕ	the empty set, null set
S'	the complement of the set S
$\cup$	the union of
$\cap$	the intersection of
$\subseteq, \subset$	is a subset of, is a proper subset of, respectively
$\backslash$	the set difference, $A \backslash B = A \cap B'$
$\mathbb{N}$	the set of positive integers and zero $\{0, 1, 2, \ldots\}$
$\mathbb{Z}$	the set of integers $\{0, \pm 1, \pm 2, \ldots\}$
$\mathbb{Z}^+$	the set of positive integers
$\mathbb{Q}$	the set of rational numbers
$\mathbb{Q}^+$	the set of positive rational numbers
$\mathbb{R}$	the set of real numbers
$\mathbb{R}^+$	the set of positive real numbers
$\mathbb{C}$	the set of complex numbers
$x \mathrm{R} y$	x has the relation R to y
$\mathrm{f}: A \to B$	the function f maps the set A into the set B

$\mathrm{f}(x)$	the function value for x
$\mathrm{f}: x \mapsto y$	f is the function under which x is mapped into y
f^{-1}	the inverse function of f

For functions f and g where domains and ranges are subsets of $\mathbb{R}$

$\mathrm{f}+\mathrm{g}$	is defined by $(\mathrm{f}+\mathrm{g})(x) = \mathrm{f}(x)+\mathrm{g}(x)$
$\mathrm{f}.\mathrm{g}$	is defined by $(\mathrm{f}.\mathrm{g})(x) = \mathrm{f}(x)\mathrm{g}(x)$
fg	is defined by $(\mathrm{fg})(x) = \mathrm{f}(\mathrm{g}(x))$
$\rightarrow$	approaches, tends to (in the context of a limit)
∞	infinity
$\lvert \; \rvert$	the unsigned part of a signed number, the modulus
$[x]$	the integer part of the number x
Σ	the sum of (precise limits may be given)
$n!$	n factorial
$\binom{n}{r}$	the binomial coefficient $n(n-1)(n-2)\ldots(n-r+1)/r!$
δx	a small increment of x
$\mathrm{f}', \mathrm{f}'', \mathrm{f}^{(3)}, \ldots \mathrm{f}^{(n)}$	the first, second, third, . . . nth derivatives, respectively, of f
$\dfrac{\mathrm{d}y}{\mathrm{d}x}$	the derivative of y with respect to x
$\int y \, \mathrm{d}x$	the indefinite integral of y with respect to x
$\int_a^b y \, \mathrm{d}x$	the definite integral of y with respect to x between the limits $x = a$ and $x = b$
$\ln x$	the natural logarithm of x
e^x	the exponential function of x
$\lvert \mathbf{r} \rvert, r$	the magnitude of the vector $\mathbf{r}$
$\mathbf{i}, \mathbf{j}, \mathbf{k}$	unit vectors in the mutually perpendicular directions Ox, Oy, Oz
$\mathbf{a}.\mathbf{b}$	the scalar product of the vectors $\mathbf{a}$ and $\mathbf{b}$
$\mathbf{a} \times \mathbf{b}$	the vector product of the vectors $\mathbf{a}$ and $\mathbf{b}$
$\mathbf{M}^{\mathrm{T}}$	the transpose of the matrix $\mathbf{M}$
$\mathbf{M}^{-1}$	the inverse of the matrix $\mathbf{M}$
$\det \mathbf{M}$	the determinant of a square matrix $\mathbf{M}$
i	square root of -1
$\lvert z \rvert$	the modulus of the complex number z
$\arg z$	the argument of the complex number z
z^*	the conjugate of the complex number z
$\mathrm{Re}(z)$	the real part of the complex number z
$\mathrm{Im}(z)$	the imaginary part of the complex number z

Introduction: Mathematical Proof

In this introduction we analyse the methods of mathematical proof. Mathematics is a logical structure and to understand mathematics it is necessary to have a clear idea of the logic involved. We do not attempt to treat the logic formally but, through an informal discussion, we attempt to explain the common patterns and structure of mathematical proof.

This section may be read as an introduction to the book but it can usefully be re-read occasionally whilst working through later chapters. In fact, this is the best way to use it since the methods of proof can only be appreciated through their applications. We give a few illustrative examples here and refer to later places in the book where the methods being discussed are actually used.

Set theory provides a language in which ambiguities may be avoided. For example, a common way of writing the solution of the equation $x^2 = 4$ is to state that '$x = 2$ or $x = -2$'. This is a satisfactory statement for the experienced mathematician who understands just what it means. But the use of 'or' is an ambiguity which can lead to many difficulties for the inexperienced. To clarify the situation,

$$\begin{array}{llll} \text{if } x = 2 & \text{then } x^2 = 4 & & \text{is true,} \\ \text{if } x = -2 & \text{then } x^2 = 4 & & \text{is true,} \\ \text{if } x^2 = 4 & \text{then } x = 2 & & \text{is false,} \\ \text{if } x^2 = 4 & \text{then } x = -2 & & \text{is false.} \end{array}$$

The unambiguous way to express the solution is to avoid the use of 'or' and to write:

'the solution set of the equation $x^2 = 4$ is $\{2, -2\}$.'

This means that two statements are being made, namely

$$\begin{array}{ll} \text{if } x^2 = 4 & \text{then } x \in \{2, -2\} \\ \text{if } x \in \{2, -2\} & \text{then } x^2 = 4. \end{array}$$

Notice that in order to provide a correct solution one must prove that if the equation is satisfied then x lies in the solution set *and* if x lies in the solution set then the equation is satisfied.

Implication

Results in mathematics are obtained by starting from some set of assumptions (the premise or hypothesis) and, using logical arguments, arriving at a result (the conclusion). We shall use bold capital letters to denote statements. Such statements may be obviously true or false or may be true only when certain extra conditions are added. If **P** and **Q** are statements and there is a logical argument which produces the conclusion **Q** from the hypothesis **P**, then we say that

$$\mathbf{P} \text{ implies } \mathbf{Q} \text{ with the notation } \mathbf{P} \Rightarrow \mathbf{Q},$$

or equivalently,

$$\text{if } \mathbf{P} \text{ then } \mathbf{Q} \text{ or } \mathbf{P} \text{ is sufficient for } \mathbf{Q}.$$

It must be remembered that $\mathbf{P} \Rightarrow \mathbf{Q}$ only means that if **P** is true then **Q** is also true. It does not say anything about the truth of **Q** if **P** is false and also it says nothing about the truth of **P**.
For example, working in the set of real numbers,

let **P** be the statement '$x = 2$'
let **Q** be the statement '$x^2 = 4$'

then clearly $\mathbf{P} \Rightarrow \mathbf{Q}$. However, this only means that **Q** is true if **P** is true. If **P** is false then **Q** may be either true or false. For example,

if $x = 3$ then **P** is false and **Q** is false,
if $x = -2$ then **P** is false and **Q** is true.

Generally, when **P** is false then $\mathbf{P} \Rightarrow \mathbf{Q}$ is true, whether **Q** is true or false. This means that from a false hypothesis you can 'prove' anything, a device used by those who would like to convert you to their point of view. This aspect of the logic may be hard to understand since it is not in line with everyday use of the words. However, it does not enter into the rest of our work since we are not concerned with the formal logic.

Negation

The statement 'not **P**' is the negation of **P** and is defined by

not **P** is true when **P** is false,
not **P** is false when **P** is true.

Since $\mathbf{P} \Rightarrow \mathbf{Q}$ means that **Q** is true whenever **P** is true, it is equivalent to the statement that if **Q** is false then so is **P**, thus

$$\mathbf{P} \Rightarrow \mathbf{Q} \text{ is equivalent to not } \mathbf{Q} \Rightarrow \text{not } \mathbf{P}.$$

As a simple example of the equivalence of the two statements,

let **P** be the statement $x \geqslant 3$
let **Q** be the statement $x > 2$

then $\mathbf{P} \Rightarrow \mathbf{Q}$ is the statement if $x \geqslant 3$ then $x > 2$

and an equivalent statement to this is

not $\mathbf{Q} \Rightarrow$ not $\mathbf{P}$ that is if $x \leqslant 2$ then $x < 3$.

See §3.5, Theorem 3.1(iii).

Warning The above equivalence of $\mathbf{P} \Rightarrow \mathbf{Q}$ and not $\mathbf{Q} \Rightarrow$ not $\mathbf{P}$ must not be confused with the negation of the statement $\mathbf{P} \Rightarrow \mathbf{Q}$, that is, not $(\mathbf{P} \Rightarrow \mathbf{Q})$ meaning $\mathbf{P}$ does not imply $\mathbf{Q}$ and denoted by $\mathbf{P} \nRightarrow \mathbf{Q}$. $\mathbf{P}$ does not imply $\mathbf{Q}$ is equivalent to the pair of statements $\mathbf{P}$ is true and $\mathbf{Q}$ is false. Thus $\mathbf{P} \nRightarrow \mathbf{Q}$ means not $(\mathbf{P} \Rightarrow \mathbf{Q})$ and is equivalent to $\mathbf{P}$ *and* not $\mathbf{Q}$. In the above example, $\mathbf{Q} \nRightarrow \mathbf{P}$, that is

$$x > 2 \nRightarrow x \geqslant 3$$

as is seen by the counter-example $x = 5/2$.

Converse

The converse of $\mathbf{P} \Rightarrow \mathbf{Q}$ is $\mathbf{Q} \Rightarrow \mathbf{P}$ and this must not be confused with the negation. Generally, a statement and its converse are not equivalent since one may be true and the other false, as was shown above. As a further example, in the integers,

n is divisible by 6 $\Rightarrow n$ is divisible by 3

but n is divisible by 3 $\nRightarrow n$ is divisible by 6

since 9 is divisible by 3 but not by 6.
See §13.3, Theorems 13.4 and 13.7.

Necessary and sufficient conditions

The statement $\mathbf{P} \Rightarrow \mathbf{Q}$ means that

$\mathbf{P}$ is a sufficient condition for $\mathbf{Q}$ to be true

and that $\mathbf{Q}$ is a necessary condition for $\mathbf{P}$ to be true,

since if $\mathbf{P}$ is true $\mathbf{Q}$ cannot be false.

When both a theorem and its converse are true, we have the special situation $\mathbf{P} \Rightarrow \mathbf{Q}$ and $\mathbf{Q} \Rightarrow \mathbf{P}$. This double implication is written $\mathbf{P} \Leftrightarrow \mathbf{Q}$. The following statements are equivalent:

$\mathbf{P} \Rightarrow \mathbf{Q}$, $\mathbf{P}$ implies $\mathbf{Q}$, if $\mathbf{P}$ then $\mathbf{Q}$,

$\mathbf{P}$ is sufficient for $\mathbf{Q}$, $\mathbf{Q}$ is necessary for $\mathbf{P}$.

The following statements are equivalent:

$\mathbf{P} \Leftrightarrow \mathbf{Q}$, $\mathbf{P}$ implies $\mathbf{Q}$ and $\mathbf{Q}$ implies $\mathbf{P}$,

$\mathbf{P}$ if and only if $\mathbf{Q}$, $\mathbf{P}$ is both necessary and sufficient for $\mathbf{Q}$.

For example, in the set of real numbers,
let **P** mean '$x > 1$', let **Q** mean '$x^2 > 1$' and let **R** mean '$x^3 > 1$'.
Then $\mathbf{P} \Rightarrow \mathbf{R}$ and $\mathbf{R} \Rightarrow \mathbf{P}$ so that $\mathbf{P} \Leftrightarrow \mathbf{R}$.
Also $\mathbf{P} \Rightarrow \mathbf{Q}$ but $\mathbf{Q} \not\Rightarrow \mathbf{P}$ as is shown by taking $x = -2$.
See §1.4, Theorem 1.4; §4.2, Theorem 4.2; §8.2, Theorem 8.1; §12.2, Theorems 12.2 and 12.4; §14.2, Theorems 14.4 and 14.6; §14.3, Theorems 14.17 and 14.18; §14.4, Theorem 14.20.

Counter-example

The last result $\mathbf{Q} \not\Rightarrow \mathbf{P}$, that is, not $(\mathbf{Q} \Rightarrow \mathbf{P})$ is proved by showing that **Q** is true and **P** is false when $x = -2$. This is expressed by saying that $x = -2$ is a counter-example to the statement $\mathbf{Q} \Rightarrow \mathbf{P}$. Generally, the negation of a statement is proved by producing a counter-example. If the statement involves a variable x, then one given value of x will be enough to provide the counter-example.
See §2.2, Example 2.

Generality

Many statements may include a variable x (or more than one variable) and may be true for some values of x and not others. The universal set of all values of x that are being considered may be stated explicitly or may be understood from the context. It will commonly be the set $\mathbb{R}$ of real numbers or the set $\mathbb{C}$ of complex numbers.

The statement $x = x$ is true for all x.

The statement $x^2 \geqslant 0$ is true for all $x \in \mathbb{R}$.

The set of values of x for which a statement is true is called its *solution set*. Thus

S is the solution set of $\mathbf{P}(x)$ means $\mathbf{P}(x)$ is true $\Leftrightarrow x \in S$.

Note that the double implication is needed since

$$\mathbf{P}(x) \Rightarrow x \in S \quad \text{and} \quad x \in S \Rightarrow \mathbf{P}(x)$$

are both required for a solution set.
When a statement $\mathbf{P}(n)$ is to be proved true for all natural numbers n, then a suitable method may be by:

Mathematical Induction

This type of proof is conveniently divided into three steps.

Step 1 The basis of induction. This is the starting point and involves proving $\mathbf{P}(n)$ for some suitable small value of n. For example, it could be a proof that $\mathbf{P}(0)$ is true, or that $\mathbf{P}(1)$ is true.

Step 2 The induction step. This involves using the hypothesis

for a given $n \in \mathbb{N}$, $\mathbf{P}(n)$ is true.

This is then used to deduce that $\mathbf{P}(n+1)$ is also true, so the step is described as a proof of the statement

$$\mathbf{P}(n) \Rightarrow \mathbf{P}(n+1), \text{ for all } n \in \mathbb{N}.$$

Step 3 The first two steps are now combined to say that

P(0) is true so P(1) is true by use of $n = 0$ in step 2,
P(1) is true so P(2) is true by use of $n = 1$ in step 2,
P(2) is true so P(3) is true by use of $n = 2$ in step 2,

and so on. Since the natural numbers $\mathbb{N}$ are generated by counting, it follows that $\mathbf{P}(n)$ is true for all $n \in \mathbb{N}$. To summarise

Proof by induction
(i) $\mathbf{P}(0)$ is true,
(ii) $\mathbf{P}(n) \Rightarrow \mathbf{P}(n+1)$, for all $n \in \mathbb{N}$,
(iii) hence, by induction, $\mathbf{P}(n)$ is true for all $n \in \mathbb{N}$.

As an example, consider the statement

$$\mathbf{P}(n) \text{ means '}n^3 - n \text{ is a multiple of 3'}.$$

(i) Basis of induction: since $0 - 0 = 3 \times 0$, $\mathbf{P}(0)$ is true.
(ii) Induction hypothesis: assume that $\mathbf{P}(n)$ is true, namely that for some natural number k, $n^3 - n = 3k$.

Then
$$\begin{aligned}(n+1)^3 - (n+1) &= n^3 + 3n^2 + 3n + 1 - n - 1\\ &= (n^3 - n) + 3(n^2 + n)\\ &= 3k + 3(n^2 + n) = 3(k + n^2 + n),\end{aligned}$$

and so $\mathbf{P}(n+1)$ is true.
(iii) Hence, by induction, $n^3 - n$ is a multiple of 3 for all $n \in \mathbb{N}$.
See §4.3, Theorems 4.4 and 4.5; §6.5, Theorem 6.6; §10.2, Theorem 10.1 and Example 7; §14.3, Theorem 14.3.

Proof by contradiction

It may be that a method of proof is constructive in the sense that the statement says that something can be done and the proof provides an algorithm for doing it. This is done in §4.2, Theorem 4.3, and §4.4, Theorem 4.6, for the echelon reduction of matrices. Another type of proof is non-constructive in the sense that the statement is proved by showing that its negation leads to a contradiction. The form of the proof is:

assume that $\mathbf{P}$ is false,
deduce that some statement is both true and false,
this is impossible so $\mathbf{P}$ must be true.

The logic of the proof relies on two ideas. First, the system of logic must be consistent so that no statement can be both true and false. Therefore, if $\mathbf{P}$ is false implies that $\mathbf{Q}$ is both true and false then it must be false that $\mathbf{P}$ is false. Second, it assumes that a statement must be either true or false, so that if it is false that **P** is false then **P** is true. An example of this type of proof is the proof that $\sqrt{2}$ is irrational, given in *Pure Mathematics* §9.5.

As another example, let **P** be the statement that there are infinitely many prime numbers. Assume that **P** is false, that is, assume that there are only k prime numbers, for some $k \in \mathbb{N}$. Let M be the product of these k primes and consider the number $M+1$. On division of $M+1$ by any prime number there is a remainder 1 and so $M+1$ has no factors. Therefore $M+1$ is a prime number. But $M+1$ is larger than each of the k primes so it is not equal to any of them. Hence, $M+1$ is another prime and it is false that there are only k primes. Therefore **P** is true.

Solving problems

When presented with a problem in mathematics there are a number of strategies which may be adopted. Begin by writing down the given information which restricts the problem to certain conditions. Note any immediate consequences which may be useful. Write down what it is that is to be proved and then look for possible methods of proof.

Experiment with various approaches and, if the problem seems too hard, try to simplify it by adding further restrictive conditions. This solution of the easier problem may suggest a plan of attack on the harder problem. Use diagrams to assist your thoughts and look for the possibility of transforming the problem into a familiar one. This may be done by a change in the variables, as is often used in integration, or by a change in the coordinate system. Look also for any recognisable patterns in the problem to see if you can use some isomorphism (§3.7) to translate the problem into a more familiar one, for example an algebraic problem may be translated into a geometric one or *vice-versa*. Sometimes a problem needs to be divided into a number of cases, each of which must be considered separately. This is referred to as proof by exhaustion!

It may well be useful to work backwards from the desired conclusion to see what intermediate results would lead to the conclusion. The approaches from front and back of the problem will, we hope, lead to some path through the logic and provide a solution. Be careful not to fall into the trap of finishing with an incomplete logical argument or one in which the logic is reversed and you have assumed the solution and proved the hypothesis.

Example of a false proof to show that

$$\frac{\cos(A-B)-\cos(A+B)}{\sin(A+B)+\sin(A-B)} = \tan B.$$

Multiply both sides by the denominator

$$\cos(A-B)-\cos(A+B) = \tan B\,(\sin(A+B)+\sin(A-B))$$

$$2\sin A\sin B = 2\tan B\sin A\cos B$$

$$= 2\sin A\sin B,$$

this is true and hence

$$\frac{\cos(A-B)-\cos(A+B)}{\sin(A+B)+\sin(A-B).}=\tan B.$$

The logic is false and all that has been 'proved' is that, given the required identity, then $2 \sin A \sin B = 2 \sin A \sin B$, a fact that we know is true and did not want to prove. In this case, the 'proof' has been written backwards and could be corrected by inserting 'because' between each line. This is because, in this case, every step can be reversed. A better approach to a proof of such an identity is to start from one side, usually the more complicated, and attempt to simplify it to produce the other side. Thus

$$\frac{\cos(A-B)-\cos(A+B)}{\sin(A+B)+\sin(A-B)}=\frac{2\sin A\sin B}{2\sin A\cos B}=\tan B.$$

In working out how to solve the problem, you may have experimented and worked from both ends as well as in the middle. All this is a preliminary to producing a final solution. In reordering the material to produce the final solution, make sure that each step follows logically from the previous one. If you use the double implication, make sure that the statements each side of it each follow from the other. Finally, make sure that what you write down is good English. If you read it out aloud, including all the mathematical symbols, it should read as a connected piece of English. All too often, students produce a list of equations with no connectives between them and if read out the 'solution' does not even make sense.

1 Matrices

1.1 Information tables

Information is normally presented to the reader in one of three forms: the written word, a diagram, or a table of figures. This chapter is concerned with developing a notation for dealing with numerical information presented in a tabular form. Such a rectangular table of numbers is called a *matrix* and we begin by looking at some examples.

Example: multiplication table Consider the multiplication table of products of the integers 1 to 4 with the integers 1 to 6 shown below.

	1	2	3	4	5	6
1	1	2	3	4	5	6
2	2	4	6	8	10	12
3	3	6	9	12	15	18
4	4	8	12	16	20	24

The 4×6 rectangular block of numbers enclosed in the vertical lines is called a *matrix*. It can be regarded as a series of four *rows*, labelled 1, 2, 3, 4 and each row is the sequence of products of the row number with the various column numbers. Thus the third row consists of 3×1, 3×2, $3 \times 3, \ldots, 3 \times 6$. Alternatively, the matrix can be thought of as a series of six columns, labelled above by 1, 2, 3, 4, 5, 6 so that the fourth column is the sequence of products 1×4, 2×4, 3×4, 4×4. The row and column headings are not part of the matrix but we will usually need them to identify the entries, thus the entry in row 3 and column 4 is 12. The matrix is said to have *order* 'four by six', written (4×6).

Definition A *matrix* **M** is a rectangular array with *a rows* and *b columns*. Such a matrix has *order* $(a \times b)$.

Example: league table Part of a football league table is shown below. The 5×5 block of numbers, between the vertical lines, is a matrix of order (5×5), called a square matrix, since it has the same number of rows as columns.

	W	D	L	F	A
Arsenal	1	4	4	7	21
Everton	3	2	4	9	28
Leeds	7	0	2	20	11
Liverpool	5	2	2	15	14
Manchester United	8	0	1	23	5

Each row of the matrix shows the performance of one team, namely the number of matches won, drawn or lost and the number of goals scored for and against. The information is presented in a compact form, organised so that it is fairly easy to compare the performance of two teams or perhaps to predict future results or the final league placings.

Example: statistical distribution The table below shows data relating to the length and breadth of a fully developed human head, collected from a survey conducted on a sample of 1300 people.

		Length in cm				
		17·4	18·2	19·0	19·8	20·6
Breadth	14·2	11	74	101	23	1
in cm	15	7	146	418	163	18
	15·8	2	20	130	136	24
	16·6	0	1	6	17	2

Numbers of people with a given head size

A quick glance at the table reveals that the average human head is about 15 cm wide and 19 cm long, but further analysis could provide much more information, useful, say, to the medical profession or to a manufacturer of hats. This might include details of the spread of the values of width and length about their mean values, and the general relationship between the width and the length of a head.

The commercial world's performance is often judged by tables of figures: sales figures for different products in different areas, stock lists and price lists, distance and weight limit charts for a transport system, or accounts and final balance sheets. This type of information is often stored in computers and organised in a file structure which reflects the type of information and the use to which it will be put. These files will often be organised in a matrix format, enabling comparable information to be extracted efficiently.

Another use of matrices is to describe the connections between the elements of some system. Such connections are often represented by means of a diagram in which the connections are shown by lines joining the points. In cases where the connection is directional, such as in a flow diagram, these directed lines are indicated by arrows. In the corresponding matrix, the rows and columns are labelled by the elements of the structure, and the entry in position row A, column B is the number of connections between A and B, or from A to B in the directed case. The matrix enables an efficient handling of the information to be achieved mathematically.

Example: networks Part of a road network is shown diagrammatically in Fig. 1.1(a) and a similar road network, in which some of the streets are one-way, is shown in Fig. 1.1(b). We construct two matrices representing these networks. The rows and columns of the two matrices will be labelled by the four points A, B, C and D.

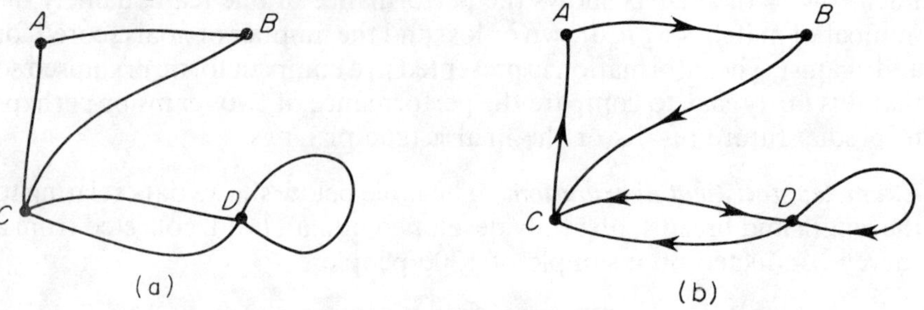

Fig. 1.1 (a) Undirected road network (b) Directed road network

(a) In the undirected case, each entry gives the number of connections between the points. There are two connections between C and D so that the (C, D) entry and the (D, C) entry are each 2. The (A, D) entry and the (D, A) entry are each 0, since there are no connections between A and D. Thus the matrix is built up, as shown in Fig. 1.2(a). This matrix is square and is *symmetric* about its *leading diagonal* (from top left to bottom right, shown by the dotted line). This symmetry reflects the fact that each route is entered twice, for example once for the route from A to B and once for the route from B to A. For a given junction, the sum of all the entries in the row labelled by that junction, which is called the row sum for that row, specifies the number of routes at that junction.

$$\begin{array}{c} \\ A \\ B \\ C \\ D \end{array} \begin{array}{c} \begin{matrix} A & B & C & D \end{matrix} \\ \begin{vmatrix} 0 & 1 & 1 & 0 \\ 1 & 0 & 1 & 0 \\ 1 & 1 & 0 & 2 \\ 0 & 0 & 2 & 2 \end{vmatrix} \end{array} \qquad \begin{array}{c} \\ A \\ B \\ C \\ D \end{array} \begin{array}{c} \begin{matrix} A & B & C & D \end{matrix} \\ \begin{vmatrix} 0 & 1 & 0 & 0 \\ 0 & 0 & 1 & 0 \\ 1 & 0 & 0 & 1 \\ 0 & 0 & 2 & 1 \end{vmatrix} \end{array}$$

(a) (b)

Fig. 1.2 Matrices corresponding to the networks of Fig. 1.1

(b) When some of the roads are one-way, the matrix is a *directed route matrix* where the entry counts the number of routes *from* the point labelling the row *to* the point labelling the column. The matrix is shown in Fig. 1.2(b) and it is still square but is no longer symmetric. The row sum of a given row specifies the number of routes *leaving* a junction whilst the column sum specifies the number of routes *arriving* at a junction. A zero row sum would imply a dead end and a zero column sum would imply a junction which cannot be reached from any other junction. Note that, in both (a) and (b), an entry on the leading diagonal indicates a loop which begins and ends at the same junction.

Other matrix representations of networks are also useful. Entries can represent the distance between junctions on a road network, the length of time it takes to travel between the junctions, the rate of flow along a pipe connecting units in a manufacturing plant, the length of wires connecting electronic units, and such matrices may often be large. Methods have been developed which analyse these large order matrices in order to find efficient distribution routes or travelling salesman circuits, (the *order* of a *square* matrix is the number of its rows, or columns). Matrix methods can provide a way of predicting the behaviour of an electrical circuit or an hydraulic network without having to build the actual network. Such matrix modelling methods are important in the design of control systems generally.

A matrix is a general purpose tool for storing information in an organised fashion and we shall develop techniques for manipulating this information by manipulating the matrices. When a method of storing the essential information about a system in a matrix has been evolved, the matrix techniques can then be used to provide new information about the system and to spot similarities between the structures of apparently different systems.

In the above examples where the matrices formed part of tables, the matrices were shown enclosed in vertical lines. When a matrix is to be considered alone as a rectangular array, it is enclosed in brackets. It is useful to use a single symbol to name the matrix and, when this is done, the symbol is printed in bold type (hand written by underlining with a wavy line), just as a vector is, in order to distinguish a matrix from a scalar. Of course, a three-dimensional column vector is a (3×1) matrix. The components of a column vector are the row entries in its column and it is sometimes useful to use suffices to indicate the components. In the same way, it can be useful to label the entries of a matrix by means of a double suffix notation. The same letter is used throughout, thus a matrix can be denoted by $\mathbf{A}$ and its entry in row i and column j is denoted by a_{ij}.

Notation The matrix $\mathbf{A}$ is a rectangular array of elements a_{ij} enclosed within brackets. If $\mathbf{A}$ is a $(m \times n)$ matrix, with m rows and n columns, the ith row of $\mathbf{A}$ is $(a_{i1}, a_{i2}, \ldots, a_{in})$ and the jth column of $\mathbf{A}$ is

$$\begin{pmatrix} a_{1j} \\ a_{2j} \\ a_{3j} \\ a_{4j} \\ \cdot \\ \cdot \\ a_{mj} \end{pmatrix}, \text{ and } \mathbf{A} = \begin{pmatrix} a_{11} & a_{12} & a_{13} & a_{14} & \cdots & a_{1n} \\ a_{21} & a_{22} & a_{23} & a_{24} & \cdots & a_{2n} \\ a_{31} & a_{32} & a_{33} & a_{34} & \cdots & a_{3n} \\ a_{41} & \cdot & \cdot & \cdot & \cdots & a_{4n} \\ \cdot & \cdot & \cdot & \cdot & \cdot & \cdot \\ \cdot & \cdot & \cdot & \cdot & \cdot & \cdot \\ a_{m1} & a_{m2} & a_{m3} & a_{m4} & \cdots & a_{mn} \end{pmatrix}$$

The $(m \times n)$ matrix **A** may also be written
$$\mathbf{A} = (a_{ij}),\ 1 \leqslant i \leqslant m,\ 1 \leqslant j \leqslant n.$$

EXERCISE 1.1

1 A milk roundsman uses the matrix below to help him to deliver the correct dairy products to five houses.

House number	1	3	5	4	2
Silver	5	8	2	0	6
Gold	0	2	0	0	1
Cream	0	1	1	0	0

(a) How many pints of gold-top milk does No. 3 have?
(b) How many pints of silver-top milk are needed in total for the five houses?
(c) The following week, No. 3 cancels the order for cream and changes to 7 silver and 3 gold, No. 4 returns from holiday and needs 2 gold and 4 silver and No. 5 doubles his order for cream. Write down the matrix for the next week's deliveries.
(d) Over the two weeks, calculate the total number of pints of silver-top, gold-top and cream delivered to each of the five houses and represent your answer as a matrix.
(e) If a pint of silver-top costs 21p, gold-top 23p and cream 37p, calculate the first week's bill for each household.

2 Using the Football League Table, given as an example on page 9, answer the following.
(a) How many matches have each of the teams played?
(b) Which team has had the greatest number of goals scored against it?
(c) If each team is allocated 2 points for a win and 1 point for a draw, calculate the points obtained by each team so far.
(d) Assuming that the set of points scored is added as an extra column to the table and that three more teams are added to the five already shown, what is the order of the new matrix?
(e) A pools punter uses his own points system in order to predict the next weeks draws. He gives 1 point for a win, no points for a draw, -2 points for each loss, 0·5 for each goal for and $-0{\cdot}3$ for each goal against. He then predicts a draw if the points scored by the two teams are separated by one point or less. Everton and Arsenal play together next week. Does the punter assume a draw?

3 Using the matrix representations of networks find the route matrices for (a) and (b) and the directed route matrices for (c) and (d), as shown in Fig. 1.3.
Calculate the total sum of all the entries for each of these four matrices and explain the significance of your answers.
Find also the matrix for the network (e) and compare this with the matrix for (a). These networks are said to be *equivalent*.

4 The position vectors of the three points $A(2, 4, -1)$, $B(3, -2, 5)$ and $C(-1, 4, 2)$ are **a**, **b** and **c**, respectively.
(a) Regarding these vectors as matrices, write down the order of **a**.
(b) Vectors may be added and multiplied by a scalar. Calculate the vectors (i) $\mathbf{a}+\mathbf{b}$, (ii) $3\mathbf{c}$, (iii) $2\mathbf{a}-\mathbf{b}-3\mathbf{c}$.

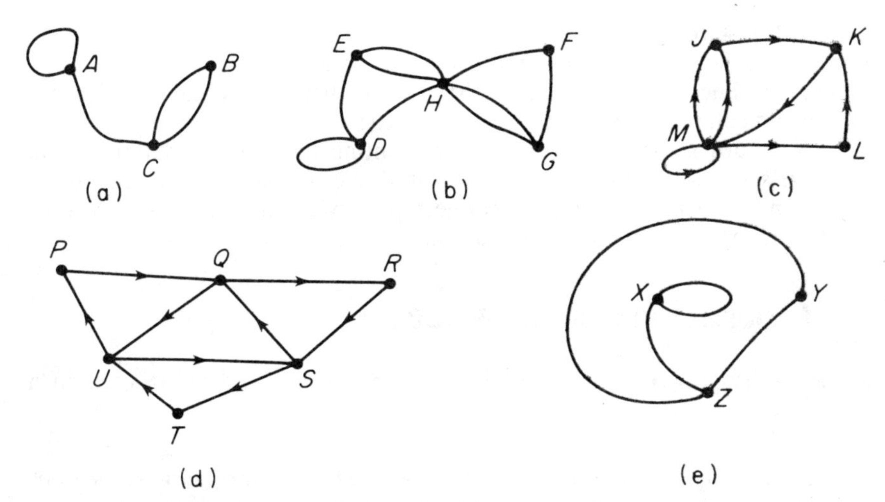

Fig. 1.3

(c) $\mathbf{P} = \begin{pmatrix} 2 & 3 \\ 4 & -2 \\ -1 & 5 \end{pmatrix}$, $\mathbf{Q} = \begin{pmatrix} -1 & 6 \\ 4 & 3 \\ 2 & 1 \end{pmatrix}$. If matrix addition is to be consistent with vector addition, write down the matrix $\mathbf{P}+\mathbf{Q}$.
(d) Similarly, write down $3\mathbf{P}-2\mathbf{Q}$.

5 Let $\mathbf{T}$ be the $m \times n$ matrix with entries t_{ij}.
(a) Write down the second entry in the third row of $\mathbf{T}$ and also the xth entry in the yth row.
(b) Write down the sum of all the entries in the fourth column of $\mathbf{T}$, using a $\sum$ notation.
(c) Write down, in $\sum$ notation, the sum of all the entries in the rth row of $\mathbf{T}$.
(d) Write down the total sum of all the mn entries in $\mathbf{T}$, using the $\sum$ notation.
(e) Let $\mathbf{T}^{\mathrm{T}}$ be the matrix with entries $t^{\mathrm{T}}_{ij} = t_{ji}$. Write down the order of $\mathbf{T}^{\mathrm{T}}$.
(f) If $n = m$, write down a condition on the entries of $\mathbf{T}$ so that $\mathbf{T}$ and $\mathbf{T}^{\mathrm{T}}$ have identical entries. Show that this implies that $\mathbf{T}$ is symmetric. Describe what the series $\sum t_{ii}$ represents.

6

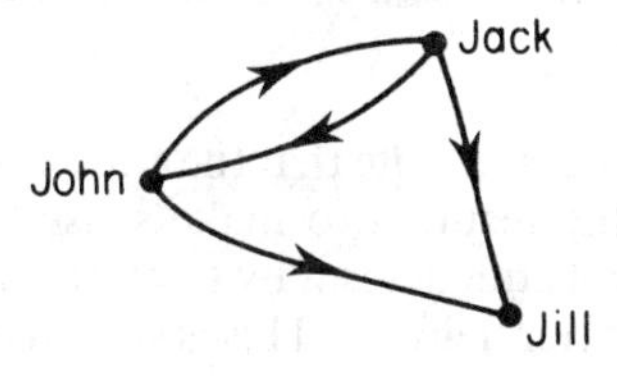

	Ja	Jo	Ji
Ja	0	1	1
Jo	1	0	1
Ji	0	0	0

Fig. 1.4

Fig. 1.4 shows the relation 'is the brother of' as an arrow diagram connecting the three children in a family, together with a matrix method of representing this relation (similar to a directed route matrix). Use a similar method to find the matrix representation of the following sets and relations.

(a) $\{2, 3, 4, 6, 8\}$ and 'is a factor of'.
(b) $\{-2, -1, 0, 1, 2, 3, 4\}$ and 'is the square of'.
(c) Four lines k, l, m, n in space with l perpendicular to k and m and with n perpendicular to l, and 'is perpendicular to'.

Write down the three matrices representing the inverse relation for the above sets and relations, namely 'is a multiple of', 'is a square root of', 'is perpendicular to'. Describe the connection between the matrix of a relation and the matrix of the inverse relation.

1.2 Operations on matrices

We begin by showing how matrices can be added and subtracted in a natural way.

EXAMPLE 1 *The owner of a small hardware shop uses a matrix to store the stock information about boxes of wood screws. Every month, a new delivery of boxes of wood screws is made and the stock list has to be updated. In Fig. 1.5, tables are shown (a) of the stock just before a delivery, (b) the delivery note and (c) the stock at the end of the month.*

Screw size	Length in inches $\frac{1}{2}$	$\frac{3}{4}$	1	$1\frac{1}{4}$	$1\frac{1}{2}$
No. 6	21	56	31	0	0
No. 8	0	13	68	115	88
No. 10	12	0	42	13	58

(a) Stock at start of month

(b) Delivery note
25 No 6, $\frac{1}{2}''$
10 No 6, $1''$
20 No 8, $\frac{3}{4}''$
40 No 8, $1\frac{1}{2}''$
50 No 10, $1\frac{1}{4}''$

(b) Delivery note

Screw size	Length in inches $\frac{1}{2}$	$\frac{3}{4}$	1	$1\frac{1}{4}$	$1\frac{1}{2}$
No. 6	32	15	26	0	0
No. 8	0	31	24	56	82
No. 10	0	0	3	14	23

(c) Stock at end of month

Fig. 1.5

Show how the stock matrix can be updated by adding a delivery matrix, the matrices being added by adding entries in corresponding positions. Show also how the month's sales can be found by subtraction of matrices.

We omit the row and column headings of the matrices but note that the entries in the matrices must be in the correct positions if the result is to make sense. A delivery matrix is created from the delivery note and added, term by term, to the initial stock list matrix to give the stock list matrix after delivery. This is shown in Figure 1.6.

$$\begin{pmatrix} 21 & 56 & 31 & 0 & 0 \\ 0 & 13 & 68 & 115 & 88 \\ 12 & 0 & 42 & 13 & 58 \end{pmatrix} + \begin{pmatrix} 25 & 0 & 10 & 0 & 0 \\ 0 & 20 & 0 & 0 & 40 \\ 0 & 0 & 0 & 50 & 0 \end{pmatrix} = \begin{pmatrix} 46 & 56 & 41 & 0 & 0 \\ 0 & 33 & 68 & 115 & 128 \\ 12 & 0 & 42 & 63 & 58 \end{pmatrix}$$

Old stock + Delivery = New Stock

Fig. 1.6

Similarly, the sales matrix, showing the sales of the boxes of various sizes of screws during the month, is found by subtracting the stock list matrix at the end of the month from the stock list matrix at the start of the month. This is shown in Fig. 1.7.

$$\underset{\text{Stock at start of month}}{\begin{pmatrix} 46 & 56 & 41 & 0 & 0 \\ 0 & 33 & 68 & 115 & 128 \\ 12 & 0 & 42 & 63 & 58 \end{pmatrix}} - \underset{\text{Stock at end of month}}{\begin{pmatrix} 32 & 15 & 26 & 0 & 0 \\ 0 & 31 & 24 & 56 & 82 \\ 0 & 0 & 3 & 14 & 23 \end{pmatrix}} = \underset{\text{Sales}}{\begin{pmatrix} 14 & 41 & 15 & 0 & 0 \\ 0 & 2 & 44 & 59 & 46 \\ 12 & 0 & 39 & 49 & 35 \end{pmatrix}}$$

Fig. 1.7

Note that, in order to add or subtract two matrices, the matrices must be of the same order since the addition or subtraction is made entry-wise.

In the above type of stock system, it would be possible to have a matrix containing the lowest stock level allowed by the shopkeeper before reordering and another matrix containing the standard size of order for each type of screw. In this way, stock control could become a series of matrix operations, ideally suited to computerisation.

Definition Let $\mathbf{A} = (a_{ij})$, $\mathbf{B} = (b_{ij})$ and $\mathbf{C} = (c_{ij})$, where $\mathbf{A}$, $\mathbf{B}$ and $\mathbf{C}$ are matrices, all of order $(m \times n)$. Then:
(i) $\mathbf{C} = \mathbf{A} + \mathbf{B}$ if $c_{ij} = a_{ij} + b_{ij}$, for all i, j, $1 \leqslant i \leqslant m$, $1 \leqslant j \leqslant n$,
(ii) $\mathbf{C} = \mathbf{A} - \mathbf{B}$ if $c_{ij} = a_{ij} - b_{ij}$, for all i, j, $1 \leqslant i \leqslant m$, $1 \leqslant j \leqslant n$.
Note that if $n = 1$ then these definitions are consistent with the definitions of addition and subtraction of vectors, see *Pure Mathematics* Chapter 12.

The multiplication of a vector by a scalar can also be naturally extended to the multiplication of a matrix by a scalar, by multiplying each entry of the matrix by the scalar. By combining this with matrix addition, it is possible to calculate linear combinations of matrices of the same order. We define the operations and demonstrate them in an example.

Definition Let $\mathbf{A} = (a_{ij})$, $\mathbf{B} = (b_{ij})$ and $\mathbf{C} = (c_{ij})$, where $\mathbf{A}$, $\mathbf{B}$ and $\mathbf{C}$ are matrices, all of order $(m \times n)$, and let p and q be scalars. Then:
(i) $\mathbf{C} = p\mathbf{A}$ if $c_{ij} = pa_{ij}$, for all i, j, $1 \leqslant i \leqslant m$, $1 \leqslant j \leqslant n$,
(ii) $\mathbf{C} = p\mathbf{A} + q\mathbf{B}$ if $c_{ij} = pa_{ij} + qb_{ij}$, for all i, j, $1 \leqslant i \leqslant m$, $1 \leqslant j \leqslant n$.

EXAMPLE 2 *Let* $\mathbf{P} = \begin{pmatrix} 3 & -1 & 2 \\ 4 & 0 & -3 \end{pmatrix}$, $\mathbf{Q} = \begin{pmatrix} 5 & 0 & -3 \\ 2 & -1 & 5 \end{pmatrix}$, $\mathbf{R} = \begin{pmatrix} 7 & -4 \\ 2 & -3 \end{pmatrix}$.

Calculate, where possible: (*i*) $3\mathbf{P}$, (*ii*) $\mathbf{P} + \mathbf{Q}$, (*iii*) $2\mathbf{P} - 3\mathbf{Q}$, (*iv*) $\mathbf{Q} - \mathbf{R}$.

(i) $3\mathbf{P} = 3\begin{pmatrix} 3 & -1 & 2 \\ 4 & 0 & -3 \end{pmatrix} = \begin{pmatrix} \mathbf{9} & \mathbf{-3} & \mathbf{6} \\ \mathbf{12} & \mathbf{0} & \mathbf{-9} \end{pmatrix}$,

(ii) $\mathbf{P} + \mathbf{Q} = \begin{pmatrix} 3 & -1 & 2 \\ 4 & 0 & -3 \end{pmatrix} + \begin{pmatrix} 5 & 0 & -3 \\ 2 & -1 & 5 \end{pmatrix} = \begin{pmatrix} \mathbf{8} & \mathbf{-1} & \mathbf{-1} \\ \mathbf{6} & \mathbf{-1} & \mathbf{2} \end{pmatrix}$,

(iii) $2\mathbf{P}-3\mathbf{Q}=2\begin{pmatrix}3 & -1 & 2\\ 4 & 0 & -3\end{pmatrix}-3\begin{pmatrix}5 & 0 & -3\\ 2 & -1 & 5\end{pmatrix}$

$$=\begin{pmatrix}6 & -2 & 4\\ 8 & 0 & -6\end{pmatrix}-\begin{pmatrix}15 & 0 & -9\\ 6 & -3 & 15\end{pmatrix}$$

$$=\begin{pmatrix}\mathbf{-9} & \mathbf{-2} & \mathbf{13}\\ \mathbf{2} & \mathbf{3} & \mathbf{-21}\end{pmatrix},$$

(iv) $\mathbf{Q}-\mathbf{R}$ **is impossible to calculate**. It has no meaning since $\mathbf{Q}$ has order (2×3) and $\mathbf{R}$ has order (2×2).

Matrix multiplication

We now consider how to define a useful multiplication operation between matrices. One possibility would be to define multiplication entry-wise, exactly like addition and multiplication by a scalar. This is not a useful process, just as the same process for vectors has no meaning.

The type of matrix multiplication that is useful has been met in questions **1** and **2** of Exercise 1.1. We consider this type of product in a very simple shopping list example which is easily generalised.

EXAMPLE 3 *Find the cost of the purchases of vegetables in week* 1, *where the quantities and unit prices are given in tabular form:*

	Quantity in kg			
	Potatoes	Carrots	Onions	Peas
Week 1 (	4	1	1	2)

	Price in p per kg			
	Potatoes	Carrots	Onions	Peas
Week 1 (	15	32	24	18)

The total cost of the purchases is obtained by multiplying the corresponding entries and adding the four amounts. In order to anticipate the generalisation, we use a row-column presentation, thus

$$(4\quad 1\quad 1\quad 2)\times\begin{pmatrix}15\\ 32\\ 24\\ 18\end{pmatrix}=(4\times 15+1\times 32+1\times 24+2\times 18)=\mathbf{(152)}$$

The notation indicates the multiplication of a (1×4) matrix by a (4×1) matrix, giving a (1×1) matrix product (in this case a scalar).

EXAMPLE 4 *Generalise Example* 3 *by considering three weeks, where the orders for weeks* 2 *and* 3 *are given by*

	Quantity in kg			
	Potatoes	Carrots	Onions	Peas
Week 2 (	3	2	1	3)
Week 3 (	5	0	2	4)

This time we find three total costs, one for each week. The calculation is displayed in matrix form

$$\begin{pmatrix} 4 & 1 & 1 & 2 \\ 3 & 2 & 1 & 3 \\ 5 & 0 & 2 & 4 \end{pmatrix} \times \begin{pmatrix} 15 \\ 32 \\ 24 \\ 18 \end{pmatrix} = \begin{pmatrix} 4\times 15+1\times 32+1\times 24+2\times 18 \\ 3\times 15+2\times 32+1\times 24+3\times 18 \\ 5\times 15+0\times 32+2\times 24+4\times 18 \end{pmatrix} = \begin{pmatrix} \mathbf{152} \\ \mathbf{187} \\ \mathbf{195} \end{pmatrix}$$

The final column vector gives the cost of the purchases in each of the three weeks.

EXAMPLE 5 *Generalise the previous examples further by including a second set of prices, offered at a different store and given by the list*

	Price in p per kg		
Potatoes	Carrots	Onions	Peas
(16	29	25	18)

The new set of prices means that we must add a second column to the price matrix and we show the results in a matrix product, in which the various rows and columns are labelled, (Pr1 and Pr2 denoting the prices at the two stores).

$$\begin{array}{c} \\ \text{Wk 1} \\ \text{Wk 2} \\ \text{Wk 3} \end{array} \begin{array}{c} \begin{array}{cccc} \text{Po} & \text{Ca} & \text{On} & \text{Pe} \end{array} \\ \begin{pmatrix} 4 & 1 & 1 & 2 \\ \boxed{3} & \boxed{2} & \boxed{1} & \boxed{3} \\ 5 & 0 & 2 & 4 \end{pmatrix} \end{array} \times \begin{array}{c} \text{Po} \\ \text{Ca} \\ \text{On} \\ \text{Pe} \end{array} \begin{array}{c} \begin{array}{cc} \text{Pr1} & \text{Pr2} \end{array} \\ \begin{pmatrix} \boxed{15} & 16 \\ \boxed{32} & 29 \\ \boxed{24} & 25 \\ \boxed{18} & 18 \end{pmatrix} \end{array} = \begin{array}{c} \\ \text{Wk 1} \\ \text{Wk 2} \\ \text{Wk 3} \end{array} \begin{array}{c} \begin{array}{cc} \text{Pr1} & \text{Pr2} \end{array} \\ \begin{pmatrix} 152 & 154 \\ \boxed{187} & 185 \\ 195 & 202 \end{pmatrix} \end{array}$$

Fig. 1.8

Each of the 3 rows of the first matrix has been multiplied by each of the 2 columns of the second matrix to produce the six products, giving the 6 total costs shown in the 3 rows and 2 columns of the final cost matrix. The position in the cost matrix of each total cost is found by placing it in the row corresponding to the row used in the first matrix and in the column corresponding to the column used in the second matrix. In Fig. 1.8, the second row of the first matrix and the first column of the second matrix have been shaded and so has their product which is the entry in row 2 and column 1 of the final matrix. The row and column headings identify the meaning of the final cost matrix and the row-column multiplication process keeps the structure readable. Notice that, in the final cost matrix, the vegetable headings have disappeared since the separate vegetable quantities and prices do not appear in the total cost of the orders. This can be expressed by the 'equation'

$$\begin{array}{ccccc} \text{(week by product)} & \times & \text{(product by cost)} & = & \text{(week by cost)} \\ (3\times 4) & \times & (4\times 2) & = & (3\times 2) \end{array}$$

The process of matrix multiplication is quite tedious and a systematic approach is helpful. In order to find the entry in the ith row and the jth column of the product matrix, use one finger of the left hand on the ith row of the first matrix and one finger of the right hand on the jth column of the second matrix. Then use two scanning motions: the left finger moving across the row and the right finger moving down the column, corresponding entries being multiplied together and the products added as the scan proceeds. Clearly, in order for this definition of a matrix product to work, it is necessary for the first matrix to have the same number of columns as the second matrix has rows and then the product matrix has the same number of rows as the first matrix and the same number of columns as the second matrix.

Note It would be possible to define a matrix product working down a column of the first matrix and along a row of the second but this is not done and the product which we have defined is universally accepted.

Definition Let $\mathbf{A} = (a_{ij})$, $\mathbf{B} = (b_{ij})$ and $\mathbf{C} = (c_{ij})$.
Then the matrix $\mathbf{C}$ is the product of the matrix $\mathbf{A}$, of order $(m \times n)$, with the matrix $\mathbf{B}$, of order $(n \times p)$, if

$$\begin{aligned} c_{ij} &= (a_{i1} \times b_{1j}) + (a_{i2} \times b_{2j}) + (a_{i3} \times b_{3j}) + \ldots + (a_{in} \times b_{nj}) \\ &= \sum_{k=1}^{n} a_{ik}b_{kj}, \text{ for } 1 \leqslant i \leqslant m \text{ and } 1 \leqslant j \leqslant p. \end{aligned}$$

Since i ranges from 1 to m and j ranges from 1 to p, the matrix $\mathbf{C}$ has m rows and p columns so it has order $(m \times p)$.

The double suffix notation gives a very neat form for the entry c_{ij}, in the product matrix $\mathbf{C}$, which arises from the multiplication of

$$\text{the row } (a_{i1}\ a_{i2}\ a_{i3}\ a_{in}) \text{ with the column } \begin{pmatrix} b_{1j} \\ b_{2j} \\ b_{3j} \\ \cdot \\ \cdot \\ b_{nj} \end{pmatrix}$$

> The multiplication of the matrices **A** and **B** can take place only if the number (n) of columns of **A** equals the number of rows of **B** or if the length of a row of **A** equals the length of a column of **B**

EXAMPLE 6 *The matrices* **W**, **X**, **Y** *and* **Z** *are given by*

$$\mathbf{W} = \begin{pmatrix} 2 & 1 \\ 0 & 1 \end{pmatrix}, \quad \mathbf{X} = \begin{pmatrix} 3 \\ 2 \end{pmatrix}, \quad \mathbf{Y} = (1 \quad 3 \quad -2), \quad \mathbf{Z} = \begin{pmatrix} 4 & 6 \\ 2 & -1 \\ 0 & -2 \end{pmatrix}$$

Calculate all possible products of **W**, **X**, **Y** *and* **Z**.

This problem is best solved by considering the orders of the four matrices, namely $\mathbf{W}\,(2 \times 2)$, $\mathbf{X}\,(2 \times 1)$, $\mathbf{Y}\,(1 \times 3)$ and $\mathbf{Z}\,(3 \times 2)$. Using the last result, we list the possible products, showing the matching of row and column lengths.

	Possible	Since	
WX	$(2\times \underbrace{2)\times(2}_{\text{match}}\times 1) = (2\times 1)$		$\begin{pmatrix}2&1\\0&1\end{pmatrix}\times\begin{pmatrix}3\\2\end{pmatrix}=\begin{pmatrix}\mathbf{8}\\\mathbf{2}\end{pmatrix}$
XY	$(2\times \underbrace{1)\times(1}_{\text{match}}\times 3) = (2\times 3)$		$\begin{pmatrix}3\\2\end{pmatrix}\times(1\quad 3\quad -2)=\begin{pmatrix}\mathbf{3}&\mathbf{9}&\mathbf{-6}\\\mathbf{2}&\mathbf{6}&\mathbf{-4}\end{pmatrix}$
YZ	$(1\times \underbrace{3)\times(3}_{\text{match}}\times 2) = (1\times 2)$		$(1\quad 3\quad -2)\times\begin{pmatrix}4&6\\2&-1\\0&-2\end{pmatrix}=(\mathbf{10}\quad\mathbf{7})$
ZW	$(3\times \underbrace{2)\times(2}_{\text{match}}\times 2) = (3\times 2)$		$\begin{pmatrix}4&6\\2&-1\\0&-2\end{pmatrix}\times\begin{pmatrix}2&1\\0&1\end{pmatrix}=\begin{pmatrix}\mathbf{8}&\mathbf{10}\\\mathbf{4}&\mathbf{1}\\\mathbf{0}&\mathbf{-2}\end{pmatrix}$
ZX	$(3\times \underbrace{2)\times(2}_{\text{match}}\times 1) = (3\times 1)$		$\begin{pmatrix}4&6\\2&-1\\0&-2\end{pmatrix}\times\begin{pmatrix}3\\2\end{pmatrix}=\begin{pmatrix}\mathbf{24}\\\mathbf{4}\\\mathbf{-4}\end{pmatrix}$

The condition on the orders of two matrices in order that they may be multiplied together immediately implies that the order of the two matrices in the product is important. It is also important because, in the product, the *rows* of the first matrix are multiplied by the *columns* of the second. A (2×3) matrix **K** can be multiplied by a (3×2) matrix **L** in two different ways. The matrix **KL** will be a (2×2) matrix but **LK** will be a (3×3), so **KL** and **LK** will have a different number of entries and so are certainly not the same. Two square matrices of the same order n can be multiplied in two ways to give matrices of order n and these two products may not be the same.

Square matrices of a given order n may be combined by means of multiplication, addition, subtraction and linear combinations to form square matrices, also of order n, so they form a special set of matrices with closure properties. We consider these sets of matrices in section 1.4.

The *transpose* of a matrix is formed by interchanging the rows and the columns of the matrix, that is, by reflecting the matrix in its leading diagonal. Since the order of the transpose of a matrix of order $(m\times n)$ is $(n\times m)$, any matrix can be multiplied by its transpose in two ways.

Definition Let **A** be an $(m\times n)$ matrix with $\mathbf{A}=(a_{ij})$, then the *transpose* matrix of **A** is the matrix $\mathbf{A}^{\mathrm{T}}$, with $\mathbf{A}^{\mathrm{T}}=(a_{ij}^{\mathrm{T}})$, defined by

$$a_{ij}^{\mathrm{T}}=a_{ji},\ \text{for } 1\leqslant i\leqslant n \text{ and } 1\leqslant j\leqslant m,$$

and so $\mathbf{A}^{\mathrm{T}}$ is an $(n\times m)$ matrix.

EXAMPLE 7 *Let* $\mathbf{A}=\begin{pmatrix}2&1\\3&0\\0&2\end{pmatrix}$. *Find* $\mathbf{A}^{\mathrm{T}}$, $\mathbf{A}\mathbf{A}^{\mathrm{T}}$ *and* $\mathbf{A}^{\mathrm{T}}\mathbf{A}$ *and show that* $\mathbf{A}\mathbf{A}^{\mathrm{T}}$ *and* $\mathbf{A}^{\mathrm{T}}\mathbf{A}$ *are square matrices.*

We begin by proving the last two results. Since the order of **A** is (3 × 2), the order of $\mathbf{A}^T$ will be (2 × 3) and so the order of $\mathbf{AA}^T$ will be (3 × 3) and the order of $\mathbf{A}^T\mathbf{A}$ will be (2 × 2), and both are square. Now

$$\mathbf{A}^T = \begin{pmatrix} 2 & 3 & 0 \\ 1 & 0 & 2 \end{pmatrix}, \quad \mathbf{AA}^T = \begin{pmatrix} 2 & 1 \\ 3 & 0 \\ 0 & 2 \end{pmatrix}\begin{pmatrix} 2 & 3 & 0 \\ 1 & 0 & 2 \end{pmatrix} = \begin{pmatrix} \mathbf{5} & \mathbf{6} & \mathbf{2} \\ \mathbf{6} & \mathbf{9} & \mathbf{0} \\ \mathbf{2} & \mathbf{0} & \mathbf{4} \end{pmatrix}$$

and $$\mathbf{A}^T\mathbf{A} = \begin{pmatrix} 2 & 3 & 0 \\ 1 & 0 & 2 \end{pmatrix}\begin{pmatrix} 2 & 1 \\ 3 & 0 \\ 0 & 2 \end{pmatrix} = \begin{pmatrix} \mathbf{13} & \mathbf{2} \\ \mathbf{2} & \mathbf{5} \end{pmatrix}$$

Notice that $\mathbf{AA}^T$ and $\mathbf{A}^T\mathbf{A}$ are also symmetric about their leading diagonals. Such a matrix is called *symmetric.*

Definition The square matrix **A**, where $\mathbf{A} = (a_{ij})$, of order n, is *symmetric* if $a_{ij} = a_{ji}$, for $1 \leqslant i \leqslant n$ and $1 \leqslant j \leqslant n$. That is, **A** is symmetric if $\mathbf{A}^T = \mathbf{A}$.

Theorem 1.1 Let **A** be an $(m \times n)$ matrix with transpose $\mathbf{A}^T$, then:
(i) $\mathbf{A}^T$ is an $(n \times m)$ matrix,
(ii) $\mathbf{AA}^T$ is square of order m and $\mathbf{A}^T\mathbf{A}$ is square of order n,
(iii) $\mathbf{AA}^T$ and $\mathbf{A}^T\mathbf{A}$ are both symmetric.

We shall prove (iii) and leave the proofs of (i) and (ii) as exercises for the reader. Let $\mathbf{A} = (a_{ij})$, $\mathbf{X} = \mathbf{AA}^T = (x_{ij})$, $\mathbf{Y} = \mathbf{A}^T\mathbf{A} = (y_{ij})$. Then $\mathbf{A}^T = (a_{ij}^T)$, with $a_{ij}^T = a_{ji}$, so that

$$x_{ij} = \sum_{k=1}^{n} a_{ik}a_{kj}^T, \qquad y_{ij} = \sum_{l=1}^{m} a_{il}^T a_{lj},$$

and $$x_{ij} = \sum_{k=1}^{n} a_{ik}a_{jk}, \qquad y_{ij} = \sum_{l=1}^{m} a_{li}a_{lj}.$$

Thus $x_{ji} = \sum_{k=1}^{n} a_{jk}a_{ik} = \sum_{k=1}^{n} a_{ik}a_{jk} = x_{ij}$ and similarly $y_{ij} = y_{ji}$ and so **X** and **Y** are symmetric.

Notice that $x_{ii} = \sum_{k=1}^{n} (a_{ik})^2$ and $y_{ii} = \sum_{l=1}^{m} (a_{li})^2$, and these are the sums of the squares of the entries of the ith row and of the ith column of the matrix **A**. This property will be looked at again in Chapter 14.

EXERCISE 1.2

1 Given that $\mathbf{R} = \begin{pmatrix} 2 & -3 & 2 \\ 4 & 0 & 6 \end{pmatrix}$, $\mathbf{S} = \begin{pmatrix} 4 & -2 & -1 \\ 0 & -3 & -5 \end{pmatrix}$, $\mathbf{T} = \begin{pmatrix} -2 & -1 & 3 \\ 4 & 0 & 0 \end{pmatrix}$, calculate: (i) $\mathbf{R}+\mathbf{S}$, (ii) $2\mathbf{R}+\mathbf{T}$, (iii) $3\mathbf{S}-2\mathbf{T}$, (iv) $\mathbf{T}-\mathbf{R}+\mathbf{S}$, (v) $2\mathbf{S}-3\mathbf{T}+4\mathbf{R}$.

2 Given that $\mathbf{O} = \begin{pmatrix} 0 & 0 \\ 0 & 0 \end{pmatrix}$, $\mathbf{P} = \begin{pmatrix} 2 & -3 \\ 4 & 0 \end{pmatrix}$, $\mathbf{Q} = \begin{pmatrix} -2 & 3 \\ -4 & 0 \end{pmatrix}$, show that:

(i) $\mathbf{P}+\mathbf{Q} = \mathbf{O}$, (ii) $\mathbf{P}+\mathbf{O} = \mathbf{P} = \mathbf{O}+\mathbf{P}$, (iii) $-\mathbf{P} = \mathbf{Q}$,

(iv) $\mathbf{OP} = \mathbf{O} = \mathbf{PO}$.

3 Let $\mathbf{W} = \begin{pmatrix} 1 & 0 \\ 0 & 0 \end{pmatrix}$, $\mathbf{X} = \begin{pmatrix} 0 & 1 \\ 0 & 0 \end{pmatrix}$, $\mathbf{Y} = \begin{pmatrix} 0 & 0 \\ 1 & 0 \end{pmatrix}$ and $\mathbf{Z} = \begin{pmatrix} 0 & 0 \\ 0 & 1 \end{pmatrix}$.

(a) Calculate: (i) $2\mathbf{W}+3\mathbf{X}-4\mathbf{Y}-\mathbf{Z}$, (ii) $3\mathbf{X}+2\mathbf{Z}$, (iii) $a\mathbf{W}-b\mathbf{Z}$.

(b) Express the following matrices as linear combinations of **W**, **X**, **Y** and **Z**:

(i) $\begin{pmatrix} 0 & 2 \\ 5 & -1 \end{pmatrix}$, (ii) $\begin{pmatrix} 3 & 0 \\ 5 & 2 \end{pmatrix}$, (iii) $\begin{pmatrix} 4 & 4 \\ 4 & 4 \end{pmatrix}$, (iv) $\begin{pmatrix} a & b \\ c & d \end{pmatrix}$.

4 Calculate the following matrix products:

(i) $\begin{pmatrix} 2 & 3 \\ 4 & -2 \end{pmatrix}\begin{pmatrix} 4 \\ 6 \end{pmatrix}$, (ii) $\begin{pmatrix} 3 \\ 4 \end{pmatrix}(4 \quad -2 \quad -1)$,

(iii) $\begin{pmatrix} 5 & 6 \\ -2 & -1 \\ 0 & 3 \end{pmatrix}\begin{pmatrix} 4 & 3 & -2 \\ 2 & 1 & -1 \end{pmatrix}$, (iv) $\begin{pmatrix} 5 & 2 \\ -1 & -2 \\ 0 & -3 \\ 5 & -1 \end{pmatrix}\begin{pmatrix} 1 & 2 & -1 \\ 5 & -2 & 0 \end{pmatrix}$,

(v) $\begin{pmatrix} 2 & 7 & 3 \\ -4 & 6 & -8 \end{pmatrix}\begin{pmatrix} 4 & 5 \\ 2 & -1 \\ -3 & 0 \end{pmatrix}$, (vi) $\begin{pmatrix} 3 & 0 & 1 \\ 0 & 2 & 1 \\ -1 & 0 & 0 \end{pmatrix}\begin{pmatrix} 4 & 0 & 2 \\ 2 & 0 & 3 \\ 0 & 1 & -1 \end{pmatrix}$

5 Let $\mathbf{A} = \begin{pmatrix} 2 & 0 \\ 0 & 3 \end{pmatrix}$, $\mathbf{B} = \begin{pmatrix} 4 & 0 & 2 \\ 1 & 3 & 6 \end{pmatrix}$, $\mathbf{C} = \begin{pmatrix} 2 & -1 & 1 \\ 0 & 3 & 0 \\ 4 & 2 & 0 \end{pmatrix}$, $\mathbf{D} = \begin{pmatrix} 3 & 4 \\ 2 & 1 \\ -1 & 0 \end{pmatrix}$.

Calculate, *where possible*, (i) **AB**, (ii) **BA**, (iii) **BC**, (iv) **BD**, (v) **DB**, (vi) **DC**, (vii) **CD**, (viii) **DA**, (ix) (**AB**)**C**, (x) **A**(**BC**).

6 Let $\mathbf{X} = (x_{ij})$, an $(m \times n)$ matrix, and let $\mathbf{Y} = (y_{ij})$, a $(p \times q)$ matrix. Write down necessary conditions in order that:

(i) $2\mathbf{X}+3\mathbf{Y}$ may be calculated,

(ii) **YX** may be calculated,

(iii) **XY** may be calculated and have the same order as **X**,

(iv) $\mathbf{X}^{\mathrm{T}}$ has the same order as **Y**,

(v) **XY** and **YX** may be calculated and have the same order,

(vi) $\mathbf{X}^{\mathrm{T}}\mathbf{Y}$ and $\mathbf{Y}^{\mathrm{T}}\mathbf{X}$ may be calculated and $\mathbf{X}^{\mathrm{T}}\mathbf{Y}$ has the same order as $(\mathbf{Y}^{\mathrm{T}}\mathbf{X})^{\mathrm{T}}$.

7 Let $\mathbf{J} = \begin{pmatrix} 3 & 1 & 0 \\ 2 & -1 & 4 \end{pmatrix}$ and $\mathbf{K} = \begin{pmatrix} 4 & 0 \\ -2 & 3 \\ 1 & -1 \end{pmatrix}$. Find $\mathbf{J}^{\mathrm{T}}$, $\mathbf{K}^{\mathrm{T}}$, $\mathbf{J}^{\mathrm{T}}\mathbf{K}^{\mathrm{T}}$, **JK**, $\mathbf{K}^{\mathrm{T}}\mathbf{J}^{\mathrm{T}}$, **KJ**. Show that $\mathbf{J}^{\mathrm{T}}\mathbf{K}^{\mathrm{T}} = (\mathbf{KJ})^{\mathrm{T}}$ and state a matrix which is the same as $(\mathbf{JK})^{\mathrm{T}}$.

8 Fig. 1.9(a) shows two islands I_1 and I_2, 3 mainland ports P_1, P_2 and P_3 and 2 towns T_1 and T_2, connected by ferry routes and railways.

(a) Calculate the directed route matrices: (i) from I_1, I_2 to P_1, P_2, P_3, (ii) from P_1, P_2, P_3 to T_1, T_2. Multiply these two matrices together and explain the meaning of the entries of the product matrix.

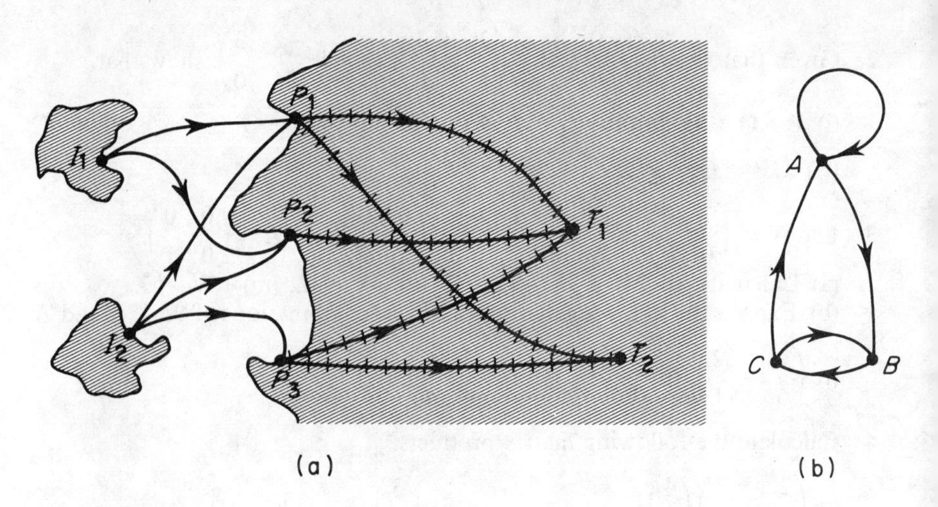

Fig. 1.9

(b) Use your results to write down the number of routes there are from: (i) I_1 to T_1, (ii) I_2 to T_1, (iii) I_2 to T_2.

9 Calculate **M**, the (3×3) directed route matrix for the network shown in Fig. 1.9(b). Multiply **M** by itself to produce $\mathbf{M}^2$. The entries in $\mathbf{M}^2$ represent the 2-stage routes in the network. For example, there is just one 2-stage route from A to C, namely $A \to B \to C$. In a similar way, describe the 2-stage routes from: (i) A to A, (ii) A to B, (iii) C to C. Now calculate $\mathbf{M}^3$. Describe the 3-stage routes from: (i) A to A, (ii) C to B, (iii) C to A, and relate the number of such routes to certain entries in $\mathbf{M}^3$. How many 4-stage routes are there from A to B? Describe the meaning of the first (top left) entry in the matrix $\mathbf{M}^n$.

10 Let $\mathbf{T} = \begin{pmatrix} 3 & 2 \\ 2 & -1 \end{pmatrix}$, $\mathbf{P} = \begin{pmatrix} 1 \\ -2 \end{pmatrix}$ and $\mathbf{X} = \begin{pmatrix} x \\ y \end{pmatrix}$.

Prove that $\mathbf{TP} = \begin{pmatrix} -1 \\ 4 \end{pmatrix}$ and $\mathbf{TX} = \begin{pmatrix} 3x+2y \\ 2x-y \end{pmatrix}$.

Solve the simultaneous equations $\begin{aligned} 3x+2y &= -1 \\ 2x-y &= 4 \end{aligned}$.

Explain the connection of the solution of these equations with **P**. Represent the following system of linear equations in matrix form:

$$\begin{aligned} 4x - y + z &= 2, \\ 3x + 2y - z &= 1, \\ x - y + 2z &= 0. \end{aligned}$$

11 The matrix $\mathbf{M} = \begin{pmatrix} a & b \\ c & d \end{pmatrix}$, and $\mathbf{I} = \begin{pmatrix} 1 & 0 \\ 0 & 1 \end{pmatrix}$. Prove that, if $a+d \neq 0$, then $\mathbf{M} = \dfrac{1}{a+d}[\mathbf{M}^2 + (ad-bc)\mathbf{I}]$.

Prove also that, if $\mathbf{M}^2 = \begin{pmatrix} p & q \\ r & s \end{pmatrix}$, then $p+s = (a+d)^2 - 2(ad-bc)$ and $ps - qr = (ad-bc)^2$. Hence, or otherwise, find four distinct matrices $\mathbf{M}$ such that $\mathbf{M}^2 = \begin{pmatrix} 2 & 1 \\ 2 & 3 \end{pmatrix}$. (C)

1.3 Vectors

A non-empty set of vectors which is closed under the operations of taking linear combinations is called a *vector space*, see Chapter **14**. The set of all position vectors in two dimensions forms a vector space. This space is called $\mathbb{R}^2$ and we make a general definition in n dimensions.

Definition $\mathbb{R}^n$ is the space of all $n \times 1$ column vectors with real entries.

Linear dependence

In *Pure Mathematics* Chapter 12, the basis for the set of all vectors in $\mathbb{R}^3$ was considered and such a basis consists of three non-zero, non-coplanar vectors. A corresponding basis in $\mathbb{R}^2$ consists of two non-zero, non-parallel vectors. A basis is a set of independent vectors in the sense that no one of them can be written as a linear combination of the rest. This is formalised in the definition of linear dependence and independence.

Definition In the space $\mathbb{R}^n$: A set of vectors is *linearly independent* if the only linear combination of the vectors in the set that is equal to zero is that with all zero multipliers. A set of vectors is *linearly dependent* if some linear combination of the vectors in the set, with not all the multipliers zero, is equal to the zero vector.

$$\text{Let } S = \{\mathbf{a}_1, \mathbf{a}_2, \ldots, \mathbf{a}_k\}, \text{ then}$$

$$S \text{ is linearly independent} \Leftrightarrow \sum_{r=1}^{k} x_r \mathbf{a}_r = \mathbf{0} \Rightarrow x_r = 0,\ 1 \leqslant r \leqslant k,$$

$$S \text{ is linearly dependent} \Leftrightarrow \sum_{r=1}^{k} x_r \mathbf{a}_r = \mathbf{0} \quad \text{for some } x_r, \text{ not all zero.}$$

EXAMPLE 1 *Determine whether the set S, of vectors in $\mathbb{R}^2$ or $\mathbb{R}^3$, is linearly independent or linearly dependent:*

(*i*) $S = \left\{\begin{pmatrix} 0 \\ 1 \end{pmatrix}, \begin{pmatrix} 1 \\ 0 \end{pmatrix}\right\}$, (*ii*) $S = \left\{\begin{pmatrix} 0 \\ 1 \end{pmatrix}, \begin{pmatrix} 1 \\ 1 \end{pmatrix}\right\}$,

(*iii*) $S = \left\{\begin{pmatrix} 0 \\ 0 \end{pmatrix}, \begin{pmatrix} 0 \\ 1 \end{pmatrix}\right\}$, (*iv*) $S = \left\{\begin{pmatrix} 1 \\ 1 \end{pmatrix}, \begin{pmatrix} 1 \\ -1 \end{pmatrix}\right\}$,

(*v*) $S = \left\{\begin{pmatrix} 2 \\ 3 \end{pmatrix}, \begin{pmatrix} 4 \\ 6 \end{pmatrix}\right\}$, (*vi*) $S = \{\mathbf{i}, \mathbf{j}, \mathbf{k}\}$, (*vii*) $S = \{\mathbf{i}, \mathbf{j}, \mathbf{j}\}$.

The method is to investigate whether or not a set of 'not all zero' multipliers can be found to give a linear combination equal to $\mathbf{0}$.

(i) $x\begin{pmatrix}0\\1\end{pmatrix}+y\begin{pmatrix}1\\0\end{pmatrix}=\begin{pmatrix}y\\x\end{pmatrix}=\begin{pmatrix}0\\0\end{pmatrix}$ only if $x=0=y$, so S is *linearly independent.*

(ii) $x\begin{pmatrix}0\\1\end{pmatrix}+y\begin{pmatrix}1\\1\end{pmatrix}=\begin{pmatrix}y\\x+y\end{pmatrix}=\begin{pmatrix}0\\0\end{pmatrix}$ only if $x=0=y$, so S is *linearly independent.*

(iii) $2\begin{pmatrix}0\\0\end{pmatrix}+0\begin{pmatrix}0\\1\end{pmatrix}=\begin{pmatrix}0\\0\end{pmatrix}$ so S is *linearly dependent.*

(iv) $x\begin{pmatrix}1\\1\end{pmatrix}+y\begin{pmatrix}1\\-1\end{pmatrix}=\begin{pmatrix}x+y\\x-y\end{pmatrix}=\begin{pmatrix}0\\0\end{pmatrix}$ only if $x+y=0=x-y$, that is, only if $x=0=y$, so S is *linearly independent.*

(v) $2\begin{pmatrix}2\\3\end{pmatrix}-\begin{pmatrix}4\\6\end{pmatrix}=\begin{pmatrix}0\\0\end{pmatrix}$ so S is *linearly dependent.*

(vi) $x\mathbf{i}+y\mathbf{j}+z\mathbf{k}=\mathbf{0}$ only if $x=y=z=0$ so S is *linearly independent.*
(vii) $0\mathbf{i}+\mathbf{j}-\mathbf{j}=\mathbf{0}$, so S is *linearly dependent.*

Spanning sets

The space $\mathbb{R}^2$ consists of all linear combinations of the vectors in the set $\left\{\begin{pmatrix}1\\0\end{pmatrix},\begin{pmatrix}0\\1\end{pmatrix}\right\}$ and it also consists of all linear combinations of the vectors in the set $\left\{\begin{pmatrix}1\\0\end{pmatrix},\begin{pmatrix}1\\1\end{pmatrix},\begin{pmatrix}0\\1\end{pmatrix}\right\}$, since the extra vector in this set does not lead to any more vectors by forming linear combinations. Such a set of vectors is said to *span* $\mathbb{R}^2$.

Definition Let V be a vector space and let S be a finite set of vectors in V. Then, if V is the set of all linear combinations of the vectors in S,

$$S \text{ is a } \textit{spanning set} \text{ of } V \qquad \text{or } S \textit{ spans } V$$
$$\text{or } V \text{ is the } \textit{span} \text{ of } S \qquad \text{or } V = \textit{span}\ (S).$$

Since the set of all linear combinations of a given set of vectors is a vector space, for any non-empty set S of vectors, span(S) is a vector space. Note that we restrict the term 'spanning set' to a finite set of vectors which spans the space, even though there will be many infinite sets which span the space.

EXAMPLE 2 *Find the span of the set S: (i)* $S=\left\{\begin{pmatrix}1\\2\end{pmatrix}\right\}$,

(ii) $S=\left\{\begin{pmatrix}1\\2\end{pmatrix},\begin{pmatrix}2\\1\end{pmatrix}\right\}$, *(iii)* $S=\left\{\begin{pmatrix}1\\-1\\0\end{pmatrix},\begin{pmatrix}0\\-1\\1\end{pmatrix}\right\}$, *(iv)* $S=\{\mathbf{i},\mathbf{j}-\mathbf{k}\}$.

(i) $\text{span}(S) = \left\{\mathbf{a}\colon \mathbf{a} = s\begin{pmatrix}1\\2\end{pmatrix}, \text{ for all } s \in \mathbb{R}\right\}$. This is the set of all vectors originating from the origin and ending at points $(s, 2s)$, that is, the **line** $y = 2x$.

(ii) $\text{span}(S) = \left\{\mathbf{a}\colon \mathbf{a} = s\begin{pmatrix}1\\2\end{pmatrix} + t\begin{pmatrix}2\\1\end{pmatrix}, \text{ for all } s, t \text{ in } \mathbb{R}\right\} = \mathbf{R}^2$, since

$$\mathbf{a} = \begin{pmatrix}x\\y\end{pmatrix} \text{ if } s = (2y - x)/3 \text{ and } t = (2x - y)/3.$$

(iii) $\text{span}(S) = \left\{\mathbf{a}\colon \mathbf{a} = s\begin{pmatrix}1\\-1\\0\end{pmatrix} + t\begin{pmatrix}0\\-1\\1\end{pmatrix} \text{for all } s, t \text{ in } \mathbb{R}\right\}$. This is the set of all vectors from the origin ending at points $(s, -s-t, t)$, that is, the **plane** $x + z = y$.

(iv) $\text{span}(S) = \{\mathbf{a}\colon \mathbf{a} = s\mathbf{i} + t(\mathbf{j} - \mathbf{k}) \text{ for all } s, t \text{ in } \mathbb{R}\}$. That is the **plane** $y = z$.

When $S = \{\mathbf{0}\}$, $\text{span}(S) = \{\mathbf{0}\}$, the zero space consisting of just the zero vector. In all other cases, $\text{span}(S)$ contains infinitely many vectors because, if $\mathbf{a} \in \text{span}(S)$ then, for all $s \in \mathbb{R}$, $s\mathbf{a} \in \text{span}(S)$.

The question may be asked whether the same span can be obtained if some of the vectors in S are omitted and, if so, is there a least number of vectors required in a spanning set.

EXAMPLE 3 *The vector space V is spanned by the set S of vectors, given by* $S = \left\{\begin{pmatrix}1\\0\\-1\end{pmatrix}, \begin{pmatrix}1\\-1\\0\end{pmatrix}, \begin{pmatrix}0\\1\\-1\end{pmatrix}, \begin{pmatrix}-3\\2\\1\end{pmatrix}\right\}$. *Prove that V can be spanned by a set of two vectors and find two sets of vectors which each span V and which have empty intersection.*

Let $\mathbf{a} = \begin{pmatrix}1\\0\\-1\end{pmatrix}$, $\mathbf{b} = \begin{pmatrix}1\\-1\\0\end{pmatrix}$, $\mathbf{c} = \begin{pmatrix}0\\1\\-1\end{pmatrix}$, $\mathbf{d} = \begin{pmatrix}-3\\2\\1\end{pmatrix}$, then $\mathbf{c} = \mathbf{a} - \mathbf{b}$ and $\mathbf{d} = -\mathbf{a} - 2\mathbf{b}$. Hence $p\mathbf{a} + q\mathbf{b} + r\mathbf{c} + t\mathbf{d} = (p + r - t)\mathbf{a} + (q - r - 2t)\mathbf{b}$ and so V is spanned by $\{\mathbf{a}, \mathbf{b}\}$. Likewise, $\mathbf{a} = (2\mathbf{c} - \mathbf{d})/3$, $\mathbf{b} = (-\mathbf{c} - \mathbf{d})/3$, so that $p\mathbf{a} + q\mathbf{b} + r\mathbf{c} + t\mathbf{d} = (2p - q + 3r)\mathbf{c}/3 + (-p - q + 3t)\mathbf{d}/3$ and so V is also spanned by $\{\mathbf{c}, \mathbf{d}\}$ and the two spanning sets obtained have empty intersection.

In Example **3**, the two spanning sets obtained were as small as possible in the sense that if a vector is removed from the set the remaining set is no longer a spanning set. This is because no set consisting of just one of the vectors **a**, **b**, **c** or **d** can span V. Such a smallest possible spanning set can be shown to be linearly independent see §14.2, and is defined to be a basis of the space. The sets of basis vectors of $\mathbb{R}^2$ and $\mathbb{R}^3$, see *Pure Mathematics* §5.5 and 12.3, are linearly independent spanning sets and this property is used to define a basis of a general vector space.

Definition A *basis* of a vector space V is a linearly independent spanning set of V.

We shall show in Chapter 14 that any basis of $\mathbb{R}^2$ contains 2 vectors and any basis of $\mathbb{R}^3$ contains 3 vectors. This is described by saying that the spaces have dimensions 2 and 3 respectively.

Definition The *dimension* of a vector space (with a finite spanning set) is the number of vectors in any basis.

EXERCISE 1.3

1 Determine whether the set S is linearly independent or linearly dependent.

(i) $S=\left\{\begin{pmatrix}2\\0\end{pmatrix},\begin{pmatrix}0\\3\end{pmatrix}\right\}$, (ii) $S=\left\{\begin{pmatrix}2\\1\end{pmatrix},\begin{pmatrix}3\\4\end{pmatrix}\right\}$,

(iii) $S=\left\{\begin{pmatrix}1\\2\end{pmatrix},\begin{pmatrix}2\\1\end{pmatrix},\begin{pmatrix}3\\4\end{pmatrix}\right\}$, (iv) $S=\left\{\begin{pmatrix}0\\0\end{pmatrix},\begin{pmatrix}1\\5\end{pmatrix}\right\}$.

2 Determine whether the set S is linearly independent or linearly dependent, given that $\mathbf{a}$ and $\mathbf{b}$ are non-zero, non-parallel vectors.
(i) $S=\{\mathbf{a},\mathbf{b}\}$, (ii) $S=\{\mathbf{a},3\mathbf{a}\}$, (iii) $S=\{\mathbf{a},\mathbf{a}+\mathbf{b}\}$,
(iv) $S=\{\mathbf{a}-\mathbf{b},\mathbf{b}-\mathbf{a}\}$, (v) $S=\{\mathbf{a}-\mathbf{b},\mathbf{b}+\mathbf{a}\}$, (vi) $S=\{2\mathbf{a}-\mathbf{b},\mathbf{a}+3\mathbf{b}\}$.

3 Prove that any set of vectors containing the zero vector is linearly dependent.

4 Prove that any set of vectors containing a repeated vector is linearly dependent.

5 Determine whether the set S is linearly independent or linearly dependent.

(i) $S=\left\{\begin{pmatrix}0\\1\\0\end{pmatrix},\begin{pmatrix}1\\0\\1\end{pmatrix}\right\}$, (ii) $S=\left\{\begin{pmatrix}0\\1\\0\end{pmatrix},\begin{pmatrix}1\\0\\1\end{pmatrix},\begin{pmatrix}2\\0\\0\end{pmatrix}\right\}$,

(iii) $S=\left\{\begin{pmatrix}1\\1\\0\end{pmatrix},\begin{pmatrix}1\\0\\-1\end{pmatrix},\begin{pmatrix}0\\3\\3\end{pmatrix}\right\}$, (iv) $S=\left\{\begin{pmatrix}2\\0\\1\end{pmatrix},\begin{pmatrix}0\\1\\2\end{pmatrix},\begin{pmatrix}-2\\-1\\-3\end{pmatrix}\right\}$,

(v) $S=\left\{\begin{pmatrix}2\\1\\3\end{pmatrix},\begin{pmatrix}1\\0\\2\end{pmatrix},\begin{pmatrix}4\\2\\6\end{pmatrix}\right\}$, (vi) $S=\left\{\begin{pmatrix}3\\1\\5\end{pmatrix},\begin{pmatrix}0\\0\\0\end{pmatrix},\begin{pmatrix}4\\0\\3\end{pmatrix}\right\}$,

(vii) $S=\left\{\begin{pmatrix}2\\1\\4\end{pmatrix},\begin{pmatrix}1\\0\\2\end{pmatrix},\begin{pmatrix}3\\1\\1\end{pmatrix},\begin{pmatrix}5\\3\\4\end{pmatrix}\right\}$, (viii) $S=\{\mathbf{i},\mathbf{i}+\mathbf{j},\mathbf{i}+\mathbf{j}+\mathbf{k}\}$,

(ix) $S=\{\mathbf{i}-\mathbf{j},\mathbf{j}-\mathbf{k},\mathbf{k}-\mathbf{i}\}$, (x) $S=\{\mathbf{i}+\mathbf{j},\mathbf{j}+\mathbf{k},\mathbf{k}+\mathbf{i}\}$.

6 For each of the cases in Question **5**, find the dimension of the space spanned by S.

7 Given that $k\neq 0$, prove that: (i) if $\{\mathbf{a},\mathbf{b},\mathbf{c}\}$ is linearly dependent then $\{k\mathbf{a},k\mathbf{b},k\mathbf{c}\}$ is linearly dependent, (ii) if $\{\mathbf{a},\mathbf{b},\mathbf{c}\}$ is linearly independent then $\{k\mathbf{a},k\mathbf{b},k\mathbf{c}\}$ is linearly independent, (iii) if $\{\mathbf{a},\mathbf{b},\mathbf{c}\}$ is linearly independent then so is $\{\mathbf{a},\mathbf{b},k\mathbf{c}\}$.

8 Given that $\{\mathbf{a}, \mathbf{b}, \mathbf{c}\}$ is a linearly independent set of vectors in $\mathbb{R}^3$, determine whether or not the following set of vectors is linearly independent:
(i) $\{\mathbf{a}, \mathbf{c}\}$, (ii) $\{\mathbf{a}, \mathbf{0}, \mathbf{c}\}$, (iii) $\{3\mathbf{a}\}$, (iv) $\{\mathbf{a}, \mathbf{a}-\mathbf{b}, \mathbf{a}+\mathbf{b}+\mathbf{c}\}$,
(v) $\{\mathbf{a}+\mathbf{b}-\mathbf{c}, \mathbf{b}+\mathbf{c}-\mathbf{a}, \mathbf{c}+\mathbf{a}-\mathbf{b}\}$,
(vi) $\{\mathbf{a}+2\mathbf{b}+3\mathbf{c}, 2\mathbf{a}-3\mathbf{b}+\mathbf{c}, 3\mathbf{a}+\mathbf{b}-2\mathbf{c}\}$.

9 Given that $\mathbf{a} = \begin{pmatrix} 1 \\ 0 \\ 1 \end{pmatrix}$, $\mathbf{b} = \begin{pmatrix} 0 \\ 1 \\ -1 \end{pmatrix}$, $\mathbf{c} = \begin{pmatrix} 1 \\ 1 \\ 1 \end{pmatrix}$, prove that $\{\mathbf{a}, \mathbf{b}, \mathbf{c}\}$ is linearly independent. Express, as a linear combination of $\mathbf{a}$, $\mathbf{b}$ and $\mathbf{c}$, the vector
(i) $\begin{pmatrix} 1 \\ 2 \\ -2 \end{pmatrix}$, (ii) $\begin{pmatrix} x \\ y \\ z \end{pmatrix}$.

10 Given that $\mathbf{a} = \begin{pmatrix} 2 \\ 0 \\ 1 \end{pmatrix}$, $\mathbf{b} = \begin{pmatrix} 1 \\ -1 \\ 2 \end{pmatrix}$, $\mathbf{c} = \begin{pmatrix} 1 \\ 1 \\ -1 \end{pmatrix}$, $\mathbf{d} = \begin{pmatrix} 1 \\ 1 \\ 1 \end{pmatrix}$, $\mathbf{e} = \begin{pmatrix} 1 \\ 1 \\ 3 \end{pmatrix}$, prove that (i) $\{\mathbf{a}, \mathbf{b}, \mathbf{c}\}$ is not a basis of $\mathbb{R}^3$, (ii) $\{\mathbf{a}, \mathbf{b}, \mathbf{d}\}$ is a basis of $\mathbb{R}^3$. Express $\mathbf{e}$ as a linear combination of the vectors $\mathbf{a}$, $\mathbf{b}$ and $\mathbf{d}$, and also as a linear combination of the vectors $\mathbf{a}$, $\mathbf{c}$ and $\mathbf{d}$.

11 In $\mathbb{R}^3$, find a set of vectors which:
(i) spans $\mathbb{R}^3$ but is not a basis of $\mathbb{R}^3$,
(ii) is linearly independent but is not a basis of $\mathbb{R}^3$,
(iii) is a basis of $\mathbb{R}^3$,
(iv) spans a proper subspace of $\mathbb{R}^3$ but is linearly dependent,
(v) is a basis of a proper subspace of $\mathbb{R}^3$.

12 The scalar k is non-zero and $\mathbf{a}$, $\mathbf{b}$ and $\mathbf{c}$ are any vectors in a vector space V. Prove that: (i) $\{\mathbf{a}, k\mathbf{a}\}$ is linearly dependent, (ii) if $\{\mathbf{a}, \mathbf{b}\}$ is linearly independent, then so is $\{k\mathbf{a}, \mathbf{b}\}$, (iii) if $\{\mathbf{a}, \mathbf{b}\}$ is linearly dependent, then so is $\{k\mathbf{a}, \mathbf{b}\}$, (iv) span $(\{\mathbf{a}, \mathbf{b}\})$ = span $(\{k\mathbf{a}, \mathbf{b}\})$, (v) if $\{\mathbf{a}, \mathbf{b}, \mathbf{c}\}$ is a basis of V then $\{\mathbf{a}, k\mathbf{b}, \mathbf{c}\}$ and $\{\mathbf{a}+k\mathbf{b}, \mathbf{b}, \mathbf{c}\}$ are also bases of V.

1.4 The algebra of square matrices

Let M_n be the set of all real square matrices, of a given order n.

Definition The set M_n is the set of all square matrices of order n with real entries.

For example, for $n = 2$, if $\mathbf{A} \in M_2$, then $\mathbf{A} = \begin{pmatrix} a & b \\ c & d \end{pmatrix}$, where a, b, c and d are real numbers. Since a (1×1) matrix is effectively a scalar, the set M_1 is in one-one correspondence with the set $\mathbb{R}$ of all real numbers. We shall be concerned with the properties of M_n under the two operations of matrix addition and multiplication. In particular, we shall be concerned with results about M_2 and M_3, but the results and their proofs will usually be true for M_n for any positive integer n.

Under addition, M_n has the same properties as the set of real numbers, since addition is performed placewise in the matrix entries. We state these properties, using the vocabulary of chapter **3**.

Theorem 1.2 Under matrix addition, the set M_n forms a commutative group, that is, the operation is closed, associative and commutative, with a zero element and with negatives.

Proof Most of the properties are self evident, as we show. Suppose that $\mathbf{A}$, $\mathbf{B}$ and $\mathbf{C}$ are real $(n \times n)$ matrices in M_n, with $\mathbf{A} = (a_{ij})$, $\mathbf{B} = (b_{ij})$, $\mathbf{C} = (c_{ij})$.

Closure $\mathbf{A} + \mathbf{B}$ is an $(n \times n)$ matrix with entries $(a_{ij} + b_{ij})$ which are real, so $\mathbf{A} + \mathbf{B} \in M_n$ and M_n is closed under addition.

Commutativity $\mathbf{A} + \mathbf{B} = \mathbf{B} + \mathbf{A}$, since $(a_{ij} + b_{ij}) = (b_{ij} + a_{ij})$.

Associativity $(\mathbf{A} + \mathbf{B}) + \mathbf{C} = \mathbf{A} + (\mathbf{B} + \mathbf{C})$, since

$$(a_{ij} + b_{ij}) + c_{ij} = a_{ij} + (b_{ij} + c_{ij}).$$

In the above, the rules for the closure, commutativity and associativity of the addition of real numbers are applied n^2 times.

Zero The zero matrix $\mathbf{0}_n$ is defined as the $(n \times n)$ matrix with every entry equal to zero. Then $\mathbf{0}_n + \mathbf{A} = \mathbf{A} = \mathbf{A} + \mathbf{0}_n$, as required, since

$$a_{ij} + 0 = a_{ij} = 0 + a_{ij}.$$

Negatives If $\mathbf{A} = (a_{ij})$, then $-\mathbf{A} = (-a_{ij})$ is the additive inverse, or negative, of $\mathbf{A}$ since $a_{ij} - a_{ij} = 0 = -a_{ij} + a_{ij}$, that is

$$\mathbf{A} + (-\mathbf{A}) = \mathbf{0}_n = (-\mathbf{A}) + \mathbf{A}.$$

Theorem 1.2 is true, in fact for the set of all real matrices of any given order, since all the properties hold when $\mathbf{A}$, $\mathbf{B}$, $\mathbf{C}$ and $\mathbf{0}$ are all $(m \times n)$ matrices, for given m and n.

Under multiplication, the set M_n possesses some, but not all, of the properties of real numbers. These are stated in the next theorem.

Theorem 1.3 Under matrix multiplication, the set M_n is closed and associative, has an identity element and a zero element, and matrix multiplication distributes over matrix addition.

The important point to notice about this theorem concerns those properties of the arithmetic of real numbers which have been omitted. Matrix multiplication is not commutative, in general, (although there are certain matrices which commute with all others in the set), and yet the operation is associative. Also, not all non-zero matrices have multiplicative inverses. The set of all those matrices which have inverses forms an important subset of M_n.

Proof of theorem 1.3: Again, let $\mathbf{A}$, $\mathbf{B}$ and $\mathbf{C}$ be real $(n \times n)$ matrices in M_n, with $\mathbf{A} = (a_{ij})$, $\mathbf{B} = (b_{ij})$, $\mathbf{C} = (c_{ij})$.

Closure $\mathbf{AB}$ is also an $(n \times n)$ matrix and its entries are the sums of products of real numbers and are also real, and so $\mathbf{AB} \in M_n$.

Identity The unit matrix in M_n is defined to be $\mathbf{I}_n$, where $\mathbf{I}_n$ is the $(n \times n)$ matrix with 1's on the leading diagonal and 0's elsewhere. That is,

$$\mathbf{I}_n = (i_{jk}) \text{ and, for all } j, k \text{ with } j \neq k,\ i_{jj} = 1 \text{ and } i_{jk} = 0.$$

Then, if $\mathbf{A} = (a_{ij})$ and $\mathbf{X} = (x_{ij}) = \mathbf{AI}_n$, for all i, j,

$$x_{ij} = \sum_{k=1}^{n} a_{ik} i_{kj} = a_{i1} i_{1j} + a_{i2} i_{2j} + \ldots + a_{in} i_{nj} = a_{ij},$$

since $i_{jj} = 1$ and, for all $k \neq j$, $i_{kj} = 0$. This means that $\mathbf{X} = \mathbf{A}$ and so $\mathbf{AI}_n = \mathbf{A}$. The proof that $\mathbf{I}_n\mathbf{A} = \mathbf{A}$ is similar. Therefore

$$\mathbf{AI}_n = \mathbf{A} = \mathbf{I}_n\mathbf{A}$$

and $\mathbf{I}_n$ is the multiplicative identity.

Zero Clearly, $\mathbf{A0}_n = \mathbf{0}_n = \mathbf{0}_n\mathbf{A}$.

As a demonstration in M_2,

$$\begin{pmatrix} 2 & -1 \\ 3 & 0 \end{pmatrix}\begin{pmatrix} 1 & 0 \\ 0 & 1 \end{pmatrix} = \begin{pmatrix} 2 & -1 \\ 3 & 0 \end{pmatrix} = \begin{pmatrix} 1 & 0 \\ 0 & 1 \end{pmatrix}\begin{pmatrix} 2 & -1 \\ 3 & 0 \end{pmatrix},$$

$$\begin{pmatrix} 2 & -1 \\ 3 & 0 \end{pmatrix}\begin{pmatrix} 0 & 0 \\ 0 & 0 \end{pmatrix} = \begin{pmatrix} 0 & 0 \\ 0 & 0 \end{pmatrix} = \begin{pmatrix} 0 & 0 \\ 0 & 0 \end{pmatrix}\begin{pmatrix} 2 & -1 \\ 3 & 0 \end{pmatrix}.$$

Distributivity We prove that $\mathbf{A}(\mathbf{B}+\mathbf{C}) = \mathbf{AB}+\mathbf{AC}$, using the sigma notation to keep the proof compact. Let $\mathbf{A} = (a_{ij})$, $\mathbf{B} = (b_{ij})$ and $\mathbf{C} = (c_{ij})$, then the (i, j) entry in $\mathbf{A}(\mathbf{B}+\mathbf{C})$ will be x_{ij} where

$$x_{ij} = \sum_{k=1}^{n} a_{ik}(b_{kj} + c_{kj}) = \sum_{k=1}^{n} a_{ik} b_{kj} + \sum_{k=1}^{n} a_{ik} c_{kj},$$

by expansion of the sums, and x_{ij} is also the (i, j) entry in $\mathbf{AB}+\mathbf{AC}$. In a similar manner, $(\mathbf{B}+\mathbf{C})\mathbf{A} = \mathbf{BA}+\mathbf{CA}$.

Associativity We have left associativity until the end because the proof is rather complicated. We give a proof in M_2 and a general proof in M_n. The reader is recommended to look at Chapter **4** where the result becomes an easy consequence of the identification of a matrix with a linear transformation set up there.

The proof for $n = 2$ will not use the sigma notation. Define matrices

$$\mathbf{X} = \begin{pmatrix} a & b \\ c & d \end{pmatrix}, \quad \mathbf{Y} = \begin{pmatrix} e & f \\ g & h \end{pmatrix}, \quad \mathbf{Z} = \begin{pmatrix} j & k \\ l & m \end{pmatrix}. \text{ Then,}$$

$$(\mathbf{XY})\mathbf{Z} = \begin{pmatrix} ae+bg & af+bh \\ ce+dg & cf+dh \end{pmatrix}\begin{pmatrix} j & k \\ l & m \end{pmatrix} = \begin{pmatrix} aej+bgj+afl+bhl & aek+bgk+afm+bhm \\ cej+dgj+cfl+dhl & cek+dgk+cfm+dhm \end{pmatrix}$$

and

$$\mathbf{X(YZ)} = \begin{pmatrix} a & b \\ c & d \end{pmatrix} \begin{pmatrix} ej+fl & ek+fm \\ gj+hl & gk+hm \end{pmatrix} = \begin{pmatrix} aej+afl+bgj+bhl & aek+afm+bgk+bhm \\ cej+cfl+dgj+dhl & cek+cfm+dgk+dhm \end{pmatrix}$$

and so $\mathbf{(XY)Z} = \mathbf{X(YZ)}$, as was to be proved.

We now give a general proof that multiplication in M_n is associative and, this time, we use the sigma notation. Given three $(n \times n)$ matrices $\mathbf{A}$, $\mathbf{B}$ and $\mathbf{C}$, with $\mathbf{A} = (a_{ij})$, $\mathbf{B} = (b_{ij})$ and $\mathbf{C} = (c_{ij})$, the (i, j) entries in $\mathbf{(AB)C}$ and in $\mathbf{A(BC)}$ are

$$\sum_{l=1}^{n} \left(\sum_{k=1}^{n} a_{ik} b_{kl} \right) c_{lj} \quad \text{and} \quad \sum_{k=1}^{n} a_{ik} \left(\sum_{l=1}^{n} b_{kl} c_{lj} \right)$$

and both these summations are equal to $\sum_{k=1}^{n} \sum_{l=1}^{n} a_{ik} b_{kl} c_{lj}$, therefore, they are equal to each other. The double summation and its manipulation is the difficult part of this proof to assimilate. The reader should spend a considerable amount of time experimenting with writing out such summations in full, for small values of n, in order to understand how the sums are manipulated. As an example, we write out the case for $n = 2$.

$$\begin{aligned} \sum_{l=1}^{2} \sum_{k=1}^{2} a_{ik} b_{kl} c_{lj} &= \sum_{l=1}^{2} (a_{i1} b_{1l} + a_{i2} b_{2l}) c_{lj} \\ &= a_{i1} b_{11} c_{1j} + a_{i2} b_{21} c_{1j} + a_{i1} b_{12} c_{2j} + a_{i2} b_{22} c_{2j}, \\ \sum_{k=1}^{2} a_{ik} \sum_{l=1}^{2} b_{kl} c_{lj} &= \sum_{k=1}^{2} a_{ik} (b_{k1} c_{1j} + b_{k2} c_{2j}) \\ &= a_{i1} b_{11} c_{1j} + a_{i1} b_{12} c_{2j} + a_{i2} b_{21} c_{1j} + a_{i2} b_{22} c_{2j}. \end{aligned}$$

Consequences Since $\mathbf{(AB)C} = \mathbf{A(BC)}$, we can omit the brackets and write the product as $\mathbf{ABC}$ with no ambiguity. Then, in the usual way, we can write $\mathbf{AA} = \mathbf{A}^2$, $\mathbf{AAA} = \mathbf{A}^3$, and so on, defining positive integer powers of $\mathbf{A}$.

Definition If matrices $\mathbf{A}$ and $\mathbf{B}$ exist in M_n, with $\mathbf{AB} = \mathbf{I}_n = \mathbf{BA}$, then $\mathbf{A}$ and $\mathbf{B}$ are defined to be inverses of each other. This is written

$$\mathbf{B} = \mathbf{A}^{-1}, \mathbf{A} = \mathbf{B}^{-1}.$$

Zero divisors In the real numbers, a product can only be zero if one of the terms in the product is zero. This is not true in matrix multiplication, which may seem rather odd. In fact, the square of a non-zero matrix can be zero and the product of two matrices can be zero whilst the product in the other order is not zero. For the following matrices $\mathbf{AB} \neq \mathbf{0}_n$, $\mathbf{BA} = \mathbf{0}_n$, $\mathbf{C} \neq \mathbf{0}_n$ and $\mathbf{C}^2 = \mathbf{0}_n$.

Let $$\mathbf{A} = \begin{pmatrix} 2 & 4 \\ 3 & 6 \end{pmatrix}, \mathbf{B} = \begin{pmatrix} 3 & -2 \\ -3 & 2 \end{pmatrix}, \mathbf{C} = \begin{pmatrix} 4 & 2 \\ -8 & -4 \end{pmatrix},$$

then $\quad \mathbf{AB} = \begin{pmatrix} -6 & 4 \\ -9 & 6 \end{pmatrix}, \; \mathbf{BA} = \begin{pmatrix} 0 & 0 \\ 0 & 0 \end{pmatrix}, \; \mathbf{C}^2 = \begin{pmatrix} 0 & 0 \\ 0 & 0 \end{pmatrix}.$

A matrix with this property is called a *zero divisor.*

Definition The matrix $\mathbf{A}$ is a zero divisor if $\mathbf{A} \neq \mathbf{0}_n$ and there exists a non-zero matrix $\mathbf{B}$ such that either $\mathbf{AB} = \mathbf{0}_n$ or $\mathbf{BA} = \mathbf{0}_n$.

Note that, in the above definition, $\mathbf{B}$ is also a zero divisor. As a result of the existence of zero divisors, cancellation is not possible in a matrix equation. If there is a matrix equation $\mathbf{AB} = \mathbf{AC}$ with $\mathbf{A} \neq \mathbf{0}_n$, then $\mathbf{AB} - \mathbf{AC} = \mathbf{0}_n$ so that $\mathbf{A}(\mathbf{B} - \mathbf{C}) = \mathbf{0}_n$ and all that can be said is that either $\mathbf{B} = \mathbf{C}$, or $\mathbf{A}$ and $\mathbf{B} - \mathbf{C}$ are zero divisors. Also, a matrix which is a zero divisor cannot have a multiplicative inverse. For suppose that $\mathbf{AB} = \mathbf{0}_n$ and $\mathbf{A}$ has an inverse $\mathbf{A}^{-1}$, then

$$\mathbf{B} = \mathbf{I}_n\mathbf{B} = (\mathbf{A}^{-1}\mathbf{A})\mathbf{B} = \mathbf{A}^{-1}(\mathbf{AB}) = \mathbf{A}^{-1}\mathbf{0}_n = \mathbf{0}_n.$$

Clearly, these results provide a warning against treating the arithmetic of matrices exactly like that of real numbers.

It should be noted in the above examples that the matrices, which are zero divisors, have the property that one row (or column) is a multiple of another row (or column). We shall see later that the property that one row (or column) of a matrix is a linear combination of some other rows (or columns) characterises those matrices which have no inverses and which are therefore zero divisors.

EXERCISE 1.4A

1 Check the associativity and the distributive properties of the arithmetic of 3×3 matrices by calculating: (i) $(\mathbf{AB})\mathbf{C}$, (ii) $\mathbf{A}(\mathbf{BC})$, (iii) $\mathbf{A}(\mathbf{B} + \mathbf{C})$, (iv) $\mathbf{AB} + \mathbf{AC}$, where

$$\mathbf{A} = \begin{pmatrix} 1 & 2 & 3 \\ 0 & 1 & -1 \\ 3 & 1 & 2 \end{pmatrix}, \; \mathbf{B} = \begin{pmatrix} -1 & 2 & 1 \\ 1 & 0 & 1 \\ 3 & 2 & -1 \end{pmatrix} \text{ and } \mathbf{C} = \begin{pmatrix} -2 & -1 & 0 \\ 1 & 0 & 0 \\ 0 & -1 & -2 \end{pmatrix}.$$

2 Let $\mathbf{A} = (a_{ij})$ be an $(n \times n)$ matrix and let $\mathbf{I}_n$ be the $(n \times n)$ identity matrix. Use the double suffix notation to prove that $\mathbf{I}_n\mathbf{A} = \mathbf{A}$.

3 If $ad - bc = 0$, $\mathbf{X} = \begin{pmatrix} a & b \\ c & d \end{pmatrix}$ and if $\mathbf{X}$ is a non-zero matrix, prove that $\mathbf{X}$ must either have one zero row and one non-zero row or one zero column and one non-zero column or that $\mathbf{X}$ must take the form $\begin{pmatrix} a & b \\ ka & kb \end{pmatrix}$, where a, b and k are non-zero.

4 Prove that $\begin{pmatrix} a & 0 \\ b & 0 \end{pmatrix}\begin{pmatrix} 0 & 0 \\ c & d \end{pmatrix} = \begin{pmatrix} 0 & 0 \\ 0 & 0 \end{pmatrix} = \begin{pmatrix} 0 & a \\ 0 & b \end{pmatrix}\begin{pmatrix} c & d \\ 0 & 0 \end{pmatrix}$

5 By constructing another non-zero matrix so that the product is the zero matrix, prove that the following matrices are zero divisors:

(i) $\begin{pmatrix} 1 & 3 \\ 0 & 0 \end{pmatrix}$ (ii) $\begin{pmatrix} 0 & -1 \\ 0 & 3 \end{pmatrix}$ (iii) $\begin{pmatrix} 2 & 4 \\ 3 & 6 \end{pmatrix}$ (iv) $\begin{pmatrix} -x & y \\ x & -y \end{pmatrix}$

Inverses

We now look for conditions under which a square matrix possesses an inverse, beginning by considering 2×2 matrices.

Let $\mathbf{X} = \begin{pmatrix} a & b \\ c & d \end{pmatrix}$, and suppose that $\mathbf{X}^{-1}$ is an inverse of $\mathbf{X}$. This means

that $\mathbf{X}\mathbf{X}^{-1} = \mathbf{I}_2$, so that if $\mathbf{X}^{-1} = \begin{pmatrix} e & f \\ g & h \end{pmatrix}$, then

$$\mathbf{X}\mathbf{X}^{-1} = \begin{pmatrix} ae+bg & af+bh \\ ce+dg & cf+dh \end{pmatrix} = \begin{pmatrix} 1 & 0 \\ 0 & 1 \end{pmatrix}$$

To find $\mathbf{X}^{-1}$, it is necessary to solve the simultaneous equations

$$ae+bg = 1 = cf+dh \qquad 1$$
$$af+bh = 0 = ce+dg \qquad 2$$

It will be seen from the equations 2 that:
the ratio of h to f is the negative of the ratio of a to b and the ratio of g to e is the negative of the ratio of c to d.

Therefore, let $h = ka$, $f = -kb$, $g = lc$ and $e = -ld$,
then $af+bh = -akb+bka = 0$ and $ce+dg = -cld+dlc = 0$,
as required. Substituting into equations 1,

$$-ald+blc = 1 = -ckb+dka,$$

that is

$$-l(ad-bc) = 1 = k(ad-bc).$$

If $ad-bc \neq 0$, then $k = 1/(ad-bc)$ and $l = -1/(ad-bc)$ and

$$\mathbf{X}^{-1} = \begin{pmatrix} e & f \\ g & h \end{pmatrix} = \begin{pmatrix} -ld & -kb \\ lc & ka \end{pmatrix} = \frac{1}{ad-bc}\begin{pmatrix} d & -b \\ -c & a \end{pmatrix}.$$

We verify that this is indeed the inverse of $\mathbf{X}$:

$$\mathbf{X}^{-1}\mathbf{X} = \frac{1}{ad-bc}\begin{pmatrix} d & -b \\ -c & a \end{pmatrix}\begin{pmatrix} a & b \\ c & d \end{pmatrix} = \frac{1}{ad-bc}\begin{pmatrix} ad-bc & 0 \\ 0 & ad-bc \end{pmatrix} = \begin{pmatrix} 1 & 0 \\ 0 & 1 \end{pmatrix} = \mathbf{I}_2,$$

$$\mathbf{X}\mathbf{X}^{-1} = \begin{pmatrix} a & b \\ c & d \end{pmatrix}\frac{1}{ad-bc}\begin{pmatrix} d & -b \\ -c & a \end{pmatrix} = \frac{1}{ad-bc}\begin{pmatrix} ad-bc & 0 \\ 0 & ad-bc \end{pmatrix} = \begin{pmatrix} 1 & 0 \\ 0 & 1 \end{pmatrix} = \mathbf{I}_2.$$

Thus, if $ad-bc \neq 0$, the matrix $\begin{pmatrix} a & b \\ c & d \end{pmatrix}$ has an inverse $\dfrac{1}{ad-bc}\begin{pmatrix} d & -b \\ -c & a \end{pmatrix}$. It remains to find out what happens if $ad-bc=0$. In this case,

$$\begin{pmatrix} a & b \\ c & d \end{pmatrix}\begin{pmatrix} d & -b \\ -c & a \end{pmatrix} = \begin{pmatrix} 0 & 0 \\ 0 & 0 \end{pmatrix} \text{ and so } \begin{pmatrix} a & b \\ c & d \end{pmatrix} \text{ is a zero divisor}$$

and hence has no inverse. The condition $ad-bc=0$ can be written $ad = bc$, that is, one of the row vectors $(a \quad b)$ and $(c \quad d)$ is a multiple of the other and the same is true of the column vectors $\begin{pmatrix} a \\ c \end{pmatrix}$ and $\begin{pmatrix} b \\ d \end{pmatrix}$. This is because:

(i) if $a \neq 0$, $d = cb/a$ and $\quad (c \quad d) = \dfrac{c}{a}(a \quad b)$, and

(ii) if $a = 0$, $ad = 0$, and therefore $bc = 0$ and then either

$$b = 0 \quad \text{and} \quad (a \quad b) = (0 \quad 0) = 0(c \quad d)$$

or else

$$b \neq 0,\ c = 0 \quad \text{and} \quad (c \quad d) = (0 \quad d) = \frac{d}{b}(0 \quad b) = \frac{d}{b}(a \quad b).$$

Two vectors are linearly dependent if one is a multiple of the other. Therefore,

if $ad-bc=0$ the rows of $\begin{pmatrix} a & b \\ c & d \end{pmatrix}$ are linearly dependent, and the columns of $\begin{pmatrix} a & b \\ c & d \end{pmatrix}$ are linearly dependent.

The converse is also true. If $l(a \quad b) + m(c \quad d) = \mathbf{0}$, with say $l \neq 0$, then $(a \quad b) = -\dfrac{m}{l}(c \quad d)$, $a = -mc/l$, $b = -md/l$ and

$$ad-bc = -\frac{mcd}{l} + \frac{mcd}{l} = 0.$$

Since the matrix $\begin{pmatrix} a & b \\ c & d \end{pmatrix}$ has an inverse if $ad-bc \neq 0$ and has no inverse if $ad-bc=0$, the condition $ad-bc \neq 0$ is both necessary and sufficient for the inverse to exist. We say that a matrix which has an inverse is non-singular and that a matrix which has no inverse is singular. We can then summarise our results in a form which generalises to $(n \times n)$ matrices.

Definition A square matrix $\mathbf{X}$ is non-singular if it has an inverse $\mathbf{X}^{-1}$, a square matrix $\mathbf{X}$ is singular if it does not possess an inverse.

Theorem 1.4 Let $\mathbf{X} = \begin{pmatrix} a & b \\ c & d \end{pmatrix}$, where $a, b, c, d \in \mathbb{R}$. Then

I the following statements are equivalent:
(i) $ad - bc \neq 0$,
(ii) $\mathbf{X}$ is non-singular,
(iii) $\mathbf{X}$ has an inverse $\mathbf{X}^{-1}$ and $\mathbf{X}^{-1} = \dfrac{1}{ad-bc}\begin{pmatrix} d & -b \\ -c & a \end{pmatrix}$,
(iv) the rows of $\mathbf{X}$ are linearly independent,
(v) the columns of $\mathbf{X}$ are linearly independent.

II the following statements are equivalent:
(i) $ad - bc = 0$,
(ii) $\mathbf{X}$ is singular,
(iii) $\mathbf{X}$ has no inverse,
(iv) the rows of $\mathbf{X}$ are linearly dependent,
(v) the columns of $\mathbf{X}$ are linearly dependent.

Clearly, the number $ad - bc$ is an important number associated with the matrix $\mathbf{X}$ and it is called the *determinant* of $\mathbf{X}$.

Definition The determinant of the 2×2 matrix $\mathbf{X}$, where $\mathbf{X} = \begin{pmatrix} a & b \\ c & d \end{pmatrix}$, is the real number $ad - bc$. This is written

$$ad - bc = \det(\mathbf{X}) = |\mathbf{X}|.$$

Theorem 1.4 shows that M_2 is partitioned into two disjoint sets of 2×2 real matrices, namely:

(i) the set of non-singular matrices which have inverses and non-zero determinants,

(ii) the set of singular matrices which have no inverses and zero determinants.

EXAMPLE 1 *Determine which of these matrices has an inverse and find the inverse matrix when it exists:*

$$(i)\ \mathbf{A} = \begin{pmatrix} 3 & 2 \\ 4 & 3 \end{pmatrix},\quad (ii)\ \mathbf{B} = \begin{pmatrix} 2 & 1 \\ -4 & -2 \end{pmatrix},\quad (iii)\ \mathbf{C} = \begin{pmatrix} 5 & 3 \\ -3 & -1 \end{pmatrix}.$$

(i) $\text{Det}\,(\mathbf{A}) = 3.3 - 4.2 = 1$, so $\mathbf{A}^{-1} = \begin{pmatrix} \mathbf{3} & \mathbf{-2} \\ \mathbf{-4} & \mathbf{3} \end{pmatrix}$, and

$$\mathbf{AA}^{-1} = \begin{pmatrix} 3 & 2 \\ 4 & 3 \end{pmatrix}\begin{pmatrix} 3 & -2 \\ -4 & 3 \end{pmatrix} = \begin{pmatrix} 1 & 0 \\ 0 & 1 \end{pmatrix}.$$

(ii) $\text{Det}(\mathbf{B}) = 2.(-2) - 1.(-4) = 0$, so $\mathbf{B}$ is singular and has no inverse. Notice that column 1 of $\mathbf{B}$ is double column 2.

(iii) $\text{Det}(\mathbf{C}) = 5.(-1) - 3.(-3) = 4$, so

$$\mathbf{C}^{-1} = \frac{1}{4}\begin{pmatrix} -1 & -3 \\ 3 & 5 \end{pmatrix} = \begin{pmatrix} \mathbf{-0{\cdot}25} & \mathbf{-0{\cdot}75} \\ \mathbf{0{\cdot}75} & \mathbf{1{\cdot}25} \end{pmatrix}$$

and $\mathbf{CC}^{-1} = \begin{pmatrix} 5 & 3 \\ -3 & -1 \end{pmatrix}\begin{pmatrix} -0{\cdot}25 & -0{\cdot}75 \\ 0{\cdot}75 & 1{\cdot}25 \end{pmatrix} = \begin{pmatrix} 1 & 0 \\ 0 & 1 \end{pmatrix}.$

The inverse of a product of two matrices

The inverse of the composition of two one-one functions is the composition of their inverse functions in the reverse order. The same structure occurs for the inverse of the product of two non-singular matrices. This is a consequence of the definition of matrix product.

Theorem 1.5 If $\mathbf{A}$ and $\mathbf{B}$ are two non-singular $(n \times n)$ matrices, then

$$(\mathbf{AB})^{-1} = \mathbf{B}^{-1}\mathbf{A}^{-1}.$$

Proof (in M_n) Let $\mathbf{A}$ and $\mathbf{B}$ have inverses $\mathbf{A}^{-1}$ and $\mathbf{B}^{-1}$, respectively and let $\mathbf{C} = \mathbf{B}^{-1}\mathbf{A}^{-1}$.

Then $(\mathbf{AB})\mathbf{C} = \mathbf{ABC} = \mathbf{A}(\mathbf{BB}^{-1})\mathbf{A}^{-1} = \mathbf{AI}_n\mathbf{A}^{-1} = \mathbf{AA}^{-1} = \mathbf{I}_n$

and $\mathbf{C}(\mathbf{AB}) = \mathbf{CAB} = \mathbf{B}^{-1}(\mathbf{A}^{-1}\mathbf{A})\mathbf{B} = \mathbf{B}^{-1}\mathbf{I}_n\mathbf{B} = \mathbf{B}^{-1}\mathbf{B} = \mathbf{I}_n.$

By the definition of the inverse of a matrix, $\mathbf{C} = \mathbf{B}^{-1}\mathbf{A}^{-1} = (\mathbf{AB})^{-1}$.

In Theorem 1.5, because **A** and **B** are non-singular, they each possess an inverse and then their product, **C**, also possesses an inverse. Thus the product of non-singular matrices is non-singular. Conversely, let **A** be a singular matrix, then **A** is a zero divisor and there exists a non-zero matrix **D** such that $\mathbf{DA} = \mathbf{0}$. For any other matrix, **B**, either $\mathbf{AB} = \mathbf{0}$ or $\mathbf{AB} \neq \mathbf{0}$ and $\mathbf{D}(\mathbf{AB}) = (\mathbf{DA})\mathbf{B} = \mathbf{0B} = \mathbf{0}$ so that **AB** is a zero divisor. In either case **AB** is singular. Similarly **BA** is singular. We have just proved the following theorem.

Theorem 1.6 A product of non-singular matrices is non-singular. A product of matrices, in which one factor is a singular matrix, is singular.

In Theorem 1.6, we are using the fact that the zero matrix, **0**, is singular. This follows from the definition.

In the next example, inverse matrices are used to solve equations.

EXAMPLE 2 *Solve the matrix equations:* (*i*) $\begin{pmatrix} 3 & 2 \\ 4 & 3 \end{pmatrix}\begin{pmatrix} x \\ y \end{pmatrix} = \begin{pmatrix} 2 \\ -1 \end{pmatrix}$,

(*ii*) $\begin{pmatrix} 2 & 1 \\ -4 & -2 \end{pmatrix}\begin{pmatrix} x \\ y \end{pmatrix} = \begin{pmatrix} 2 \\ -1 \end{pmatrix}$, (*iii*) $\begin{pmatrix} 5 & 3 \\ -3 & -1 \end{pmatrix}\begin{pmatrix} w & x \\ y & z \end{pmatrix} = \begin{pmatrix} 2 & 1 \\ 4 & -2 \end{pmatrix}$.

(i) From example 1, the matrix **A**, where $\mathbf{A} = \begin{pmatrix} 3 & 2 \\ 4 & 3 \end{pmatrix}$, has an inverse $\begin{pmatrix} 3 & -2 \\ -4 & 3 \end{pmatrix}$, so $\begin{pmatrix} 3 & -2 \\ -4 & 3 \end{pmatrix}\begin{pmatrix} 3 & 2 \\ 4 & 3 \end{pmatrix}\begin{pmatrix} x \\ y \end{pmatrix} = \begin{pmatrix} 3 & -2 \\ -4 & 3 \end{pmatrix}\begin{pmatrix} 2 \\ -1 \end{pmatrix}$, on pre-multiplying both sides of the equation by $\mathbf{A}^{-1}$. Therefore $\begin{pmatrix} x \\ y \end{pmatrix} = \begin{pmatrix} 8 \\ -11 \end{pmatrix}$ so the solution is $\boldsymbol{x = 8, y = -11}$.

(ii) The matrix **B**, where $\mathbf{B} = \begin{pmatrix} 2 & 1 \\ -4 & -2 \end{pmatrix}$ has no inverse, since det $(\mathbf{B}) = 0$.

The equation can be written $\begin{pmatrix} 2x + y \\ -4x - 2y \end{pmatrix} = \begin{pmatrix} 2 \\ -1 \end{pmatrix}$ and, if $2x + y = 2$ then $-4x - 2y = -4$ and not -1, so there is **no solution**. If the vector on the right hand side of the equation had been $\begin{pmatrix} 2 \\ -4 \end{pmatrix}$, then there would have been an infinite number of solutions of the form $x = \alpha$, $y = 2 - 2\alpha$.

(iii) From Example 1, the inverse of the matrix **C**, where $\mathbf{C} = \begin{pmatrix} 5 & 3 \\ -3 & -1 \end{pmatrix}$, is $\begin{pmatrix} -0{\cdot}25 & -0{\cdot}75 \\ 0{\cdot}75 & 1{\cdot}25 \end{pmatrix}$, so

$$\begin{pmatrix} w & x \\ y & z \end{pmatrix} = \begin{pmatrix} -0{\cdot}25 & -0{\cdot}75 \\ 0{\cdot}75 & 1{\cdot}25 \end{pmatrix}\begin{pmatrix} 5 & 3 \\ -3 & -1 \end{pmatrix}\begin{pmatrix} w & x \\ y & z \end{pmatrix}$$

$$= \begin{pmatrix} -0{\cdot}25 & -0{\cdot}75 \\ 0{\cdot}75 & 1{\cdot}25 \end{pmatrix}\begin{pmatrix} 2 & 1 \\ 4 & -2 \end{pmatrix}$$

$$= \begin{pmatrix} \mathbf{-3{\cdot}5} & \mathbf{1{\cdot}25} \\ \mathbf{6{\cdot}5} & \mathbf{-1{\cdot}75} \end{pmatrix}.$$

Elementary matrices

There are certain very useful matrices which are obtained from simple operations of the unit matrix. The operations concerned are called *elementary* operations and consist of

(i) multiplying a row (or a column) by a non-zero constant,
(ii) interchanging two rows (or columns),
(iii) adding a multiple of one row (or column) to another row (or column).

The corresponding matrices are called *elementary* matrices. If a matrix is multiplied by an elementary matrix on the left the result is the same as performing upon the matrix the same elementary row operation as was performed upon the unit matrix to form the elementary matrix. Similarly, multiplication on the right corresponds to column operations.

Illustrating these operations for (3×3) matrices, let $\mathbf{A} = \begin{pmatrix} a & b & c \\ d & e & f \\ g & h & k \end{pmatrix}$.

(i) Let $\mathbf{P} = \begin{pmatrix} p & 0 & 0 \\ 0 & 1 & 0 \\ 0 & 0 & 1 \end{pmatrix}$, and let $\mathbf{I} = \begin{pmatrix} 1 & 0 & 0 \\ 0 & 1 & 0 \\ 0 & 0 & 1 \end{pmatrix}$, so that **P** is the elementary matrix obtained from **I** by multiplying its first row (or column) by p, which we assume to be non-zero. Then

$$\mathbf{PA} = \begin{pmatrix} pa & pb & pc \\ d & e & f \\ g & h & k \end{pmatrix} \quad \text{and} \quad \mathbf{AP} = \begin{pmatrix} pa & b & c \\ pd & e & f \\ pg & h & k \end{pmatrix}.$$

Similarly,

$$\begin{pmatrix} 1 & 0 & 0 \\ 0 & p & 0 \\ 0 & 0 & 1 \end{pmatrix}\begin{pmatrix} a & b & c \\ d & e & f \\ g & h & k \end{pmatrix} = \begin{pmatrix} a & b & c \\ pd & pe & pf \\ g & h & k \end{pmatrix}, \begin{pmatrix} a & b & c \\ d & e & f \\ g & h & k \end{pmatrix}\begin{pmatrix} 1 & 0 & 0 \\ 0 & p & 0 \\ 0 & 0 & 1 \end{pmatrix} = \begin{pmatrix} a & pb & c \\ d & pe & f \\ g & ph & k \end{pmatrix},$$

$$\begin{pmatrix} 1 & 0 & 0 \\ 0 & 1 & 0 \\ 0 & 0 & p \end{pmatrix}\begin{pmatrix} a & b & c \\ d & e & f \\ g & h & k \end{pmatrix} = \begin{pmatrix} a & b & c \\ d & e & f \\ pg & ph & pk \end{pmatrix}, \begin{pmatrix} a & b & c \\ d & e & f \\ g & h & k \end{pmatrix}\begin{pmatrix} 1 & 0 & 0 \\ 0 & 1 & 0 \\ 0 & 0 & p \end{pmatrix} = \begin{pmatrix} a & b & pc \\ d & e & pf \\ g & h & pk \end{pmatrix}.$$

If p is replaced by $1/p$ in **P**, then the corresponding row or column operations will consist of multiplying the appropriate row or column by $1/p$. This means that the corresponding elementary matrix is $\mathbf{P}^{-1}$ and the operation is the inverse of the operation for **P**. In particular, **P** is non-singular. The same will be true for the other two matrices, obtained by replacing 1 by p in the second, or third, row of **I**.

(ii) Let $\mathbf{Q} = \begin{pmatrix} 0 & 1 & 0 \\ 1 & 0 & 0 \\ 0 & 0 & 1 \end{pmatrix}$, which is obtained from **I** by interchanging the first two rows (or the first two columns). Then multiplying **A** by **Q** on the left will interchange its first two rows and multiplying **A** by **Q** on the right will interchange its first two columns. Other elementary matrices, similar to **Q** are obtained by interchanging other pairs of rows, or columns, of **I**. Thus

$$\begin{pmatrix} 0 & 1 & 0 \\ 1 & 0 & 0 \\ 0 & 0 & 1 \end{pmatrix}\begin{pmatrix} a & b & c \\ d & e & f \\ g & h & k \end{pmatrix} = \begin{pmatrix} d & e & f \\ a & b & c \\ g & h & k \end{pmatrix}, \begin{pmatrix} a & b & c \\ d & e & f \\ g & h & k \end{pmatrix}\begin{pmatrix} 0 & 1 & 0 \\ 1 & 0 & 0 \\ 0 & 0 & 1 \end{pmatrix} = \begin{pmatrix} b & a & c \\ e & d & f \\ h & g & k \end{pmatrix},$$

$$\begin{pmatrix} 1 & 0 & 0 \\ 0 & 0 & 1 \\ 0 & 1 & 0 \end{pmatrix}\begin{pmatrix} a & b & c \\ d & e & f \\ g & h & k \end{pmatrix} = \begin{pmatrix} a & b & c \\ g & h & k \\ d & e & f \end{pmatrix}, \begin{pmatrix} a & b & c \\ d & e & f \\ g & h & k \end{pmatrix}\begin{pmatrix} 1 & 0 & 0 \\ 0 & 0 & 1 \\ 0 & 1 & 0 \end{pmatrix} = \begin{pmatrix} a & c & b \\ d & f & e \\ g & k & h \end{pmatrix},$$

$$\begin{pmatrix} 0 & 0 & 1 \\ 0 & 1 & 0 \\ 1 & 0 & 0 \end{pmatrix}\begin{pmatrix} a & b & c \\ d & e & f \\ g & h & k \end{pmatrix} = \begin{pmatrix} g & h & k \\ d & e & f \\ a & b & c \end{pmatrix}, \begin{pmatrix} a & b & c \\ d & e & f \\ g & h & k \end{pmatrix}\begin{pmatrix} 0 & 0 & 1 \\ 0 & 1 & 0 \\ 1 & 0 & 0 \end{pmatrix} = \begin{pmatrix} c & b & a \\ f & e & d \\ k & h & g \end{pmatrix}.$$

All these elementary matrices are self-inverses since the square of each is the unit matrix **I**.

(iii) Let $\mathbf{R} = \begin{pmatrix} 1 & r & 0 \\ 0 & 1 & 0 \\ 0 & 0 & 1 \end{pmatrix}$, which is obtained from **I** by adding either r times the second row to the first row, or r times the first column to the second column. Multiplying **A** by **R** on the left (right) gives a matrix obtained from **A** by adding r times its second row (first column) to its first row (second column). Thus

$$\begin{pmatrix} 1 & r & 0 \\ 0 & 1 & 0 \\ 0 & 0 & 1 \end{pmatrix}\begin{pmatrix} a & b & c \\ d & e & f \\ g & h & k \end{pmatrix} = \begin{pmatrix} a+rd & b+re & c+rf \\ d & e & f \\ g & h & k \end{pmatrix},$$

$$\begin{pmatrix} a & b & c \\ d & e & f \\ g & h & k \end{pmatrix}\begin{pmatrix} 1 & r & 0 \\ 0 & 1 & 0 \\ 0 & 0 & 1 \end{pmatrix} = \begin{pmatrix} a & rb+b & c \\ d & re+e & f \\ g & rh+h & k \end{pmatrix}.$$

If r is replaced in **R** by $-r$, then the effect of multiplication on the left (right) is to subtract r times the second row (first column) from the first row (second column), which is the inverse of the previous operation. Thus **R** is non-singular with an inverse matrix $\mathbf{R}^{-1}$, given by

$$\mathbf{R}^{-1} = \begin{pmatrix} 1 & -r & 0 \\ 0 & 1 & 0 \\ 0 & 0 & 1 \end{pmatrix}.$$

There are six elementary matrices of the form of **R**, namely **R** and five others. They are obtained from **I** by replacing one of the six zero entries by a non-zero real number r. Each has an inverse obtained by changing the sign of r.

EXERCISE 1.4B

1 Show that the set $\left\{\begin{pmatrix} 1 & 0 \\ 0 & 1 \end{pmatrix}, \begin{pmatrix} 0 & 1 \\ 1 & 0 \end{pmatrix}, \begin{pmatrix} -1 & 0 \\ 0 & -1 \end{pmatrix}, \begin{pmatrix} 0 & -1 \\ -1 & 0 \end{pmatrix}\right\}$ of 2×2 matrices is closed under multiplication, has an identity and that each matrix is its own inverse.

2 Let $S = \left\{\begin{pmatrix} 1 & 0 \\ 0 & 1 \end{pmatrix}, \begin{pmatrix} 0 & -1 \\ 1 & 0 \end{pmatrix}, \begin{pmatrix} -1 & 0 \\ 0 & -1 \end{pmatrix}\right\}$. Find the matrix **M** which must be adjoined to S to make the set $S \cup \{\mathbf{M}\}$ a closed set under matrix multiplication. Find the inverse matrix of each of these 4 matrices.

3 Multiply each of the following matrices by itself and use your result to find its inverse matrix, given that $a^2 + bc \neq 0$.

(i) $\begin{pmatrix} 4 & -5 \\ 3 & -4 \end{pmatrix}$, (ii) $\begin{pmatrix} 10 & -7 \\ 14 & -10 \end{pmatrix}$, (iii) $\begin{pmatrix} -2 & 0 \\ -4 & 2 \end{pmatrix}$, (iv) $\begin{pmatrix} a & b \\ c & -a \end{pmatrix}$.

4 Prove that each of the following are pairs of inverse matrices:

(i) $\begin{pmatrix} 3 & 2 \\ -1 & 0 \end{pmatrix}, \begin{pmatrix} 0 & -1 \\ 1/2 & 3/2 \end{pmatrix}$, (ii) $\begin{pmatrix} 1 & 0 & 2 \\ 0 & 3 & 0 \\ 2 & 0 & 1 \end{pmatrix}, \begin{pmatrix} -1/3 & 0 & 2/3 \\ 0 & 1/3 & 0 \\ 2/3 & 0 & -1/3 \end{pmatrix}$,

(iii) $\begin{pmatrix} 1 & -1 & 0 \\ 0 & 2 & 1 \\ 0 & 0 & 1 \end{pmatrix}, \begin{pmatrix} 1 & 1/2 & -1/2 \\ 0 & 1/2 & -1/2 \\ 0 & 0 & 1 \end{pmatrix}$,

(iv) $\begin{pmatrix} 2 & 0 & 1 \\ 3 & -2 & -1 \\ -4 & 1 & -1 \end{pmatrix}, \begin{pmatrix} 3 & 1 & 2 \\ 7 & 2 & 5 \\ -5 & -2 & -4 \end{pmatrix}$.

Hence solve the sets of simultaneous equations:

(i) $3x + 2y = 5,\quad -x = 2,$

(ii) $x + 2z = -1,\quad 3y = 4,\quad 2x + z = 5,$

(iii) $x - y = -1,\quad 2y + z = 7,\quad z = 3,$

(iv) $2x + z = -3,\quad 3x - 2y - z = 4,\quad -4x + y - z = 1.$

5 Solve the matrix equations, where possible, by using an inverse matrix:

(i) $\begin{pmatrix} 2 & 5 \\ 3 & 8 \end{pmatrix}\begin{pmatrix} w & x \\ y & z \end{pmatrix} = \begin{pmatrix} 3 & 5 \\ -2 & 2 \end{pmatrix}$, (ii) $\begin{pmatrix} 1 & -3 \\ 1 & 2 \end{pmatrix}\begin{pmatrix} x \\ y \end{pmatrix} = \begin{pmatrix} -1 \\ 3 \end{pmatrix}$,

(iii) $\begin{pmatrix} 2 & -4 \\ 3 & -6 \end{pmatrix}\begin{pmatrix} w & x \\ y & z \end{pmatrix} = \begin{pmatrix} 0 & 2 \\ -1 & 3 \end{pmatrix}$, (iv) $\begin{pmatrix} 1 & 3 \\ -2 & -6 \end{pmatrix}\begin{pmatrix} x \\ y \end{pmatrix} = \begin{pmatrix} -4 \\ 8 \end{pmatrix}$.

6 In the set M of (2×2) real matrices show that the set of elementary matrices consists of the matrices:

(i) $\begin{pmatrix} p & 0 \\ 0 & 1 \end{pmatrix}$ and $\begin{pmatrix} 1 & 0 \\ 0 & p \end{pmatrix}$, for all non-zero real numbers p

(ii) $\begin{pmatrix} 0 & 1 \\ 1 & 0 \end{pmatrix}$, (iii) $\begin{pmatrix} 1 & p \\ 0 & 1 \end{pmatrix}$ and $\begin{pmatrix} 1 & 0 \\ p & 1 \end{pmatrix}$, for all real p.

Write down the inverse of each of these elementary matrices. Using $p = 2$, verify the elementary row and column operations which are performed on the matrix $\begin{pmatrix} 3 & -1 \\ 2 & 5 \end{pmatrix}$ by means of multiplication of the left and on the right by each of the elementary matrices.

7 Let $\mathbf{A} = \begin{pmatrix} 3 & 5 \\ 2 & 4 \end{pmatrix}$. Let **P**, **Q**, **R**, **S** be the elementary matrices such that multiplication of a (2×2) matrix by:

(i) **P** on the left multiplies the first row by 1/3,
(ii) **Q** on the left subtracts 2 times the first row from the second,
(iii) **R** on the right subtracts 5/3 times the first column from the second,
(iv) **S** on the right divides the second column by 2/3.

Show that $\mathbf{QPARS} = \mathbf{I}_2$ and show that $\mathbf{A}$ can be written as a product of elementary matrices $\mathbf{P}^{-1}$, $\mathbf{Q}^{-1}$, $\mathbf{S}^{-1}$, $\mathbf{R}^{-1}$ in some order.
(v) Hence prove that $\mathbf{A}^{-1} = \mathbf{RSQP}$ and find $\mathbf{A}^{-1}$.

8 Find, in a manner similar to question 7, four elementary matrices $\mathbf{P}, \mathbf{Q}, \mathbf{R}, \mathbf{S}$, such that $\mathbf{QPARS} = \mathbf{I}_2$, where $\mathbf{A}$ is the matrix:

$$\text{(i)} \begin{pmatrix} 2 & 1 \\ 3 & 4 \end{pmatrix}, \quad \text{(ii)} \begin{pmatrix} -2 & 1 \\ 3 & 5 \end{pmatrix}, \quad \text{(iii)} \begin{pmatrix} 6 & 2 \\ -1 & 4 \end{pmatrix}, \quad \text{(iv)} \begin{pmatrix} a & 1 \\ 1 & 2/a \end{pmatrix}, \ a \neq 0.$$

9 The real matrix $\mathbf{A}$, where $\mathbf{A} = \begin{pmatrix} a & b \\ c & d \end{pmatrix}$, satisfies the equation $\mathbf{A}^2 - p\mathbf{A} + q\mathbf{I} = \mathbf{0}$, where p, q are real constants, $\mathbf{I}$ is the 2×2 unit matrix and $\mathbf{0}$ is the 2×2 zero matrix. (i) Given that $bc \neq 0$, obtain the values of p and q in terms of the elements of $\mathbf{A}$, and show that if the matrix $\mathbf{A} - x\mathbf{I}$ is singular then x satisfies $x^2 - px + q = 0$. (ii) Obtain two distinct matrices $\mathbf{A}$ with $a, b, c, d \in \mathbb{N}$, such that $\mathbf{A}^2 - 3\mathbf{A} + \mathbf{I} = \mathbf{0}$. (*C*)

10 For each of the following assertions concerning 2×2 matrices $\mathbf{A}$ and $\mathbf{B}$, state whether it is true or false, justifying your answer.
(i) If $\mathbf{A}^2 = \mathbf{A}$ then $\mathbf{A}$ is a singular matrix.
(ii) For all $\mathbf{A}$, $\mathbf{A}^2 - 5\mathbf{A} + 6\mathbf{I} = (\mathbf{A} - 3\mathbf{I})(\mathbf{A} - 2\mathbf{I})$, where $\mathbf{I}$ is the unit 2×2 matrix.
(iii) For all $\mathbf{A}$ and $\mathbf{B}$, $\mathbf{A}^2 - 5\mathbf{AB} + 6\mathbf{B}^2 = (\mathbf{A} - 3\mathbf{B})(\mathbf{A} - 2\mathbf{B})$.
(iv) If $\mathbf{A}$ and $\mathbf{B}$ are symmetric matrices then $\mathbf{AB}$ is also a symmetric matrix. (A symmetric matrix $\mathbf{A}$ satisfies $\mathbf{A}^{\mathrm{T}} = \mathbf{A}$.)
(v) If $\mathbf{A} = \begin{pmatrix} \cos \pi/6 & -\sin \pi/6 \\ \sin \pi/6 & \cos \pi/6 \end{pmatrix}$, then $\mathbf{A}^{12} = \mathbf{I}$. (*C*)

11 (i) The matrices $\mathbf{A}$ and $\mathbf{B}$ are such that $\mathbf{AB} = \mathbf{BA}$. Show that

$$(\mathbf{A} + \mathbf{B})^3 = \mathbf{A}^3 + 3\mathbf{A}^2\mathbf{B} + 3\mathbf{AB}^2 + \mathbf{B}^3.$$

(ii) Let $\mathbf{C}$ be a given matrix, where $\mathbf{C} \neq \mathbf{0}$. Find two different matrices $\mathbf{P}$ satisfying both the equations $\mathbf{CP} = \mathbf{PC}$ and $\mathbf{P}^2 - \mathbf{PC} - 6\mathbf{C}^2 = 0$.
(iii) Given that $\mathbf{D} = \begin{pmatrix} 1 & 1 \\ 1 & 0 \end{pmatrix}$, show that all the solutions of the equation $\mathbf{QD} = \mathbf{DQ}$ are of the form $\mathbf{Q} = \alpha\mathbf{I} + \beta\mathbf{D}$ where α and β are scalars. (*C*)

12 It is given that $\mathbf{M}$ is the matrix $\begin{pmatrix} a & b \\ c & d \end{pmatrix}$ where $b \neq 0$, $c \neq 0$.
(i) Given that $\mathbf{A}$ is a matrix of the form $\begin{pmatrix} \lambda a + \mu & \lambda b \\ x & y \end{pmatrix}$ and is such that $\mathbf{MA} = \mathbf{AM}$, obtain expressions for x and y in terms of λ, μ and the elements of $\mathbf{M}$. Hence express $\mathbf{A}$ in terms of $\mathbf{M}$ and the unit matrix $\mathbf{I}$. (ii) Deduce that any matrix which commutes with $\mathbf{M}$ under the operation of matrix multiplication can be expressed in the form $\alpha\mathbf{M} + \beta\mathbf{I}$. (iii) Prove the converse of (ii). (iv) Obtain two singular matrices each of which commutes with $\begin{pmatrix} 4 & 5 \\ 2 & 1 \end{pmatrix}$ under the operation of matrix multiplication, and such that neither is a scalar multiple of the other. (*C*)

1.5 Determinants

The determinant of a (2×2) matrix was defined in §1.4. The matrix is singular (or non-singular) if its determinant is zero (or non zero). In the case of an $(n \times n)$ matrix, a similar scalar, its determinant, can be defined with the same properties. We shall investigate the (3×3) case.

Consider the problem of finding the inverse, $\mathbf{X}$, of the matrix $\mathbf{M}$, given by $\mathbf{M} = \begin{pmatrix} a_1 & b_1 & c_1 \\ a_2 & b_2 & c_2 \\ a_3 & b_3 & c_3 \end{pmatrix}$, so that $\mathbf{MX} = \mathbf{I}_3$. This matrix equation corresponds to nine scalar equations, one for each matrix entry, and we look at three of these equations, corresponding to one column.

Suppose that $\mathbf{r} = \begin{pmatrix} x \\ y \\ z \end{pmatrix}$ and $\mathbf{d} = \begin{pmatrix} d_1 \\ d_2 \\ d_3 \end{pmatrix}$, and that $\mathbf{r}$ and $\mathbf{d}$ are corresponding columns of $\mathbf{X}$ and $\mathbf{I}_3$. The matrix equation, corresponding to these columns is $\mathbf{Mr} = \mathbf{d}$ and this corresponds to the simultaneous equations

$$
\begin{aligned}
a_1x + b_1y + c_1z &= d_1, \\
a_2x + b_2y + c_2z &= d_2, \\
a_3x + b_3y + c_3z &= d_3.
\end{aligned}
$$

The solution of these equations by elimination is very tedious. We indicate a fairly short method but omit much of the tedious algebra, which the reader should check for himself.

We eliminate y and z by multiplying the three equations respectively by $(b_2c_3 - b_3c_2)$, $(b_3c_1 - b_1c_3)$ and $(b_1c_2 - b_2c_1)$ and adding. In the resulting sum, the coefficients of y and z are respectively

$$b_1(b_2c_3 - b_3c_2) + b_2(b_3c_1 - b_1c_3) + b_3(b_1c_2 - b_2c_1)$$

and $$c_1(b_2c_3 - b_3c_2) + c_2(b_3c_1 - b_1c_3) + c_3(b_1c_2 - b_2c_1)$$

and, on expansion, each of these sums is seen to be zero. The coefficient of x is Δ and the constant term is D_1, where

$$\Delta = a_1(b_2c_3 - b_3c_2) + a_2(b_3c_1 - b_1c_3) + a_3(b_1c_2 - b_2c_1)$$

and $$D_1 = d_1(b_2c_3 - b_3c_2) + d_2(b_3c_1 - b_1c_3) + d_3(b_1c_2 - b_2c_1).$$

Note the symmetry of these expressions. Repeating the process after cyclically permuting (x, y, z) and (a, b, c), first z and x and then x and y can be eliminated in turn and we obtain three equations

$$\Delta x = D_1, \quad \Delta y = D_2, \quad \Delta z = D_3,$$

where D_1 is obtained by replacing a_i by d_i in Δ, $1 \leqslant i \leqslant 3$,
D_2 is obtained by replacing b_i by d_i in Δ, $1 \leqslant i \leqslant 3$,
and D_3 is obtained by replacing c_i by d_i in Δ, $1 \leqslant i \leqslant 3$,

We now take as the vector **d** the three columns of $\mathbf{I}_3$ in turn and let **X** and **D** be the (3×3) matrices whose corresponding columns are $\begin{pmatrix} x \\ y \\ z \end{pmatrix}$ and $\begin{pmatrix} D_1 \\ D_2 \\ D_3 \end{pmatrix}$. That is, on putting $d_1 = 1, d_2 = 0, d_3 = 0$,

$$D_1 = b_2c_3 - b_3c_2, \quad D_2 = b_3c_1 - b_1c_3, \quad D_3 = b_1c_2 - b_2c_1.$$

Similarly, if $d_1 = 0, d_2 = 1, d_3 = 0$, then

$$D_1 = c_2a_3 - c_3a_2, \quad D_2 = c_3a_1 - c_1a_3, \quad D_3 = c_1a_2 - c_2a_1,$$

and, if $d_1 = 0, d_2 = 0, d_3 = 1$,

$$D_1 = a_2b_3 - a_3b_2, \quad D_2 = a_3b_1 - a_1b_3, \quad D_3 = a_1b_2 - a_2b_1.$$

This gives us nine equations which can be written as one matrix equation

$$\Delta\mathbf{X} = \mathbf{D} = \begin{pmatrix} b_2c_3 - b_3c_2 & c_2a_3 - c_3a_2 & a_2b_3 - a_3b_2 \\ b_3c_1 - b_1c_3 & c_3a_1 - c_1a_3 & a_3b_1 - a_1b_3 \\ b_1c_2 - b_2c_1 & c_1a_2 - c_2a_1 & a_1b_2 - a_2b_1 \end{pmatrix}.$$

We have therefore shown that, if $\mathbf{MX} = \mathbf{I}_3$ then $\Delta\mathbf{X} = \mathbf{D}$. By means of a tedious matrix multiplication, of which we omit the details, the reader will be able to show that, given the above matrix **D**,

$$\mathbf{MD} = \Delta\mathbf{I}_3.$$

This means that

if $\Delta \neq 0$, then **M** is non-singular and $\mathbf{M}^{-1} = \dfrac{1}{\Delta}\mathbf{D}$,

if $\Delta = 0$, then $\mathbf{MD} = \mathbf{0}$, **M** is a zero divisor and hence singular.

As in the (2×2) case, Δ is called the determinant of **M**.

Definition Let $\mathbf{M} = \begin{pmatrix} a_1 & b_1 & c_1 \\ a_2 & b_2 & c_2 \\ a_3 & b_3 & c_3 \end{pmatrix}$, then the determinant of **M** is the scalar det (**M**), also written as $|\mathbf{M}|$, given by,

$$\det(\mathbf{M}) = \Delta = a_1b_2c_3 - a_1b_3c_2 + a_2b_3c_1 - a_2b_1c_3 + a_3b_1c_2 - a_3b_2c_1,$$

or

$$\Delta = a_1(b_2c_3 - c_2b_3) + b_1(c_2a_3 - a_2c_3) + c_1(a_2b_3 - b_2a_3).$$

Calculation of a determinant The symmetry of the definition of a determinant can be used to lay out the entries in a manner which assists the calculation. We avoid the use of suffices and consider the (2×2) and (3×3) determinants, given by

$$\begin{vmatrix} a & b \\ c & d \end{vmatrix} = ad - bc, \quad \begin{vmatrix} a & b & c \\ d & e & f \\ g & h & j \end{vmatrix} = aej - afh + bfg - bdj + cdh - ceg.$$

In the second and third rows, we repeat the terms cyclically on the left and right and pick out terms of the determinant by using diagonals forwards and backwards. Since the terms of the determinant each contain just one entry from each row and one from each column, we obtain all the terms in this way, those from a forwards diagonal being positive and those from a backwards diagonal being negative.

This is shown in Fig. 1.10.

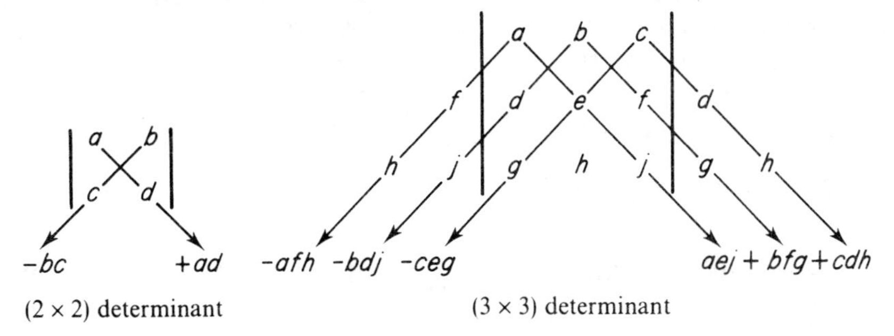

Fig. 1.10

We shall not be concerned with the evaluation of determinants of order greater than 3, when the use of forwards and backwards diagonals no longer applies. The determinant is a sum of all possible products of elements, taken one from each row and column with signs determined by an associated permutation.

EXAMPLE 1 *Write out the determinant of the square matrix* **A**, *given by* $\mathbf{A} = (a_{ij})$, *given that the order of* **A** *is* (*i*) 2, (*ii*) 3.

(i) $\begin{vmatrix} a_{11} & a_{12} \\ a_{21} & a_{22} \end{vmatrix} = a_{11}a_{22} - a_{12}a_{21}.$

(ii) $\begin{vmatrix} a_{11} & a_{12} & a_{13} \\ a_{21} & a_{22} & a_{23} \\ a_{31} & a_{32} & a_{33} \end{vmatrix} = \begin{matrix} a_{11}(a_{22}a_{33} - a_{23}a_{32}) \\ -a_{12}(a_{21}a_{33} - a_{23}a_{31}) \\ +a_{13}(a_{21}a_{32} - a_{22}a_{31}) \end{matrix}$

$$= \mathbf{a_{11}} \begin{vmatrix} a_{22} & a_{23} \\ a_{32} & a_{33} \end{vmatrix} - \mathbf{a_{12}} \begin{vmatrix} a_{21} & a_{23} \\ a_{31} & a_{33} \end{vmatrix} + \mathbf{a_{13}} \begin{vmatrix} a_{21} & a_{22} \\ a_{31} & a_{32} \end{vmatrix}.$$

EXAMPLE 2 *Find the determinant of the matrix:*

(*i*) $\begin{pmatrix} 3 & 2 \\ -2 & 1 \end{pmatrix}$, (*ii*) $\begin{pmatrix} 1 & 2 & 3 \\ 3 & 2 & 1 \\ -1 & -2 & 1 \end{pmatrix}$, (*iii*) $\begin{pmatrix} 1 & 2 & 3 \\ 2 & 4 & 6 \\ 0 & 1 & 2 \end{pmatrix}$.

(i) $\begin{vmatrix} 3 & 2 \\ -2 & 1 \end{vmatrix} = 3(1) - 2(-2) = \mathbf{7}.$

$$\text{(ii)} \begin{vmatrix} 1 & 2 & 3 \\ 3 & 2 & 1 \\ -1 & -2 & 1 \end{vmatrix} = \begin{matrix} (1)(2)(1)-(1)(1)(-2) \\ +(2)(1)(-1)-(2)(3)(1) \\ +(3)(3)(-2)-(3)(2)(-1) \end{matrix} = 2+2-2-6-18+6 = \mathbf{-16}.$$

$$\text{(iii)} \begin{vmatrix} 1 & 2 & 3 \\ 2 & 4 & 6 \\ 0 & 1 & 2 \end{vmatrix} = \begin{matrix} 1\{(4)(2)-(6)(1)\} \\ -2\{(2)(2)-(6)(0)\} \\ +3\{(2)(1)-(0)(4)\} \end{matrix} = 8-6-8+0+6-0 = \mathbf{0}.$$

Notice that, in the last matrix, rows 1 and 2 are dependent. In fact, whenever the rows (or columns) of a matrix are linearly dependent the determinant of the matrix is zero, see Chapter **14**.

EXAMPLE 3 *Let* $\mathbf{A} = \begin{pmatrix} 2 & 1 \\ 3 & -1 \end{pmatrix}$, $\mathbf{B} = \begin{pmatrix} 4 & -3 \\ 1 & 2 \end{pmatrix}$. *Show that*
(*i*) $|\mathbf{AB}| = |\mathbf{A}|\,|\mathbf{B}|$, (*ii*) $|\mathbf{A}|\,|\mathbf{A}^{-1}| = 1$, (*iii*) $|\mathbf{A}^{\mathrm{T}}| = |\mathbf{A}|$.

(i) $|\mathbf{A}| = -5$, $|\mathbf{B}| = 11$, $\mathbf{AB} = \begin{pmatrix} 9 & -4 \\ 11 & -11 \end{pmatrix}$, $|\mathbf{AB}| = -99+44 = -55 = |\mathbf{A}|\,|\mathbf{B}|$.

(ii) $\mathbf{A}^{-1} = \begin{pmatrix} 1/5 & 1/5 \\ 3/5 & -2/5 \end{pmatrix}$, $|\mathbf{A}^{-1}| = -2/25 - 3/25 = -1/5$, so

$|\mathbf{A}|\,|\mathbf{A}^{-1}| = (-5)\,(-1/5) = 1$.

(iii) $\mathbf{A}^{\mathrm{T}} = \begin{pmatrix} 2 & 3 \\ 1 & -1 \end{pmatrix}$, $|\mathbf{A}^{\mathrm{T}}| = -5 = |\mathbf{A}|$.

The results indicated in these examples are true in general. The determinant of the product of two matrices is equal to the product of their determinants. Since the determinant of the unit matrix is 1, the product of the determinants of a non-singular matrix and its inverse is 1. The determinant of the transpose of a matrix is equal to its determinant. Also, singular matrices are characterised by the vanishing of their determinants. These results are summarised in the next two theorems. Theorem 1.7 follows, for $n = 2$ and $n = 3$, from our investigation into the existence of inverses. It is true in general, but we give no proof since the determinant has not been defined for $n > 3$.

Theorem 1.7 Let $\mathbf{A}$ be an $n \times n$ matrix. Then
(i) $\mathbf{A}$ is non-singular $\Leftrightarrow \det(\mathbf{A}) \neq 0$,
(ii) $\mathbf{A}$ is singular $\Leftrightarrow \det(\mathbf{A}) = 0$.

Theorem 1.8 Let $\mathbf{A}$ and $\mathbf{B}$ be $(n \times n)$ matrices. Then
(i) $\det(\mathbf{AB}) = \det(\mathbf{A})\det(\mathbf{B})$,
(ii) $\mathbf{A}$ is non-singular $\Leftrightarrow$ the rows of $\mathbf{A}$ are linearly independent $\Leftrightarrow$ the columns of $\mathbf{A}$ are linearly independent.
(iii) $\mathbf{A}$ is singular $\Leftrightarrow$ the rows of $\mathbf{A}$ are linearly dependent $\Leftrightarrow$ the columns of $\mathbf{A}$ are linearly dependent.

Again, these results are generally true but we only prove them for $n \leqslant 3$.

The proof of (i) is given in Theorem 4.4. Parts (ii) and (iii) were proved for $n = 2$ in Theorem 1.4 and, for $n = 3$, they are proved in Chapter 14.

Some other important results, needed in future work, are given in Theorem 1.9 which states the effect on the determinant of a matrix of taking its transpose and of elementary row operations. The results are true for any square matrix but we only prove them for the (3×3) case.

Theorem 1.9

(i) Interchanging the rows with the columns of a matrix, to form the transpose matrix, leaves the value of its determinant unaltered.
(ii) Interchanging two rows (columns) of a matrix changes the sign of its determinant.
(iii) Multiplication of all the elements of one row (column) of a matrix by a constant k will multiply the determinant by k.
(iv) The value of the determinant of a matrix is unchanged if a constant multiple of one row (column) of the matrix is added to another row (column).

Proof Let $\mathbf{M} = \begin{pmatrix} a_1 & b_1 & c_1 \\ a_2 & b_2 & c_2 \\ a_3 & b_3 & c_3 \end{pmatrix}$, then we use the expansion

$$\det(\mathbf{M}) = a_1(b_2c_3 - c_2b_3) + b_1(c_2a_3 - a_2c_3) + c_1(a_2b_3 - b_2a_3).$$

(i) By rearranging the terms in the expansion,

$$\begin{aligned}\det(\mathbf{M}) &= a_1b_2c_3 - a_1b_3c_2 + a_2b_3c_1 - a_2b_1c_3 + a_3b_1c_2 - a_3b_2c_1 \\ &= a_1(b_2c_3 - b_3c_2) + a_2(b_3c_1 - b_1c_3) + a_3(b_1c_2 - b_2c_1) = \det(\mathbf{M}^\mathrm{T}).\end{aligned}$$

(ii) Interchanging rows 1 and 2 transforms $\mathbf{M}$ to $\mathbf{N}$ where

$$\begin{aligned}\det(\mathbf{N}) &= a_2(b_1c_3 - c_1b_3) + b_2(c_1a_3 - a_1c_3) + c_2(a_1b_3 - b_1a_3) \\ &= a_2b_1c_3 - a_2c_1b_3 + b_2c_1a_3 - b_2a_1c_3 + c_2a_1b_3 - c_2b_1a_3 \\ &= -a_1b_2c_3 + a_1b_3c_2 - a_3b_1c_2 + a_2b_1c_3 - a_2b_3c_1 + a_3b_2c_1 \\ &= -a_1(b_2c_3 - c_2b_3) - b_1(c_2a_3 - a_2c_3) - c_1(a_2b_3 - b_2a_3) \\ &= -\det(\mathbf{M}).\end{aligned}$$

Similarly, interchanging any other pair of rows, or, by (i), interchanging any pair of columns, has the same effect.

(iii) Multiplication by k of all the elements of row 1 of $\mathbf{M}$ gives a matrix $\mathbf{N}$, where

$$\begin{aligned}\det(\mathbf{N}) &= ka_1(b_2c_3 - c_2b_3) + kb_1(c_2a_3 - a_2c_3) + kc_1(a_2b_3 - b_2a_3) \\ &= k\det(\mathbf{M}),\end{aligned}$$

and similarly for the multiplication by k of any other row. By (i), the same is true for the multiplication by k of any column.

(iv) Let $\mathbf{L}$ be the matrix formed by adding k times the second row of $\mathbf{M}$ to the first row. Then

$$\begin{aligned}\det(\mathbf{L}) &= (a_1+ka_2)(b_2c_3-c_2b_3)+(b_1+kb_2)(c_2a_3-a_2c_3)\\ &\quad +(c_1+kc_2)\,(a_2b_3-b_2a_3)\\ &= a_1(b_2c_3-c_2b_3)+b_1(c_2a_3-a_2c_3)+c_1(a_2b_3-b_2a_3)\\ &\quad +k\{a_2(b_2c_3-c_2b_3)+b_2(c_2a_3-a_2c_3)+c_2(a_2b_3-b_2a_3)\}\\ &= \det(\mathbf{M})+k.0=\det(\mathbf{M}).\end{aligned}$$

Similarly, the determinant is unaltered if a multiple of any row is added to another row and, on using (i), the same is true for columns.

The operations described in (ii), (iii) and (iv) of Theorem 1.9 are elementary operations on the matrix, and can be performed by multiplication by elementary matrices.

These are of three types and we evaluate typical determinants in the 3×3 case. $\begin{vmatrix} p & 0 & 0\\ 0 & 1 & 0\\ 0 & 0 & 1\end{vmatrix}=p,\quad \begin{vmatrix} 0 & 1 & 0\\ 1 & 0 & 0\\ 0 & 0 & 1\end{vmatrix}=-1,\quad \begin{vmatrix} 1 & r & 0\\ 0 & 1 & 0\\ 0 & 0 & 1\end{vmatrix}=1$, using the notation of §1.4. Using these results and Theorem 1.9, Theorem 1.10 follows for 3×3 matrices. The theorem is also true for square matrices of any order. The proof is similar and we omit it.

Theorem 1.10 If $\mathbf{M}$ is a square matrix and $\mathbf{E}$ is an elementary matrix of the same order, then $\det(\mathbf{EM})=\det(\mathbf{E})\det(\mathbf{M})$.

These properties are demonstrated in the next example.

EXAMPLE 4 *Let the rows of the* (3×3) *matrix* $\mathbf{A}$ *be labelled* $\mathbf{r}_1$, $\mathbf{r}_2$ *and* $\mathbf{r}_3$, *where* $\mathbf{A}=\begin{pmatrix} 2 & 1 & -1\\ 0 & 1 & 2\\ 3 & -1 & 2\end{pmatrix}$. *Show that: (i)* $\left|\begin{pmatrix}\mathbf{r}_2\\ \mathbf{r}_1\\ \mathbf{r}_3\end{pmatrix}\right|=-|\mathbf{A}|$,

(ii) $\left|\begin{pmatrix}2\mathbf{r}_1\\ \mathbf{r}_2\\ \mathbf{r}_3\end{pmatrix}\right|=2|\mathbf{A}|$, *(iii)* $\left|\begin{pmatrix}\mathbf{r}_1-3\mathbf{r}_3\\ \mathbf{r}_2\\ \mathbf{r}_3\end{pmatrix}\right|=|\mathbf{A}|$.

(i) $\left|\begin{pmatrix}\mathbf{r}_2\\ \mathbf{r}_1\\ \mathbf{r}_3\end{pmatrix}\right|=\begin{vmatrix} 0 & 1 & 2\\ 2 & 1 & -1\\ 3 & -1 & 2\end{vmatrix}=0-0+(-3)-4+(-4)-6=-17$, and

$$|\mathbf{A}|=4-(-4)+6-0+0-(-3)=17, \text{ as required.}$$

(ii) $\left|\begin{pmatrix}2\mathbf{r}_1\\ \mathbf{r}_2\\ \mathbf{r}_3\end{pmatrix}\right|=\begin{vmatrix} 4 & 2 & -2\\ 0 & 1 & 2\\ 3 & -1 & 2\end{vmatrix}=8-(-8)+12-0+0-(-6)=34=2|\mathbf{A}|.$

(iii) $\left|\begin{pmatrix}\mathbf{r}_1-3\mathbf{r}_3\\ \mathbf{r}_2\\ \mathbf{r}_3\end{pmatrix}\right|=\begin{vmatrix} -7 & 4 & -7\\ 0 & 1 & 2\\ 3 & -1 & 2\end{vmatrix}=-14-14+24-0+0-(-21)=17=|\mathbf{A}|.$

These properties can also be used in the evaluation of the determinant of a matrix in cases when its rows and columns have certain symmetry properties. This is demonstrated in the next example.

EXAMPLE 5 *Evaluate the determinant* Δ, *given by*

$$\Delta = \begin{vmatrix} x & x-y & x+y \\ x-y & x+y & x \\ x+y & x & x-y \end{vmatrix}.$$

Use this result in order to show that

$$\begin{vmatrix} 4 & 11 & -3 \\ 11 & -3 & 4 \\ -3 & 4 & 11 \end{vmatrix} = \begin{vmatrix} 4 & -3 & 11 \\ -3 & 11 & 4 \\ 11 & 4 & -3 \end{vmatrix}.$$

At each stage of the work, we label the rows of the determinant $\mathbf{r}_1$, $\mathbf{r}_2$ and $\mathbf{r}_3$ and the columns $\mathbf{c}_1$, $\mathbf{c}_2$ and $\mathbf{c}_3$. We use elementary operations which either leave the value of the determinant unchanged or multiply it by a constant. It is useful to record what is being done at each step and we do this by borrowing the notation $\leftarrow$ from computer programs, where '$\leftarrow$' means 'becomes'.

$$\Delta = \begin{vmatrix} x & x-y & x+y \\ x-y & x+y & x \\ x+y & x & x-y \end{vmatrix} = \begin{vmatrix} x & x-y & x+y \\ -y & 2y & -y \\ y & y & -2y \end{vmatrix} \qquad \begin{matrix} (\mathbf{r}_1 \leftarrow \mathbf{r}_1) \\ (\mathbf{r}_2 \leftarrow \mathbf{r}_2 - \mathbf{r}_1) \\ (\mathbf{r}_3 \leftarrow \mathbf{r}_3 - \mathbf{r}_1) \end{matrix}$$

$$= -y^2 \begin{vmatrix} x & x-y & x+y \\ 1 & -2 & 1 \\ 1 & 1 & -2 \end{vmatrix}, \text{ if } y \neq 0, \qquad \begin{matrix} (\mathbf{r}_1 \leftarrow \mathbf{r}_1) \\ (\mathbf{r}_2 \leftarrow (-1/y)\mathbf{r}_2) \\ (\mathbf{r}_3 \leftarrow (1/y)\mathbf{r}_3) \end{matrix}$$

$$= -y^2 \begin{vmatrix} 3x & x-y & x+y \\ 0 & -2 & 1 \\ 0 & 1 & -2 \end{vmatrix} \qquad \begin{matrix} (\mathbf{c}_1 \leftarrow \mathbf{c}_1 + \mathbf{c}_2 + \mathbf{c}_3) \\ (\mathbf{c}_2 \leftarrow \mathbf{c}_2) \\ (\mathbf{c}_3 \leftarrow \mathbf{c}_3) \end{matrix}$$

$$= -3xy^2 \begin{vmatrix} -2 & 1 \\ 1 & -2 \end{vmatrix} = -3xy^2(4-1) = \mathbf{-9xy^2}.$$

In the last line above, we use the fact that the two zeros in the first column mean that there are only two non-zero terms in the expansion of the determinant, namely the terms of the (2×2) determinant. Although we assumed that $y \neq 0$, it is easily checked that the result is still true when $y = 0$.

Now let $x = 4$, $x + y = -3$ and $x - y = 11$, so that $y = -7$, then $\begin{vmatrix} 4 & 11 & -3 \\ 11 & -3 & 4 \\ -3 & 4 & 11 \end{vmatrix} = -9(4)(49)$. Alternatively, let $x=4$, $x+y=11$, $x-y=-3$, so that $y = 7$, then $\begin{vmatrix} 4 & -3 & 11 \\ -3 & 11 & 4 \\ 11 & 4 & -3 \end{vmatrix} = -9(4)(49)$, as was to be shown.

EXERCISE 1.5

1 Find the determinants of the following matrices:

(i) $\begin{pmatrix} 3 & 2 \\ 4 & 6 \end{pmatrix}$, (ii) $\begin{pmatrix} -5 & 2 \\ -3 & -1 \end{pmatrix}$, (iii) $\begin{pmatrix} 2 & -1 & 0 \\ 1 & -2 & 3 \\ 4 & 6 & -2 \end{pmatrix}$, (iv) $\begin{pmatrix} 3 & -2 & 1 \\ 1 & -3 & -1 \\ 0 & 2 & -1 \end{pmatrix}$.

2 Prove that the following determinants are zero, without evaluating them:

(i) $\begin{vmatrix} 2 & 1 & -3 \\ 3 & 2 & -1 \\ 4 & 2 & -6 \end{vmatrix}$, (ii) $\begin{vmatrix} 1 & 2 & 3 \\ -1 & 3 & 2 \\ 0 & 1 & 1 \end{vmatrix}$, (iii) $\begin{vmatrix} 2 & 4 & 3 & 1 \\ 1 & -2 & -3 & 5 \\ 7 & 8 & 2 & 1 \\ 0 & 0 & 0 & 0 \end{vmatrix}$.

3 Prove that theorem 1.9 is true for a general (3×3) determinant.

4 Evaluate the determinant $\begin{vmatrix} 1 & 3 & 4 \\ 13 & 12 & 10 \\ 27 & 27 & 24 \end{vmatrix}$. *(L)*

5 Factorise completely $\begin{vmatrix} 1 & x & x^3+1 \\ 1 & y & y^3+1 \\ 1 & 1 & 2 \end{vmatrix}$. *(JMB)*

6 Evaluate the determinant $\begin{vmatrix} \lambda & \lambda+1 & \lambda-1 \\ \lambda+2 & \lambda & \lambda-2 \\ \lambda-3 & \lambda+3 & \lambda \end{vmatrix}$ showing that its value is a numerical multiple of λ. Evaluate $\begin{vmatrix} 7 & 8 & 6 \\ 4 & 10 & 7 \\ 9 & 7 & 5 \end{vmatrix}$. *(JMB)*

7 Show that $(\alpha+\beta+\gamma)$ is a factor of the determinant $\Delta = \begin{vmatrix} \alpha & \beta & \gamma \\ \beta & \gamma & \alpha \\ \gamma & \alpha & \beta \end{vmatrix}$ and find the other factor. Hence find the value of Δ in terms of k when α, β and γ are the roots of the equation $x^3 - kx^2 - k^2x + 1 = 0$. *(JMB)*

8 Show that $a-1$ is a factor of the determinant $\begin{vmatrix} 1 & a+bc & a^2+b^2c^2 \\ 1 & b+ca & b^2+c^2a^2 \\ 1 & c+ab & c^2+a^2b^2 \end{vmatrix}$. Hence, or otherwise, factorise the determinant completely. *(JMB)*

9 Show that $x+a+y$ is a factor of the determinant $\begin{vmatrix} a & x & y \\ x & a & y \\ x & y & a \end{vmatrix}$. Express the determinant as a product of three factors. Hence find all the values of θ in the range $0 \leqslant \theta \leqslant \pi$ which satisfy the equation $\begin{vmatrix} 1 & \cos\theta & \cos 2\theta \\ \cos\theta & 1 & \cos 2\theta \\ \cos\theta & \cos 2\theta & 1 \end{vmatrix} = 0$. *(JMB)*

MISCELLANEOUS EXERCISE 1

1 Given that $\mathbf{A} = \begin{pmatrix} a & 0 & a \\ 0 & a & 0 \\ a & 0 & a \end{pmatrix}$, $\mathbf{B} = \begin{pmatrix} 0 & b & 0 \\ b & b & b \\ 0 & b & 0 \end{pmatrix}$, find (a) $\mathbf{A}+\mathbf{B}$, (b) $\mathbf{AB}$. *(L)*

2 Given the matrices $\mathbf{A} = \begin{pmatrix} 1 & 2 \\ 0 & 1 \end{pmatrix}$, $\mathbf{I} = \begin{pmatrix} 1 & 0 \\ 0 & 1 \end{pmatrix}$, obtain constants p and q such that $\mathbf{A}^2 = p\mathbf{A}+q\mathbf{I}$. *(L)*

3 Show that, if $\mathbf{A} = \begin{pmatrix} 1 & -1 \\ 2 & -1 \end{pmatrix}$ and $\mathbf{B} = \begin{pmatrix} 1 & 1 \\ 4 & -1 \end{pmatrix}$, then $(\mathbf{A}+\mathbf{B})^2 = \mathbf{A}^2+\mathbf{B}^2$. *(L)*

4 Given that $\mathbf{A} = \begin{pmatrix} 1 & -2 & 0 \\ 2 & -5 & -4 \\ 3 & -6 & 1 \end{pmatrix}$, $\mathbf{B} = \begin{pmatrix} 29 & -2 & -8 \\ 14 & -1 & -4 \\ -3 & 0 & 1 \end{pmatrix}$, find $\mathbf{AB}$ and $\mathbf{BA}$.
Hence, or otherwise, solve the equations:

(a) $x-2y = 3$,
$2x-5y-4z = 5$,
$3x-6y+z = 9$.

(b) $29x-2y-8z = -3$,
$14x-y-4z = 1$,
$-3x+z = -1$. *(L)*

5 The matrices $\mathbf{A}$, $\mathbf{B}$ and $\mathbf{A}+\mathbf{B}$ are non-singular. Show that $\{\mathbf{A}(\mathbf{A}+\mathbf{B})^{-1}\mathbf{B}\}^{-1} = \mathbf{A}^{-1}+\mathbf{B}^{-1}$. *(L)*

6 The transpose of any matrix $\mathbf{M}$ is denoted by $\mathbf{M}^{\mathrm{T}}$, square matrices such that $\mathbf{M}^{\mathrm{T}} = \mathbf{M}$ are called symmetric, and square matrices such that $\mathbf{M}^{\mathrm{T}} = -\mathbf{M}$ are called skew-symmetric. Determine the truth or falsity of each of the following statements about (2×2) matrices, giving the proofs or counter-examples as appropriate.
(i) No matrix, other than the null matrix, is both symmetric and skew-symmetric.
(ii) $\mathbf{M}$ is skew-symmetric $\Rightarrow \mathbf{M}^2$ is symmetric.
(iii) $\mathbf{A}$ and $\mathbf{B}$ are both symmetric $\Rightarrow \mathbf{AB}$ is symmetric.
(iv) Any matrix $\mathbf{M}$ can be expressed as $\mathbf{P}+\mathbf{Q}$ where $\mathbf{P}$ is symmetric and $\mathbf{Q}$ is skew-symmetric. *(C)*

7 Given that $\mathbf{a}$, $\mathbf{b}$, $\mathbf{c}$ are linearly independent vectors, determine whether the following vectors are linearly independent: (i) $\{\mathbf{a}, \mathbf{0}\}$ (ii) $\{\mathbf{a}+\mathbf{b}, \mathbf{b}+\mathbf{c}, \mathbf{c}+\mathbf{a}\}$ (iii) $\{\mathbf{a}+2\mathbf{b}+\mathbf{c}, \mathbf{a}-\mathbf{b}-\mathbf{c}, 5\mathbf{a}+\mathbf{b}-\mathbf{c}\}$. *(JMB)*

8 If $\mathbf{M} = \begin{pmatrix} 1 & a \\ 0 & 1 \end{pmatrix}$, prove by induction that, when n is a positive integer,

$$\mathbf{M}^n = \begin{pmatrix} 1 & na \\ 0 & 1 \end{pmatrix}.$$ *(L)*

9 If $\mathbf{A} = \begin{pmatrix} 1 & 1 \\ 0 & 2 \end{pmatrix}$, find $\mathbf{A}^2$, $\mathbf{A}^3$ and conjecture a form for $\mathbf{A}^n$ where n is any positive integer. Prove the truth of your conjecture by mathematical induction.

10 (i) A matrix $\mathbf{A}$ is said to be symmetric if and only if $\mathbf{A}^{\mathrm{T}} = \mathbf{A}$. It is said to be skew-symmetric if and only if $\mathbf{A}^{\mathrm{T}} = -\mathbf{A}$. Given the matrix $\mathbf{A} = \begin{pmatrix} a & b \\ c & d \end{pmatrix}$,

$abcd \neq 0$, obtain the matrix $\mathbf{AA}^{\mathrm{T}} - \mathbf{A}^{\mathrm{T}}\mathbf{A}$. Given that $\mathbf{AA}^{\mathrm{T}} = \mathbf{A}^{\mathrm{T}}\mathbf{A}$, show that either $\mathbf{A}$ is symmetric or $\mathbf{A} - a\mathbf{I}$ is skew-symmetric, where $\mathbf{I}$ is the unit matrix.
(ii) $\mathbf{P}$ and $\mathbf{Q}$ are non-singular square matrices.
Prove that $(\mathbf{PQ})^{-1} = \mathbf{Q}^{-1}\mathbf{P}^{-1}$. (*L*)

11 The vectors **a**, **b**, **c**, **d** are given by

$$\mathbf{a} = \begin{pmatrix} 1 \\ 2 \\ -1 \end{pmatrix}, \quad \mathbf{b} = \begin{pmatrix} 0 \\ 3 \\ 4 \end{pmatrix}, \quad \mathbf{c} = \begin{pmatrix} 1 \\ 2 \\ 0 \end{pmatrix}, \quad \mathbf{d} = \begin{pmatrix} 3 \\ -3 \\ -14 \end{pmatrix}.$$

(i) Show that $\{\mathbf{a}, \mathbf{b}, \mathbf{c}\}$ is a basis for the space of 3-dimensional vectors
(ii) Express **d** as a linear combination of **a**, **b** and **c**. (*JMB*)

12 (i) Given that $\mathbf{A} = \begin{pmatrix} 1 & -1 & 1 \\ -1 & 0 & 0 \\ 2 & 0 & -1 \end{pmatrix}$, show that $\mathbf{A}^3 - 4\mathbf{A} - \mathbf{I} = \mathbf{0}$, where $\mathbf{I}$ is the unit matrix and $\mathbf{0}$ is the null matrix. Deduce that $\mathbf{A}^{-1} = \mathbf{A}^2 - 4\mathbf{I}$ and hence find $\mathbf{A}^{-1}$.
(ii) A matrix $\mathbf{A}$ is said to be symmetric if $\mathbf{A} = \mathbf{A}^{\mathrm{T}}$. For any 2×2 matrix $\mathbf{B}$, show that both $\mathbf{B} + \mathbf{B}^{\mathrm{T}}$ and $\mathbf{BB}^{\mathrm{T}}$ are symmetric. (*L*)

13 Given that $\mathbf{A} = \begin{pmatrix} -1 & 2 & -3 \\ 2 & -1 & 4 \\ 3 & 4 & 1 \end{pmatrix}$, $\mathbf{B} = \begin{pmatrix} -17 & -14 & 5 \\ 10 & 8 & -2 \\ 11 & 10 & -3 \end{pmatrix}$,

$\mathbf{I} = \begin{pmatrix} 1 & 0 & 0 \\ 0 & 1 & 0 \\ 0 & 0 & 1 \end{pmatrix}$, show that $\mathbf{AB}$ is a scalar multiple of $\mathbf{I}$ and obtain $\mathbf{A}^{-1}$ and $\mathbf{B}^{-1}\mathbf{A}^{-1}$. Hence, or otherwise, solve the equations

$$\begin{aligned} -x + 2y - 3z &= -8, \\ 2x - y + 4z &= 17, \\ 3x + 4y + z &= 22, \end{aligned}$$

and express the vector $(-8\mathbf{i} + 17\mathbf{j} + 22\mathbf{k})$ as a linear combination of the vectors $(-\mathbf{i} + 2\mathbf{j} + 3\mathbf{k})$, $(2\mathbf{i} - \mathbf{j} + 4\mathbf{k})$, $(-3\mathbf{i} + 4\mathbf{j} + \mathbf{k})$. (*L*)

14 Evaluate the determinant $\begin{vmatrix} 1 & \sin A & \cos A \\ \cos B & \cos C & \sin C \\ \sin B & -\sin C & \cos C \end{vmatrix}$, where A, B and C are the angles of a triangle. (*JMB*)

15 Show that the set of vectors $\mathbf{x}_1 = \begin{pmatrix} -1 \\ 2 \\ 1 \end{pmatrix}$, $\mathbf{x}_2 = \begin{pmatrix} 1 \\ 1 \\ 2 \end{pmatrix}$ and $\mathbf{x}_3 = \begin{pmatrix} -1 \\ 8 \\ 7 \end{pmatrix}$

spans the same space as the set of vectors $\mathbf{y}_1 = \begin{pmatrix} -5 \\ 4 \\ -1 \end{pmatrix}$, $\mathbf{y}_2 = \begin{pmatrix} 1 \\ 7 \\ 8 \end{pmatrix}$ and

$\mathbf{y}_3 = \begin{pmatrix} -11 \\ 1 \\ -10 \end{pmatrix}$. (*JMB*)

16 Prove that $(a + b + c)$ is a factor of the determinant Δ, where $\Delta = \begin{vmatrix} a & b & c \\ c & a & b \\ b & c & a \end{vmatrix}$,

and determine the other factor. Show that $\Delta = 0$ only if $a + b + c = 0$ or

$a = b = c$. Hence, or otherwise, solve the equations

(i) $\begin{vmatrix} x^2 & -x & 1 \\ 1 & x^2 & -x \\ -x & 1 & x^2 \end{vmatrix} = 0$, (ii) $\begin{vmatrix} x & 1 & -x \\ -x & x & 1 \\ 1 & -x & x \end{vmatrix} + \begin{vmatrix} i & 1 & -i \\ -i & i & 1 \\ 1 & -i & i \end{vmatrix} = 0$,

$\{$where $i = \sqrt{(-1)}\}$. (*JMB*)

17 (a) Given that the roots of the equation $x^3 + qx + r = 0$ are α, β, γ, express the value of the determinant $\begin{vmatrix} 1+\alpha & 1 & 1 \\ 1 & 1+\beta & 1 \\ 1 & 1 & 1+\gamma \end{vmatrix}$ in terms of q and r.

(b) Show that $x = -6$ is the only real root of the equation $\begin{vmatrix} x & 4 & 2 \\ 1 & x & 5 \\ 3 & 3 & x \end{vmatrix} = 0$.

(*JMB*)

18 (a) Show that the set of vectors $\left\{\begin{pmatrix} 1 \\ 0 \\ 1 \end{pmatrix}, \begin{pmatrix} -1 \\ 1 \\ 1 \end{pmatrix}, \begin{pmatrix} 0 \\ -2 \\ 0 \end{pmatrix}\right\}$ in $\mathbb{R}^3$ is both a linearly independent set and a spanning set of $\mathbb{R}^3$.

(b) For each of the following sets of vectors in $\mathbb{R}^3$, either extend the set to form a basis for $\mathbb{R}^3$, by adding a suitable vector to the set, or give a reason why it cannot be so extended to form a basis.

(i) $\left\{\begin{pmatrix} 1 \\ 1 \\ 2 \end{pmatrix}, \begin{pmatrix} 0 \\ 0 \\ 0 \end{pmatrix}\right\}$ (ii) $\left\{\begin{pmatrix} 0 \\ 1 \\ 1 \end{pmatrix}, \begin{pmatrix} 1 \\ 0 \\ 0 \end{pmatrix}\right\}$ (iii) $\left\{\begin{pmatrix} 1 \\ 0 \\ 3 \end{pmatrix}, \begin{pmatrix} 2 \\ 0 \\ 6 \end{pmatrix}\right\}$.

(c) For each of the following sets of vectors in $\mathbb{R}^3$, either select a subset of vectors which will form a basis for $\mathbb{R}^3$, or give a reason why it is not possible to do this.

(i) $\left\{\begin{pmatrix} 1 \\ 0 \\ 0 \end{pmatrix}, \begin{pmatrix} 0 \\ 1 \\ 0 \end{pmatrix}, \begin{pmatrix} 1 \\ 1 \\ 1 \end{pmatrix}, \begin{pmatrix} 1 \\ 1 \\ 0 \end{pmatrix}\right\}$ (ii) $\left\{\begin{pmatrix} 0 \\ 0 \\ 0 \end{pmatrix}, \begin{pmatrix} 1 \\ 0 \\ 0 \end{pmatrix}, \begin{pmatrix} 0 \\ 0 \\ 1 \end{pmatrix}, \begin{pmatrix} 1 \\ 0 \\ 1 \end{pmatrix}\right\}$

(iii) $\left\{\begin{pmatrix} 1 \\ 2 \\ 1 \end{pmatrix}, \begin{pmatrix} 0 \\ 1 \\ 2 \end{pmatrix}, \begin{pmatrix} 0 \\ -1 \\ -2 \end{pmatrix}, \begin{pmatrix} 1 \\ 3 \\ 3 \end{pmatrix}\right\}$. (*C*)

19 (i) Given that $D = \begin{vmatrix} a^2 & a & 1 \\ b^2 & b & 1 \\ c^2 & c & 1 \end{vmatrix}$, express D in a factorised form.

(ii) Given the matrix $\mathbf{A}$, where $\mathbf{A} = \begin{pmatrix} 0 & 0 & 0 \\ 1 & 0 & 0 \\ 0 & 1 & 0 \end{pmatrix}$, find $\mathbf{A}^2$ and $\mathbf{A}^3$. Hence show that $\mathbf{I} + \mathbf{A}t + \mathbf{A}^2t^2/2! + \ldots + \mathbf{A}^n t^n/n! + \ldots = \begin{pmatrix} 1 & 0 & 0 \\ t & 1 & 0 \\ t^2/2 & t & 1 \end{pmatrix}$, where $\mathbf{I}$ is the 3×3 unit matrix and t is a real number. (*L*)

2 Inequalities

2.1 Revision

Inequalities were studied in Chapter **13** of *Pure Mathematics*. We now extend this work and begin with a revision exercise. To remind the reader of some of the methods, Examples 1 and 2 illustrate the treatment of inequalities involving rational functions and modulus signs.

EXAMPLE 1 *Solve the inequalities* $0 \leqslant \dfrac{x^2-x-6}{2x-2} \leqslant 1$.

For the left hand inequality, factorise numerator and denominator giving $0 \leqslant \dfrac{(x-3)(x+2)}{2(x-1)}$. This means that either two or none of the three factors $(x-3)$, $(x+2)$ and $(x-1)$ can be negative. The points where these factors change sign are marked on the number line together with the signs of the factors on each side of the points, as is shown in Fig. 2.1. It is then possible to read off the figure that the inequality is satisfied by $-2 \leqslant x < 1$ and by $3 \leqslant x$. For the right hand inequality, bring all the terms to the right hand side and factorise.

$$0 \leqslant 1 - \frac{x^2-x-6}{2x-2} = \frac{2x-2-x^2+x+6}{2x-2} = \frac{-x^2+3x+4}{2x-2} = -\frac{(x-4)(x+1)}{2(x-1)}.$$

This inequality is satisfied if one or three of the factors $(x+1)$, $(x-1)$, $(x-4)$ are negative. By marking the signs of these factors for various values of x on

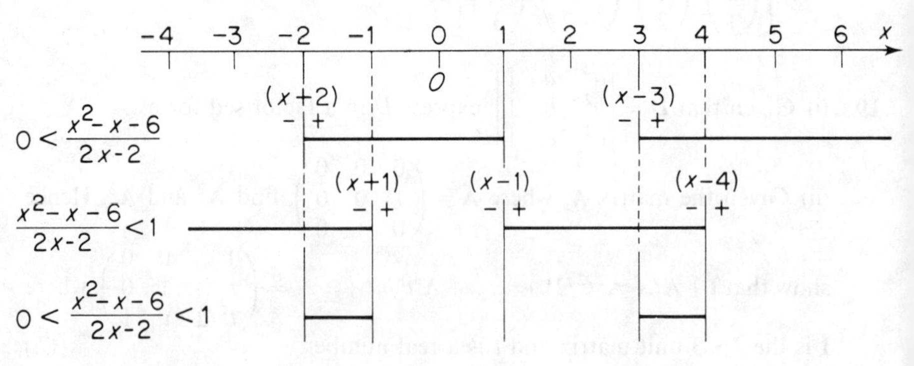

Fig. 2.1 $0 \leqslant \dfrac{x^2-x-6}{2x-2} \leqslant 1$ is satisfied by $-2 \leqslant x \leqslant -1$ and by $3 \leqslant x \leqslant 4$

the number line, the inequality is seen to be satisfied when $x \leqslant -1$ and when $1 < x \leqslant 4$, see Fig. 2.1. The combined inequalities are then satisfied by values of x in the intersection of the two sets previously obtained. From the diagram, the pair of inequalities is satisfied by $-2 \leqslant x \leqslant -1$ and by $3 \leqslant x \leqslant 4$. The solution set is $\{\mathbf{x}: -\mathbf{2} \leqslant \mathbf{x} \leqslant -\mathbf{1}\} \cup \{\mathbf{x}: \mathbf{3} \leqslant \mathbf{x} \leqslant \mathbf{4}\}$.

EXAMPLE 2 *Solve the inequality* $|x+1|+|3-2x| < 5$.

First consider the points where the arguments of the two moduli functions change sign. These are at $x = -1$ and at $x = 3/2$ so there are three ranges of values of x to consider.

If $x < -1$, then $-(x+1)+(3-2x) < 5$, so $-x-1+3-2x < 5$, that is $-3x < 3$ or $x > -1$, which is false. So the inequality is never satisfied for $x < -1$.

If $-1 \leqslant x < 3/2$, then $x+1+(3-2x) < 5$, so $-x+4 < 5$, $-x < 1, x > -1$. The inequality is satisfied provided that $-1 < x < 3/2$.

If $3/2 \leqslant x$, then $x+1-(3-2x) < 5$, so $3x < 7$, $x < 7/3$. Combining these results, the solution is $\mathbf{-1 < x < 7/3}$.

An alternative method is to use a graph. In Fig. 2.2, the graphs of $y = |x+1|$ and $y = |3-2x|$ are drawn. Each graph consists of two half lines and the graph of $y = |x+1|+|3-2x|$ is therefore made up of three line segments. The gradients of these line segments are -3, -1 and 3 respectively and they meet at $(-1, 5)$ and $(3/2, 5/2)$. The line $y = 5$ is also drawn on the graph and it can then be seen that the required inequality is satisfied for $\mathbf{-1 < x < 7/3}$.

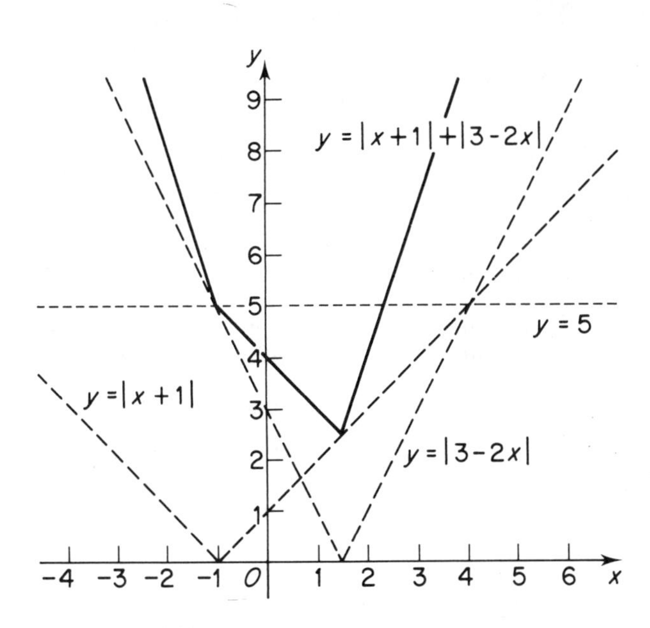

Fig. 2.2 $|x+1|+|3-2x| < 5$ for $-1 < x < 7/3$

EXERCISE 2.1

1 Solve the inequalities: (i) $(x+1)(x-2)(x-3) > 0$, (ii) $(x^2-1)(x+4) < 0$, (iii) $\dfrac{x(x-2)}{x^2-1} > 0$, (iv) $\dfrac{x^2+2x-3}{x^2-6x+8} < 0$.

2 Find the set of values of x for which: (i) $\dfrac{x}{x-4} > \dfrac{1}{x-3}$, (ii) $\dfrac{x+1}{x-2} > \dfrac{x}{x-1}$, (iii) $0 < \dfrac{x-1}{x(x+1)} < 1$, (iv) $0 < \dfrac{(x+3)(x-2)}{2x} < 1$.

3 Find the set of values of x for which $|2x-5| < |x|$. (*L*)

4 Find those values of x which satisfy: (i) $|x-1| < x^2+1$, (ii) $|x| < |3x-2|$, (iii) $|x+1| < 2|x-3|$, (iv) $|x^2-3| > |x^2-5|$.

5 Find the set of values of x for which $|x^2-3| > |x^2-2|$. (*L*)

6 (i) The geometric mean and the arithmetic mean of the real numbers x and y are G and A respectively. Given that $x > y > 0$, prove that $A > G$.
(ii) Solve the inequality $\dfrac{1}{x-4} > \dfrac{x}{x-6}$. (*L*)

7 (a) Sketch the graph of $y = \dfrac{4x-1}{x+5}$. Hence, or otherwise, find the set of values of x for which $-1 < \dfrac{4x-1}{x+5} < 1$.
(b) Find the set of values of x for which $3x^3+2x^2+6 \geqslant 19x$. (*C*)

8 Find the set of values of x for which (a) $x+4 > x(x-2)$, (b) $\dfrac{1}{x-2} > \dfrac{x}{x+4}$, (c) $|x+4| > 4|x-2|$. (*L*)

2.2 Inequalities arising from squares

The fact that a square is non-negative gives rise to some interesting inequalities. For example, if a and b are any real numbers,

$$(a+b)^2 \geqslant 0 \text{ so } a^2+2ab+b^2 \geqslant 0 \quad \text{and} \quad a^2+b^2 \geqslant -2ab.$$

Also $$(a-b)^2 \geqslant 0 \text{ so } a^2-2ab+b^2 \geqslant 0 \quad \text{and} \quad a^2+b^2 \geqslant 2ab.$$

This means that $a^2+b^2 \geqslant 2|ab|$ and the equality only occurs when either $a = b$ or $a = -b$, that is, only when $|a| = |b|$.

EXAMPLE 1 *For all real numbers a and b, prove that $a^2+ab+b^2 \geqslant 0$. Deduce that $a^4+b^4 \geqslant a^3b+ab^3$ and show that equality occurs only when $a = b$.*

Completing the square, $a^2+ab = (a+\frac{1}{2}b)^2 - \frac{1}{4}b^2$, so that

$$a^2+ab+b^2 = (a+\tfrac{1}{2}b)^2 + \tfrac{3}{4}b^2 \geqslant 0, \text{ (sum of two squares)}.$$

For the other inequality, it may be helpful to see how the values of a^4+b^4 and a^3b+ab^3 vary for different values of a and b.

a =	0	1	1	1	−1	1
b =	0	1	−1	2	−1	−2
a^4+b^4 =	0	2	2	17	2	17
a^3b+ab^3 =	0	2	−2	10	2	−10

The results indicate that the inequality is satisfied and that equality occurs only when $a = b$. This is not a proof, however, and the general result is proved algebraically. Bring everything to the left hand side, and note that $a^4 + b^4 - a^3b - ab^3 = 0$ when $a = b$, so $(a-b)$ is a factor. On factorising,

$$a^4 + b^4 - a^3b - ab^3 = (a-b)(a^3 - b^3) = (a-b)(a-b)(a^2 + ab + b^2).$$

Now $(a-b)^2 \geqslant 0$ and $(a^2 + ab + b^2) \geqslant 0$ and hence $a^4 + b^4 - a^3b - ab^3 \geqslant 0$ so that $a^4 + b^4 \geqslant a^3b + ab^3$. For equality, either $a - b = 0$ or $a^2 + ab + b^2 = 0$ and this last equality only occurs for $a = 0 = b$. Hence, the condition for equality is $a = b$.

EXAMPLE 2 *Prove that, if a and b are distinct positive numbers, then* $a^3 + b^3 > a^2b + ab^2$. *Show that the result may be false if one of a or b is negative.*

As in Example 1, bring all the terms to the L.H.S. and factorise.

$$a^3 + b^3 - a^2b - ab^2 = (a-b)(a^2 - b^2) = (a-b)^2(a+b).$$

Since a and b are positive, $a + b > 0$ and, since a and b are distinct, $(a-b)^2 > 0$. Hence $a^3 + b^3 - a^2b - ab^2 > 0$ and so $a^3 + b^3 > a^2b + ab^2$. Again, as in Example 1, try some values of a and b to obtain some idea of the values of the two expressions.

a	$= 0$	1	1	1	1	-1	-1
b	$= 0$	1	-1	-2	2	-1	-2
$a^3 + b^3$	$= 0$	2	0	-7	9	-2	-9
$a^2b + ab^2$	$= 0$	2	0	2	6	-2	-6

The inequality is false for $a = 1$, $b = -2$, or for $a = -1$, $b = -2$.

EXERCISE 2.2

1 Prove that $a^4 + b^4 \geqslant 2a^2b^2$ and deduce that $a^4 + b^4 + c^4 + d^4 \geqslant 4abcd$.

2 Given that $0 < a < b$, prove that $a^3 + 2b^3 > 3ab^2$.

3 Use the fact that the squares of $a-b$, $b-c$, $c-a$ are non-negative to show that $a^2 + b^2 + c^2 \geqslant ab + bc + ca$.

4 Prove that if $a > b$ then $a^5 > b^5$ and deduce that, for all real numbers a and b, if $a \neq b$ then $(a-b)$ and $(a^5 - b^5)$ have the same sign. Prove that, if $a \neq b$, then $a^4 + a^3b + a^2b^2 + ab^3 + b^4 > 0$. Find a counter-example to show that $a^3 + a^2b + ab^2 + b^3$ is not always positive.

2.3 Inequalities in two variables

When a set of inequalities in two variables x and y is to be satisfied, the situation can be illustrated in the Cartesian plane. Each inequality corresponds to points lying on one side of some curve, given by the corresponding equality. The solution of the inequalities will be some area of the plane which can be indicated by shading the regions which are not required. The boundaries of the area may correspond to strict inequalities

'$<$' or to weak inequalities '$\leqslant$' and the convention is to denote the former by a dotted line and the latter by a full line.

EXAMPLE 1 *Indicate on a diagram, by shading the regions not required, the region of the plane $O(x, y)$ for which $y^2 \leqslant 4x$ and $x^2 < 4y$.*

The boundaries are the two parabolas $y^2 = 4x$ and $x^2 = 4y$ and the first step is to find their points of intersection. If $y^2 = 4x$ and $x^2 = 4y$, then $y^4 = 16x^2 = 64y$ so that $y = 0$ or $y^3 = 64$ and $y = 4$. The corresponding values of x are 0 and 4 and the curves intersect at (0, 0) and (4, 4). The bounding curves are shown in Fig. 2.3 and the area outside them is shaded, leaving the required area between them unshaded. The boundary $y^2 = 4x$ is a solid line since $y^2 \leqslant 4x$ and the boundary $x^2 = 4y$ is dotted since $x^2 < 4y$.

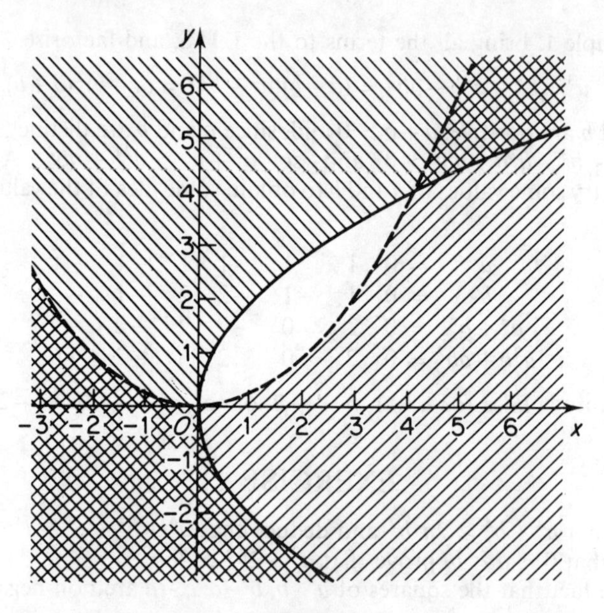

Fig. 2.3

Problems arise involving inequalities when a number of variables have to satisfy certain constraints and it is necessary to minimise or to maximise some function of the variables. Thus a company may wish to maximise its profits or to minimise its costs and the firing of the engine of a space vehicle must be programmed so as to minimise the expenditure of fuel. When there are only two variables involved, the problem is to determine the area of the plane in which the inequalities are satisfied and then to search within this area for the coordinates which give the appropriate minimum or maximum.

EXAMPLE 2 *A small factory produces two products A and B. The finished products are packed either in a combined pack containing one unit of A and one unit of B or else in a smaller pack containing just one unit of B. In order to make the* *products, each unit of A requires 4 hours of machine time and each unit of B requires 1 hour of machine time. The total machine time available in one week is 200 hours. Each week, the completed packs are stored and a pack containing one unit of A and one unit of B occupies 8 m^3 of storage space whilst a pack containing one unit of B needs 5 m^3 of space. The storage space is limited to 600 m^3. The profit on a unit of A is £15 and the profit on a unit of B is £5. Write down the inequalities relating x and y, where x units of A and y units of B are produced each week and indicate the allowable points (x, y) by shading on a diagram. Find the number of units of A and of B which should be produced each week in order to maximise the profit.*

The number of units produced each week must be positive so $x \geqslant 0$ and $y \geqslant 0$. The machine hours needed to produce x units of A and y units of B is $4x + y$ and so $4x + y \leqslant 200$. Similarly the restriction on the storage space requires that $8x + 5(y - x) \leqslant 600$, and the profit each week is $15x + 5y$. Also, since the units of A are only packed together with units of B it is necessary to produce at least as many units of B as of A, otherwise the excess units of A cannot be used. The problem may now be expressed: find the values of x and y which give a maximum

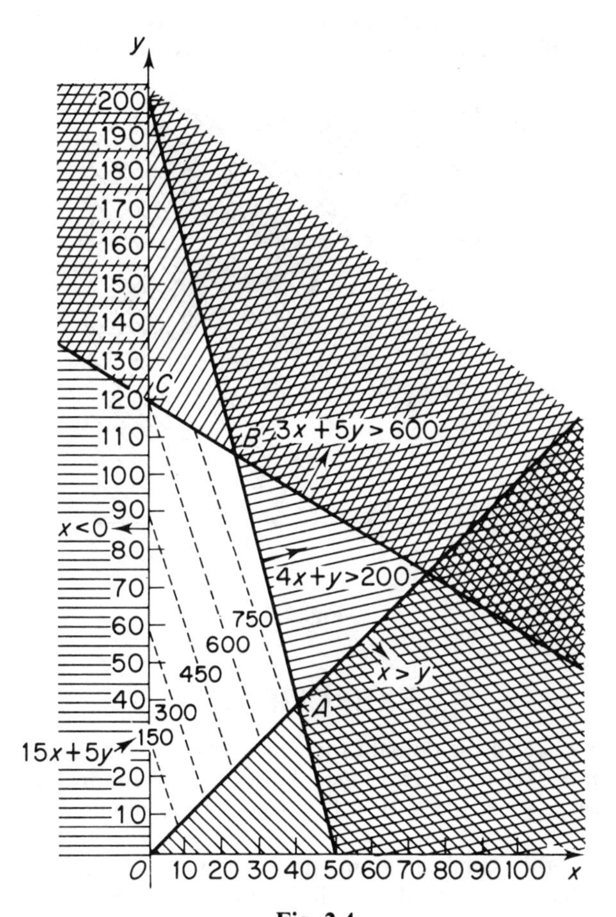

Fig. 2.4

value of $15x+5y$ subject to the conditions

$$x \geqslant 0,\ y \geqslant 0,\ 4x+y \leqslant 200,\ 3x+5y \leqslant 600,\ x \leqslant y.$$

In Fig. 2.4, the boundaries corresponding to these inequalities are drawn and the regions excluded are shaded. This leaves an unshaded region, bounded by a quadrilateral $OABC$, where O is the origin, A is the point of intersection of $y = x$ and $y+4x = 200$, B is the point of intersection of $y+4x = 200$ and $3x+5y = 600$, C is the point of intersection of $x = 0$ and $3x+5y = 600$. Note that the inequality $y \geqslant 0$ does not bound the required region. Solving the above equations for the coordinates of A, B and C, it is found that A is the point (40, 40), B is the point (400/17, 1800/17) and C is the point (0, 120).

Now the profit P is given by $P = 15x+5y$ so, on the diagram, dotted lines are drawn showing constant values of P, namely for P in the set $\{150, 300, 450, 600, 750\}$. It is then clear, from the diagram, that P is a maximum at B. Therefore the optimum values of x and y are 400/17 and 1800/17, that is **23·53 and 105·88.**

The factory may need to plan to construct a whole number of units of A and B each week. In this case, the solution must be in the set of integers, so we need to look at an enlarged diagram, near to the point B. This is drawn in Fig. 2.5 and it is then seen that the optimum solution, in integers, is $\mathbf{x = 24,\ y = 104}$.

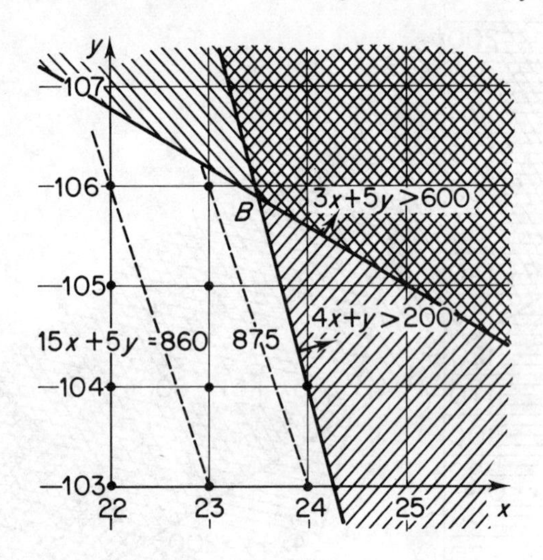

Fig. 2.5

EXERCISE 2.3

1 Shade on a sketch the domain for which $y^2 - x < 0$ and $x^2 - y < 0$. (*L*)

2 Shade on a sketch the region(s) of the $x-y$ plane in which the three inequalities $x+y < 1$, $x-2y < 1$, $y-2x < 1$ are simultaneously satisfied. (*L*)

3 Sketch on the same diagram the line $x+y = 1$ and the parabola $y^2 = 4x$. Shade on your sketch the regions of the $x-y$ plane for which $(x+y-1)(y^2-4x) > 0$. (*L*)

4 Sketch the three lines whose equations are $y-x-6=0$, $2y+x-18=0$, $2y-x=0$. Shade on your diagram the domain defined by $y-x-6<0$, $2y+x-18<0$, $2y-x>0$. (*L*)

5 Shade a diagram so as to leave unshaded the region in the plane $O(x, y)$ where all the inequalities are satisfied:
(a) (i) $x^2+y^2 \leqslant 9$ (ii) $x+y<1$,
(b) (i) $x^2+y^2 \leqslant 4x$ (ii) $x+y>1$ (iii) $y \geqslant x-1$,
(c) (i) $x \geqslant 0$ (ii) $y \geqslant 0$ (iii) $x+y \leqslant 5$ (iv) $x+2y \geqslant 2$.

6 Find the coordinates of the point nearest to the origin such that:
(a (i) $2x+y \geqslant 1$ (ii) $y \geqslant 1+x$ (iii) $x+y \leqslant 3$,
(b) (i) $x^2+y^2 \leqslant 4$ (ii) $2x+y-2 \geqslant 0$ (iii) $x+2y-2 \geqslant 0$,
(c) (i) $x+y \leqslant 2$ (ii) $x+3y \geqslant 3$ (iii) $4x+y \geqslant 4$.

7 On a diagram, indicate the region satisfied by the three inequalities $x^2+y^2 \leqslant 9$, $3x+y \leqslant 6$, $3y \leqslant x+6$. On the same diagram, draw the family of lines $x+y=k$ for k in the set $\{0, \pm 1, \pm 2, \pm 3\}$. Find the greatest value of $x+y$ satisfying the above three inequalities.

8 Find the maximum value of $x+y$ such that (i) $y^2 \leqslant 8(18-x)$ (ii) $5y \leqslant 2x+20$ and (iii) $y+2x \geqslant 12$.

9 A diet is to include products A and B. Each packet of A costs 50 p and contains 1 unit of carbohydrate, 3 units of vitamin and 3 units of protein. The corresponding units in one packet of B are 3, 4 and 1, and the packet costs 25 p. The minimum daily requirement to be supplied by A and B is 9 units of carbohydrate, 18 units of vitamin and 6 units of protein. Show that, if the daily consumption uses x packets of A and y packets of B then, to satisfy the requirements, $x+3y \geqslant 9$, $3x+4y \geqslant 18$, $3x+y \geqslant 6$, and represent these inequalities on a diagram. Draw a family of curves $50x+25y=c$, of constant cost, and find the number of packets of A and B which will be needed in June.

10 A gardener wishes to use 9 units of chemical A, 8 units of B and 12 of C on his garden. He can purchase packets of products X and Y at £2 and £3 respectively. One packet of X contains 3 units of A, 1 unit of B and 1 unit of C and a packet of Ycontains 1 unit of A, 2 units of B and 4 units of C. Determine how many packets of X and Y he should buy in order to satisfy his needs and minimise his costs.

MISCELLANEOUS EXERCISE 2

1 Show that, for all positive values of a and b, $a^3+2b^3 \geqslant 3ab^2$. (*L*)

2 Indicate on a diagram of the $x-y$ plane the regions satisfied by the inequality $(x+y)/(x-1)<1$. (*JMB*)

3 In one diagram, sketch the three lines $x+y-3=0$, $y-2x-4=0$, $y-x/2+2=0$. Indicate, by shading on your diagram, the region in which the following three inequalities are all satisfied, marking this region with a large X: $x+y-3>0$, $y-2x-4>0$, $y-x/2+2>0$. (*L*)

4 Given that $x^2+x(y-2)+(y^2-2y+1)=0$, where x and y are real, show that $0 \leqslant y \leqslant 4/3$. (*L*)

5 In the same diagram sketch the graphs of $3y+4x=12$, $\dfrac{x^2}{25}+\dfrac{y^2}{16}=1$.
Indicate, by shading, the region for which both the inequalities $3y+4x \geqslant 12$,

$\frac{x^2}{25}+\frac{y^2}{16} \leqslant 1$ are satisfied. Write down the coordinates of the point in this region for which $(x-3)^2+y^2$ is greatest. (*L*)

6 Show that there is no value of x for which $|x|+|x-2| = x^2-2x+4$ and sketch on the same axes the graphs of (i) $y = |x|+|x-2|$ and (ii) $y = x^2-2x+4$. Shade the region satisfying the four inequalities (a) $y > 2$, (b) $y < x^2-2x+4$, (c) $y > 2x-2$ and (d) $y > 2-2x$. (*JMB*)

7 (i) Find the set of values of x for which $x(x-4)(x^2-4) < 0$.
(ii) Given that p, q, r and s are positive and unequal, show that
(a) $(pq+rs)^2 \leqslant (p^2+r^2)(q^2+s^2)$, (b) $p^3q+pq^3 < p^4+q^4$.
Show also that $p^4+q^4+r^4+s^4 > 4pqrs$. (*L*)

8 Find, graphically or otherwise, the set of values of x for which $|2x-1|+|4-x| > 5$. (*L*)

9 Sketch on the same diagram the curves whose equations are $x^2+y^2 = 16$ and $x^2 = 6y$. (You need not find the coordinates of the points of intersection.) Shade in your diagram the regions of the plane for which $(x^2+y^2-16)(x^2-6y) < 0$. (*L*)

10 Solve the inequalities: (i) $\frac{1}{x+6} \leqslant \frac{2}{2-3x}$, (ii) $|5-3x| \leqslant |x+1|$.

11 Show, by shading on a sketch of the xy-plane, the region for which $x^2+y^2 \leqslant 1$, $y \geqslant x$ and $y \leqslant x+1$. Hence find (i) the greatest value of y, (ii) the least value of $x+y$, for which these inequalities hold.

12 (i) Show that the arithmetic mean of two positive real numbers is greater than or equal to their geometric mean. Hence show that, when a, b, c, d are real, $a^4+b^4+c^4+d^4 \geqslant 4abcd$. (ii) Use a carefully labelled sketch to obtain the point (x, y) whose coordinates satisfy simultaneously the inequalities $y^2 \geqslant 4x$, $x+y \geqslant 3$, and for which x^2+y^2 is least. (*L*)

13 Sketch, using the same axes, the two curves whose equations are $(x-2)^2+(y-4)^2 = 1$, $2x+y = 9$. Shade the region R of the x-y plane giving the solution set of the simultaneous inequalities $(x-2)^2+(y-4)^2 \leqslant 1$, $2x+y \leqslant 9$. By considering the family of lines whose equations are of the form $x+y = k$, where k is a real parameter, find the greatest value of $x+y$ for points (x, y) in the region R. Find also the greatest value of (x^2+y^2) in the region R. (*L*)

14 Find a and b such that $\{x: a < x < b\} = \{x: |2x-3|-|x-5| < 10\}$. (*L*)

3 Abstract Algebra

3.1 Unary operations or functions

In mathematics, we often perform operations on the elements of a set to produce elements of another set (or the same set). When we consider the set together with its operation, we have a mathematical structure. We may have quite different sets and operations which nevertheless have the same structure. Then any results which we prove about the structure of one set can, at once, be applied to the other set.

Some operations involve only one element of the set so that they are functions having the set as their domain. They are called *unary operations.* Let the function f have the set S as its domain. Then f is a unary operation on the set S. We can denote the structure consisting of the set S with its unary operation f by the notation (S, f). For example, $(\mathbb{C}, *)$ is the set of complex numbers with the operation $*$ which maps each complex number z to its conjugate z^*. It will usually be preferable, however, to use the function notation.

Closure Let f be a unary operation on the set S, so that f is a function with domain S, then S is said to be *closed* under f if the range of f is a subset of S, that is, if $\mathrm{f}(S) \subseteq S$.

Definition (S, f) is closed if, for all x in S, $\mathrm{f}(x)$ is in S.

EXAMPLE *Determine whether the operation on the set is closed:*
(i) the operation of taking the square on the set of integers,
(ii) the operation of taking the square root on the set of natural numbers.

(i) Let n be any integer, then n^2 is also an integer, so $(\mathbb{Z}, \text{square})$ is **closed.**
(ii) 2 is a natural number but $\sqrt{2}$ is not (*Pure Mathematics* §9.5), so $(N, \sqrt{\ })$ is **not closed.**

EXERCISE 3.1

1 In the following, the set S is named first and then the unary operation f is described. State whether (S, f) is closed or not:

(i) real numbers, square,
(ii) the rational numbers, square root,
(iii) integers, modulus,
(iv) positive real numbers, reciprocal,
(v) integers, sign change,
(vi) real numbers, integer part,
(vii) $\{0, 1, 2\}$, square,
(viii) complex numbers, argument,

(ix) $\{0, 1, -1\}$, square, (x) a set of vectors, modulus,
(xi) $\{0, 1\}$, square root, (xii) $\{-1, 0, 1\}$, square root,
(xiii) subsets of $\{a, b\}$ so that $S = \{\phi, \{a\}, \{b\}, \{a, b\}\}$, complement,
(xiv) all subsets of a set U, complement.

2 Given a unary operation f, which is closed on a set S, show how we can define on S the unary operation f^2.

3.2 Binary operations

When an operation on a set S involves two elements of S, it is called a *binary* operation on S. Such an operation is a function whose domain consists of the set of all ordered pairs of elements of S.

Definition The binary operation $*$ on the set S is some rule which defines a unique element z for every x and y in S, where $z = x*y$.

Notation We refer to the mathematical object consisting of the set S together with its binary operation $*$ by the notation $(S, *)$.

EXAMPLE 1 *Prove that the following are examples of a binary operation $*$ on a set S:*

Elements of the set S	*Binary operation* $*$	
(*i*) *integers*	*multiplication*	$m*n = mn$,
(*ii*) *positive integers*	*division*	$m*n = m/n$,
(*iii*) *vectors in space*	*scalar product*	$a*b = a.b$,
(*iv*) 2×2 *real matrices*	*matrix addition*	$M*N = M+N$,
(*v*) *rational numbers*	*subtraction*	$p*q = p-q$.

In each case, the proof consists of showing that, for every pair of elements x, y in S, $x*y$ is properly defined. This is true for:
(i) the product of integers, (ii) the quotient of positive integers, (iii) the scalar product of two vectors, (iv) the addition of 2×2 matrices, (v) the subtraction of rational numbers.

EXAMPLE 2 *Prove that the following are examples which are not binary operations because $a*b$ is not always defined:*

Elements of the set S	*Operation* $*$
(*i*) *integers*	*division*,
(*ii*) *vectors and scalars*	*addition*.

To show that $*$ is not a binary operation, we give a counter-example in which x and y are elements of S but $x*y$ is not defined:
(i) $x = 1$, $y = 0$, (ii) $x = \mathbf{a}$, $y = 3$.

Closure An important property of a binary operation $*$ on a set S is the property that the result is always a member of S. As in the case of a unary operation, we then say that $(S, *)$ is closed.

Definition The set S is closed under the binary operation $*$ if the range of $*$ is a subset of S, that is

$$(S, *) \text{ is closed if } x*y \text{ is in } S \text{ for all } x \text{ and } y \text{ in } S.$$

Note. Some authors include the property of closure within the definition of a binary operation. In this case, they use the words binary operation to mean what we define as a closed binary operation.

The table of a binary operation A binary operation $*$ on a set S may be described by means of its table, which is a square array, i.e. a matrix, in which the rows and the columns are labelled by the elements of S, and the entry in the position row x and column y is $x*y$. This is an agreed definition which is a matter of convention.

Definition The table of $(S, *)$ is a matrix in which: the rows are labelled by the elements of S and the columns are labelled by the elements of S and the entry in row x and column y is $x*y$.

EXAMPLE 3 *Draw the table of* $(S, *)$, *where* $S = \{0, 1, 2\}$ *and where:* (*i*) $x*y = x+y$, (*ii*) $x*y = xy$, (*iii*) $x*y = 2x-y$.

We indicate the operation in the top left hand corner:

(i)

$+$	0	1	2
0	0	1	2
1	1	2	3
2	2	3	4

(ii)

xy	0	1	2
0	0	0	0
1	0	1	2
2	0	2	4

(iii)

$2x-y$	0	1	2
0	0	-1	-2
1	2	1	0
2	4	3	2

Note that, in this example, none of the operations is closed because each table contains at least one entry which is not an element of S.

EXERCISE 3.2

1 In each of (i), (ii), (iii), and (v) of Example 1, check that the operation is an example of a binary operation by showing that $x*y$ is properly defined for all pairs of elements x and y in S. Prove that in (i), (iv) and (v) the operation is closed. Prove that the binary operations in (ii) and (iii) are not closed by finding, in each case, a pair of elements x and y in S with $x*y$ not in S. (This last method is called finding a counter-example.)

2 Verify that each of the following is a binary operation on the set and state whether or not it is closed:
(i) $(\mathbb{N}, +)$, (ii) $(\mathbb{N}, -)$, (iii) $(\mathbb{N}, \times)$,
(iv) $(\mathbb{Z}, -)$, (v) $(\mathbb{R}^+, \div)$, (vi) $(\{-1, 0, 1\}, +)$,
(vii) $(\{-1, 0, 1\}, \times)$.

3 Draw the table of:
(i) $\{2, 4, 6, 8, 10\}$ with operation 'add',
(ii) $\{15, 18, 21, 22\}$ with operation 'highest common factor',

(iii) $\{1, 3, 5, 7, 9, 11\}$ with operation 'largest',
(iv) $\{0, 1, 2, 3, 4, 5, 6\}$ with operation 'modulus of the difference'.

4 Let A be a set. Draw the table for $(\{A, \phi\}$, intersection).

5 Draw the table of the binary operation of set union for the set S, where S is the set of all subsets of the set $\{a, b\}$, so that $S = \{A, B, C, D\}$ and $A = \phi, B = \{a\}$, $C = \{b\}, D = \{a, b\}$.

6 Let S be the set of all subsets of a universal set E. State which of the following operations on S are unary and which are binary:
(i) union, (ii) complementation,
(iii) intersection, (iv) set difference.
Verify that all the operations are closed.

3.3 Clock arithmetic

A fruitful source of examples of the binary operations of addition and multiplication is in clock arithmetic. When measuring time in hours, using the hour hand on a 12 hour clock, we identify 12 with 0, 13 with 1, 14 with 2 and so on. So, when adding hours, we find that $8+5=1, 9+7=4$ and $3-8=7$. In this clock addition, the addition of an integer n is obtained by moving the hour hand through an angle $n\pi/6$ clockwise. The rotation through a negative angle clockwise is a rotation through the modulus of that angle anticlockwise. Clock multiplication is defined in a similar manner. We define $m \times n$ to be the result of adding m n's together so that $2 \times 3 = 3+3 = 6$, $2 \times 8 = 8+8 = 4$, $3 \times 5 = 5+5+5 = 5+10 = 3$, and so on. In this clock arithmetic, we are ignoring all multiples of 12 since we equate 12 with 0. The arithmetic is called *arithmetic modulo 12*, or with base 12, and the set of integers $\{0, 1, 2, 3, \ldots, 10, 11\}$ is referred to as $\mathbb{Z}_{12}$, the integers modulo 12.

Any other natural number, greater than 1, can be used as the base for modular (or clock) arithmetic, so we shall look at the bases 2 and 3 as particular examples.

Arithmetic modulo 2

Imagine a clock, with one hand, and with only two figures on the dial, namely 0 and 1 placed diametrically opposite each other, (Fig. 3.1). We shall represent addition of a number by an appropriate rotation of the hand, clockwise for a positive number and anticlockwise for a negative number. The addition of zero does not move the hand and so it leaves the numbers 0 and 1 unaltered. The addition of 1 corresponds to a rotation π of the hand and so this changes 0 to 1 and 1 to 0. Thus we find that the rules of addition are: $0+0=0, 0+1=1+0=1, 1+1=0$. The corresponding multiplication is very simple since any number multiplied by zero becomes zero and so the only non-zero product is 1×1 which equals 1. In this way, we obtain the set $\mathbb{Z}_2$ of integers modulo 2, where $\mathbb{Z}_2 = \{0, 1\}$, and

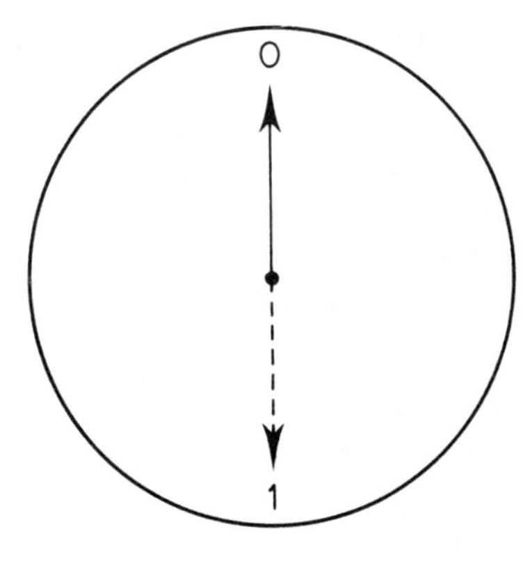

Fig. 3.1

$\mathbb{Z}_2$ is a set with the two binary operations of addition and multiplication denoted by $+$ and $\times$. The tables of these operations are:

$(\mathbb{Z}_2, +)$	0	1
0	0	1
1	1	0

$(\mathbb{Z}_2, \times)$	0	1
0	0	0
1	0	1

Arithmetic modulo 3

For the corresponding situation with base 3, we have a clock, with one hand, and with three figures, 0, 1 and 2, equally spaced round the dial, (Fig. 3.2). Addition of 1 corresponds to a clockwise rotation of the hand through 1/3 of a full turn, that is, through $2\pi/3$, and addition of 2 corresponds to a clockwise rotation through 2/3 of a full turn, that is, through $4\pi/3$. Thus $1+1=2$ and $1+2=2+1=0$, since this corresponds to a rotation through one whole turn which leaves the hand

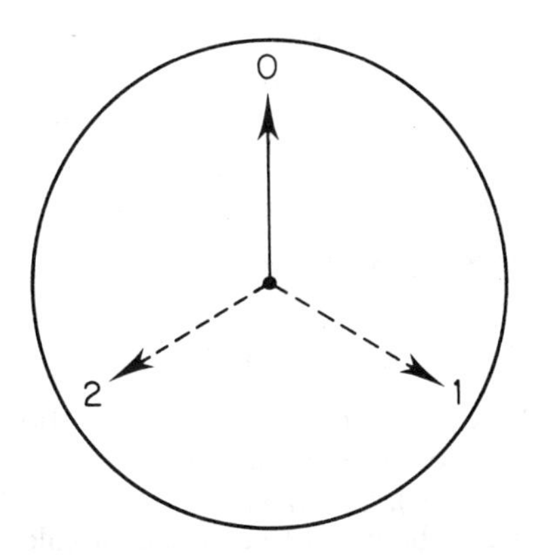

Fig. 3.2

unmoved. Also $2+2=1$, since this corresponds to a rotation through 4/3 of a full turn clockwise, which is the same as a rotation through 1/3 of a turn. As far as multiplication is concerned, multiplication by zero always gives zero, multiplication by 1 leaves any number unaltered. Therefore the only new product is 2×2, which is the same as $2+2$, that is, 1. Thus, for arithmetic modulo 3, we have the set $\mathbb{Z}_3$, given by $\mathbb{Z}_3 = \{0, 1, 2\}$, with the operations of addition and multiplication given by the tables:

$(\mathbb{Z}_3, +)$	0	1	2
0	0	1	2
1	1	2	0
2	2	0	1

$(\mathbb{Z}_3, \times)$	0	1	2
0	0	0	0
1	0	1	2
2	0	2	1

Arithmetic modulo *n*

The above can be generalised to a clock arithmetic with a base n, for any natural number n greater than 1. The integers modulo n are a set $\mathbb{Z}_n$, given by $\mathbb{Z}_n = \{0, 1, 2, \ldots, n-1\}$, corresponding to n positions evenly spaced round a clock face. Any number m in $\mathbb{Z}_n$ corresponds to a clockwise rotation of the hand through an angle $2\pi m/n$ and sums and products are worked out by successive rotations.

Thus in $\mathbb{Z}_{11}$, $2 \times 7 = 7+7 = 3$ and $7 \times 8 = 1$ because $56 = 5 \times 11 + 1$. We shall return to modulo n arithmetic in Exercise 3.5, Question **9**.

EXAMPLE *Draw the table of $\mathbb{Z}_5$ under the operation of subtraction.*

The subtraction of a smaller number from a larger number in $\mathbb{Z}_5$ gives the same result as in $\mathbb{Z}$. To subtract a larger number from a smaller, we first add 5 to the smaller and then subtract the larger. The addition and subtraction are done in $\mathbb{Z}$ and the result is then in $\mathbb{Z}_5$. Thus, in $\mathbb{Z}_5$, $2-4 = 2+5-4 = 3$. This allows us to complete the table.

$(\mathbb{Z}_5, -)$	0	1	2	3	4
0	0	4	3	2	1
1	1	0	4	3	2
2	2	1	0	4	3
3	3	2	1	0	4
4	4	3	2	1	0

EXERCISE 3.3

1 Verify that each of the following is a binary operation on the given set. Find the table of the operation and check that it is closed:
(i) $(\mathbb{Z}_4, +)$, (ii) $(\mathbb{Z}_4, \times)$, (iii) $(\mathbb{Z}_4, -)$, (iv) $(\mathbb{Z}_3, -)$.

2 Two instructions are to be given to a squad of soldiers:
(i) face as you are, that is, remain facing the same direction, denoted by F,

(ii) about turn, that is, turn through 180° to face the opposite direction, denoted by A.
Let $S = \{F, A\}$ and let $\circ$ be the operation of composition of instructions. Draw the table of $(S, \circ)$.

3 Refer to question **2** and add two more instructions:
(iii) right turn, denoted by R, (iv) left turn, denoted by L.
Let $T = \{F, A, R, L\}$ and draw the table of $(T, \circ)$.

4 Draw the tables of:
(i) $(\{0, 1\}, +)$, (ii) $(\{0, 1\}, \times)$, (iii) $(\{1, -1\}, +)$,
(iv) $\{1, -1\}, \times)$, (v) $(1, i, -1, -i\}, \times)$, (vi) $(\mathbb{Z}_2, +)$,
(vii) $(\mathbb{Z}_2, \times)$.
Determine whether any of these tables have a similar structure. Determine also whether there is a correspondence in structure between these tables and the tables you obtained in Questions **1**, **2** and **3** above, and Questions **4** and **5** of Exercise 3.2.

3.4 Binary relations

Sometimes we are interested in a given connection between pairs of elements in a set. For example, in the set $\mathbb{Z}$ of integers, we have such connections as 'is less than' or 'is a divisor of', and, in the set of members of a family, we have 'is a cousin of' or 'has the same grandfather as'. Such a connection between two elements is called a binary relation, or just a relation. If the relation is denoted by R, then the fact that two elements x and y are connected by R is expressed by saying that x and y are *related* by the *relation* R, and this is denoted by $x\mathrm{R}y$.

Notation $x\mathrm{R}y$ means that x and y (in that order) are related by R.

For example, on the set $\mathbb{Z}$ we can define the three relations $<$, M, D by means of

$x < y$ means x is less than y,
$x\mathrm{M}y$ means x has the same modulus as y,
$x\mathrm{D}y$ means x is a divisor of y.

Relations may also be defined on sets which are not sets of numbers. For example, on the set of all the living members of a family, we could define the relations B and P by

$x\mathrm{B}y$ means x is a brother of y,
$x\mathrm{P}y$ means x has the same parents as y.

For a given relation R on a set S, the relation $x\mathrm{R}y$ may be true or false for a given ordered pair (x, y) of elements of S. Thus

$2 < 3$ is true, $3 < 2$ and $2 < 2$ are both false,
$2\,\mathrm{M}\,2$ and $2\,\mathrm{M}\,-2$ are both true, $2\,\mathrm{M}\,3$ and $3\,\mathrm{M}\,2$ are both false,
$2\,\mathrm{D}\,4$ and $2\,\mathrm{D}\,2$ are both true, $2\,\mathrm{D}\,3$ and $3\,\mathrm{D}\,2$ are both false,
if George B Fred is true then Fred B George is true and George P Fred is true,
but if Fred B Mary is true then Mary P Fred is true and Mary B Fred is false.

We classify relations according to their properties and now consider three important properties of binary relations: reflexivity, symmetry and transitivity.

Definition Let R be a binary relation on a set S, then:
(i) R is reflexive if, for all x in S, $x\mathrm{R}x$,
(ii) R is symmetric if, for x, y in S, $x\mathrm{R}y \Rightarrow y\mathrm{R}x$,
(iii) R is transitive if, for x, y, z in S, $(x\mathrm{R}y$ and $y\mathrm{R}z) \Rightarrow x\mathrm{R}z$.

EXERCISE 3.4

1 For the examples of binary relations, given above, prove that:
(i) $<$ is transitive but not reflexive or symmetric,
(ii) M is reflexive, symmetric and transitive,
(iii) D is reflexive and transitive but not symmetric,
(iv) B is not symmetric, not reflexive and not transitive,
(v) P is reflexive, symmetric and transitive.

2 Test the following binary relations for the three properties of reflexivity, symmetry and transitivity:
(i) the relation 'is greater than' on the set $\mathbb{R}$,
(ii) the relation 'has the same argument as' on the set $\mathbb{C}$,
(iii) the relation 'differs by 2 from' on the set $\mathbb{Z}$,
(iv) the relation 'has no common factor with' on the set $\mathbb{N}$,
(v) the relation 'equality' on any set.

3.5 Equivalence relations

Relations such as equality or 'has the same modulus as' or 'has the same argument as' or 'has the same parents as' possess all three of the properties of reflexivity, symmetry and transitivity and are called *equivalence relations* (or *equivalences*).

Definition A relation R on a set S is an equivalence relation (or an equivalence) if it satisfies all three of the following axioms:

(i) E1 for all a in S,	$a\mathrm{R}a$,	(R is reflexive)
(ii) E2 for all a, b in S,	$a\mathrm{R}b \Rightarrow b\mathrm{R}a$,	(R is symmetric)
(iii) E3 for all a, b, c in S,	$(a\mathrm{R}b$ and $b\mathrm{R}c) \Rightarrow a\mathrm{R}c$,	(R is transitive)

Suppose that R is an equivalence relation on the set S. For any element a in S, we define a subset $S(a)$ of S, called the *equivalence class* of a, consisting of all those elements of S which are equivalent to a.

Definition Given an equivalence relation R on a set S and an element a in S, the equivalence class $S(a)$ of a is given by:

$$S(a) = \{x: x \in S \text{ and } x\mathrm{R}a\}.$$

In Theorem 3.1 we prove some of the properties of equivalence classes.

Theorem 3.1 Let R be an equivalence on S and, for any a in S, let $S(a)$ be the equivalence class of a. Then, for all a and b in S,
(i) $a \in S(a)$,
(ii) if $a \in S(b)$ then $S(a) = S(b)$,
(iii) if $a \notin S(b)$ then $S(a) \cap S(b) = \phi$,
(iv) S is the disjoint union of all the equivalence classes of R.

Proof (i) By E1, $a\mathrm{R}a$ and so $a \in S(a)$.
(ii) Suppose that $a \in S(b)$, then $a\mathrm{R}b$ and, by E2, $b\mathrm{R}a$. We prove that $S(a) = S(b)$ by showing that any element in one of the two sets also lies in the other.
$x \in S(a) \Rightarrow x\mathrm{R}a \Rightarrow x\mathrm{R}b \Rightarrow x \in S(b)$, using $a\mathrm{R}b$ and E3.
$x \in S(b) \Rightarrow x\mathrm{R}b \Rightarrow x\mathrm{R}a \Rightarrow x \in S(a)$, using $b\mathrm{R}a$ and E3.
Hence $S(a) \subseteq S(b)$ and $S(b) \subseteq S(a)$ and so $S(a) = S(b)$.
(iii) If $S(a) \cap S(b) \neq \phi$, suppose that $c \in S(a) \cap S(b)$. Then $c\mathrm{R}a$ and $c\mathrm{R}b$, so that $a\mathrm{R}c$, by E2, and $a\mathrm{R}b$, by E3. Thus $a \in S(b)$ and, by (ii), $S(a) = S(b)$ $(= S(c))$.
This is equivalent to the statement: $a \notin S(b) \Rightarrow S(a) \cap S(b) = \phi$.
(iv) Let T be the set of all equivalence classes,

$$T = \{S(a) \colon a \in S\}.$$

Then, for all a in S, $a \in S(a)$ so S is the union of the sets in T. Also, for any two sets $S(a)$ and $S(b)$ in T,

either $a \in S(b)$ when $S(a) = S(b)$ by (ii),
or else $a \notin S(b)$ when $S(a)$ and $S(b)$ are disjoint by (iii).

Therefore, S is the union of the set T of disjoint classes.

Partition of a set

Let T be a set of disjoint (non-empty) sets. That is $T = \{A, B, C, \ldots\}$ and for any A, B in T, $A \neq \phi$, $B \neq \phi$ and, if $A \neq B$, $A \cap B = \phi$. Then, if S is the union of all the sets in T, we say that S is *partitioned* into the set T of disjoint sets.

For example, the following are partitions of a set:

$$\{1,2,3,4,5,6\} = \{1\} \cup \{2,3\} \cup \{4,5,6\},$$

$$\mathbb{Z} = E \cup O,\ E = \{2n \colon n \in \mathbb{Z}\},\ O = \{2n+1 \colon n \in \mathbb{Z}\}$$

$$\mathbb{Z} = \{3n \colon n \in \mathbb{Z}\} \cup \{3n+1 \colon n \in \mathbb{Z}\} \cup \{3n+2 \colon n \in \mathbb{Z}\}.$$

Theorem 3.1 states that an equivalence relation R on a set S partitions S into its disjoint equivalence classes. There is a converse to this result. Any partition of a set S defines an equivalence on S whose classes give the partition.

Theorem 3.2 Let S be partitioned into a set T of disjoint subsets of S. Define the relation R on S by

$$\text{for } x, y \text{ in } S, x\text{R}y \text{ if, for some } A \in T,\ x \in A \text{ and } y \in A.$$

Then R is an equivalence relation on S and T is the set of equivalence classes.

Proof Since the sets in T form a partition of S, every element a in S lies in just one set in T. Call this set $S(a)$ and then the definition of R may be expressed in the form $a\text{R}b$ if $S(a) = S(b)$. It is then straightforward to check that R is an equivalence relation and that $S(a)$ is the equivalence class containing a.

EXAMPLE *Let S be the set* $\{1, 2, 3, 4, 5, 6, 7, 8, 9, 10\}$. *Define relations* Q *and* R *by:*

$$a\text{Q}b \text{ if and only if } |a-b| \text{ is divisible by } 2,$$
$$a\text{R}b \text{ if and only if } |a-b| \text{ is divisible by } 3.$$

Prove that Q *and* R *are equivalence relations and identify the equivalence classes. A third relation* P *is defined by*

$$a\text{P}b \text{ if and only if } a\text{Q}b \text{ or } a\text{R}b.$$

Show that P *is not an equivalence relation.*

We prove first that Q is an equivalence relation.
Reflexive: $a\text{Q}a$ since 0 is divisible by 2.
Symmetric: if $a\text{Q}b$ then $|a-b|$ is divisible by 2, but $|b-a| = |a-b|$, so $b\text{Q}a$.
Transitive: if $a\text{Q}b$ and $b\text{Q}c$ then $a-b = 2m$ and $b-c = 2n$, $m, n \in \mathbb{Z}$, so $a-c = 2(m+n)$ and $m+n \in \mathbb{Z}$, therefore $a\text{Q}c$.

The proof that R is an equivalence relation follows in exactly the same way.

Under Q the elements of S in the same equivalence class differ by a multiple of 2, so the classes are $\{1, 3, 5, 7, 9\}$ and $\{2, 4, 6, 8, 10\}$. Under R the elements of S in the same equivalence class differ by a multiple of 3, so the classes are $\{1, 4, 7, 10\}$, $\{2, 5, 8\}$ and $\{3, 6, 9\}$. Since Q (or R) is reflexive, P must be reflexive. Suppose that $a\text{P}b$ then either $a\text{Q}b$ or $a\text{R}b$ and so either $b\text{Q}a$ or $b\text{R}a$ and therefore $b\text{P}a$. Hence P is symmetric and to show that P is not an equivalence relation we must find a counter-example to transitivity. Such an example is: 1P3 and 3P6, since 1Q3 and 3R6, but 1P6 is false since neither 1Q6 nor 1R6. Hence P is not transitive and thus not an equivalence relation.

EXERCISE 3.5

1 Check that the following are equivalence relations and identify the equivalence classes: (i) $S = \{1, 2, 3, 4, 5\}$, $a\text{R}b \Leftrightarrow a+b$ is even, (ii) $S = \mathbb{N}$, $a\text{R}b \Leftrightarrow a = b \pmod 5$, (iii) $S = \{-4, -2, -1, 1, 2, 4\}$, $a\text{R}b \Leftrightarrow |a/b| = 1$.

2 For each of the following relations, find which of the three properties of reflexivity, symmetry and transitivity hold and give an example of the failure of the others: (i) $S = \{0, 1, 2, 3\}$, $a\text{R}b \Leftrightarrow a > b$, (ii) $S = \{3, 6, 9, 12\}$, $a\text{R}b \Leftrightarrow a+b$ is odd, (iii) $S = \mathbb{N}$, $a\text{R}b \Leftrightarrow a$ and b have a common divisor greater than 1.

3 Define the relation R on $\mathbb{Z}$ by

$$\text{for } x, y \text{ in } \mathbb{Z}, x\text{R}y \text{ if } (x-y) \text{ is even.}$$

Prove that R is an equivalence relation and that there are two equivalence classes, namely the set E of even integers and the set O of odd integers.

4 Given any two sets X and Y of numbers, define two new sets, the sum and product of X and Y, by

$$X+Y=\{x+y: x\in X \text{ and } y\in Y\},\ X\times Y=\{xy: x\in X \text{ and } y\in Y\}.$$

Write down $X+Y$ and $X\times Y$, given that,
(i) $X=\{0,1\}$, $Y=\{0,1\}$, (ii) $X=\{-1,1\}$, $Y=\{0,1\}$,
(iii) $X=E$, $Y=\mathbb{Z}$, (iv) $X=E$, $Y=E$,
(v) $X=E$, $Y=O$, (vi) $X=O$, $Y=O$.

5 Using the results of the Question 4, write down the tables of $(\{E,O\}, +)$ and $(\{E,O\}, \times)$. Show that the two binary operations are closed and commutative. Prove that $\{E,O\}$ has the same structure as $\mathbb{Z}_2$ under: (a) addition, (b) multiplication.

6 In an electronic computer, the operations of an AND gate and of an EXCLUSIVE OR gate can be described in terms of the output voltages xANDy and xORy which arise from given input voltages x and y.
Two voltages are used, a high voltage denoted by 1 and a low voltage denoted by 0. For the AND gate the output xANDy is high only if both the inputs x and y are high. For the EXCLUSIVE OR gate the output xORy is high only if the two inputs x and y are different, that is only if one is high and one is low. Draw the tables for AND and OR, showing the output in terms of the two inputs. Prove that the AND gate corresponds to $(\mathbb{Z}_2, \times)$ and the OR gate corresponds to $(\mathbb{Z}_2, +)$.

7 Write out the tables of $(\mathbb{Z}_5, +)$ and $(\mathbb{Z}_5, -)$. Solve the equations in the set $\mathbb{Z}_5$ of integers modulo 5:
(i) $x+3=2$, (ii) $3-x=4$, (iii) $x^2=1$,
(iv) $x^2=4$, (v) $x^2=2$, (vi) $x^2+x+1=0$.

8 Prove that the relation R on $\mathbb{Z}$, given by:

$$\text{for } x, y \text{ in } \mathbb{Z}, x\text{R}y \text{ if } (x-y) \text{ is a multiple of 3,}$$

is an equivalence. Prove that the equivalence classes are:

$$\mathbf{0}=\{3k \quad : k\in\mathbb{Z}\},$$
$$\mathbf{1}=\{3k+1 : k\in\mathbb{Z}\},$$
$$\mathbf{2}=\{3k+2 : k\in\mathbb{Z}\}.$$

Define addition and multiplication on the set $T=\{\mathbf{0},\mathbf{1},\mathbf{2}\}$, as described in question **4** above, and prove that $(T, +)$ and $(T, \times)$ are closed. Write down the tables of $(T, +)$ and $(T, \times)$ and show that they have the same structure as $\mathbb{Z}_3$ under addition and multiplication.

9 Given a fixed integer n, greater than 1, extend question **8** to the set S of equivalence classes of the equivalence relation R on $\mathbb{Z}$, given by,

for $x, y\in\mathbb{Z}$, $x\text{R}y$ if $(x-y)=nk, k\in\mathbb{Z}$.
Let $S=\{\mathbf{0}, \mathbf{1}, \mathbf{2}, \ldots, \boldsymbol{n-1}\}$, where $\mathbf{r}=\{nk+r, k\in\mathbb{Z}\}$, $0\leqslant r<n$.

Prove that S is closed under the addition and product of sets, defined in question **4** by checking that

$$\mathbf{r}+\mathbf{s}=\mathbf{t}, \text{ where } t=r+s, \text{ modulo } n,$$
$$\mathbf{r}\times\mathbf{s}=\mathbf{t}, \text{ where } t=rs, \text{ modulo } n.$$

Show that the above procedure is, in fact, a construction of $\mathbb{Z}_n$, the integers modulo n, by proving that S has the same structure as $\mathbb{Z}_n$ under the operations of addition and multiplication.

10 Let $S=\{(x, y)\colon\ x, y\in\mathbb{N}\}$, where (x, y) is an ordered pair of natural numbers. Define the relation R on S by

$$(x, y)\ \mathrm{R}\ (a, b) \text{ if } (x+b)=(y+a).$$

Prove that R is an equivalence relation on S. Prove that there is a one-one correspondence between the equivalence classes of R and the integers, in which the class of (x, y) corresponds to $(x-y)$.

11 Prove that the relation R on the set S of ordered pairs of non-zero integers, given by $(x, y)\ \mathrm{R}\ (a, b)$ if $xa=yb$ is an equivalence. Describe the equivalence classes in terms of a well known set of numbers.

12 The relations P, Q, R are defined on the real numbers by: $a\mathrm{P}b$ if $a\leqslant b$, $a\mathrm{Q}b$ if $|a-b|<2$, $a\mathrm{R}b$ if $a\equiv b \pmod 6$. Determine which one of these relations is an equivalence relation. For each of the other two relations give a numerical example to show the failure of one of the properties required for an equivalence relation. (*JMB*)

13 A relation R is defined on a set S and is known to be both symmetric and transitive. Given that a and b are distinct elements of S such that $a\mathrm{R}b$, prove that $a\mathrm{R}a$. Explain why R is not necessarily a reflexive relation. (*JMB*)

3.6 Plane symmetries

A symmetry of a geometrical figure is a rigid transformation of the figure, in which points and lines are transformed into points and lines with no change in distances, and in which the figure is unaltered as a whole. The identity transformation is always a symmetry of any figure and we shall call this identity symmetry E. We combine symmetries by the operation of composition of transformations and refer to such a composition as a product of symmetries. Thus if A and B are symmetries of some figure, AB is the symmetry obtained by making the transformation B first and then following this by the transformation A. Note that the order of making the transformations is the reverse of the order of writing them down, as is standard when we compose transformations or functions. If S is the set of all symmetries of a given figure, then the product of symmetries is a closed binary operation on S.

As an example, consider the symmetries of the letter Z. We label the corners of the letter Z by A, B, C and D, and the centre of the figure by O, (Fig. 3.3).

The figure has two symmetries. First the identity symmetry E and second the symmetry R of rotation through one half turn about O. E

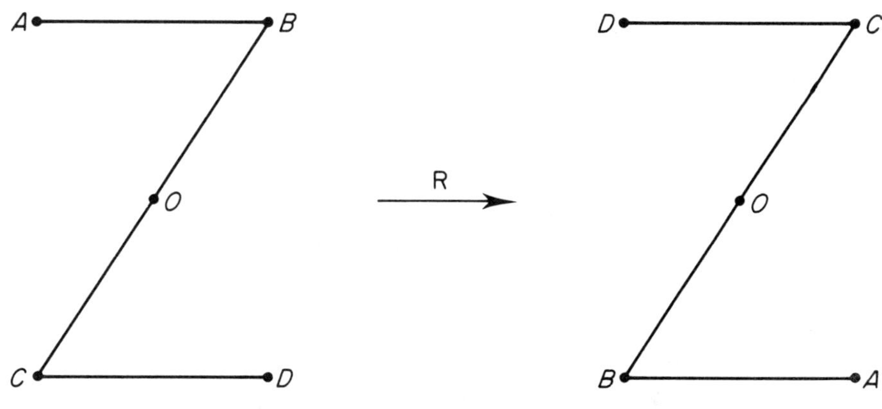

Fig. 3.3

leaves all the points unchanged and R interchanges A and D and also B and C. A repetition of R will bring the figure back to its starting point and so RR = E. The set S of symmetries of the letter Z is thus given by $S = \{\mathrm{E}, \mathrm{R}\}$ and its table under the operation of product (that is, composition) of symmetries is

$(S, \circ)$	E	R
E	E	R
R	R	E

Note the correspondence between this table and the table for $(\mathbb{Z}_2, +)$. Of course, a clock face with just two marks on it, at the ends of a diameter has the same set of symmetries as the letter Z.

EXAMPLE *Find the set of symmetries of a rectangle and display its table under the operation of composition.*

Let $ABCD$ be a rectangle with centre O. Let Ox and Oy be axes, fixed in the plane, such that initially AB is parallel to Ox and DA is parallel to Oy. (Fig. 3.4). Like the letter Z, considered above, the rectangle will have the identity symmetry E and also the rotational symmetry R, in which A and C are interchanged and also B and D are interchanged. The symmetry R is shown in the top half of Fig. 3.4. The axes Ox and Oy are lines of symmetry of the rectangle and so the rectangle has corresponding mirror symmetries M and N, given by reflections in Ox and Oy respectively. These two symmetries are shown in the lower half of Fig. 3.4. The symmetry M interchanges A and D and also B and C whilst N interchanges A and B and also C and D. The set S of all the symmetries of the rectangle is given by $S = \{\mathrm{E}, \mathrm{R}, \mathrm{M}, \mathrm{N}\}$. In order to discuss the composition of the symmetries in S, we note that each symmetry is defined by the position of A after applying the symmetry to the rectangle in the initial position. The four positions of A being A, C, D and B, corresponding respectively to E, R, M and N. The position of the rectangle is determined by the position of A since, once the position of A is fixed, the positions of the other three vertices are also fixed.

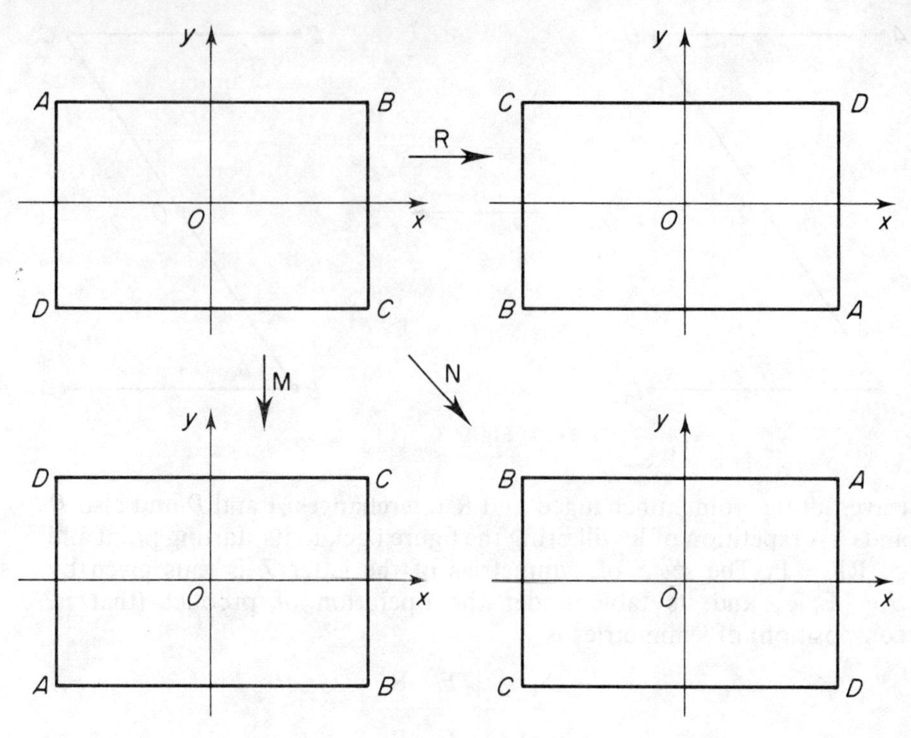

Fig. 3.4

Remember that, in representing a composition of symmetries X and Y, the product symmetry XY is the symmetry Y followed by X. The square of R is the identity and so is the square each of the mirror symmetries M and N, thus E = RR = MM = NN. The product of E with any other symmetry leaves it unaltered so we now need to find the product of any two different symmetries in the set {R, M, N}. Consider what happens to *A* under these products.

M sends *A* to *D* and R then sends it to *B* so RM = N,
R sends *A* to *C* and M then sends it to *B* so MR = N.

Note that we have to check the two products RM and MR which, in this case, both equal N. Proceeding in a similar way for the other products, we find that the product of any two of R, M and N in either order equals the third. We now have all the information needed to complete the table of *S* under composition.

$(S, \circ)$	E	R	M	N
E	E	R	M	N
R	R	E	N	M
M	M	N	E	R
N	N	M	R	E

EXERCISE 3.6

1 Using the symmetries E, R, M, N of the rectangle, as defined above, write down the table for the set of all symmetries of the letter: (i) A, (ii) N, (iii) F, (iv) O, (v) E.
2 The equilateral triangle ABC has three symmetries of rotation: E through the zero angle, R through $2\pi/3$ and P through $4\pi/3$. Let $S = \{E, R, P\}$ and write down the table for $(S, \circ)$.
3 Referring to question **2**, let L, M, N be the mirror symmetries of reflection of the triangle in the altitudes AF, BG, CH. Prove that the set T of all the symmetries of the triangle is given by $T = \{E, R, P, L, M, N\}$ and find the table of $(T, \circ)$.
4 A swastika (see §3.10, Example 1 (iii)) has rotational symmetries through angles of $\pi/2$, π, $3\pi/2$ and 0. Give names to these symmetries and write down the table of their composition.

3.7 Isomorphism

In Question **4** of Exercise 3.3, you will have discovered that there are cases of binary operations on sets where the structure is the same and the two systems differ only in the names of the elements, the sets and the operation. When this happens we say that the two systems are *isomorphic*, the word coming from the Greek, meaning a like (iso) shape (morph). Obviously, the sets of two isomorphic structures must contain the same number of elements and so there must be a one–one mapping between them. More than this, the operations must correspond so that there must be a one–one map between the two sets such that the same element is obtained from a pair of elements of one set whether we apply the operation of that set first and then map into the other set or whether we first map each element into the other set and then apply the operation of the second set. This property is described by saying that the operation is *preserved* by the mapping.

Definition The two systems $(S, *)$ and $(T, \circ)$ are isomorphic if both the following conditions hold:

(i) there is a one-one function f with domain S and range T,
that is f: $S \to T$, f is one-one and $f(S) = T$;
(ii) f preserves the binary operation,
that is for all x and y in S, $f(x * y) = f(x) \circ f(y)$.

Notation $(S, *) \cong (T, \circ)$ means that $(S, *)$ is isomorphic to $(T, \circ)$.

When the binary operations on the two sets S and T are given by their tables, then the two systems are isomorphic if the elements of the two sets can be arranged in such a way that the two tables look exactly alike except for the names of the elements. If the two tables do not look exactly alike, this does not prove that the systems are not isomorphic because it is possible that the tables can be made to look alike by a rearrangement of the elements of one of the sets.

EXAMPLE 1 *Prove that the set of symmetries of the letter Z under the operation of composition is isomorphic to* $(\mathbb{Z}_2, +)$.

We draw the tables of the two systems, which we have previously obtained in §3.6 and 3.3.

$(S, \circ)$	E	R
E	E	R
R	R	E

$(\mathbb{Z}_2, +)$	0	1
0	0	1
1	1	0

It can be seen that these two tables are the same apart from the names of the elements. Let $f: S \to \mathbb{Z}_2$ be defined by $f(E) = 0$, $f(R) = 1$, then the map f changes one table into the other and clearly gives an isomorphism between $(S, \circ)$ and $(\mathbb{Z}_2, +)$.

EXAMPLE 2 *Find the set S of symmetries of the figure shown in the diagram, (Fig. 3.5). Prove that* $(S, \circ) \cong (\mathbb{Z}_3, +)$.

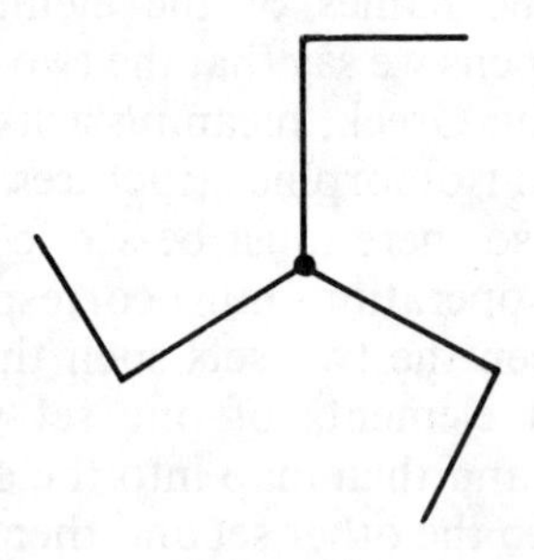

Fig. 3.5

In any symmetry of the figure, its centre must remain fixed and the three legs must be mapped on to each other. Thus, the figure has just three symmetries consisting of the identity symmetry E, the rotational symmetry R of clockwise rotation about the centre through $2\pi/3$ and the rotational symmetry RR, i.e. R^2, of clockwise rotation through $4\pi/3$ (or anticlockwise rotation through $2\pi/3$). Clearly $RR^2 = R^3 = E$, since this corresponds to a rotation through 2π. Thus $\mathbf{S} = \{\mathbf{E}, \mathbf{R}, \mathbf{R}^2\}$ and the tables of $(S, \circ)$ and $(\mathbb{Z}_3, +)$ may be compared. We shall write the table of $(\mathbb{Z}_3, +)$ in two ways.

$(S, \circ)$	E	R	R^2
E	E	R	R^2
R	R	R^2	E
R^2	R^2	E	R

$(\mathbb{Z}_3, +)$	0	1	2
0	0	1	2
1	1	2	0
2	2	0	1

$(\mathbb{Z}_3, +)$	0	2	1
0	0	2	1
2	2	1	0
1	1	0	2

These tables display two isomorphisms f and g between $(S, \circ)$ and $(\mathbb{Z}_3, +)$, given by

$$f(E) = 0, \quad f(R) = 1, \quad f(R^2) = 2,$$
$$g(E) = 0, \quad g(R) = 2, \quad g(R^2) = 1.$$

Note also that there is a isomorphism h of $(\mathbb{Z}_3, +)$ on to $(\mathbb{Z}_3, +)$, given by

$$h(0) = 0, \quad h(1) = 2, \quad h(2) = 1.$$

When two sets with binary operations are infinite sets, it is no longer possible to use their tables in order to prove that they are isomorphic by displaying an isomorphism between them. It is then necessary to find an appropriate function of one set on to the other set and then to prove that this function is an isomorphism.

EXAMPLE 3 *Show that* $(\mathbb{R}^+, \times) \cong (\mathbb{R}, +)$.

We use the function

$$f\colon \mathbb{R}^+ \to \mathbb{R}, \text{ defined by } f(x) = \ln x.$$

This is a one-one function, with inverse

$$f^{-1} \text{ given by } f^{-1}(a) = e^a, \text{ for all } a \text{ in } \mathbb{R}.$$

To complete the proof that f is an isomorphism, we must show that it preserves the operations, that is, we must prove that,

$$\text{for } x > 0 \text{ and } y > 0,\ f(xy) = f(x) + f(y).$$

Now this is true because

$$f(xy) = \ln(xy) = \ln x + \ln y = f(x) + f(y).$$

EXERCISE 3.7

1 By comparing their tables and finding one-one maps which preserve the operations between the sets, prove that the following sets with binary operations are isomorphic, (A is any non-empty set):
(i) $(\mathbb{Z}_2, \times)$, (ii) $(\{A, \phi\}, \cup)$,
(iii) $(\{A, \phi\}, \cap)$, (iv) $(\{2, 4\}$, larger number).

2 Prove that the following are isomorphic: (i) $(\mathbb{Z}_3, +)$,
(ii) $(\{2, 4, 8\}$, multiplication modulo 14),
(iii) $(\{0, 4, 8\}$, addition modulo 12),
(iv) $(S, \circ)$, given in Example 2 above.

3 Prove that the following are isomorphic: (i) $(\mathbb{Z}_4, +)$
(ii) $(T, \circ)$, given in Question 3 of Exercise 3.3,
(iii) $(\{1, 2, 4, 8\}$, multiplication modulo 15),
(iv) $(\{1, -1, i, -i\}$, multiplication of complex numbers),
(v) $(\{1, 3, 7, 9\}$, multiplication modulo 10).

4 Prove that the following are isomorphic: (i) $(\mathbb{Z}_3, \times)$,
(ii) $(\{0, 2, 4\}$, multiplication modulo 6),
(iii) $(\{1, 4, 7\}$, multiplication modulo 12).

5 Prove that the following are isomorphic: (i) $(\mathbb{Z}_2, +)$, (ii) $(\{1, -1\}, \times)$, (iii) $(\{1, 2\}$, multiplication modulo 3), (iv) $(S, \circ)$, S is the set of symmetries of the letter A.

6 Prove that if $(S, *) \cong (T, \circ)$ and $(T, \circ) \cong (U, \times)$ then $(S, *) \cong (U, \times)$.

7 Let E be the set of all even integers. Prove that

$$\text{f}: \mathbb{Z} \to E, \text{ given by } \text{f}(n) = 2n,$$

is an isomorphism of $(\mathbb{Z}, +)$ on to $(E, +)$. Prove also that

$$\text{g}: \mathbb{Z} \to E, \text{ given by } \text{g}(n) = -2n,$$

is an isomorphism of $(\mathbb{Z}, +)$ on to $(E, +)$.

8 Using the notation of Question 7, prove that f is not an isomorphism of $(\mathbb{Z}, \times)$ on to $(E, \times)$, by considering the image under f of the product 1×1 in $\mathbb{Z}$.

3.8 Properties of binary operations

Commutativity

If, in a binary operation, the order of the two elements does not matter, then we say that the two elements *commute*,

i.e. if $b * a = a * b$ then a and b commute under $*$.

If, in $(S, *)$, all pairs of elements of S commute, then we say that $*$ is a commutative operation on S.

Definition The binary operation $*$ on the set S is commutative if for all a, b in S, $a * b = b * a$. The binary operation $*$ on the set S is non-commutative if there is one pair of elements a, b in S with $a * b \neq b * a$.

Clearly, an operation is commutative when its table is symmetrical about its leading diagonal—the diagonal running down and to the right from the top left hand corner.

Most of the binary operations which we have considered so far are commutative. The only non-commutative ones which we have considered are (a) subtraction, $4 - 3 \neq 3 - 4$, (b) division, $4/3 \neq 3/4$ and (c) $2x - y$ used in Example 3 of §3.2. Check that the operations of ll the other examples and exercises are commutative.

Neutral element

Under the operation of addition, the number zero leaves the other number unchanged, since $0 + x = x = x + 0$. The number 1 acts in the same way under multiplication: $1 \times x = x = x \times 1$. Such elements are called *neutral* elements of the binary operation concerned.

Definition The element e is a neutral element of the binary operation $*$ on the set S if $e \in S$ and, for all x in S, $e * x = x = x * e$.

When the operation is either addition or multiplication, it is usual to use special notations.

Addition: write $*$ as $+$ and call the neutral element *zero* or 0.

Multiplication: write $*$ as $\times$ and call the neutral element *unity* or 1 or the *unit* element or the *identity* element.

A neutral element can be spotted from the table of a binary operation on a set since in its corresponding row and column all the elements of the set occur unchanged. In the next example, we prove that there is only one neutral element and so we refer to it as *the* neutral element. In the case of addition we refer to *the* zero and in the case of multiplication we refer to *the* identity.

EXAMPLE 1 *If* $(S, *)$ *contains a neutral element* e, *prove that this is unique.*

Suppose that the two elements e and f are neutral elements in $(S, *)$. Then $e*f = e$ and $e*f = f$. Hence $e = f$.

Inverse elements

In the case when a neutral element e exists in $(S, *)$, it may happen that two elements a and b exist in S and that $a * b = e = b * a$. In this case we say that a is the *inverse* of b and that b is the inverse of a.

Definition If e is the neutral element of $(S, *)$ and $a * b = e = b * a$, then a is the inverse element of b and b is the inverse of a.

EXAMPLE 2 *Prove that in* $(\mathbb{Z}, +)$ *every element has an inverse but that in* $(\mathbb{Z}, \times)$ *only two elements have inverses.*

In $(\mathbb{Z}, +)$ the neutral element is 0 and, for any $n \in \mathbb{Z}$, $n + (-n) = 0 = (-n) + n$. Therefore $-n$ is the inverse of n. Note that 0 is the only self-inverse element.

In $(\mathbb{Z}, \times)$ the neutral element is 1 and, for any integer n, the equations $n \times x = 1 = x \times n$ have a solution for x only when $n \in \{1, -1\}$. Thus the only elements with inverses are 1 and -1, and each is its own inverse.

EXAMPLE 3 *Given that* $\mathrm{f}: S \to T$ *is an isomorphism of* $(S, *)$ *on to* $(T, \circ)$ *and that* e *is the neutral element of* $(S, *)$, *prove that* $\mathrm{f}(e)$ *is the neutral element of* $(T, \circ)$. *If, furthermore,* a *and* b *are inverse elements in* $(S, *)$, *prove that* $\mathrm{f}(a)$ *and* $\mathrm{f}(b)$ *are inverse elements in* $(T, \circ)$.

Let e be the neutral element of $(S, *)$. Then for all x in S,

$$e * x = x = x * e.$$

Now $\mathrm{f}: S \to T$ is an isomorphism so, for every y in T, there is an element x in S, given by $x = \mathrm{f}^{-1}(y)$, such that $\mathrm{f}(x) = y$.

Then $\quad \mathrm{f}(e) \circ y = \mathrm{f}(e) \circ \mathrm{f}(x) = \mathrm{f}(e * x) = \mathrm{f}(x) = y,$

and $\quad y \circ \mathrm{f}(e) = \mathrm{f}(x) \circ \mathrm{f}(e) = \mathrm{f}(x * e) = \mathrm{f}(x) = y.$

Hence $f(e)$ is the neutral element in $(T, \circ)$. If $a*b = e = b*a$,

then $$f(a) \circ f(b) = f(a*b) = f(e) = f(b*a) = f(b) \circ f(a),$$

and so $f(a)$ and $f(b)$ are inverse elements in $(T, \circ)$.

It follows from Example 3 that, if two structures are isomorphic, then the neutral elements must map on to one another and inverse pairs must map on to inverse pairs, under any isomorphism. This provides a useful guide when trying to set up a suitable mapping between the sets in order to prove the isomorphism. It also may help to provide a counter-example when trying to prove that two structures are not isomorphic.

EXERCISE 3.8A

1 For the binary operation given by the table, (a) state whether or not it is commutative, (b) state whether or not it possesses a neutral element, (c) if it possesses a neutral element, state those elements which possess inverses:

(i)

	a	b	c
a	c	a	a
b	a	b	c
c	a	c	b

(ii)

	a	b	c
a	b	c	a
b	c	b	b
c	a	a	b

(iii)

	a	b	c	d
a	a	b	c	d
b	b	c	a	b
c	c	a	b	c
d	d	b	c	a

2 The operation $*$ in $(\mathbb{R}, *)$ is defined by $x*y = xy + x + y$. Show that 0 is the neutral element for this operation. Prove that every real number except -1 has an inverse under $*$ and, if $x \neq -1$, write down the inverse of x.

3 In the following set with binary operation, find if it possesses a neutral element. If a neutral element exists, find the subset of all elements which possess inverses in the set:
(i) $(\mathbb{R}, \times)$ (ii) $(\mathbb{R}, +)$ (iii) $(\mathbb{Z}, -)$ (iv) $(\mathbb{Q}^+, \div)$ (v) $(\mathbb{Z}_4, \times)$.

Associativity

There is a very important property of binary operations which involves three elements. Let a, b and c be three elements of the set S and let $*$ be a closed binary operation on S. We can operate on a, b, c, in that order, in two ways:

(i) we can first form $a*b$ and then form $(a*b)*c$,
(ii) we can first form $b*c$ and then form $a*(b*c)$.

If these two results are always the same for all a, b, c in S then we say that the operation $*$ on S is *associative*. In this case, the brackets may be omitted because we obtain the same product for either position of the brackets.

Definition $(S, *)$ is associative if, for all a, b, c in S,

$$(a*b)*c = a*(b*c),$$

and then the product is written $a*b*c$, without brackets.

EXAMPLE 4 *Prove that the operation $*$ of Exercise 3.8A, question* **2**, *is associative.*

In this case, for any real numbers a, b, c, $a*b = ab+a+b$.

Therefore
$$\begin{aligned}(a*b)*c &= (ab+a+b)c+(ab+a+b)+c\\ &= abc+ac+bc+ab+a+b+c\\ &= a(bc+b+c)+a+(bc+b+c)\\ &= a*(b*c),\end{aligned}$$
and so (R, $*$) is associative.

Note that we re careful not to change the order of the three elements in dealing with associativity. In the above example, the operation was also commutative so we could have changed the order of elements in any product if we so wished.

When $(S, *)$ is associative, we can define powers of an element. We let $a*a = a^2$, and then $a*a*a = a^3$, and similarly for higher powers. This would not be possible if $(a*a)*a$, $(= a^2*a)$ was not the same element as $a*(a*a)$, $(= a*a^2)$. In the same way, we can attach no meaning to a^4 if $a*(a*a)*a$ and $(a*a)*(a*a)$ are different, or if any two ways of bracketing the four a's give different elements.

EXAMPLE 5 *Test* $(\mathbb{R}, *)$ *for associativity, where* $x * y = x^2y$.

If we begin by trying to prove that $*$ is associative, we take three real numbers x, y, z. Then we evaluate the two triple products:

$$(x*y)*z = (x^2y)*z = x^4y^2z;\ x*(y*z) = x*(y^2z) = x^2y^2z.$$

These two numbers will be different provided that $x^2 \neq x^4$, that is, for $x \notin \{0, 1, -1\}$. Therefore we can provide a counter-example to associativity by choosing $x = 2$, $y = 1$, $z = 1$.

Then $$(2*1)*1 = 4*1 = 16 \quad \text{but} \quad 2*(1*1) = 2*1 = 4,$$

and hence $*$ is **not associative**.

It is difficult to test a binary operation for associativity by using its table. If the set S has n elements, then an ordered triplet $\{a, b, c\}$ of elements of S can be chosen in n^3 ways and so we have to test $2n^3$ products $a*(b*c)$ and $(a*b)*c$. Even for a three element set, this means 27 tests in order to prove associativity. Of course, if we want to prove that $(S, *)$ is non-associative, we only need one suitable triple (a, b, c) to provide a counter-example.

We usually use known properties of the sets and operations concerned in order to prove associativity. For real and complex numbers, $(a+b)+c = a+(b+c)$ and $(ab)c = a(bc)$, so any set which is a subset of $\mathbb{C}$ will be associative under the operations of addition and multiplication. The same is true of $\mathbb{Z}_n$ since, in $\mathbb{Z}_n$, addition and multiplication modulo n derive from addition and multiplication in $\mathbb{Z}$ followed by the removal of multiples of n. We shall also be able to quote the fact that addition and multiplication of matrices are both associative and apply this to sets of square $n \times n$ matrices (Chapter 1).

Another operation which we know to be associative is that of the composition of functions or mappings and we will usually quote this property without proof. We remind the reader of the proof: let f, g, h be functions with domain and codomain equal to D. Then, for any x in D,

$$(\mathrm{f}(\mathrm{gh}))(x) = \mathrm{f}(\mathrm{gh}(x)) = \mathrm{f}(\mathrm{g}(\mathrm{h}(x))) = (\mathrm{fg})(\mathrm{h}(x)) = ((\mathrm{fg})\mathrm{h})(x)$$

and so $\mathrm{f}(\mathrm{gh}) = (\mathrm{fg})\mathrm{h}$.

A consequence of this is that, if S is a set of symmetries of a geometrical figure then $(S, \circ)$ is associative.

EXERCISE 3.8B

1 Test the following binary operation, given by a table, for closure, commutativity, associativity, the existence of a neutral element and the existence of inverses:

(i)

	a	b	c
a	a	a	a
b	b	b	b
c	c	c	c

(ii)

	a	b	c
a	a	a	a
b	a	b	b
c	a	b	c

(iii)

	a	b	c	d
a	a	b	c	d
b	b	c	d	b
c	c	d	b	a
d	d	b	a	b

2 Test, for the properties mentioned in question **1**, the following binary operation:
(i) $(\mathbb{R}, *)$, $x * y = \text{minimum}\{x, y\}$ (ii) $(\mathbb{Z}, *)$, $m * n = \text{H.C.F.}\{m, n\}$
(iii) $(\mathbb{Z}^+, *)$, $m * n = |m - n|$ (iv) $(\mathbb{Z}, *)$, $m * n = |m + n|$
(v) $(\mathbb{Z}^+, *)$, $m * n = \text{L.C.M.}\{m, n\}$ (vi) $(\{A, B\}, \cup)$, A, B are sets, $A \subseteq B$
(vii) $(S, \cap)$, S is the set of all subsets of a set U
(viii) $(S, +)$, S is the set of all real 2×2 matrices (ix) $(S, \times)$, S is the set of all real non-singular 3×3 matrices (x) $(S, \times)$, S is the set of all 2×2 complex matrices.

3 The binary operation $*$ on the set S of all real numbers is defined by $a * b = |a - b|$. Investigate whether $*$ is (i) commutative, (ii) associative. The relation R on S is defined by $a\mathrm{R}b \Leftrightarrow a * b = 1$. State, with reasons, whether or not R is (i) reflexive, (ii) symmetric, (iii) transitive. (*JMB*)

3.9 Groups

Permutations

We begin with a definition of a permutation of a set.

Definition A permutation of *a set* X is a one-one map of X n to X.

In *Pure Mathematics* §17.7, we considered a permutation of a sequence, which is a one-one mapping of the sequence on to another sequence, consisting of the same terms in a different order. The elements of each sequence form the same set X and so the permutation of the sequence is also a permutation of X. A symmetry of a geometrical figure is a permutation of the set of points of the figure, with the extra property that the distance between any two points is the same as the distance between their images. This type of permutation is sometimes called an *isometry*.

A permutation f of a set X is described when we know the image $\mathrm{f}(x)$ of every element x of the set X. A convenient way of representing f in the case when X has n elements is to use a $2 \times n$ matrix, in which the first row contains the elements of X and the second their images under f. For each x in X, $\mathrm{f}(x)$ is placed under x. Thus, if $X = \{a, b, c\}$ and f is the permutation given by $\mathrm{f}(a) = b$, $\mathrm{f}(b) = c$, $\mathrm{f}(c) = a$, then we write

$$\mathrm{f} = \begin{pmatrix} a & b & c \\ \mathrm{f}(a) & \mathrm{f}(b) & \mathrm{f}(c) \end{pmatrix} = \begin{pmatrix} a & b & c \\ b & c & a \end{pmatrix}.$$

Briefly, when we do not wish to list all the elements of X, we could write

$$\mathrm{f} = \begin{pmatrix} x \\ \mathrm{f}(x) \end{pmatrix}, \; x \in X.$$

This notation can also be used when the set X is infinite.

We define the product fg of two permutations f and g of a set X as the composition of f and g, so that fg is the permutation g followed by the permutation f.

Definition Let f and g be permutations of the set X, then fg is the permutation of X given by $\mathrm{fg}(x) = \mathrm{f}(\mathrm{g}(x))$, for all $x \in X$.

Note that the permutations are performed in the reverse order to the order in which they are written. For example, if $X = \{a, b, c\}$,

$$\mathrm{f} = \begin{pmatrix} a & b & c \\ b & c & a \end{pmatrix} \text{ and } \mathrm{g} = \begin{pmatrix} a & b & c \\ a & c & b \end{pmatrix}, \text{ then } \mathrm{fg} = \begin{pmatrix} a & b & c \\ b & a & c \end{pmatrix}$$

since $\mathrm{fg}(a) = \mathrm{f}(a) = b$, $\mathrm{fg}(b) = \mathrm{f}(c) = a$ and $\mathrm{fg}(c) = \mathrm{f}(b) = c$.

EXAMPLE 1 *Describe the set S of symmetries of a rectangle $ABCD$ in terms of permutations of the set $\{A, B, C, D\}$.*

We refer to the Example of §3.6, where we obtained the set S of the four symmetries of the rectangle, with the notation $S = \{\mathrm{E}, \mathrm{R}, \mathrm{M}, \mathrm{N}\}$. We describe these symmetries as permutations of $\{A, B, C, D\}$, that is, as functions with this set as domain:

$$\begin{array}{llll} \mathrm{E}(A) = A, & \mathrm{E}(B) = B, & \mathrm{E}(C) = C, & \mathrm{E}(D) = D, \\ \mathrm{R}(A) = C, & \mathrm{R}(B) = D, & \mathrm{R}(C) = A, & \mathrm{R}(D) = B, \\ \mathrm{M}(A) = D, & \mathrm{M}(B) = C, & \mathrm{M}(C) = B, & \mathrm{M}(D) = A, \\ \mathrm{N}(A) = B, & \mathrm{N}(B) = A, & \mathrm{N}(C) = D, & \mathrm{N}(D) = C. \end{array}$$

Thus, in the notation for permutations of $\{A, B, C, D\}$:

$$\mathrm{E}=\begin{pmatrix} A & B & C & D \\ A & B & C & D \end{pmatrix}, \mathrm{R}=\begin{pmatrix} A & B & C & D \\ C & D & A & B \end{pmatrix}, \mathrm{M}=\begin{pmatrix} A & B & C & D \\ D & C & B & A \end{pmatrix}, \mathrm{N}=\begin{pmatrix} A & B & C & D \\ B & A & D & C \end{pmatrix}.$$

The table, which we obtained in §3.6, shows that, for this set S of symmetries of $\{A, B, C, D\}$, $(S, \circ)$ is closed, commutative, has an identity E, and every element in S is its own inverse. We also know that $(S, \circ)$ is associative, because the operation $\circ$ of composition of functions is associative and permutations are functions.

Note that, in general, a set of permutations will not be commutative because the composition of functions is not commutative except in special cases.

Let us consider the set S of all permutations of a given set X, then $(S, \circ)$ has the following properties:

(i) it is closed because we include all the permutations of X

(ii) it is associative as composition of permutations is associative

(iii) it possesses a neutral (identity) element e, given by

$$\mathrm{e}(x) = x \text{ for all } x \text{ in } X,$$

because, if f is any permutation in S, for all x in X,

$$\mathrm{fe}(x) = \mathrm{f}(\mathrm{e}(x)) = \mathrm{f}(x) = \mathrm{e}(\mathrm{f}(x)) = \mathrm{ef}(x), \text{ and so fe} = \mathrm{f} = \mathrm{ef};$$

(iv) every permutation f in S has an inverse f^{-1} in S and f^{-1} is the inverse of f in S, because, for every x in X,

$$\mathrm{ff}^{-1}(x) = \mathrm{f}(\mathrm{f}^{-1}(x)) = x = \mathrm{e}(x) = \mathrm{f}^{-1}(\mathrm{f}(x)) = \mathrm{f}^{-1}\mathrm{f}(x)$$

and so $\mathrm{ff}^{-1} = \mathrm{e} = \mathrm{f}^{-1}\mathrm{f}$.

These four properties (i)–(iv), of the set S of all permutations of X with the binary operation of composition, occur in many algebraic structures and when all four properties hold, the structure is called a *group*. An abstract group is a set with a binary operation satisfying the four axioms of closure, associativity, existence of identity and existence of inverses. We now give a formal definition.

Definition Let S be a set with a binary operation $*$, $G = (S, *)$. Then G is a group if the following four axioms are all satisfied:

G1 closure: for all a and b in S, $a * b$ is in S;

G2 associativity: for all a, b and c in S, $(a * b) * c = a * (b * c)$;

G3 identity: there exists e in S and, for all a in S, $e * a = a = a * e$;

G4 inverses: for all a in S, there exists a^{-1} in S and $a^{-1} * a = e = a * a^{-1}$.

When there is no ambiguity concerning the operation $*$, we shall identify G with S and say that S is a group, omitting reference to $*$.

In general, a group is not commutative. If a and b are any elements in a group G, then $a * b$ and $b * a$ may be different elements. This means that it is very important not to change the order of the elements in a group product.

Abelian groups If the group is commutative, then the group table is symmetrical about its leading diagonal and, for any two elements a and b of the group, $a * b = b * a$. It is common to refer to a commutative group as an *Abelian group.*

Order of a group The number of elements in a group G is called the order of the group and is denoted by $n(G)$.

Definition If G has n elements, $n \in \mathbb{N}$, then $n(G) = n$ and G is a finite group. If G has infinitely many elements, then $n(G) = \infty$ and G is an infinite group.

Note In the infinite case, the equation $n(G) = \infty$ simply means that the group is infinite, that is that its order is infinite. It must not be interpreted to mean that there is such a number as ∞.

EXAMPLE 2 *Show that* $(\mathbb{Z}_3, +)$ *is a group of order* 3.

Under addition, $\mathbb{Z}_3$ is closed and associative because $(\mathbb{Z}, +)$ is associative. Of the three elements in $\mathbb{Z}_3$, 0 is the identity under addition and is its own inverse, and 1 and 2 are inverses of each other. Therefore $(\mathbb{Z}_3, +)$ satisfies the axioms G1, G2, G3 and G4 and so it is a group.

$(\mathbb{Z}_3, +)$	0	1	2
0	0	1	2
1	1	2	0
2	2	0	1

Another way of showing that $(\mathbb{Z}_3, +)$ satisfies axioms G1, G3 and G4 is to use its table. From the table we can see that all entries lie in $\mathbb{Z}_3$ so G1 is satisfied. Also the first row and column leave all the elements unaltered so G3 is satisfied. Finally, in every row and column the identity 0 occurs just once and its positions are symmetrically placed relative to the leading diagonal, so G4 is satisfied. Notice, however, that we can not see from the table that the operation is associative.

EXAMPLE 3 *Show that* $(\mathbb{Z}_3, \times)$ *is not a group.*

Under multiplication, $\mathbb{Z}_3$ is closed, associative and has an identity element, namely 1. However, the first row of the table of $(\mathbb{Z}_3, \times)$ does not contain 1 and so 0 has no inverse. Therefore $(\mathbb{Z}_3, \times)$ is not a group since it does not satisfy axiom G4.

Theorem 3.3 The set $S(X)$ of all permutations of a given set X is a group under composition.

Proof We have already proved that $(S(X), \circ)$ is closed, associative, has an identity e and that, in $S(X)$, every permutation f has an inverse f^{-1}. Hence $(S(X), \circ)$ is a group.

EXAMPLE 4 *Prove that the set S of symmetries of a rectangle is a group under composition.*

We reproduce the table of $(S, \circ)$ from the Example of §3.6.

$(S, \circ)$	E	R	M	N
E	E	R	M	N
R	R	E	N	M
M	M	N	E	R
N	N	M	R	E

The table is closed and the operation of composition is associative. The symmetry E is the identity and every symmetry is its own inverse. Hence $(S, \circ)$ is a group.

We may generalise this result and show that, as in the case of all the permutations of a set, the set of all symmetries of a geometrical figure is a group under composition.

Theorem 3.4 Let X be a geometrical figure with a set S of all symmetries of X. Then $(S, \circ)$ is a group called the symmetry group of X.

Proof We have already proved that the composition of two symmetries of X is also a symmetry, so $(S, \circ)$ is closed. The operation is associative and S contains the identity symmetry E. Also, if A is any symmetry of X, then A is a one-one permutation of the points of X which preserves distance and so A has an inverse map A^{-1} which also preserves distance and so is a symmetry. Hence $(S, \circ)$ is a group.

It should be noted that, in the table for $(S, \circ)$ of Example 4, every element of S occurs exactly once in each row and in each column of the table. In other words, each row and each column of the table is a permutation of the elements of S. Such a table is called a *Latin square.* Latin squares are of great importance in the design of statistical experiments.

It is necessary that the table of a group is a Latin square, as we prove in Theorem 3.6. However it is not true that every Latin square is the table of a group, as we show in Example 5. We first prove that it is possible to cancel elements on the right (and on the left) in an equation relating group elements. For a group which is not commutative, it is not possible to cancel an element which is on the right on one side of the equation and on the left on the other side of the equation.

Theorem 3.5 (Cancellation on the right, and on the left, in a group)
Let $(G, *)$ be a group and let a, b and x be elements of G, then

$$a * x = b * x \Rightarrow a = b,$$
$$x * a = x * b \Rightarrow a = b.$$

Proof The results follow at once by multiplying the equation on the appropriate side by x^{-1}, the element inverse to x, and using associativity. If $a*x=b*x$, then

$$\begin{aligned} a = a*e = a*(x*x^{-1}) &= (a*x)*x^{-1} = (b*x)*x^{-1} \\ &= b*(x*x^{-1}) = b*e = b. \end{aligned}$$

If $x*a=x*b$, then

$$\begin{aligned} a = e*a = (x^{-1}*x)*a &= x^{-1}*(x*a) = x^{-1}*(x*b) \\ &= (x^{-1}*x)*b = e*b = b, \end{aligned}$$

as was to be proved.

Theorem 3.6 In the table of a group $(G, *)$, each row and each column consists of a permutation of the elements of G, i.e. the table is a Latin square.

Proof Consider the row labelled by the element a of G. For any element x of G, let $y=a^{-1}*x$, then the element in column y of row a is $a*y$ and $a*y=a*(a^{-1}*x)=(a*a^{-1})*x=e*x=x$, and so x occurs in row a. Similarly, x also occurs in column b, for any b in G, because the element in row $x*b^{-1}$ and column b is $x*b^{-1}*b$, which is equal to x. So far, we have proved that an element x of G occurs in each row and in each column of the table. To prove that each row and each column is a permutation of the elements of G, we must also show that an element x only occurs once in each row and in each column. Suppose that x occurs in columns b and c of row a. Then $x=a*b=a*c$, and then, by cancellation of a on the left, $b=c$. Hence x occurs only once in row a. Similarly, if x occurs in both rows a and c of column b, $x=a*b=c*b$ and, by cancellation of b on the right, $a=c$, and x only occurs in one row in column b.

EXAMPLE 5 *Prove that $(S, *)$, with the given table, is not a group even though each row and each column of the table is a permutation of S.*

G	e	a	b	c	d
e	e	a	b	c	d
a	a	b	e	d	c
b	b	c	d	e	a
c	c	d	a	b	e
d	d	e	c	a	b

The table is closed and e is the identity element, so $(S, *)$ satisfies axioms G1 and G3. However it does not satisfy axioms G2 and G4: $a*(b*c)=a*e=a$ and $(a*b)*c=e*c=c$, so the operation is not associative; also the element a has different inverses on the right and on the left, since $a*b=e=d*a$, and so a has no inverse. The same is true for b, c and d. Therefore $(S, *)$ is not a group.

As well as the results of Theorems 3.5 and 3.6 there are some other immediate consequences of the group axioms which are very useful. We give them in the next theorem.

Theorem 3.7 Let $(G, *)$ be a group with an identity element e and an inverse element a^{-1} for each element a in G. Then
(i) the identity element e in G is unique,
(ii) for each element a in G, the inverse a^{-1} is unique,
(iii) for a, b in G, $(a * b)^{-1} = b^{-1} * a^{-1}$,
(iv) for a, b in G, the equation $a * x = b$ has the unique solution

$$x = a^{-1} * b,$$

(v) for a, b in G, the equation $x * a = b$ has the unique solution

$$x = b * a^{-1}.$$

Proof (i) The uniqueness of the neutral element of a binary operation was proved in §3.8, Example 1.

(ii) Suppose that a^{-1} and b are two inverses of the element a of G. Then, using associativity,

$$a^{-1} = a^{-1} * e = a^{-1} * (a * b) = (a^{-1} * a) * b = e * b = b.$$

Hence the inverse of a is unique. Note that the method of proving that an element is unique is to assume that two elements exist with the required property and then to prove that the two elements are equal.

(iii) Let $b^{-1} * a^{-1} = y$, then

$$(a * b) * y = a * (b * y) = a * (b * (b^{-1} * a^{-1})) = a * ((b * b^{-1}) * a^{-1})$$
$$= a * (e * a^{-1}) = a * a^{-1} = e,$$

and so $\quad y = (a * b)^{-1}$, as was to be proved.

(iv) If $a * x = b$, then $x = e * x = (a^{-1} * a) * x = a^{-1} * (a * x) = a^{-1} * b$. Conversely, if $x = a^{-1} * b$ then
$a * x = a * (a^{-1} * b) = (a * a^{-1}) * b = e * b = b$, as required.

(v) The proof is similar to the proof of (iv) above, with the orders of the elements reversed in each of the products.

Multiplicative and additive notations

There are two notations which are very common in group theory. The group can be written either multiplicatively or additively.

Notation

(a) In a multiplicative group G the operation is written xy the identity is either e or 1 and the inverse of x is x^{-1}.

(b) In an additive group G the operation is written $x + y$ the identity is 0 and the inverse of x is $-x$.

EXAMPLE 6 *Write the results of Theorems 3.5 and 3.7 in (a) the multiplicative notation and (b) the additive notation.*

Let a, b, x be any elements in a group G. Then if G is

(a) a multiplicative group	(b) an additive group
$ax = bx \Rightarrow a = b$	$a+x = b+x \Rightarrow a = b$
$xa = xb \Rightarrow a = b$	$x+a = x+b \Rightarrow a = b$
the identity 1 is unique	the identity 0 is unique
the inverse a^{-1} of a is unique	the inverse $-a$ of a is unique
$(ab)^{-1} = b^{-1}a^{-1}$	$-(a+b) = -b-a$
$xa = b \Leftrightarrow x = ba^{-1}$	$x+a = b \Leftrightarrow x = b-a$
$ax = b \Leftrightarrow x = a^{-1}b$	$a+x = b \Leftrightarrow x = -a+b$.

EXERCISE 3.9

1 In the multiplicative group G, the inverse of any element a is a^{-1}. Prove that, for all a, b, c in G:
(i) $(a^{-1})^{-1} = a$, (ii) $(abc)^{-1} = c^{-1}b^{-1}a^{-1}$.

2 The element e is the identity element in the group G and $G = \{e, a\}$. Find the inverse of the element a.

3 Prove that the following are groups:
(i) $(\{1, -1, \mathrm{i}, -\mathrm{i}\}, \times)$, (ii) $(\mathbb{Z}, *)$, where $m * n = m + n - mn$,
(iii) $(\mathbb{Z}_2\backslash\{0\}, \times)$, (iv) $(\mathbb{Z}_2, +)$.

4 Prove that the following are not groups and state which of the axioms G1–G4 are not satisfied in each: (i) $(\{\phi, \{x\}, \{y\}, \{x, y\}\}, \cap)$
(ii) $(\mathbb{Z}$, subtraction), (iii) $(\{1, 2, 3\}$, product modulo 4), (iv) $(\mathbb{Q}$, division),
(v) $(\mathbb{R}, *)$, $a * b = a$, (vi) $(\mathbb{R}, *)$, $a * b = a^b$, (vii) $(\mathbb{Z}_2$, product modulo 2),
(viii) $(\mathbb{Z}_4, -)$, (ix) $(\mathbb{Z}_4, *)$, $m * n = |m - n|$.

5 Determine whether or not the following are groups: (i) $(\mathbb{Q}, +)$, (ii) $(\mathbb{C}, +)$,
(iii) $(\mathbb{C}^*, \times)$, (iv) $(\mathbb{R}^+, \times)$, (v) $(C, \times)$, $C = \{z\colon z \in \mathbb{C}$ and $|z| = 1\}$,
(vi) $(\mathbb{Z}_3\backslash\{0\}, \times)$.
(Remember that $C^* = \mathbb{C}\backslash\{0\}$.)

6 In a group G, the square of every element is equal to the identity. Prove that, for all a and b in G,
(i) $aba = b^{-1}$, (ii) $aba = b$, (iii) $ab = ba$.
Deduce that a group, in which the square of every element is equal to the identity, is a commutative group.

7 For all elements a and b in a group G, $(ab)^2 = a^2b^2$. Prove that, G is commutative.

8 Let V be the set of all real column vectors in three dimensions. Prove that $(V, +)$ is a group. State the identity element in this group and the inverse of the vector $\begin{pmatrix} 1 \\ -2 \\ 3 \end{pmatrix}$.

9 Let S be the set of all real 2×2 matrices and let T be the set of all real, non-singular 2×2 matrices. Prove that:
(i) $(S, +)$ is a group (ii) $(S, \times)$ is not a group (iii) $(T, +)$ is not a group
(iv) $(T, \times)$ is a group.

10 By rearranging the order of the elements of S, prove that $(S, *)$ is a group, given that the table of $(S, *)$ is:

(i) $(S, *)$	a	b	c
a	c	a	b
b	a	b	c
c	b	c	a

(ii) $(S, *)$	a	b	c	d
a	b	a	d	c
b	a	b	c	d
c	d	c	b	a
d	c	d	a	b

(ii) $(S, *)$	a	b	c	d
a	d	c	b	a
b	c	a	d	b
c	b	d	a	c
d	a	b	c	d

11 Draw the table of $(S, *)$ and determine whether $(S, *)$ is a group:

(i) $S = \mathbb{Z}_4$ and $*$ is product modulo 4,
(ii) $S = \mathbb{Z}_4 \backslash \{0\}$ and $*$ is product modulo 4,
(iii) $S = \mathbb{Z}_5 \backslash \{0\}$ and $*$ is product modulo 5,
(iv) $S = \mathbb{Z}_6 \backslash \{0\}$ and $*$ is product modulo 6.

12 Given two isomorphic groups, $(G, \times)$ and $(H, *)$, with identity elements e in G and d in H, and with inverses a^{-1} of a in G and x' of x in H, let $\mathrm{f}: G \to H$ be a group isomorphism. Prove that (i) $\mathrm{f}(e) = d$, (ii) $\mathrm{f}(a^{-1}) = (\mathrm{f}(a))'$.

13 Prove that the following are groups and that they are all isomorphic:
(i) $\{0, 4, 8\}$ with the operation of addition modulo 12,
(ii) $\{1, 2, 4\}$ with the operation of multiplication modulo 7,
(iii) $\{1, 4, 16\}$ with the operation of multiplication modulo 63,
(iv) $\{1, 3, 9\}$ with the operation of multiplication modulo 26.

14 Prove that the following are groups and that they are all isomorphic:
(i) $\{1, 5, 7, 11\}$ with the operation of multiplication modulo 12,
(ii) $\{1, 3, 5, 7\}$ with the operation of multiplication modulo 8,
(iii) $\{1, 4, 11, 14\}$ with the operation of multiplication modulo 15,
(iv) the symmetry group of a rectangle.

15 Show that the set $A = \{1, 2, 3, 4, 5, 6\}$ forms a group under the operation of multiplication modulo 7. State the inverse of each element. An element a of the set A is said to have order n if n is the smallest positive integer such that $a^n = 1 \pmod 7$. Find the order of each of the six elements of A and write 5 as the product of two elements of order 2 and 3. *(L)*

16 The multiplication table for the set $\{e, a, b, c, d\}$ is given below.

	e	a	b	c	d
e	e	a	b	c	d
a	a	e	c	d	b
b	b	d	e	a	c
c	c	b	d	e	a
d	d	c	a	b	e

Using the above table, determine $(ab)c$, $a(bc)$, $(bc)d$, $b(cd)$. Ascertain which group axioms are satisfied by the given set under the given multiplication. Find two subsets from the above set which form a group under the given multiplication. *(L)*

3.10 Cyclic groups

Let a be an element in a group $(G, *)$ with identity e. Then we define the powers of a in the usual manner.

Definition $a^1 = a, \quad a^2 = a * a, \quad a * a^2 = a^3,$

$$a^{n+1} = a^n * a, \text{ for all } n \in \mathbb{N},$$
$$a^{-2} = a^{-1} * a^{-1} = (a^{-1})^2 = (a^2)^{-1},$$
$$a^{-n} = (a^{-1})^n = (a^n)^{-1},$$
$$a^0 = e.$$

These definitions are consistent since the powers of a commute and $(a^m * a^n)^{-1} = a^{-m} * a^{-n}$. It follows that a^n is defined for all $n \in \mathbb{Z}$ and the usual laws of indices hold, namely

$$a^m * a^n = a^{(m+n)} \text{ and } (a^m)^n = a^{mn}.$$

Theorem 3.8 Let A be the set of all powers of a, under an associative operation $*$, so that $A = \{a^n: n \in \mathbb{Z}\}$. Then $(A, *)$ is a group.

Proof The operation $*$ is associative and $(A, *)$ is closed since, if $x, y \in A$, $x = a^m$ and $y = a^n$ for some $m, n \in \mathbb{Z}$, and $x * y = a^m * a^n = a^{m+n}$ which is in A. $(A, *)$ contains the identity e since $e = a^0$. Also if $x \in A$ and $x = a^n$ then $x^{-1} = a^{-n}$ and this element is in A, so $(A, *)$ satisfies axiom G4. Therefore $(A, *)$ is a group and A is called the *cyclic* group generated by a and a is called a *generator* of A.

Definition The group A is cyclic if, for some $a \in A$, $A = \{a^n: n \in \mathbb{Z}\}$. Then A is generated by a and a is a generator of A.

The name, cyclic, arises from cyclic groups associated with rotational symmetries. In the next example, the generator of the symmetry group rotates the arms of the figure cyclically in a cyclic permutation of these arms.

EXAMPLE 1 *For each of the three shapes, shown in Fig. 3.6, prove that the symmetry group of the shape is a cyclic group.*

(i) We previously drew the table of the set S of symmetries of the letter Z in §3.6 and we repeat it here.

$(S, \circ)$	E	R
E	E	R
R	R	E

The elements of $(S, \circ)$ are E and R, with $\text{E} = \text{R}^2$, hence $(S, \circ)$ is a cyclic group, generated by R.

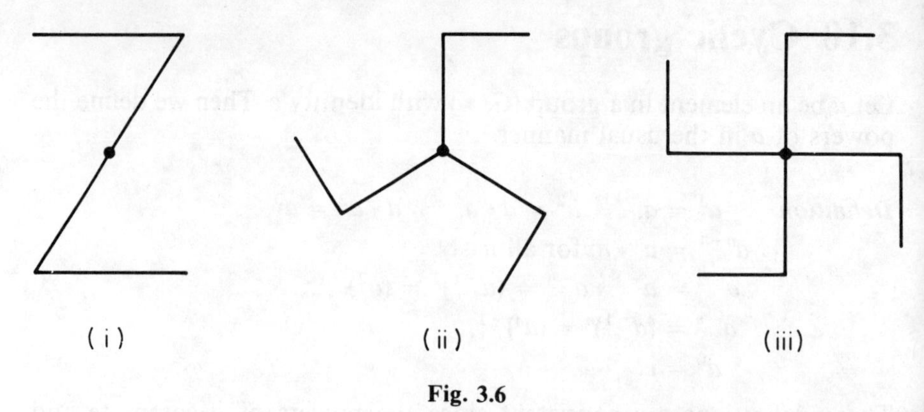

Fig. 3.6

(ii) We reproduce the table of $(S, \circ)$ from Example 2 of §3.7.

$(S,\circ)$	E	R	R^2
E	E	R	R^2
R	R	R^2	E
R^2	R^2	E	R

The elements of $(S, \circ)$ are R, R^2 and E which is equal to R^3, therefore $(S, \circ)$ is a cyclic group generated by R.
(iii) For the swastika, let R be the symmetry of rotation through $\pi/2$, so that R^2 is the symmetry of rotation through π and R^3 is the symmetry of rotation through $3\pi/2$. Then $R^4 = E$ and $S = \{E, R, R^2, R^3\}$, therefore $(S, \circ)$ is a cyclic group generated by R. The table for $(S, \circ)$ is

$(S, \circ)$	E	R	R^2	R^3
E	E	R	R^2	R^3
R	R	R^2	R^3	E
R^2	R^2	R^3	E	R
R^3	R^3	E	R	R^2

Finite and infinite cyclic groups

Let A be the cyclic group generated by the element a, so that

$$A = \{a^n : n \in \mathbb{Z}\}.$$

Then either

I any two different powers of a are different elements in A,

or II there are two different powers of a which are the same element in A.

We show that in the first case A is an infinite set and in the second case A is a finite set.

Case I Since all the powers of a are distinct, there must be an infinite number of distinct elements of A and therefore A is an infinite cyclic group. Consider the mapping

$$\mathrm{f}: A \to (\mathbb{Z}, +), \text{ given by } \mathrm{f}(a^r) = r.$$

Then, if r and s are distinct integers, $a^r \neq a^s$ and so f is one-one. Also $\mathrm{f}(a^r a^s) = \mathrm{f}(a^{r+s}) = r + s = \mathrm{f}(a^r) + \mathrm{f}(a^s)$ and therefore f is an isomorphiwm. Obviously, the same is true for any other infinite cyclic group and so all infinite cyclic groups are isomorphic. We use the notation C_∞ for the typical infinite cyclic group.

Definition The infinite cyclic group C_∞ is given by

$$C_\infty = \{x^n : n \in \mathbb{Z}\}, \text{ with } x^n = e \text{ only if } n = 0.$$

Then $A \cong C_\infty \cong (\mathbb{Z}, +)$.

Case II Not all the powers of a are distinct so that,

$$\text{for some } p \text{ and } q,\ p \neq q \text{ and } a^p = a^q.$$

Without loss of generality, we assume that $p < q$, then

$$a^{q-p} = a^q a^{-p} = a^p a^{-p} = e \text{ and } q - p > 0.$$

It follows that some positive power of a is equal to e and hence there must be a least positive power of a which is equal to e. Suppose that this power is a^n, that is,

$$a^n = e,\ n > 0 \text{ and, for } 0 < r < n,\ a^r \neq e.$$

Then, if $0 < p < q < n$, $0 < q - p < n$ and so $a^{q-p} \neq e$ and hence $a^q \neq a^p$, (for if $a^q = a^p$ then $a^{q-p} = a^q a^{-p} = a^p a^{-p} = e$). Therefore the set $\{a^r : 0 \leqslant r < n\}$ contains n distinct powers of a. Now consider any element x in A, so that $x = a^p$ for some p in $\mathbb{Z}$. If we divide p by n to give a quotient q and a remainder r, we find that

$$p = qn + r, \text{ with } 0 \leqslant r < n.$$

Then $$x = a^p = a^{qn+r} = a^{qn}a^r = (a^n)^q a^r = e^q a^r = ea^r = a^r.$$

This means that each element of A is one of the n elements a^r, with $0 \leqslant r < n$, and each of these n elements is in A. Therefore

$$A = \{a^r : 0 \leqslant r < n\} = \{e, a, a^2, \ldots a^{n-1}\}, \text{ with } a^n = e.$$

A contains n distinct elements and is the cyclic group of order n, generated by a. The number n is called the *period* of a.

Definition I $a^n = e$ and $a^r \neq e$ for $0 < r < n$, then a has period n.

Now consider the mapping

$$\mathrm{f}: A \to (\mathbb{Z}_n, +), \text{ given by } \mathrm{f}(a^r) = r,\ 0 \leqslant r < n.$$

Clearly f is one-one and

$$f(a^r a^s) = f(a^{r+s}) = f(a^t) = t$$
$$= r + s = f(a^r) + f(a^s), \text{ where } r + s = t \text{ in } \mathbb{Z}_n.$$

Therefore, f is an isomorphism. Obviously, the same is true for any other finite cyclic group, of order n, and so all finite cyclic groups of a given order are isomorphic. We use the notation C_n for the typical finite cyclic group of order n.

Definition The finite cyclic group C_n is given by

$$C_n = \{x^r : 0 \leqslant r < n, \; x^n = e\}.$$
$$= \{e, x, x^2, \ldots, x^{n-1}\}.$$

Then $A \cong C_n \cong (\mathbb{Z}_n, +)$.

EXAMPLE 2 *Prove that (i)* $(\mathbb{Z}_3, +) \cong C_3$; *(ii)* $(\mathbb{Z}, +) \cong C_\infty$.

i) Let f: $\mathbb{Z}_3 \to C_3$ be given by $f(0) = e$, $f(1) = x$, $f(2) = x^2$. The tables of $(\mathbb{Z}_3, +)$ and C_3 can be written

$(\mathbb{Z}_3, +)$	0	1	2
0	0	1	2
1	1	2	0
2	2	0	1

C_3	e	x	x^2
e	e	x	x^2
x	x	x^2	e
x^2	x^2	e	x

and this shows that f is an isomorphism.
(ii) Let f: $(\mathbb{Z}, +) \to C_\infty$ be given by $f(n) = x^n$ for all n in $\mathbb{Z}$. Then f is one-one and $f(m+n) = x^{m+n} = x^m x^n = f(m)f(n)$, so f is an isomorphism.

EXAMPLE 3 *Given a cyclic group A, generated by a, prove that* a^{-1} *is also a generator of A.*

We write the group multiplicatively. Let $a^{-1} = b$, and suppose that x is any element of A. Then, for some integer n, $x = a^n$, since A is generated by a. Let $m = -n$ so that m is an integer and

$$x = a^n = a^{-m} = (a^{-1})^m = b^m.$$

Hence every element in A is a power of b and so A is generated by b, that is, by a^{-1}.

Note that when A is a finite cyclic group, generated by a, with $a^n = e$, then $a^{-1} = a^n . a^{-1} = a^{n-1}$ and it is then possible to define $A = \{a^r : r \in \mathbb{N}\}$. This is not so in the case of an infinite cyclic group since, in case I, no positive power of the generator can equal the identity. When A is an infinite cyclic group, the set $\{a^r : r \in \mathbb{N}\}$ only contains half of the elements of A. (It is, in fact, called a semigroup.)

EXAMPLE 4 *Find the number of generators in the cyclic group:* (*i*) $C_\infty = \{x^n : n \in \mathbb{Z}\}$; (*ii*) $C_4 = \{e, x, x^2, x^3\}$; (*iii*) $C_5 = \{e, x, x^2, x^3, x^4\}$.

In each case, we know that x and x^{-1} are generators, from Example 3. (i) If x^n is a generator, then, for some m, $x = (x^n)^m = x^{mn}$. Since all powers of x are distinct, $mn = 1$ and m and n are integers. Hence either $m = 1 = n$ or $m = -1 = n$ and so x and x^{-1} are the only generators.
(ii) $(x^2)^2 = x^4 = e$ and so the cyclic group generated by x^2 is $\{e, x^2\}$. Therefore x^2 does not generate C_4 which has only two generators x and $x^3\,(=x^{-1})$.
(iii) $(x^2)^2 = x^4$, $(x^2)^3 = x^6 = x, (x^2)^4 = x^8 = x^3$, $(x^2)^5 = x^{10} = e$, so x^2 is a generator of C_5. Now $xx^4 = x^5 = e$ and $x^2x^3 = x^5 = e$ and so $x^4 = x^{-1}$ and $x^3 = (x^2)^{-1}$. Therefore, on using Example 3, C_5 has four generators, x, x^2, x^3 and x^4.

Notice that C_5 is generated by each of its non-identity elements. This is because 5 is a prime number and has no proper divisors. On the other hand, C_4 is not generated by x^2 because 2 is a proper divisor of 4 and so $(x^2)^2 = x^4 = e$. Therefore x^2 generates a group, $\{e, x^2\}$, which is inside C_4 and this group has order 2.

Cycles

A permutation of a set X containing n elements, which cyclically permutes m of the elements of X and which leaves the other $(n-m)$ elements unaltered, is called a *cycle* of length m. Thus, if $X = \{1, 2, 3, 4\}$ and $\mathrm{a} = \begin{pmatrix} 1 & 2 & 3 & 4 \\ 2 & 4 & 3 & 1 \end{pmatrix}$, then a is a cycle of length 3, since it cyclically permutes (1, 2, 4) and leaves 3 unaltered. In this case a useful shorthand notation for the cycle is to write

$$\mathrm{a} = (1\ \ 2\ \ 4).$$

Notation If the cycle c is a permutation of X which cyclically permutes $(x_1, x_2, \ldots, x_m)$ and leaves the other elements of X unaltered, then we write

$$\mathrm{c} = (x_1, x_2, \ldots x_m).$$

The above notation for c means that, for all elements x_i in X,

$$\begin{aligned} \mathrm{c}(x_i) &= x_{i+1}, \quad \text{for } 1 < i < m, \\ \mathrm{c}(x_m) &= x_1, \\ \mathrm{c}(x_i) &= x_i, \qquad \text{for } m < i. \end{aligned}$$

Clearly, the period of a cycle of length m is m and the cycle generates a cyclic group of order m.

Thus if $\mathrm{a} = \begin{pmatrix} 1 & 2 & 3 & 4 \\ 2 & 4 & 3 & 1 \end{pmatrix} = (1\ \ 2\ \ 4)$, then

$$\mathrm{a}(1) = 2, \quad \mathrm{a}(2) = 4, \quad \mathrm{a}(3) = 3, \quad \mathrm{a}(4) = 1,$$

and so
$$a^2(1) = a(2) = 4, \qquad a^2(4) = a(1) = 2,$$
$$a^2(2) = a(4) = 1, \qquad a^2(3) = a(3) = 3.$$

Hence $a^2 = \begin{pmatrix} 1 & 2 & 3 & 4 \\ 4 & 1 & 3 & 2 \end{pmatrix} = (1\ 4\ 2)$ and, of course, $a^3 = e$.

The cycles a and a^2 can be represented graphically as permutations of (1, 2, 4), which may be of help in understanding their action as mappings. This representation is shown in Fig. 3.7.

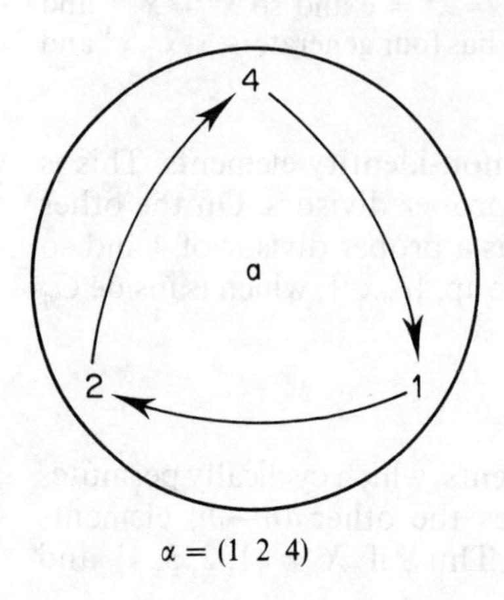

$\alpha = (1\ 2\ 4)$

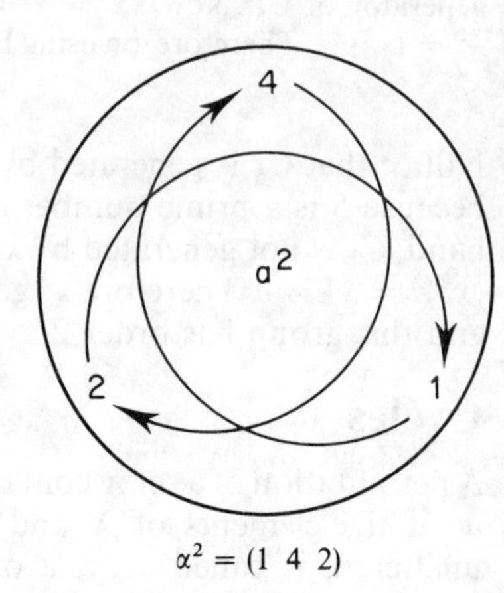

$\alpha^2 = (1\ 4\ 2)$

Fig. 3.7

EXAMPLE 5 *Using the cycle notation, prove that:*
(*i*) $(1\ \ 2)(1\ \ 3) = (1\ \ 3\ \ 2)$, (*ii*) $(1\ \ 2\ \ 3\ \ 4)(2\ \ 3\ \ 4\ \ 5) = (1\ \ 2\ \ 4\ \ 5\ \ 3)$.

(i) Let $a = (1\ \ 2)$ and $b = (1\ \ 3)$, then
$ab(1) = a(b(1)) = a(3) = 3,$
$ab(3) = a(b(3)) = a(1) = 2,$
$ab(2) = a(b(2)) = a(2) = 1.$
Therefore $ab = (1\ \ 3\ \ 2)$.
(ii) Let $c = (1\ \ 2\ \ 3\ \ 4)$ and $d = (2\ \ 3\ \ 4\ \ 5)$, then $cd(1) = c(1) = 2$, $cd(2) = c(3) = 4$, $cd(4) = c(5) = 5$, $cd(5) = c(2) = 3$, $cd(3) = c(4) = 1$, and so $cd = (1\ \ 2\ \ 4\ \ 5\ \ 3)$.

EXERCISE 3.10

1 The instruction 'cross the road' is denoted by 'C' and the instruction 'remain on the same side of the road' is denoted by 'R'. Show that, under the composition of instructions, the set S, given by $S = \{C, R\}$, is a group isomorphic to C_2.

2 The complex number ω is given by $\omega = \cos 2\pi/3 + i \sin 2\pi/3$, and $S = \{1, \omega, \omega^2\}$. Show that $\omega^3 = 1$. Prove that, under the multiplication of complex numbers, S is a group isomorphic to C_3. Give a geometrical interpretation of this group as a group of transformations of the plane.

3 The transformation T of the plane $O(x, y)$ is the translation through one unit, parallel to Ox and the transformation E leaves the plane unaltered.

Describe the transformations T^{-1} and T^2. Prove that the cyclic group G, generated by T, is isomorphic to C_∞.

4 The transformation R of the plane $O(x, y)$ is a rotation, in the positive sense, through an angle $2\pi/5$ about O. In the cyclic group $(S, \circ)$, where $S = \{R^n: n \in \mathbb{Z}\}$, state the inverses of the elements R and R^2 and prove that $(S, \circ) \cong (\mathbb{Z}_5, +)$.

5 Prove that $(\mathbb{R}, \times)$ is not a group.
Given a real number a, let $S = \{a^n: n \in \mathbb{Z}\}$. Prove that $(S, \times)$ is a cyclic group.
In each of the cases
(i) $a = 0$, (ii) $a = 1$, (iii) $a = -1$, (iv) $a(a^2 - 1) \neq 0$:
 (a) find the order of the group $(S, \times)$,
 (b) state whether the group is isomorphic to $(\mathbb{Z}, +)$ or to $(\mathbb{Z}_n, +)$ for some particular value of n,
 (c) list all possible generators of $(S, \times)$:

6 The group G contains an identity element e and one other element a. Prove that a generates G and that $G \cong C_2$.

7 The group $(G, *)$ contains three elements, consisting of the identity e and two other elements a and b.
Prove that (i) $a * b \neq a$, (ii) $a * b \neq b$, (iii) $a * a \neq a$, (iv) $a * a \neq e$.
Deduce that (v) $a^2 = b$, (vi) $a^{-1} = b$, (vii) $G \cong C_3$.

8 Prove that: (i) $(\mathbb{Z}_4, +) \cong C_4$.
 (ii) $(\{1, 2, 4\}$, product modulo 7$) \cong C_3$,
 (iii) $(\{1, 2, 3, 4\}$, product modulo 5$) \cong C_4$,

9 Let a and b be any two generators of a cyclic group G. Define $f: G \to G$, by $f(a^r) = b^r$ for all r in $\mathbb{Z}$. Prove that f is a group isomorphism. (Such an isomorphism of a group G on to G is called an automorphism of G.)

10 Using the notation for cycles, prove that:
 (i) $(1\ 2\ 3)(1\ 3\ 2) = e$, (ii) $(1\ 2\ 3)(1\ 2) = (1\ 3)$,
 (iii) $(1\ 2)(1\ 2\ 3) = (2\ 3)$, (iv) $(1\ 2)(1\ 3) = (1\ 3\ 2)$,
 (v) $(1\ 3)(1\ 2) = (1\ 2\ 3)$, (vi) $(1\ 2\ 3\ 4)^2 = (1\ 3)(2\ 4)$,
 (vii) $(1\ 2\ 3\ 4\ 5)^2 = (1\ 3\ 5\ 2\ 4)$,
 (viii) $(1\ 2\ 3\ 4\ 5)^3 = (1\ 4\ 2\ 5\ 3)$,
 (ix) $(1\ 2\ 3\ 4\ 5)^4 = (5\ 4\ 3\ 2\ 1)$.

11 A group G of order 6 has identity e, and the other five elements are a, b, a^2, ab, a^2b. (i) Given that $ab^2a = a^2$, prove that $b^2 = e$. (ii) Given in addition tjat $bab = a$, prove that $ab = ba$. (iii) Given the above conditions, explain why G is commutative. (iv) Deduce from (iii) whether or not G is cyclic. (*SMP*)

3.11 Subgroups

Not all groups are cyclic, as we see from the symmetry group G of the rectangle. We copy the table of G from §3.9, Example 4.

G	E	R	M	N
E	E	R	M	N
R	R	E	N	M
M	M	N	E	R
N	N	M	R	E

Since the square of every element is the identity E, the group is not cyclic. However, within this table, we can spot some smaller tables which are also the tables of groups. They are all cyclic groups of order 2, isomorphic to $(\mathbb{Z}_2, +)$, and we list them.

A	E	R
E	E	R
R	R	E

B	E	M
E	E	M
M	M	E

C	E	N
E	E	N
N	N	E

We therefore have three subsets *A*, *B* and *C* of *G*,each of which is a group under the same operation as *G*. Such subsets are called *subgroups* of *G*.

Definition A subgroup of a group *G* is a subset of *G* which is also a group under the same operation as *G*.

Notation $H \leqslant G$ means that *H* is a subgroup of *G*,
$H < G$ means that $H \leqslant G$ and $H \neq G$.

Special cases
For any group *G*, *G* and $\{e\}$ are both subgroups of *G*.
$\{e\}$ is called the trivial subgroup of *G* and is denoted by *E*.
Therefore, for any group *G*, $E \leqslant G$ and $G \leqslant G$.
The following are trivial groups, that is groups of order 1: (i) $(\{1\}, \times)$, (ii) $(\{0\}, +)$, (iii) $(\{0\}, \times)$, (iv) $(\{3\}, \times$ modulo 6). Therefore, these four groups are all isomorphic to *E*.
A subgroup of *G*, which is neither *E* nor *G* is called a *proper* subgroup of *G*.

Definition *H* is a proper subgroup of *G* if $E < H < G$.

In the group *G* of symmetries of the rectangle, discussed above, $E = \{E\}$, $A = \{E, R\}$, $B = \{E, M\}$, $C = \{E, N\}$, and *A*, *B* and *C* are all proper subgroups of *G*, each isomorphic to C_2.

EXAMPLE 1 *Let E be the set of all even integers and let O be the set of odd integers, so that* $\mathbb{Z} = E \cup O$. *Prove that* $(E, +)$ *is a subgroup of* $(\mathbb{Z}, +)$ *but that* $(O, +)$ *is not a subgroup of* $(\mathbb{Z}, +)$. *Prove also that* $(E, +)$ *is isomorphic to* $(\mathbb{Z}, +)$.

Let $(\mathbb{Z}, +) = G$, $(E, +) = H$, $(O, +) = K$. Then the neutral element of *G* is 0 which lies in *E*, but does not lie in *O*. Therefore *K* is not a group so it is not a subgroup of *G*.
For any *x* and *y* in *E*, $x = 2m$ and $y = 2n$ with *m* and *n* in $\mathbb{Z}$.

$$(x+y) = 2m + 2n = 2(m+n) \text{ and } -x = -2m = 2(-m),$$

so that both $(x+y)$ and $-x$ lie in *E*. Addition in *E* is associative since it is associative in $\mathbb{Z}$. Therefore H is a group and $H < G$. Now f is a one-one function of $\mathbb{Z}$ on to E under the mapping

$$\mathrm{f}: \mathbb{Z} \to E, \text{ given by } \mathrm{f}(n) = 2n.$$

Also $$\mathrm{f}(m+n) = 2(m+n) = 2m+2n = \mathrm{f}(m)+\mathrm{f}(n)$$

and, therefore, f is an isomorphism.

Note that the above example shows that a group can be isomorphic to one of its own proper subgroups.

EXAMPLE 2 *Prove that a finite group can not be isomorphic to a proper subgroup. Deduce that, if a group is isomorphic to a proper subgroup, then the group is infinite.*

In an isomorphism of two groups G and H, the isomorphic mapping is one-one and so $\mathrm{n}(H) = \mathrm{n}(G)$. If H is a proper subgroup of a finite group G, then $\mathrm{n}(H) < \mathrm{n}(G)$ and, therefore, H can not be isomorphic to G. Hence, if a group G is isomorphic to a proper subgroup H, then $\mathrm{n}(H) = \mathrm{n}(G) = \infty$.

EXAMPLE 3 *Find all the subgroups of C_{12}, where $C_{12} = G = \{x^r : 0 \leqslant r \leqslant 11, x^{12} = x^0 = e\}$.*

From Example 3 of §3.8, we know that x^{-1} $(= x^{11})$ generates G, as well as x. If, for some r and for some element y of G, $x = y^r$, then, for any s in $\mathbb{Z}$, $x^s = y^{rs}$ and so y is a generator of G.
Now $(x^5)^5 = x^{25} = x$, $(x^7)^7 = x^{49} = x$ and $(x^{11})^{11} = x^{121} = x$, so that x^5, x^7 and x^{11} each generate G.
In a similar way, it is seen that:

(i) x^2 and x^{10} each generate the subgroup A, given by $A = \{e, x^2, x^4, x^6, x^8, x^{10}\}$, which is isomorphic to C_6,
(ii) x^3 and x^9 each generate the subgroup B, given by $B = \{e, x^3, x^6, x^9\}$, which is isomorphic to C_4,
(iii) x^4 and x^8 each generate the subgroup C, given by $C = \{e, x^4, x^8\}$, which is isomorphic to C_3,
(iv) x^6 generates the subgroup D, given by $D = \{e, x^6\}$, which is isomorphic to C_2.

The subgroups of G are therefore **E, A, B, C, D and G.**

EXAMPLE 4 *The functions* e, f, g *and* h *all have the same domain, $R\backslash\{0\}$, and are given by:*

$$\mathrm{e}(x) = x,\ \mathrm{f}(x) = -x,\ \mathrm{g}(x) = 1/x,\ \mathrm{h}(x) = -1/x.$$

Let $S = \{\mathrm{e}, \mathrm{f}, \mathrm{g}, \mathrm{h}\}$ and let $G = (S, \circ)$. Prove that G is a group and list its proper subgroups.

We know that the binary operation of G is associative, since it is the composition of functions. We next calculate the table of G. Since e is the identity function, ek = k = ke for all k $\in S$. Each of f, g and h has a square equal to e since

$$f^2(x) = ff(x) = f(-x) = -(-x) = x = e(x),$$

$$g^2(x) = gg(x) = g(1/x) = \frac{1}{1/x} = x = e(x),$$

$$h^2(x) = hh(x) = h\left(\frac{-1}{x}\right) = \frac{-1}{-1/x} = x = e(x).$$

Also

$$fg(x) = f(1/x) = -1/x = h(x) = gf(x),$$
$$gh(x) = g(-1/x) = -x = f(x) = hg(x),$$
$$fh(x) = f(-1/x) = 1/x = g(x) = hf(x).$$

G	e	f	g	h
e	e	f	g	h
f	f	e	h	g
g	g	h	e	f
h	h	g	f	e

The above table for G can now be drawn. The operation is closed and associative, with identity e, and every element is self-inverse. Hence G is a group. Its proper subgroups are **{e, f}**, **{e, g}** and **{e, h}**.

EXAMPLE 5 *Draw the table of the group of all permutations of* {1, 2, 3} *under composition. List the proper subgroups of G.*

We list the permutations in G and give them names.

$$e = \begin{pmatrix} 1 & 2 & 3 \\ 1 & 2 & 3 \end{pmatrix} \quad a = \begin{pmatrix} 1 & 2 & 3 \\ 2 & 3 & 1 \end{pmatrix} \quad b = \begin{pmatrix} 1 & 2 & 3 \\ 3 & 1 & 2 \end{pmatrix}$$

$$p = \begin{pmatrix} 1 & 2 & 3 \\ 1 & 3 & 2 \end{pmatrix} \quad q = \begin{pmatrix} 1 & 2 & 3 \\ 3 & 2 & 1 \end{pmatrix} \quad r = \begin{pmatrix} 1 & 2 & 3 \\ 2 & 1 & 3 \end{pmatrix}$$

Since e is the identity mapping, ex = x = xe, for all x in S. Each of the permutations p, q and r interchanges two of the numbers so the square of each permutation will be the identity e, since two numbers are interchanged and then interchanged back again. Therefore e, p, q and r are all self inverses. Now consider the images of 1, 2 and 3 under a and b.

$$\begin{array}{lll} a: 1 \to 2, & a: 2 \to 3, \text{ so } a^2: & 1 \to 3, \quad \text{and hence} \quad a^2 = b. \\ 2 \to 3, & 3 \to 1, & 2 \to 1 \\ 3 \to 1, & 1 \to 2, & 3 \to 2 \end{array}$$

$$\begin{array}{lll} a: 1 \to 2, & b: 2 \to 1, \text{ so } ba: & 1 \to 1, \quad \text{and hence} \quad a^3 = ba = e = ab \\ 2 \to 3, & 3 \to 2, & 2 \to 2 \\ 3 \to 1, & 1 \to 3, & 3 \to 3 \end{array}$$

Therefore, a and b are inverses of each other and

$$b^2 = (a^2)^2 = a^4 = a^3a = ea = a.$$

We now draw a partial table for G, using the above results.

G	e	a	b	p	q	r
e	e	a	b	p	q	r
a	a	b	e			
b	b	e	a			
p	p			e		
q	q				e	
r	r					e

Consider some further images of 1, 2 and 3.

$$p: \begin{matrix} 1 \to 1, \\ 2 \to 3, \\ 3 \to 2, \end{matrix} \quad a: \begin{matrix} 1 \to 2, \\ 3 \to 1, \\ 2 \to 3, \end{matrix} \quad \text{so } ap: \begin{matrix} 1 \to 2, \\ 2 \to 1 \\ 3 \to 3 \end{matrix} \quad \text{and hence} \quad ap = r.$$

$$p: \begin{matrix} 1 \to 1, \\ 2 \to 3, \\ 3 \to 2, \end{matrix} \quad b: \begin{matrix} 1 \to 3, \\ 3 \to 2, \\ 2 \to 1, \end{matrix} \quad \text{so } bp: \begin{matrix} 1 \to 3, \\ 2 \to 2 \\ 3 \to 1 \end{matrix} \quad \text{and hence} \quad bp = q.$$

We can now use the above results to find all the remaining products in the table. There are many ways of working out these results, try and find some alternative ones for yourself.

$$aq = a(bp) = (ab)p = ep = p, \qquad ar = a(ap) = (aa)p = bp = q,$$
$$bq = b(bp) = (bb)p = ap = r, \qquad br = b(ap) = (ba)p = ep = p,$$
$$pq = (aq)q = a(qq) = ae = a, \qquad qr = (ar)r = a(rr) = ae = a,$$
$$rq = (bq)q = b(qq) = be = b, \qquad pr = (br)r = b(rr) = be = b.$$

In the next line we use associativity to rearrange the brackets in a product of four elements,

$$qp = e(qp) = (ba)(qp) = b(aq)p = b(pp) = be = b.$$

Then $rp = (bq)p = b(qp) = bb = a, \qquad pa = p(pq) = (pp)q = eq = q,$
$$pb = p(pr) = (pp)r = er = r, \qquad qa = q(qr) = (qq)r = er = r,$$
$$qb = q(qp) = (qq)p = ep = p, \qquad ra = (bq)a = b(qa) = br = p,$$
$$rb = r(rq) = (rr)q = eq = q.$$

We have now calculated all the products, which we display in the table for G.

G	e	a	b	p	q	r
e	e	a	b	p	q	r
a	a	b	e	r	p	q
b	b	e	a	q	r	p
p	p	q	r	e	a	b
q	q	r	p	b	e	a
r	r	p	q	a	b	e

We note that a and b have period 3 and that p, q and r have period 2. Any subgroup containing a will also contain a^2 and hence will contain e, a and b. If it also contains p, then it will contain q and r since ap = r and bp = q. Similarly, if it contains any of p, q or r, it will contain the other two and will consist of the whole group. Therefore there is just one proper subgroup containing a, namely $A = \{e, a, b\}$. The subgroup A is the cyclic subgroup generated by a (and by b) and it is isomorphic to C_3.

Any subgroup containing two of the three elements p, q, r, will contain their products a and b and will therefore be the whole group. Hence there are only three more proper subgroups, $P = \{e, p\}$, $Q = \{e, q\}$ and $R = \{e, r\}$. These are the cyclic subgroups generated by p, q and r respectively and are all isomorphic to C_2.

The proper subgroups of G are therefore ***A, P, Q* and *R***.

It should be noted that, since the group G in Example 5 consists of all the permutations of (1, 2, 3), it must be isomorphic to the symmetry group of an equilateral triangle. See Exercise 3.6, Question **3**. Any symmetry of the triangle will be a permutation of the vertices. The elements a and b in G correspond to rotations of the triangle through the angles $2\pi/3$ and $4\pi/3$, that is, to cyclic permutations of the vertices. The elements p, q and r in G correspond to the symmetries of reflection of the triangle in its three altitudes.

EXERCISE 3.11

1 Find the number of proper subgroups of the group:
(i) C_4, (ii) C_5, (iii) C_6,
(iv) C_{15}, (v) C_{16}, (vi) C_{28}.

2 Prove that:
(i) C_7 and C_{11} have no proper subgroups,
(ii) C_9 and C_{25} have one proper subgroup,
(iii) C_8, C_{10}, C_{14} and C_{21} have two proper subgroups,
(iv) C_{81} has three proper subgroups,
(v) C_{20}, and C_{18} have four proper subgroups.

3 Prove that $(\mathbb{Z}, +) < (\mathbb{Q}, +)$.

4 Prove that $(\mathbb{Q}^+, \times) < (\mathbb{R}^+, \times)$.

5 Refer to question **3** of Exercise 3.3, where $T = \{F, R, A, L\}$. Write down the table of T and prove that T is a group.
Prove that $T \cong C_4$. State the periods of the elements of T and show that T has just one proper subgroup.

6 The symmetry group of the rectangle is G, where $G = \{E, R, M, N\}$. State the periods of the elements of G and show that G has three proper subgroups. Prove that G is not isomorphic to the group T of question 5.

7 Given two subgroups S and T of a group G, with $S \cap T = R$, prove that R is a subgroup of G, and of S and of T.

8 Use as an example the symmetry group of the rectangle to show that the union of two subgroups of a group may not be a subgroup.

9 Let $G = (\mathbb{C}\backslash\{0\}, \times)$ and let A be the set of complex numbers of unit modulus. Prove that $(\{1, -1\}, \times) < (\{1, -1, i, -i\}, \times) < (A, \times) < G$.

10 The set S is the set of all polynomials in x, with integer coefficients and of degree less than or equal to 2. The set T is the subset of S of polynomials of degree less than or equal to 1. The set U is the subset of S of polynomials which have $(x-2)$ as a factor. The set V is the cyclic subgroup of S generated by $(x-2)$. Prove that:

(i) $(S, +)$ is a group, (ii) $(T, +) < (S, +)$,
(iii) $(U, +) < (S, +)$, (iv) $T \cap U < V$,
(v) $(T, +)$ is not cyclic.

11 Let G be the group of all non-singular 2×2 real matrices, under the operation of matrix product. Let A be the subset of G of those matrices with determinant equal to 1 or -1. Let K be the subset of G of those matrices with determinant equal to 1. Prove that $K < H < G$.

MISCELLANEOUS EXERCISE 3

1 Prove that the following are groups and that they are all isomorphic:
(i) $\{1, 3, 7, 9\}$ under multiplication modulo 10,
(ii) $\{1, 2, 3, 4\}$ under multiplication modulo 5,
(iii) $\{1, 4, 7, 13\}$ under multiplication modulo 15,
(iv) the symmetry group of a swastika.

2 The binary system $(S, *)$ is closed and associative. For any pair of elements a, b in S, there exist unique elements x, y in S, such that $a * x = b$ and $y * a = b$. Prove that $(S, *)$ is a group.

3 Prove that the group of all symmetries of an equilateral triangle is isomorphic to the group of all permutations of $\{1, 2, 3\}$.

4 Let $S = \{\mathrm{e}, \mathrm{a}, \mathrm{b}, \mathrm{p}, \mathrm{q}, \mathrm{r}\}$, where e, a, b, p, q and r are functions, with the same domain $\mathbb{R}\backslash\{0, 1\}$, given by:

$$\mathrm{e}(x) = x, \quad \mathrm{a}(x) = 1/(1-x), \quad \mathrm{b}(x) = (x-1)/x,$$
$$\mathrm{p}(x) = 1/x, \quad \mathrm{q}(x) = 1-x, \quad \mathrm{r}(x) = x/(x-1).$$

Draw the table of $(S, \circ)$ and prove that, if $G = (S, \circ)$, then:
(i) G is a group, (ii) G has one subgroup of order 3, (iii) G is isomorphic to the group of all permutations of $\{1, 2, 3\}$.

5 Let a and b be two different non-zero integers and let

$$S = \{xa + yb \colon x, y \in \mathbb{Z}\}.$$

Prove that $(S, +)$ is a group. Prove also that:
(i) if $a = 2$, $b = 3$, then $(S, +) = (\mathbb{Z}, +)$,
(ii) if $a = 4$, $b = 6$, then $(S, +)$ is the cyclic subgroup of $(\mathbb{Z}, +)$ generated by 2,
(iii) if $a = 18$, $b = 24$, then $(S, +)$ is the cyclic subgroup of $(\mathbb{Z}, +)$ generated by 6.

6 Prove that if $\mathrm{f}\colon G \to H$ is a group isomorphism then, for any element a in G, a and $\mathrm{f}(a)$ have the same period. (Consider separately the cases when the period of a is (i) finite, (ii) infinite).

7 Find the order of the multiplicative group of matrices, which is the cyclic group generated by:

(i) $\begin{pmatrix} 1 & 0 \\ 0 & -1 \end{pmatrix}$ (ii) $\begin{pmatrix} 0 & 1 \\ 1 & 0 \end{pmatrix}$ (iii) $\begin{pmatrix} 0 & 1 \\ -1 & 0 \end{pmatrix}$

(iv) $\begin{pmatrix} 1 & 1 \\ 0 & 1 \end{pmatrix}$ (v) $\begin{pmatrix} 1 & -1 \\ 0 & -1 \end{pmatrix}$ (vi) $\begin{pmatrix} 0 & -1 \\ 1 & -1 \end{pmatrix}$

(vii) $\begin{pmatrix} 1/\sqrt{2} & -1/\sqrt{2} \\ 1/\sqrt{2} & 1/\sqrt{2} \end{pmatrix}$ (viii) $\begin{pmatrix} i & 27 \\ 0 & -i \end{pmatrix}$

8 Prove that, if S is a non-empty subset of a group $(G, *)$ and if, for all x and y in S, $x * y^{-1}$ lies in S, then $(S, *) \leqslant (G, *)$.

9 Given a fixed element a in a multiplicative group G, define:
(i) $C(a) = \{x: x \in G \text{ and } xa = ax\}$,
(ii) $Z = \{z: z \in G \text{ and, for all } y \text{ in } G,\ yz = zy\}$.
Prove that $C(a) \leqslant Z \leqslant G$. Prove further that Z is the intersection of all subgroups $C(a)$, for all a in G.

10 Given that S is the set of elements $\{a, b, \ldots\}$ and $*$ is a binary operation, state in terms of $a, b, \ldots$ and $*$, four conditions which must be satisfied for $(S, *)$ to form a group. State also the condition under which the group is commutative. Given that $S = \{R, T, V, W\}$ where $R = \{p, q\}$, $T = \{p\}$, $V = \{q\}$, $W = \phi$, draw up tables for the binary operations of (a) union and (b) intersection, stating the identity element in each case. State, with reasons, (i) whether or not $(S, \cup)$ is a group, (ii) whether or not $(S, \cap)$ is a group. (*L*)

11 The operation $*$ is defined on the set of vectors in two dimensions by $\begin{pmatrix} a \\ b \end{pmatrix} * \begin{pmatrix} c \\ d \end{pmatrix} = \begin{pmatrix} ac \\ bc + d \end{pmatrix}$. Show that the operation $*$ is associative. (*JMB*)

12 For each of the relations $R_1, R_2, \ldots, R_5$ defined below, state whether it is (a) reflexive, (b) symmetric, (c) transitive. For each relation which you consider to be an equivalence relation, give the equivalence classes.
(i) For $p, q \in \mathbb{N}$, $pR_1q \Leftrightarrow |p - q|$ is odd.
(ii) For $p, q \in \mathbb{N}$, $pR_2q \Leftrightarrow p + q$ is even.
(iii) For $p, q \in \mathbb{N}$, $pR_3q \Leftrightarrow p, q$ have a common divisor greater than unity.
(iv) For $p, q \in \mathbb{Q}$, $pR_4q \Leftrightarrow p, q$ have the same denominator when in their lowest terms with positive denominator.
(v) For $p, q \in \{a, b, c, d, e, f\}$, $pR_5q \Leftrightarrow p * q = f$, where the operation $*$ is defined by the following table:

*	a	b	c	d	e	f
a	f	e	f	e	e	e
b	e	f	e	f	f	e
c	f	e	f	e	e	e
d	e	f	e	f	f	e
e	e	f	e	f	f	e
f	e	e	e	e	e	f

(*C*)

13 A binary operation (denoted by $*$) is defined on the set of real numbers by $a * b = ab/(a - b)$, $(a \neq b)$. Investigate whether or not this binary operation is associative. (*SMP*)

14 Given that the set $G = \{0, 1, 2\}$, show that G under the binary operation 'addition modulo 3' forms a group P, but that under the binary operation 'multiplication modulo 3' G does not form a group. Let H be the rotation group of an equilateral triangle. Is H isomorphic to P? Give reasons for your answer. Examine whether or not the set G forms a group under the binary operation $*$, where $A * B = |A^2 - B^2|$, (mod 3). (*L*)

15 Given that $z = r(\cos \pi/3 + \mathrm{i} \sin \pi/3)$, where $r > 0$, and that $S = \{1, z, z^2, z^3, z^4, z^5\}$, show that $r = 1$ is a necessary and sufficient condition for S to form a group under multiplication. (The associative property for the multiplication of complex numbers may be assumed.) If $r = 1$, list the proper subgroups of S. Show that this group is isomorphic to the group $G = \{1, 2, 3, 4, 5, 6\}$ of residue classes modulo 7 under multiplication. (*JMB*)

16 A group G has the binary operation $\circ$. The elements $a, b, p \in G$. State which of the following statements follow from the group axioms:
(i) $p \circ a = p \circ b \Rightarrow a = b$, (ii) $p \circ a = b \circ p \Rightarrow a = b$,
(iii) $a \circ p = b \circ p \Rightarrow a = b$. Give a detailed proof of one of these statements, stating at each step which of the group axioms you have used. A group S has exactly three distinct elements e, a, b, where e is the identity element. Explain why $a \circ b \neq a$ and $a \circ b \neq b$. What can you deduce about the elements a and b? Hence, or otherwise, prove that S is a cyclic group. Sketch a figure whose group of symmetries is isomorphic to S, and a second figure whose group of symmetries has a non-trivial subgroup isomorphic to S. (*SMP*)

17 Given the set S and the operation intersection ($\cap$), where $S = \{\phi, \{k\}, \{l\}, \{k, l\}\}$, find the solutions, if any, of each of the following equations, (i) $\{k, l\} \cap Q = \{l\}$, (ii) $\{l\} \cap Q = \{l\}$, (iii) $\{k\} \cap Q = \{l\}$, for the set Q, a subset of S. Given the set T and the operation $\otimes_6$, where $\otimes_6$ represents multiplication modulo 6, and $T = \{1, 2, 3, 4, 5\}$, find the values of r in T for which the equation $2 \otimes_6 q = r$ has two solutions for q in T. State the properties which any set A with operation $*$ must possess, sufficient to ensure that $(A, *)$ shall be a group. (*L*)

18 Write out combination tables for the given operation on each of the following sets, and prove that each forms a group of order 4: (i) $S = \{1, 3, 5, 7\}$, $a * b$ denotes the remainder when ab is divided by 8, (ii) $T = \{2, 4, 6, 8\}$, $a \circ b$ denotes the remainder when ab is divided by 10, (iii) $U = \{1, -1, j, -j\}$, $a \times b$ denotes the product ab, and $j \times j = -1$. For each group, state which pairs of elements (if any) are inverse to each other, and which elements (if any) are their own inverses. Two of the three groups are isomorphic to one another. State which, and give *two* different isomorphisms between the groups concerned. How many subgroups of order 2 can be found within (i) S, (ii) T, (iii) U? (*SMP*)

19 The set $\{1, 2, 3, 4, 5, 6\}$ of residue classes modulo 7 forms a group G under multiplication. State the order of each element of the group, and name the elements which are generators. List the proper subgroups of G. A group S, isomorphic to G, is generated by the permutation $p = \begin{pmatrix} a & b & c & d & e & f \\ b & c & d & e & f & a \end{pmatrix}$, under the operation of successive application. State another generator of S. (*JMB*)

20 In the set G of symmetry transformations of the square $ABCD$, let R denote an anticlockwise rotation of $\pi/2$ radians about the centre O of the square, let I denote the identity transformation, let L denote a reflection in the axis AC and let M denote a reflection in the axis BD. Form the combination table for the elements I, R^2, L, M. Deduce that these elements form a group of order 4. (Associativity may be assumed.) State another set of elements of G which forms a group of order 4. (*L*)

21 Show (i) that the set of symmetry transformations of the rhombus forms a group, (ii) the set of matrices $\{\mathbf{A}, \mathbf{B}, \mathbf{C}, \mathbf{D}\}$, where

$$\mathbf{A} = \begin{pmatrix} 1 & 0 \\ 0 & 1 \end{pmatrix}, \mathbf{B} = \begin{pmatrix} 0 & 1 \\ -1 & 0 \end{pmatrix}, \mathbf{C} = \begin{pmatrix} -1 & 0 \\ 0 & -1 \end{pmatrix}, \mathbf{D} = \begin{pmatrix} 0 & -1 \\ 1 & 0 \end{pmatrix},$$

forms a group under matrix multiplication. In each case state (a) the identity element, (b) the inverse of each of the other elements. State whether or not the groups are isomorphic and justify your answer. (Associativity may be assumed in each case.) *(L)*

22 Show that H, the set of all non-singular 2×2 real matrices, forms a group under matrix multiplication. (It may be assumed that matrix multiplication is associative and that $\det \mathbf{A} \times \det \mathbf{B} = \det \mathbf{AB}$.) Given that S is the set of all non-singular 2×2 matrices which commute with the matrix $\begin{pmatrix} 1 & 0 \\ 1 & 1 \end{pmatrix}$, show that each member of S must be of the form $\begin{pmatrix} a & 0 \\ b & a \end{pmatrix}$, where $a \neq 0$. Show that under the operation of matrix multiplication, S is a commutative group. *(L)*

23 T is the set of all 2×2 real matrices $\begin{pmatrix} a & b \\ c & d \end{pmatrix}$ with $ad - bc = 1$. Prove that T is a group under matrix multiplication. U is the set of 2×2 matrices: $U = \left\{ \begin{pmatrix} p & -q \\ q & p \end{pmatrix} : p, q \text{ real but not both zero} \right\}$. Show that U is also a group under matrix multiplication. Establish whether either T or U is abelian. Four more sets of matrices are defined as follows:

$$V = \left\{ \begin{pmatrix} p & 0 \\ 0 & p \end{pmatrix} : p \neq 0, p \text{ real} \right\}, \; W = \left\{ \begin{pmatrix} 0 & -q \\ q & 0 \end{pmatrix} : q \neq 0, q \text{ real} \right\},$$

$$X = \left\{ \begin{pmatrix} r & 0 \\ 0 & 1/r \end{pmatrix} : r \neq 0, r \text{ real} \right\}, \; Y = \left\{ \begin{pmatrix} p & -q \\ q & p \end{pmatrix} : p^2 + q^2 = 1, p, q \text{ real} \right\}.$$

For each of V, W, X and Y, under matrix multiplication, state whether (a) it is a group, (b) it is a subgroup of T, (c) it is a subgroup of U. *(SMP)*

24 Find 2×2 matrices $\mathbf{I}$, $\mathbf{J}$ independent of a and b such that the matrix $\mathbf{M} = \begin{pmatrix} a & b \\ -b & a \end{pmatrix}$ can be written in the form $a\mathbf{I} + b\mathbf{J}$ and show that $\mathbf{J}^2 = -\mathbf{I}$ (a and b are real numbers). Prove also that $\det \mathbf{M} = 0$ if and only if $\mathbf{M}$ is the zero matrix. Hence prove that the set S of all matrices of the form $\mathbf{M}$, excluding the zero matrix, forms a group under matrix multiplication. (The associativity of matrix multiplication may be assumed.) Deduce that if $\mathrm{i}^2 = -1$ and $\mathrm{f}\begin{pmatrix} a & b \\ -b & a \end{pmatrix} = a + \mathrm{i}b$, then f is an isomorphism of this set of matrices on to the set of non-zero complex numbers under multiplication. Show that the set of matrices of the form $\begin{pmatrix} \cos\theta & \sin\theta \\ -\sin\theta & \cos\theta \end{pmatrix}$ forms a subgroup of S. To what set of complex numbers is this subgroup isomorphic? Write down a cyclic subgroup of S of order 3. (*JMB*)

25 Let S be the set of permutations of the numbers 1, 2 and 3. The elements of S may be listed as follows:

$$p_1 = \begin{pmatrix} 1 & 2 & 3 \\ 1 & 2 & 3 \end{pmatrix} \quad p_2 = \begin{pmatrix} 1 & 2 & 3 \\ 3 & 1 & 2 \end{pmatrix} \quad p_3 = \begin{pmatrix} 1 & 2 & 3 \\ 2 & 3 & 1 \end{pmatrix} \quad p_4 = \begin{pmatrix} 1 & 2 & 3 \\ 2 & 1 & 3 \end{pmatrix}$$

$$p_5 = \begin{pmatrix} 1 & 2 & 3 \\ 3 & 2 & 1 \end{pmatrix} \quad p_6 = \begin{pmatrix} 1 & 2 & 3 \\ 1 & 3 & 2 \end{pmatrix},$$

and $p_1 * p_2$ is defined to mean 'p_2 followed by p_1'. *It is given that* the elements of S under $*$ form a group. State the order of each element of S. Construct the group table for $(S, *)$. Show that the group $(S, *)$ is not commutative. Write down a cyclic subgroup of $(S, *)$ of order three. Name a generator of this subgroup. List all other non-trivial subgroups of $(S, *)$. (*JMB*)

26 It is given that the set S contains six elements each of which is a function with domain $\{x: x \in \mathbb{R}, x \neq 0, 1\}$. Three of the elements of S are e, f, g, given by $\mathrm{e}(x) = x, \mathrm{f}(x) = x/(x-1), \mathrm{g}(x) = (x-1)/x$. It is further given that S forms a group under functional composition. Find the orders of the elements f and g. Determine the other three functions which are elements of S. List the elements of all non-trivial subgroups. (*JMB*)

27 The mappings $\mathrm{f}_1, \mathrm{f}_2, \mathrm{f}_3, \mathrm{f}_4$ of $D \to D$, under composition of mappings are $\mathrm{f}_1(x) = (x-1)/(x+1), \mathrm{f}_2(x) = \mathrm{f}_1(\mathrm{f}_1(x)), \mathrm{f}_3(x) = \mathrm{f}_1(\mathrm{f}_2(x)), \mathrm{f}_4(x) = \mathrm{f}_1(\mathrm{f}_3(x))$, where $D = \{x: x \in \mathbb{R},\ x \neq 0,\ x \neq 1,\ x \neq -1\}$. Determine $\mathrm{f}_2(x)$, $\mathrm{f}_3(x)$ and show that $\mathrm{f}_4(x) = \mathrm{I}(x)$, where I is the identity mapping. Prove that the four mappings $\mathrm{I}, \mathrm{f}_1, \mathrm{f}_2, \mathrm{f}_3$ form a group with composition as the group operation, and show that this group is isomorphic with the group consisting of $\{1, \mathrm{i}, -1, -\mathrm{i}\}$, with multiplication as the group operation. (*L*)

28 Show, from geometrical considerations, or otherwise, that if $\mathbf{A} = \begin{pmatrix} \cos\theta & -\sin\theta \\ \sin\theta & \cos\theta \end{pmatrix}$ then $\mathbf{A}^n = \begin{pmatrix} \cos n\theta & -\sin n\theta \\ \sin n\theta & \cos n\theta \end{pmatrix}$ where n is a positive integer. Show that if $\theta = \pi/2$ there are only 4 distinct matrices of the above kind whatever the value of n, and prove that they form a group under the operation of matrix multiplication. (Associativity of matrix multiplication may be assumed.) (*JMB*)

29 Consider the six matrices $\mathbf{A} = \begin{pmatrix} 1 & 0 \\ 0 & 1 \end{pmatrix}$, $\mathbf{B} = \begin{pmatrix} 0 & 1 \\ 1 & 0 \end{pmatrix}$, $\mathbf{C} = \begin{pmatrix} 0 & \omega^2 \\ \omega & 0 \end{pmatrix}$,

$\mathbf{D} = \begin{pmatrix} 0 & \omega \\ \omega^2 & 0 \end{pmatrix}$, $\mathbf{E} = \begin{pmatrix} \omega & 0 \\ 0 & \omega^2 \end{pmatrix}$, $\mathbf{F} = \begin{pmatrix} \omega^2 & 0 \\ 0 & \omega \end{pmatrix}$, where $\omega^3 = 1,\ \omega \neq 1$.

Complete the multiplication table given below, where the product is the ordinary matrix product with the factor taken from the left hand column preceding that taken from the top row. Hence show that the six matrices form a non-Abelian (non-commutative) group with respect to matrix multiplication. (Associativity of matrix multiplication may be assumed.) Find a subgroup of order 3 and all subgroups of order 2. Obtain a mapping of the group onto itself by the substitution $\omega \to \omega^2, \omega^2 \to \omega$ and show that this mapping is an isomorphism.

	A	B	C	D	E	F
A	A	B	C	D	E	F
B	B	A	E	F	C	D
C	C	.	.	.	D	B
D	D	.	.	.	B	C
E	E	D	B	C	F	A
F	F	C	D	B	A	E

(*JMB*)

30 The set of all 2×2 matrices is denoted by S. The relations $\rho_1, \rho_2, \rho_3, \rho_4$ are defined on S as follows: $\mathbf{A}\rho_1\mathbf{B} \Leftrightarrow \mathbf{A}^{\mathrm{T}} = \mathbf{B}$, $\mathbf{A}\rho_2\mathbf{B} \Leftrightarrow \mathbf{AB} = \mathbf{BA}$, $\mathbf{A}\rho_3\mathbf{B} \Leftrightarrow \mathbf{AB} = \mathbf{I}$, (where $\mathbf{I}$ is the 2×2 identity matrix), $\mathbf{A}\rho_4\mathbf{B} \Leftrightarrow$ non-singular 2×2 matrices $\mathbf{P}$ and $\mathbf{Q}$ can be found such that $\mathbf{B} = \mathbf{PAQ}$. Show, by counter-examples, that ρ_1, ρ_2 nd ρ_3 are not equivalence relations. Prove that ρ_4 is an equivalence relation and that all non-singular matrices in S belong to the same equivalence class. (*C*)

31 A group, with identity e, is defined in terms of generators a and b by $a^4 = b^2 = e$, $ba = a^3b$. Part of the group table is given below. Copy and complete the table.

	e	a	a^2	a^3	b	ab	a^2b	a^3b
e					b	ab	a^2b	a^3b
a					ab	a^2b	a^3b	b
a^2					a^2b	a^3b	b	ab
a^3					a^3b	b	ab	a^2b
b	b	a^3b	a^2b	ab				
ab	ab	b	a^3b	a^2b				
a^2b	a^2b	ab	b	a^3b				
a^3b	a^3b	a^2b	ab	b				

State the order of each element and the number of subgroups of order 2. Find (i) a subgroup which is isomorphic to the group $(\{1, 2, 3, 4\}, \times \pmod 5)$, (ii) a subgroup which is isomorphic to the group of symmetries of the rectangle. (*C*)

32 Let G be the set of all matrices of the form $\begin{pmatrix} 1 & x \\ 0 & 1 \end{pmatrix}$ with x a real number.

(i) Show that G does *not* form a group under matrix addition. (ii) Show that G does form a group under matrix multiplication. (You may assume associativity of matrix addition and multiplication.) (*JMB*)

33 A binary operation $*$ is defined on a set G so that

I $a * b$ is in G for all a, b in G,
II $a * (b * c) = (a * b) * c$ for all a, b, c in G,
III G contains an element e such that $a * e = a$ for all a in G,
IV given any a in G, there exists an element a^{-1} in G such that $a * a^{-1} = e$.

Justifying each step in your argument, prove that, for all a in G, (i) $(a^{-1} * a) * a^{-1} = a^{-1}$, (ii) $a^{-1} * a = (a^{-1} * a) * \{a^{-1} * (a^{-1})^{-1}\} = e$, (iii) $e * a = a$. (*JMB*)

4 Linear transformations

4.1 Definitions and examples

In this chapter, the algebra of square matrices is linked with certain geometrical transformations. The theory applies to $(n \times n)$ matrices and spaces of n dimensions but we shall restrict ourselves to matrices of order 2 and 3 and to spaces of dimension 2 and 3. A *linear transformation* of space can be defined as a transformation of coordinates in which the new coordinates are a (homogeneous) linear combination of the old coordinates. This means that the origin is always unchanged and that lines (or planes) through the origin are transformed into lines (or planes) through the origin, or possibly to the origin itself.

Definition A *linear transformation* of $\mathbb{R}^2$ is a mapping of points, given by a change in coordinates $(x, y) \mapsto (x', y')$, where $x' = ax + by$, $y' = cx + dy$, with $a, b, c, d \in \mathbb{R}$.

This definition can easily be extended to other dimensions: in a 1 dimensional space, $x \mapsto x'$, where $x' = ax$, $a \in \mathbb{R}$, in a 3 dimensional space, $(x, y, z,) \mapsto (x', y', z',)$, where

$$\left.\begin{aligned} x' &= ax + by + cz, \\ y' &= dx + ey + fz, \\ z' &= gx + hy + iz, \end{aligned}\right\} \quad a, b, \ldots, i \in \mathbb{R}.$$

The term, *linear transformation*, is also applied to vectors by transforming the coordinates of a vector in the above manner. This explains the reason for the word *linear* since a linear transformation is seen to preserve linearity, that is, the transformation of any linear combination of vectors is the same linear combination of their transforms. Thus, in a vector space (Chapter 14), if $\mathbf{u}$ and $\mathbf{v}$ are any vectors and a and b are any scalars, and T is a linear transformation (or mapping), $\mathrm{T}(a\mathbf{u} + b\mathbf{v}) = a\mathrm{T}(\mathbf{u}) + b\mathrm{T}(\mathbf{v})$.

Any transformation of coordinates can be regarded in two ways: either (i) as a transformation of position vectors or (ii) as a change of axes. Both of these ways can be useful in certain situations in the case of linear transformations.

EXAMPLE 1 *Describe geometrically the transformation* T *of* $\mathbb{R}^2$, *given by* $x' = \mathrm{T}(x) = 2x + y$, $y' = \mathrm{T}(y) = 3y$.

The effect of T on a position vector is given by

$$\begin{pmatrix} x' \\ y' \end{pmatrix} = \mathrm{T}\begin{pmatrix} x \\ y \end{pmatrix} = \begin{pmatrix} 2x+y \\ 3y \end{pmatrix}$$

Consider the images of the four corners of a unit square, Fig. 4.1(i):

$$\mathrm{T}\begin{pmatrix} 0 \\ 0 \end{pmatrix} = \begin{pmatrix} 0 \\ 0 \end{pmatrix}, \quad \mathrm{T}\begin{pmatrix} 1 \\ 0 \end{pmatrix} = \begin{pmatrix} 2 \\ 0 \end{pmatrix}, \quad \mathrm{T}\begin{pmatrix} 0 \\ 1 \end{pmatrix} = \begin{pmatrix} 1 \\ 3 \end{pmatrix}, \quad \mathrm{T}\begin{pmatrix} 1 \\ 1 \end{pmatrix} = \begin{pmatrix} 3 \\ 3 \end{pmatrix}.$$

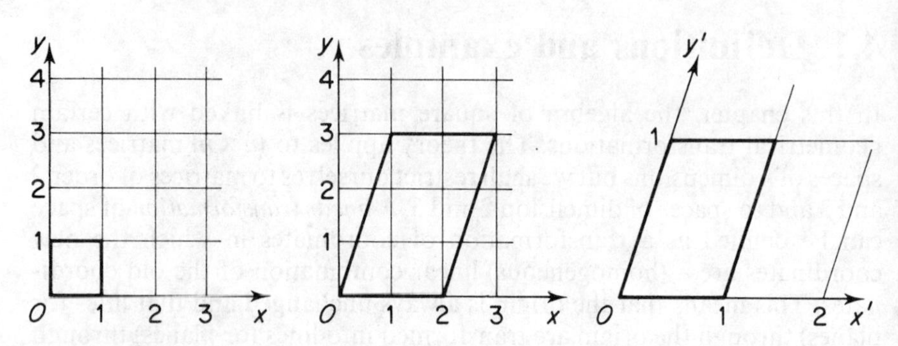

Fig. 4.1 (i) The unit square (ii) Transformed square (iii) Transformed grid

The transformed position of the unit square is shown in Fig. 4.1 (ii). It is also useful to see what happens to the system of grid lines which are parallel to the coordinate axes and unit distance apart. The line $y = n$ consists of points with coordinates (s, n) which are transformed to $(2s + n, 3n)$, that is to the line $y = 3n$. Similarly, the line $x = m$ becomes the set of points $\{(2m + t, 3t) : t \in \mathbb{R}\}$, that is the line $y = 3(x - 2m)$. These lines are shown in Fig. 4.1 (iii).

Matrix of a transformation

Let the linear transformation T of $\mathbb{R}^2$ be given by the equations

$$x' = ax + by,$$
$$y' = cx + dy,$$

Then this may be written in matrix form

$$\begin{pmatrix} x' \\ y' \end{pmatrix} = \begin{pmatrix} a & b \\ c & d \end{pmatrix}\begin{pmatrix} x \\ y \end{pmatrix} = \begin{pmatrix} ax+by \\ cx+dy \end{pmatrix}.$$

Let $\mathbf{M} = \begin{pmatrix} a & b \\ c & d \end{pmatrix}$ and let $\mathbf{r}$ and $\mathbf{r}'$ be the position vectors of the points $P\ (x, y)$ and $P'\ (x', y')$ respectively, where $\mathrm{T}(P) = P'$. Then the matrix form of the transformation equations is $\mathbf{r}' = \mathrm{T}(\mathbf{r}) = \mathbf{Mr}$. We say that $\mathbf{M}$ is the matrix corresponding to the linear transformation T, or that $\mathbf{M}$ is the matrix of T.

EXAMPLE 2 *Find the matrix of the linear transformation* T *of Example* 1.

The transformation T is given by $x' = \mathrm{T}(x) = 2x + y$, $y' = \mathrm{T}(y) = 3y$. In matrix form, these equations become $\begin{pmatrix} x' \\ y' \end{pmatrix} = \begin{pmatrix} 2 & 1 \\ 0 & 3 \end{pmatrix}\begin{pmatrix} x \\ y \end{pmatrix}$, so the matrix of T is $\begin{pmatrix} \mathbf{2} & \mathbf{1} \\ \mathbf{0} & \mathbf{3} \end{pmatrix}$.

Consider the effect of a general linear transformation T of $\mathbb{R}^2$ on the unit square with one vertex at the origin and with sides **i** and **j**. Let the matrix corresponding to T be **M**, given by

$$\mathbf{M} = \begin{pmatrix} a & b \\ c & d \end{pmatrix}, \text{ then } \mathrm{T}\begin{pmatrix} 1 \\ 0 \end{pmatrix} = \mathbf{M}\begin{pmatrix} 1 \\ 0 \end{pmatrix} = \begin{pmatrix} a & b \\ c & d \end{pmatrix}\begin{pmatrix} 1 \\ 0 \end{pmatrix} = \begin{pmatrix} a \\ c \end{pmatrix}$$

and $\mathrm{T}\begin{pmatrix} 0 \\ 1 \end{pmatrix} = \mathbf{M}\begin{pmatrix} 0 \\ 1 \end{pmatrix} = \begin{pmatrix} a & b \\ c & d \end{pmatrix}\begin{pmatrix} 0 \\ 1 \end{pmatrix} = \begin{pmatrix} b \\ d \end{pmatrix}$. Thus the columns of **M** are the images of the unit vectors **i** and **j**. Conversely, given any (2×2) matrix **M**, a transformation T of $\mathbb{R}^2$ can be defined by the equation $\mathrm{T}(\mathbf{r}) = \mathbf{Mr}$, for any position vector **r** in $\mathbb{R}^2$. If $\mathbf{M} = \begin{pmatrix} a & b \\ c & d \end{pmatrix}$ and $\mathbf{r} = \begin{pmatrix} x \\ y \end{pmatrix}$,

$\mathrm{T}(\mathbf{r}) = \begin{pmatrix} x' \\ y' \end{pmatrix} = \begin{pmatrix} a & b \\ c & d \end{pmatrix}\begin{pmatrix} x \\ y \end{pmatrix} = \begin{pmatrix} ax + by \\ cx + dy \end{pmatrix}$ so that T is, in fact, a linear transformation and has **M** as its matrix. This shows that there is a one-one correspondence between linear transformations of $\mathbb{R}^2$ and 2×2 matrices. The result generalises to higher dimensions, for example, there is a one-one correspondence between linear transformations of $\mathbb{R}^3$ and 3×3 matrices in which the matrix corresponding to a given transformation T has columns equal to the images under T of the unit Cartesian vectors **i**, **j** and **k**.

Images of the unit square

One tool which is useful in describing a linear transformation is to consider the image of the unit square, as considered in Example 1. Let the linear transformation T of $\mathbb{R}^2$ have matrix **M**, given by $\mathbf{M} = \begin{pmatrix} a & b \\ c & d \end{pmatrix}$.

Then the image of the point (x, y) is (x', y'), where

$$\begin{aligned} \begin{pmatrix} x' \\ y' \end{pmatrix} &= \mathbf{M}\begin{pmatrix} x \\ y \end{pmatrix} = \mathbf{M}\left(x\begin{pmatrix} 1 \\ 0 \end{pmatrix} + y\begin{pmatrix} 0 \\ 1 \end{pmatrix}\right), \\ &= x\mathbf{M}\begin{pmatrix} 1 \\ 0 \end{pmatrix} + y\mathbf{M}\begin{pmatrix} 0 \\ 1 \end{pmatrix} = x\begin{pmatrix} a \\ c \end{pmatrix} + y\begin{pmatrix} b \\ d \end{pmatrix}. \end{aligned}$$

The transformation is thus characterised by the images of the two Cartesian unit vectors, that is, the columns of **M**. The image of the origin is always the origin and the image of the point (1, 1) is the point with position vector $\mathbf{M}\begin{pmatrix}1\\1\end{pmatrix}$, that is, $\begin{pmatrix}a & b\\c & d\end{pmatrix}\begin{pmatrix}1\\1\end{pmatrix}=\begin{pmatrix}a+b\\c+d\end{pmatrix}$.

If **M** is non-singular the vectors formed by its columns are not parallel. In this case, the unit square is transformed into a parallelogram with vertices at (0, 0), (a, c), $(a+b, c+d)$, (b, d) and this is shown in Fig. 4.2.

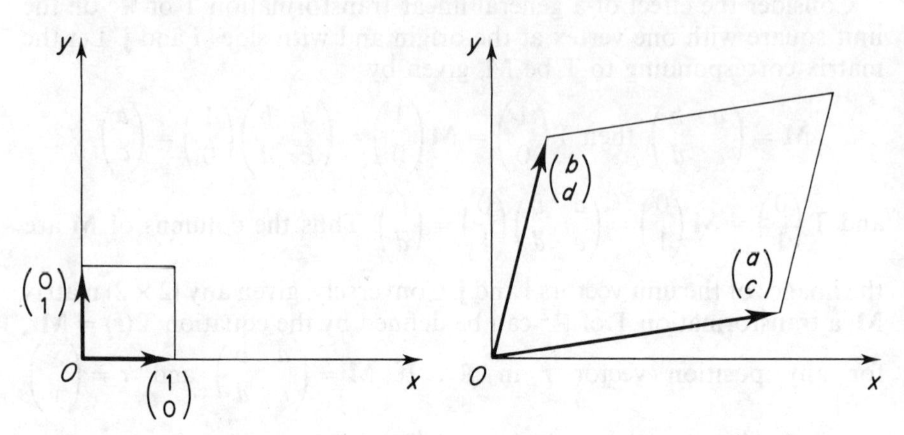

Fig. 4.2 The transformation of the unit square by the matrix $\begin{pmatrix}a & b\\c & d\end{pmatrix}$

If **M** is singular, but non-zero, then its column vectors are parallel and the image of every point (x, y) will lie on one line. The transformation squashes the plane on to a line, a space of dimension 1. If **M** is the zero matrix, then every point in the plane is mapped on to the origin and we call this a null transformation.

EXAMPLE 3 *Describe the transformation: (i) of* $\mathbb{R}^2$ *represented by the matrix* $\begin{pmatrix}3 & 0\\0 & 3\end{pmatrix}$, *(ii) of* $\mathbb{R}^3$ *represented by* $\begin{pmatrix}0 & 1 & 0\\1 & 0 & 0\\0 & 0 & 1\end{pmatrix}$.

(i) $\begin{pmatrix}1\\0\end{pmatrix}\mapsto\begin{pmatrix}3\\0\end{pmatrix}$ and $\begin{pmatrix}0\\1\end{pmatrix}\mapsto\begin{pmatrix}0\\3\end{pmatrix}$, so the unit square is enlarged to 3 times its size.

Every vector is similarly enlarged and the matrix $\begin{pmatrix}3 & 0\\0 & 3\end{pmatrix}$ represents an enlargement, scale factor 3, centred at the origin, see Fig. 4.3 (i).

(ii) $\begin{pmatrix}1\\0\\0\end{pmatrix}\mapsto\begin{pmatrix}0\\1\\0\end{pmatrix}$, $\begin{pmatrix}0\\1\\0\end{pmatrix}\mapsto\begin{pmatrix}1\\0\\0\end{pmatrix}$, $\begin{pmatrix}0\\0\\1\end{pmatrix}\mapsto\begin{pmatrix}0\\0\\1\end{pmatrix}$, so this transformation of the

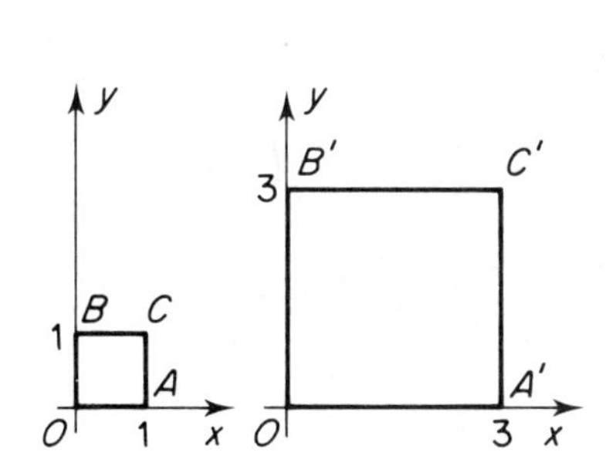

(i) Enlargement, scale factor 3,

centre (0, 0), $\mathbf{M} = \begin{pmatrix} 3 & 0 \\ 0 & 3 \end{pmatrix}$

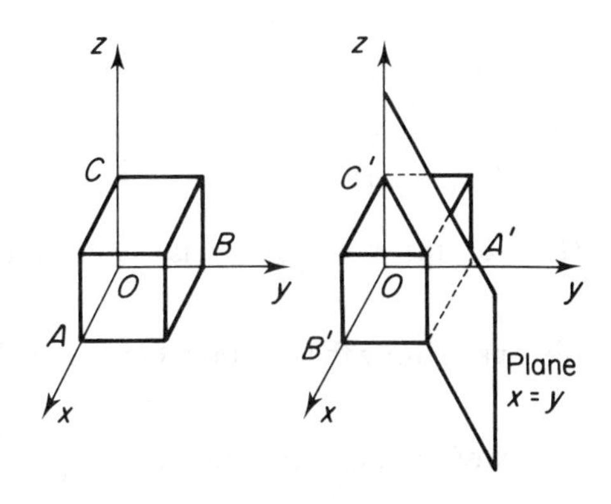

(ii) Reflection in the plane $x = y$,

matrix $\begin{pmatrix} 0 & 1 & 0 \\ 1 & 0 & 0 \\ 0 & 0 & 1 \end{pmatrix}$

Fig. 4.3

unit cube leaves the z-axis fixed pointwise but interchanges the x and y coordinates. This is a reflection in the plane $x = y$, shown in Fig. 4.3 (ii). The transformation can also be interpreted as that change of the coordinate system in which the Ox and Oy axes are interchanged, since $\begin{pmatrix} x \\ y \\ z \end{pmatrix} \mapsto \begin{pmatrix} y \\ x \\ z \end{pmatrix}$.

Since the columns of the matrix of the transformation are the images of the Cartesian unit vectors, the unit cube (square) may also be used to find the matrix of a given linear transformation.

EXAMPLE 4 *Find the matrix corresponding to the transformation of* $\mathbb{R}^2$:
(*i*) *the rotation of* $\pi/2$ *anticlockwise about* (0, 0),
(*ii*) *the one-way stretch, scale factor* $-1/2$, *in the direction* Ox.

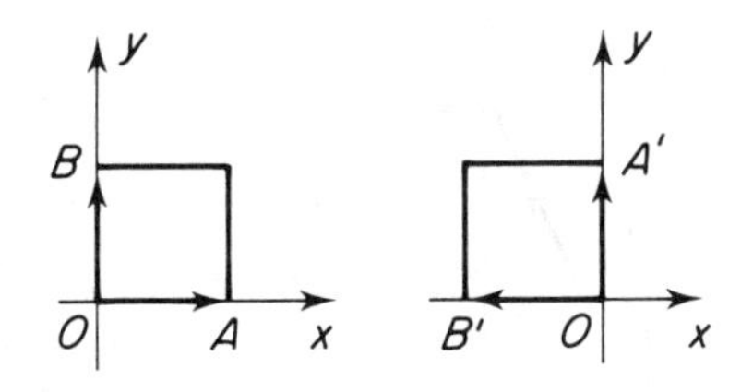

(i) Anticlockwise rotation
π about O,

$\begin{pmatrix} 1 \\ 0 \end{pmatrix} \mapsto \begin{pmatrix} 0 \\ 1 \end{pmatrix}, \begin{pmatrix} 0 \\ 1 \end{pmatrix} \mapsto \begin{pmatrix} -1 \\ 0 \end{pmatrix}$

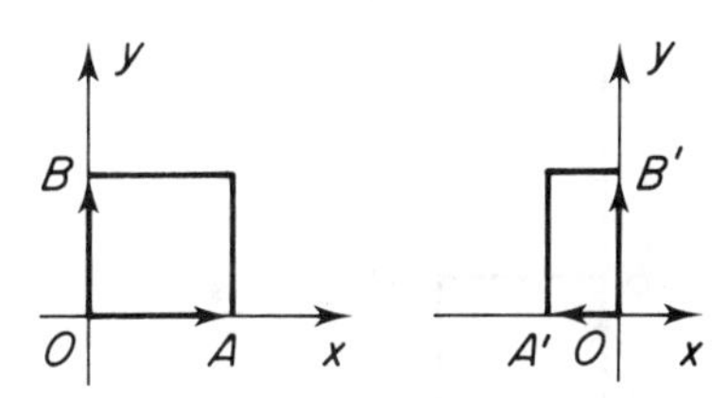

(ii) one way stretch,
scale factor $-1/2$, along Ox,

$\begin{pmatrix} 1 \\ 0 \end{pmatrix} \mapsto \begin{pmatrix} -1/2 \\ 0 \end{pmatrix}, \begin{pmatrix} 0 \\ 1 \end{pmatrix} \mapsto \begin{pmatrix} 0 \\ 1 \end{pmatrix}$

Fig. 4.4

(i) Let the vertices (1, 0) and (0, 1) of the unit square be A and B, respectively, as shown in Fig. 4.4 (i). The rotation maps these points to the points A' (0, 1) and B' (−1, 0), respectively. Therefore, the columns of the corresponding matrix are $\begin{pmatrix} 0 \\ 1 \end{pmatrix}$ and $\begin{pmatrix} -1 \\ 0 \end{pmatrix}$ and the matrix is $\begin{pmatrix} \mathbf{0} & \mathbf{-1} \\ \mathbf{1} & \mathbf{0} \end{pmatrix}$. We may check that the fourth vertex (1, 1) of the unit square is mapped to (−1, 1): $\begin{pmatrix} 0 & -1 \\ 1 & 0 \end{pmatrix}\begin{pmatrix} 1 \\ 1 \end{pmatrix} = \begin{pmatrix} -1 \\ 1 \end{pmatrix}$.

(ii) Again, using the diagram, Fig. 4.4 (ii), $\begin{pmatrix} 1 \\ 0 \end{pmatrix} \mapsto \begin{pmatrix} -1/2 \\ 0 \end{pmatrix}, \begin{pmatrix} 0 \\ 1 \end{pmatrix} \mapsto \begin{pmatrix} 0 \\ 1 \end{pmatrix}$, and the matrix of the transformation is $\begin{pmatrix} \mathbf{-1/2} & \mathbf{0} \\ \mathbf{0} & \mathbf{1} \end{pmatrix}$.

If the matrix of a linear transformation of $\mathbb{R}^2$ is singular and non-zero then all points are mapped on to a line, as shown in the next example.

EXAMPLE 5 *Describe the transformation given by the matrix* $\begin{pmatrix} 1 & 2 \\ 2 & 4 \end{pmatrix}$.

The images of the vectors $\begin{pmatrix} 1 \\ 0 \end{pmatrix}$ and $\begin{pmatrix} 0 \\ 1 \end{pmatrix}$ are $\begin{pmatrix} 1 \\ 2 \end{pmatrix}$ and $\begin{pmatrix} 2 \\ 4 \end{pmatrix}$, respectively. Every point (x, y) is mapped to $(x+2y,\ 2x+4y)$ on the line $y = 2x$, so that the transformation can be described as a *squash* on to $y = 2x$. This is shown in Fig. 4.5.

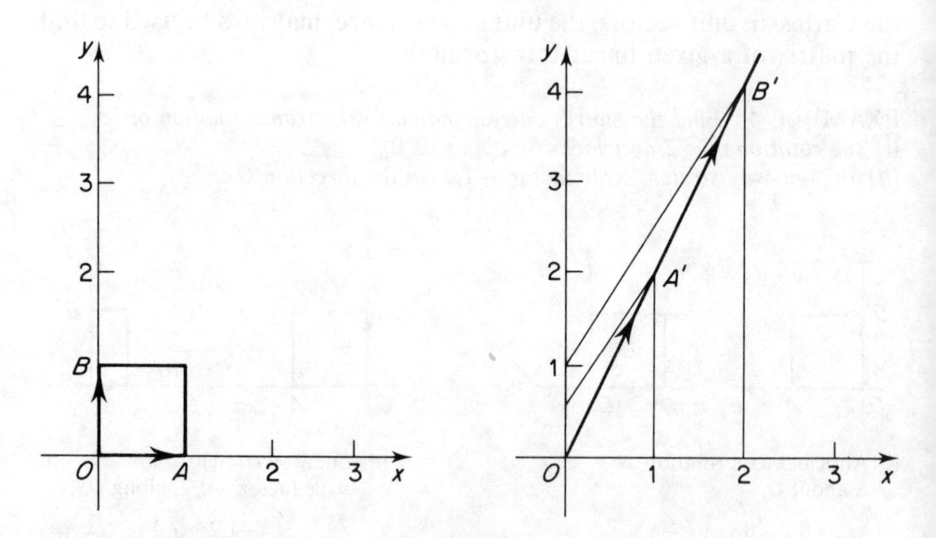

Fig. 4.5 The mapping given by the matrix $\begin{pmatrix} 1 & 2 \\ 2 & 4 \end{pmatrix}, \begin{pmatrix} 1 \\ 0 \end{pmatrix} \mapsto \begin{pmatrix} 1 \\ 2 \end{pmatrix}, \begin{pmatrix} 0 \\ 1 \end{pmatrix} \mapsto \begin{pmatrix} 2 \\ 4 \end{pmatrix}$

Standard transformations

We now list some standard transformations of the plane, together with their corresponding matrices. We shall show in §4.2 how other transformations may be constructed by the composition of these standard transformations.

Zero $\begin{pmatrix} 0 & 0 \\ 0 & 0 \end{pmatrix}$ represents the zero (or null) transformation in which the whole plane is squashed into the origin.

Projections $\begin{pmatrix} 1 & 0 \\ 0 & 0 \end{pmatrix}$ and $\begin{pmatrix} 0 & 0 \\ 0 & 1 \end{pmatrix}$ represent the projections on to the x-axis and y-axis respectively. These transformations squash the whole plane on to a line.

Identity $\begin{pmatrix} 1 & 0 \\ 0 & 1 \end{pmatrix}$ represents the identity transformation under which every point (or vector) remains fixed, Fig. 4.6 (i).

Enlargements $\begin{pmatrix} c & 0 \\ 0 & c \end{pmatrix}$, $c \in \mathbb{R}\backslash\{0, 1\}$, represents an enlargement, centre the origin, with scale factor c. If $c > 1$, this is a true enlargement, if $0 < c < 1$ it is a shrinking, if $c < 0$ the image is on the opposite side of the origin. If $c = -1$, the transformation is a reflection in the origin which is the same as a rotation of the plane about the origin through π. The case $c = -2$ is shown in Fig. 4.6 (ii).

One way and two way stretches $\begin{pmatrix} c & 0 \\ 0 & d \end{pmatrix}$, $c, d \in \mathbb{R}\backslash\{0\}$, $c \neq d$, represents a stretch. If $d = 1$, it is a one way stretch, with scale factor c, in the Ox direction, keeping the y-axis fixed. If $c = 1$, it is a one way stretch, with scale factor d, in the Oy direction, keeping the x-axis fixed. If neither c nor d is 1, it is a two way stretch with scale factor c in the x-direction and scale factor d in the y-direction. The two way stretch, given by the matrix $\begin{pmatrix} -1 & 0 \\ 0 & 3 \end{pmatrix}$ is shown in Fig. 4.6 (iii).

Shears The matrices $\begin{pmatrix} 1 & c \\ 0 & 1 \end{pmatrix}$ and $\begin{pmatrix} 1 & 0 \\ d & 1 \end{pmatrix}$, $c, d \in \mathbb{R}\backslash\{0\}$, represent shears which keep fixed the x-axis and the y-axis, respectively. All other points in the plane are moved parallel to the fixed axis through a distance c (or d) times its distance from the fixed axis. This is best understood by referring to Fig. 4.6 (iv) and (v) which show the shears given by $c = 2$ and $d = -1$.

Reflections The matrix $\begin{pmatrix} \cos 2\alpha & \sin 2\alpha \\ \sin 2\alpha & -\cos 2\alpha \end{pmatrix}$ represents a reflection in the

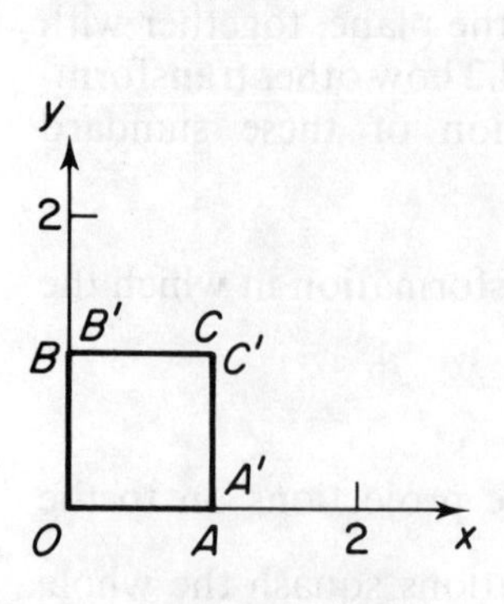

(i) Identity, $\mathbf{M} = \begin{pmatrix} 1 & 0 \\ 0 & 1 \end{pmatrix}$

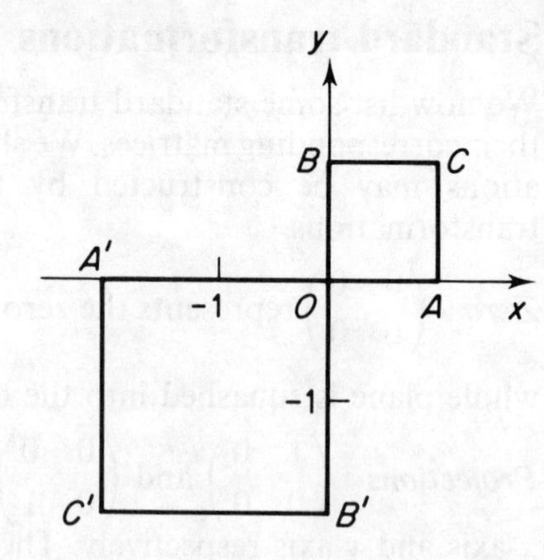

(ii) Enlargement, $\mathbf{M} = \begin{pmatrix} -2 & 0 \\ 0 & -2 \end{pmatrix}$

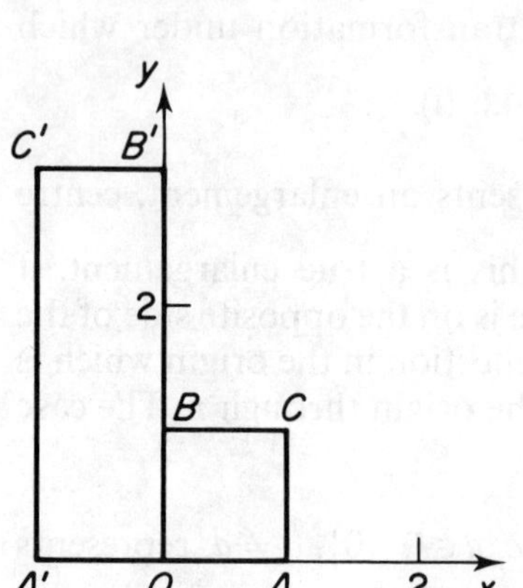

(iii) 2 way stretch, $\mathbf{M} = \begin{pmatrix} -1 & 0 \\ 0 & 3 \end{pmatrix}$

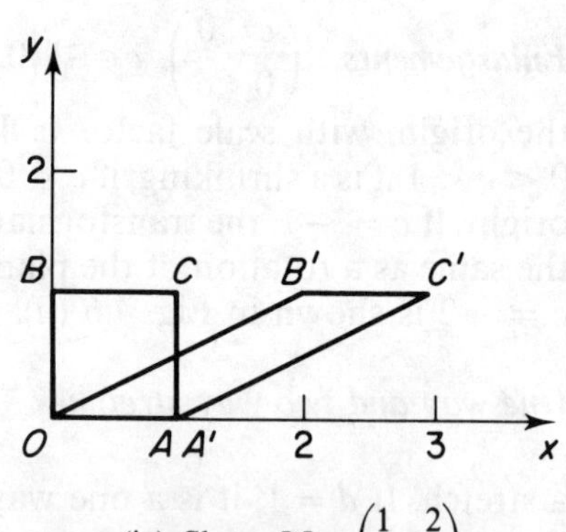

(iv) Shear, $\mathbf{M} = \begin{pmatrix} 1 & 2 \\ 0 & 1 \end{pmatrix}$

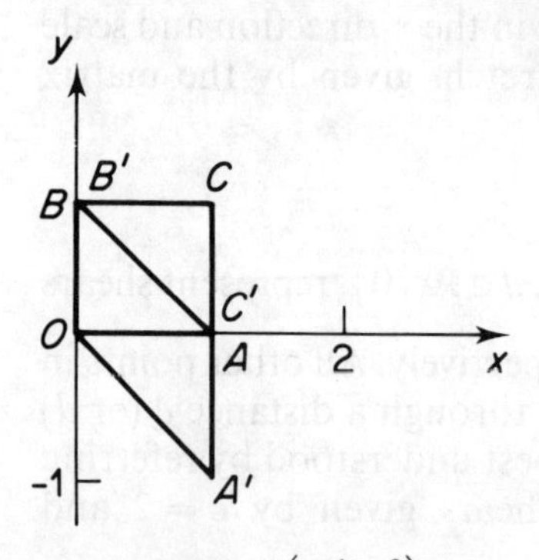

(v) Shear, $\mathbf{M} = \begin{pmatrix} 1 & 0 \\ -1 & 1 \end{pmatrix}$

Fig. 4.6

line $y = x \tan\alpha$, $-\pi/2 < \alpha < \pi/2$. This is the most general reflection since, if a reflection is a linear transformation, the mirror line must pass through the origin.

Proof The situation for $0 < \alpha < \pi/4$ is shown in Fig. 4.7 (i). The image of A $(1, 0)$ is the point A' $(\cos 2\alpha, \sin 2\alpha)$ and $|OA'| = 1 = |OA|$. Since $1 + \cos 2\alpha = 2\cos^2\alpha$ and $\sin 2\alpha = 2\cos\alpha\sin\alpha$, the midpoint of AA' is the point $(\cos^2\alpha, \cos\alpha\sin\alpha)$ which lies on the line $y = x\tan\alpha$. Hence A' is the reflection of A in this line. Similarly, the image of B $(0, 1)$ is B' $(\sin 2\alpha, -\cos 2\alpha)$, $|OB'| = 1 = |OB|$ and the midpoint of BB' is $(\sin\alpha\cos\alpha, \sin^2\alpha)$ which also lies on the line $y = x\tan\alpha$ and so B' is the reflection of B in the same line.

We list some particular reflections, as a table:

α	matrix	mirror line	α	matrix	mirror line
0	$\begin{pmatrix} 1 & 0 \\ 0 & -1 \end{pmatrix}$	axis Ox	$\pi/2$	$\begin{pmatrix} -1 & 0 \\ 0 & 1 \end{pmatrix}$	axis Oy
$\pi/4$	$\begin{pmatrix} 0 & 1 \\ 1 & 0 \end{pmatrix}$	line $y = x$	$-\pi/4$	$\begin{pmatrix} 0 & -1 \\ -1 & 0 \end{pmatrix}$	line $y = -x$

Rotations The matrix $\begin{pmatrix} \cos\alpha & -\sin\alpha \\ \sin\alpha & \cos\alpha \end{pmatrix}$ represents a rotation about the origin through an angle α, measured positively in the anticlockwise sense, the convention used in our early definitions of circular functions. Again,

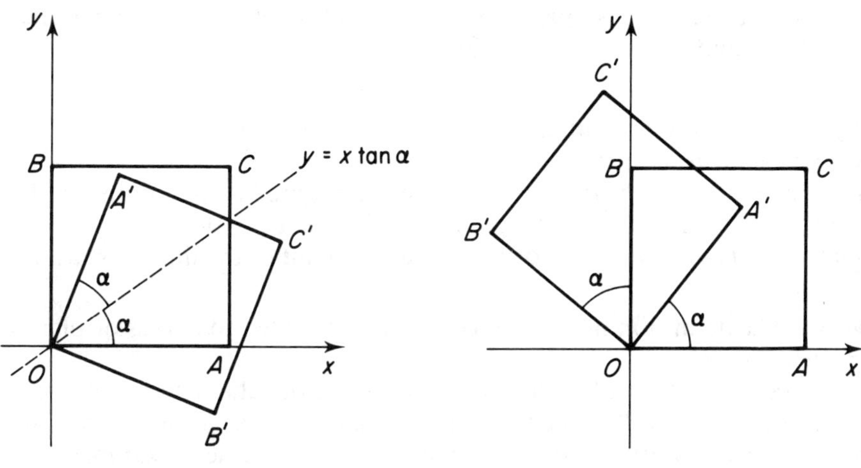

(i) Reflection in $y = x\tan\alpha$, $\mathbf{M} = \begin{pmatrix} \cos 2\alpha & \sin 2\alpha \\ \sin 2\alpha & -\cos 2\alpha \end{pmatrix}$

(ii) Rotation about O, angle α, $\mathbf{M} = \begin{pmatrix} \cos\alpha & -\sin\alpha \\ \sin\alpha & \cos\alpha \end{pmatrix}$

Fig. 4.7

the rotation must be about the origin for a linear transformation since the origin remains fixed.

Proof The situation for an acute angle α is shown in Fig. 4.7 (ii). The images A' and B' of the points $A(1, 0)$ and $B(0, 1)$ have coordinates $(\cos\alpha, \sin\alpha)$ and $(-\sin\alpha, \cos\alpha)$ respectively which shows that the unit square is rotated about O through α. As for reflections, we list some particular rotations:

matrix	α = angle of rotation	matrix	α = angle of rotation
$\begin{pmatrix} 0 & -1 \\ 1 & 0 \end{pmatrix}$	$\pi/2$ (a right-angle)	$\begin{pmatrix} 0 & 1 \\ -1 & 0 \end{pmatrix}$	$-\pi/2$ (or $+3\pi/2$)
$\begin{pmatrix} \frac{1}{\sqrt{2}} & \frac{-1}{\sqrt{2}} \\ \frac{1}{\sqrt{2}} & \frac{1}{\sqrt{2}} \end{pmatrix}$	$\pi/4$	$\begin{pmatrix} -1 & 0 \\ 0 & -1 \end{pmatrix}$	π (or an enlargement scale factor -1)

As with other more complicated transformations, the rotations can be expressed in terms of the composition of the earlier standard transformations. However, the rotation is given here as a standard transformation because it is an important basic isometry.

Images and invariants

A linear transformation T of $\mathbb{R}^n$ is a function with domain $\mathbb{R}^n$. It is often important to find the image under T of a point, or a line, or a curve, or, in the case of three dimensions, of a plane. A set of points may, as a set, be left fixed by T and the set is then said to be *invariant*.

Definition Let T be a linear transformation of some space.
A point is *invariant* if it is left fixed by T.
A line is *invariant* if every point of the line is mapped on to a point on the line (possibly the same point, possibly a different point).
A line is *pointwise invariant*, or *fixed*, if every point of the line is invariant.

Note The definition of invariance is easily extended to curves or planes or surfaces.

The consideration of invariants is useful in the characterisation of a transformation as will be seen in §§4.3 and 4.5. Some techniques in dealing with linear transformations are demonstrated in the next examples.

EXAMPLE 6 *The linear transformation* T *of the plane is the reflection in the line* $y = x$. *Prove that the line* $y = x$ *is the only set of points which is pointwise invariant and that the other invariant lines are those perpendicular to* $y = x$.

The matrix of T is $\begin{pmatrix} 0 & 1 \\ 1 & 0 \end{pmatrix}$ and T maps the point (s, t) on to (t, s).

The point (s, t) is invariant if and only if $s = t$, so the only invariant set of points is the line $y = x$. The line given parametrically by $x = a + su$, $y = b + sv$, is invariant if and only if, for some t, $a + tu = b + sv$ and $b + tv = a + su$. By adding these two equations, it is seen that this is so only if $(t - s)(u + v) = 0$ and, since $t = s$ gives pointwise invariance, the line is invariant only if $u = -v$, that is if the line has slope -1 and is perpendicular to $y = x$. Clearly every line with slope -1 is reflected into itself.

EXAMPLE 7 *Find the images of the points P* (3, 4) *and Q* (−2, 1) *under the given transformation:* (i) *a rotation* R *of* 30° *about O,* (ii) *the shear* S *with the matrix* $\begin{pmatrix} 0 & -2 \\ 1/2 & 2 \end{pmatrix}$.

(i) Using our standard form, the matrix of the rotation of 30° is given by

$\mathbf{R} = \begin{pmatrix} \sqrt{3}/2 & -1/2 \\ 1/2 & \sqrt{3}/2 \end{pmatrix}$, since $\cos 30° = \sqrt{3}/2$ and $\sin 30° = 1/2$.

The position vectors of the images P' and Q' of P and Q are found by premultiplying the position vectors of P and Q by $\mathbf{R}$.

$$\overrightarrow{OP'} = \mathbf{R}(\overrightarrow{OP}) = \begin{pmatrix} \sqrt{3}/2 & -1/2 \\ 1/2 & \sqrt{3}/2 \end{pmatrix}\begin{pmatrix} 3 \\ 4 \end{pmatrix} = \begin{pmatrix} (3\sqrt{3}-4)/2 \\ (3+4\sqrt{3})/2 \end{pmatrix}.$$

$$\overrightarrow{OQ'} = \mathbf{R}(\overrightarrow{OQ}) = \begin{pmatrix} \sqrt{3}/2 & -1/2 \\ 1/2 & \sqrt{3}/2 \end{pmatrix}\begin{pmatrix} -2 \\ 1 \end{pmatrix} = \begin{pmatrix} -\sqrt{3}-1/2 \\ -1+\sqrt{3}/2 \end{pmatrix}.$$

Thus the coordinates of P' and Q' are respectively $\boldsymbol{\left(\frac{3\sqrt{3}-4}{2}, \frac{3+4\sqrt{3}}{2}\right)}$ and $\boldsymbol{(-\sqrt{3}-1/2, -1+\sqrt{3}/2)}$.

(ii) Using the same technique, with the given matrix,

$$\overrightarrow{OP'} = \mathrm{S}(\overrightarrow{OP}) = \begin{pmatrix} 0 & -2 \\ 1/2 & 2 \end{pmatrix}\begin{pmatrix} 3 \\ 4 \end{pmatrix} = \begin{pmatrix} -8 \\ 19/2 \end{pmatrix}.$$

$$\overrightarrow{OQ'} = \mathrm{S}(\overrightarrow{OQ}) = \begin{pmatrix} 0 & -2 \\ 1/2 & 2 \end{pmatrix}\begin{pmatrix} -2 \\ 1 \end{pmatrix} = \begin{pmatrix} -2 \\ 1 \end{pmatrix}.$$

Thus P' is the point **(−8, 19/2)** and Q is **fixed** under this shear.
Note that this means that Q is on the invariant line, which is therefore given by $2y + x = 0$.

EXAMPLE 8 *Find the image of the line* $y = 3x$ *under the transformation represented by the matrix* $\begin{pmatrix} 1 & 2 \\ 0 & 1 \end{pmatrix}$. *Find the set of points which remain the same distance from the origin under this transformation.*

The first step is to represent the position vector of a general point on the line $y = 3x$ and this can be done in terms of a scalar parameter t. If the x-coordinate of a point P on the line is t then the y-coordinate will be $3t$. Therefore

$\overrightarrow{OP} = \begin{pmatrix} t \\ 3t \end{pmatrix}$ and the position vector of the image of P is $\begin{pmatrix} 1 & 2 \\ 0 & 1 \end{pmatrix}\begin{pmatrix} t \\ 3t \end{pmatrix} = \begin{pmatrix} 7t \\ 3t \end{pmatrix}$.
The image of P is the point $(7t,\ 3t)$ and, since t can take all real values, the image of the line $y = 3x$ is the line $\mathbf{7y = 3x}$. The distance d of a general point $P(x, y)$ from the origin is given by $d = \sqrt{(x^2+y^2)}$. P is transformed to the point $P'(x+2y, y)$ whose distance from the origin is $\sqrt{\{(x+2y)^2+y^2\}}$. This must equal d if the distance is invariant and so:

The distance from O is invariant $\Leftrightarrow \sqrt{(x^2+y^2)} = \sqrt{\{(x+2y)^2+y^2\}}$

$$\Leftrightarrow x^2+y^2 = x^2+4xy+5y^2$$
$$\Leftrightarrow 0 = 4xy+4y^2$$
$$\Leftrightarrow y(x+y) = 0.$$

Thus the required locus is the line pair $\{(x, y): y = 0 \text{ or } y+x = 0\}$, that is, the set $\{(\mathbf{x}, \mathbf{0}): \mathbf{x} \in \mathbf{R}\} \cup \{(\mathbf{x}, -\mathbf{x}): \mathbf{x} \in \mathbf{R}\}$.

EXAMPLE 9 *The curve C has Cartesian equation* $x^2+y^2-2xy-2x-2y = 0$. *Let* $\mathbf{R}$ *be the* 2×2 *matrix which represents the rotation of the coordinate axes through* $\pi/4$ *from* $O(x, y)$ *to* $O(X, Y)$, *such that* $\mathbf{R}\begin{pmatrix} x \\ y \end{pmatrix} = \begin{pmatrix} X \\ Y \end{pmatrix}$. *Determine the equation of C in terms of X and Y. Sketch the two sets of axes and the curve C.*

Using the standard form for a rotation matrix, $\mathbf{R} = \begin{pmatrix} \frac{1}{\sqrt{2}} & \frac{-1}{\sqrt{2}} \\ \frac{1}{\sqrt{2}} & \frac{1}{\sqrt{2}} \end{pmatrix}$ so

$$\begin{pmatrix} X \\ Y \end{pmatrix} = \mathbf{R}\begin{pmatrix} x \\ y \end{pmatrix} = \begin{pmatrix} \frac{1}{\sqrt{2}} & \frac{-1}{\sqrt{2}} \\ \frac{1}{\sqrt{2}} & \frac{1}{\sqrt{2}} \end{pmatrix}\begin{pmatrix} x \\ y \end{pmatrix} = \begin{pmatrix} \frac{x-y}{\sqrt{2}} \\ \frac{x+y}{\sqrt{2}} \end{pmatrix}.$$

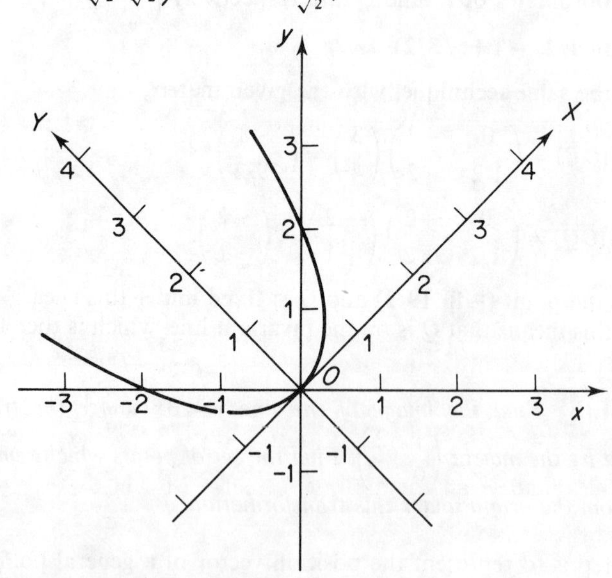

Fig. 4.8 The curve with equation $x^2+y^2-2xy-2x-2y = 0$ for axes $O(x, y)$ and with equation $Y = X^2/\sqrt{2}$ for axes $O(X, Y)$

Rather than solve for x and y in terms of X and Y, we rearrange the equation of C in the form $(x-y)^2-2(x+y)=0$, so $(\sqrt{2}X)^2-2(\sqrt{2}Y)=0$ or $\mathbf{Y=X^2/\sqrt{2}}$. The curve C is a parabola, with axis $X=0$ and is shown, together with the two sets of axes, in Fig. 4.8.

EXERCISE 4.1

1 Write the change of coordinates in matrix form and hence describe the transformation geometrically: (i) $x'=x$, $y'=-y$, (ii) $x'=2x$, $y'=-3y$, (iii) $x'=y$, $y'=x$, (iv) $x'=y$, $y'=-x$, (v) $x'=x+y$, $y'=y$, (vi) $x'=z$, $y'=y$, $z'=x$, (vii) $x'=2x$, $y'=2y$, $z'=2z$, (viii) $x'=x$, $y'=y$, $z'=2y+z$.

2 Find the matrix for each of the transformations, of $\mathbb{R}^2$ in cases (i)–(iv) and of $\mathbb{R}^3$ in cases (v)–(viii):

(i) a one way stretch, scale factor 3, in the Ox direction,
(ii) a reflection in the line $y=2x$,
(iii) a rotation through $3\pi/4$ about $(0, 0)$,
(iv) a shear keeping $y=0$ fixed and mapping $\begin{pmatrix}0\\1\end{pmatrix}$ to $\begin{pmatrix}-2\\1\end{pmatrix}$
(v) a two way stretch keeping Oy fixed, S.F. 2 in the Ox direction and S.F. 5 in the Oz direction,
(vi) a reflection in the plane $y+z=0$,
(vii) a rotation of $\pi/2$ about the axis Oz,
(viii) a rotation of $2\pi/3$ about the line $x=y=z$.

3 The point $P(x, y)$ is mapped into the point $Q(X, Y)$ by the transformation $\begin{pmatrix}X\\Y\end{pmatrix}=\begin{pmatrix}0{\cdot}6 & -0{\cdot}8\\0{\cdot}8 & 0{\cdot}6\end{pmatrix}\begin{pmatrix}x\\y\end{pmatrix}$. Show that the distance of P from the origin in the x-y plane is equal to the distance of Q from the origin in the X-Y plane. (*L*)

4 If 2-dimensional space is transformed by $\begin{pmatrix}x\\y\end{pmatrix}\mapsto\begin{pmatrix}4 & -1\\6 & -3\end{pmatrix}\begin{pmatrix}x\\y\end{pmatrix}$ find the equations of the straight lines which are mapped on to themselves. (*L*)

5 When column vectors, representing points in a plane with the usual axes, are premultiplied by the matrix $\mathbf{M}=\begin{pmatrix}3 & -2\\6 & k\end{pmatrix}$, a transformation of the plane results. (i) If $k=3$, find (a) the point whose image is $(8, 9)$, (b) the image of the line $y=x$. (ii) If $k=-4$, find (a) the set of points which map on to the origin, (b) the image of the line $2y=3x-1$. (*C*)

6 The point $P(x, y)$ in the plane of the Cartesian axes Ox, Oy is rotated in an anticlockwise sense through an angle α about an axis through O perpendicular to the plane to a point $P'(x', y')$. Show that $\begin{pmatrix}x'\\y'\end{pmatrix}=\begin{pmatrix}\cos\alpha & -\sin\alpha\\\sin\alpha & \cos\alpha\end{pmatrix}\begin{pmatrix}x\\y\end{pmatrix}$. Given that $\tan\alpha=3/4$, where α is acute, and that P lies on the curve whose equation is $12x^2+7xy-12y^2=25$, find, in terms of x' and y', an equation of the curve on which P' lies. (*L*)

7 The transformation with matrix $\mathbf{T}$, where $\mathbf{T}=\begin{pmatrix}2 & 1\\2 & -2\end{pmatrix}$, maps the point (x, y) into the point (x', y') so that $\mathbf{T}\begin{pmatrix}x\\y\end{pmatrix}=\begin{pmatrix}x'\\y'\end{pmatrix}$. Find the equation of the image of the line $y = 3x$ under this transformation. Find also the equations of the lines through the origin which are turned through a right angle about the origin under this transformation. *(JMB)*

8 The point (x, y) is transformed to the point (x', y') by means of the transformation $\begin{pmatrix} x' \\ y' \end{pmatrix} = \begin{pmatrix} 3 & 0 \\ 0 & 4 \end{pmatrix}\begin{pmatrix} x \\ y \end{pmatrix} + \begin{pmatrix} 1 \\ 1 \end{pmatrix}$. Find the image of the line $y = 2x$ under this transformation. *(L)*

9 (i) The position vector of a point P with respect to the origin is $\begin{pmatrix} x \\ y \end{pmatrix}$. The transformation T_1 reflects P in the y-axis. The transformation T_2 is an anticlockwise rotation about the origin through an angle θ. Obtain the matrices corresponding to T_1 and to T_2. Show that, if $x_1 > 0$, $y_1 > 0$ and $\tan\theta/2 = x_1/y_1$, then the point P_1 with coordinates (x_1, y_1) is transformed to the same new point P_2 by both transformations.
(ii) Show that the transformation

$$\mathrm{T}\begin{pmatrix} x \\ y \end{pmatrix} = \begin{pmatrix} 1 \\ 0 \end{pmatrix} + \begin{pmatrix} \cos\theta & \sin\theta \\ -\sin\theta & \cos\theta \end{pmatrix}\begin{pmatrix} x-1 \\ y \end{pmatrix}$$ is a rotation of P about Q through an angle θ in a clockwise sense, P and Q being the points (x, y) and $(1, 0)$ respectively. *(L)*

10 Multiply out $(a^2+b^2)(x^2+y^2)-(ax+by)^2$, and deduce that, if a, b, x, y are real numbers, then $(ax+by)^2 \leqslant (a^2+b^2)(x^2+y^2)$. The points of a plane are transformed by a matrix $\mathbf{M} = \begin{pmatrix} a & b \\ c & d \end{pmatrix}$, so that a general point P with position vector $\mathbf{r} = \begin{pmatrix} x \\ y \end{pmatrix}$ is transformed to a point P' with position vector $\mathbf{r}' = \mathbf{Mr}$. Write down an expression for OP'^2 in terms of a, b, c, d, x, y. Using the above inequality, or otherwise, deduce that the ratio of the length OP' to the length OP cannot be greater than $\sqrt{(a^2+b^2+c^2+d^2)}$. *(SMP)*

11 It is required to find the position vector $\begin{pmatrix} x \\ y \end{pmatrix}$ of a point which, as a result of a transformation of the plane whose matrix is $\mathbf{A}$, is moved to the same place as it would be by a displacement $\begin{pmatrix} 4 \\ -2 \end{pmatrix}$.

(i) If $\mathbf{A} = \begin{pmatrix} 0 & 1 \\ 1 & 2 \end{pmatrix}$, write down an equation for $\begin{pmatrix} x \\ y \end{pmatrix}$ and solve it.

(ii) If $\mathbf{A} = \begin{pmatrix} 1 & 2 \\ 0 & 1 \end{pmatrix}$, write down an equation for $\begin{pmatrix} x \\ y \end{pmatrix}$ and show that it has no solution. State in this case what kind of transformation is represented by $\mathbf{A}$, and give a geometrical explanation why there is no solution. *(SMP)*

12 A transformation matrix $\mathbf{M}$ is given by $\mathbf{M} = \begin{pmatrix} 1 & 0 & 4 \\ 0 & 5 & 4 \\ 4 & 4 & 3 \end{pmatrix}$. If O is the origin and A is the point $(1, 2, 2)$, what are the coordinates of the image of A under the transformation M? If A' is the image of A, what is the ratio of the lengths of OA and OA'? A second point B has its image B' under $\mathbf{M}$ such that $\overrightarrow{OB'} = 3\overrightarrow{OB}$. Find a set of possible coordinates of B. A third point C, not in the plane OAB, is related to its image C' under M by $\overrightarrow{OC'} = k\overrightarrow{OC}$, where k is a scalar. Find the value of k and a set of possible coordinates of C.

13 Describe geometrically the transformation with matrix $\begin{pmatrix} 1 & 0 \\ 0 & k \end{pmatrix}$, where $0 < k < 1$. Find the image of the circle $x^2 + y^2 = a^2$ under this transformation when $k = b/a$, where $0 < b < a$. Deduce the area within the ellipse $\frac{x^2}{a^2} + \frac{y^2}{b^2} = 1$, and state the lengths cut off by the ellipse on its two axes of symmetry. The rear hold of a ship is of depth 20 ft. Horizontal sections are half ellipses, similar to each other, and the linear dimensions of the sections decrease uniformly with depth, see Fig. 4.9(a). If the top of the hold is a half ellipse with semi-axes of lengths a, b, and the bottom is a half ellipse with semi-axes of lengths $a/2$, $b/2$, show that the semi-axes of the half ellipse at depth x are of lengths p, q, where $p = a(40-x)/40$ and $q = b(40-x)/40$. Deduce that, if the top of the hold has area 4800 sq. ft., then the cross-sectional area at depth x, denoted by $A(x)$, is given by $A(x) = 3(40-x)^2$. The hold is filled with grain. Find the volume of the grain and the depth of the centre of mass of the grain. (*SMP*)

14 The symmetry S of the regular octahedron $ABCDEF$ shown in Fig. 4.9(b), takes C to B, D to E and A to F. By taking axes $Oxyz$ as shown, find a matrix for S and hence decide whether S is direct or opposite. (*SMP*)

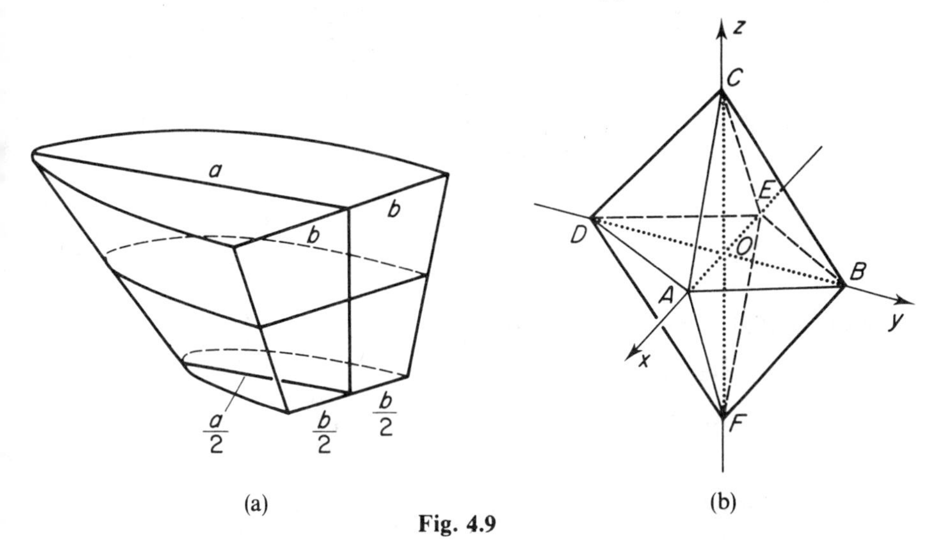

Fig. 4.9

4.2 Combining transformations

A transformation is a function so there is a natural way of combining transformations by composition. Let the transformation S map the point P to its image P' and let the transformation T map P' to its image P''. Then the combined transformation S followed by T, which maps P to P'' is the composite map TS. Thus:

$$S(P) = P', \ T(P') = P'', \text{ so } TS(P) = T(S(P)) = T(P') = P''.$$

The operation is associative, being a composition of functions and so:

$$T(SR(P)) = T(S(R(P))) = (TS)(R(P)) \text{ so } T(SR) = (TS)R = TSR.$$

EXAMPLE 1 *Describe a single transformation of the plane equivalent to a rotation of 90° about the origin followed by a reflection in $y = 0$. Write down the corresponding matrices of the transformations and show their effect on the position vector of $P(2, 1)$.*

The transformations are shown in a diagram by drawing the images of an asymmetrical object, like a flag. It is then clear that the combined transformation is a reflection in the line $y + x = 0$, see Fig. 4.10.

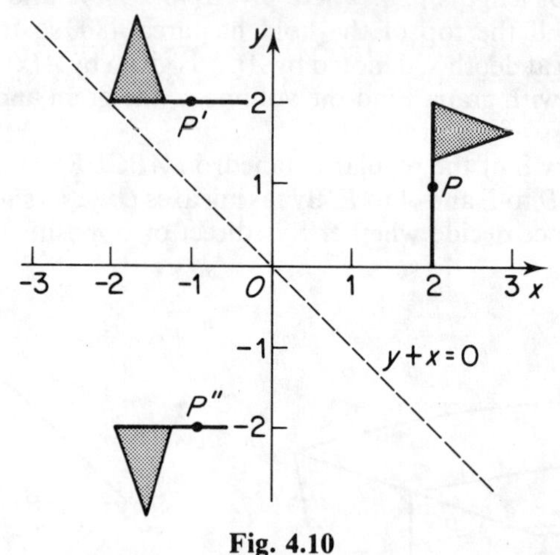

Fig. 4.10

This is verified by an algebraic proof: $\begin{pmatrix} x \\ y \end{pmatrix} \mapsto \begin{pmatrix} -y \\ x \end{pmatrix} \mapsto \begin{pmatrix} -y \\ -x \end{pmatrix}$.

The corresponding matrices are:

for rotation 90° about O, $\begin{pmatrix} 0 & -1 \\ 1 & 0 \end{pmatrix}$,

for reflection in $y = 0$, $\begin{pmatrix} 1 & 0 \\ 0 & -1 \end{pmatrix}$,

for reflection in $y + x = 0$, $\begin{pmatrix} 0 & -1 \\ -1 & 0 \end{pmatrix}$.

Then $\begin{pmatrix} 0 & -1 \\ 1 & 0 \end{pmatrix}\begin{pmatrix} 2 \\ 1 \end{pmatrix} = \begin{pmatrix} -1 \\ 2 \end{pmatrix}$, $\begin{pmatrix} 1 & 0 \\ 0 & -1 \end{pmatrix}\begin{pmatrix} -1 \\ 2 \end{pmatrix} = \begin{pmatrix} -1 \\ -2 \end{pmatrix}$,

and $\begin{pmatrix} 1 & 0 \\ 0 & -1 \end{pmatrix}\begin{pmatrix} 0 & -1 \\ 1 & 0 \end{pmatrix}\begin{pmatrix} 2 \\ 1 \end{pmatrix} = \begin{pmatrix} 0 & -1 \\ -1 & 0 \end{pmatrix}\begin{pmatrix} 2 \\ 1 \end{pmatrix} = \begin{pmatrix} -1 \\ -2 \end{pmatrix}$.

Note that the operation S followed by T for transformations S and T gives the transformation TS (in the reverse order) and is equivalent to the matrix multiplication **TS** of the corresponding matrices **S** and **T**, (again in the reverse order). The order of the matrix product is the same as the order of the product of the transformations.

The associativity of the composition of transformations can be used to prove that matrix multiplication is associative. This is shown in Theorem 4.1 and should be compared with the proof shown in Theorem 1.4.

Theorem 4.1 Matrix multiplication is associative.

Proof The proof is given for square matrices but it also applies to non-square matrices if transformations of spaces of appropriate dimension are used.
Let **A**, **B** and **C** be three $(n \times n)$ matrices with A, B and C their corresponding transformations of $\mathbb{R}^n$. Then A (BC) = (AB) C so **A (BC)** = **(AB) C**.

Matrix multiplication may now be used for the solution of problems involving the combination of transformations.

EXAMPLE 2 *Find the* (2×2) *matrix* **M** *of the transformation which reflects each point P of the plane in the line* $y = x$. *Find also the matrix* **R** *of the transformation which rotates OP anticlockwise through* 90° *about the origin O. Find the matrix* **S** *of the transformation which is obtained by following the reflection by the rotation. Show that, under this transformation, the image of the line* $y = 3x + 4$ *is the line* $y + 3x = 4$. *Find also the image of the line* $y = 3x + 4$ *under the transformation obtained by following the rotation by the reflection and show that the two images are parallel.*

Using the unit square method, $\mathbf{M} = \begin{pmatrix} 0 & 1 \\ 1 & 0 \end{pmatrix}$ and $\mathbf{R} = \begin{pmatrix} 0 & -1 \\ 1 & 0 \end{pmatrix}$. Then $\mathbf{S} = \mathbf{RM} = \begin{pmatrix} 0 & -1 \\ 1 & 0 \end{pmatrix}\begin{pmatrix} 0 & 1 \\ 1 & 0 \end{pmatrix} = \begin{pmatrix} -1 & 0 \\ 0 & 1 \end{pmatrix}$. A point on the line $y = 3x + 4$ has position vector $\begin{pmatrix} t \\ 3t+4 \end{pmatrix}$, where t is a scalar parameter, and this transforms to the point with position vector $\begin{pmatrix} -1 & 0 \\ 0 & 1 \end{pmatrix}\begin{pmatrix} t \\ 3t+4 \end{pmatrix} = \begin{pmatrix} -t \\ 3t+4 \end{pmatrix}$ so the image of the line is the line $y = 3(-x) + 4$, that is $y + 3x = 4$. The second transformation, reflection then rotation, has matrix **MR** and $\mathbf{MR} = \begin{pmatrix} 0 & 1 \\ 1 & 0 \end{pmatrix}\begin{pmatrix} 0 & -1 \\ 1 & 0 \end{pmatrix} = \begin{pmatrix} 1 & 0 \\ 0 & -1 \end{pmatrix}$, $\begin{pmatrix} 1 & 0 \\ 0 & -1 \end{pmatrix}\begin{pmatrix} t \\ 3t+4 \end{pmatrix} = \begin{pmatrix} t \\ -3t-4 \end{pmatrix}$ so the second image line is $y = -3x - 4$, or $\mathbf{y + 3x = -4}$. Both image lines have gradient -3 and therefore they are parallel.

The results of Chapter 1 on the set of real $n \times n$ matrices under multiplication can be applied to transformations. If the matrix $\mathbf{M}$ of a linear transformation T is non-singular, $\mathbf{M}$ has an inverse matrix $\mathbf{M}^{-1}$ and $\mathbf{M}^{-1}\mathbf{M} = \mathbf{I}_n$. So, if T^{-1} is the transformation with matrix $\mathbf{M}^{-1}$, then $\mathrm{T}^{-1}\mathrm{T}$ is the identity mapping and T^{-1} is the inverse transformation of T. Conversely, if a transformation T has an inverse T^{-1}, then $\mathrm{T}^{-1}\mathrm{T}$ is the identity map with matrix $\mathbf{I}_n$ so the matrix of T is non-singular with an inverse which is the matrix of T^{-1}. We apply the words *singular* and *non-singular* to linear transformations and then write the above results as a theorem.

Table 4.1 Standard transformations and their matrices

Transformation	Matrix	Inverse transformation	Inverse matrix
zero	$\begin{pmatrix} 0 & 0 \\ 0 & 0 \end{pmatrix}$	none	none
projection on to Ox	$\begin{pmatrix} 1 & 0 \\ 0 & 0 \end{pmatrix}$	none	none
projection on to Oy	$\begin{pmatrix} 0 & 0 \\ 0 & 1 \end{pmatrix}$	none	none
identity	$\begin{pmatrix} 1 & 0 \\ 0 & 1 \end{pmatrix}$	identity	$\begin{pmatrix} 1 & 0 \\ 0 & 1 \end{pmatrix}$
enlargement, $c \in \mathbb{R}\setminus\{0, 1\}$	$\begin{pmatrix} c & 0 \\ 0 & c \end{pmatrix}$	enlargement	$\begin{pmatrix} 1/c & 0 \\ 0 & 1/c \end{pmatrix}$
stretch, $c, d \in \mathbb{R}\setminus\{0\}$	$\begin{pmatrix} c & 0 \\ 0 & d \end{pmatrix}$	stretch	$\begin{pmatrix} 1/c & 0 \\ 0 & 1/d \end{pmatrix}$
shear, $c \in \mathbb{R}\setminus\{0\}$	$\begin{pmatrix} 1 & c \\ 0 & 1 \end{pmatrix}$	shear	$\begin{pmatrix} 1 & -c \\ 0 & 1 \end{pmatrix}$
shear, $d \in \mathbb{R}\setminus\{0\}$	$\begin{pmatrix} 1 & 0 \\ d & 1 \end{pmatrix}$	shear	$\begin{pmatrix} 1 & 0 \\ -d & 1 \end{pmatrix}$
reflection in $y = x \tan \alpha$	$\begin{pmatrix} \cos 2\alpha & \sin 2\alpha \\ \sin 2\alpha & -\cos 2\alpha \end{pmatrix}$	the same: a reflection is its own inverse	
rotation α	$\begin{pmatrix} \cos \alpha & -\sin \alpha \\ \sin \alpha & \cos \alpha \end{pmatrix}$	rotation $-\alpha$	$\begin{pmatrix} \cos \alpha & \sin \alpha \\ -\sin \alpha & \cos \alpha \end{pmatrix}$

Definition A linear transformation T of $\mathbb{R}^n$ is:
(i) non-singular if it is one-one, that is, if it has an inverse T^{-1},
(ii) singular if it is not one-one, that is, it has no inverse.

Theorem 4.2 Let T be a linear transformation of $\mathbb{R}^n$ with matrix $\mathbf{M}$. Then (i) T is non-singular if and only if $\mathbf{M}$ is non-singular and, in this case the inverse transformation T^{-1} has matrix $\mathbf{M}^{-1}$,
(ii) T if singular if and only if $\mathbf{M}$ is singular, that is $\det(\mathbf{M}) = 0$.

In the case when a linear transformation T (and its matrix) is singular, T is not one-one, so the image of T will be a space of dimension less than n so that T will have some squashing effect.

The list of standard transformations, given in §4.1, can now be expanded by including the inverse transformations, where they exist. We display them in Table 4.1.

Elementary matrices and transformations

Elementary matrices were defined in §1.4 and we define an *elementary transformation* to be a linear transformation whose matrix is elementary. In two dimensions, the elementary matrices (and their corresponding transformations) are

$\begin{pmatrix} p & 0 \\ 0 & 1 \end{pmatrix}$, a one way stretch, $\begin{pmatrix} 1 & 0 \\ 0 & p \end{pmatrix}$, a one way stretch, $(p \neq 0)$

$\begin{pmatrix} 1 & r \\ 0 & 1 \end{pmatrix}$, a shear, $\begin{pmatrix} 1 & 0 \\ r & 1 \end{pmatrix}$, a shear, $(r \neq 0)$

$\begin{pmatrix} 0 & 1 \\ 1 & 0 \end{pmatrix}$, a reflection in $y = x$.

Row reduction of a matrix Any non-singular linear transformation (matrix) can be expressed as a product of elementary transformations (matrices). We shall demonstrate how this may be done in the two dimensional case. Let $\mathbf{M}$ be a general non-singular (2×2) matrix, given by

$\mathbf{M} = \begin{pmatrix} a & b \\ c & d \end{pmatrix}$, and let M be the corresponding linear transformation.

Suppose that $\mathbf{M} = \mathbf{E}_k \ldots \mathbf{E}_2 \mathbf{E}_1$, where $\mathbf{E}_1, \mathbf{E}_2, \ldots, \mathbf{E}_k$ are elementary matrices. Note that this means that, as transformations, the order of operations of the elementary matrices is $\mathbf{E}_1$ first, then $\mathbf{E}_2$, and lastly $\mathbf{E}_k$.

Then, for each elementary matrix $\mathbf{E}_r$, we know that $\mathbf{E}_r$ is non-singular and, moreover, we know its inverse $\mathbf{E}_r^{-1}$. Therefore,

$$\mathbf{M}^{-1} = \mathbf{E}_1^{-1}\mathbf{E}_2^{-1} \ldots \mathbf{E}_k^{-1} \quad \text{and} \quad \mathbf{E}_1^{-1}\mathbf{E}_2^{-1} \ldots \mathbf{E}_k^{-1}\mathbf{M} = \mathbf{I}_2.$$

This gives us a clue to a method. Premultiply $\mathbf{M}$ by a succession of elementary matrices (the inverses of $\mathbf{E}_r$, $1 \leqslant r \leqslant k$), to produce the unit matrix. But this is the same as using elementary row operations on $\mathbf{M}$ (see §1.3) in order to convert it to the unit matrix. This we now proceed to do assuming that $a \neq 0$. If $a = 0$ then $c \neq 0$ since $\mathbf{M}$ is non-singular, so interchange the rows by multiplying by the elementary matrix $\begin{pmatrix} 0 & 1 \\ 1 & 0 \end{pmatrix}$, that is, by the transformation of reflection in $y = x$. We may now assume that $a \neq 0$. The next four steps are shown in Fig. 4.11. Since we shall later take inverses, we label the elementary matrices from $\mathbf{E}_4$ to $\mathbf{E}_1$.

Step 1 Make the top left-hand corner entry 1 by dividing the first row by a,

$$\mathbf{E}_4^{-1}\mathbf{M} = \begin{pmatrix} 1 & b/a \\ c & d \end{pmatrix} = \begin{pmatrix} 1/a & 0 \\ 0 & 1 \end{pmatrix}\begin{pmatrix} a & b \\ c & d \end{pmatrix}, \quad \mathbf{E}_4^{-1} = \begin{pmatrix} 1/a & 0 \\ 0 & 1 \end{pmatrix}.$$

Step 2 Make the first entry in row 2 zero by subtracting c times row 1 from row 2,

$$\mathbf{E}_3^{-1}\mathbf{E}_4^{-1}\mathbf{M} = \begin{pmatrix} 1 & b/a \\ 0 & \Delta/a \end{pmatrix} = \begin{pmatrix} 1 & 0 \\ -c & 1 \end{pmatrix}\begin{pmatrix} 1 & b/a \\ c & d \end{pmatrix}, \quad \mathbf{E}_3^{-1} = \begin{pmatrix} 1 & 0 \\ -c & 1 \end{pmatrix}.$$

where $\Delta = ad - bc = \det(\mathbf{M})$.

Step 3 Make the second entry in row 2 equal to 1 by dividing row 2 by Δ/a which is non-zero, since $\mathbf{M}$ is non-singular.

$$\mathbf{E}_2^{-1}\mathbf{E}_3^{-1}\mathbf{E}_4^{-1}\mathbf{M} = \begin{pmatrix} 1 & b/a \\ 0 & 1 \end{pmatrix} = \begin{pmatrix} 1 & 0 \\ 0 & a/\Delta \end{pmatrix}\begin{pmatrix} 1 & b/a \\ 0 & \Delta/a \end{pmatrix}, \mathbf{E}_2^{-1} = \begin{pmatrix} 1 & 0 \\ 0 & a/\Delta \end{pmatrix}.$$

Step 4 Make the second entry in row 1 equal to 0 by subtracting b/a times row 2 from row 1

$$\mathbf{E}_1^{-1}\mathbf{E}_2^{-1}\mathbf{E}_3^{-1}\mathbf{E}_4^{-1}\mathbf{M} = \begin{pmatrix} 1 & 0 \\ 0 & 1 \end{pmatrix} = \begin{pmatrix} 1 & -b/a \\ 0 & 1 \end{pmatrix}\begin{pmatrix} 1 & b/a \\ 0 & 1 \end{pmatrix},$$

$$\mathbf{E}_1^{-1} = \begin{pmatrix} 1 & -b/a \\ 0 & 1 \end{pmatrix}.$$

Then $\mathbf{M} = \mathbf{E}_4\mathbf{E}_3\mathbf{E}_2\mathbf{E}_1 = \begin{pmatrix} a & 0 \\ 0 & 1 \end{pmatrix}\begin{pmatrix} 1 & 0 \\ c & 1 \end{pmatrix}\begin{pmatrix} 1 & 0 \\ 0 & \Delta/a \end{pmatrix}\begin{pmatrix} 1 & b/a \\ 0 & 1 \end{pmatrix}$, on using our table above for the inverses of the elementary matrices.

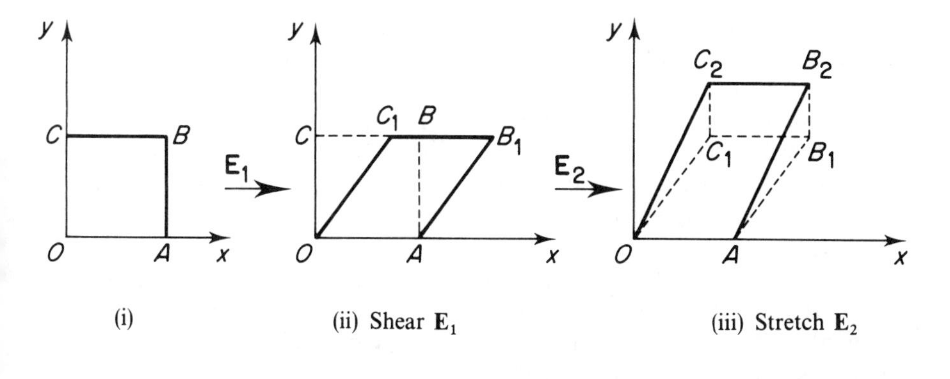

(i) (ii) Shear $\mathbf{E}_1$ (iii) Stretch $\mathbf{E}_2$

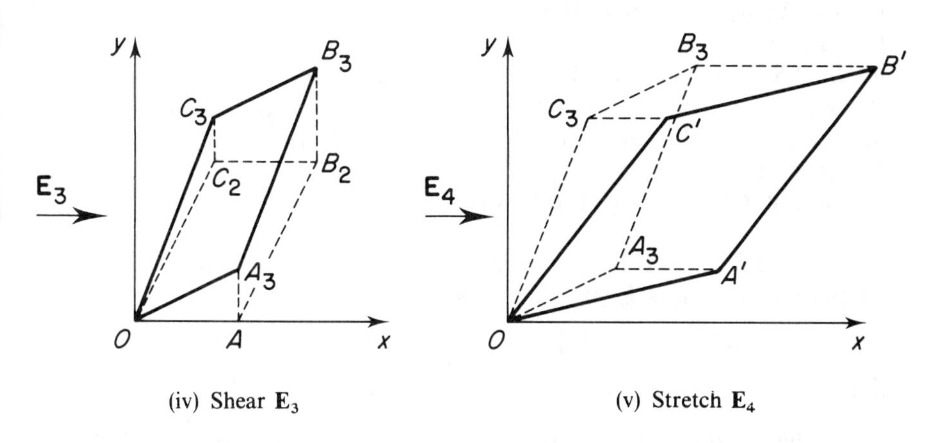

(iv) Shear $\mathbf{E}_3$ (v) Stretch $\mathbf{E}_4$

Fig. 4.11 Four elementary transformations equivalent to $\begin{pmatrix} a & b \\ c & d \end{pmatrix}$, $a \neq 0$

The expression for **M** as a product of elementary matrices can be used to show how the transformation M is built up from elementary transformations, using the unit square method. We start with the unit square $OABC$ and transform this into the parallelogram $OA'B'C'$, where $\overrightarrow{OA'} = \begin{pmatrix} a \\ c \end{pmatrix}$ and $\overrightarrow{OC'} = \begin{pmatrix} b \\ d \end{pmatrix}$, using the steps indicated above. The process is shown in Fig. 4.11.

The initial square $OABC$ is shown in (i).

Step 1, E_1 is a shear moving C to C_1 $(b/a, 1)$ shown in (ii).

Step 2, E_2 is a stretch moving C_1 to C_2 $(b/a, \Delta/a)$ shown in (iii).

Step 3, E_3 is a shear moving C_2 to C_3 $(b/a, d)$ and A to A_3 $(1, c)$ shown in (iv).

Step 4, E_4 is a stretch moving C_3 to C' (b, d) and A_3 to A' (a, c) shown in (v).

The technique of row reduction of non-singular matrices can be extended to $n \times n$ matrices. The process can be described as an algorithm in a series of steps, as follows.

Take in turn, each term, a_{ii}, of the main diagonal, convert this to unity and then make all other entries in the same column zero.
For $i = 1$ to n:

Step 1 If $a_{ii} = 0$, make this entry non-zero by interchanging rows.
Step 2 Make a_{ii} equal 1 by dividing row i by a_{ii}.
Step 3 Make all other entries in column i zero by subtracting multiples of row i from the other rows.

Next i:

The above algorithm is a constructive way of transforming any non-singular square matrix to the unit matrix by row operations, which are equivalent to premultiplying the matrix by a succession of elementary matrices. Step 1 is justified, when dealing with a_{ii}, for the following reason. At this stage the entries in the first $i-1$ columns are all zero, except for the main diagonal entries which are 1, and these columns are unaffected by steps 2 and 3. Therefore there is always a non-zero entry a_{ji} for some j with $i \leqslant j \leqslant n$, otherwise column i of $\mathbf{A}$ would be linearly dependent on the previous columns and $\mathbf{A}$ would be singular, by Theorem 1.8.

The algorithm proves the theorem:

Theorem 4.3 If $\mathbf{M}$ is a non-singular $n \times n$ matrix, then there exist elementary matrices $\mathbf{E}_i$, $1 \leqslant i \leqslant k$, such that

$$\mathbf{M} = \mathbf{E}_k \ldots \mathbf{E}_2\mathbf{E}_1, \quad \mathbf{M}^{-1} = \mathbf{E}_1^{-1}\mathbf{E}_2^{-1} \ldots \mathbf{E}_k^{-1}$$

and
$$\mathbf{E}_1^{-1}\,\mathbf{E}_2^{-1} \ldots \mathbf{E}_k^{-1}\,\mathbf{M} = \mathbf{I}_n.$$

The technique of row reduction is central to this chapter and may be used to find the inverse of a non-singular matrix, to solve sets of simultaneous equations, and to determine the rank (the number of linearly independent rows) of a matrix. The calculation of an inverse matrix is demonstrated in the next example and the other uses occur in later sections.

EXAMPLE 3 *Use the method of row reduction to find the inverse of the matrix* $\mathbf{A}$, *where* $\mathbf{A} = \begin{pmatrix} 0 & 2 & 3 \\ 1 & -1 & -1 \\ 4 & -2 & 1 \end{pmatrix}$ *and hence find the solution of the equations*

$$\begin{aligned} 2y + 3z &= 2 \\ x - y - z &= -1 \\ 4x - 2y + z &= 4 \end{aligned}$$

If $\mathbf{A} = \mathbf{E}_k \ldots \mathbf{E}_2\mathbf{E}_1$, then $\mathbf{A}^{-1} = \mathbf{E}_1^{-1}\mathbf{E}_2^{-1} \ldots \mathbf{E}_k^{-1}$. The matrix $\mathbf{A}^{-1}$ is constructed by performing the same row operations on the unit matrix as are performed on $\mathbf{A}$ in order to convert it to the unit matrix. Begin by interchanging rows 1 and 2 to get a non-zero entry in the top left-hand corner (see Table 4.2). The successive steps are explained on each line. To help the reader follow the work and to recognise the various steps of the algorithm, at each stage the value of i is given as well as the step number from the algorithm.

Table 4.2

$$\left(\begin{array}{rrr|rrr} 0 & 2 & 3 & 1 & 0 & 0 \\ 1 & -1 & -1 & 0 & 1 & 0 \\ 4 & -2 & 1 & 0 & 0 & 1 \end{array}\right)$$

Initial position, the matrix **A** together with the unit matrix, interchange $\mathbf{r}_1$ and $\mathbf{r}_2$,

$i = 1$, step 1, step 2,

$$\left(\begin{array}{rrr|rrr} 1 & -1 & -1 & 0 & 1 & 0 \\ 0 & 2 & 3 & 1 & 0 & 0 \\ 4 & -2 & 1 & 0 & 0 & 1 \end{array}\right)$$

make first term in row 3 zero, make second term in row 2 unity, $\mathbf{r}_3 \leftarrow \mathbf{r}_3 - 4\mathbf{r}_1$, $\mathbf{r}_2 \leftarrow \mathbf{r}_2/2$,

$i = 1$, step 3, $i = 2$, step 2,

$$\left(\begin{array}{rrr|rrr} 1 & -1 & -1 & 0 & 1 & 0 \\ 0 & 1 & 3/2 & 1/2 & 0 & 0 \\ 0 & 2 & 5 & 0 & -4 & 1 \end{array}\right)$$

make second terms in rows 1 and 3 both zero, $\mathbf{r}_1 \leftarrow \mathbf{r}_1 + \mathbf{r}_2$, $\mathbf{r}_3 \leftarrow \mathbf{r}_3 - 2\mathbf{r}_2$,

$i = 2$, step 3,

$$\left(\begin{array}{rrr|rrr} 1 & 0 & 1/2 & 1/2 & 1 & 0 \\ 0 & 1 & 3/2 & 1/2 & 0 & 0 \\ 0 & 0 & 2 & -1 & -4 & 1 \end{array}\right)$$

make third term in row 3 unity, $\mathbf{r}_3 \leftarrow \mathbf{r}_3/2$,

$i = 3$, step 2,

$$\left(\begin{array}{rrr|rrr} 1 & 0 & 1/2 & 1/2 & 1 & 0 \\ 0 & 1 & 3/2 & 1/2 & 0 & 0 \\ 0 & 0 & 1 & -1/2 & -2 & 1/2 \end{array}\right)$$

make third terms in rows 1 and 2 both zero, $\mathbf{r}_1 \leftarrow \mathbf{r}_1 - \mathbf{r}_3/2$, $\mathbf{r}_2 \leftarrow \mathbf{r}_2 - 3\mathbf{r}_3/2$,

$i = 3$, step 3,

$$\left(\begin{array}{rrr|rrr} 1 & 0 & 0 & 3/4 & 2 & -1/4 \\ 0 & 1 & 0 & 5/4 & 3 & -3/4 \\ 0 & 0 & 1 & -1/2 & -2 & 1/2 \end{array}\right)$$

this is now the matrix $(\mathbf{I} \mid \mathbf{A}^{-1})$.

Therefore $\mathbf{A}^{-1} = \begin{pmatrix} \mathbf{3/4} & \mathbf{2} & \mathbf{-1/4} \\ \mathbf{5/4} & \mathbf{3} & \mathbf{-3/4} \\ \mathbf{-1/2} & \mathbf{-2} & \mathbf{1/2} \end{pmatrix}$ and the solution of the equations is given by

$$\begin{pmatrix} x \\ y \\ z \end{pmatrix} = \mathbf{A}^{-1} \begin{pmatrix} 2 \\ -1 \\ 4 \end{pmatrix} = \begin{pmatrix} 3/4 & 2 & -1/4 \\ 5/4 & 3 & -3/4 \\ -1/2 & -2 & 1/2 \end{pmatrix} \begin{pmatrix} 2 \\ -1 \\ 4 \end{pmatrix} = \begin{pmatrix} \mathbf{-3/2} \\ \mathbf{-7/2} \\ \mathbf{3} \end{pmatrix}.$$

EXERCISE 4.2

1 Find the elements a, b, c, d in order that the transformation $\begin{pmatrix} x' \\ y' \end{pmatrix} = \begin{pmatrix} a & b \\ c & d \end{pmatrix}\begin{pmatrix} x \\ y \end{pmatrix}$ may represent (a) an anticlockwise rotation through an angle α, (b) a reflection in the line $y = x \tan(\beta/2)$. Hence prove that a reflection in the line $y = x \tan(\beta/2)$ followed by a reflection in the line $y = x \tan(\gamma/2)$ is equivalent to a rotation. (*L*)

2 Find the coefficients a, b, c, d such that the transformation $\begin{pmatrix} x_1 \\ y_1 \end{pmatrix} = \begin{pmatrix} a & b \\ c & d \end{pmatrix}\begin{pmatrix} x_0 \\ y_0 \end{pmatrix}$ may represent a reflection in the line $y = x$. Find also the coefficients a, b, c, d such that this transformation may represent a reflection in the line $y = -x$. The point $P_1(x_1, y_1)$ is the image of $P_0(x_0, y_0)$ when it is reflected in the line $y = x$. The point $P_2(x_2, y_2)$ is the reflection of P_1 when it is reflected in the line $y = -x$. Show that the combined effect of these successive reflections is equivalent to a rotation. Find, in terms of x_0, y_0, the radius of the circle passing through P_0, P_1 and P_2. (*L*)

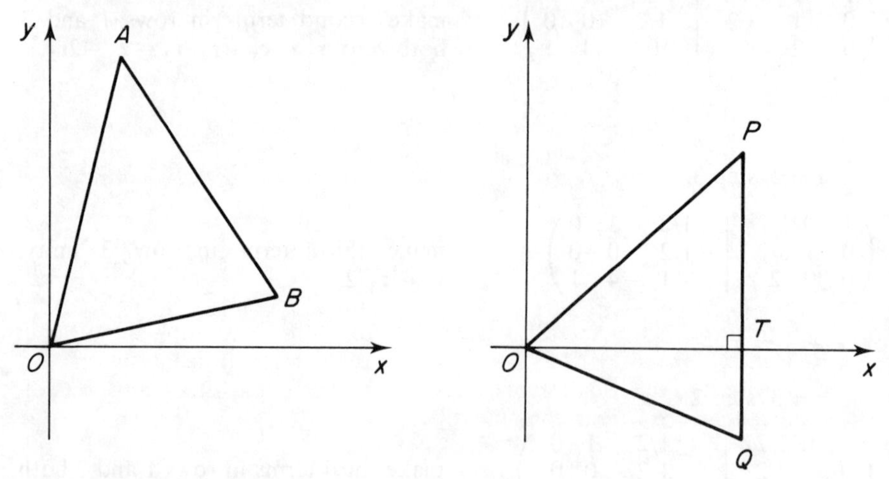

Fig. 4.12

3 Let A, B be the points in the x-y plane with position vectors $\begin{pmatrix} 1 \\ 5 \end{pmatrix}$ and $\begin{pmatrix} 4 \\ 1 \end{pmatrix}$ respectively. Write down, in terms of θ, the position vectors of the images P, Q of A, B under the transformation $\mathbf{r}' = \mathbf{Mr}$, where $\mathbf{M}$ is the rotation matrix $\begin{pmatrix} \cos\theta & \sin\theta \\ -\sin\theta & \cos\theta \end{pmatrix}$. Show that if PQ is perpendicular to the x-axis, as shown in Fig. 4.12, then $\tan\theta = 3/4$. Determine the position vector of the point T where PQ cuts the x-axis. Find the position vector of the point C on AB which is mapped on to T by the above transformation. (*JMB*)

4 Use row reduction on the matrix $\mathbf{M}$ to (a) find the inverse of $\mathbf{M}$ and (b) express $\mathbf{M}$ as a product of elementary matrices, where $\mathbf{M}$ is: (i) $\begin{pmatrix} 1 & 3 \\ -2 & -5 \end{pmatrix}$, (ii) $\begin{pmatrix} 2 & 4 \\ 1 & 3 \end{pmatrix}$, (iii) $\begin{pmatrix} 0 & 2 \\ -1 & 5 \end{pmatrix}$.

5 In each case of question 4, draw the transformations of the unit square corresponding to the sequence of elementary matrices in the expression for **M** and verify that the final parallelogram is the image of the unit square under **M**.

6 Show that the transformation of the plane given by the matrix **S**, where $\mathbf{S} = \begin{pmatrix} 1 & 0 \\ 0 & -1 \end{pmatrix}$, is a reflection in the x-axis. Show also that the transformation of the plane given by the matrix $\mathbf{R}_\alpha$, where $\mathbf{R}_\alpha = \begin{pmatrix} \cos\alpha & -\sin\alpha \\ \sin\alpha & \cos\alpha \end{pmatrix}$, is a rotation about the origin through an angle α. Form the product $\mathbf{R}_\alpha\mathbf{S}\mathbf{R}_{-\alpha}$ and show that the transformation of the plane given by this matrix is a reflection in the line $y = x\tan\alpha$. (*L*)

7 **M** denotes the matrix $\begin{pmatrix} 1{\cdot}36 & 0{\cdot}48 \\ 0{\cdot}48 & 1{\cdot}64 \end{pmatrix}$. A transformation T of points of the x-y plane is such that $\mathrm{T}\begin{pmatrix} x \\ y \end{pmatrix} = \mathbf{M}\begin{pmatrix} x \\ y \end{pmatrix}$. Show that each point on the line $3x + 4y = 0$ is invariant under T. Verify, showing your working, that $\mathbf{M} = \mathbf{R}^{-1}\mathbf{D}\mathbf{R}$, where $\mathbf{D} = \begin{pmatrix} 2 & 0 \\ 0 & 1 \end{pmatrix}$ and $\mathbf{R} = \begin{pmatrix} 0{\cdot}6 & 0{\cdot}8 \\ -0{\cdot}8 & 0{\cdot}6 \end{pmatrix}$. By considering the effect on $\begin{pmatrix} 1 \\ 0 \end{pmatrix}$ and $\begin{pmatrix} 0 \\ 1 \end{pmatrix}$, or otherwise, describe the geometrical transformations which result when column vectors are multiplied by **D** and **R**. Hence, by considering T as a sequence of three transformations, or otherwise, find the equation of a line l which is such that T maps any point of l, other than the origin, to some different point of l. (*C*)

8 The matrices **A** and **B** are given by $\mathbf{A} = \begin{pmatrix} -3/5 & 4/5 \\ 4/5 & 3/5 \end{pmatrix}$ and $\mathbf{B} = \begin{pmatrix} 3 & -1 \\ 6 & -2 \end{pmatrix}$.

Transformations $\mathrm{T_A}$, $\mathrm{T_B}$ of points of the x-y plane are obtained by premultiplication of the position vectors of the points by **A**, **B** respectively. Prove that the points which are invariant under $\mathrm{T_A}$ are also invariant under $\mathrm{T_B}$. Show that the transformation $\mathrm{T_A}$ is self-inverse, and describe the transformation in geometrical terms. Verify that the image of the line $y = 3x + c$ under the transformation $\mathrm{T_{AB}}$ (corresponding to the matrix product **AB**) is the same as the image of the line $9y = 13x - 5c$ under the transformation $\mathrm{T_{BA}}$. (*C*)

9 (a) Prove that under the transformation $\begin{pmatrix} x \\ y \end{pmatrix} \mapsto \begin{pmatrix} -5 & 3 \\ 2 & -1 \end{pmatrix}\begin{pmatrix} x \\ y \end{pmatrix}$ no point, other than the origin, is its own image. Find the set of points which is the image of $\{(x, y): 2y - 3x = 1\}$ under this transformation.

(b) Find the matrices A and B such that (i) the transformation $\begin{pmatrix} x \\ y \end{pmatrix} \mapsto \mathbf{A}\begin{pmatrix} x \\ y \end{pmatrix}$ gives an anti-clockwise rotation about the origin through an angle α, (ii) the transformation $\begin{pmatrix} x \\ y \end{pmatrix} \mapsto \mathbf{B}\begin{pmatrix} x \\ y \end{pmatrix}$ gives a reflection in the line $y = x \tan \beta$. Obtain the matrix **AB**, expressing the elements in simplified form, and describe the transformation given by this matrix. (*C*)

10 A point P lies in the plane of the cartesian axes Ox, Oy and has coordinates (x, y). The line OP is rotated, through an angle α in an anticlockwise sense about an axis through the origin perpendicular to the plane, to a position OP' where P' has coordinates (x', y'). Show that $\begin{pmatrix} x' \\ y' \end{pmatrix} = \begin{pmatrix} \cos\alpha & -\sin\alpha \\ \sin\alpha & \cos\alpha \end{pmatrix}\begin{pmatrix} x \\ y \end{pmatrix}$. Obtain the inverse of this rotation matrix, verifying that it represents a clockwise rotation through an angle α. The line OP' is similarly rotated in an anticlockwise direction through an angle β to a position OP'' where P'' has coordinates (x'', y''). By making use of matrix multiplication, find a single matrix which represents the rotation of OP to OP''. Hence verify that $\cos(\alpha + \beta) = \cos\alpha\cos\beta - \sin\alpha\sin\beta$. (*L*)

11 Find the inverse matrix, $\mathbf{A}^{-1}$, where $\mathbf{A} = \begin{pmatrix} 2 & 0 & 1 \\ 3 & -2 & -1 \\ -4 & 1 & -1 \end{pmatrix}$. (*L*)

12 Calculate the inverse of the matrix $\begin{pmatrix} 1 & 2 & 3 \\ -3 & 1 & 0 \\ 2 & 0 & 1 \end{pmatrix}$. (*SMP*)

13 Find the inverse of the matrix $\begin{pmatrix} 1 & 0 & 0 \\ a & 1 & 0 \\ b & c & 1 \end{pmatrix}$. (*SMP*)

14 Let **A** be the matrix $\begin{pmatrix} 4 & 8 & 3 \\ 3 & 5 & 1 \\ 1 & 4 & 3 \end{pmatrix}$. By performing elementary row operations on the matrix $(\mathbf{A} \mid \mathbf{I})$ find $\mathbf{A}^{-1}$. Hence, or otherwise, find the inverses of the following matrices:

(i) $\begin{pmatrix} 4 & 3 & 1 \\ 8 & 5 & 4 \\ 3 & 1 & 3 \end{pmatrix}$, (ii) $\begin{pmatrix} 8 & 16 & 6 \\ 3 & 5 & 1 \\ 1 & 4 & 3 \end{pmatrix}$, (iii) $\begin{pmatrix} -22 & 24 & 14 \\ 16 & -18 & -10 \\ -14 & 16 & 8 \end{pmatrix}$. (*C*)

15 Find the inverse of the matrix $\begin{pmatrix} 1 & 1 & 1 \\ 1 & 2 & 3 \\ 3 & -2 & 2 \end{pmatrix}$ and hence solve the equations

$$\begin{cases} x + y + z = 3 \\ x + 2y + 3z = -5 \\ 3x - 2y + 2z = 4 \end{cases}$$

(*SMP*)

16 You are given the matrix $\mathbf{M} = \begin{pmatrix} 0 & 6 \\ 1 & 3 \end{pmatrix}$. Three matrices $\mathbf{E}_1, \mathbf{E}_2, \mathbf{E}_3$ are to be found such that the product $\mathbf{E}_3\mathbf{E}_2\mathbf{E}_1\mathbf{M}$ is the identity matrix. These three matrices are to be 'elementary matrices', i.e. matrices having one of the forms

(i) $\begin{pmatrix} 0 & 1 \\ 1 & 0 \end{pmatrix}$, (ii) $\begin{pmatrix} a & 0 \\ 0 & 1 \end{pmatrix}$ or $\begin{pmatrix} 1 & 0 \\ 0 & a \end{pmatrix}$ for some number a,

(iii) $\begin{pmatrix} 1 & b \\ 0 & 1 \end{pmatrix}$ or $\begin{pmatrix} 1 & 0 \\ b & 1 \end{pmatrix}$ for some number b.

Find suitable matrices $\mathbf{E}_1$, $\mathbf{E}_2$, $\mathbf{E}_3$. Describe in words three transformations of the plane which, applied in succession in the correct order, will map the origin on to itself, the point (0, 1) on to (1, 0) and the point (6, 3) on to (0, 1). *(SMP)*

17 When column vectors representing points in a plane with the usual axes are premultiplied by the matrix $\mathbf{M}$, where $\mathbf{M} = \begin{pmatrix} 2 & -1 \\ 1 & 0 \end{pmatrix}$, a transformation of the plane results. Show that, under this transformation, (i) all points on the line $y = x$ are invariant points, (ii) any point on the line $y = x + c$ is transformed to another point on that line. Prove by induction that if $n \in \mathbb{Z}^+$, then $\mathbf{M}^n = \begin{pmatrix} n+1 & -n \\ n & 1-n \end{pmatrix}$. Hence show that, if the transformation represented by $\mathbf{M}$ is applied n times in succession, the line $y + x = 0$ is transformed to the line $y = \dfrac{2n-1}{2n+1}x$. *(C)*

4.3 Determinants and scale factors

The determinant of the matrix of a linear transformation of $\mathbb{R}^n$ has an important geometrical significance. In two dimensions, it is the scale factor for change in area in the plane. In three dimensions it is the scale factor for change in volume. In order to prove this, we first prove part (i) of Theorem 1.8, which we number Theorem 4.4. We shall then look at the determinant of an elementary matrix and the scale factor of an elementary transformation.

Theorem 4.4 Given any $n \times n$ matrices $\mathbf{A}$ and $\mathbf{B}$,

$$\det(\mathbf{AB}) = \det(\mathbf{A}) \det(\mathbf{B}).$$

Proof From Theorem 1.6, $\mathbf{AB}$ is singular if and only if either $\mathbf{A}$ or $\mathbf{B}$ is singular. Now, by Theorem 1.7, a matrix is singular if and only if its determinant is zero. This deals with the zero case since:

$$\begin{aligned} \det(\mathbf{AB}) = 0 &\Leftrightarrow \mathbf{AB} \text{ is singular} \Leftrightarrow \text{either } \mathbf{A} \text{ or } \mathbf{B} \text{ is singular} \\ &\Leftrightarrow \text{either } \det(\mathbf{A}) = 0 \text{ or } \det(\mathbf{B}) = 0. \end{aligned}$$

Hence, if either $\mathbf{A}$ or $\mathbf{B}$ is singular, $\det(\mathbf{AB}) = \det(\mathbf{A})\det(\mathbf{B})$ $(= 0)$. Now assume that both $\mathbf{A}$ and $\mathbf{B}$ are non-singular.

Then, by Theorem 4.3, $\mathbf{A} = \mathbf{E}_k \ldots \mathbf{E}_2\mathbf{E}_1$, a product of elementary matrices, so $\mathbf{AB} = \mathbf{E}_k \ldots \mathbf{E}_2\mathbf{E}_1\mathbf{B}$. Now, by Theorem 1.10, $\det(\mathbf{E}_1\mathbf{B}) = \det(\mathbf{E}_1)\det(\mathbf{B})$, and this forms the basis of an induction proof. As the inductive hypothesis, assume that for some r, $1 \leqslant r < k$, $\det(\mathbf{E}_r \ldots \mathbf{E}_2\mathbf{E}_1\mathbf{B}) = \det(\mathbf{E}_r \ldots \mathbf{E}_2\mathbf{E}_1)\det(\mathbf{B})$.

Then $\det(\mathbf{E}_{r+1}\mathbf{E}_r \ldots \mathbf{E}_2\mathbf{E}_1\mathbf{B}) = \det(\mathbf{E}_{r+1}\{\mathbf{E}_r \ldots \mathbf{E}_2\mathbf{E}_1\mathbf{B}\})$

$$= \det(\mathbf{E}_{r+1})\det(\mathbf{E}_r \ldots \mathbf{E}_2\mathbf{E}_1\mathbf{B})$$
(Theorem 1.10)

$$= \det(\mathbf{E}_{r+1})\det(\mathbf{E}_r \ldots \mathbf{E}_2\mathbf{E}_1)\det(\mathbf{B})$$
(hypothesis)

$$= \det(\mathbf{E}_{r+1}\mathbf{E}_r \ldots \mathbf{E}_2\mathbf{E}_1)\det(\mathbf{B}).$$
(Theorem 1.10)

By mathematical induction, we obtain the result,

$$\det(\mathbf{AB}) = \det(\mathbf{E}_k \ldots \mathbf{E}_2\mathbf{E}_1\mathbf{B}) = \det(\mathbf{E}_k \ldots \mathbf{E}_2\mathbf{E}_1)\det(\mathbf{B}) = \det(\mathbf{A})\det(\mathbf{B}).$$

Elementary transformations and their matrices are of three types. In two dimensions, these are listed in §4.2. They consist of:

1 A one way stretch, with scale factor p ($\neq 0$), in the direction of the ith Cartesian axis. The corresponding elementary matrix is the unit matrix $\mathbf{I}_2$ with 1 replaced by p in the (i, i) entry. The determinant of this matrix is p and the elementary transformation changes area by the factor p, since there is a change of scale in just one direction.

2 A shear with matrix equal to the unit matrix with one of the zero entries, off the main diagonal, changed to r. The matrix has unit determinant and the transformation has no effect on area since the scales in the directions of the Cartesian axes are unchanged.

3 A reflection obtained by interchanging two of the Cartesian axes with matrix equal to the unit matrix with the corresponding rows interchanged. The matrix has determinant -1 and the transformation has no effect on area but the sense (left and right handedness) is changed.

The corresponding situation in three dimensions is just the same except that the word 'area' must everywhere be replaced by 'volume'. The matrices are shown in §1.4. A generalisation to n dimensions is also possible but then 'volume' must mean an 'n-dimensional volume'.

Now consider a linear transformation A, with matrix $\mathbf{A}$, which has the effect of changing 'volume' by a scale factor α. Suppose that an elementary row operation is performed on $\mathbf{A}$ by premultiplying it by the elementary matrix $\mathbf{E}$. Then the transformation EA changes 'volume' first by α and then by $\det(\mathbf{E})$ so the total change is by a scale factor $\det(\mathbf{E})\alpha$. An induction argument, similar to that used in the proof of Theorem 4.4, shows that the scale change due to a succession of elementary transformations, with matrices $\mathbf{E}_k, \ldots, \mathbf{E}_2, \mathbf{E}_1$, is $\det(\mathbf{E}_k \ldots \mathbf{E}_2\mathbf{E}_1)$. Now by Theorem 4.3, any non-singular matrix $\mathbf{M}$ can be written as a product of elementary matrices, that is, $\mathbf{M} = \mathbf{E}_k \ldots \mathbf{E}_2\mathbf{E}_1$. Therefore the matrix $\mathbf{M}$ has the effect of changing 'volume' by a scale factor $\det(\mathbf{M})$ and we have proved the theorem:

Theorem 4.5 If $\mathbf{M}$ is a non-singular $n \times n$ matrix, then the linear transformation with matrix $\mathbf{M}$ changes the 'volume' of $\mathbb{R}^n$ by a scale factor $\det(\mathbf{M})$.

This theorem is illustrated by considering the basic transformations of the plane, given in §4.1, and studying the way in which they change the shape of the unit square.

Reflections and rotations are both isometries so they keep the shape of an object invariant. A rotation is a *direct* isometry, preserving sense, and a reflection is an *opposite* isometry, flipping the shape over and reversing the sense. In both cases, lengths, angles and areas are preserved and this is indicated by the determinants of their matrices.

The general rotation through the angle α has matrix $\mathbf{M}$, given by

$$\mathbf{M} = \begin{pmatrix} \cos\alpha & -\sin\alpha \\ \sin\alpha & \cos\alpha \end{pmatrix}, \text{ and } \det(\mathbf{M}) = \cos^2\alpha + \sin^2\alpha = 1.$$

The general reflection in the line $y = x\tan\alpha$ has matrix $\mathbf{M}$, where

$$\mathbf{M} = \begin{pmatrix} \cos 2\alpha & \sin 2\alpha \\ \sin 2\alpha & -\cos 2\alpha \end{pmatrix} \quad \text{and} \quad \det(\mathbf{M}) = -\cos^2 2\alpha - \sin^2 2\alpha = -1,$$

indicating no change in area but a change in sense.

A shear preserves lengths parallel to the fixed line whilst all other lengths and angles are changed. However the area of any shape is preserved since there is no scale change perpendicular to the fixed line. The general shears have matrices $\begin{pmatrix} 1 & r \\ 0 & 1 \end{pmatrix}$ or $\begin{pmatrix} 1 & 0 \\ r & 1 \end{pmatrix}$, with determinant 1.

A one way stretch preserves length parallel to the fixed line and changes the linear scale in the stretch direction by the scale factor p. Area is enlarged by the same scale factor. The general one way stretches have matrices $\begin{pmatrix} p & 0 \\ 0 & 1 \end{pmatrix}$ or $\begin{pmatrix} 1 & 0 \\ 0 & p \end{pmatrix}$, with determinant p. The matrices of enlargements and two way stretches are of the forms $\begin{pmatrix} p & 0 \\ 0 & p \end{pmatrix}$ and $\begin{pmatrix} p & 0 \\ 0 & q \end{pmatrix}$. Their determinants, p^2 and pq, give the scale factors for change in area.

EXAMPLE *The transformation* T *of the plane has matrix* $\mathbf{M}$, *given by* $\mathbf{M} = \begin{pmatrix} a & b \\ c & d \end{pmatrix}$ *and* $\det(\mathbf{M}) \neq 0$. *Show on a diagram the transformed position of the unit square and use this to show that the area scale factor is* $\det(\mathbf{M})$.

Let the unit square $OABC$ be transformed to the parallelogram $OA'B'C'$ and draw this on a diagram. In Fig. 4.13, lines are drawn through the vertices of the parallelogram, parallel to the Cartesian axes, and the lengths of the various line segments are indicated. The large rectangle is made up from the parallelogram, two small rectangles and four triangles so that its area is given by

$$(a+b)(c+d) = S + 2bc + 2(ac/2) + 2(bd/2)$$

where S is the area of the parallelogram, that is, the area scale factor. Hence $S = (a+b)(c+d) - 2bc - ac - bd = ad - bc = \det(\mathbf{M})$.

Note The diagram in Fig. 4.13 has been drawn with all four constants a, b, c and d positive. If any of these constants are negative or zero, the same argument holds but a different diagram is needed.

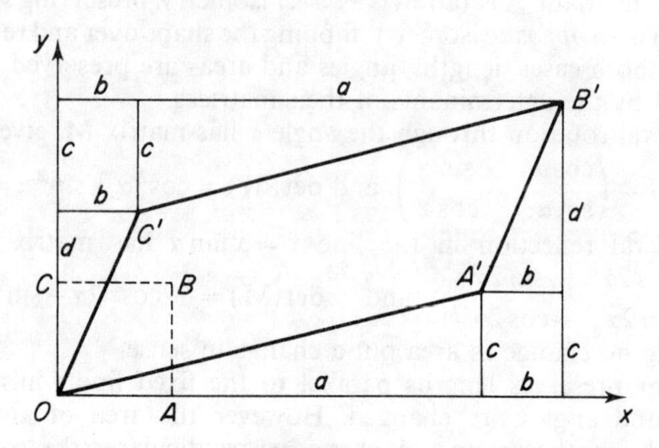

Fig. 4.13

EXERCISE 4.3

1 Calculate the determinant of the matrices and hence determine which reverse sense and which preserve sense:

(i) $\begin{pmatrix} \frac{3}{5} & \frac{4}{5} \\ \frac{4}{5} & -\frac{3}{5} \end{pmatrix}$, (ii) $\begin{pmatrix} \frac{4}{5} & -\frac{3}{5} \\ \frac{3}{5} & \frac{4}{5} \end{pmatrix}$, (iii) $\begin{pmatrix} \frac{1}{2} & -\frac{\sqrt{3}}{2} & 0 \\ \frac{\sqrt{3}}{2} & \frac{1}{2} & 0 \\ 0 & 0 & 1 \end{pmatrix}$,

(iv) $\begin{pmatrix} \frac{1}{\sqrt{2}} & 0 & \frac{1}{\sqrt{2}} \\ 0 & 1 & 0 \\ \frac{1}{\sqrt{2}} & 0 & -\frac{1}{\sqrt{2}} \end{pmatrix}$, (v) $\begin{pmatrix} 0 & -1 & 0 \\ -1 & 0 & 0 \\ 0 & 0 & 1 \end{pmatrix}$, (vi) $\begin{pmatrix} 0 & 0 & 1 \\ 0 & -1 & 0 \\ -1 & 0 & 0 \end{pmatrix}$,

(vii) $\begin{pmatrix} 0 & -1 & 0 \\ 0 & 0 & 1 \\ -1 & 0 & 0 \end{pmatrix}$, (viii) $\begin{pmatrix} 0 & 0 & -1 \\ 0 & -1 & 0 \\ -1 & 0 & 0 \end{pmatrix}$.

2 For the following pairs of matrices, check that $\det(\mathbf{AB}) = \det(\mathbf{A})\det(\mathbf{B}) = \det(\mathbf{BA})$:

(i) $\begin{pmatrix} 2 & 1 \\ 3 & 2 \end{pmatrix}, \begin{pmatrix} 5 & -1 \\ -8 & 2 \end{pmatrix}$, (ii) $\begin{pmatrix} 0 & 3 \\ 5 & 2 \end{pmatrix}, \begin{pmatrix} 1 & 0 \\ 1 & -2 \end{pmatrix}$,

(iii) $\begin{pmatrix} 2 & -1 & 3 \\ -1 & 1 & 0 \\ 0 & 1 & 2 \end{pmatrix}, \begin{pmatrix} 0 & -1 & 2 \\ -4 & 2 & 1 \\ 3 & 0 & 1 \end{pmatrix}$.

3 The determinant of a shear is 1 and the determinant of a reflection is -1. The determinant of a one way stretch is p (the scale factor). Use the decomposition of a non-singular matrix into a product of elementary matrices, given in §4.2, to prove that $\det\begin{pmatrix} a & b \\ c & d \end{pmatrix} = ad - bc$.

4 Draw a new diagram for the example above in the case where a and c are negative and b and d are positive. Verify that the proof is still valid.

5 The matrix $\mathbf{M}$, where $\mathbf{M} = \begin{pmatrix} a & b \\ c & d \end{pmatrix}$, ($a, b, c, d$ real and non-zero) and its transpose $\mathbf{M}^{\mathrm{T}}$ satisfy $\mathbf{M}^{\mathrm{T}}\mathbf{M} = \begin{pmatrix} k^2 & 0 \\ 0 & k^2 \end{pmatrix}$, ($k$ real and non-zero). If $\begin{pmatrix} X \\ Y \end{pmatrix} = \mathbf{M}\begin{pmatrix} x \\ y \end{pmatrix}$, show that $X^2 + Y^2 = k^2(x^2 + y^2)$. Show also that (i) $a^2 - d^2 = b^2 - c^2$, (ii) $\dfrac{a^2}{d^2} = \dfrac{c^2}{b^2}$, and deduce that $a^2 = d^2$ and $b^2 = c^2$.

Write down the two matrices $\mathbf{M}_1$, $\mathbf{M}_2$ of the above form for which $a = 3$, $b = 4$. Describe the geometric transformation represented by either $\mathbf{M}_1$ or $\mathbf{M}_2$, interpreting the significance of the determinant of the matrix.

(*JMB*)

4.4 Solution of simultaneous linear equations

In Chapter 1, a system of n linear equations in n unknown variables, $x_1, x_2, \ldots, x_n$, was represented in a matrix form

$$\begin{pmatrix} a_{11} & a_{12} & \cdots & a_{1n} \\ a_{21} & a_{22} & \cdots & a_{2n} \\ \cdot & \cdot & \cdots & \cdot \\ \cdot & \cdot & \cdots & \cdot \\ \cdot & \cdot & \cdots & \cdot \\ a_{n1} & a_{n2} & \cdots & a_{nn} \end{pmatrix} \begin{pmatrix} x_1 \\ x_2 \\ \cdot \\ \cdot \\ \cdot \\ x_n \end{pmatrix} = \begin{pmatrix} b_1 \\ b_2 \\ \cdot \\ \cdot \\ \cdot \\ b_n \end{pmatrix},$$

that is, $\mathbf{Ax} = \mathbf{b}$, where $\mathbf{A}$ is an $n \times n$ matrix of coefficients and $\mathbf{x}$ and $\mathbf{b}$ are $n \times 1$ column vectors. The method of solution, given in Chapter 1, was to use the inverse matrix $\mathbf{A}^{-1}$, when this existed, and to premultiply each side of the equation by $\mathbf{A}^{-1}$ to obtain the solution $\mathbf{x} = \mathbf{A}^{-1}\mathbf{Ax} = \mathbf{A}^{-1}\mathbf{b}$.

In this section, we develop a more satisfactory method of handling such systems of equations which will work even when $\mathbf{A}$ is singular and has no inverse. The method is easily extended to systems of m equations in n unknowns when $\mathbf{A}$ is an $m \times n$ matrix, $\mathbf{x}$ is an $n \times 1$ vector and $\mathbf{b}$ is an $m \times 1$ vector. The solution for $\mathbf{x}$ in $\mathbb{R}^n$ may not always be a unique point but may represent a line or a plane or, alternatively, there may be no solution at all. The method relies on row reduction of a matrix by means of multiplication by elementary matrices and the fact that the set of linear equations so produced are equivalent to the original set. This is because any elementary matrix $\mathbf{E}$ has an inverse $\mathbf{E}^{-1}$ and so every elementary row operation is reversible.

$$\mathbf{EAx} = \mathbf{Eb} \Leftrightarrow \mathbf{Ax} = \mathbf{b}\,(\Leftrightarrow \mathbf{E}^{-1}\mathbf{EAx} = \mathbf{E}^{-1}\mathbf{Eb}),$$

so the solution sets of the two equations are the same.

In §4.2, the inverse of a matrix was found by reducing a matrix to the unit matrix. In the current work, we only use half of that process and reduce the matrix to upper triangular form.

Definition A matrix $\mathbf{A}$, (a_{ij}), is *reduced upper triangular* if all the terms below the main diagonal are zero and the first non-zero term in any row is 1, that is the two conditions are satisfied:
(i) $a_{ij} = 0$ for $i > j$.
(ii) if $a_{ij} \neq 0$ and, for $1 \leqslant k < j$, $a_{ik} = 0$, then $a_{ij} = 1$.

We begin by taking a close look at the two dimensional case when the matrix $\mathbf{A}$ is singular.

EXAMPLE 1 *Solve the equations:* (i) $\begin{pmatrix} 2 & 6 \\ 1 & 4 \end{pmatrix}\begin{pmatrix} x \\ y \end{pmatrix} = \begin{pmatrix} 2 \\ 2 \end{pmatrix}$,

(ii) $\begin{pmatrix} 2 & 6 \\ 1 & 3 \end{pmatrix}\begin{pmatrix} x \\ y \end{pmatrix} = \begin{pmatrix} 4 \\ 2 \end{pmatrix}$, *(iii)* $\begin{pmatrix} 2 & 6 \\ 1 & 3 \end{pmatrix}\begin{pmatrix} x \\ y \end{pmatrix} = \begin{pmatrix} 5 \\ 2 \end{pmatrix}$.

(i) Det $\begin{pmatrix} 2 & 6 \\ 1 & 4 \end{pmatrix} = 2 \times 4 - 6 = 2$, so the inverse matrix $\mathbf{A}^{-1} = \begin{pmatrix} 2 & -3 \\ -1/2 & 1 \end{pmatrix}$ and

the solution is $\begin{pmatrix} x \\ y \end{pmatrix} = \begin{pmatrix} 2 & -3 \\ -1/2 & 1 \end{pmatrix}\begin{pmatrix} 2 \\ 2 \end{pmatrix} = \begin{pmatrix} -2 \\ 1 \end{pmatrix}$. The transformation with matrix $\mathbf{A}$ has an inverse and the point $(-2, 1)$ has image $(2, 2)$. The row reduction method is to reduce the matrix $(\mathbf{A}\,|\,\mathbf{b})$ to a reduced upper triangular form by means of elementary row operations. The matrix $(\mathbf{A}\,|\,\mathbf{b})$ is the matrix $\mathbf{A}$ augmented by adding $\mathbf{b}$ as an extra column. The method is now shown:

$(\mathbf{A}\,|\,\mathbf{b}) = \left(\begin{array}{cc|c} 2 & 6 & 2 \\ 1 & 4 & 2 \end{array}\right)$, $\mathbf{r}_1 \leftarrow \mathbf{r}_1/2$ gives $\left(\begin{array}{cc|c} 1 & 3 & 1 \\ 1 & 4 & 2 \end{array}\right)$, $\mathbf{r}_2 \leftarrow \mathbf{r}_2 - \mathbf{r}_1$ gives $\left(\begin{array}{cc|c} 1 & 3 & 1 \\ 0 & 1 & 1 \end{array}\right)$, so the equations become $x + 3y = 1$ and $y = 1$, and so $x = 1 - 3 = -2$, giving the solution $\mathbf{x = -2,\ y = 1}$.

(ii) $\begin{pmatrix} 2 & 6 \\ 1 & 3 \end{pmatrix}$ has zero determinant and hence no inverse. The point (x, y) is mapped to $(2x + 6y, x + 3y)$ which lies on the line $2y = x$. Thus the transformation is a squash with the plane being squashed on to the line $2y = x$. Since $\begin{pmatrix} 4 \\ 2 \end{pmatrix}$ is the position vector of the point $(4, 2)$ which lies on the line on to which the plane is squashed, there is a solution, in fact, there is an infinite number of solutions all satisfying the equation $x + 3y = 2$. The reason is that the other equation $2x + 6y = 4$ is a multiple of this equation. There is a line of solutions which can be given in terms of a parameter t by $y = t$, $x = 2 - 3t$. Using the row reduction method:

$\left(\begin{array}{cc|c} 2 & 6 & 4 \\ 1 & 3 & 2 \end{array}\right)$ reduces to $\left(\begin{array}{cc|c} 1 & 3 & 2 \\ 0 & 0 & 0 \end{array}\right)$, giving a row of zeros. So there is only one equation

and the solution is given by $\begin{pmatrix} x \\ y \end{pmatrix} = \begin{pmatrix} 2-3t \\ t \end{pmatrix} = \begin{pmatrix} 2 \\ 0 \end{pmatrix} + t\begin{pmatrix} -3 \\ 1 \end{pmatrix}$. This is a line

which maps on to the point (4, 2) which lies on the squashing line $2y = x$.
(iii) As in case (ii), the range is the squashing line $2y = x$. Since the point (5, 2)

does not lie on the line, there is **no solution**. By row reduction, $\left(\begin{array}{cc|c} 2 & 6 & 5 \\ 1 & 3 & 2 \end{array}\right)$ leads to

$\left(\begin{array}{cc|c} 1 & 3 & 5/2 \\ 0 & 0 & 1 \end{array}\right)$ which gives a contradiction $0 = 1$ from the second row. This corresponds to the fact that the two original equations $2x + 6y = 5$ and $x + 3y = 2$ represent a pair of parallel lines with no intersection.

The various types of solutions in the 2×2 case are summarised below. In the matrices $*$ is used for an entry which can be any number.

Geometrical interpretation Row reduction of the augmented matrix $(\mathbf{A}|\mathbf{b})$ (to solve the pair of linear equations $\mathbf{Ax} = \mathbf{b}$) can lead to three possible cases. The form of the reduced matrix may be:

(i) $\left(\begin{array}{cc|c} 1 & * & * \\ 0 & 1 & * \end{array}\right)$, in which case $\mathbf{A}$ is non-singular, there is a single point solution, the transformation A with matrix $\mathbf{A}$ is one-one and the geometrical interpretation is that the two equations represent non-parallel lines which intersect at just one point, the solution point,

(ii) $\left(\begin{array}{cc|c} 1 & * & * \\ 0 & 0 & 0 \end{array}\right)$, in which case $\mathbf{A}$ is singular, there is a line of solutions, the transformation A with matrix $\mathbf{A}$ squashes the plane on to a line, and the point with position vector $\mathbf{b}$ lies on this line. The geometrical interpretation is that the two equations represent the same line which is the line of solutions,

(iii) $\left(\begin{array}{cc|c} 1 & * & * \\ 0 & 0 & 1 \end{array}\right)$, and, as in (ii), $\mathbf{A}$ is singular, the transformation A squashes the plane on to a line and the point with position vector $\mathbf{b}$ does not lie on this line so there is no solution. The geometrical interpretation is that the two equations represent parallel distinct lines which have no common point.

The row reduction method is easily extended to the 3×3 case as is shown in the next example.

EXAMPLE 2 *Solve the equations*: (*i*) $\begin{pmatrix} 2 & -1 & 3 \\ 1 & 3 & -2 \\ 4 & -2 & 1 \end{pmatrix}\begin{pmatrix} x \\ y \\ z \end{pmatrix} = \begin{pmatrix} 1 \\ 4 \\ -8 \end{pmatrix}$,

(*ii*) $\begin{pmatrix} 2 & -1 & 3 \\ 1 & 3 & -2 \\ 4 & -2 & 6 \end{pmatrix}\begin{pmatrix} x \\ y \\ z \end{pmatrix} = \begin{pmatrix} 1 \\ 4 \\ 2 \end{pmatrix}$, (*iii*) $\begin{pmatrix} 2 & -1 & 3 \\ 1 & 3 & -2 \\ 4 & -2 & 6 \end{pmatrix}\begin{pmatrix} x \\ y \\ z \end{pmatrix} = \begin{pmatrix} 1 \\ 4 \\ 0 \end{pmatrix}$.

We show the row reduction on all three augmented matrices at once since the same elementary row operations are used.

(i) (ii) (iii)

$$\left(\begin{array}{ccc|c} 2 & -1 & 3 & 1 \\ 1 & 3 & -2 & 4 \\ 4 & -2 & 1 & -8 \end{array}\right), \quad \left(\begin{array}{ccc|c} 2 & -1 & 3 & 1 \\ 1 & 3 & -2 & 4 \\ 4 & -2 & 6 & 2 \end{array}\right), \quad \left(\begin{array}{ccc|c} 2 & -1 & 3 & 1 \\ 1 & 3 & -2 & 4 \\ 4 & -2 & 6 & 0 \end{array}\right).$$

$\mathbf{r}_1 \leftarrow \mathbf{r}_2$, $\mathbf{r}_2 \leftarrow \mathbf{r}_1$,

$$\left(\begin{array}{ccc|c} 1 & 3 & -2 & 4 \\ 2 & -1 & 3 & 1 \\ 4 & -2 & 1 & -8 \end{array}\right), \quad \left(\begin{array}{ccc|c} 1 & 3 & -2 & 4 \\ 2 & -1 & 3 & 1 \\ 4 & -2 & 6 & 2 \end{array}\right), \quad \left(\begin{array}{ccc|c} 1 & 3 & -2 & 4 \\ 2 & -1 & 3 & 1 \\ 4 & -2 & 6 & 0 \end{array}\right).$$

$\mathbf{r}_2 \leftarrow \mathbf{r}_2 - 2\mathbf{r}_1$, $\mathbf{r}_3 \leftarrow \mathbf{r}_3 - 4\mathbf{r}_1$,

$$\left(\begin{array}{ccc|c} 1 & 3 & -2 & 4 \\ 0 & -7 & 7 & -7 \\ 0 & -14 & 9 & -24 \end{array}\right), \quad \left(\begin{array}{ccc|c} 1 & 3 & -2 & 4 \\ 0 & -7 & 7 & -7 \\ 0 & -14 & 14 & -14 \end{array}\right), \quad \left(\begin{array}{ccc|c} 1 & 3 & -2 & 4 \\ 0 & -7 & 7 & -7 \\ 0 & -14 & 14 & -16 \end{array}\right).$$

$\mathbf{r}_2 \leftarrow \mathbf{r}_2/(-7)$, $\mathbf{r}_3 \leftarrow \mathbf{r}_3 - 2\mathbf{r}_2$,

$$\left(\begin{array}{ccc|c} 1 & 3 & -2 & 4 \\ 0 & 1 & -1 & 1 \\ 0 & 0 & -5 & -10 \end{array}\right), \quad \left(\begin{array}{ccc|c} 1 & 3 & -2 & 4 \\ 0 & 1 & -1 & 1 \\ 0 & 0 & 0 & 0 \end{array}\right), \quad \left(\begin{array}{ccc|c} 1 & 3 & -2 & 4 \\ 0 & 1 & -1 & 1 \\ 0 & 0 & 0 & -2 \end{array}\right).$$

$\mathbf{r}_3 \leftarrow \mathbf{r}_3/(-5)$, $\mathbf{r}_3 \leftarrow \mathbf{r}_3/(-2)$,

$$\left(\begin{array}{ccc|c} 1 & 3 & -2 & 4 \\ 0 & 1 & -1 & 1 \\ 0 & 0 & 1 & 2 \end{array}\right), \quad \left(\begin{array}{ccc|c} 1 & 3 & -2 & 4 \\ 0 & 1 & -1 & 1 \\ 0 & 0 & 0 & 0 \end{array}\right), \quad \left(\begin{array}{ccc|c} 1 & 3 & -2 & 4 \\ 0 & 1 & -1 & 1 \\ 0 & 0 & 0 & 1 \end{array}\right).$$

Now deal with each case.

(i) Back substitution gives, from $\mathbf{r}_3$, $z = 2$, from $\mathbf{r}_2$, $y - z = 1$ so $y = z + 1 = 3$, from $\mathbf{r}_1$, $x + 3y - 2z = 4$ so $x = 4 + 2z - 3y = -1$. The solution is unique, $\begin{pmatrix} x \\ y \\ z \end{pmatrix} = \begin{pmatrix} -1 \\ 3 \\ 2 \end{pmatrix}$, a point with coordinates $(-1, 3, 2)$.

(ii) The row of zeros implies that only two of the three original equations are independent. The three equations are consistent and, from the first two rows of the reduced matrix,

$$x+3y-2z = 4, \; y-z = 1.$$

In terms of a parameter t,

$$z = t, \; y = 1+t, \; x = 4+2z-3y, \text{ so } x = 1-t.$$

The solution is $\begin{pmatrix} x \\ y \\ z \end{pmatrix} = \begin{pmatrix} 1-t \\ 1+t \\ t \end{pmatrix} = \begin{pmatrix} 1 \\ 1 \\ 0 \end{pmatrix} + t\begin{pmatrix} -1 \\ 1 \\ 1 \end{pmatrix}$, a line with Cartesian equations $1-x = y-1 = z$.

(iii) The third row gives the equation $0x+0y+0z = 1$, which is inconsistent, so the original three equations are inconsistent and there is **no solution**.

Notes 1 After the reduction to reduced upper triangular form, the solution can be read from the matrix by back substitution, first z, then y, then x.

2 It may be necessary to interchange rows in order to get a non-zero entry in the top left position and it can also be useful to interchange rows to get a 1 in this position (to save fractions).

3 It is not essential to make the first non-zero entry in each row equal to 1 and it may be better to leave it as a number larger than 1 in order to avoid fractions during the reduction process.

4 The presence of the first three zeros in the third row, in cases (ii) and (iii), indicates that $\det(\mathbf{A}) = 0$ and **A** is a squashing (singular) transformation.

Geometrical interpretation The equation $\mathbf{Ax} = \mathbf{b}$ corresponds to three scalar equations $\mathbf{r}_i\mathbf{x} = b_i$, $1 \leqslant i \leqslant 3$, where $\mathbf{r}_i$ is the ith row of **A** and b_i is the ith entry of **b**. Each of these equations represents a plane with normal vector $\mathbf{r}_i$. The solution of the equations corresponds to finding the intersection of the three planes. By inspection of the normal vectors (the rows of **A**) it is easy to see if any of the planes are parallel.

As in the 2×2 case, we list the different possibilities that can occur and, in the reduced upper triangular form of the augmented matrix, indicate arbitrary constants by $*$. The augmented matrix is a 3×4 matrix, obtained from $(\mathbf{A}|\mathbf{b})$ by elementary row operations, and its reduced form can be:

(i) $\left(\begin{array}{ccc|c} 1 & * & * & * \\ 0 & 1 & * & * \\ 0 & 0 & 1 & * \end{array}\right)$, in which case **A** is non-singular, there is just one point solution, found by back substitution, as in example 2 (i). The transformation A with matrix **A** is one-one and the geometrical interpretation is that the three equations represent non-parallel planes which intersect at just one point, the solution point, as shown in Fig. 4.14 (i).

(ii) $\left(\begin{array}{ccc|c} 1 & * & * & * \\ 0 & 1 & * & * \\ 0 & 0 & 0 & 0 \end{array}\right)$, in which case **A** is singular, there is a line of solutions. The transformation A with matrix **A** squashes $\mathbb{R}^3$ on to a plane, and the point with position vector **b** lies on this plane. The three rows of the augmented matrix are dependent so that the three normals are coplanar. Either (a) the three planes meet in the solution line forming a sheaf of planes, or (b) two planes are identical and meet the third in a line. The two alternatives are distinguished by reference to the original equations, Fig. 4.14 (ii).

(iii) $\left(\begin{array}{ccc|c} 1 & * & * & * \\ 0 & 1 & * & * \\ 0 & 0 & 0 & 1 \end{array}\right)$, and again **A** is singular. However, there is no solution, since the third row gives an inconsistency $0 = 1$. The transformation A with matrix **A** squashes $\mathbb{R}^3$ on to a plane, and the point with position vector **b** does not lie on this plane. The three rows of the matrix **A** are dependent so that the three normals are coplanar. There are two sub-cases, shown in Fig. 4.14 (iii):

(a) the normals to the three planes are all distinct and the planes meet in pairs, forming a prism,

(b) two of the normals are parallel, two of the planes are parallel and each meets the third plane in a line,

(iv) $\left(\begin{array}{ccc|c} 1 & * & * & * \\ 0 & 0 & 0 & 0 \\ 0 & 0 & 0 & 0 \end{array}\right)$, **A** is singular with only one independent row in the augmented matrix. The transformation A with matrix **A** squashes $\mathbb{R}^3$ on to a line, and the point with position vector **b** lies on this line. The three equations all represent the same plane which is the plane of solutions, Fig. 4.14 (iv),

(v) $\left(\begin{array}{ccc|c} 1 & * & * & * \\ 0 & 0 & 0 & 1 \\ 0 & 0 & 0 & 0 \end{array}\right)$, **A** is singular and there is no solution since the second row gives an inconsistency $0 = 1$. The transformation A with matrix **A** squashes $\mathbb{R}^3$ on to a line, and the point with position vector **b** does not lie on this line. The three planes are parallel, but not identical, and either (a) there are three distinct parallel planes or (b) two planes are identical and parallel to the third. The two alternatives may be distinguished by considering the original equations. See Fig. 4.14 (v) (a) and (b).

Reduction of a matrix to echelon form

The methods of solving simultaneous equations, described above, can be generalised to any number of equations in any number of unknowns. The algorithm used to prove Theorem 4.3 can easily be modified to deal with non-singular matrices and also with non-square matrices. The augmented matrix is reduced to an upper triangular form in which all entries below the leading diagonal are zero. Each row begins with more zero entries than

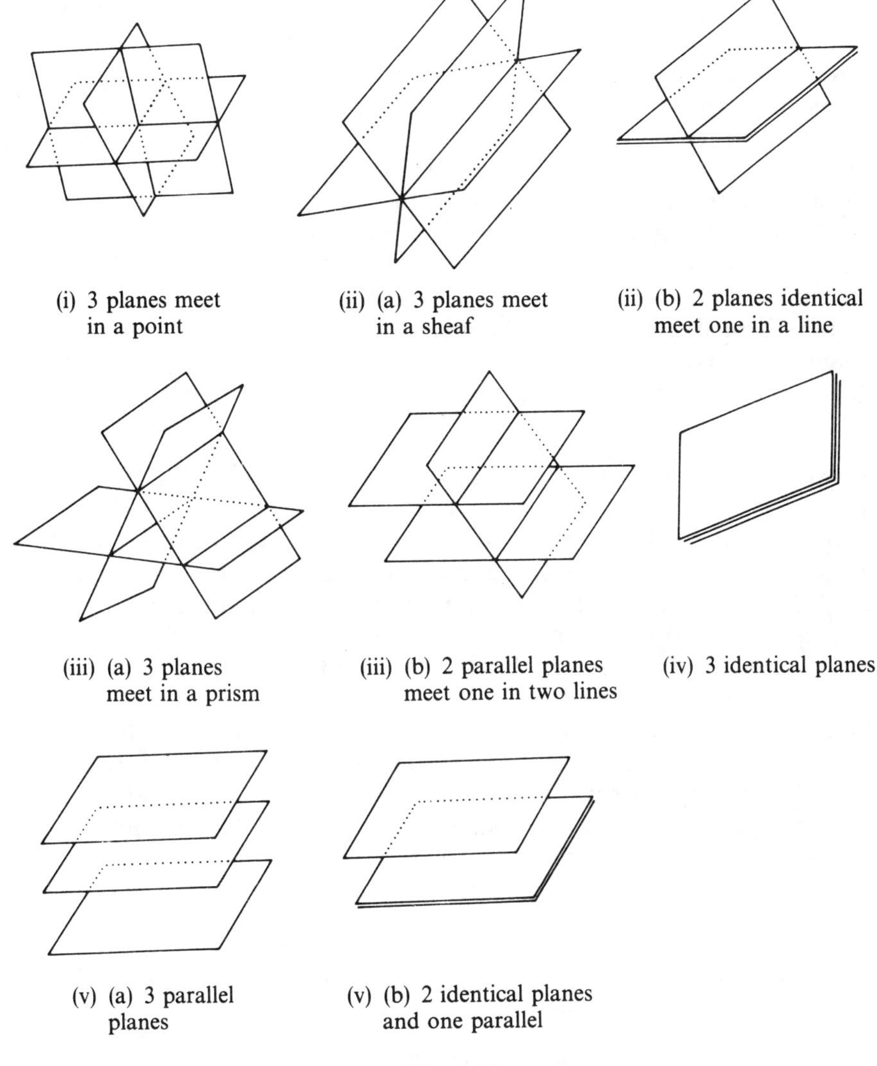

(i) 3 planes meet in a point

(ii) (a) 3 planes meet in a sheaf

(ii) (b) 2 planes identical meet one in a line

(iii) (a) 3 planes meet in a prism

(iii) (b) 2 parallel planes meet one in two lines

(iv) 3 identical planes

(v) (a) 3 parallel planes

(v) (b) 2 identical planes and one parallel

Fig. 4.14

the previous row, except for the first row and for those rows lying below full rows of zeros. The form of the matrix is then called an *echelon* form and, after a formal definition, we give the modified algorithm and the theorem.

Definition An $m \times n$ matrix $\mathbf{A}$, $\mathbf{A} = (a_{ij})$ is in echelon form means that: if a_{ij} is the first non-zero entry in row i so that

$$a_{ij} \neq 0 \text{ and, for } 1 \leqslant t < j,\ a_{it} = 0,$$

$$\text{then } a_{st} = 0 \text{ for } 1 \leqslant t \leqslant j \text{ and } i+1 \leqslant s \leqslant m.$$

This means that, if $\mathbf{A}$ is in echelon form, and a_{ij} is the first non-zero entry in row i, then all the entries of $\mathbf{A}$ which lie to the left and/or below the entry a_{ij} are zero. In particular, $\mathbf{A}$ is an upper diagonal matrix, with possibly some zeros on the leading diagonal.

Let $\mathbf{A}$ be any $m \times n$ matrix. Then the following algorithm converts $\mathbf{A}$ into echelon form by means of elementary row operations. Construct a sequence of matrices $\mathbf{A}_1, \mathbf{A}_2, \ldots, \mathbf{A}_k$, such that $\mathbf{A}_1 = \mathbf{A}$ and $\mathbf{A}_{t+1}$ is constructed from $\mathbf{A}_t$ as follows: if the first column of $\mathbf{A}_t$ contains a non-zero entry a, bring this to the top left hand corner by interchanging rows and make every other term in the first column zero by subtracting multiples of the first row. Then $\mathbf{A}_{t+1}$ is the submatrix of the resulting matrix formed by removing its first row and first column. If the first column of $\mathbf{A}_t$ is a zero vector then $\mathbf{A}_{t+1}$ is the submatrix of $\mathbf{A}_t$ consisting of $\mathbf{A}_t$ less its first column. The process ends when $\mathbf{A}_t$ has only one row.

Formal Algorithm Until the matrix $\mathbf{A}_t$ has just one row, let $\mathbf{B} = \mathbf{A}_t$, $\mathbf{B} = (b_{ij})$:
Step 1 If the first column of $\mathbf{B}$ is the zero vector, go to step 5.
Step 2 If $b_{11} = 0$, make this entry non-zero by interchanging rows.
Step 3 Make all other entries in the first column of $\mathbf{B}$ zero by subtracting multiples of row 1 from the other rows.
Step 4 Form a new matrix $\mathbf{B}$ (with zero first column) by removing the first row from $\mathbf{B}$.
Step 5 If $\mathbf{B}$ has more than one row then $\mathbf{A}_{t+1}$ is the matrix formed by removing the first column (a zero column) from $\mathbf{B}$, $\mathbf{B} = \mathbf{A}_{t+1}$, go to Step 1.
Step 6 $\mathbf{B}$ is now a row vector, end.

The above algorithm is a constructive way of transforming any matrix to echelon form by row operations, that is by premultiplying the matrix by a succession of elementary matrices. The algorithm proves the theorem:

Theorem 4.6 If $\mathbf{M}$ is any $m \times n$ matrix, then there exist elementary $m \times m$ matrices $\mathbf{E}_i$, $1 \leqslant i \leqslant k$, such that

$$\text{if } \mathbf{E} = \mathbf{E}_k \ldots \mathbf{E}_2 \mathbf{E}_1, \text{ then } \mathbf{EM} \text{ is in echelon form.}$$

The algorithm is best understood by following through an example.

EXAMPLE 3 *Use the method of row reduction to reduce the matrix* $\mathbf{A}$ *to echelon form, where* $\mathbf{A} = \begin{pmatrix} 0 & 2 & 3 & 3 \\ 1 & -1 & 0 & -1 \\ 4 & -2 & 3 & 1 \end{pmatrix}$.

As in Example 3 of §4.2, in order to help the reader follow the work and to recognise the various steps of the algorithm, at each stage we give the value of t, the matrix $\mathbf{B}(= \mathbf{A}_t)$ and the step number from the algorithm.

Reduced matrix	t	$\mathbf{B}\,(=\mathbf{A}_t)$	
$\begin{pmatrix} 0 & 2 & 3 & 3 \\ 1 & -1 & 0 & -1 \\ 4 & -2 & 3 & 1 \end{pmatrix}$	1	$\begin{pmatrix} 0 & 2 & 3 & 3 \\ 1 & -1 & 0 & -1 \\ 4 & -2 & 3 & 1 \end{pmatrix}$	Step 2 interchange rows 1 and 2
$\begin{pmatrix} 1 & -1 & 0 & -1 \\ 0 & 2 & 3 & 3 \\ 4 & -2 & 3 & 1 \end{pmatrix}$	1	$\begin{pmatrix} 1 & -1 & 0 & -1 \\ 0 & 2 & 3 & 3 \\ 4 & -2 & 3 & 1 \end{pmatrix}$	Step 3 subtract 4 times row 1 from row 3
$\begin{pmatrix} 1 & -1 & 0 & -1 \\ 0 & 2 & 3 & 3 \\ 0 & 2 & 3 & 5 \end{pmatrix}$	1	$\begin{pmatrix} 1 & -1 & 0 & -1 \\ 0 & 2 & 3 & 3 \\ 0 & 2 & 3 & 5 \end{pmatrix}$	Step 4 delete first row of **B**
$\begin{pmatrix} 1 & -1 & 0 & -1 \\ 0 & 2 & 3 & 3 \\ 0 & 2 & 3 & 5 \end{pmatrix}$	1	$\begin{pmatrix} 0 & 2 & 3 & 3 \\ 0 & 2 & 3 & 5 \end{pmatrix}$	Step 5 delete first column of **B**, to form $\mathbf{A}_2$
$\begin{pmatrix} 1 & -1 & 0 & -1 \\ 0 & 2 & 3 & 3 \\ 0 & 2 & 3 & 5 \end{pmatrix}$	2	$\begin{pmatrix} 2 & 3 & 3 \\ 2 & 3 & 5 \end{pmatrix}$	Step 3 subtract the first row of **B** from the second
$\begin{pmatrix} 1 & -1 & 0 & -1 \\ 0 & 2 & 3 & 3 \\ 0 & 0 & 0 & 2 \end{pmatrix}$	2	$\begin{pmatrix} 2 & 3 & 3 \\ 0 & 0 & 2 \end{pmatrix}$	Step 4 delete first row of **B**
$\begin{pmatrix} 1 & -1 & 0 & -1 \\ 0 & 2 & 3 & 3 \\ 0 & 0 & 0 & 2 \end{pmatrix}$	2	$\begin{pmatrix} 0 & 0 & 2 \end{pmatrix}$	Step 6 **B** has only one row, $\mathbf{A}_2$ is in echelon form.

Note When writing out a solution, only the left hand column of matrices is needed, and then, only those matrices which differ from the line above. However, it is also useful to keep a note of the row operations used, as given in the right hand column.

EXAMPLE 4 *For what value of k does the set of simultaneous equations*

$$\begin{aligned} 2y + 3z &= 3, \\ x - y \quad &= -1, \\ 4x - 2y + 3z &= k, \end{aligned}$$

have a solution. Solve the equations for this value of k.

The augmented matrix is $\mathbf{A}$, where $\mathbf{A} = \left(\begin{array}{ccc|c} 0 & 2 & 3 & 3 \\ 1 & -1 & 0 & -1 \\ 4 & -2 & 3 & k \end{array}\right)$ and, except for the entry k, this is the matrix of example 3. Therefore, we use the same row operations to reduce it to echelon form, thus.

$$\left(\begin{array}{ccc|c} 0 & 2 & 3 & 3 \\ 1 & -1 & 0 & -1 \\ 4 & -2 & 3 & k \end{array}\right) \to \left(\begin{array}{ccc|c} 1 & -1 & 0 & -1 \\ 0 & 2 & 3 & 3 \\ 4 & -2 & 3 & k \end{array}\right) \to \left(\begin{array}{ccc|c} 1 & -1 & 0 & -1 \\ 0 & 2 & 3 & 3 \\ 0 & 2 & 3 & 4+k \end{array}\right)$$

$$\to \left(\begin{array}{ccc|c} 1 & -1 & 0 & -1 \\ 0 & 2 & 3 & 3 \\ 0 & 0 & 0 & 1+k \end{array}\right).$$

In this echelon form, it is seen that the third row gives an equation which is inconsistent unless $1+k=0$. Therefore, the equations have a solution only when $k=-1$. When this is so, the echelon form shows that there is a line of solutions. If $z=t$, a parameter, then by back substitution, $2y+3z=3$, so $\mathbf{y=3(1-t)/2}$, $x-y=-1$, so $\mathbf{x=(1-3t)/2}$.

EXAMPLE 5 *Show that there are 2 values of λ for which the equations*

$$\left\{\begin{array}{l} 2x+\lambda y+\lambda^2 z = -4 \\ x+y+2z = 1 \\ x-y = -3 \end{array}\right\}$$

do not have a unique solution and find these values of λ. When the equations have a unique solution express this solution in terms of λ. Interpret geometrically the cases $\lambda=0$, $\lambda=2$ and $\lambda=-1$.

Row reduce the augmented matrix to echelon form taking care to note any assumptions about the value of λ needed in order to proceed.

$$\left(\begin{array}{ccc|c} 2 & \lambda & \lambda^2 & -4 \\ 1 & 1 & 2 & 1 \\ 1 & -1 & 0 & -3 \end{array}\right) \begin{array}{l} \mathbf{r}_1 \leftarrow \mathbf{r}_3 \\ \\ \mathbf{r}_3 \leftarrow \mathbf{r}_1 \end{array} \left(\begin{array}{ccc|c} 1 & -1 & 0 & -3 \\ 1 & 1 & 2 & 1 \\ 2 & \lambda & \lambda^2 & -4 \end{array}\right) \begin{array}{l} \mathbf{r}_2 \leftarrow \mathbf{r}_2 - \mathbf{r}_1 \\ \mathbf{r}_3 \leftarrow \mathbf{r}_3 - 2\mathbf{r}_1 \end{array}$$

$$\left(\begin{array}{ccc|c} 1 & -1 & 0 & -3 \\ 0 & 2 & 2 & 4 \\ 0 & \lambda+2 & \lambda^2 & 2 \end{array}\right) \mathbf{r}_2 \leftarrow \mathbf{r}_2/2 \left(\begin{array}{ccc|c} 1 & -1 & 0 & -3 \\ 0 & 1 & 1 & 2 \\ 0 & \lambda+2 & \lambda^2 & 2 \end{array}\right) \mathbf{r}_3 \leftarrow \mathbf{r}_3 - (\lambda+2)\mathbf{r}_2$$

$$\left(\begin{array}{ccc|c} 1 & -1 & 0 & -3 \\ 0 & 1 & 1 & 2 \\ 0 & 0 & \lambda^2-\lambda-2 & -2\lambda-2 \end{array}\right) = \left(\begin{array}{ccc|c} 1 & -1 & 0 & -3 \\ 0 & 1 & 1 & 2 \\ 0 & 0 & (\lambda+1)(\lambda-2) & -2(\lambda+1) \end{array}\right).$$

This matrix is now in echelon form and the third row depends on λ. If $\lambda=-1$, the third row is the zero row and there is a one parameter solution. If $\lambda=2$, then the third row gives $0=-2$ and the equations are inconsistent, with no solution. For other values of λ there will be a unique solution. We can now answer the questions. The equations do not have a unique solution for $\boldsymbol{\lambda \in \{-1, 2\}}$. When $\lambda \notin \{-1, 2\}$, the solution is obtained by back substitution. From row 3, $\mathbf{z=-2/(\lambda-2)}$, from row 2, $\mathbf{y=2-z=2(\lambda-1)/(\lambda-2)}$ and, from row 1, $\mathbf{x=y-3=(-\lambda+4)/(\lambda-2)}$.

When $\lambda = 0$, the equations represent three planes meeting in one point $(-2, 1, 1)$. When $\lambda = 2$, the first two equations represent parallel planes, perpendicular to $\mathbf{i}+\mathbf{j}+2\mathbf{k}$ and the third equation represents a plane which meets the other two in parallel lines.

When $\lambda = -1$, there is a line of solutions, given by the equations $x = y-3$ and $z = 2-y$, and the three equations represent three planes which form a sheaf, meeting in this line.

If the image, under a linear transformation T, of some geometrical object is required, it is best to use parametric equations to describe the object. It could be a line or a plane or a curve or a surface but, in each case, the object may be described as the set of all points whose coordinates are given in terms of (i) a parameter t in the case of a line or a curve or (ii) two parameters s and t in the case of a plane or other surface. The set of image points under T is then also given in terms of the same parameters. This device is illustrated in Example 6.

EXAMPLE 6 *The linear transformation* T: $\mathbb{R}^3 \to \mathbb{R}^3$ *has matrix* $\mathbf{A}$, *given by* $\mathbf{A} = \begin{pmatrix} 1 & 2 & 1 \\ 3 & 3 & 1 \\ 4 & -1 & -2 \end{pmatrix}$. *Show that the image space of* T *is the plane* $5x-3y+z=0$. *Find the images under* T *of (i) the line* $x-1 = \dfrac{y-2}{-2} = \dfrac{z+1}{3}$, *(ii) the plane* $x-y-z=1$.

It is important to avoid any confusion between the coordinates of the original point and the transformed point. Therefore, let the point (x, y, z) be transformed by T to the point (X, Y, Z). This means that

$$\begin{pmatrix} X \\ Y \\ Z \end{pmatrix} = \mathrm{T}\begin{pmatrix} x \\ y \\ z \end{pmatrix} = \mathbf{A}\begin{pmatrix} x \\ y \\ z \end{pmatrix} = \begin{pmatrix} 1 & 2 & 1 \\ 3 & 3 & 1 \\ 4 & -1 & -2 \end{pmatrix}\begin{pmatrix} x \\ y \\ z \end{pmatrix} = \begin{pmatrix} x+2y+z \\ 3x+3y+z \\ 4x-y-2z \end{pmatrix}.$$

Then $5X-3Y+Z = 5x+10y+5z-9x-9y-3z+4x-y-2z = 0$, so every image point lies in the plane $5x-3y+z=0$. Since the first two columns of $\mathbf{A}$ are independent, the images of $\mathbf{i}$ and $\mathbf{j}$ are not parallel and so the image space is the required plane.

(i) In terms of a parameter t, the points on the line are given by $x = 1+t$, $y = 2-2t$, $z = -1+3t$. These points are mapped under T to points with coordinates (X, Y, Z), given by

$$\begin{pmatrix} X \\ Y \\ Z \end{pmatrix} = \begin{pmatrix} 1+t+4-4t-1+3t \\ 3+3t+6-6t-1+3t \\ 4+4t-2+2t+2-6t \end{pmatrix} = \begin{pmatrix} 4 \\ 8 \\ 4 \end{pmatrix}$$

So the whole line is mapped on to a single point $(4, 8, 4)$.

(ii) In terms of parameters s and t, a general point in the plane $x-y-z=1$ has coordinates given by $x = 1+s+t$, $y = s$, $z = t$. Under T, this becomes the point (X, Y, Z), given by

$$\begin{pmatrix} X \\ Y \\ Z \end{pmatrix} = \begin{pmatrix} 1+s+t+2s+t \\ 3+3s+3t+3s+t \\ 4+4s+4t-s-2t \end{pmatrix} = \begin{pmatrix} 1+3s+2t \\ 3+6s+4t \\ 4+3s+2t \end{pmatrix}$$

Clearly, $3s+2t = X-1 = \dfrac{Y-3}{2} = Z-4$, so (X, Y, Z) lies on the line $x-1 = \dfrac{y-3}{2} = z-4$, and this line is the image of the plane $x-y-z=1$.

EXERCISE 4.4

1 Solve, for x, y and z, the equations:

(i) $\begin{aligned} 2x+6y+z &= 0 \\ -x+2y-z &= 10, \\ 4x+3y+z &= 1 \end{aligned}$ (ii) $\begin{aligned} 3x+y-z &= -5 \\ x-2y+2z &= 3. \\ 2x+3y+z &= 12 \end{aligned}$

2 Find the solution set of the system of equations $\begin{aligned} x+y+z &= -1, \\ x+2y+3z &= -2, \\ 3x-2y-7z &= 2. \end{aligned}$

If x, y, z represent coordinates of points in three dimensional space, state
(i) what is represented geometrically by the solution set,
(ii) what is represented geometrically by each of the three equations,
(iii) how the geometrical figures represented by the equations are related to each other. (*SMP*)

3 Find the value of λ such that the equations $\begin{aligned} x+y-3z &= 1 \\ 3x-y-z &= 7 \\ 5x-3y+\lambda z &= 13 \end{aligned}$ do *not* have a unique solution and obtain the general solution in this case. (*JMB*)

4 Express in matrix form the transformation of coordinates given by the equations

$$\begin{aligned} x' &= 7x+5y+6z, \\ y' &= 4x+3y+3z, \\ z' &= 10x+7y+\lambda z. \end{aligned}$$

When $\lambda = 8$, obtain three linear equations expressing x, y, z each in terms of x', y', z', and find the point (x, y, z) which corresponds to $(x', y', z') = (6, 2, 9)$.

Obtain the value of λ for which there is no inverse transformation and, in this case, show that all the points (x, y, z) are transformed into points of the plane $2x'-y'-z'=0$. (*L*)

5 Find the solution sets of the equations

$$\begin{aligned} x-2y+2z &= 1, \\ 2x-y+z &= 2, \\ x-5y+kz &= 1, \end{aligned}$$

in the cases (*a*) $k=-1$, (b) $k=5$. Give geometrical interpretations of your solutions. (*SMP*)

6 Given that $\mathbf{A} = \begin{pmatrix} 2 & a & -3 \\ 3 & 1 & -2 \\ 1 & 4 & 1 \end{pmatrix}$ and that $\mathbf{B}$ is a given vector in $\mathbb{R}^3$, find the value of a for which the equation $\mathbf{AX} = \mathbf{B}$ does not have exactly one solution. Using this value of a, find the solutions, if any, of the equation $\mathbf{AX} = \mathbf{B}$ in the cases

(i) $\mathbf{B} = \begin{pmatrix} 4 \\ 2 \\ -2 \end{pmatrix}$, (ii) $\mathbf{B} = \begin{pmatrix} 0 \\ 2 \\ -2 \end{pmatrix}$. (*C*)

7 Find a vector equation for the line of intersection of the planes $2x+y-z=3$ and $x+2y+4z=0$. Hence, or otherwise, solve the equations $\begin{aligned} 2x+\ y-\ z &= 3, \\ x+2y+4z &= 0, \\ 3x+\lambda y+6z &= 2 \end{aligned}$ for *all* real values of λ, interpreting your results geometrically. (*JMB*)

8 Using elementary row operations, reduce the matrix $\begin{pmatrix} 1 & 2 & -1 & 1 \\ 2 & 5 & a & 3 \\ 3 & a & 6 & 4 \end{pmatrix}$ to echelon form. Hence, or otherwise, determine the values of a so that the equation $\begin{pmatrix} 1 & 2 & -1 \\ 2 & 5 & a \\ 3 & a & 6 \end{pmatrix}\begin{pmatrix} x \\ y \\ z \end{pmatrix} = \begin{pmatrix} 1 \\ 3 \\ 4 \end{pmatrix}$ has (i) no solution, (ii) more than one solution, (iii) a unique solution, and solve the equation completely in case (ii). (*C*)

9 By reducing the equation $\begin{pmatrix} 1 & 2 & -3 \\ 2 & 6 & -11 \\ 1 & -2 & 7 \end{pmatrix}\begin{pmatrix} x \\ y \\ z \end{pmatrix} = \begin{pmatrix} a \\ b \\ c \end{pmatrix}$ to echelon form, or otherwise, prove that the equation is soluble only if $c+2b-5a=0$. Hence show that the planes $\begin{aligned} x+2y-\ 3z &= 1 \\ 2x+6y-11z &= 2 \\ x-2y+\ 7z &= 1 \end{aligned}$ intersect in a line. Find, in terms of t, the coordinates of the point in which this line meets the plane $z=t$. (*JMB*)

10 The point $P(x, y, z)$ is transformed to the point $Q(X, Y, Z)$ by the relation $\begin{pmatrix} X \\ Y \\ Z \end{pmatrix} = \mathbf{M}\begin{pmatrix} x \\ y \\ z \end{pmatrix}$. (a) If $\mathbf{M} = \begin{pmatrix} 6 & 8 & 4 \\ 9 & 12 & 6 \\ 4 & -1 & 3 \end{pmatrix}$, show that for all P the corresponding point Q lies on a plane and give an equation of this plane.
(b) If $\mathbf{M} = \begin{pmatrix} 1 & 2 & -1 \\ 3 & 6 & -3 \\ 5 & 10 & -5 \end{pmatrix}$, show that for all P the corresponding point Q lies on a line and give equations for this line. (c) If $\mathbf{M} = \begin{pmatrix} 0 & -1 & 0 \\ 1 & 0 & 0 \\ 0 & 0 & 1 \end{pmatrix}$, show that for all P the corresponding point Q is in the position P would reach if it were rotated through 90° about Oz and state the inverse matrix in this case. (*L*)

11 A transformation T of three-dimensional space is defined by $\mathbf{r}' = \mathbf{Mr}$, where $\mathbf{r}' = \begin{pmatrix} x' \\ y' \\ z' \end{pmatrix}$, $\mathbf{r} = \begin{pmatrix} x \\ y \\ z \end{pmatrix}$ and $\mathbf{M} = \begin{pmatrix} 1 & 3 & 2 \\ -3 & 6 & 4 \\ 5 & -3 & -2 \end{pmatrix}$. Prove that T transforms the plane $x=a$ into a line, and show that the direction of this line is independent of the value of a. Given that this line passes through the point $(5, -5, 13)$, determine the value of a. (*JMB*)

12 Given that $\mathbf{A} = \begin{pmatrix} 1 & 2 & 3 \\ 5 & 4 & a \\ -5 & a & 11 \end{pmatrix}$ and that $\mathbf{B}$ is a 3×1 matrix, find the values of a for which the equation $\mathbf{AX} = \mathbf{B}$ does not have exactly one solution. Using each of these values of a in turn, find the solutions, if any of the equation $\mathbf{AX} = \begin{pmatrix} 1 \\ 4 \\ -3 \end{pmatrix}$. *(C)*

13 Reduce the matrix equation $\begin{pmatrix} 1 & 2 & -1 \\ 2 & 1 & \alpha \\ \alpha & 1 & 2 \end{pmatrix}\begin{pmatrix} x \\ y \\ z \end{pmatrix} = \begin{pmatrix} \beta \\ 0 \\ 0 \end{pmatrix}$ $(\alpha, \beta \text{ real})$ (1)

to the form $\begin{pmatrix} * & * & * \\ 0 & * & * \\ 0 & 0 & * \end{pmatrix}\begin{pmatrix} x \\ y \\ z \end{pmatrix} = \begin{pmatrix} * \\ * \\ * \end{pmatrix}$ (where the asterisks denote real numbers). Hence show that equation (1) has a unique solution if $\alpha^2 \neq 4$, has a non-empty solution set if $\alpha = 2$ whatever the value of β, but has no solution when $\alpha = -2$ unless $\beta = 0$. Find the solution set when (i) $\alpha = -1, \beta = 1$, (ii) $\alpha = -2, \beta = 0$. *(SMP)*

14 Show that the planes

$$\begin{aligned} \pi_1&: \ x+2y+\ z = 3 \\ \pi_2&: \ x+\ y+3z = 4 \\ \pi_3&: \ 2x+7y-4z = 3 \end{aligned}$$

have a line L in common and find a vector equation for L. Obtain the Cartesian equation of the plane π_4 which is perpendicular to the plane π_3 and which also contains the line L. Find a unit vector which is perpendicular to π_4 and hence, or otherwise, find the perpendicular distance from the origin to π_4. *(JMB)*

15 Show that the simultaneous equations

$$\begin{aligned} kx+\ y+\ z &= \ \ 1, \\ 2x+ky-2z &= -1, \\ x-2y+kz &= -2 \end{aligned}$$

have a unique solution except for three values of k which are to be found. Show that, when $k = 1$, the planes represented by the equations meet at a point P which lies in the plane $y = 1$, and that, when $k = -1$, the planes meet in a line L which also lies in the plane $y = 1$. Find the perpendicular distance of P from L. *(JMB)*

MISCELLANEOUS EXERCISE 4

1 Let $\mathbf{A}$ be the matrix $\begin{pmatrix} 1 & 2 & 7 \\ 1 & 3 & 0 \\ 0 & -1 & 8 \end{pmatrix}$. By performing elementary row operations on the matrix $(\mathbf{A}|\mathbf{I})$ find $\mathbf{A}^{-1}$. Solve the equation $\mathbf{AX} = \mathbf{K}$, when (i) $\mathbf{K} = \begin{pmatrix} 0 \\ 0 \\ 0 \end{pmatrix}$, (ii) $\mathbf{K} = \begin{pmatrix} -1 \\ 1 \\ 0 \end{pmatrix}$, (iii) $\mathbf{K} = \begin{pmatrix} 1 \\ 2 \\ 3 \end{pmatrix}$. *(C)*

2 If $\mathbf{M} = \begin{pmatrix} 0 & 1 & 0 \\ -1 & 0 & 0 \\ 0 & 0 & 1 \end{pmatrix}$, find $\mathbf{M}^{-1}$. Interpret geometrically the transformations of x-y-z space defined by $\mathbf{M}$ and $\mathbf{M}^2$: (*L*)

3 State the condition necessary in order that a square matrix $\mathbf{A}$ shall have an inverse $\mathbf{A}^{-1}$. Prove that, if $\mathbf{A}$ and $\mathbf{B}$ are two $(n \times n)$ non-singular matrices, then $(\mathbf{AB})^{-1} = \mathbf{B}^{-1}\mathbf{A}^{-1}$.
Find $\mathbf{A}^{-1}$ given that

$\mathbf{A} = \begin{pmatrix} 3 & 2 & -1 \\ 1 & -1 & 2 \\ 2 & 4 & -1 \end{pmatrix}$. Solve the system of equations

$$\begin{aligned} 3x + 2y - z &= -5, \\ x - y + 2z &= 11, \\ 2x + 4y - z &= -10. \end{aligned}$$ (*L*)

4 Write down the matrices for (i) a rotation R through 45° in the positive sense, (ii) a shear S of angle 45° to the right, as shown in Fig. 4.15.

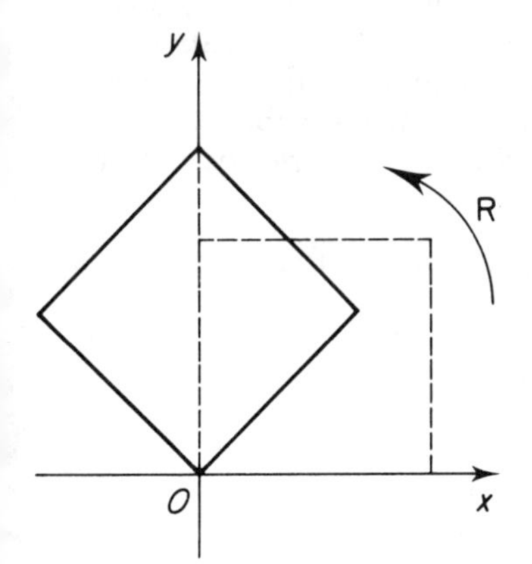

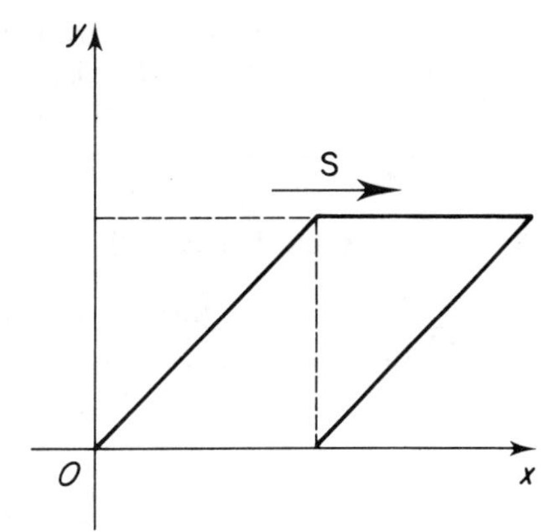

Fig. 4.15

If $\mathrm{T} = \mathrm{R}^{-1}\mathrm{SR}$, show that $\begin{pmatrix} 1 \\ -1 \end{pmatrix}$ is unchanged by T, and find the effect of T on $\begin{pmatrix} 1 \\ 1 \end{pmatrix}$. State whether T is (a) a rotation, (b) a reflection, (c) a shear, (d) an enlargement, (e) a one-way stretch. (*SMP*)

5 (i) If $\mathbf{X} = \begin{pmatrix} x \\ y \end{pmatrix}$, $\mathbf{X}_1 = \begin{pmatrix} x_1 \\ y_1 \end{pmatrix}$, $\mathbf{A} = \begin{pmatrix} -5 & 0 \\ 0 & 5 \end{pmatrix}$, find the image of $y = x^2 + 2x + 1$ under the transformation $\mathbf{X}_1 = \mathbf{AX}$. Describe the effect of this transformation on the curve in geometrical terms.

(ii) Given that $\begin{aligned} x + y + z &= a, \\ 2x - y + 2z &= b, \\ 2x + 2y - z &= c, \end{aligned}$ find x, y, z in terms of a, b and c. Hence obtain the inverse of the matrix $\begin{pmatrix} 1 & 1 & 1 \\ 2 & -1 & 2 \\ 2 & 2 & -1 \end{pmatrix}$. (*L*)

6 Find non-singular matrices **P** and **Q** such that **PA** and **QB** are echelon matrices, where **A** and **B** are given by $\mathbf{A} = \begin{pmatrix} -2 & 0 & 3 \\ 1 & 3 & 1 \\ -1 & 9 & 9 \end{pmatrix}$, $\mathbf{B} = \begin{pmatrix} 1 & -2 & 1 \\ 3 & -1 & 5 \\ -1 & 4 & 0 \end{pmatrix}$. For each of **A** and **B** find its inverse if it exists. Find numbers α, β, γ, not all zero, such that
$\alpha(-2, 0, 3) + \beta(1, 3, 1) + \gamma(-1, 9, 9) = (0, 0, 0)$. (*C*)

7 The transformation T_1 of three-dimensional space is given by $\mathbf{r}' = \mathbf{Mr}$, $\mathbf{r} = \begin{pmatrix} x \\ y \\ z \end{pmatrix}$, $\mathbf{r}' = \begin{pmatrix} x' \\ y' \\ z' \end{pmatrix}$ and $\mathbf{M} = \begin{pmatrix} 1 & 0 & 0 \\ 0 & 0 & -1 \\ 0 & 1 & 0 \end{pmatrix}$. Describe the transformation geometrically. Two other transformations are defined as follows: T_2 is a reflection in the x-z plane and T_3 is a rotation through 180° about the line $x = 0, y + z = 0$. By considering the image under each transformation of the points with position vectors **i, j, k**, or otherwise, find the matrix for each of T_2 and T_3. Determine the matrix for the combined transformations T_3T_1 and T_1T_3 and describe each of these transformations geometrically. (*JMB*)

8 Given that the matrix $\begin{pmatrix} 1/2 & \sqrt{3}/2 & 0 \\ 0 & 0 & 1 \\ \sqrt{3}/2 & -1/2 & 0 \end{pmatrix}$ corresponds to a rotation about a fixed axis through the origin, determine this axis.

9 The three planes π_1, π_2, π_3 have the equations

$$\pi_1: 2x - z = 0,$$
$$\pi_2: x + y + 2z = 0,$$
$$\pi_3: x - 5y + 2z = 0.$$

Show that each of these planes is perpendicular to the other two planes. Verify that, for all values of the parameters s and t, the point $P(2s, 2t, -s-t)$ lies in the plane π_2. Find a vector equation of the line L which is perpendicular to π_2 and contains the point P. The transformation $T: \mathbb{R}^3 \to \mathbb{R}^3$ is given by $T(\mathbf{r}) = \mathbf{r}'$, where $\mathbf{r}' = \mathbf{Mr}$, and $\mathbf{r} = \begin{pmatrix} x \\ y \\ z \end{pmatrix}$, $\mathbf{r}' = \begin{pmatrix} x' \\ y' \\ z' \end{pmatrix}$,
$\mathbf{M} = \frac{1}{6}\begin{pmatrix} 5 & -1 & -2 \\ -1 & 5 & -2 \\ -2 & -2 & 2 \end{pmatrix}$.

Show that each point of the line L is transformed by T to the point P. Show also that the image of the plane π_1 under the transformation T is a line and find a vector equation of this line. (*JMB*)

10 The matrix **A** is given by $\mathbf{A} = \begin{pmatrix} 8 & 5 & 3 \\ 5 & 3 & 1 \\ 3 & 2 & 1 \end{pmatrix}$. By performing elementary row operations on the matrix $(\mathbf{A} \mid \mathbf{I})$ find $\mathbf{A}^{-1}$.

(i) The linear transformation L: $\mathbb{R}^3 \to \mathbb{R}^3$ is defined by L: $\begin{pmatrix} x \\ y \\ z \end{pmatrix} \mapsto \mathbf{A}\begin{pmatrix} x \\ y \\ z \end{pmatrix}$.

Find the element in $\mathbb{R}^3$ whose image under L is $\begin{pmatrix} 1 \\ -1 \\ 2 \end{pmatrix}$.

(ii) Solve the equation $\mathbf{XA} = \mathbf{K}$, where $\mathbf{K} = (1 \quad -1 \quad 2)$. (*C*)

11 The equations $\begin{aligned} x+5y+2z &= 6, \\ x+3y+\ z &= 4, \\ 2x+4y+\ z &= 6 \end{aligned}$ represent three planes π_1, π_2 and π_3 respectively. By reducing the system of equations to echelon form, or otherwise, show that the planes belong to a sheaf (i.e. a set of planes with a line in common). Explain why, for any value of λ, the equation $x+3y+z-4+\lambda(2x+4y+z-6) = 0$ represents a plane belonging to the sheaf. By making a suitable choice for λ, find the plane π_4 of the sheaf that is perpendicular to π_1. Write down a unit vector normal to π_4, and hence show that the perpendicular distance from (0, 0, 0) on to π_4 is $2/\sqrt{5}$. (*SMP*)

12 (i) If **A** and **B** are any square matrices for which $\mathbf{AB} = \mu\mathbf{A}$ and $\mathbf{BA} = \lambda\mathbf{B}$, where λ and μ are non-zero scalars, prove that $\mathbf{A}^2 = \lambda\mathbf{A}$ and $\mathbf{B}^2 = \mu\mathbf{B}$.
(ii) If **A** is any 2×2 matrix such that $\mathbf{A}^2 = \lambda\mathbf{A}$ where λ is a non-zero scalar, prove that either **A** is singular or $\mathbf{A} = \lambda\mathbf{I}$, where **I** is the unit matrix.
(iii) If $\mathbf{A} = \begin{pmatrix} 3 & 6 \\ 1 & 2 \end{pmatrix}$ and $\mathbf{B} = \begin{pmatrix} 5 & 0 \\ 0 & 5 \end{pmatrix}$, describe in geometrical terms the transformations of points of the *x*-*y* plane given by each of **A** and **AB**, explaining clearly the difference between the two transformations. (*C*)

13 (i) Show that the determinant $\begin{vmatrix} 1 & 1 & 1 \\ x & y & z \\ yz & zx & xy \end{vmatrix} = (x-y)(y-z)(z-x)$.

(ii) The equations $\begin{aligned} 2\lambda x - 3y + \lambda - 3 &= 0, \\ 3x - 2y + 1 &= 0, \\ 4x - \lambda y + 2 &= 0, \end{aligned}$ represent three straight lines in the *x*-*y* plane. Find the values of λ for which the lines are concurrent. For each of these values of λ, find the coordinates of the point at which the lines are concurrent. (*L*)

14 (i) The matrix **M** is given by $\mathbf{M} = \begin{pmatrix} a & b \\ c & d \end{pmatrix}$, where $a+d = -1$ and $ad-bc = 1$. Show that $\mathbf{M}^2 = \begin{pmatrix} d & -b \\ -c & a \end{pmatrix}$, and deduce that $\mathbf{M}^2 + \mathbf{M} + \mathbf{I} = \mathbf{0}$. (ii) A transformation is given by $\mathbf{X} \mapsto \mathbf{AX}$, where **A** is a 2×2 matrix and $\mathbf{X} = \begin{pmatrix} x \\ y \end{pmatrix}$. The points (1, 0) and (0, 1) are mapped to (1, 0) and (*p*, 1) respectively. Write down the matrix **A**. A second transformation is given by $\mathbf{X} \mapsto \mathbf{BX}$ and is an anticlockwise rotation about the origin through an obtuse angle θ, where $\cos\theta = -4/5$. Write down the matrix **B**. Given that the matrix **M** in (i) above is such that $\mathbf{M} = \mathbf{BA}$, find the value of *p*. (*C*)

15 (a) Prove that, when n is a positive integer, $\begin{pmatrix} 1 & 0 \\ a & 1 \end{pmatrix}^n = \begin{pmatrix} 1 & 0 \\ na & 1 \end{pmatrix}$. Prove that the result also holds when n is a negative integer.

(b) The transformation T of the $x-y$ plane is defined by T: $\begin{pmatrix} x \\ y \end{pmatrix} \mapsto \begin{pmatrix} a & b \\ c & d \end{pmatrix}\begin{pmatrix} x \\ y \end{pmatrix}$. Show that T leaves all lines through the origin invariant if and only if $b = c = 0$ and $a = d \neq 0$. In this case, describe T geometrically. (*C*)

16 Given that $\mathbf{a} = \begin{pmatrix} -1 \\ 3 \\ 13 \end{pmatrix}$, $\mathbf{b} = \begin{pmatrix} 1 \\ 2 \\ 2 \end{pmatrix}$, $\mathbf{c} = \begin{pmatrix} 1 \\ 3 \\ 5 \end{pmatrix}$, express $\mathbf{a}$ as a linear combination of $\mathbf{b}$ and $\mathbf{c}$. Hence, or otherwise, evaluate $\begin{vmatrix} 1 & 2 & 2 \\ 1 & 3 & 5 \\ -1 & 3 & 13 \end{vmatrix}$.

A transformation of three dimensional space is defined by $\mathbf{r}' = \mathbf{M}\mathbf{r}$ where $\mathbf{r}' = \begin{pmatrix} x' \\ y' \\ z' \end{pmatrix}$, $\mathbf{r} = \begin{pmatrix} x \\ y \\ z \end{pmatrix}$, $\mathbf{M} = \begin{pmatrix} 1 & 2 & 2 \\ 1 & 3 & 5 \\ -1 & 3 & 13 \end{pmatrix}$. Show that for all $\mathbf{r}$ the point with position vector $\mathbf{r}'$ lies in a plane and find an equation of this plane. (*JMB*)

17 Express the determinant $\begin{vmatrix} bc & b+c & a \\ ca & c+a & b \\ ab & a+b & c \end{vmatrix}$ as a product of linear factors. Also show that $\begin{vmatrix} b+c & a & a^3 \\ c+a & b & b^3 \\ a+b & c & c^3 \end{vmatrix} = (b-c)(c-a)(a-b)(a+b+c)^2$. Given that a, b and c are constants, no two of which are equal and whose sum is not zero, show that the planes

$$\begin{aligned} bcx + (b+c)y + az + a^3 &= 0 \\ cax + (c+a)y + bz + b^3 &= 0 \\ abx + (a+b)y + cz + c^3 &= 0 \end{aligned}$$

meet in a point. Find the x-coordinate of the point of intersection of the planes. (*JMB*)

18 The transformation T: $\mathbf{r} \to \mathbf{r}'$ of a three-dimensional space is given by $\mathbf{r}' = \mathbf{M}\mathbf{r}$ where $\mathbf{r} = \begin{pmatrix} x \\ y \\ z \end{pmatrix}$, $\mathbf{r}' = \begin{pmatrix} x' \\ y' \\ z' \end{pmatrix}$ and $\mathbf{M} = \begin{pmatrix} 1 & 2 & 3 \\ 2 & 0 & -2 \\ 3 & -2 & -7 \end{pmatrix}$. Show that T maps the whole space on to the plane π whose equation is $x - 2y + z = 0$. Show that $\mathbf{e} = \begin{pmatrix} 1 \\ 1 \\ 1 \end{pmatrix}$ and $\mathbf{f} = \begin{pmatrix} 2 \\ 1 \\ 0 \end{pmatrix}$ form a basis in π. Find the image under T of (i) the line $x = -y = z/2$, (ii) the plane $x - y - z = 0$, (iii) the plane $x - z = 0$. If the answers to (i) and (iii) are expressed in the form $\lambda\mathbf{e} + \mu\mathbf{f}$, where λ and μ are scalars, find the relation between λ and μ in each case. (*JMB*)

5 Hyperbolic Functions

5.1 Definitions

We remind the reader of the definition of even and odd functions, (*Pure Mathematics* §8.7). Let f be a real function with domain D. Then

$$\text{f is even if } f(-x) = f(x), \text{ for all } x \text{ in } D,$$
$$\text{f is odd if } f(-x) = -f(x), \text{ for all } x \text{ in } D.$$

Any real function f can be expressed as the sum of an even function g and an odd function h by defining

$$2g(x) = f(x) + f(-x), \text{ and } 2h(x) = f(x) - f(-x).$$

Clearly $f(x) = g(x) + h(x)$, and also

$$g(-x) = \tfrac{1}{2}\{f(-x) + f(x)\} = g(x), \text{ so g is even,}$$
$$h(-x) = \tfrac{1}{2}\{f(-x) - f(x)\} = -h(x), \text{ so h is odd.}$$

When f is a polynomial function, g will consist of the even terms of f and h will consist of the odd terms. Thus, if $f(x) = x^2 + 3x + 2$, $f(-x) = x^2 - 3x + 2$, $g(x) = x^2 + 2$ and $h(x) = 3x$.

If the domain of f is restricted, for example, if f is not defined at $x = a$, then $f(-x)$ is not defined at $x = -a$ so that g and h will not be defined at both $x = a$ and $x = -a$. The equation $f = g + h$ is then only true for a restricted domain and not for the whole domain of f. Consider the function f given by $f(x) = (x+1)/(x-1)$ so that $f(-x) = (-x+1)/(-x-1)$, $g(x) = (x^2+1)/(x^2-1)$ and $h(x) = 2x/(x^2-1)$. Then f has domain $\mathbb{R}\setminus\{1\}$ but g and h have domain $\mathbb{R}\setminus\{1, -1\}$. The reader may find it instructive to draw the graphs of f, g and h for the above two examples, and for others, in order to see what happens graphically.

In the case when f is the exponential function so that, $f(x) = e^x$, the functions g and h are called the hyperbolic cosine and sine functions, denoted by cosh and sinh.

Definition The functions cosh and sinh have domain $\mathbb{R}$ and are defined by:

$$\cosh x = \tfrac{1}{2}(e^x + e^{-x}) \text{ and } \sinh x = \tfrac{1}{2}(e^x - e^{-x}), \text{ for all } x \text{ in } \mathbb{R}.$$

Cosh is an even function and sinh is an odd function and

$$e^x = \cosh x + \sinh x, \text{ and } e^{-x} = \cosh x - \sinh x.$$

Then

$$\cosh^2 x - \sinh^2 x = (\cosh x + \sinh x)(\cosh x - \sinh x) = e^x e^{-x} = 1,$$

and this may be compared with a similar formula for circular functions, namely

$$\cos^2 x + \sin^2 x = 1.$$

From these two equations, we find an explanation of the names of the functions. Consider the locus of a point (x, y) given, in terms of a parameter t, in each of the two cases:
(i) $x = \cos t$, $y = \sin t$, $x^2 + y^2 = 1$, and the locus is the unit circle,
(ii) $x = \cosh t$, $y = \sinh t$, $x^2 - y^2 = 1$, and the locus is one branch of a rectangular hyperbola. The two curves are shown in Fig. 5.1. It should be noted that, since $e^x > 0$ and $e^{-x} > 0$, $\cosh x > 0$ and hence, since $\cosh^2 x = 1 + \sinh^2 x$, $\cosh x \geq 1$. However, $\sinh x$ can take all real values, positive and negative. Therefore, only the one branch of the hyperbola occurs in Fig. 5.1.

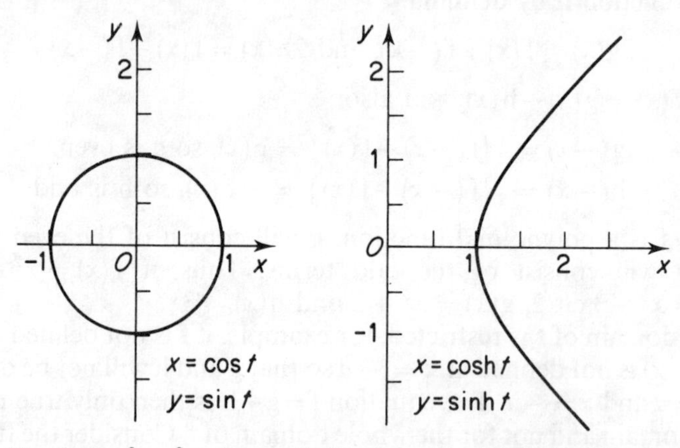

(a) The circle $x^2 + y^2 = 1$. (b) One branch of the hyperbola $x^2 - y^2 = 1$.

Fig. 5.1

Many properties of hyperbolic functions are similar to corresponding properties of circular functions, apart from a change of sign of any term which involves a product of two sinh's instead of two sines. These are illustrated in Exercise 5.1.

Corresponding to other trigonometric functions, we define further hyperbolic functions.

Definition

$$\tanh x = \frac{\sinh x}{\cosh x}, \quad \operatorname{sech} x = \frac{1}{\cosh x},$$

$$\coth x = \frac{\cosh x}{\sinh x}, \quad \operatorname{cosech} x = \frac{1}{\sinh x}.$$

EXERCISE 5.1

1 Using the definitions of cosh and sinh in terms of the exponential function, prove that:

(i) $\cosh(x+y) = \cosh x \cosh y + \sinh x \sinh y$,
(ii) $\sinh(x+y) = \sinh x \cosh y + \cosh x \sinh y$,
(iii) $\sinh 2x = 2\sinh x \cosh x$,
(iv) $\cosh 2x = \cosh^2 x + \sinh^2 x = 2\cosh^2 x - 1 = 1 + 2\sinh^2 x$,
(v) $\cosh(x-y) = \cosh x \cosh y - \sinh x \sinh y$,
(vi) $\sinh(x-y) = \sinh x \cosh y - \cosh x \sinh y$,
(vii) $\operatorname{sech}^2 y + \tanh^2 y = 1$.

2 (i) Given that $\cosh x = \frac{5}{4}$, find the two possible pairs of values for $\sinh x$ and $\tanh x$,
(ii) Given that $\tanh x = \frac{12}{13}$, find $\operatorname{cosech} x$ and $\operatorname{sech} x$.

3 By expressing $\cosh x$ and $\sinh x$ in terms of u, where $u = e^x$, solve the equations:
(i) $3\cosh x + 3\sinh x = 5$, (ii) $3\cosh x - \sinh x = 3$.

4 Solve the equations:
(i) $\sinh x + 4 = 4\cosh x$, (ii) $2\tanh x - \operatorname{sech} x = 1$,
(iii) $9\coth x = \tanh x$, (iv) $2\sinh x + 6\cosh x = 9$,
(v) $10\operatorname{cosech} x + \coth x = 5$.

5 Solve for x the equation $2\cosh x - 2\sinh x = 3$, leaving your answer as a natural logarithm. (*L*)

6 Prove that:
(i) $\cosh 3x = 4\cosh^3 x - 3\cosh x$,
(ii) $2\cosh x \cosh y = \cosh(x+y) + \cosh(x-y)$,
(iii) $2\sinh x \cosh y = \sinh(x+y) + \sinh(x-y)$,

(iv) $\cosh x - \cosh y = 2\sinh\left(\dfrac{x+y}{2}\right)\sinh\left(\dfrac{x-y}{2}\right)$,

(v) $\sinh x - \sinh y = 2\cosh\left(\dfrac{x+y}{2}\right)\sinh\left(\dfrac{x-y}{2}\right)$.

7 Find corresponding expressions to those in question **6** for:
(i) $\sinh 3x$, (ii) $2\sinh x \sinh y$, (iii) $2\cosh x \sinh y$, (iv) $\cosh x + \cosh y$,
(v) $\sinh x + \sinh y$.

8 Prove that:
(i) $\tanh(x+y) = (\tanh x + \tanh y)/(1 + \tanh x \tanh y)$,
(ii) $\tanh 2x = 2\tanh x/(1 + \tanh^2 x)$,

(iii) $\tanh x = \dfrac{\cosh 2x + \sinh 2x - 1}{\cosh 2x + \sinh 2x + 1}$.

(iv) $2\cos^2 x \cosh^2 x + 2\sin^2 x \sinh^2 x = \cos 2x + \cosh 2x$.

9 Using the fact that $\cosh x + \sinh x = e^x$, prove that:
(i) $(\cosh x + \sinh x)^2 = \cosh 2x + \sinh 2x$,
(ii) $(\cosh x + \sinh x)^n = \cosh nx + \sinh nx$, for all n in $\mathbb{N}$,
(iii) $(\cosh x + \sinh x)^{-1} = \cosh x - \sinh x$,
(iv) $(\cosh x + \sinh x)^n = \cosh nx + \sinh nx$, for all n in $\mathbb{Z}$.

10 Prove that: (i) $\cosh 0 = 1$, $\sinh 0 = 0$, $\tanh 0 = 0$,

(ii) as $x \to \infty$, $\cosh x$ and $\sinh x$ approximate to $\frac{1}{2}e^x$, and $\tanh x$ approaches the value 1,
(iii) as $x \to -\infty$, $\cosh x$ approximates to $\frac{1}{2}e^{-x}$, $\sinh x$ approximates to $-\frac{1}{2}e^{-x}$, and $\tanh x$ approaches the value -1,
(iv) for all real x, $\sinh x < \frac{1}{2}e^x < \cosh x$ and $|\tanh x| < 1$.

5.2 Derivatives and integrals of hyperbolic functions

The derivatives of cosh and sinh follow from their definitions:

$$\frac{d}{dx}\cosh x = \frac{d}{dx}\frac{1}{2}(e^x + e^{-x}) = \frac{1}{2}(e^x - e^{-x}) = \sinh x,$$

$$\frac{d}{dx}\sinh x = \frac{d}{dx}\frac{1}{2}(e^x - e^{-x}) = \frac{1}{2}(e^x + e^{-x}) = \cosh x,$$

$$\frac{d}{dx}\tanh x = \frac{d}{dx}\frac{\sinh x}{\cosh x} = \frac{\cosh^2 x - \sinh^2 x}{\cosh^2 x} = \frac{1}{\cosh^2 x} = \operatorname{sech}^2 x.$$

The derivatives of $\sinh x$ and $\tanh x$ are always positive. This means that the graphs of these two functions are monotonic increasing and have no stationary points. In particular, each function is one-one and takes the same sign as x.

The derivative of $\cosh x$ has the same sign as x and vanishes when $x = 0$. Thus, as x increases $\cosh x$ decreases for negative values of x, but increases for positive values of x. It has a minimum at $(0, 1)$. The above properties of the hyperbolic functions are illustrated in their graphs which are shown in Fig. 5.2.

EXAMPLE 1 *Prove that* $\frac{d}{dx}\operatorname{sech} x = -\operatorname{sech} x \tanh x$.

$$\frac{d}{dx}\operatorname{sech} x = \frac{d}{dx}(\cosh x)^{-1} = \frac{-\sinh x}{\cosh^2 x} = -\operatorname{sech} x \tanh x.$$

EXAMPLE 2 *Find the coordinates of the maximum and of the minimum points of the curve with equation* $y = 10\sinh 2x - 6\cosh 2x - 65x$.

To find the turning points, we put $y' = 0$, namely

$$y' = 20\cosh 2x - 12\sinh 2x - 65 = 0.$$

In terms of exponentials, this equation is

$$10(e^{2x} + e^{-2x}) - 6(e^{2x} - e^{-2x}) - 65 = 0,$$

that is, $4e^{2x} - 65 + 16e^{-2x} = 0$.

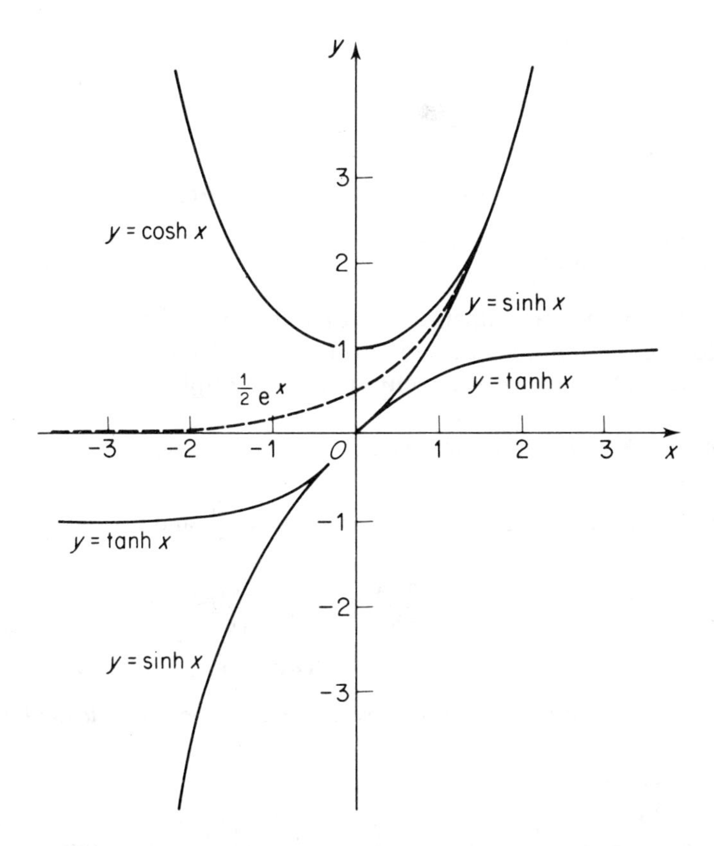

Fig. 5.2 Graphs of $\cosh x$, $\sinh x$ and $\tanh x$

Put $e^{2x} = z$ and multiply by z, which is non-zero,

$$4z^2 - 65z + 16 = (4z - 1)(z - 16) = 0,$$

so that $e^{2x} = z = \frac{1}{4}$ or $e^{2x} = 16$.
When $e^{2x} = \frac{1}{4}$, $e^x \in \{\frac{1}{2}, -\frac{1}{2}\}$ and, since $0 < e^x$, $e^x = \frac{1}{2}$.

Then $x = \ln\frac{1}{2} = -\ln 2$, so that
$\cosh 2x = \left(\frac{1}{4} + 4\right)\Big/2 = \frac{17}{8}$, $\sinh 2x = \left(\frac{1}{4} - 4\right)\Big/2 = -\frac{15}{8}$ and

$$y = -\frac{150}{8} - \frac{102}{8} + 65\ln 2 = -\frac{63}{2} + 65\ln 2.$$

Similarly when $e^{2x} = 16$, $e^x = 4$, since it is positive, and $x = \ln 4$.

Then $\quad \cosh 2x = \left(16 + \frac{1}{16}\right)\Big/2 = \frac{257}{32}$, $\quad \sinh 2x = \left(16 - \frac{1}{16}\right)\Big/2 = \frac{255}{32}$, $\quad$ and

$$y = \frac{1275}{16} - \frac{771}{16} - 65\ln 4 = \frac{63}{2} - 130\ln 2.$$

Now $\qquad y'' = 40\sinh 2x - 24\cosh 2x = 8e^{2x} - 32e^{-2x}$.

Thus, for $e^{2x} = \frac{1}{4}$, $e^{-2x} = 4$ and $y'' = 2 - 128 < 0$, while for $e^{2x} = 16$, $e^{-2x} = \frac{1}{16}$ and $y'' = 128 - 2 > 0$.

Therefore the curve has a **maximum at $\left(-\ln 2, -\frac{63}{2} + 65\ln 2\right)$** and a **minimum at $\left(2\ln 2, \frac{63}{2} - 130\ln 2\right)$**.

In order to integrate functions involving hyperbolic functions, we may need to use the methods of substitution and of integration by parts.

We illustrate some of the techniques in examples.

EXAMPLE 3 *Find (i)* $\int \tanh x\,dx$, *(ii)* $\int \operatorname{sech} x\,dx$.

(i) Use the substitution $\cosh x = u$, then $\sinh x = \frac{du}{dx}$ and

$$\int \tanh x\,dx = \int \frac{\sinh x}{\cosh x}dx = \int \frac{1}{u}\frac{du}{dx}dx = \int \frac{1}{u}du$$
$$= \ln u + c = \mathbf{\ln(\cosh x) + c}.$$

(ii) By the substitution $u = e^x$, $\operatorname{sech} x = 2/(e^x + e^{-x}) = 2u/(u^2+1)$, and $\frac{du}{dx} = e^x = u$. Therefore

$$\int \operatorname{sech} x\,dx = \int \frac{2u}{u^2+1}\frac{1}{u}du = \int \frac{2}{u^2+1}du = 2\tan^{-1} u + c = \mathbf{2\tan^{-1} e^x + c}.$$

EXAMPLE 4 *Evaluate* $\int_0^1 \operatorname{sech}^3 x\,dx$.

Since $\frac{d}{dx}\tanh x = \operatorname{sech}^2 x$, we integrate by parts.

$$\int_0^1 \operatorname{sech}^3 x\,dx = \int_0^1 \operatorname{sech} x \frac{d}{dx}\tanh x\,dx$$
$$= \Big[\operatorname{sech} x \tanh x\Big]_0^1 - \int_0^1 \tanh x \frac{d}{dx}\operatorname{sech} x\,dx$$
$$= \frac{2(e-e^{-1})}{(e+e^{-1})^2} + \int_0^1 \operatorname{sech} x \tanh^2 x\,dx$$
$$= \frac{2(e-e^{-1})}{(e+e^{-1})^2} + \int_0^1 \operatorname{sech} x\,(1-\operatorname{sech}^2 x)dx$$
$$= \frac{2(e-e^{-1})}{(e+e^{-1})^2} + \int_0^1 \operatorname{sech} x\,dx - \int_0^1 \operatorname{sech}^3 x\,dx.$$

Hence $$2\int_0^1 \operatorname{sech}^3 x\,dx = \frac{2(e-e^{-1})}{(e+e^{-1})^2} + \int_0^1 \operatorname{sech} x\,dx$$

$$= \frac{2(e-e^{-1})}{(e+e^{-1})^2} + 2\Big[\tan^{-1} e^x\Big]_0^1, \text{ on using example 3.}$$

So $$\int_0^1 \operatorname{sech}^3 x\,dx = \frac{(e-e^{-1})}{(e+e^{-1})^2} + \tan^{-1} e - \frac{\pi}{4}$$

$$= \mathbf{\frac{(e^3-e)}{(e^2+1)^2} + \tan^{-1} e - \frac{\pi}{4}}.$$

EXERCISE 5.2

1 Prove that:

(i) $\frac{d}{dx}\coth x = -\operatorname{cosech}^2 x$, (ii) $\frac{d}{dx}\operatorname{cosech} x = -\operatorname{cosech} x \coth x$,

(iii) $\frac{d}{dx}\tanh x = \operatorname{sech}^2 x$, (iv) $\frac{d}{dx}(\cosh^2 x - \sinh^2 x) = 0$.

2 By expressing the hyperbolic functions in terms of e^x and e^{-x}, prove that: (i) $\int \cosh x\,dx = \sinh x + c$, (ii) $\int \sinh x\,dx = \cosh x + c$.

3 Find (i) $\int \operatorname{cosech}^2 2x\,dx$, (ii) $\int \operatorname{sech} 3x \tanh 3x\,dx$.

4 Find the derivatives of: (i) $\cosh 2x$, (ii) $3\sinh x \cosh x$, (iii) $\sinh(3x-2)$, (iv) $x\cosh x$, (v) $e^x \sinh x$, (vi) $(\tanh x)/x$, (vii) $\ln(\cosh x)$, (viii) $\cos(\cosh x)$, (ix) $\tan^{-1}(\sinh x)$, (x) $\sinh(\ln x)$.

5 Find the indefinite integral of:

(i) $\sinh^2 x$, by expressing it in terms of $\cosh 2x$,

(ii) $\dfrac{\sinh x}{\cosh^2 x}$, by using the substitution $\cosh x = u$,

(iii) $\sinh^3 x$, by using the substitution $\cosh x = u$,

(iv) $\tanh^2 x$, by using the result of Question **1**, (iii),

(v) $\tanh x \operatorname{sech}^3 x$, by using the substitution $\operatorname{sech} x = u$,

(vi) $e^x \sinh x$, by expressing $\sinh x$ in terms of exponentials,

(vii) $x\cosh x$, by integrating by parts,

(viii) $\sinh 2x \cosh 4x$, by writing it as a sum of sinh functions.

6 Find the equations of the tangent and the normal to the curve with equation $y = \cosh(3x-1)$ at the point $(1, \cosh 2)$.

7 The point P is given by $t = p$ on the curve with parametric equations

$$x = a\cosh t,\ y = b\sinh t.$$

Find the equations of the tangent and the normal to the curve at P. Find the area enclosed by the curve and the line $x = a\cosh p$.

8 The particle P moves on the line Ox and, at time t, its distance from O is x. If $x = 2\cosh t + \sinh 3t$, find the speed and the acceleration of P when $t = 1$.

9 (i) Given that $y = A\cosh nx + B\sinh nx$, prove that

$$\frac{d^2y}{dx^2} = n^2 y,$$

(ii) Given that $y = A\cos nx + B\sin nx$, prove that

$$\frac{d^2y}{dx^2} = -n^2y.$$

10 Find the indefinite integral of:
(i) $\coth x$, (ii) $\tanh^3 x \operatorname{sech}^2 x$,
(iii) $\tanh^4 x$, (iv) $\sinh x \cosh^3 x$,
(v) $\sinh^3 x \cosh^2 x$, (vi) $x^2 \sinh 3x$.

11 Prove that the minimum value of $5\cosh x + 3\sinh x$ is equal to 4.

12 Find the volume formed by rotating the area between the lines $x = 0$, $y = 0$, $x = 2$ and the curve $y = \sinh x$ through four right-angles about the axis Ox.

5.3 Inverse hyperbolic functions

The function sinh is a one-one function with both domain and range equal to $\mathbb{R}$. Hence, it has an inverse function $\sinh^{-1}$ (or arsinh) which also has domain and range equal to $\mathbb{R}$. The function tanh is also one-one and has domain $\mathbb{R}$ and range $\{x: -1 < x < 1\}$. Thus it has an inverse function $\tanh^{-1}$ (or artanh) with domain equal to $\{x: -1 < x < 1\}$ and range equal to $\mathbb{R}$.

We have to be more careful when defining an inverse function to the function cosh, since it is not one-one, for $\cosh(-x) = \cosh x$. We produce a one-one function, with range $\{x: x \geqslant 1\}$, by restricting the domain to $\{x: x \geqslant 0\}$. Then we have an inverse function $\cosh^{-1}$ (or arcosh) with domain $\{x: x \geqslant 1\}$ and range $\{x: x \geqslant 0\}$.

The graphs of $\cosh^{-1}$, $\sinh^{-1}$ and $\tanh^{-1}$ may be obtained from the graphs of $\cosh_{(x \geqslant 0)}$, sinh and tanh (Fig. 5.2) by reflection in the line $y = x$. We show these graphs in Fig. 5.3.

Inverse hyperbolic functions in terms of logarithms

Let $y = \cosh^{-1} x$ so that $x = \cosh y$ and $x^2 - 1 = \sinh^2 y$. Now $x \geqslant 1$ and $y \geqslant 0$, by definition, so $\sinh y \geqslant 0$ and the positive square root of the equation gives $\sinh y = \sqrt{(x^2 - 1)}$.

Then $$e^y = \cosh y + \sinh y = x + \sqrt{(x^2 - 1)},$$
so that $$y = \cosh^{-1} x = \ln\{x + \sqrt{(x^2 - 1)}\},\ x \geqslant 1.$$

Let $y = \sinh^{-1} x$, so that $x = \sinh y$, $x^2 + 1 = \cosh^2 y$ and, since $\cosh y > 0$, $\cosh y = \sqrt{(x^2 + 1)}$.

Then $$e^y = \sinh y + \cosh y = x + \sqrt{(x^2 + 1)},$$
so that $$y = \sinh^{-1} x = \ln\{x + \sqrt{(x^2 + 1)}\},\ x \in \mathbb{R}.$$

Let $y = \tanh^{-1} x$, $|x| < 1$, then

$$x = \tanh y = \frac{e^y - e^{-y}}{e^y + e^{-y}} = \frac{e^{2y} - 1}{e^{2y} + 1}.$$

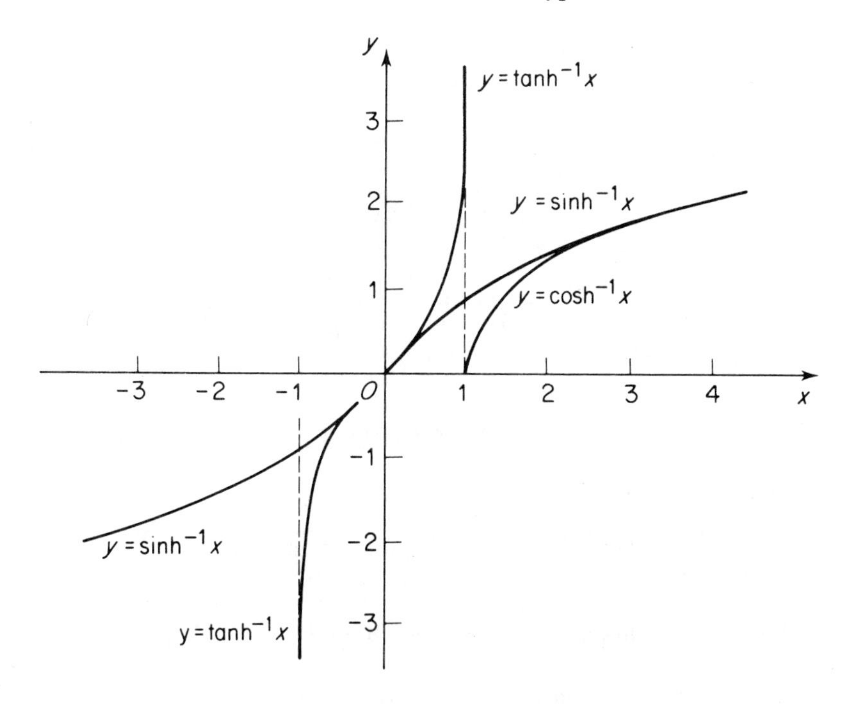

Fig. 5.3 Graphs of $\cosh^{-1} x$, $\sinh^{-1} x$ and $\tanh^{-1} x$

Solving for e^{2y} we find that $e^{2y} = \dfrac{1+x}{1-x}$, so that $y = \dfrac{1}{2}\ln\left(\dfrac{1+x}{1-x}\right)$. Thus

$$\tanh^{-1} x = \ln\sqrt{\left(\frac{1+x}{1-x}\right)}, \quad \text{for} \quad |x| < 1.$$

Note that, for $|x| < 1$, $\dfrac{1+x}{1-x} > 0$, so there is no need for a modulus sign inside the square root.

In the following examples, the solutions require the expression of inverse hyperbolic functions in terms of logarithms.

EXAMPLE 1 *Solve the equation* $1 + 7\sinh x = 4\cosh^2 x$.

Let $y = \sinh x$, then $1 + 7y = 4(y^2 + 1)$, so $4y^2 - 7y + 3 = 0$. On factorising, we find that $(4y-3)(y-1) = 0$, so that $y = \frac{3}{4}$ or $y = 1$. Now $x = \sinh^{-1} y = \ln\{y + \sqrt{(y^2+1)}\}$ and so $\mathbf{x = \ln 2}$ **or** $\mathbf{\ln(1+\sqrt{2})}$.

EXAMPLE 2 *Show that* $\displaystyle\int_0^1 \sqrt{(1+x^2)}\,dx = (\sqrt{2} + \ln(1+\sqrt{2}))/2$.

Use the substitution $x = \sinh u$, and let $p = \operatorname{arsinh} 1 = \ln(1+\sqrt{2})$. Then $\sinh p = 1$ and $\cosh p = \sqrt{2}$, and

$$\int_0^1 \sqrt{(1+x^2)}\,dx = \int_0^p \cosh u \cosh u \, du$$

$$= \frac{1}{2}\int_0^p (\cosh 2u + 1)\,du$$

$$= \frac{1}{4}\Big[\sinh 2u + 2u\Big]_0^p = \frac{1}{2}(\sinh p \cosh p + p)$$

$$= [\sqrt{2} + \ln(1+\sqrt{2})]/2.$$

EXERCISE 5.3

1 Given that $a > 0$, prove that:

(i) $\cosh^{-1}\dfrac{x}{a} = \ln\left\{\dfrac{x+\sqrt{(x^2-a^2)}}{a}\right\}$, for $x \geqslant a$,

(ii) $\sinh^{-1}\dfrac{x}{a} = \ln\left\{\dfrac{x+\sqrt{(x^2+a^2)}}{a}\right\}$, for $x \in \mathbb{R}$,

(iii) $\tanh^{-1}\dfrac{x}{a} = \ln\sqrt{\left(\dfrac{a+x}{a-x}\right)}$, for $|x| < a$.

2 Evaluate: (i) $\cosh^{-1} 2$, (ii) $\sinh^{-1}(-3)$, (iii) $\tanh^{-1}\frac{1}{2}$.

3 Given that $f(x) = 3\sinh^{-1} x + 4\cosh^{-1} x$, evaluate, where possible: (i) $f(2)$, (ii) $f(1)$, (iii) $f(\frac{1}{2})$.

4 Evaluate: (i) $\Big[\sinh^{-1} x\Big]_1^2$, (ii) $\Big[\cosh^{-1} x\Big]_1^{3/2}$.

5 Given that $\operatorname{arsinh} x = \ln\sqrt{6}$, prove that $\operatorname{arcosh} x = \ln\sqrt{(3/2)}$.

6 Solve the equations in terms of logarithms: (i) $\cosh x - 1 = \sinh^2 x$, (ii) $7 + \sinh^2 x = 5\cosh x$, (iii) $2\cosh x = 5\sinh x$.

7 Evaluate: (i) $\int_0^2 \sqrt{(4+x^2)}\,dx$, (ii) $\int_1^2 \sqrt{(x^2-1)}\,dx$.

5.4 Derivatives of inverse hyperbolic functions

Let $y = \cosh^{-1} x$, $x > 1$, so that $x = \cosh y$, $y > 0$, and then

$$\frac{dx}{dy} = \sinh y = \sqrt{(x^2-1)} \text{ and } \frac{dy}{dx} = \frac{1}{\dfrac{dx}{dy}} = \frac{1}{\sinh y} = \frac{1}{\sqrt{(x^2-1)}}.$$

Let $y = \sinh^{-1} x$, so that $x = \sinh y$ and then

$$\frac{dx}{dy} = \cosh y = \sqrt{(1+x^2)} \text{ and } \frac{dy}{dx} = \frac{1}{\sqrt{(x^2+1)}}.$$

Let $y = \tanh^{-1} x$, so that $x = \tanh y$ and then

$$\frac{dx}{dy} = \operatorname{sech}^2 y = 1 - x^2 \text{ and } \frac{dy}{dx} = \frac{1}{1-x^2}.$$

Derivatives of hyperbolic and circular functions and their inverses

We tabulate all the basic results obtained so far, comparing them with the corresponding results for trigonometric functions:

$\frac{d}{dx}\cosh x = \sinh x,$	$\frac{d}{dx}\cos x = -\sin x,$
$\frac{d}{dx}\sinh x = \cosh x,$	$\frac{d}{dx}\sin x = \cos x,$
$\frac{d}{dx}\tanh x = \operatorname{sech}^2 x,$	$\frac{d}{dx}\tan x = \sec^2 x,$
$\frac{d}{dx}\operatorname{sech} x = -\operatorname{sech} x \tanh x,$	$\frac{d}{dx}\sec x = \sec x \tan x,$
$\frac{d}{dx}\operatorname{cosech} x = -\operatorname{cosech} x \coth x,$	$\frac{d}{dx}\operatorname{cosec} x = -\operatorname{cosec} x \cot x,$
$\frac{d}{dx}\coth x = -\operatorname{cosech}^2 x,$	$\frac{d}{dx}\cot x = -\operatorname{cosec}^2 x,$
$\frac{d}{dx}\cosh^{-1} x = \frac{1}{\sqrt{(x^2-1)}},$	$\frac{d}{dx}\cos^{-1} x = \frac{-1}{\sqrt{(1-x^2)}},$
$\frac{d}{dx}\sinh^{-1} x = \frac{1}{\sqrt{(x^2+1)}},$	$\frac{d}{dx}\sin^{-1} x = \frac{1}{\sqrt{(1-x^2)}},$
$\frac{d}{dx}\tanh^{-1} x = \frac{1}{1-x^2},$	$\frac{d}{dx}\tan^{-1} x = \frac{1}{1+x^2}.$

These last three equations may be written in the form of standard integrals:

$$\int \frac{1}{\sqrt{(x^2-1)}}\,dx = \cosh^{-1} x + c = \ln\{x + \sqrt{(x^2-1)}\} + c,$$

$$\int \frac{1}{\sqrt{(x^2+1)}}\,dx = \sinh^{-1} x + c = \ln\{x + \sqrt{(x^2+1)}\} + c,$$

$$\int \frac{1}{1-x^2}\,dx = \tanh^{-1} x + c = \ln\sqrt{\left(\frac{1+x}{1-x}\right)} + c.$$

The last of these results can also be obtained directly by using partial fractions.

For the purpose of making use of hyperbolic functions in integration, it is useful to extend the integral forms, given above, by changing the variable to x/a. This uses the result that if $\int f(x)\,dx = F(x)+c$, then $\int \frac{1}{a} f\left(\frac{x}{a}\right) dx = F\left(\frac{x}{a}\right)+c$ which follows from the equation

$$\frac{d}{dx} F\left(\frac{x}{a}\right) = \frac{1}{a} f\left(\frac{x}{a}\right).$$

This substitution gives the standard forms:

$$\int \frac{1}{\sqrt{(x^2-a^2)}}\,dx = \cosh^{-1}\frac{x}{a}+c = \ln\{x+\sqrt{(x^2-a^2)}\} - \ln a + c,$$

$$\int \frac{1}{\sqrt{(x^2+a^2)}}\,dx = \sinh^{-1}\frac{x}{a}+c = \ln\{x+\sqrt{(x^2+a^2)}\} - \ln a + c,$$

$$\int \frac{a}{a^2-x^2}\,dx = \tanh^{-1}\frac{x}{a}+c = \ln\sqrt{\left(\frac{a+x}{a-x}\right)}+c.$$

In order to integrate a function of the type $\dfrac{1}{\sqrt{(Ax^2+Bx+C)}}$, it is necessary to transform it into a standard form by a change of variable. The quadratic expression in the square root is manipulated by completing the square and then the change of variable is made.

EXAMPLE 1 *Evaluate:* (*i*) $\displaystyle\int_0^1 \frac{1}{\sqrt{(x^2+4x+13)}}\,dx$; (*ii*) $\displaystyle\int_2^4 \frac{1}{\sqrt{(x^2+4x)}}\,dx$.

(i) Completing the square of the quadratic expression in the denominator

$$x^2+4x+13 = (x+2)^2+9 = t^2+9$$

on substituting $t = x+2$. Then

$$\int_0^1 \frac{1}{\sqrt{(x^2+4x+13)}}\,dx = \int_2^3 \frac{1}{\sqrt{(t^2+9)}}\,dt = \left[\operatorname{arsinh}\frac{t}{3}\right]_2^3$$

$$= \operatorname{arsinh}1 - \operatorname{arsinh}\frac{2}{3} = \ln(1+\sqrt{2}) - \ln\left(\frac{2+\sqrt{13}}{3}\right)$$

$$= \ln\left(\frac{3+3\sqrt{2}}{2+\sqrt{13}}\right).$$

(ii) Since $x^2+4x=(x+2)^2-4$, we use the same substitution, $x+2=t$, as we used in (i).

$$\int_2^4 \frac{1}{\sqrt{(x^2+4x)}}\,\mathrm{d}x = \int_4^6 \frac{1}{\sqrt{(t^2-4)}}\,\mathrm{d}t = \left[\operatorname{arcosh}\frac{t}{2}\right]_4^6 = \operatorname{arcosh} 3 - \operatorname{arcosh} 2$$

$$= \ln(3+\sqrt{8}) - \ln(2+\sqrt{3}) = \mathbf{ln}\left(\frac{\mathbf{3+\sqrt{8}}}{\mathbf{2+\sqrt{3}}}\right).$$

EXAMPLE 2 *Find the area of the finite region of the plane bounded by the curve* $y = \operatorname{arsinh} x$, *the* x-*axis and the line* $x = 1$.

Let the area be A. Then, using integration by parts,

$$A = \int_0^1 \operatorname{arsinh} x\,\mathrm{d}x = \Big[x \operatorname{arsinh} x\Big]_0^1 - \int_0^1 \frac{x}{\sqrt{(1+x^2)}}\,\mathrm{d}x$$

$$= \operatorname{arsinh} 1 - \left[\sqrt{(1+x^2)}\right]_0^1$$

$$= \mathbf{ln(1+\sqrt{2})-\sqrt{2}+1}.$$

Sometimes, an equation is to be solved which contains both hyperbolic and polynomial functions. In this case, a solution cannot be given in algebraic form but an approximate numerical solution may be found by using graphical and numerical methods.

EXAMPLE 3 *Solve for* x, *correct to two places of decimals, the equation:*

$$\int_0^x (3-\cosh t)\,\mathrm{d}t = 0.$$

We begin by evaluating the definite integral.

$$\int_0^x (3-\cosh t)\,\mathrm{d}t = \Big[3t - \sinh t\Big]_0^x = 3x - \sinh x.$$

We need to solve the equation $f(x) = 0$, where $f(x) = 3x - \sinh x$. The roots of $f(x) = 0$ are given by the points of intersection of the two graphs, with equations $y = 3x$ and $y = \sinh x$ and, as is seen in Fig. 5.4, there are three solutions, namely 0, α and $-\alpha$, where $f(\alpha) = 0$ and $0 < \alpha$. The solution $x = \alpha$ cannot be obtained exactly and we use linear interpolation to find an approximation to α. The graphs of Fig. 5.4 suggest that $2 < \alpha < 3$ and this gives us the starting point.

$f(2) = 2{\cdot}37$, $f(3) = -1{\cdot}02$, and $2{\cdot}37/(2{\cdot}37+1{\cdot}02) = 0{\cdot}7$,
$f(2{\cdot}7) = 0{\cdot}69$, $f(2{\cdot}8) = 0{\cdot}21$, so α does not lie between $2{\cdot}7$ and $2{\cdot}8$)

$f(2{\cdot}8) = 0{\cdot}21$, $f(2{\cdot}9) = -0{\cdot}36$, and $0{\cdot}21/(0{\cdot}21+0{\cdot}36) = 0{\cdot}4$.
$f(2{\cdot}84) = -0{\cdot}009$, $f(2{\cdot}83) = 0{\cdot}047$, so we obtain the solution set $\{\mathbf{2{\cdot}84, 0, -2{\cdot}84}\}$.

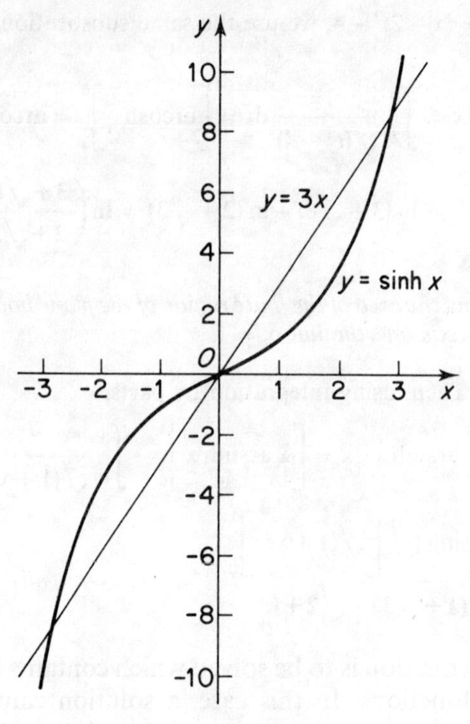

Fig. 5.4

EXERCISE 5.4

1 Verify that:

(i) $\displaystyle\int \frac{1}{\sqrt{(x^2-a^2)}}\,dx = \cosh^{-1}\frac{x}{a} + c$, (use the substitution $x = a\cosh u$),

(ii) $\displaystyle\int \frac{1}{\sqrt{(x^2+a^2)}}\,dx = \sinh^{-1}\frac{x}{a} + c$, (use the substitution $x = a\sinh u$),

(iii) $\displaystyle\int \frac{a}{a^2-x^2}\,dx = \tanh^{-1}\frac{x}{a} + c$, (use the substitution $x = a\tanh u$),

(iv) $\displaystyle\int \frac{a}{a^2-x^2}\,dx = \ln\sqrt{\left(\frac{a+x}{a-x}\right)} + c$, (use partial fractions).

In (iii) and (iv) above, assume that $|x| < a$. Deduce that, for $|x| < a$, $\tanh^{-1}\dfrac{x}{a} = \dfrac{1}{2}\ln\left(\dfrac{a+x}{a-x}\right)$.

2 Differentiate, with respect to x:

(i) $\sinh^{-1}(3x-4)$, (ii) $\tanh^{-1}(2x-1)$, (iii) $\cosh^{-1}(x^2)$,
(iv) $\text{arcosh}\,(1/x)$, (v) $\text{arsinh}\,(x^2+2)$, (vi) $x\sinh^{-1}x$,
(vii) $\sinh^{-1}(\cosh x)$.

3 Find the indefinite integral of:
(i) $1/\sqrt{(x^2-4)}$, by using the substitution $x = 2\cosh u$,
(ii) $1/\sqrt{(x^2+9)}$, by using the substitution $x = 3\sinh u$,
(iii) $1/(x^2-9)$, by using the substitution $x = 3\tanh u$,
(iv) $1/(x^2+9)$, by using the substitution $x = 3\tan u$,
(v) $1/\sqrt{(9-x^2)}$, by using the substitution $x = 3\cos u$.

4 Integrate, with respect to x:
(i) $1/\sqrt{(x^2-4x+5)}$, by using the substitution $x = 2+\sinh t$,
(ii) $1/\sqrt{(x^2-4x+3)}$, by using the substitution $x = 2+\cosh t$,
(iii) $1/\sqrt{(4x^2+12x)}$, by completing the square of the quadratic expression.

5 Integrate with respect to x: (i) $\dfrac{1}{\sqrt{(4+x^2)}}$, (ii) $\dfrac{1}{\sqrt{(9x^2-1)}}$,

(iii) $\dfrac{1}{\sqrt{\{(x+1)^2+1\}}}$, (iv) $\dfrac{1}{\sqrt{(x^2+4x)}}$, (v) $\dfrac{1}{4x^2-9}$, (vi) $\dfrac{1}{\sqrt{(4x^2+9)}}$.

6 Evaluate: (i) $\displaystyle\int_0^1 \frac{1}{\sqrt{(1+x^2)}}\,dx$, (ii) $\displaystyle\int_2^3 \frac{1}{\sqrt{(x^2-1)}}\,dx$, (iii) $\displaystyle\int_a^b \frac{1}{\sqrt{(x^2+4)}}\,dx$,

(iv) $\displaystyle\int_0^1 \frac{1}{\sqrt{(x^2+2x+2)}}\,dx$, (v) $\displaystyle\int_0^1 \frac{1}{\sqrt{(x^2+4x+5)}}\,dx$,

(vi) $\displaystyle\int_0^1 \frac{1}{\sqrt{(x^2+4x+3)}}\,dx$, (vii) $\displaystyle\int_0^1 \frac{2x+4}{\sqrt{(x^2+4x+3)}}\,dx$,

(viii) $\displaystyle\int_0^1 \frac{x}{\sqrt{(x^2+4x+3)}}\,dx$.

7 Calculate the area between the lines $x = 0$, $x = 1$ and the two parts of the curve given by the equation $y^2(x^2+1) = 1$.

MISCELLANEOUS EXERCISE 5

1 Solve, for real x, the equation $5\cosh x - 3\sinh x = 5$. (*L*)

2 Find the possible values of $\sinh x$ if $\begin{vmatrix} \cosh x & -\sinh x \\ \sinh x & \cosh x \end{vmatrix} = 2$. (*L*)

3 Differentiate with respect to x (a) $\ln(\tanh x)$, (b) $\cosh(\sin 2x)$. (*L*)

4 The function f is defined by $f(x) = 6\cosh 2x + 10\sinh 2x - 65x$, $x \in \mathbb{R}$. Find the values of x for which $f'(x)$ is zero. (*JMB*)

5 Sketch, on the same diagram, the curves $y = 5e^{-2x}$ and $y = \cosh x$. Verify that the curves intersect where $y = 5/4$. Find the area of the finite region enclosed by the two curves and the axis of y. This region is rotated completely about the axis of y. Show that the volume V of the solid of revolution so formed is given by

$$V = 2\pi \int_0^{\ln 2} x(5e^{-2x} - \cosh x)\,dx.$$

Hence determine V, leaving your result in terms of π and natural logarithms. (*C*)

6 Find (a) $\displaystyle\int \frac{1}{\sqrt{(1+9x^2)}}\,dx$, (b) $\displaystyle\int \sinh^2 3x\,dx$. (*L*)

7 By means of the substitution $u = 1 + \cosh x$, or otherwise, evaluate

$$\int_0^{\text{arcosh}\,2} \frac{\tanh x}{1+\cosh x}\,dx.$$ (*L*)

8 (i) Find, in logarithmic form, the solutions of the equation

$$9\cosh x - 6\sinh x = 7.$$

(ii) If $y = \cosh(2\,\text{arsinh}\,x)$, show that $(x^2+1)\dfrac{d^2y}{dx^2} + x\dfrac{dy}{dx} - 4y = 0.$ (*L*)

9 Find positive numbers R and a such that

$$5\cosh x - 4\sinh x = R\cosh(x - \ln a)$$

for all real x. Hence, or otherwise,
(i) find the minimum value of $5\cosh x - 4\sinh x$,
(ii) solve the equation $5\cosh x - 4\sinh x = 5$ $(x > 0)$, giving your answer in the form $x = \ln b$ for some real number b. (*JMB*)

10 Sketch on the same diagram the graphs of $y = \cosh x$ and $y = \tanh x$. Evaluate $\cosh x$ when $x = \ln(2+\sqrt{3})$. If $z = \cosh x - \tanh x$, evaluate z and $\dfrac{dz}{dx}$ when $x = 0$ and show that when $x = \ln(2+\sqrt{3})$ then $z > 1$. Deduce that z has a minimum value for some value of x between 0 and $\ln(2+\sqrt{3})$. (*L*)

11 Prove that the curve whose equation is $y = \sinh^{-1}(x+1)$ has a point of inflection at $(-1, 0)$. Find, in its simplest form, the equation of the normal to the curve at this point. (*JMB*)

12 Show that $\dfrac{1+\tanh^2 x}{1-\tanh^2 x} = \cosh 2x$. By means of the substitution $t = \tanh x$, or otherwise, find the indefinite integral $\int \text{sech}\,2x\,dx$. (*JMB*)

13 Sketch, for all real values of x, the curve whose equation is $y = \text{arsinh}\,x$. Find the angle between the x axis and the tangent to this curve at the origin, and draw this tangent on your sketch. Find the volume of the solid generated when the finite region bounded by the curve and the lines $y = 0$ and $x = 1$ is rotated through one complete revolution about the y-axis. (*L*)

14 Prove that $\dfrac{d}{dx}(\tanh^{-1} x) = \dfrac{1}{1-x^2}$. Hence, or otherwise, prove that

$$\int_0^{1/2} \tanh^{-1} x\,dx = \frac{3}{4}\ln 3 - \ln 2.$$ (*C*)

15 By using the substitution $y = \sinh x$, or otherwise,

(i) find $\displaystyle\int \frac{y}{\sqrt{(y^2+1)}} \ln\{y+\sqrt{(y^2+1)}\}\,dy$,

(ii) show that $\displaystyle\int_0^{4/3} y \ln\{y+\sqrt{(y^2+1)}\}\,dy = \frac{41\ln 3 - 20}{36}$. (*C*)

16 Prove that $\sinh^{-1} x = \ln\{x+\sqrt{(x^2+1)}\}$. Show that $\dfrac{d}{dx}(\sinh^{-1} x) = \dfrac{1}{\sqrt{(x^2+1)}}$. Evaluate $\displaystyle\int_1^8 \frac{1}{\sqrt{(x^2-2x+2)}}\,dx$, expressing your answer as a natural logarithm. Show that $\displaystyle\int_1^2 \frac{3}{\sqrt{(x^2-2x+2)}}\,dx = \int_1^8 \frac{1}{\sqrt{(x^2-2x+2)}}\,dx$. (*JMB*)

17 (a) Starting from the definition of $\sinh x$ in terms of e^x, prove that $\sinh^{-1} x = \ln\{x+\sqrt{(x^2+1)}\}$. Using the same axes, sketch the graphs of $\sinh x$ and of $\sinh^{-1} x$, labelling your graphs carefully.
(b) Prove that the derivative of $\sinh^{-1} x$ is $(1+x^2)^{-1/2}$. Show that the equation $(1+x^2)\dfrac{d^2y}{dx^2} + x\dfrac{dy}{dx} - 2 = 0$ is satisfied when $y = (\sinh^{-1} x)^2$.
(c) Obtain $\displaystyle\int_0^1 x \sinh x\,dx$, leaving your answer in terms of e. (*L*)

18 Show that $\sinh 3\theta = 3\sinh\theta + 4\sinh^3\theta$. Use the substitution $x = \sinh\theta$ to find, in terms of $2^{1/3}$, the real root of the equation

$$16x^3 + 12x - 3 = 0.$$ (*JMB*)

19 By substituting $u = e^x$, or otherwise, evaluate

$$\int_0^{\ln 2} \frac{1}{4+5\cosh x - 3\sinh x}\,dx.$$ (*JMB*)

20 Show that there exists a real number x such that $\cosh x = \sqrt{a} + \sqrt{b}$ and $\sinh x = \sqrt{a} - \sqrt{b}$ if and only if a and b are positive real numbers such that $16ab = 1$. (*JMB*)

21 Show that the derivative of $\tan^{-1}(e^{2x})$ is $\operatorname{sech} 2x$. Show also that the derivative of $\tan^{-1}(\tanh x)$ is $\operatorname{sech} 2x$. Deduce that $\tan^{-1}(e^{2x}) = \tan^{-1}(\tanh x) + c$, where c is a constant. (*JMB*)

22 From the definition of the hyperbolic cosine in terms of exponentials, prove that $\cosh^{-1}\left(\dfrac{1}{x}\right) = \ln\left\{\dfrac{1+\sqrt{(1-x^2)}}{x}\right\}$ $(0 < x \leqslant 1)$. (*JMB*)

23 Given that $\operatorname{artanh} x = \ln 2$, show that $\operatorname{arsinh} x = \ln\left(\dfrac{3+\sqrt{34}}{5}\right)$. (*L*)

24 If $y = (\operatorname{arcosh} x)^2$ prove that $(x^2-1)\left(\dfrac{dy}{dx}\right)^2 = 4y$. (*L*)

25 By substituting $\sinh u = \tan\theta$, show that $\displaystyle\int_0^a \operatorname{sech} u\,du = \arctan(\sinh a)$. (*L*)

6 Power Series

6.1 Approximations to a function

In *Pure Mathematics* we considered various methods of approximating to functions. In Chapter **14**, we used approximations by a linear function in order to solve equations. In graphical terms, both a chord and a tangent were used as approximations to the function. The same linear approximations were employed in Chapter **18** to estimate the gradient of a function and then to define the derived function, or derivative. Finally, in Chapter **20**, linear approximation was used for the trapezium rule and quadratic approximation was used for Simpson's rule, both rules being used to calculate the value of a definite integral.

The above approximations fall into two types. In the case of numerical integration, the function is approximated by a polynomial (linear or quadratic) *over an interval.* In the case of the derivative, the linear approximation used is an approximation *close to a point*, the point where the derivative is to be found. The present chapter is concerned with this second type of approximation, namely the approximation to a function by a polynomial near to (or in the neighbourhood of) a point.

The two types of quadratic approximation to the function e^x are shown in Fig. 6.1. In (a) the approximating polynomial $p_1(x)$ near

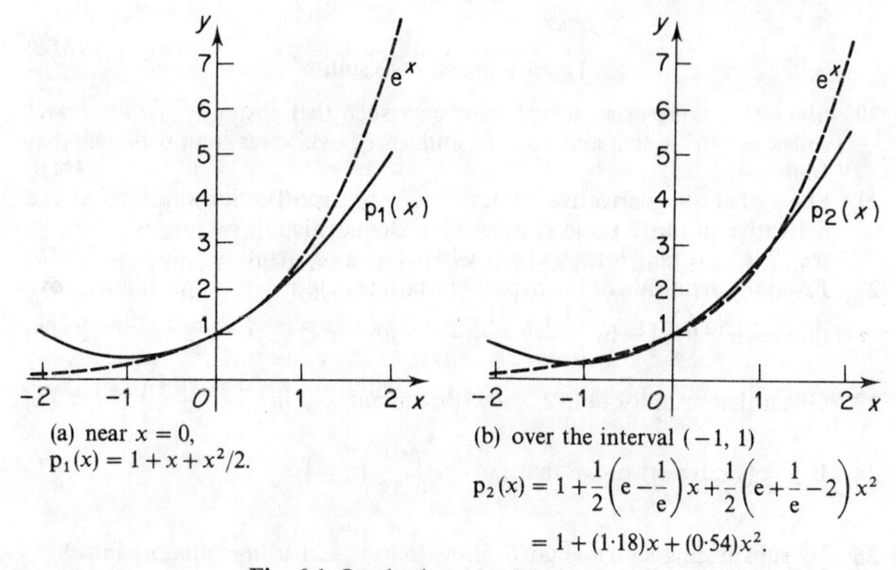

(a) near $x = 0$,
$p_1(x) = 1 + x + x^2/2$.

(b) over the interval $(-1, 1)$

$$p_2(x) = 1 + \frac{1}{2}\left(e - \frac{1}{e}\right)x + \frac{1}{2}\left(e + \frac{1}{e} - 2\right)x^2$$
$$= 1 + (1{\cdot}18)x + (0{\cdot}54)x^2.$$

Fig. 6.1 Quadratic approximations to e^x.

to the origin is given by $p_1(x) = 1 + x + x^2/2$ and in (b) the approximating polynomial $p_2(x)$ over the interval $(-1, 1)$ is given by $p_2(x) = 1 + \frac{1}{2}\left(e - \frac{1}{e}\right)x + \frac{1}{2}\left(e + \frac{1}{e} - 2\right)x^2 = 1 + (1{\cdot}18)x + (0{\cdot}54)x^2$. The values of $p_1(x)$ are only correct for one value of x, that is $x = 0$, but $p_1(x)$ gives a very good approximation for values of x which are close to this point, see the second and third rows of table 6.1.

Table 6.1

x	$-0{\cdot}2$	$-0{\cdot}1$	0	$0{\cdot}1$	$0{\cdot}2$
e^x	0·819	0·905	1·000	1·105	1·221
$1 + x + x^2/2$	0·820	0·905	1·000	1·105	1·220
$1 + 1{\cdot}18x + 0{\cdot}54x^2$	0·786	0·877	1·000	1·123	1·258

The values of $p_2(x)$ are correct for $x = -1$, 0 and 1 but are not as good near to $x = 0$. This is because the first and second derivatives of $p_1(x)$, as well as the value of the function, are matched to e^x at $x = 0$, which gives an almost identical shape to the graphs near to $x = 0$.

The approximation to $f(x)$ may be needed near to any point, such as $(a, f(a))$, but in many cases it is needed near to $(0, f(0))$, as in the example above. Then, in the neighbourhood of $x = 0$, x will be small and, if f is continuous, $f(x)$ will not differ much from $f(0)$. The approximation to $f(x)$ by means of a polynomial $p(x)$ of degree n, where

$$p(x) = a_0 + a_1x + a_2x^2 + \ldots + a_nx^n = \sum_{r=0}^{n} a_rx^r,$$

is often a good approximation for small values of $|x|$. It may be used in complicated mathematical models in order to simplify the mathematics. This is because such a polynomial can be differentiated and integrated easily and the derivative and integral are good approximations to the derivative and integral of the function f. For a particular value of x, the value of the polynomial is easily calculated and this is how function values are often calculated in a digital computer. A calculator, or computer, with facilities for calculating mathematical functions does not store large tables of values of the functions since this would use up far too much memory. Instead, it stores programs which calculate the values of the functions to the degree of accuracy displayed by the calculator and these programs will usually use polynomial approximations.

The method is useful when we have an infinite series expansion for the function, such as

$$f(x) = \sum a_rx^r = a_0 + a_1x + a_2x^2 + \ldots,$$

such that the terms decrease rapidly in magnitude. In this case, a small number of terms, say $n+1$, of the series is used and this is effectively using a polynomial approximation of degree n.

EXAMPLE *Use the series* $\sin x = x - x^3/3! + x^5/5! - x^7/7! + \ldots$ *to calculate* $\sin(\pi/6)$, *correct to 5 decimal places.*

Substitute $x = \pi/6$ and work to 6 decimal places for the required accuracy,

$$\begin{aligned}\sin(\pi/6) &= 0{\cdot}523599 - 0{\cdot}023925 + 0{\cdot}000328 - 0{\cdot}000002 + \ldots \\ &= \mathbf{0{\cdot}500000 +} \ldots .\end{aligned}$$

Thus, by using only the first four terms of the series an accuracy of 5 decimal places is obtained, which means that the polynomial $p(x)$, of degree 7, given by

$$p(x) = x - x^3/3! + x^5/5! - x^7/7!$$

is a fairly accurate approximation to $\sin x$ in the interval

$$-\pi/6 < x < \pi/6.$$

Note that, if x is much larger, then the series converges much more slowly and more terms of the series would be needed, (see Question **1** of Exercise 6.1).

EXERCISE 6.1

1 Use the series expansion for $\sin x$, given in the Example to evaluate: (i) $\sin 0{\cdot}1$, (ii) $\sin 1$, (iii) $\sin 10$. State the number of terms required in order to achieve an accuracy of 3 decimal places in each case.

2 A formula for the rth term of the sine series is $(-1)^{r+1}\dfrac{x^{2r-1}}{(2r-1)!}$. Use this formula to estimate how many terms of the series will be required to find $\sin \pi$, accurate to 3 decimal places. Find also how many terms will be required to obtain an accuracy of 8 decimal places.

6.2 Convergence of power series

The idea of an infinite series was introduced in *Pure Mathematics*, Chapter 26. Given a real sequence (u_r), the formal series $\sum u_r$ is obtained by placing + signs between the terms of the sequence,

$$\sum u_r = u_1 + u_2 + u_3 + \ldots .$$

The nth partial sum, S_n, of the series is the sum of the first n terms

$$S_n = \sum_{r=1}^{n} u_r = u_1 + u_2 + u_3 + \ldots + u_n.$$

As n tends to infinity, S_n may, or may not, tend to a limit. If S_n tends to a limit S, as n tends to infinity, then the series is said to converge to a limit S, called the sum (or the sum to infinity) of the series. Otherwise, the series is said to diverge.

In the case of a power series, we change the notation slightly and define its nth partial sum as the sum of the first $(n+1)$ terms so that this is a polynomial of degree n.

Definition A *power series* is an infinite series of the form

$$\sum a_r x^r = a_0 + a_1 x + a_2 x^2 + \ldots + a_r x^r + \ldots,$$

Its nth partial sum is $S_n(x)$, given by

$$S_n(x) = \sum_{r=0}^{n} a_r s^r = a_0 + a_1 x + a_2 x^2 + \ldots + a_n x^n.$$

If, for a given value of x, $\lim_{n \to \infty} S_n(x) = S(x)$, then the series converges to $S(x) = \sum a_r x^r$.

The convergence or divergence of a power series may well depend upon the value of x and it can be shown that, if the series converges for $x = a$, then it also converges for all x such that $|x| < |a|$. The proof is beyond the scope of this book but we state the consequences as a theorem.

Theorem 6.1 Given a power series $\sum a_r x^r$, then just one of the following three situations holds:

either (i) $\sum a_r x^r$ converges for all real values of x,

or (ii) $\sum a_r x^r$ converges for all x such that $|x| < R$ and diverges for all x such that $|x| > R$,

or (iii) $\sum a_r x^r$ converges only for $x = 0$.

The number R is called the *radius of convergence* of the power series and in (i) the series has infinite radius of convergence,
in (ii) the series has finite radius of convergence, R,
and in (iii) the series has zero radius of convergence.
An important property of power series, which again we are unable to prove here, is that, if the series is differentiated (or integrated) term by term the series thus obtained has the same radius of convergence as the original series and, within that radius of convergence, the sum of the differentiated (or integrated) series is the derivative (or integral) of the sum of the original series. This makes it useful to use partial sums of power series as polynomial approximations to functions in mathematical models.

The series for $\sin x$, given in §6.1, has infinite radius of convergence, although the rate of convergence varies widely, depending upon the value of x. We now show a power series with a finite radius of convergence.

EXAMPLE *Expand $1/(1+x)^2$ as a power series in x, using the binomial theorem and state the range of values of x for which the series converges. Demonstrate this by considering $1/(1{\cdot}2)^2$ and $1/(3{\cdot}2)^2$.*

By the binomial theorem,

$$(1+x)^{-2} = 1+(-2)x+\frac{(-2)(-3)}{2!}x^2+\frac{(-2)(-3)(-4)}{3!}x^3+\ \dots$$

$$\mathbf{= 1-2x+3x^2-4x^3+\ \dots .}$$

This infinite series expansion converges only for $|x| < 1$, so the power series has radius of convergence equal to 1.

$$\begin{aligned} 1/(1{\cdot}2)^2 &= 1/(1+0{\cdot}2)^2 = 1-2(0{\cdot}2)+3(0{\cdot}2)^2-4(0{\cdot}2)^3+\ \dots \\ &= 1-0{\cdot}4+0{\cdot}12-0{\cdot}032+0{\cdot}008-0{\cdot}00192+\ \dots \\ &\approx 0{\cdot}6941. \end{aligned}$$

(Compare this with the true value 0·69444 . . .).

$$\begin{aligned} 1/(3{\cdot}2)^2 &= 1/(1+2{\cdot}2)^2 = 1-2(2{\cdot}2)+3(2{\cdot}2)^2-4(2{\cdot}2)^3+\ \dots \\ &= 1-4{\cdot}4+14{\cdot}52-\ \dots, \end{aligned}$$

and the values of the partial sums oscillate and diverge.

EXERCISE 6.2

1 Evaluate the partial sums S_0, S_1, S_2, S_3 of the power series for the given values of x and state whether you think that they will converge for all real values of x:

(a) $f(x) = 1+x+x^2+x^3+\ \dots,$ $\quad x = 0,\ 1/2,\ 1,\ -5,\ 10,$

(b) $f(x) = 1+x/1+x^2/2+x^3/3+\ \dots,$ $\quad x = 0,\ 1/2,\ 1,\ -2,\ 8,$

(c) $f(x) = 1-x/2+x^2/4-x^3/8+\ \dots,$ $\quad x = 0,\ 1/2,\ 1,\ -2,\ 8,$

(d) $f(x) = 1+x^2/2!+x^4/4!+x^6/6!+\ \dots,$ $\quad x = 0,\ 1/2,\ 1,\ -4,$

(e) $f(x) = x-x^3/3!+x^5/5!-x^7/7!+\ \dots,$ $\quad x = 0,\ -1/2,\ 1,\ -2.$

2 Use the power series $\sinh x = x+x^3/3!+x^5/5!+x^7/7!+\ \dots$, to find sinh (1) accurate to 6 decimal places.

3 State whether the power series expansion for $(1+x)^n$ converges for the given values of x and n:
(i) $x = 1/2,\ n = 1/2$, (ii) $x = 2,\ n = 3$, (iii) $x = 1/2,\ n = -1$,
(iv) $x = 2,\ n = 1/2$, (v) $x = 1/2,\ n = 3$, (vi) $x = 2,\ n = -1$.

4 Let $f(x)$ be the polynomial $3+4x-5x^2$. Find the derivatives $f'(x)$, $f''(x)$ and evaluate $f(0)$, $f'(0)$ and $f''(0)$. Verify that $f(x) = f(0)+f'(0)x+\frac{1}{2}f''(0)x^2$.

5 Let $f(x)$ be the cubic polynomial $1+x+x^2+x^3$. Find the values of a, b, c and d such that

$$f(x) = af(0)+bf'(0)x+cf''(0)x^2+df'''(0)x^3.$$

6.3 Linear approximation at a point

In order to find a linear approximation to a function f such that its values closely approximate to the values of f near to $x = 0$, the obvious candidate is the function whose graph is the tangent to the graph of f at the origin, *Pure Mathematics* §18.4. The function must take the value f(0) when $x = 0$ and must have the gradient f′(0) at $x = 0$. Therefore, the linear approximation to f(x) at $x = 0$ is given by

$$f(x) \approx f(0) + xf'(0).$$

EXAMPLE 1 *Find linear approximations, at $x = 0$, to:*
(*i*) $(x-1)^3+2$, (*ii*) $\ln(2+x)$.

(i) Let $f(x) = (x-1)^3+2$, then $f'(x) = 3(x-1)^2$. On putting $x = 0$, $f(0) = 1$, $f'(0) = 3$, and the linear approximation to f(x) near to $x = 0$ is $\mathbf{1+3x}$.
(ii) Let $g(x) = \ln(2+x)$, then $g'(x) = 1/(2+x)$, and $g(0) = \ln 2$, $g'(0) = 1/2$. The linear approximation to g(x) near $x = 0$ is $\mathbf{\ln 2 + x/2}$.

The graphs of the functions f and g, of Example 1, together with their linear approximations are shown in Fig. 6.2.

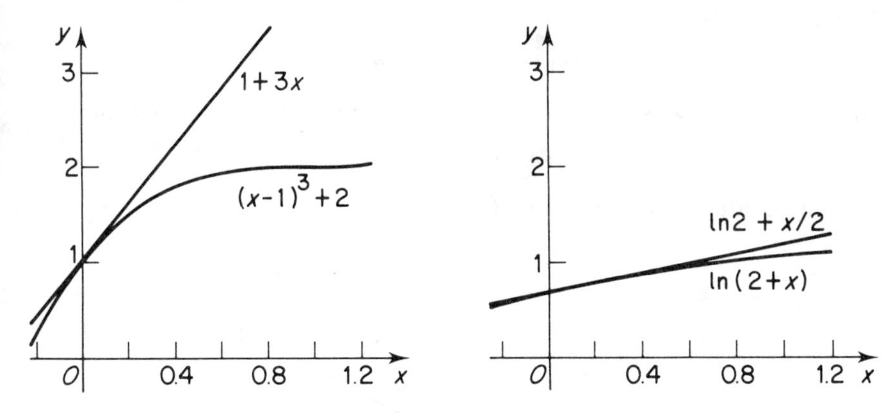

Fig. 6.2 Linear approximations at $x = 0$.

It is seen from Fig. 6.2 that the linear approximation $\ln 2 + x/2$ is likely to be a good approximation to $\ln(2+x)$ for a wide range of values of x but that the approximation $1+3x$ is a good approximation to $(x-1)^3+2$ over a much smaller range of values of x. This is shown in the following tables of values of the functions.

x	0	0·1	0·2	0·4	0·8
$(x-1)^3+2$	1	1·271	1·488	1·784	1·992
$1+3x$	1	1·3	1·6	2·2	3·4
error (1 s.f.)	0	0·03	0·1	0·4	1·4
$\ln(2+x)$	0·693	0·742	0·788	0·875	1·03
$\ln 2 + x/2$	0·693	0·743	0·793	0·893	1·09
error (1 s.f.)	0	0·001	0·005	0·02	0·06

The approximation to g is much better than the approximation to f and this is so because the graph of g does not curve away so quickly as does the graph of f. The rate at which the graph of a function curves away can be found by considering the rate at which the gradient changes, that is the second derivative.

For the function f, $f''(x) = 6(x-1)$ and $f''(0) = -6$,
for the function g, $g''(x) = -1/(2+x)^2$ and $g''(0) = -1/4$.

It therefore seems very likely that the error involved in using a linear (or first order) approximation can be estimated by considering the size of the second derivative of the function concerned. This idea will be developed later.

The linear approximation process may be carried out near to a non-zero value of x, such as $x = a$. The tangent to the graph of f at $(a, f(a))$ is the graph of the linear function $f(a)+(x-a)f'(a)$. This can be obtained from the approximation near $x = 0$ by moving the origin to $(a, 0)$, giving the approximation

$$f(x) \approx f(a)+(x-a)f'(a).$$

Alternatively, if h is the x-displacement from $x = a$, $h = x-a$, then

$$f(a+h) \approx f(a)+hf'(a).$$

This situation is shown in Fig. 6.3.

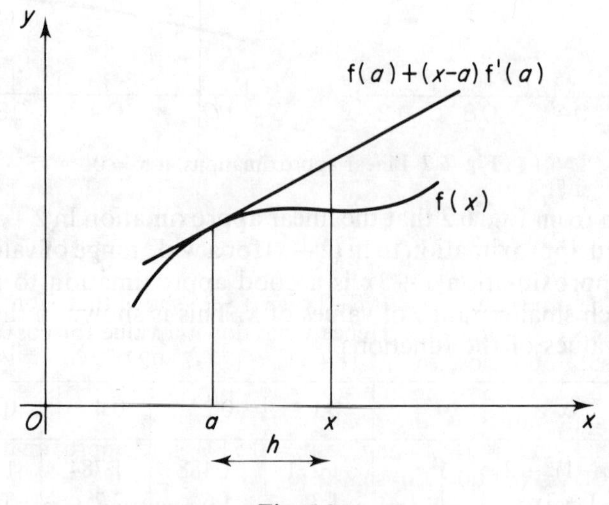

Fig. 6.3

EXAMPLE 2 *Find a linear approximation to the function x^3-2x at $x = 2$.*

Let $f(x) = x^3-2x$, then $f'(x) = 3x^2-2$, $f(2) = 2^3-4 = 4$, $f'(2) = 10$.

Then the required approximation is given by

$$f(x) \approx 4+(x-2).10 = \mathbf{-16+10x},$$

or, using the alternative notation,

$$f(2+h) \approx 4+h.10 = 4+10h.$$

EXAMPLE 3 *Use the value* $\sqrt{(16)} = 4$ *and a linear approximation to* $\sqrt{x}$ *at* $x = 16$ *to find* $\sqrt{(15{\cdot}9)}$.

Let $f(x) = \sqrt{x}$, then $f'(x) = 1/(2\sqrt{x})$, $f(16) = 4$, $f'(16) = 1/8$. Then, using the alternative notation, with $h = -0{\cdot}1$,

$$\begin{aligned}\sqrt{(15{\cdot}9)} = f(16+{}^{-}0{\cdot}1) &\approx f(16)+({}^{-}0{\cdot}1)f'(16)\\ &= 4-(0{\cdot}1)/8 = \mathbf{3{\cdot}9875}.\end{aligned}$$

The result obtained in this example compares quite favourably with the value 3·987480, correct to 6 decimal places, and this shows the value of a linear approximation to a function with a numerically small second derivative. In this case, $f''(x) = -\frac{1}{4}x^{-3/2}$ and $f''(16) \approx -0{\cdot}004$. This method has been used extensively to produce, and later to verify, tables of values of certain functions.

EXERCISE 6.3

1 Find a linear approximation to the function at $x = 0$:
(i) x^3-4x, (ii) $2x-x^2$, (iii) e^x, (iv) $\sin x$.

2 Use a linear approximation to the given function, at $x = 0$, to find the values listed: (i) e^x, $e^{0{\cdot}1}$, $e^{0{\cdot}5}$, e^2, (ii) $\ln(1+x)$, $\ln(1{\cdot}1)$, $\ln(1{\cdot}5)$, $\ln 2$, (iii) $\tan x$, $\tan 0{\cdot}1$, $\tan 0{\cdot}5$, $\tan 1$, (iv) $\sqrt{(x+1)}$, $\sqrt{(1{\cdot}21)}$, $\sqrt{(1{\cdot}44)}$, $\sqrt{(1{\cdot}96)}$, (v) $\sqrt[3]{(x+1)}$, $\sqrt[3]{(1{\cdot}331)}$, $\sqrt[3]{(1{\cdot}728)}$, $\sqrt[3]{8}$, (vi) $\cos x$, $\cos 0{\cdot}1$, $\cos 0{\cdot}5$, $\cos 1$, (vii) $1/(x+1)$, $1/1{\cdot}1$, $1/1{\cdot}5$, $1/2$.

3 Using the formula $f(x) = f(a)+(x-a)f'(a)$, show that, for x near $\pi/6$, $\cos x \approx \pi/12+\sqrt{3}/2-x/2$. Hence write down a value for $\cos 0{\cdot}5$.

4 Show that, for x close to 2, $1/(1+x)^2 \approx (7-2x)/27$.

5 Use a linear approximation to $\sqrt{(64+h)}$, where h is small, to find approximate values of (i) $\sqrt{65}$, (ii) $\sqrt{60}$.

6 Given that $\sin 1 = 0{\cdot}8415$ and $\cos 1 = 0{\cdot}5403$, find approximate values for: (i) $\sin 1{\cdot}01$, (ii) $\cos 1{\cdot}01$, (iii) $\sin 0{\cdot}9$.

7 Use the values of $\sin 1$ and $\cos 1$, given in Question 6, and apply linear approximation repeatedly to complete the table:

x	1	1·01	1·02	1·03	1·04	1·05
$\sin x$	0·8415					
$\cos x$	0·5403					

Compare with more exact values $\sin 1{\cdot}05 = 0{\cdot}8674$, $\cos 1{\cdot}05 = 0{\cdot}4976$.

6.4 Quadratic and higher approximations at a point

In order to improve upon a linear approximation to a function at a point, the next step is obviously to try a quadratic approximation.

EXAMPLE 1 *Find a quadratic approximation* q(*x*) *to the function* f, where f(*x*) = cos *x*, *near to the origin by matching the value at* $x = 0$ *of the function, its gradient* f′ *and the rate of change* f″ *of its gradient. Tabulate the values of* f(*x*) *and* q(*x*) *at intervals of* 0·2 *for* $0 \leqslant x \leqslant 1$ *and sketch the graphs of* f *and* q *in the interval* $0 \leqslant x \leqslant 1{\cdot}5$.

The linear approximation to cos (*x*) near $x = 0$ is the constant function p, where $p(x) = 1$. Assume that the quadratic approximation q is given by $q(x) = a + bx + cx^2$ and find the coefficients by equating the values of f and q and their first and second derivatives at the origin.

$$f(x) = \cos x, \quad f(0) = 1, \quad q(x) = a + bx + cx^2, \; q(0) = a, \text{ so } a = 1,$$
$$f'(x) = -\sin x, \quad f'(0) = 0, \quad q'(x) = b + 2cx, \quad q'(0) = b, \text{ so } b = 0,$$
$$f''(x) = -\cos x, \quad f''(0) = -1, \quad q''(x) = 2c, \quad q''(0) = 2c, \text{so } c = -\tfrac{1}{2}.$$

Therefore $q(x) = \mathbf{1} - \tfrac{1}{2}\,\mathbf{x^2}$ and the table of values may be calculated. The graphs are shown in Fig. 6.4 (on page 183).

x	0	0·2	0·4	0·6	0·8	1·0
$f(x) = \cos x$	1	0·980	0·921	0·825	0·697	0·540
$q(x) = 1 - \frac{1}{2}x^2$	1	0·98	0·92	0·82	0·68	0·5

Note that, although the quadratic approximation has been found by matching values of the function and its derivatives at the origin, the approximation is very good for quite a wide band of values of *x* about the origin.

The above process of finding a quadratic approximation by matching the function and its first two derivatives at the origin can be used for any function which can be differentiated twice. We state this result as a theorem. The proof will follow from the next theorem, which is a more general result.

Theorem 6.2 The quadratic approximation q to the function f at $x = 0$ is given by

$$q(x) = f(0) + f'(0)x + \tfrac{1}{2}f''(0)x^2,$$

provided that f can be differentiated twice at $x = 0$.

It is possible to continue this process and to find an approximation to a given function by means of a polynomial of higher degree by matching the values of the function and its derivatives at $x = 0$.

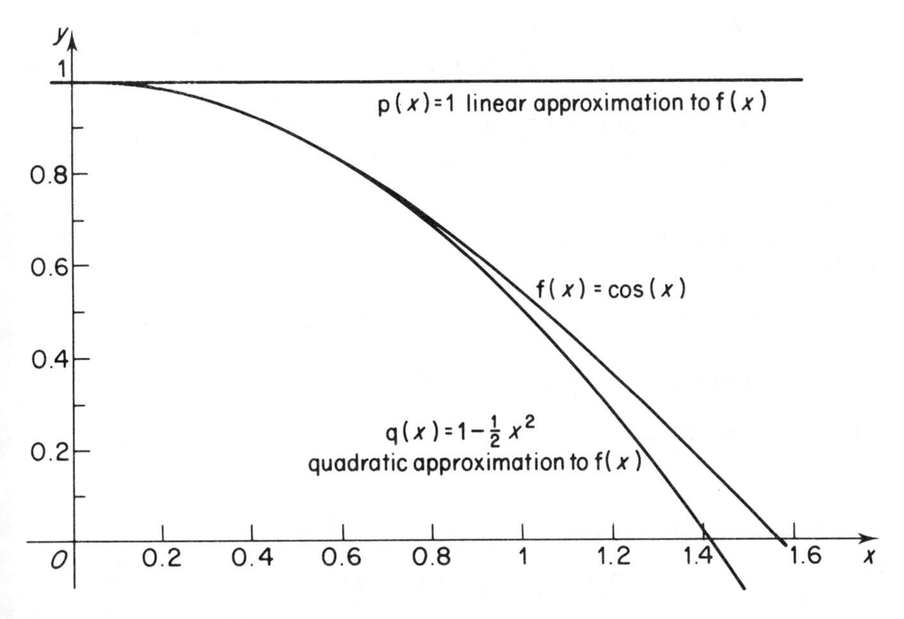

Fig. 6.4 Approximations to $\cos x$ near $x = 0$

Theorem 6.3 Let f be a function which can be differentiated n times at $x = 0$. Then the nth degree polynomial approximation $P_n(x)$ to f at $x = 0$ is given by

$$P_n(x) = f(0) + f'(0)x + \tfrac{1}{2}f''(0)x^2 + \frac{1}{3!}f'''(0)x^3 + \ldots \frac{1}{n!}f^{(n)}(0)x^n.$$

Proof Suppose that $P_n(x) = \sum_{r=0}^{n} a_r x^r = a_0 + a_1 x + a_2 x^2 + \ldots + a_n x^n$

then $P_n'(x) = a_1 + 2a_2 x + 3a_3 x^2 + \ldots + na_n x^{n-1}$, $P_n'(0) = a_1$. With successive differentiations,

$$P_n^{(2)}(x) = 2a_2 + 2.3a_3 x + 3.4a_4 x^2 + \ldots + (n-1)na_n x^{n-2},\ P_n^{(2)}(0) = 2a_2,$$
$$P_n^{(3)}(x) = 2.3a_3 + 2.3.4a_4 x + \ldots + (n-2)(n-1)na_n x^{n-3},\ P_n^{(3)}(0) = 6a_3,$$
$$P_n^{(r)}(x) = 2.3 \ldots ra_r + 2.3 \ldots (r+1)a_{r+1}x + \ldots$$
$$+ (n-r+1) \ldots (n-1)na_n x^{n-r},$$

$P_n^{(r)}(0) = r!\,a_r$, for $1 \leqslant r \leqslant n$, so that $a_r = \dfrac{1}{r!}P_n^{(r)}(0)$.

Since we are matching $P_n(x)$ and $f(x)$ at $x = 0$, together with their n derivatives, $P_n(0) = f(0)$, $P_n'(0) = f'(0)$, and in general $P_n^{(r)}(0) = f^{(r)}(0)$, and therefore $a_r = \dfrac{1}{r!}f^{(r)}(0)$.

Theorem 6.3 produces an nth degree polynomial $P_n(x)$ which fits f and its first n derivatives at $x = 0$. The question of most concern is the accuracy of the approximation to $f(x)$. Some measure is required of the expected error in using such a polynomial as an approximation to the function and a general result about this error is stated in §6.5.

EXAMPLE 2 *Find a fourth degree polynomial approximation,* $P_4(x)$, *to* $\ln(x+1)$ *at* $x = 0$. *Calculate the error in using this approximation to find:* (*i*) $\ln(1{\cdot}1)$, (*ii*) $\ln(0{\cdot}8)$, (*iii*) $\ln(1{\cdot}5)$, (*iv*) $\ln(8)$.

Let $f(x) = \ln(x+1)$, then $f'(x) = 1/(x+1)$, $f''(x) = -1/(x+1)^2$, $f'''(x) = 2/(x+1)^3$ and $f''''(x) = -6/(x+1)^4$.
So the coefficients are calculated as follows:

$$f(0) = \ln 1 = 0,\ f'(0) = \frac{1}{(0+1)} = 1,\ \frac{1}{2}f''(0) = \frac{1}{2}\frac{-1}{(0+1)^2} = -\frac{1}{2},$$

$$\frac{1}{3!}f'''(0) = \frac{1}{6}\frac{2}{(0+1)^3} = \frac{1}{3},\ \frac{1}{4!}f''''(0) = \frac{1}{24}\frac{-6}{(0+1)^4} = -\frac{1}{4}.$$

Thus $$P_4(x) = x - \tfrac{1}{2}x^2 + \tfrac{1}{3}x^3 - \tfrac{1}{4}x^4.$$

Now use $P_4(x)$ as an approximation to $\ln(x+1)$. In order to find $\ln(1{\cdot}1)$, put $x = 0{\cdot}1$ so that

$$P_4(0{\cdot}1) = (0{\cdot}1) - (0{\cdot}1)^2/2 + (0{\cdot}1)^3/3 - (0{\cdot}1)^4/4 = 0{\cdot}09530.$$

Using a calculator, the value of $\ln(1{\cdot}1)$ is 0·0953101 so the error in the above approximation is of the order of $\mathbf{1 \times 10^{-5}}$.
To find $\ln(0{\cdot}8)$, put $x = -0{\cdot}2$ so that

$$P_4(-0{\cdot}2) = (-0{\cdot}2) - (-0{\cdot}2)^2/2 + (-0{\cdot}2)^3/3 - (-0{\cdot}2)^4/4 = -0{\cdot}223067$$

and, compared with the calculator value $-0{\cdot}2231435$, the error is of the order $\mathbf{0{\cdot}8 \times 10^{-4}}$.
To find $\ln(1{\cdot}5)$, put $x = 0{\cdot}5$ so that

$$P_4(0{\cdot}5) = (0{\cdot}5) - (0{\cdot}5)^2/2 + (0{\cdot}5)^3/3 - (0{\cdot}5)^4/4 = 0{\cdot}4010416$$

and, compared with the calculator value 0·4054651, the error is of the order $\mathbf{0{\cdot}5 \times 10^{-3}}$.
To find $\ln(8)$, put $x = 7$ so that

$$P_4(7) = (7) - (7)^2/2 + (7)^3/3 - (7)^4/4 = -503{\cdot}406$$

and, compared with the calculator value 2·0794, the error is of the order $\mathbf{5 \times 10^2}$.

The last result gives no useful information about the value of $\ln(8)$ and demonstrates the limited usefulness of this process. Informally, we can see that the series $P_4(x) = x - \frac{1}{2}x^2 + \frac{1}{3}x^3 - \frac{1}{4}x^4 + \ldots$ does not appear to converge for $x = 7$ since the terms become larger and larger and the partial sums oscillate between large positive and large negative values. The method will not be useful for a series which does not converge and it seems reasonable to assume that the polynomial approximation process will only give good results when the power series in x converges rapidly.

EXERCISE 6.4

1 Show that the cubic polynomial approximation to $\sin x$ at $x = 0$ is given by $P_3(x) = x - \frac{1}{6}x^3$. Calculate the error in using this approximation for: (i) $x = \pi/12$, (ii) $x = \pi/6$, (iii) $x = \pi/3$, (iv) $x = \pi/2$, (v) $x = \pi$.

2 Prove that the quartic approximation to $\cos(x/2)$ at $x = 0$ is given by $P_4(x) = 1 - \dfrac{x^2}{8} + \dfrac{x^4}{384}$. Find the percentage error in using $P_4(x)$ as an approximation to $\cos(x/2)$ at: (i) $x = 0{\cdot}1$, (ii) $x = 0{\cdot}5$, (iii) $x = 1$, (iv) $x = 10$.

3 Find a quadratic approximation $P_2(x)$ for the function f at $x = 0$ and compare the values it gives at $x = 0{\cdot}1$, $-0{\cdot}2$, $0{\cdot}5$ and 1 with the exact value: (i) $f(x) = \sqrt{(1-x)}$, (ii) $f(x) = 2/(x+1)$, (iii) $f(x) = \sqrt{(1-x^2)}$.

4 Differentiate the function f, given by $f(x) = e^x$, and find $f^{(n)}(x)$. Hence show that $P_n(x)$, the nth degree polynomial approximation to e^x at $x = 0$, is given by $P_n(x) = 1 + x + x^2/2! + x^3/3! + \dots + x^n/n!$. Find the least value of n needed to calculate e^x accurate to 3 D.P. when: (i) $x = 0{\cdot}1$, (ii) $x = -0{\cdot}2$, (iii) $x = 0{\cdot}5$, (iv) $x = 1$, (v) $x = 2$.

5 Making use of previous results, where appropriate, list polynomial approximations of: (i) $\cos x$, (ii) $\sin x$, (iii) e^x, (iv) $\ln(1+x)$.

6 For the given function f, find the 2nd, 3rd and 4th degree approximating polynomials, $P_2(x)$, $P_3(x)$ and $P_4(x)$, at $x = 0$: (i) $f(x) = \tan x$, (ii) $f(x) = x \sin x$, (iii) $f(x) = e^{-x^2}$.

7 Translate the result for the polynomial approximation at $x = 0$ to the polynomial approximation $P_n(x)$ at $x = a$ to show that $P_n(x) = f(a) + (x-a)f'(a) + \frac{1}{2}(x-a)^2 f''(a) + \dots + \dfrac{1}{n!}(x-a)^n f^{(n)}(a)$. Use this result to find a cubic approximation to $\sin x$ which is valid near to $x = \pi$. Find the percentage error in using this approximation to find $\sin(4\pi/3)$.

8 Find a quartic approximation to $1/x$ at $x = 1$ and calculate the error in using the approximation to find: (i) 1/2, (ii) 1/3, (iii) 2.

6.5 Maclaurin Expansions

The idea of a polynomial approximation to a function at a point naturally extends to the idea of representing a function by a convergent power series. A power series has a radius of convergence, R, which may be infinite, finite or zero, as was stated in Theorem 6.1. The following typical examples show that all three cases are possible.

Case I The power series for e^x is given by

$$e^x = 1 + x + \frac{x^2}{2!} + \frac{x^3}{3!} + \dots + \frac{x^n}{n!} + \dots$$

and this series converges for any given value of x. Thus it has an infinite radius of convergence, $R = \infty$.

Case II The power series for $\dfrac{1}{1+x}$ is given by

$$\frac{1}{1+x} = 1 - x + x^2 - x^3 + \dots + (-1)^n x^n + \dots$$

and this series converges for $|x| < 1$. It has radius of convergence R, given by $R = 1$, and it diverges for $|x| \geqslant 1$.

Case III The power series

$$\sum_{r=0}^{\infty} r!x^r = 1 + x + 2x^2 + 3!x^3 + \ldots + r!x^r + \ldots$$

diverges for all non-zero values of x so it has zero radius of convergence.

We now state a theorem concerning power series and their differentiation and integration term by term, which is derived from Theorem 6.1. We cannot prove the theorem in this book but the earlier work of this chapter makes it plausible and we shall apply it directly.

Theorem 6.4 Let a power series, with radius of convergence R and sum $f(x)$ for $|x| < R$, be given by

$$f(x) = \sum a_r x^r = a_0 + a_1 x + a_2 x^2 + \ldots + a_r x^r + \ldots.$$

Then, for $|x| < R$,

(i) the power series for $f(x)$ is unique,

(ii) $f'(x) = \sum r a_r x^{r-1} = a_1 + 2a_2 x + 3a_3 x^2 + \ldots + r a_r x^{r-1} + \ldots,$

(iii) $\displaystyle\int_0^x f(t)\,dt = \sum \frac{a_r}{r+1} x^{r+1} = a_0 x + \tfrac{1}{2} a_1 x^2 + \ldots + \frac{1}{r+1} a_r x^{r+1} + \ldots.$

We shall assume that, for the functions with which we deal, a function which has derivatives of all orders can be expanded in a power series. In fact, more conditions are required to ensure that a function can be expanded in a power series but, again, we are not able to deal with this problem in this book. As an example of the difficulties that may arise, the reader should try to apply the next theorem, Maclaurin's theorem, to the function $e^{-\frac{1}{2}x^2}$.

Theorem 6.5 (The Maclaurin expansion) Let the function f have derivatives of all orders at $x = 0$ and let $f(x)$ have a power series expansion with radius of convergence R, then

$$f(x) = \sum \frac{1}{r!} f^{(r)}(0) x^r, \text{ for } |x| < R.$$

Theorem 6.5 can be seen to be the natural development of the work on nth degree polynomial approximations to f as we let n tend to infinity. As an outline of the proof of the Maclaurin expansion:

(a) the function f has a power series which converges for $|x| < R$,

(b) the series may be differentiated term by term to give the power series for $f'(x)$ which also converges for $|x| < R$,

(c) the coefficients of the power series for f can then be found in exactly the same way as were the coefficients of $P_n(x)$ in §6.4.

In much of our work, the radius of convergence of the series will be infinite and problems of convergence will not trouble us. However, when dealing with binomial type expansions, and with logarithmic expansions, the radius of convergence will be finite and we need to remember to keep within the radius of convergence in order to calculate useful results (outside the radius of convergence the results will be nonsense). We demonstrate by means of some examples.

EXAMPLE 1 *Express* $\ln\left(\dfrac{1+2x}{1-2x}\right)$ *as a series of terms in ascending powers of* x, *up to and including the term in* x^5. *Use the series to obtain an approximate value for* ln (5/3), *giving your answer to 5 decimal places.*

Let $f(x) = \ln\left(\dfrac{1+2x}{1-2x}\right) = \ln(1+2x) - \ln(1-2x)$. Then

$$f^{(1)}(x) = \frac{2}{1+2x} + \frac{2}{1-2x}, \quad f^{(2)}(x) = \frac{-4}{(1+2x)^2} + \frac{4}{(1-2x)^2},$$

$$f^{(3)}(x) = \frac{16}{(1+2x)^3} + \frac{16}{(1-2x)^3}, \quad f^{(4)}(x) = \frac{-96}{(1+2x)^4} + \frac{96}{(1-2x)^4},$$

$$f^{(5)}(x) = \frac{768}{(1+2x)^5} + \frac{768}{(1-2x)^5}.$$

On substituting $x = 0$, $f(0) = 0$, $f^{(1)}(0) = 4$, $f^{(2)}(0) = 0$, $f^{(3)}(0) = 32$, $f^{(4)}(0) = 0$, $f^{(5)}(0) = 1536$, and so

$$\ln\left(\frac{1+2x}{1-2x}\right) \approx 4x + \frac{32}{3!}x^3 + \frac{1536}{5!}x^5 = \mathbf{4x + \frac{16}{3}x^3 + \frac{64}{5}x^5}.$$

In order to find ln (5/3), put $\dfrac{1+2x}{1-2x} = \dfrac{5}{3}$, that is $3+6x = 5-10x$ or $x = 1/8$. Substituting in the series,

$$\ln\left(\frac{1+\frac{1}{4}}{1-\frac{1}{4}}\right) = \ln\left(\frac{5}{3}\right) = 0{\cdot}5 + 0{\cdot}0104167 + 0{\cdot}000390625 = 0{\cdot}5108072$$

$$= \mathbf{0{\cdot}51081}.$$

Note: The radius of convergence of the power series for $\ln(1+x)$ is 1 and so the series for $\ln(1+2x)$ converges for $|2x| < 1$, that is for $|x| < 1/2$. Thus the series which we obtained has radius of convergence equal to 1/2. However, it may be used to calculate the natural logarithm of any positive number since a value of x less than 1/2 can be chosen so that $\dfrac{1+2x}{1-2x}$ is as large as is needed.

EXAMPLE 2 *Differentiate* $\ln(\sec x + \tan x)$ *with respect to x. Express* $\sec x$ *as a series in ascending powers of x, up to and including the term in* x^4. *Hence, show that, as far as the term in* x^5, *the series for* $\ln(\sec x + \tan x)$ *is* $x + x^3/6 + x^5/24$.

$$\frac{d}{dx}\ln(\sec x + \tan x) = \frac{1}{\sec x + \tan x}\cdot(\sec x \tan x + \sec^2 x)$$

$$= \sec x \frac{\sec x + \tan x}{\sec x + \tan x} = \mathbf{\sec x}.$$

Let $f(x) = \sec x$, then $f^{(1)}(x) = \sec x \tan x$,

$$f^{(2)}(x) = \sec x \tan^2 x + \sec^3 x = 2\sec^3 x - \sec x,$$
$$f^{(3)}(x) = 6\sec^3 x \tan x - \sec x \tan x,$$
$$f^{(4)}(x) = 18\sec^3 x \tan^2 x + 6\sec^5 x - 2\sec^3 x + \sec x.$$

At $x = 0$, $f(0) = 1$, $f^{(1)}(0) = 0$, $f^{(2)}(0) = 1$,
$f^{(3)}(0) = 0$, $f^{(4)}(0) = 5$, and so

$$\sec x \approx 1 + x^2/2! + 5x^4/4! = \mathbf{1 + x^2/2 + 5x^4/24}.$$

Within the radius of convergence, this approximation may be integrated term by term (Theorem 6.4), and so

$$\int_0^x \sec t \, dt \approx \int_0^x (1 + t^2/2 + 5t^4/24)dt,$$

and $\ln(\sec x + \tan x) \approx x + x^3/6 + x^5/24.$

An alternative way of completing the last step is to use an indefinite integral $\ln(\sec x + \tan x) \approx x + x^3/6 + x^5/24 + c$, and then to evaluate the arbitrary constant c of integration by putting $x = 0$, so that $c = \ln(\sec 0 + \tan 0) = \ln(1) = 0$.

Standard Maclaurin Expansions

We now state some standard expansions, together with their radii of convergence. We make no attempt to justify the value of the radius of convergence in each case.

$\cos x = 1 - x^2/2! + x^4/4! - x^6/6! + \ldots + (-1)^r x^{2r}/(2r)! + \ldots, R = \infty$
$\sin x = x - x^3/3! + x^5/5! - x^7/7! + \ldots + (-1)^r x^{2r+1}/(2r+1)! + \ldots, R = \infty$
$\cosh x = 1 + x^2/2! + x^4/4! + x^6/6! + \ldots + x^{2r}/(2r)! + \ldots, R = \infty$
$\sinh x = x + x^3/3! + x^5/5! + x^7/7! + \ldots + x^{2r+1}/(2r+1)! + \ldots, R = \infty$
$e^x = 1 + x + x^2/2! + x^3/3! + \ldots + x^r/r! + \ldots, R = \infty$
$\ln(1+x) = x - x^2/2 + x^3/3 + \ldots + (-1)^{r+1}x^r/r + \ldots, R = 1$
$(1+x)^n = 1 + nx + \binom{n}{2}x^2 + \binom{n}{3}x^3 + \ldots + \binom{n}{r}x^r + \ldots, R = 1$

EXAMPLE 3 *Find the first two non-zero terms in the expansion of* $(1-x)(1-e^x)+\ln(1+x)$ *in ascending powers of* x*, when* x *is small.*

Using the power series expansions of e^x and $\ln(1+x)$ from earlier work,

$$\begin{aligned}(1-x)(1-e^x) &= (1-x)\{1-(1+x+x^2/2+x^3/3!+\ldots)\}\\ &= (1-x)(-x-x^2/2-x^3/3!-\ldots)\\ &= -x+x^2-x^2/2+x^3/2-x^3/3!+x^4/3!-x^4/4!+\ldots\\ &= -x+x^2/2+x^3(3-1)/3!+x^4(4-1)/4!\ldots\\ &= -x+x^2/2+x^3/3+x^4/8+\ldots\end{aligned}$$

and $\qquad \ln(1+x) = x-x^2/2+x^3/3-x^4/4+\ldots$

so that

$$(1-x)(1-e^x)+\ln(1+x) = \mathbf{2x^3/3-x^4/8}+\ldots.$$

EXAMPLE 4 *Calculate the first two terms in the expansions of* $\sin x$ *and* $\sinh x$ *in ascending powers of* x*. Given that* x *is sufficiently small for* x^5 *and higher powers of* x *to be neglected, show that*

$$\ln\left(\frac{1+\sinh x}{1+\sin x}\right) \approx ax^3+bx^4,$$

where a *and* b *are constants to be determined.*

Again use the previously quoted expansions, $\sin x = x-x^3/3!+x^5/5!-\ldots$, $\sinh x = x+x^3/3!+x^5/5!+\ldots$.
Neglect x^5 and higher powers of x as the calculation proceeds.

$$\begin{aligned}\ln(1+\sinh x) &\approx \ln(1+x+x^3/3!)\\ &\approx (x+x^3/6)-(x+x^3/6)^2/2+(x+x^3/6)^3/3-(x+x^3/6)^4/4\\ &\approx x+x^3/3!-\{x^2+2xx^3/3!\}/2+x^3/3-x^4/4,\end{aligned}$$

$$\begin{aligned}\ln(1+\sin x) &\approx \ln(1+x-x^3/3!)\\ &\approx (x-x^3/6)-(x-x^3/6)^2/2+(x-x^3/6)^3/3-(x-x^3/6)^4/4\\ &\approx x-x^3/3!-\{x^2-2xx^3/3!\}/2+x^3/3-x^4/4,\end{aligned}$$

so

$$\begin{aligned}\ln\left(\frac{1+\sinh x}{1+\sin x}\right) &= \ln(1+\sinh x)-\ln(1+\sin x)\\ &\approx 2x^3/3!-2x^4/3!,\end{aligned}$$

and $\boldsymbol{a=1/3}$, $\boldsymbol{b=-1/3}$.

Comment on the above methods We are treating the series expansions in a rather cavalier manner, taking the series expansions of expressions from the first few terms of other series expansions. This is valid only as long as x is sufficiently small for the convergence of all the series concerned to be assured. For example, in example 4, we have assumed that both $|x+x^3/3!|<1$ and $|x-x^3/3!|<1$ in order that the logarithmic expansion is valid.

In order to find the Maclaurin expansion of a function f, it is necessary to differentiate $f(x)$ any number of times. Sometimes, this may be a formidable task which can be made easier by using the theorem of Leibnitz in order to differentiate a product.

Theorem 6.6 (Leibnitz)

$$\frac{d^n}{dx^n}[f(x)g(x)] = \sum_{r=0}^{n}\binom{n}{r}f^{(n-r)}(x)g^{(r)}(x)$$

$$= f^{(n)}(x)g(x)+nf^{(n-1)}(x)g(x)+\binom{n}{2}f^{(n-2)}(x)g^{(2)}(x)+\ldots+f(x)g^{(n)}(x).$$

This theorem is particularly useful when $g(x)$ is a polynomial of low degree. For example, if $g(x)$ is a quadratic polynomial, then $g^{(r)}(x) = 0$ for $r \geqslant 3$ so that, for $n \geqslant 3$,

$$\frac{d^n}{dx^n}[f(x)g(x)] = f^{(n)}(x)g(x)+nf^{(n-1)}(x)g'(x)+\tfrac{1}{2}n(n-1)f^{(n-2)}(x)g''(x).$$

Proof of Theorem 6.6

It should be noted that the coefficients in the expansion are binomial coefficients and the theorem can be proved by induction in the same way as the binomial theorem.

Basis of induction The result is true for $n = 1$ since

$$\frac{d}{dx}[f(x)g(x)] = f'(x)g(x)+f(x)g'(x).$$

Induction step Assume that the result is true for $n = k$ so that

$$\frac{d^k}{dx^k}[f(x)g(x)] = \sum_{r=0}^{k}\binom{k}{r}f^{(k-r)}(x)g^{(r)}(x).$$

Then

$$\frac{d^{k+1}}{dx^{k+1}}[f(x)g(x)] = \sum_{r=0}^{k}\binom{k}{r}[f^{(k+1-r)}(x)g^{(r)}(x)+f^{(k-r)}(x)g^{(r+1)}(x)]$$

$$= \sum_{r=0}^{k}\binom{k}{r}f^{(k+1-r)}(x)g^{(r)}(x)+\sum_{r=1}^{k+1}\binom{k}{r-1}f^{(k+1-r)}(x)g^{(r)}(x)$$

$$= f^{(k+1)}(x)g(x)+\sum_{r=1}^{k}\left[\binom{k}{r}+\binom{k}{r-1}\right]f^{(k+1-r)}(x)g^{(r)}(x)+f(x)g^{(k+1)}(x)$$

$$= \sum_{r=0}^{k+1}\binom{k+1}{r}f^{(k+1-r)}(x)g^{(r)}(x), \text{ since } \binom{k+1}{r}=\binom{k}{r}+\binom{k}{r-1}.$$

Thus, if the result is true for $n = k$, $k \geqslant 1$, then it is also true for $n = k+1$. Since it is true for $n = 1$, it follows, by induction, that the result is true for all natural numbers n.

EXAMPLE 5 *Given that* $f(x) = \arctan x$, *prove that* $(x^2+1)f'(x) = 1$. *By differentiating this equation n times, show that, for* $n > 1$,

$$(x^2+1)f^{(n+1)}(x) + 2nxf^{(n)}(x) + n(n-1)f^{(n-1)}(x) = 0,$$

and deduce that $f^{(n+1)}(0) = -n(n-1)f^{(n-1)}(0)$.
Find the expansion of $\arctan x$, *as a powers series in x, giving the first four terms and the general term.*

$$\frac{d}{dx}f(x) = f^{(1)}(x) = \frac{d}{dx}\arctan x = \frac{1}{1+x^2}, \text{ hence } (1+x^2)f^{(1)}(x) = 1.$$

Differentiating this equation n times, using Leibnitz Theorem. For $n \geqslant 2$,

$$\frac{d^n}{dx^n}[(1+x^2)f^{(1)}(x)] = (1+x^2)f^{(n+1)}(x) + n\frac{d}{dx}(1+x^2)f^{(n)}(x)$$

$$+\binom{n}{2}\frac{d^2}{dx^2}(1+x^2)f^{(n-1)}(x),$$

since $\frac{d^3}{dx^3}(1+x^2) = 0$, and, therefore,

$$(1+x^2)f^{(n+1)}(x) + 2nxf^{(n)}(x) + 2\frac{n(n-1)}{2}f^{(n-1)}(x) = 0.$$

On putting $x = 0$, we find that, for $n \geqslant 2$,

$$f^{(n+1)}(0) = -n(n-1)f^{(n-1)}(0).$$

Now $f(0) = \arctan(0) = 0$ and $f^{(1)}(0) = 1$, and also $f^{(2)}(0) = 0$,

since $\frac{d}{dx}\{(1+x^2)f^{(1)}(x)\} = (1+x^2)f^{(2)}(x) + 2xf^{(1)}(x) = 0$.

Then $f^{(4)}(0) = 0$ and, similarly, $f^{(2n)}(0) = 0$, for all natural numbers n.
Also $f^{(3)}(0) = -2.1.f^{(1)}(0) = -2$, $f^{(5)}(0) = -4.3.f^{(3)}(0) = 4!$ and similarly $f^{(2n+1)}(0) = (-1)^n(2n)!$, for all natural numbers n. Thus, in the power series expansion of $f(x)$, the coefficients of the even powers of x are all zero and the coefficient of $x^{(2n+1)}$ is

$$(-1)^n(2n)!/(2n+1)!, \text{ that is, } (-1)^n/(2n+1).$$

Finally,

$$\arctan x = \mathbf{x - x^3/3 + x^5/5 - x^7/7 + \ldots + (-1)^n x^{2n+1}/(2n+1) + \ldots .}$$

EXERCISE 6.5

1 Use Theorem 6.5 to obtain the standard Maclaurin Expansions:

(i) $\cos x = \sum_{r=0}^{\infty}(-1)^r x^{2r}/(2r)!$, (ii) $\sin x = \sum_{r=0}^{\infty}(-1)^r x^{(2r+1)}/(2r+1)!$,

(iii) $\cosh x = \sum_{r=0}^{\infty} x^{2r}/(2r)!$, (iv) $\sinh x = \sum_{r=0}^{\infty} x^{(2r+1)}/(2r+1)!$,

(v) $e^x = \sum_{r=0}^{\infty} x^r/(r)!$, (vi) $\ln(1+x) = -\sum_{r=1}^{\infty}(-x)^r/r$,

(vii) $(1+x)^n = \sum_{r=0}^{\infty} \binom{n}{r} x^r$.

2 Write down the expansion of $\ln(1+x)$ in powers of x, giving the first three terms and the general term. Calculate $\ln(0{\cdot}97)$ to five significant figures.

3 Find non-zero constants a and b such that when x is small

$$(x+\sin x)\cos x \approx ax+bx^3.$$

4 Show that, when x is small, $e^x \ln(1+x)$ is approximately equal to $x+\frac{1}{2}x^2$. (*L*)

5 Find the first three non-zero terms in the expansion of $\ln\{(1+2x)(1-3x)\}$ in a series of ascending powers of x. (*L*)

6 When x is so small that x^3 and higher powers of x can be neglected, find non-zero constants a and b such that $e^{-x} = (1+ax)/(1+bx)$. (*L*)

7 When x is so small that x^3 and higher powers of x can be ignored, show that $\ln\left(\dfrac{1-2x}{1-x}\right) = e^x - e^{2x}$. (*L*)

8 Expand in ascending powers of x up to and including the term in x^3

$$\text{(i) } (1-x)^{1/2}, \quad \text{(ii) } \ln(1-ax).$$

Given that $(1-x)^{1/2} - \frac{1}{4}\ln(1-ax) = 1+px^2+qx^3+\ldots$, find the numerical values of a, p and q. (*AEB*)

9 State the first three terms and the nth term in the expansion of $\sin x$ in ascending powers of x. If x is sufficiently small for x^6 and higher powers of x to be neglected show that

$$\frac{3\sin x}{2+\cos x} \approx x - \frac{x^5}{180}. \quad (JMB)$$

10 Given that x is small enough for its fifth and higher powers to be negligible, show that $\ln(1+\sin x) \approx x - x^2/2 + x^3/6 - x^4/12$. Find also the corresponding approximation for $\ln(1-\sin x)$ and hence, or otherwise, show that $\ln(\cos x) \approx -x^2/2 - x^4/12$. (*JMB*)

11 By using the power series expansions for $\cosh x$ and $\cos x$ show that, if powers of x higher than the fifth are negligible, (i) $\cosh x - \cos x \approx x^2$, (ii) $\operatorname{sech} x - \sec x \approx -x^2$. (*JMB*)

12 Assuming the series expansion for $\ln(1+x)$, show that, for $m > n > 0$,

$$\ln\left(\frac{m}{n}\right) = 2\left\{\frac{m-n}{m+n} + \frac{1}{3}\left(\frac{m-n}{m+n}\right)^3 + \frac{1}{5}\left(\frac{m-n}{m+n}\right)^5 + \ldots\right\}.$$

Hint: $\dfrac{m}{n} = \dfrac{(m+n)+(m-n)}{(m+n)-(m-n)}$.

Hence calculate $\ln(13/12)$, correct to six decimal places.

13 Obtain the coefficient of x^n for $n \geqslant 2$ in the expansion of $(1-x^2)e^{-x}$ in ascending powers of x. *(L)*

14 Find the first three non-zero terms of the expansion, in ascending powers of x, of $\ln(1-xe^x)$. *(L)*

15 Express $(1+x)^9 e^{-3x}$ in ascending powers of x, neglecting powers of x greater than the first, and hence obtain an approximation to one significant figure of the small positive value of x which satisfies $(1+x)^9 e^{-3x} = 1{\cdot}03$.

16 Find the first non-zero term in the expansion of $(e^{-x}-1) - \ln(1-x)$ in ascending powers of x. *(L)*

17 Given that $f(x) = \tanh^{-1} x$, prove that $(1-x^2)f'(x) = 1$. Use Leibnitz's theorem to deduce that $f^{(n+1)}(0) - n(n-1)f^{(n-1)}(0) = 0$ for all $n \geqslant 2$. Evaluate $f'(0)$ and $f''(0)$. Hence, or otherwise, obtain the first three non-zero terms, and the general term, in the Taylor expansion of $f(x)$ centred on the point $x = 0$.

18 Expand $\dfrac{\sin x + \sinh x}{\cos x + \cosh x}$ in ascending powers of x, neglecting terms in x^9 and of higher order. *(L)*

19 (i) Find the coefficients of x, x^2 and x^3 in the expansion of $e^{nx}/(1+x)^n$ in a series of ascending powers of x.
(ii) Assuming the expansion of $\ln(1+x)$ in ascending powers of x, obtain the expansion of $f(x)$, where

$$f(x) = \frac{1}{2}\ln\left(\frac{1+x}{1-x}\right),$$

giving the general term. State the set of values of x for which this expansion is valid, and sketch the graph of $f(x)$. *(L)*

20 Find the first three non-zero terms and the coefficients of x^{2n} and x^{2n+1} in the expansion of $\ln\left(\dfrac{1+x}{1-x}\right)$ in ascending powers of x. State the values of x for which this expansion is valid. The function f is given by $f(x) = \sin^{-1} x$, for $-1 < x < 1$. Prove that $(1-x^2)f''(x) = xf'(x)$. Use Leibnitz's theorem to deduce that, for $n \geqslant 1$, $f^{(n+2)}(0) = n^2 f^{(n)}(0)$. Express $f(x)$ as a Taylor series about the origin, giving terms up to and including the term in x^7. Prove that the coefficient of x^{2n+1} in this series is less than the coefficient of x^{2n+1} in the expansion, in ascending powers of x, of $\ln\sqrt{\left(\dfrac{1+x}{1-x}\right)}$. *(JMB)*

6.6 More techniques concerning series expansions

Theorem 6.4 can be used in order to derive one series from another by means of differentiation or integration. We now demonstrate how this can be used to find the series for the inverse trigonometric function $\sin^{-1} x$.

EXAMPLE 1 *Expand $1/\sqrt{(1-x^2)}$ as a power series in x. By integrating the series term by term, show that*

$$\sin^{-1} x \approx x + x^3/6 + 3x^5/40 + 5x^7/112,$$

when x is so small that terms in x^9 and higher powers are negligible. Taking $x = 1/2$, use this result to calculate π to three significant figures.

The function $(1-x^2)^{-1/2}$ can be expanded by Maclaurin's theorem or by the binomial theorem and we use the latter.

$$(1-x^2)^{-1/2} = 1 + \frac{-\frac{1}{2}}{1}(-x^2) + \frac{-\frac{1}{2}\cdot-\frac{3}{2}}{1.2}(-x^2)^2 + \frac{-\frac{1}{2}\cdot-\frac{3}{2}\cdot-\frac{5}{2}}{1.2.3}(-x^2)^3 + o(x^8)$$

$$= 1 + x^2/2 + 3x^4/8 + 5x^6/16 + o(x^8),$$

where the notation $o(x^8)$ means the (infinite) sum of those terms which are negligible in comparison with x^7.

Now
$$\begin{aligned}\sin^{-1} x &= \int (1-x^2)^{-1/2}\,dx \\ &= \int \{1 + x^2/2 + 3x^4/8 + 5x^6/16 + o(x^8)\}\,dx \\ &= x + x^3/6 + 3x^5/40 + 5x^7/112 + c + o(x^9).\end{aligned}$$

Since $\sin^{-1}(0) = 0$, $c = 0$ and the result is proved. Taking $x = 1/2$,

$$\begin{aligned}\pi/6 = \sin^{-1}(\tfrac{1}{2}) &\approx \tfrac{1}{2} + (\tfrac{1}{2})^3/6 + 3(\tfrac{1}{2})^5/40 + 5(\tfrac{1}{2})^7/112 \\ &\approx 0{\cdot}5 + 0{\cdot}020833 + 0{\cdot}002344 + 0{\cdot}000349 \\ &= 0{\cdot}523526\end{aligned}$$

accurate to four significant figures, and so $\pi \approx 3{\cdot}14116 \approx$ **3·14** to three significant figures.

When calculating derivatives to be used in a Maclaurin expansion, it is sometimes helpful to use a differential equation. This is demonstrated in the next example.

EXAMPLE 2 *Given that $y = \tan^2 x$, find an expression for $\dfrac{dy}{dx}$ in terms of* $\tan x$ *and show that*

$$\frac{d^2y}{dx^2} = 2 + 8y + 6y^2.$$

Differentiate this result further to show that, if powers of x above x^6 are neglected, then $\tan^2 x = x^2 + 2x^4/3 + 17x^6/45$.

Since $y = \tan^2 x$, $\dfrac{dy}{dx} = 2\tan x\sec^2 x = 2\tan x + 2\tan^3 x$,

$$\begin{aligned}\frac{d^2y}{dx^2} &= 2\sec^2 x + 6\tan^2 x\sec^2 x \\ &= 2\sec^2 x(1 + 3\tan^2 x) \\ &= 2(1 + \tan^2 x)(1 + 3\tan^2 x) \\ &= 2(1+y)(1+3y) = 2 + 8y + 6y^2.\end{aligned}$$

On differentiating this equation

$$\frac{d^3y}{dx^3} = 8\frac{dy}{dx} + 12y\frac{dy}{dx}, \ \frac{d^4y}{dx^4} = 8\frac{d^2y}{dx^2} + 12\left(\frac{dy}{dx}\right)^2 + 12y\frac{d^2y}{dx^2},$$

$$\frac{d^5y}{dx^5} = (8+12y)\frac{d^3y}{dx^3} + 12\left(\frac{dy}{dx}\right)\frac{d^2y}{dx^2} + 24\left(\frac{dy}{dx}\right)\frac{d^2y}{dx^2}$$

$$= (8+12y)\frac{d^3y}{dx^3} + 36\left(\frac{dy}{dx}\right)\frac{d^2y}{dx^2}$$

$$\frac{d^6y}{dx^6} = (8+12y)\frac{d^4y}{dx^4} + 12\left(\frac{dy}{dx}\right)\frac{d^3y}{dx^3} + 36\left(\frac{d^2y}{dx^2}\right)^2 + 36\left(\frac{dy}{dx}\right)\frac{d^3y}{dx^3},$$

so, when $x = 0$, $y = 0$, $\frac{dy}{dx} = 0$, $\frac{d^2y}{dx^2} = 2$, $\frac{d^3y}{dx^3} = 0$, $\frac{d^4y}{dx^4} = 16$, $\frac{d^5y}{dx^5} = 0$,

$\frac{d^6y}{dx^6} = 8.16 + 36.4 = 2^4.17$, and

$$\tan^2 x \approx 0 + 0x + 2x^2/2! + 0x^3 + 16x^4/4! + 0x^5 + 2^4 17x^6/6!$$

$$\approx x^2 + \frac{2}{3}x^4 + \frac{17}{45}x^6, \text{ as required.}$$

Power series also provide a means of summing series by using a known series and then integrating or differentiating to obtain the desired series. This is illustrated in the next example.

EXAMPLE 3 *Express the function* f, *given by*

$$f(x) = x + \frac{x^3}{3} + \frac{x^5}{5} + \dots + \frac{x^{2r-1}}{2r-1} + \dots, \ |x| < 1,$$

in terms of logarithms and hence sum the series

$$2x + \frac{4x^3}{3} + \frac{6x^5}{5} + \dots + \frac{2rx^{2r-1}}{2r-1} + \dots, \quad |x| < 1.$$

The standard logarithmic expansion is

$$\ln(1+x) = x - \frac{x^2}{2} + \frac{x^3}{3} - \frac{x^4}{4} + \dots + \frac{(-1)^{r+1}x^r}{r} + \dots, \text{ valid for } |x| < 1.$$

It should be noticed that, in the series for $f(x)$, only the odd powers of x appear. This means that it will be valuable to reverse the signs of alternate terms by changing the sign of x, thus

$$\ln(1-x) = -x - \frac{(-x)^2}{2} + \frac{(-x)^3}{3} + \dots + \frac{(-1)^{r+1}(-x)^r}{r} + \dots,$$

is still valid for $|x| < 1$. On subtracting the two logarithmic series,

$$\ln(1+x) - \ln(1-x) = 2\left(x + \frac{x^3}{3} + \frac{x^5}{5} + \ldots + \frac{x^{2r-1}}{2r-1} + \ldots\right)$$

and so $\mathrm{f}(x) = \frac{1}{2}\ln\left(\frac{1+x}{1-x}\right) = \ln\sqrt{\left(\frac{1+x}{1-x}\right)}$.

The required series has the same denominator in each term as has the series for $\mathrm{f}(x)$ but the coefficient of x^{2r-1} is multiplied by $2r$. This can be obtained by multiplying the series by x and then differentiating, since

$$x\mathrm{f}(x) = x^2 + \frac{x^4}{3} + \frac{x^6}{5} + \ldots + \frac{x^{2r}}{2r-1} + \ldots,$$

and $$\frac{\mathrm{d}}{\mathrm{d}x}(x\mathrm{f}(x)) = 2x + \frac{4x^3}{3} + \frac{6x^5}{5} + \ldots + \frac{2rx^{2r-1}}{2r-1} + \ldots.$$

It then follows that

$$\frac{\mathrm{d}}{\mathrm{d}x}(x\mathrm{f}(x)) = \frac{\mathrm{d}}{\mathrm{d}x}\frac{x}{2}\{\ln(1+x) - \ln(1-x)\}$$

$$= \{\ln(1+x) - \ln(1-x)\}/2 + x\left(\frac{1}{1+x} + \frac{1}{1-x}\right)\Big/2$$

$$= \ln\sqrt{\left(\frac{1+x}{1-x}\right)} + \frac{x}{1-x^2}.$$

The Maclaurin expansion is designed for use when x is small in magnitude. It is sometimes useful to have a power series expansion which is valid near to some non-zero value of x, say $x = a$. In order to have convergence, it is then necessary to expand the function in a series of powers of $(x-a)$. The series is then called a Taylor series and the theorem corresponding to Theorem 6.5 is called Taylor's theorem, which we now state.

Theorem 6.7 (The Taylor expansion) Let the function f have derivatives of all orders at $x = a$ and let $\mathrm{f}(x)$ have a power series expansion, in $(x-a)$, with radius of convergence R, then, for $|x-a| < R$,

$$\mathrm{f}(x) = \sum \frac{(x-a)^r}{r!}\mathrm{f}^{(r)}(a)$$

$$= \mathrm{f}(a) + (x-a)\mathrm{f}^{(1)}(a) + \frac{(x-a)^2}{2!}\mathrm{f}^{(2)}(a) + \ldots + \frac{(x-a)^r}{r!}\mathrm{f}^{(r)}(a) + \ldots.$$

It is sometimes useful to rewrite the Taylor expansion in another form, obtained by replacing $x-a$ by h, thus:

$$\mathrm{f}(a+h) = \mathrm{f}(a) + h\mathrm{f}'(a) + \frac{h^2}{2!}\mathrm{f}^{(2)}(a) + \ldots + \frac{h^r}{r!}\mathrm{f}^{(r)}(a) + \ldots.$$

EXAMPLE 4 *Find the Taylor series expansion for* $\sin x$, *valid near to* $x = \pi$.

Let $f(x) = \sin x$, then $f'(x) = \cos x$, $f''(x) = -\sin x$, $f'''(x) = -\cos x$. Let $a = \pi$ so that $f(a) = 0$, $f'(a) = -1$, $f''(a) = 0$, $f'''(a) = 1$, and
$f(x) = 0 + (x-\pi)(-1) + 0 + (x-\pi)^3(1)/3! + 0 + \dots$, that is

$$f(x) = -(x-\pi) + \frac{(x-\pi)^3}{3!} - \frac{(x-\pi)^5}{5!} + \dots + \frac{(-1)^r(x-\pi)^{2r-1}}{(2r-1)!} + \dots,$$

or

$$f(\pi + h) = -h + h^3/3! - h^5/5! + \dots + (-1)^r h^{2r-1}/(2r-1)! + \dots$$

which can be seen to be the expansion for $-\sin h$, as expected.

EXERCISE 6.6

1 If $f(x) = e^{\cos x}$, prove that $f''(x) + \sin x f'(x) + \cos x f(x) = 0$. Assuming that $e^{\cos x}$ can be expanded in a series of ascending powers of x, show that, neglecting powers of x above the fourth,

$$e^{\cos x} = e(1 - \tfrac{1}{2}x^2 + \tfrac{1}{6}x^4). \qquad (C)$$

2 Given that $y = e^{\sin^{-1} x}$, prove that

$$(1-x^2)\frac{d^2y}{dx^2} - x\frac{dy}{dx} - y = 0.$$

Hence, or otherwise, expand $e^{\sin^{-1} x}$ in ascending powers of x up to and including the term in x^3. *(C)*

3 State the series expansions for e^x and $\cos x$ in ascending powers of x. Hence, or otherwise, determine the series expansions in ascending powers of x up to and including the term in x^4 for (i) $\cosh x$, (ii) $\cos^2 x$, (iii) $e^{\cos x}$.
If x is such that x^n can be neglected for $n > 4$ find values of a and b such that $e^{\cos x} + a\cos^2 x + b\cosh x + 5ex^4/72 = 0$. *(AEB)*

4 (i) The functions f and g are defined by

$$f(x) = 1 + x^2/2! + x^4/4! + \dots + x^{2n}/(2n)! + \dots,$$
$$g(x) = x + x^3/3! + x^5/5! + \dots + x^{2n+1}/(2n+1)! + \dots.$$

Express $f(x)$ and $g(x)$ in terms of e^x, and hence obtain the expansion of $\{f(x)\}^2 + \{g(x)\}^2$ in a series of powers of x, giving the coefficient of x^{2n}.
(ii) Find the first four terms in the expansion of $\ln\{(1+2x)/(1-x)\}$ in a series of ascending powers of x, and state the set of values of x for which this expansion is valid. If a_n and a_{n+1} are the coefficients of x^n and x^{n+1} respectively in the expansion, show that the value of $(n+1)a_{n+1} + 2na_n$ is independent of n. *(L)*

5 Find a second degree Taylor approximation for the function f, given by $f(x) = (\ln x)^2$, at $x = p + \alpha$, where α is small and $p > 0$. *(SMP)*

6 State the expansion of $\cosh x$ as a series of ascending powers of x and hence sum the series $x/2! + x^2/4! + \dots + x^n/(2n)! + \dots$. *(L)*

7 Using the expansions for e^x and $\cos x$, show that

$$\tfrac{1}{4}(e^x + e^{-x} + 2\cos x) = \sum_{n=0}^{\infty} \frac{x^{4n}}{(4n)!}.$$

Find the sum of the series

$$x + \frac{x^5}{5!} + \frac{x^9}{9!} + \ldots + \frac{x^{4n+1}}{(4n+1)!} + \ldots,$$

stating the range of values of x for which the series converges. Neglecting terms of higher order than x^8, expand $2/(e^x + e^{-x} + 2\cos x)$ in ascending powers of x, and show that the coefficient of x^8 in the expansion is $23/26880$. (*L*)

MISCELLANEOUS EXERCISE 6

1 Find the expansion of $\tan x$ in ascending powers of x as far as the term in x^3. Given that $\tan(0{\cdot}1) = 0{\cdot}1003347$ and $\tan(1) = 1{\cdot}5574077$, correct to 7 decimal places, estimate the percentage error in using the above series expansion as an approximation in each case and comment on the results.

2 Using the approximation $\ln(10) \approx 2{\cdot}302585$ and the Taylor series expansion of $\ln(1+x)$, evaluate $\ln(11)$ to 5 significant figures.

3 Use the expansion of $\ln(1+x)$ to find to four decimal places the value of $\ln(0.95)$. (*L*)

4 Expand $e^{-x}\ln(1+x)$ as a series in ascending powers of x up to and including the term in x^2. (*L*)

5 Obtain the first three non-zero terms in the expansion of $\tan x$ as a series in ascending powers of x. Write a program to evaluate to five decimal places the sum of these three terms for $x = 0(0{\cdot}001)0{\cdot}5$. (*L*)

6 Find the constants a and b such that, when x is small,

$$\frac{\sin 2x}{\sqrt{(1+x^2)}} \approx ax + bx^3. \qquad (L)$$

7 Given that the expansions of e^x and $(1+ax)/(1+bx)$, in ascending powers of x, are the same up to and including the term in x^2, find the values of the constants a and b. (*JMB*)

8 (a) If x is so small that terms in x^n, $n \geqslant 3$, can be neglected, find values of a and b for which $(1-4x)^{1/2} = (1+ax)/(1+bx)$. Hence find an approximation to $\sqrt{92}$ in the form p/q, where p and q are integers.
(b) State the first five terms of the expansions of $\ln(1+x)$ and $\ln(1-x)$. Deduce that, when $n > 1$,

$$\ln\left\{\frac{n+1}{n-1}\right\} = 2\left\{\frac{1}{n} + \frac{1}{3n^3} + \frac{1}{5n^5} + \ldots\right\}.$$

Hence evaluate $\ln(1.25)$ correct to four places of decimals. (*AEB*)

9 Expand $e^{2x}/(1+2x)$ as a series of ascending powers of x as far as the term in x^3 and give the set of values of x for which the expansion is valid. (*L*)

10 Write down the coefficient of x^{2n} in the expansion of $e^{x^2} + \ln(1+2x)$ in ascending powers of x, and give the set of values of x for which this expansion is valid. (*L*)

11 If $y = (\sin^{-1}x)^2$, prove that $(1-x^2)\dfrac{d^2y}{dx^2} - x\dfrac{dy}{dx} = 2$. By repeated differentiation of this result and use of Maclaurin's theorem, or otherwise, prove that the first three non-zero terms in the expansion of $(\sin^{-1}x)^2$ are $x^2+x^4/3+8x^6/45$. *(C)*

12 Find the coefficient of x^3 in the expansion, in ascending powers of x, of the function $f(x)=(1-x)^2e^{2x}$. Find also the coefficient of x^n in this expansion of $f(x)$ and deduce that the only values of n for which the coefficient of x^n is zero are 1 and 4. *(JMB)*

13 Find the coefficient of x^n, where $n>2$, in the expansion, in ascending powers of x, of $(a+bx+cx^2)e^x$. Given that, for $n \geqslant 0$, the coefficient of x^n is $\dfrac{(n+1)(n+2)}{n!}$, find a, b and c. *(JMB)*

14 Find the coefficient of x^n, where $n>1$, in the expansion of $(1+2x)\ln(1+2x)$ as a series of ascending powers of x, where $|x|<1/2$. *(L)*

15 Given that $1-px+x^2 \equiv (1-\alpha x)(1-\beta x)$, obtain expressions for $\alpha^2+\beta^2$ and $\alpha^3+\beta^3$ in terms of p. Hence, or otherwise, find the first three terms in the expansion of $\ln(1-px+x^2)$ in ascending powers of x. *(JMB)*

16 Expand $(e^x+e^{-x})/2$ as a series of ascending powers of x. Deduce the general term in the expansion of $(e^{\sqrt{x}}+e^{-\sqrt{x}})/2$. *(L)*

17 Given that $|2x|<1$, find the first two non-zero terms in the expansion of $\ln\{(1+x)^2(1-2x)\}$ in a series of ascending powers of x. *(L)*

18 Show that, in the expansion in ascending powers of x of $\ln(1+x)+3\ln(3+x)-3\ln(3+2x)$, the first non-zero term is that in x^3, and find this term. Obtain an expression (which need not be simplified) for the coefficient of x^n for $n>0$. *(JMB)*

19 Find the expansion of $\ln\left\{\dfrac{\sqrt{(1+x)}}{1-x}\right\}$, where $|x|<1$, in ascending powers of x up to and including the term in x^3. Given that
$\dfrac{1}{1-ax}\ln\left\{\dfrac{\sqrt{(1+x)}}{1-x}\right\} \approx \dfrac{3}{2}x+kx^3$ when terms in x^4 and higher powers of x are neglected, find the values of the constants a and k. *(JMB)*

20 If $E(x)=\{(1+x)^{1/2}+(1-x)^{-1/2}\}/(1-x^2)^{1/4}$ and if x is so small that its cube may be neglected, prove that $E(x)=2+x+3x^2/4$. Find the values of constants a and b if the expansions of $E(x)$ and of $2e^{ax}/(1-bx^2)$ are identical up to and including the term in x^2. *(L)*

21 Write down the first three terms in the expansion of $(1+y)^{1/2}$ in ascending powers of y. Given that x is so small that its cube and higher powers are negligible, find the constants a and b in the approximate formula $(1-x-2x^2)^{1/2} \approx 1+ax+bx^2$.
Prove also that $\ln(1-x-2x^2)^{1/2} \approx -x/2-5x^2/4$. *(JMB)*

22 Given that $f(x)=\{x+\sqrt{(x^2+1)}\}^3$, show that $(x^2+1)\{f'(x)\}^2=9\{f(x)\}^2$. Deduce that $(x^2+1)f''(x)+xf'(x)=9f(x)$. Use Leibnitz's theorem to deduce that, for $n \geqslant 0$, $f^{(n+2)}(0)+(n^2-9)f^{(n)}(0)=0$. Find the Taylor series for f about $x=0$, up to and including the term in x^4. State the coefficient of x^{2n+1} for $n \geqslant 2$. By putting $x=\sinh\theta$ in the series, or otherwise, show that the function g given by $g(\theta)=e^{3\theta}-3\sinh\theta-4\sinh^3\theta$ is an even function. *(JMB)*

23 Write down the first four non-zero terms in the expansions of $\cosh x$ and $\cos x$ in ascending powers of x. Given that x is sufficiently small for x^7 and higher powers of x to be neglected, show that $\ln\left(\dfrac{\cosh x}{\cos x}\right) \approx ax^2+bx^6$, where a and b are constants to be found. *(JMB)*

24 Given that x is so small that its fourth and higher powers may be neglected, show that $\dfrac{x \sin 3x}{\cos x - \cos 3x} \approx a + bx + cx^2 + dx^3$, where a, b, c, d are constants whose values are to be found. *(JMB)*

25 Given that $y = (1+x^2)^{1/2} \sinh^{-1} x$ show that $(1+x^2)\dfrac{dy}{dx} = 1 + x^2 + xy$. (1)

Hence express $(1+x^2)\dfrac{d^2y}{dx^2} + x\dfrac{dy}{dx}$ in terms of x and y, and deduce that

$(1+x^2)\dfrac{d^3y}{dx^3} + 3x\dfrac{d^2y}{dx^2} = 2$. If $y = \sum_{n=0}^{\infty} a_n x^n$, use the above results to find the values of a_n for $n = 0, 1, 2, 3$. Using Leibnitz's theorem, differentiate both sides of equation (1) n times, where $n \geqslant 3$, and hence show that $(n+1)a_{n+1} = -(n-2)a_{n-1}$. Calculate the values of a_n for $n = 4, 5, 6$ and 7. *(JMB)*

26 (a) Write down the first four terms in the expansion in ascending powers of x of $(1+4x)^{1/2}$, and simplify the coefficients. Hence, by putting $x = -1/100$, calculate $\sqrt{6}$ correct to four decimal places.
(b) Write down, as far as the terms in x^4 inclusive, the expansions in ascending powers of x of (i) $\ln(1+x)$, (ii) e^x. If $|x| < 1$ and $y = x + x^2/2 + x^3/3 + x^4/4 + \ldots$, find an expression, in a form not involving an infinite series, for y in terms of x. Hence find the expansion of x in powers of y as far as the term in y^4. *(JMB)*

27 A function f, where $y = f(x)$, satisfies the differential equation $\dfrac{dy}{dx} = x^2 - y$, and $y = 2$ when $x = 1$. Obtain the first five terms of the Taylor series expansion for $f(x)$, centred on $x = 1$, and use it to obtain an approximation to the value of $f(1{\cdot}1)$, working to 6 decimal places. *(JMB)*

28 Given that $y = e^x \cos x$ and $y_n = \dfrac{d^n y}{dx^n}$, $n > 0$, evaluate y_1 and y_2 at $x = 0$. Prove that, for all values of x, $y_{n+2} = 2(y_{n+1} - y_n)$. Hence, or otherwise, obtain an expansion of $e^x \cos x$ as a power series as far as the term in x^5 and use this to estimate the value of y to six places of decimals when $x = 0{\cdot}1$. *(L)*

29 Given that $y = f(x) = \sinh^{-1} x$, prove that $(1+x^2)\dfrac{d^2y}{dx^2} + x\dfrac{dy}{dx} = 0$. Hence prove that (i) $f^{(n+2)}(0) + n^2 f^{(n)}(0) = 0$, (ii) $f^{(2k)}(0) = 0$ for all $k \geqslant 1$, (iii) $f^{(2k+1)}(0)$ is non-zero for all $k \geqslant 1$. Prove that $f^{(4k+1)}(0)$ is a perfect square for all $k \geqslant 1$. *(JMB)*

30 Find, to the nearest integer, the positive root of the equation $e^x - x - 5 = 0$. Use the iterative formula $x = \ln(x+5)$ to obtain the root correct to four significant figures. Make the substitution $x = y + 2$, expand e^{y+2} $(\equiv e^2 e^y)$ in ascending powers of y, neglect terms of higher order than the first and so obtain another approximation, to four significant figures, for the root. (You are reminded that all steps in the calculation are to be shown.) *(L)*

7 Polar Coordinates

7.1 Definitions and some simple loci

In the Cartesian plane, the position of any point P, relative to the origin O, is given in terms of the coordinates (x, y). The position of P is also given by its distance r from O and the angle θ between Ox and OP. These are the *polar coordinates* of P and are defined with respect to an origin O, called the *pole*, and a fixed line Ox, called the *initial line*.

Definition Let the Cartesian coordinates of a point P in the plane be (x, y), then the polar coordinates of P are (r, θ), where r is the distance $|OP|$ of P from O, θ is the angle POx, measured positively (anticlockwise) from Ox to OP, the origin O is called the pole, the reference axis Ox is called the initial line.

The polar coordinates of P are illustrated in Fig. 7.1 and the following equations between the coordinates are obtained:

$$x = r\cos\theta, \qquad y = r\sin\theta,$$
$$r^2 = x^2 + y^2, \qquad r = \sqrt{(x^2 + y^2)},$$
$$\cos\theta = x/r, \quad \sin\theta = y/r, \quad \tan\theta = y/x.$$

Note that we choose r to be non-negative. When sketching a curve, given by a polar equation $r = \mathrm{f}(\theta)$, negative values of $\mathrm{f}(\theta)$ may occur for some values of θ. If $\mathrm{f}(\alpha) < 0$, then we use the corresponding positive value,

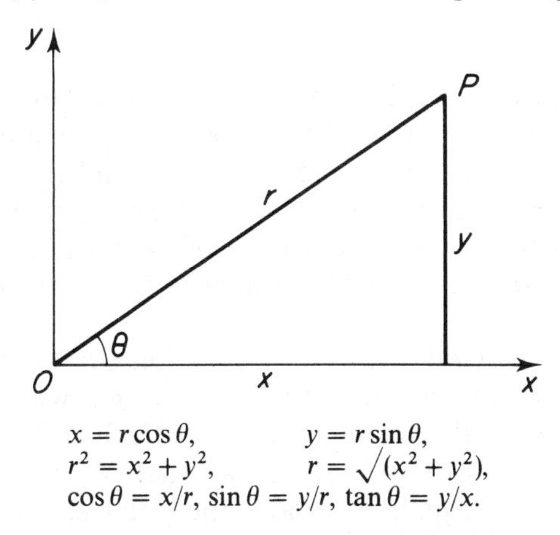

$$x = r\cos\theta, \qquad y = r\sin\theta,$$
$$r^2 = x^2 + y^2, \qquad r = \sqrt{(x^2 + y^2)},$$
$$\cos\theta = x/r, \sin\theta = y/r, \tan\theta = y/x.$$

Fig. 7.1 Polar coordinates (r, θ)

$-f(\alpha)$, and add π to (or subtract π from) α so that the point (r, θ) will be given by $r = -f(\alpha)$, $\theta = \alpha + \pi$. This is valid since the same point P is given by the polar coordinates (r, θ) and $(-r, \theta + \pi)$. The addition of π to θ gives a rotation of the plane through π about O, which is equivalent to a reflection in the origin. The addition of $2n\pi$ to θ, for any integer n, leaves the position of P unaltered and we choose a principal value of θ in the range $-\pi < \theta \leqslant \pi$.

The polar coordinates of P may be compared with the modulus $|z|(=r)$ and the argument $\arg z (=\theta)$ of the complex number z which is represented by P in the complex plane, see Pure Mathematics §18.3.

The straight line in polar coordinates

The equation $\theta = \alpha$ is the equation of the half line through the origin making an angle α with Ox. The full line through the origin with gradient $\tan\alpha$ is given by the pair of equations $\theta = \alpha$ and $\theta = \alpha - \pi$, that is, by the equation $(\theta - \alpha)(\theta - \alpha + \pi) = 0$, see Fig. 7.2.

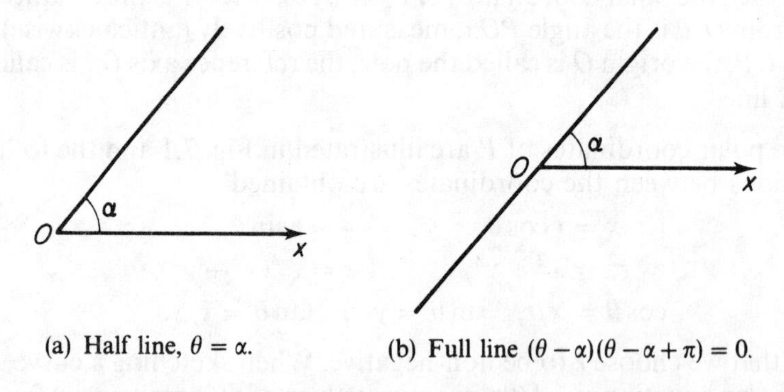

Fig. 7.2 Polar equation of a line through the origin

The polar equation of a line, given by a Cartesian equation, may be obtained by substituting for x and y in polar coordinates. Thus the lines parallel to the Cartesian axes, given by $x = a$ and by $y = b$, have polar equations $r\cos\theta = a$ and $r\sin\theta = b$. These lines are shown in Fig. 7.3 and it will be seen that the constant values of $r\cos\theta$ and of $r\sin\theta$ are the (constant) projections of the displacement OP on to the perpendicular to the line.

Now consider a general line in the plane. Let ON be the perpendicular from the origin on to the line and let $\overrightarrow{ON} = p\mathbf{n}$, where $p = |\overrightarrow{ON}|$ and $\mathbf{n}$ is a unit vector. Then the vector equation of the line is $\mathbf{r}\,.\,\mathbf{n} = p$ and, if $\overrightarrow{ON}$ makes an angle α with the initial line, so that N has polar coordinates (p, α), the angle PON equals $\theta - \alpha$ and the polar equation of the line is $r\cos(\theta - \alpha) = p$, see Fig. 7.4.

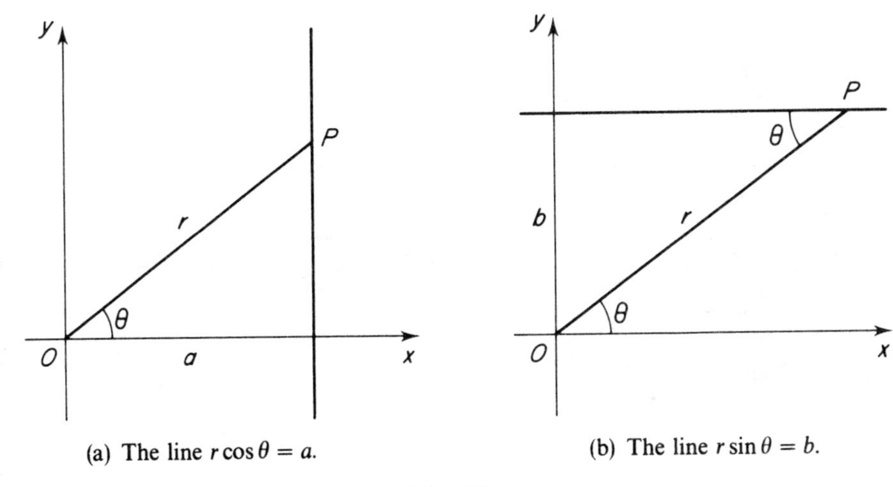

(a) The line $r\cos\theta = a$.

(b) The line $r\sin\theta = b$.

Fig. 7.3

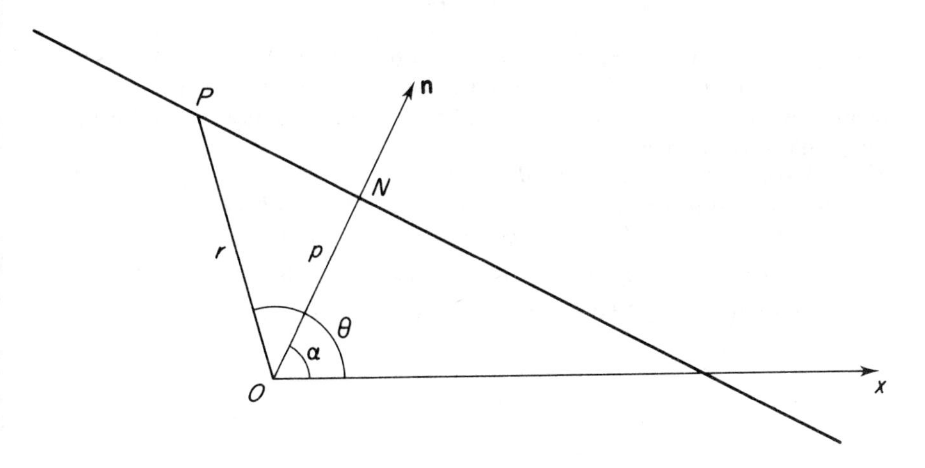

Fig. 7.4 The line $r\cos(\theta - \alpha) = p$

Spirals

The locus, given by an equation of the form $r = \mathrm{f}(\theta)$, where f is an increasing function, is a spiral. Two typical spirals are those with equations $r = a\theta$, $\theta \geqslant 0$, (the Archimedian spiral) and $r = \mathrm{e}^{a\theta}$ (the equiangular spiral), see Fig. 7.5.

EXAMPLE 1 *Sketch the loci given by the three polar equations and find the polar coordinates of their points of intersection:* (*a*) $2r = \theta$, $0 \leqslant \theta \leqslant 2\pi$, (*b*) $r\cos\theta = 1, -\pi/2 < \theta < \pi/2$, (*c*) $r(3\cos\theta - 4\sin\theta) = 6$.

The three loci are shown in Fig. 7.6, (the line with equation (c) has Cartesian equation $3x - 4y = 6$). We calculate the three intersection points.

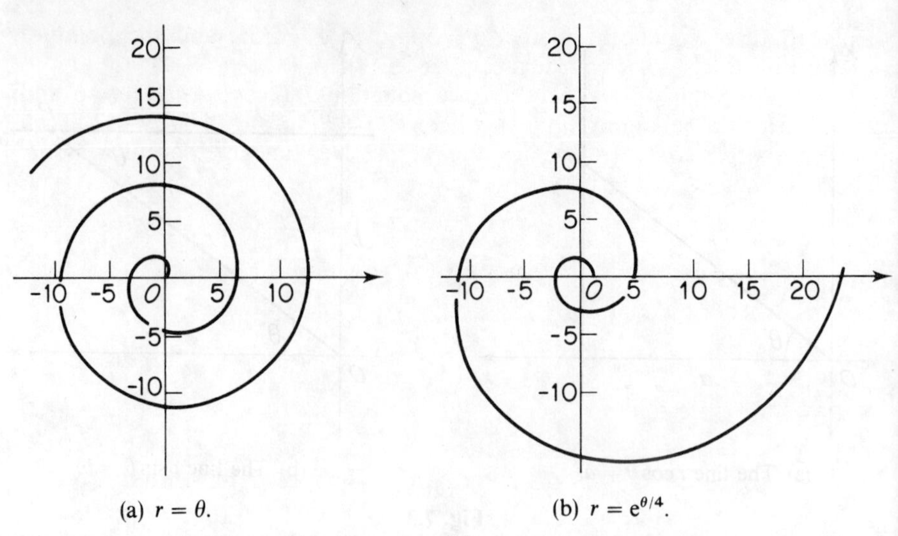

(a) $r = \theta$. (b) $r = e^{\theta/4}$.

Fig. 7.5

(b) and (c). Eliminate r, $r(3\cos\theta - 4\sin\theta) = 6r\cos\theta$, so that $4\sin\theta + 3\cos\theta = 0$, $\tan\theta = -3/4$. Since the lines cross in the fourth quadrant, we find that, to three significant figures, $\theta = 5{\cdot}64$ and $r = 1{\cdot}25$. The lines intersect at A with polar coordinates **(1·25, 5·64)**.

(a) and (b). If $2r = \theta$ and $r\cos\theta = 1$, then $\theta\cos\theta = 2$ and we have to solve numerically. From the graph, we see that the root lies between $\theta = 3\pi/2$ and $\theta = 2\pi$. Let $f(\theta) = \theta\cos\theta - 2$ and use linear interpolation beginning with $\theta_1 = 5$.

$$f(5) = -0{\cdot}58,\ f(6) = 3{\cdot}76,\ \theta_2 = 5 + 0{\cdot}58/(0{\cdot}58 + 3{\cdot}76) = 5{\cdot}13.$$

$$f(5{\cdot}13) = 0{\cdot}08,\ f(5{\cdot}12) = 0{\cdot}03,\ f(5{\cdot}11) = -0{\cdot}02.$$

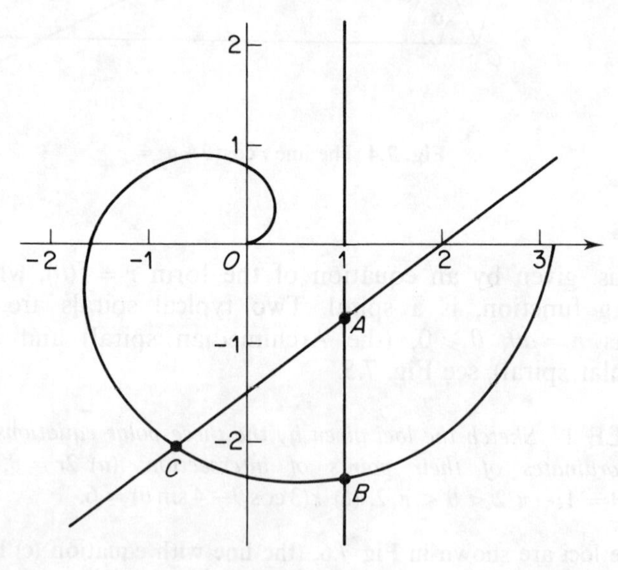

Fig. 7.6

Thus, to three significant figures $\theta = 5{\cdot}11$ and $r = 2{\cdot}56$, and the point of intersection B has polar coordinates **(2·56, 5·11)**.
(a) and (c). Eliminating r between the equations $r(3\cos\theta - 4\sin\theta) = 6$ and $2r = \theta$ gives an equation $\mathrm{f}(\theta) = 0$, where $\mathrm{f}(\theta) = \theta(3\cos\theta - 4\sin\theta) - 12$. Then $\mathrm{f}'(\theta) = (3 - 4\theta)\cos\theta - (4 + 3\theta)\sin\theta$ and we use Newton's method of solution. The graph suggests a starting point $\theta_1 = 4$. Then

$$\theta_2 = 4 - \mathrm{f}(4)/\mathrm{f}'(4) = 4{\cdot}37,\ \theta_3 = 4{\cdot}37 - \mathrm{f}(4{\cdot}37)/\mathrm{f}'(4{\cdot}37) = 4{\cdot}37.$$

So $\theta = 4{\cdot}37$, $r = 2{\cdot}19$, and the point C of intersection is **(2·19, 4·37)**.

Circles

The locus $r = a$, $0 \leqslant \theta < 2\pi$, is the circle of radius a with centre at the origin. Circles which pass through the origin also have simple polar equations.

Let C be the circle with centre $(a, 0)$ and radius a and let A be the point $(2a, 0)$ so that OA is a diameter of the circle. Then, for any point P on the circle, angle OPA is a right angle and so $OP = OA\cos\theta$ and the equation of the circle is $r = 2a\cos\theta$, $-\pi/2 \leqslant \theta \leqslant \pi/2$. This circle is shown in Fig. 7.7 (a).

In a similar manner, the circle of radius b with centre at $(b, \pi/2)$ is given by the polar equation $r = 2b\sin\theta$, $0 \leqslant \theta < \pi$, Fig. 7.7(b).

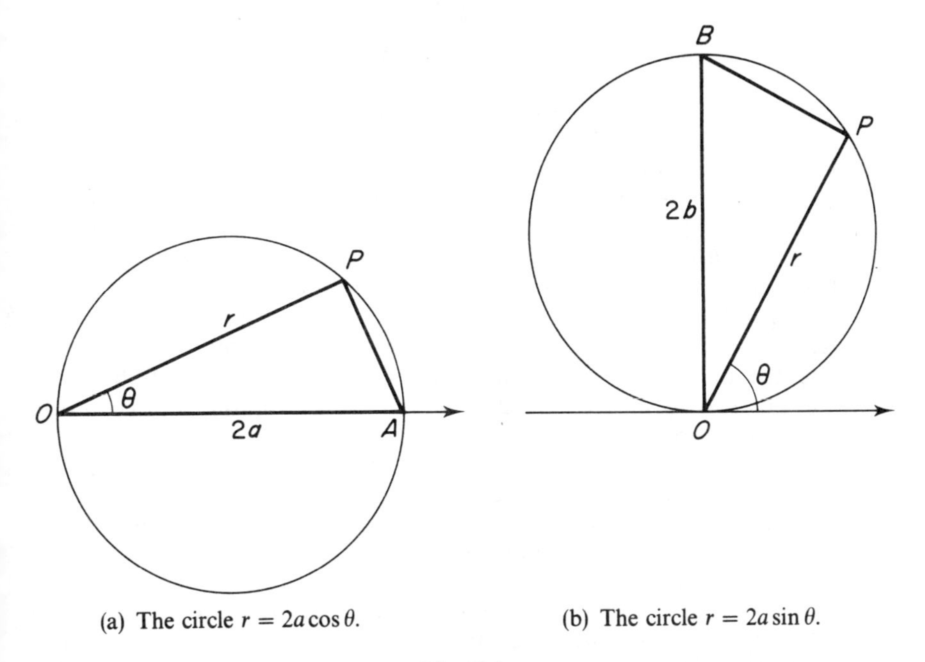

(a) The circle $r = 2a\cos\theta$.

(b) The circle $r = 2a\sin\theta$.

Fig. 7.7

EXAMPLE 2 *Find the polar equations of the tangents to the two circles $r = 2\cos\theta$ and $r = 2\sin\theta$ at their point of intersection, other than the origin.*

The circles are shown in Fig. 7.8. They intersect at the point where $\cos\theta = \sin\theta$, that is where $\theta = \pi/4$ and $r = \sqrt{2}$. Their tangents at this point are the two lines parallel, and perpendicular, to the initial line and these lines have equations $\boldsymbol{r\cos\theta = 1}$ and $\boldsymbol{r\sin\theta = 1}$.

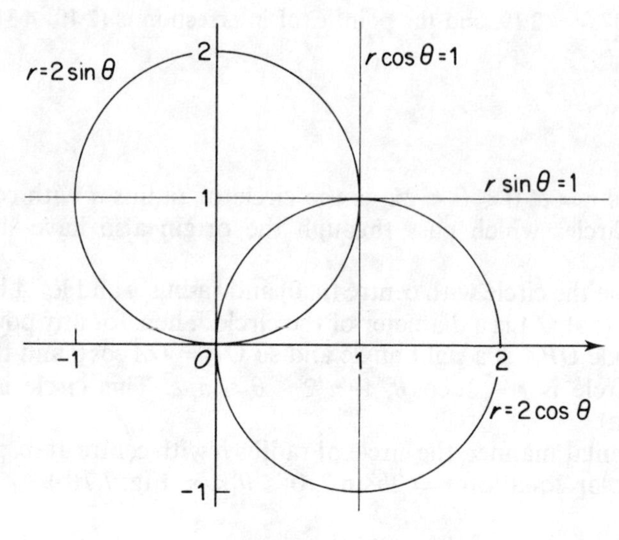

Fig. 7.8

EXERCISE 7.1

1 Sketch the locus of the point P, with polar coordinates (r, θ), given by the equation: (i) $\theta = \pi/4$, (ii) $\theta = 5\pi/4$, (iii) $\theta = -3\pi/4$, (iv) $r\cos\theta = 2$, (v) $r\sin\theta = -1$, (vi) $r(2\cos\theta + 3\sin\theta) = 6$, (vii) $r\cos(\theta - \pi/6) = 2$, (viii) $r\cos(\theta + \pi/4) = 3$, (ix) $r\sin(3\pi/4 - \theta) = 1$, (x) $r\sin(\theta + 2\pi/3) = 4$.

2 Sketch the two loci given by the polar equations (i) $r\cos\theta = -1$, $\pi/2 < \theta < 3\pi/2$ and (ii) $r = \theta$, $0 \leqslant \theta \leqslant \pi$, and find the polar coordinates of their point of intersection, correct to three significant figures.

3 State the radius and the polar coordinates of the centre of the circle with polar equation: (i) $r = 2\cos\theta$, (ii) $r = 3\sin\theta$, (iii) $2r + \cos\theta = 0$, (iv) $r = \sin\theta + \cos\theta$.

4 Find the polar equation of the tangent to the circle $r = 2\cos\theta$ at the point: (i) $(2, 0)$, (ii) $(\sqrt{2}, \pi/4)$, (iii) $(\sqrt{2}, -\pi/4)$.

5 Find the polar coordinates of the point of intersection of the two tangents to the circle $r = 2\sin\theta$, at the points where $r = 1$.

6 Find the points of intersection of the two circles $r = a$ and $r = 2a\cos\theta$. Write down the polar equation of the common chord of these two circles.

7 Sketch the curve whose polar equation is $r = 2a\cos\theta$, where $0 \leqslant \theta \leqslant \pi/2$ and a is a positive constant. The point Q has polar coordinates $(a\sqrt{2}, -\pi/4)$. Give the polar coordinates of the point P on the given curve for which the distance PQ is greatest. (*L*)

8 Sketch on the same diagram the circle with polar equation $r = 4\cos\theta$ and the line with polar equation $r = 2\sec\theta$. State polar coordinates for their points of intersection. (*L*)

9 A curve is given in polar coordinates by the equation $r = 2a\cos\theta$, where $a > 0$ and $-\pi/2 < \theta \leqslant \pi/2$. Calculate the mean value with respect to θ of the length of the straight line segment OP, where O is the pole and P is a point on the curve. (*L*)

7.2 Further polar curves

The cardioid

Consider the locus of a point P, on the circumference of a moving circle, of radius a, which rolls without slipping on a fixed circle of radius a and centre A. Let O be the initial position of P and take AO produced as the initial line. When P has polar coordinates (r, θ), let the circles touch at C and let the centre of the moving circle be D, Fig. 7.9. Since the arcs OC and PC on the two circles are equal and the circles are reflections of each other

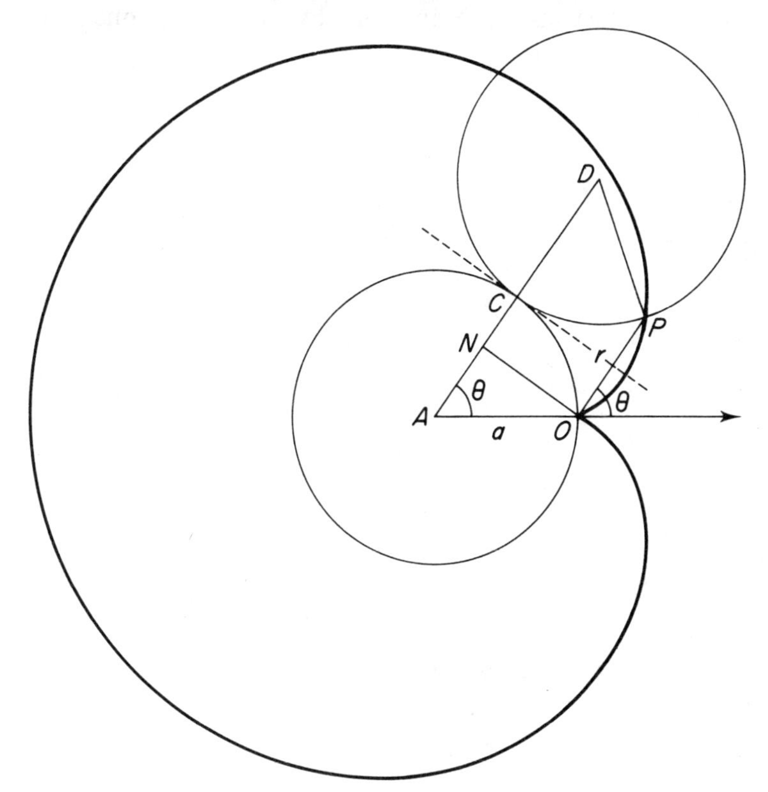

Fig. 7.9 The cardioid

in the common tangent at C, the points P and D are reflections of O and A in that tangent. Thus OP is parallel to AD and angles OAC, PDC and POx are all equal to θ. If ON is the perpendicular from O on to AC, then $AN = a\cos\theta$ and therefore $r = OP = 2CN = 2(a - a\cos\theta)$. Thus the locus of P is given by the polar equation $r = 2a(1 - \cos\theta)$. The locus is shown in Fig. 7.9 and is called a *cardioid*, since it is heart shaped.

Conics

Let L be the line $r\cos\theta = p$, where $p > 0$, and let the locus of P, (r, θ), be such that its distance from O is e times its distance from L, where e is a positive constant. In Fig. 7.10, PN is the perpendicular from P on to L and PM is the perpendicular from P on to the initial line Ox, which meets L at K. Then $OP = ePN = eMK$ and

$$r = OP = e(OK - OM) = e(p - r\cos\theta).$$

Therefore the polar equation of the locus of P is $r(1 + e\cos\theta) = ep$. This locus is a conic and

if $e < 1$, the locus is an ellipse with one focus at O,
if $e = 1$, the locus is a parabola with focus at O,
if $e > 1$, the locus is one branch of a hyperbola with one focus at O.

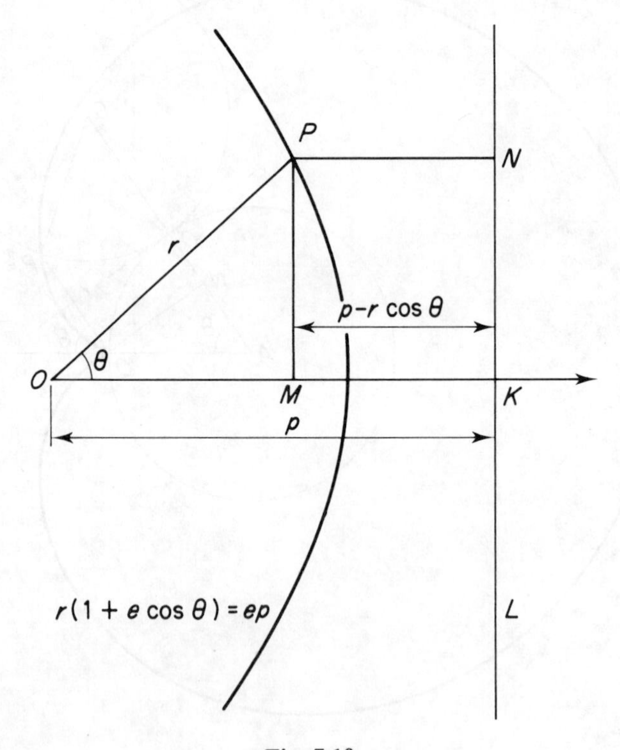

Fig. 7.10

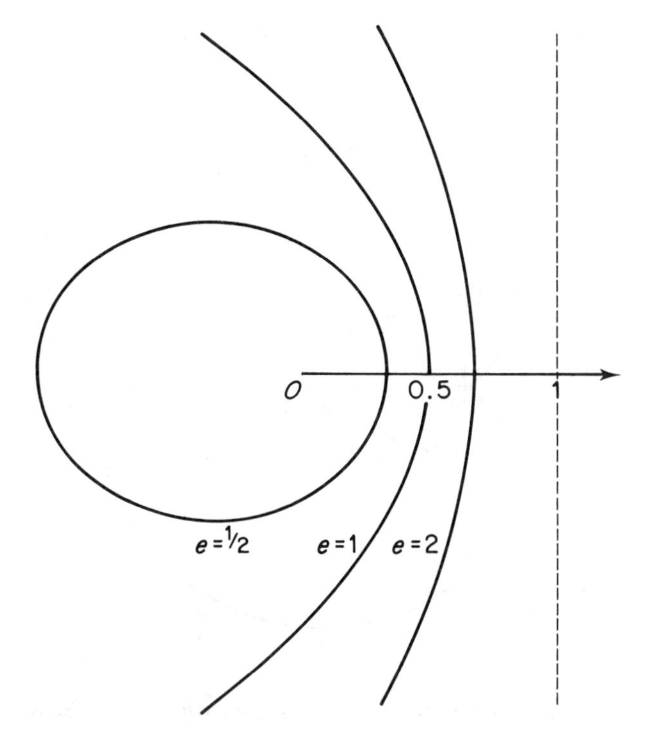

Fig. 7.11 The ellipse $r(2+\cos\theta) = 1$, the parabola $r(1+\cos\theta) = 1$ and the hyperbola $r(1+2\cos\theta) = 2$

The constant e is called the *eccentricity* of the conic and the loci for $e = 1/2$, $e = 1$ and $e = 2$ are shown in Fig. 7.11.

Note that, for the hyperbola $r(1+2\cos\theta) = 2$, where $e = 2$, the equation gives only one branch of the hyperbola. The other branch corresponds to those values of θ for which $2\pi/3 < \theta < 4\pi/3$ and these values of θ give rise to negative values of r. With our restriction that $r \geqslant 0$, the second branch does not occur on the diagram.

EXAMPLE 1 *Sketch the curves given by the polar equations*

$$r = 1+\cos\theta \text{ and } 4r\cos\theta = 3.$$

The line segment AB is the common chord of these two curves. Find (i) the length AB, (ii) the polar equation of the circle which passes through O and touches AB at its midpoint.

For the cardioid with equation $r = 1+\cos\theta$, we calculate the values of r for a few values of θ and plot the corresponding points (r, θ).

θ	0	$\pi/6$	$\pi/4$	$\pi/3$	$\pi/2$	$2\pi/3$	$3\pi/4$	$5\pi/6$	π
	2π	$11\pi/6$	$7\pi/4$	$5\pi/3$	$3\pi/2$	$4\pi/3$	$5\pi/4$	$7\pi/6$	
$r = 1+\cos\theta$	2	1·9	1·7	1·5	1	0·5	0·3	0·1	0

Joining these points by a smooth curve gives the sketch of the cardioid, as shown in Fig. 7.12.

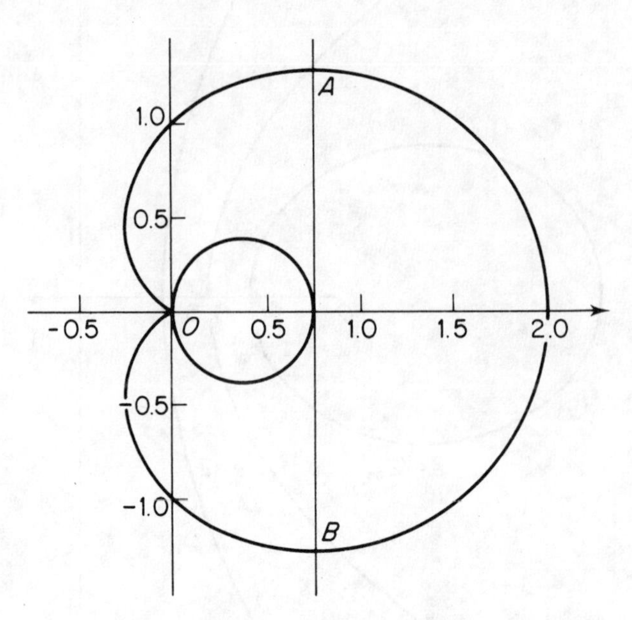

Fig. 7.12 The curves $r = 1 + \cos\theta$, $4r\cos\theta = 3$, $4r = 3\cos\theta$

The second equation, $4r\cos\theta = 3$, is the equation of the line, perpendicular to the initial line and distant 3/4 from O. This line is shown on the diagram, intersecting the cardioid at A and B. To find the coordinates of A and B, we solve the simultaneous equations

$$r = 1 + \cos\theta,\ 4r\cos\theta = 3.$$

Eliminating $\cos\theta$

$$4r^2 = 4r + 4r\cos\theta = 4r + 3.$$

Completing the square for the quadratic expression in r,

$$4r^2 - 4r + 1 = 3 + 1, \text{ that is } (2r - 1)^2 = 4,$$

which gives $2r - 1 = 2$ or $2r - 1 = -2$, that is $r = 3/2$ or $r = -1/2$. The negative solution does not satisfy the original equations so we conclude that $r = 3/2$ and then $\cos\theta = 1/2$ and $\theta = \pi/3$ or $\theta = -\pi/3$. Therefore the points A and B have polar coordinates $(3/2, \pi/3)$ and $(3/2, -\pi/3)$, and $|AB| = \mathbf{3\sqrt{3}/2}$. The midpoint of AB is the point C with coordinates $(3/4, 0)$, lying on the line $4r\cos\theta = 3$. The required circle has diameter OC and its polar equation is $\mathbf{4r = 3\cos\theta}$.

The curve given by the polar equation $r = \mathrm{f}(\theta)$, for some function f, is not usually the graph of a Cartesian function. That is, the Cartesian equation of the curve cannot be written in the form $y = \mathrm{g}(x)$ for some function g. We have seen many examples where the curve consists of a loop so that two values of y correspond to one value of x and the curve is the graph of a relation between x and y rather than the graph of a function. In fact, the curve may well contain several loops. In the next example, the curve consists of two loops.

EXAMPLE 2 *Sketch the graph with polar equation $r = 1 + \cos 2\theta$.*

The maximum value of r is 2 and this occurs when $\theta = 0$ and when $\theta = \pi$. As θ increases from 0 to $\pi/2$, r decreases from 2 to 0. It then increases again to 2 as θ increases from $\pi/2$ to π. The above changes in r are then repeated as θ increases from π to 2π. This is enough to indicate the general shape of the curve which consists of two loops, touching each other at the origin and each symmetrical about the initial line. The curve has tangents perpendicular to the initial line at the pole and at points with polar coordinates $(2, 0)$ and $(2, \pi)$. In order to obtain a reasonably accurate sketch of the curve, we calculate the values of r corresponding to values of θ between $0°$ and $90°$ at intervals of $10°$. We use degrees, instead of radians, for these calculations to facilitate the use of a protractor in plotting points. These values give the coordinates for the first quarter of the curve and the remaining three quarters are obtained by using the symmetry of the curve about the lines $\theta = 0°$, $\theta = 90°$, $\theta = 180°$ and $\theta = 270°$. We tabulate the calculated values and show the graph in Fig. 7.13.

θ (in degrees)	0	10	20	30	40	50	60	70	80	90
$r = 1 + \cos 2\theta$	2	1·94	1·77	1·5	1·17	0·83	0·05	0·23	0·06	0

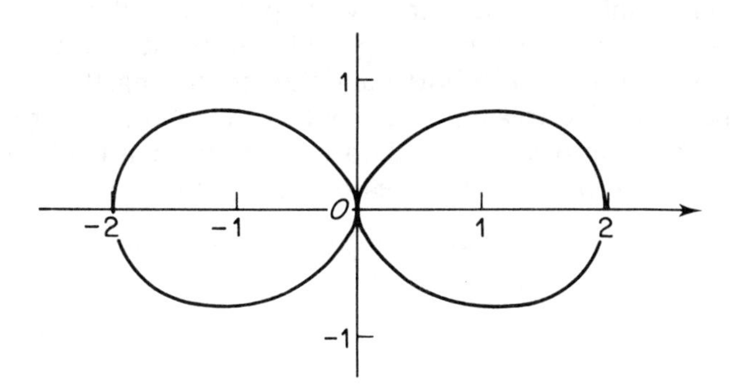

Fig. 7.13 The curve $r = 1 + \cos 2\theta$

EXERCISE 7.2

1 Sketch the curve with polar equation $r = a(1 - \cos\theta)$, where $a > 0$, stating the polar coordinates of the point on the curve at which r has its maximum value. (*L*)

2 Sketch the curve $r = a(1 + \cos\theta)$ for $0 < \theta < \pi$ where $a > 0$. Sketch also the line $r = 2a\sec\theta$ for $-\pi/2 < \theta < \pi/2$ on the same diagram. The half line $\theta = \alpha$, $0 < \alpha < \pi/2$, meets the curve at A and the line $r = 2a\sec\theta$ at B. If O is the pole, find the value of $\cos\alpha$ for which $OB = 2OA$. (*L*)

3 For the locus $r(1+e\cos\theta)=ep, e\neq 1$, define a by the equation $ep=a(1-e^2)$. Choose Cartesian coordinates (x, y), with origin at the point with polar coordinates (ea, π), so that $x=ea+r\cos\theta$ and $y=r\sin\theta$, and show that the locus has Cartesian equation

$$\frac{x^2}{a^2}+\frac{y^2}{a^2(1-e^2)}=1.$$

4 For the locus $r(1+\cos\theta)=p$, choose Cartesian coordinates with the origin at K $(p/2, 0)$ so that $x=p/2-r\cos\theta$, $y=r\sin\theta$, show that the Cartesian equation of the parabola is $y^2=2px$.

5 Transform the polar equation $r^2\cos 2\theta=a^2$ into Cartesian form and hence sketch and identify the curve.

6 Sketch the lemniscates (i) $r^2=a^2\cos 2\theta$, (ii) $r^2=a^2\sin 2\theta$, on the same axes and prove that the two curves meet on the line $(\theta-\pi/8)(\theta-9\pi/8)=0$.

7 Sketch the trifolium $r=a\cos 3\theta$. Prove that the curve has a Cartesian equation $(x^2+y^2)^2=a(x^3-3y^2x)$.

8 Sketch the trefoil $r^2=a^2\sin 3\theta$ and prove that, in Cartesian form, its equation is $(x^2+y^2)^{5/2}=a(3x^2y-y^3)$.

9 Sketch the limaçon $r=1+2\cos\theta$.

7.3 Areas in polar coordinates

Consider a curve with the polar equation $r=f(\theta)$ and let P and P' be adjacent points on the curve with polar coordinates (r, θ) and $(r+\delta r, \theta+\delta\theta)$. This is shown in Fig. 7.14, where the curve is drawn so that, on moving from P to P', both r and θ are increasing, that is, δr and $\delta\theta$ are both positive. The area δS, swept out by the radius vector in moving from $\overrightarrow{OP}$ to $\overrightarrow{OP'}$, is approximately the area of the triangle OPP' and so, (on using the formula $\frac{1}{2}ab\sin C$ for area of a triangle),

$$\delta S\approx\tfrac{1}{2}OP.OP'\sin\delta\theta\approx\tfrac{1}{2}r(r+\delta r)\delta\theta.$$

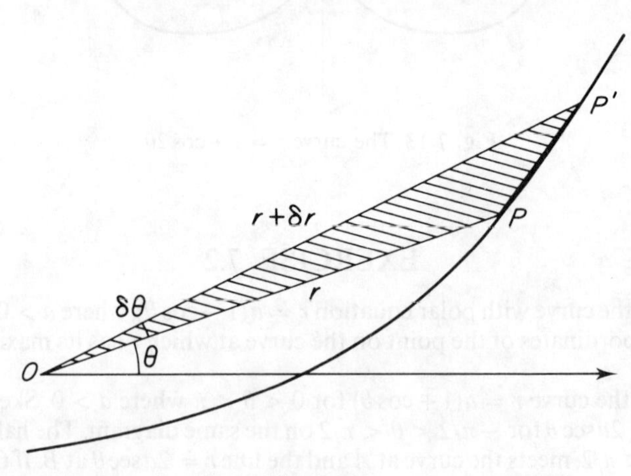

Fig. 7.14

If we ignore the products of small quantities, for small values of δr and $\delta\theta$,

$$\delta S \approx \tfrac{1}{2}r^2\,\delta\theta,$$

and, in the limit, as $P' \to P$ and both δr and $\delta\theta$ tend to zero,

$$\frac{\mathrm{d}S}{\mathrm{d}\theta} = \tfrac{1}{2}r^2.$$

This means that the area swept out, as θ increases from α to β is

$$\int_\alpha^\beta \tfrac{1}{2}r^2\,\mathrm{d}\theta = \int_\alpha^\beta \tfrac{1}{2}\{\mathrm{f}(\theta)\}^2\,\mathrm{d}\theta.$$

EXAMPLE 1 *Find the points of intersection of the circles $r = 1$ and $r = 2\cos\theta$ and find the common area enclosed by these two circles.*

The two circles are shown in Fig. 7.15. They meet at points A and B where $\boldsymbol{r=1} = 2\cos\theta$, so that $\cos\theta = 1/2$ and $\boldsymbol{\theta = \pm\pi/3}$. The area between the two circles is the sum of that area in the first circle between $\theta = -\pi/3$ and $\theta = \pi/3$ and the two areas in the second circle between $\theta = -\pi/2$ and $\theta = -\pi/3$ and between $\theta = \pi/3$ and $\theta = \pi/2$. Appealing to the symmetry of the figure, this area is given by

$$\begin{aligned}2\int_0^{\pi/3} \tfrac{1}{2}.1.\mathrm{d}\theta + 2\int_{\pi/3}^{\pi/2} \tfrac{1}{2}4\cos^2\theta\,\mathrm{d}\theta &= \pi/3 + 2\int_{\pi/3}^{\pi/2} (1+\cos 2\theta)\,\mathrm{d}\theta \\ &= \pi/3 + 2\pi/6 + \Big[\sin 2\theta\Big]_{\pi/3}^{\pi/2} \\ &= \boldsymbol{2\pi/3 - \sqrt{3}/2}.\end{aligned}$$

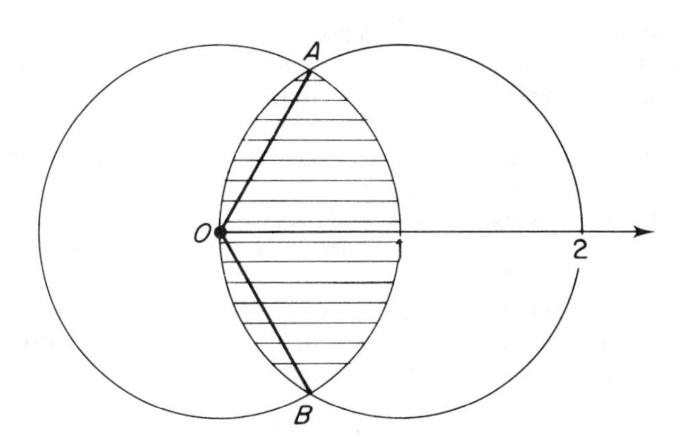

Fig. 7.15

EXAMPLE 2 *Sketch the curve with polar equation $r = 3 + 2\sin\theta$ and find the area of the region enclosed by this curve.*

On this kidney-shaped curve, r has a minimum value 1 when $\theta = 3\pi/2$ and a maximum value 5 when $\theta = \pi/2$. When θ is zero or π, $r = 3$ and, from this

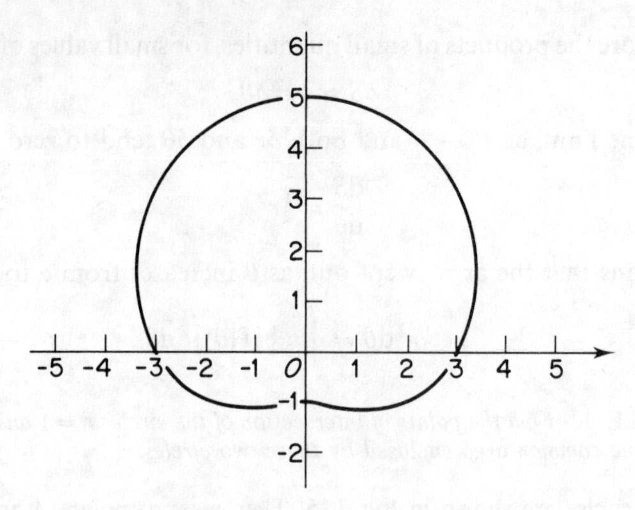

Fig. 7.16 The curve $r = 3 + 2\sin\theta$

information, we sketch the curve in Fig. 7.16. The area S, enclosed by the curve, is calculated as an integral.

$$S = \int_0^{2\pi} \tfrac{1}{2}r^2\,d\theta = \int_0^{2\pi} \tfrac{1}{2}(9 + 12\sin\theta + 4\sin^2\theta)\,d\theta$$

$$= \int_0^{2\pi} (9/2 + 6\sin\theta + 1 - \cos 2\theta)\,d\theta = \left[11\,\theta/2 - 6\cos\theta - \tfrac{1}{2}\sin 2\theta\right]_0^{2\pi}$$

$$= 11\pi.$$

EXERCISE 7.3

1 Sketch the curve $r = a\theta$, where a is a positive constant and $0 \leqslant \theta \leqslant \pi$, and find the area of the region enclosed by this curve and the half line $\theta = \pi$. *(L)*

2 Sketch, on the same diagram, the circles given by the polar equations $r = 2\cos\theta$, where $-\pi/2 \leqslant \theta \leqslant \pi/2$, and $r = 2\sin\theta$, where $0 \leqslant \theta \leqslant \pi$. Show that the circles intersect at the origin, and find the polar coordinates of the other point, P, at which the circles intersect. Find the polar equations of the tangents to the two circles at P and the polar equation of the line joining P to the origin. Calculate the area of the total region enclosed by the circles. *(L)*

3 Sketch on the same diagram the curves whose equations in polar coordinates are $r = 2a(1 + \cos\theta)$ and $r = a(3 + 2\cos\theta)$, where a is a positive constant. Find the area of the region within which

$$2a(1 + \cos\theta) < r < a(3 + 2\cos\theta). \qquad (L)$$

4 Sketch the curve whose polar equation is $r = 3\cos 2\theta$. Calculate the area of one of its loops. *(L)*

5 Sketch the curve whose polar equation is $r^2 = a^2\cos 2\theta$, where $a > 0$, $r > 0$, indicating the coordinates of any points where the curve cuts the initial line. At the point P of the curve, $\theta = \pi/6$. Prove that the tangent at P is parallel to the initial line. Find the area of the finite region enclosed by the initial line, the curve and the straight line joining the origin to P. *(L)*

MISCELLANEOUS EXERCISE 7

1 Sketch, for $0 \leqslant \theta \leqslant 4\pi$, the curve C_1 whose equation in polar coordinates is $r = 3\theta$, marking on your sketch the polar coordinates of the points of intersection with the initial line and the half lines $\theta = \pi/2$, $\theta = \pi$, $\theta = 3\pi/2$. Calculate the area of the region enclosed by the portion of the curve C_1 for which $0 \leqslant \theta \leqslant \pi/2$, and the half line $\theta = \pi/2$. On the same diagram sketch the curve C_2 whose equation is $r = (3\pi\sin\theta)/2$, $0 \leqslant \theta \leqslant \pi$, and calculate the area of the region enclosed between the curves C_1 and C_2 for the range $0 \leqslant \theta \leqslant \pi/2$. {If required, you may assume that $\sin\theta > 2\theta/\pi$ for $0 < \theta < \pi/2$. *(L)*

2 Sketch the curve $r = 2a(1+\cos\theta)$. Find the polar coordinates of the points in which the curve meets the line $2r\cos\theta + a = 0$.

3 Sketch the curve $r = 5+2\cos\theta$ and find the polar coordinates of the two points A and B in which this curve meets the straight line $r\cos\theta = 3$. Prove that the area of the triangle OAB, where O is the pole, is $9\sqrt{3}$. Find also the area of the finite region which is enclosed between the curve and the line $r\cos\theta = 3$ and which does not contain the pole. *(L)*

4 Using the same initial line and scale, sketch the curve whose polar equation is $r^2\cos 2\theta = a^2$ and the line whose polar equation is $r\cos\theta = a\sqrt{(1{\cdot}5)}$. (i) Find the general solution of the equation $3\cos 2\theta = 2\cos^2\theta$. (ii) Prove that the curve and the line meet at the points where $r = a\sqrt{2}$. (iii) Find the area of the finite region contained between the curve and the line. *(AEB)*

5 Shade the region C for which the polar coordinates r, θ satisfy $r < 4\cos 2\theta$ for $-\pi/4 < \theta < \pi/4$. Find the area of C. State the equations of the tangents to the curve $r = 4\cos 2\theta$ at the pole. *(L)*

6 Find a polar equation of the curve $x^2+y^2 = 4x$ and calculate the polar coordinates of the two points P and Q where the curve intersects the line $r = 2\sqrt{2}\sec(\pi/4-\theta)$. Find the polar equations of the two half lines from the origin which are tangents to the circle which has PQ as diameter. *(L)*

7 Calculate the values of $6/(2+\cos 2\theta)$ when $\theta = 45°$ and $\theta = 60°$. Write down the greatest and least values of this expression as θ varies. For what values of θ (between $0°$ and $360°$ inclusive) does the expression take these values? Without further calculation, sketch the complete curve whose polar equation is $r = 6/(2+\cos 2\theta)$. *(SMP)*

8 Sketch the curves with polar equations $r = a(1+\cos\theta)$ and $r = 3a/2$, where a is a positive constant and $0 \leqslant \theta \leqslant 2\pi$. These curves intersect at P and Q. Find the polar coordinates of P and Q. Obtain the polar equations of the half lines OP, OQ and the line PQ, where O is the pole. *(L)*

9 Sketch, on the same diagram, the curves $r = 3\cos\theta$ and $r = 1+\cos\theta$ and find the area of the region lying inside both curves. *(L)*

10 Find polar equations for the tangents T_1 and T_2 to the circle $r = a$ which are perpendicular to the initial line. Find also polar equations for the tangents T_3 and T_4 to the circle $r = a$ which are parallel to the half line $\theta = \alpha$. Calculate the area of the rhombus enclosed by T_1, T_2, T_3 and T_4. *(L)*

11 Sketch the curve with equation $r = a\cos 2\theta$, where $a > 0$, for $-\pi/4 \leqslant \theta \leqslant \pi/4$, and find the area of the region for which $0 \leqslant r \leqslant a\cos 2\theta$, $-\pi/4 \leqslant \theta \leqslant \pi/4$. By considering the stationary values of $\sin\theta\cos 2\theta$, or otherwise, find the polar coordinates of the points on the curve where the tangent is parallel to the initial line. *(L)*

8 Polynomials and Rational Functions

8.1 Division of a polynomial by a linear polynomial (Revision *Pure Mathematics* §7.4)

When a polynomial $p(x)$, of degree n, is divided by a linear polynomial $(ax+b)$, $a \neq 0$, there is a quotient polynomial $q(x)$, of degree $(n-1)$, and a constant remainder r. The values of the coefficients of $q(x)$ and of r may be determined by means of long division or by the loop method, *Pure Mathematics* §7.4. The required identity will be

$$p(x) \equiv (ax+b)q(x)+r. \qquad 8.1$$

EXAMPLE 1 *Find the quotient* $q(x)$ *and the remainder* r *on dividing* $2x^4+3x^3-4x+1$ *by* $2x-1$.

$$2x^4+3x^3-4x+1 = (2x-1)(x^3+ \ldots \quad) + \ldots$$

$$2x^4+3x^3-4x+1 = (2x-1)(x^3+2x^2+ \ldots \quad) + \ldots$$

$$2x^4+3x^3-4x+1 = (2x-1)(x^3+2x^2+x+ \ldots \quad) + \ldots$$

$$2x^4+3x^3-4x+1 = (2x-1)(x^3+2x^2+x-3/2) \quad + \ldots$$

$$2x^4+3x^3-4x+1 = (2x-1)(x^3+2x^2+x-3/2) \quad -1/2.$$

Therefore $\quad \mathbf{q(x)=x^3+2x^2+x-3/2} \quad$ and $\quad \mathbf{r=-1/2}.$

The substitution $x=-b/a$ into the identity 8.1, leads to the Remainder Theorem, which in turn gives rise to the Factor Theorem (*Pure Mathematics* §7.5).

Remainder Theorem
The remainder on dividing the polynomial $p(x)$ by $(ax+b)$ is $p(-b/a)$.

Factor Theorem
$(ax+b)$ is a factor of the polynomial $p(x) \Leftrightarrow p(-b/a) = 0$.

EXAMPLE 2 *Find the remainder on dividing x^3+1 by $x+2$.*

Using the remainder theorem, the remainder is $(-2)^3 + 1 = \mathbf{-7}$. This agrees with the division
$$x^3+1 = (x+2)(x^2-2x+4)-7.$$

EXAMPLE 3 *Solve the equation* $x^3+3x^2-4x-12=0$.

Since the coefficient of x^3 is unity, we look for a factor $(x-a)$ of $f(x)$, where $f(x) = x^3+3x^2-4x-12$.
We try the factors of 12 as possible values of a:
try $a = 1$, $f(1) = 1+3-4-12 = -12$, so $(x-1)$ is not a factor of $f(x)$,
try $a = -1$, $f(-1) = -1+3+4-12 = -6$, so $(x+1)$ is not a factor of $f(x)$,
try $a = 2$, $f(2) = 8+12-8-12 = 0$, so $(x-2)$ is a factor of $f(x)$.
If x^2+ax+b is the quotient, on dividing $f(x)$ by $(x-2)$, then
$$x^3+3x^2-4x-12 = (x-2)(x^2+ax+b).$$
Comparing coefficients of x^2 and of x, $3 = a-2$ and $-4 = b-2a$, so $a = 5$, and $b = 6$. Check: the constant term agrees, $-12 = -2b$. Thus
$$f(x) = (x-2)(x^2+5x+6) = (x-2)(x+2)(x+3),$$
on factorising the quadratic factor by inspection. Finally, by the factor theorem, the solution set of the equation $f(x) = 0$ is $\{\mathbf{-3, -2, 2}\}$.

Division of the identity 8.1 by $(ax+b)$ makes it possible to express a rational function with a linear denominator as a polynomial plus a rational function of the type $\dfrac{r}{ax+b}$, since $\dfrac{p(x)}{ax+b} = q(x) + \dfrac{r}{ax+b}$. This process can be useful when differentiating or integrating a rational function with linear denominator.

EXAMPLE 4 *Find (i) the derivative of* $\dfrac{x^3+1}{x+2}$ *(ii)* $\displaystyle\int_0^1 \frac{x^3+1}{x+2}\,dx$.

We use the result of Example 2 in each part.

(i) $$\frac{d}{dx}\frac{x^3+1}{x+2} = \frac{d}{dx}(x^2-2x+4) - \frac{d}{dx}\frac{7}{x+2} = \mathbf{2x-2+\frac{7}{(x+2)^2}}.$$

(ii) $$\int_0^1 \frac{x^3+1}{x+2}\,dx = \int_0^1\left(x^2-2x+4-\frac{7}{x+2}\right)dx$$
$$= \left[\tfrac{1}{3}x^3 - x^2 + 4x - 7\ln(x+2)\right]_0^1$$
$$= \tfrac{1}{3} - 1 + 4 - 7\ln 3 + 7\ln 2 = \mathbf{10/3 - 7\ln(3/2)}.$$

EXERCISE 8.1 (Revision)

1 Express as a single rational function:
(i) $\dfrac{x-1}{x+2}-\dfrac{x-2}{x+1}$, (ii) $\dfrac{x}{x^2-9}+\dfrac{2x+1}{x+3}$, (iii) $\dfrac{1}{x+2}-\dfrac{1}{x-2}$,
(iv) $3x+2-\dfrac{1}{x+1}$.

2 Find the quotient and the remainder on dividing:
(i) x^2-3x+2 by x, (ii) x^3+x^2 by $x+1$,
(iii) x^3+2x^2+3x+4 by $x+2$, (iv) x^5-3x^3+2 by $2x-1$.

3 Use the remainder theorem to find the remainder on dividing:
(i) x^3+2x^2-3x+4 by $x-1$, (ii) x^4-2 by $x+2$,
(iii) $4x^3+2x^2+x$ by $2x+3$, (iv) x^6 by $x+1$,
(v) $2x^5-3x^3+8x-12$ by $x-2$, (vi) x^2+3x+2 by $2x-1$.

4 Factorise completely into linear factors (some of which may have complex coefficients):
(i) x^4-1, (ii) $x^3+4x^2-17x-60$, (iii) $x^3-6x^2+14x-12$,
(iv) x^3-1, (v) $4t^3-4t^2-11t+6$, (vi) $x^4+y^4+2x^2y^2$.

5 Express the rational function as a polynomial plus a proper rational function and, hence, differentiate:

(i) $\dfrac{x^2-2x-1}{x+1}$, (ii) $\dfrac{2x^2-11}{x-2}$, (iii) $\dfrac{x^4-x^2}{2x+3}$.

Check your results by differentiating the original rational function, using the rule for the derivative of a quotient.

6 Evaluate:

(i) $\displaystyle\int_0^1 \frac{x^2}{x+1}\,dx$, (ii) $\displaystyle\int_3^4 \frac{x^3-3}{x-2}\,dx$, (iii) $\displaystyle\int_0^1 \frac{x^4+3x}{2x+5}\,dx$.

8.2 Division by a quadratic polynomial

When a polynomial $p(x)$, of degree n, $(n>1)$, is divided by a quadratic polynomial $d(x)$, there is a quotient polynomial $q(x)$, of degree $(n-2)$, and a remainder $r(x)$, of degree 1 or 0. Thus $r(x)=cx+d$ and, as in §8.1, the coefficients of $q(x)$ and of $r(x)$ may be found by long division or by equating the coefficients of x in the identity

$$p(x)=d(x)q(x)+r(x) \qquad 8.2$$

using the loop method.

EXAMPLE 1 *Find the quotient and the remainder on dividing*

$$2x^4-3x^3-4x^2+7x-6 \textit{ by } x^2+x-2.$$

The dividend $p(x)$ is of degree 4, the divisor $d(x)$ is of degree 2, so the quotient $q(x)$ is of degree 2 and the remainder $r(x)$ is of degree 1. We lay out the working of the loops method in a series of rows to aid comprehension, even though in practice only one row is needed.

$$2x^4 - 3x^3 - 4x^2 + 7x - 6 = (x^2 + x - 2)(2x^2 + \,.\, + \,.\,) + \,.\, + \,.\,,$$

$$2x^4 - 3x^3 - 4x^2 + 7x - 6 = (x^2 + x - 2)(2x^2 - 5x + \,.\,) + \,.\, + \,.\,,$$

$$2x^4 - 3x^3 - 4x^2 + 7x - 6 = (x^2 + x - 2)(2x^2 - 5x + 5) \quad + \,.\, + \,.\,,$$

$$2x^4 - 3x^3 - 4x^2 + 7x - 6 = (x^2 + x - 2)(2x^2 - 5x + 5) \quad - 8x + \,.\,,$$

$$2x^4 - 3x^3 - 4x^2 + 7x - 6 = (x^2 + x - 2)(2x^2 - 5x + 5) \quad - 8x + 4.$$

Thus the quotient $q(x) = \mathbf{2x^2 - 5x + 5}$ and the remainder $r(x) = \mathbf{-8x + 4}$.

Both long division and the loop method used in Example 1 can be long and tedious when $p(x)$ is of high degree. If we require only the remainder $r(x)$, then we can use shorter methods, given by the next theorem. We use ideas which are similar to those used in the proof of the Remainder Theorem.

Theorem 8.1 Let the remainder $r(x)$, after dividing the polynomial $p(x)$ by a quadratic polynomial $d(x)$, be given by $r(x) = cx + d$. Then
I If $d(x) = 0$ has distinct roots α and β, (real or complex), then

$$c = \frac{p(\alpha) - p(\beta)}{\alpha - \beta} \quad \text{and} \quad d = \frac{\alpha p(\beta) - \beta p(\alpha)}{\alpha - \beta}.$$

II If $d(x) = 0$ has a repeated root α, (real or complex), then

$$c = p(\alpha) \quad \text{and} \quad d = p(\alpha) - \alpha p(\alpha).$$

Proof I Let $d(x) = k(x - \alpha)(x - \beta)$ then, by the identity 8.2,

$$p(x) = k(x - \alpha)(x - \beta)q(x) + cx + d.$$

Hence $p(\alpha) = c\alpha + d$ and $p(\beta) = c\beta + d$.
Eliminating d by subtraction,

$$p(\alpha) - p(\beta) = (\alpha - \beta)c \text{ and, since } \alpha \neq \beta,\ c = \frac{p(\alpha) - p(\beta)}{\alpha - \beta}.$$

Finally $d = p(\alpha) - c\alpha = \dfrac{\alpha p(\beta) - \beta p(\alpha)}{\alpha - \beta}$.

II Let $d(x) = k(x - \alpha)^2$ then, by the identity 8.2,

$$p(x) = k(x - \alpha)^2 q(x) + cx + d.$$

On differentiating this equation

$$p'(x) = 2k(x-\alpha)q(x) + k(x-\alpha)^2 q'(x) + c.$$

Hence $p(\alpha) = c\alpha + d$, $p'(\alpha) = c$, and so $d = p(\alpha) - \alpha p'(\alpha)$.

When finding the remainder in a particular case, it is better to use the method of Theorem 8.1, rather than just quote the result.

EXAMPLE 2 *Use the method of Theorem* 8.1 *to find the remainder on dividing: (i)* $2x^4 - 3x^3 - 4x^2 + 7x - 6$ *by* $x^2 + x - 2$, *(ii)* $2x^3 - 3x^2 - 4x - 1$ *by* $x^2 + 2x + 1$.

(i) Write $2x^4 - 3x^3 - 4x^2 + 7x - 6 = (x+2)(x-1)q(x) + cx + d$.

Then, on putting $x = 1$, $\quad c + d = 2 - 3 - 4 + 7 - 6 = -4$

and, on putting $x = -2$, $\quad -2c + d = 32 + 24 - 16 - 14 - 6 = 20$.

On subtracting these equations, $3c = -24$, and so $c = -8$. Then $d = -4 - (-8) = 4$ and the remainder is $\mathbf{-8x+4}$, as we found in Example 1.

(ii) The divisor $x^2 + 2x + 1$ is a square, namely $(x+1)^2$, so we write

$$2x^3 - 3x^2 - 4x - 1 = (x+1)^2 q(x) + cx + d.$$

On differentiating, $6x^2 - 6x - 4 = 2(x+1)q(x) + (x+1)^2 q(x) + c$.

Then, on putting $x = -1$ into the two equations, $-c + d = -2 - 3 + 4 - 1 = -2$, $c = 6 + 6 - 4 = 8$, and $d = -2 + c = 6$. The remainder is therefore $\mathbf{8x+6}$. Check, by long division, $(2x^3 - 3x^2 - 4x - 1) = (x^2 + 2x + 1)(2x - 7) + 8x + 6$.

Corresponding to the Factor Theorem of §8.1, Theorem 8.2 may be used to provide a Factor Theorem for repeated factors.

Theorem 8.2 If $p(x)$ is a polynomial function,

$$(x-\alpha)^2 \text{ is a factor of } p(x) \Leftrightarrow p(\alpha) = 0 \text{ and } p'(\alpha) = 0.$$

Proof Let $(cx + d)$ be the remainder on dividing $p(x)$ by $(x-\alpha)^2$, so that $p(x) = (x-\alpha)^2 q(x) + cx + d$ and $p'(x) = 2(x-\alpha)q(x) + (x-\alpha)^2 q(x) + c$. Then $p'(\alpha) = c$ and $p(\alpha) = c\alpha + d$, so that $c = p'(\alpha)$ and $d = p(\alpha) - \alpha p'(\alpha)$. Hence $(x-\alpha)^2$ is a factor of $p(x) \Leftrightarrow cx + d = 0$

$$\Leftrightarrow c = 0 \text{ and } d = 0$$

$$\Leftrightarrow p'(\alpha) = 0 \text{ and } p(\alpha) = 0.$$

EXAMPLE 3 *Given that* $p(x) = x^4 + 2x^3 + 5x^2 + 4x + 4$ *and that* $p(x)$ *has a repeated linear factor, solve the equation* $p(x) = 0$.

Let $(x-\alpha)$ be the repeated factor of $p(x)$, then $p(\alpha) = 0$ and $p'(\alpha) = 0$.

That is, $\quad \alpha^4 + 2\alpha^3 + 5\alpha^2 + 4\alpha + 4 = 0 \quad$ 1.

and $\quad 4\alpha^3 + 6\alpha^2 + 10\alpha + 4 = 0$,

or, on dividing by 2, $\quad 2\alpha^3 + 3\alpha^2 + 5\alpha + 2 = 0$, $\quad$ 2.

We eliminate the higher powers of α in turn by subtracting appropriate multiples of the equations from each other. Subtract α times equation 2 from 2 times equation 1

$$2\alpha^4 + 4\alpha^3 + 10\alpha^2 + 8\alpha + 8 - (2\alpha^4 + 3\alpha^3 + 5\alpha^2 + 2\alpha) = 0$$

so that
$$\alpha^3 + 5\alpha^2 + 6\alpha + 8 = 0. \qquad 3.$$

Subtract equation 2 from 2 times equation 3,

$$2\alpha^3 + 10\alpha^2 + 12\alpha + 16 - (2\alpha^3 + 3\alpha^2 + 5\alpha + 2) = 0$$

so that
$$7\alpha^2 + 7\alpha + 14 = 0,$$

or, on dividing by 7,
$$\alpha^2 + \alpha + 2 = 0.$$

Solving this quadratic equation, either $\alpha = -(1 + i\sqrt{7})/2$ or $\alpha = -(1 - i\sqrt{7})/2$. In each case, α is complex and, since the coefficients of $p(x)$ are all real, the conjugate α^* of α is also a root of $p(x) = 0$. This means that both α and α^* are repeated roots of $p(x) = 0$ and so the solution set of this equation is $\{-(1-i\sqrt{7})/2, -(1+i\sqrt{7})/2\}$.

EXERCISE 8.2

1 Use the loop method to find the quotient and the remainder on dividing:
(i) $x^3 + 3x$ by $x^2 + 2$. (ii) $2x^3 + 4x^2$ by $x^2 - 4$,
(iii) $x^3 + 2x^2 - 5x + 1$ by $x^2 + x + 1$, (iv) $x^4 + x^2 + 2$ by $x^2 - 3x$,
(v) $3x^4 + 4x^3 - 5$ by $x^2 - 2x + 3$, (vi) x^4 by $2x^2 - 3x + 1$.

2 Find the quotient and the remainder on dividing:
(i) $x^3 + x^2 - x - 2$ by $x^2 + x - 2$, (ii) $2x^3 - 3x^2 + x + 4$ by $x^2 - 2x + 3$,
(iii) $x^4 + x^3 - x^2 + 2x - 5$ by $x^2 - 2x + 1$, (iv) $x^4 + 1$ by $x^2 + 1$,
(v) $x^4 - 3x + 2$ by $x^2 + 2x$, (vi) $x^4 + 2x^3 - 3x$ by $3x^2 + 7$.

3 Use the method of Example 2 to find the remainder on dividing:
(i) $x^3 - 2x^2$ by $(x-1)(x-2)$, (ii) $3x - x^3$ by $(x+2)(x-3)$,
(iii) $3x^3 - 2x - 2$ by $(2x-1)(x-1)$, (iv) $x^4 - x^3$ by $(2x+1)(x-2)$,
(v) $x^4 + 2x^3 - 3x + 1$ by $(2x+3)(3x+2)$, (vi) $2x^4 + 4$ by $x(x-3)$.

4 Use the method of Example 2 to find the remainder on dividing:
(i) $x^3 + 2$ by $(x-1)^2$, (ii) $(x+1)^3$ by $(x+3)^2$,
(iii) $2x^3 + 4x$ by $(2x-1)^2$, (iv) $x^3 - 3x^2 + 2x + 1$ by $x^2 - 6x + 9$,
(v) $x^4 + 2x^2 + 3x$ by $4x^2 + 4x + 1$, (vi) $x^4 - x^3 + x^2 - 1$ by $4x^2 + 12x + 9$.

5 Find the remainder on dividing:
(i) $2x^3 + x^2 - x + 2$ by $(x-3)(x+2)$, (ii) $x^4 + 5x - 2$ by $x^2 - 1$,
(iii) $x^4 + 5x - 2$ by $(x-1)^2$, (iv) $2x^5 + 4x^3$ by $x^2 + 6x + 9$,
(v) x^9 by $(x+2)(x-1)$, (vi) $x^6 - 1$ by $x^2 + 3x + 2$.

6 Given that $f(x)$ has a repeated linear factor, factorise $f(x)$ completely, where:
(i) $f(x) = x^3 - 3x - 2$, (ii) $f(x) = 4x^3 - 27x + 27$,
(iii) $f(x) = 4x^4 - 27x^2 - 25x - 6$, (iv) $f(x) = x^4 + 2x^3 + 3x^2 + 2x + 1$,
(v) $f(x) = 4x^4 - 20x^3 + 33x^2 - 20x + 4$, (vi) $f(x) = x^4 - 4x^3 + 8x^2 - 8x + 4$.

7 When the expression $x^3 + ax^2 + bx + c$ is divided by $x^2 - 4$ the remainder is $18 - x$ and when it is divided by $x + 3$ the remainder is 21. Find the remainder when the expression is divided by $x + 1$. *(AEB)*

8 The polynomials $P(x)$ and $Q(x)$ are of degree n and $n - 2$ respectively, and $P(x) = (x-2)^2 Q(x) + rx + s$. Prove that $r = P'(2)$, $s = P(2) - 2P'(2)$. Given that $P(x)$ has remainders 4 and 1 when divided by $x - 2$ and $x - 1$ respectively, and that $P'(x)$ has remainder 3 when divided by $x - 2$, find the remainder when $Q(x)$ is divided by $x - 1$. *(JMB)*

9 When the polynomial $f(x)$ is divided by $(x-a)^2$ the remainder may be written in the form $A(x-a)+B$, where A and B are constants. Show that $A = f'(a)$ and $B = f(a)$. Deduce that $(x-a)$ is a repeated factor of $f(x)$ if $f'(a) = f(a) = 0$. Solve the equation $16x^3 - 20x^2 - 32x - 9 = 0$ given that it has a repeated root. (*JMB*)

8.3 Roots and coefficients of a polynomial equation

Symmetric properties of the roots of a quadratic equation (Revision)

The connection between the roots and the coefficients of a quadratic equation was considered in *Pure Mathematics* §2.6. Let $f(x) = x^2 + ax + b$ and let α and β be the zeros of f. Then, by the Factor Theorem, $(x-\alpha)$ and $(x-\beta)$ are factors of $f(x)$, so that

$$f(x) = x^2 + ax + b = (x-\alpha)(x-\beta) = x^2 - (\alpha+\beta)x + \alpha\beta.$$

Comparing coefficients in this equation, we find that

$$a = -\alpha - \beta, \; b = \alpha\beta.$$

Symmetric properties of the roots of a cubic equation

Let $f(x) = x^3 + ax^2 + bx + c$ and suppose that $f(x) = 0$ has roots α, β and γ. Then, on using the same method as was used above for a quadratic equation,

$$\begin{aligned} f(x) &= x^3 + ax^2 + bx + c = (x-\alpha)(x-\beta)(x-\gamma) \\ &= x^3 - (\alpha+\beta+\gamma)x^2 + (\alpha\beta+\beta\gamma+\gamma\alpha)x - \alpha\beta\gamma, \end{aligned}$$

so that $\quad a = -\alpha - \beta - \gamma, \; b = \alpha\beta + \beta\gamma + \gamma\alpha, \; c = -\alpha\beta\gamma.$

EXAMPLE 1 *Given that the roots of the equation $x^3 + ax^2 + bx + c = 0$ are in arithmetical progression, prove that $27c = 9ab - 2a^3$.*

Let the roots of the equation be α, $\alpha - d$ and $\alpha + d$. Then

$$\begin{aligned} \alpha + (\alpha-d) + (\alpha+d) &= -a \quad \text{so that} \quad & 3\alpha &= -a, \\ \alpha(\alpha-d) + \alpha(\alpha+d) + (\alpha-d)(\alpha+d) &= b \quad \text{so that} \quad & 3\alpha^2 - d^2 &= b, \\ \alpha(\alpha-d)(\alpha+d) &= -c \quad \text{so that} \quad & \alpha(\alpha^2-d^2) &= -c. \end{aligned}$$

Then $\quad \alpha^2 - d^2 = 3\alpha^2 - d^2 - 2\alpha^2 = b - 2a^2/9,$

so that $\quad -c = \alpha(\alpha^2 - d^2) = -a(9b - 2a^2)/27$

and hence $\quad 27c = 9ab - 2a^3.$

EXAMPLE 2 *Two of the roots of the equation $x^3+ax^2+bx+c=0$ are the reciprocals of one another. Prove that $b=1+c(a-c)$.*

Let the roots of the equation be α, $1/\alpha$ and β. Then

$$-a=\alpha+1/\alpha+\beta,$$
$$b=\alpha/\alpha+\alpha\beta+\beta/\alpha=1+\beta(\alpha+1/\alpha),$$
$$-c=\alpha\beta/\alpha=\beta.$$

Thus $\qquad \alpha+1/\alpha=-a-\beta=-a+c$

and $\qquad \beta(\alpha+1/\alpha)=b-1$

and so $\qquad b-1=-c(\alpha+1/\alpha)=ac-c^2,$

and $\qquad b=1+ac-c^2=1+c(a-c).$

EXAMPLE 3 *Solve the equation $2x^3-13x^2-26x+16=0$, given that its roots are in geometric progression.*

Divide the equation by 2 so that the coefficient of x^3 is unity, giving the equation $x^3-\dfrac{13}{2}x^2-13x+8=0$.

Let the roots of the equation be α, $r\alpha$ and $r^2\alpha$, so that they are in geometric progression, with common ratio r. Then

$$\alpha+r\alpha+r^2\alpha=13/2,$$
$$\alpha(r\alpha)+(r\alpha)(r^2\alpha)+\alpha(r^2\alpha)=-13,$$
$$\alpha(r\alpha)(r^2\alpha)=-8.$$

Thus $\alpha(1+r+r^2)=13/2$, $\alpha^2(r+r^2+r^3)=-13$ and $\alpha^3r^3=-8$, so $\alpha r=-2$. Eliminating α, $\alpha=-2/r$, $-2(1+r+r^2)/r=13/2$, so $-4-4r-4r^2=13r$. Then $4r^2+17r+4=(4r+1)(r+4)=0$, so that $r=-4$ or $r=-1/4$ and $\alpha=1/2$ or $\alpha=8$. $r\alpha=-2$ in each case and $r^2\alpha=8$ or $r^2\alpha=1/2$. Therefore the solution set for the equation is $\{\mathbf{1/2, -2, 8}\}$.

EXAMPLE 4 *Given that α, β and γ are the roots of the equation $x^3+2x^2+3x+5=0$, find an equation whose roots are $(\alpha+\beta)$, $(\beta+\gamma)$ and $(\gamma+\alpha)$.*

Since $\alpha+\beta+\gamma=-2$, $\alpha\beta+\beta\gamma+\gamma\alpha=3$ and $\alpha\beta\gamma=-5$, $(\alpha+\beta+\gamma)^2=\alpha^2+\beta^2+\gamma^2+2(\alpha\beta+\beta\gamma+\gamma\alpha)=4$ and so $\alpha^2+\beta^2+\gamma^2=4-2\times 3=-2$. We now calculate the sum of the roots of the new equation, the sum of the products of pairs of these roots and the product of these roots.

$$(\alpha+\beta)+(\beta+\gamma)+(\gamma+\alpha)=2(\alpha+\beta+\gamma)=-4,$$
$$(\alpha+\beta)(\beta+\gamma)+(\gamma+\alpha)(\alpha+\beta)+(\beta+\gamma)(\gamma+\alpha)=\alpha^2+\beta^2+\gamma^2+3\alpha\beta+3\beta\gamma+3\gamma\alpha$$
$$=-2+3\times 3=7,$$
$$(\alpha+\beta)(\beta+\gamma)(\gamma+\alpha)=2\alpha\beta\gamma+\alpha^2\beta+\alpha^2\gamma+\beta^2\gamma+\beta^2\alpha+\gamma^2\alpha+\gamma^2\beta.$$

Now $\qquad (\alpha+\beta+\gamma)(\alpha\beta+\beta\gamma+\gamma\alpha)=3\alpha\beta\gamma+\alpha^2\beta+\alpha^2\gamma+\beta^2\gamma+\beta^2\alpha+\gamma^2\alpha+\gamma^2\beta$

so that $\alpha^2\beta+\alpha^2\gamma+\beta^2\gamma+\beta^2\alpha+\gamma^2\alpha+\gamma^2\beta = (\alpha+\beta+\gamma)(\alpha\beta+\beta\gamma+\gamma\alpha) - 3\alpha\beta\gamma$

$$= (-2)(3) - 3(-5) = 9$$

and so $(\alpha+\beta)(\beta+\gamma)(\gamma+\alpha) = 2(-5)+9 = -1.$

The required equation is therefore $\mathbf{x^3+4x^2+7x+1=0}$.

Symmetric properties of the roots of a polynomial equation

The above results for a quadratic and a cubic equation generalise to any polynomial. We assume that the polynomial is of degree n and that the coefficient of x^n is unity.

Let $f(x) = x^n + a_{n-1}x^{n-1} + a_{n-2}x^{n-2} + \ldots + a_1x + a_0$

Then, if the roots of the equation $f(x) = 0$ are $\alpha_1, \alpha_2, \ldots, \alpha_n$,

$$f(x) = (x-\alpha_1)(x-\alpha_2)\ldots(x-\alpha_n).$$

Comparing the coefficients of the powers of x, we find that

$-a_{n-1}$ is the sum of the roots,
a_{n-2} is the sum of products of the roots taken 2 at a time,
$-a_{n-3}$ is the sum of products of the roots taken 3 at a time,
$(-1)^r a_{n-r}$ is the sum of products of the roots taken r at a time,
$(-1)^n a_0$ is the product of the roots.

EXERCISE 8.3

1 Find the zeros of the following polynomials, given that the zeros are in arithmetic progression: (i) x^3-3x^2-6x+8, (ii) $9x^3+9x^2-13x-5$.

2 Find the solutions of the following polynomial equations, given that the solutions are in geometric progression: (i) $2x^3-3x^2-3x+2=0$, (ii) $16x^3-26x^2-39x+54=0$.

3 The roots of the equation $x^3+px^2+qx+30=0$ are in the ratio 2:3:5. Find the values of p and q. *(L)*

4 The product of two roots of the equation $x^3+qx-r=0$ is equal to r. Prove that $r=q+1$. *(L)*

5 The roots of the equation $x^3-26x^2+156x+p=0$ are in geometric progression. Find p. *(L)*

6 Prove that (i) $\alpha^2+\beta^2+\gamma^2 = (\alpha+\beta+\gamma)^2 - 2(\alpha\beta+\beta\gamma+\gamma\alpha)$, (ii) $\alpha^2\beta^2+\beta^2\gamma^2+\gamma^2\alpha^2 = (\alpha\beta+\beta\gamma+\gamma\alpha)^2 - 2\alpha\beta\gamma(\alpha+\beta+\gamma)$. Hence prove that the equation $x^3+(2q-p^2)x^2+(q^2-2pr)x-r^2=0$ has roots equal to the squares of the roots of the equation $x^3+px^2+qx+r=0$.

7 Given that two roots of the equation $x^4+bx^3+cx^2+dx+e=0$ are such that their sum is zero and also that b, c, d and e are all non-zero, prove that the product of these two roots is d/b and that the product of the other two roots is be/d. Hence, or otherwise, prove that $b^2e+d^2=bcd$.

Solve the equations (i) $x^4+x^3-2x^2-3x-3=0$, (ii) $x^4-x^3+2x^2-3x-3=0$, assuming that in each case two of the roots, not necessarily real, are such that their sum is zero. *(JMB)*

8 The equation $2x^4-3x^3-x^2+2x-3=0$ has roots $\alpha, \beta, \gamma, \delta$. Prove that $\frac{1}{\alpha}+\frac{1}{\beta}+\frac{1}{\gamma}+\frac{1}{\delta}=\frac{2}{3}$ and $\frac{1}{\alpha\beta}+\frac{1}{\alpha\gamma}+\frac{1}{\alpha\delta}+\frac{1}{\beta\gamma}+\frac{1}{\beta\delta}+\frac{1}{\gamma\delta}=\frac{1}{3}$. Hence write down an equation whose roots are $\frac{1}{\alpha}, \frac{1}{\beta}, \frac{1}{\gamma}, \frac{1}{\delta}$.

9 (a) Solve the equation $6x^3+25x^2-62x+24=0$, given that the product of two of the roots is -8. (b) The roots of the equation $36x^3+72x^2+23x-6=0$ are α, β, γ. Find a cubic equation, with numerical coefficients, which has roots $6\alpha+1$, $6\beta+1$, $6\gamma+1$. *(C)*

10 The equation $x^2-2x+a+bx^{-1}+4x^{-2}=0$ has roots $\alpha, \beta, \gamma, \delta$. Given that $\alpha^2+\beta^2+\gamma^2+\delta^2=12$ and $\frac{1}{\alpha}+\frac{1}{\beta}+\frac{1}{\gamma}+\frac{1}{\delta}=1$, find the values of a and b. Show that, if $y=x+2x^{-1}$, the given equation can be written as a quadratic equation in y. Hence, or otherwise, solve the given equation. *(C)*

11 The cubic equation $x^3+px+q=0$ has roots α, β and γ. Calculate $\alpha^2+\beta^2+\gamma^2$ and $\beta^2\gamma^2+\gamma^2\alpha^2+\alpha^2\beta^2$ in terms of p and q. Hence, or otherwise, show that the equation whose roots are $\beta^2\gamma^2$, $\gamma^2\alpha^2$ and $\alpha^2\beta^2$ is $\mathrm{f}(x)=0$ where $\mathrm{f}(x)=x^3-p^2x^2-2pq^2x-q^4$. Further, show that the equation $\mathrm{f}(x)=0$ has at least one root in the interval $0<x<1$ if p and q are real and $|p+q^2|<1$. *(L)*

8.4 Location of roots of a polynomial equation

Let $\mathrm{f}(x)$ be a polynomial of degree n, with real coefficients. Then we know from *Pure Mathematics* §15.6 that any complex roots of the equation $\mathrm{f}(x)=0$ that is, the complex zeros of $\mathrm{f}(x)$, must occur in conjugate pairs. Information about the real zeros of f can be obtained by considering the graph of f, its maxima and minima, and the sign of the derivative $\mathrm{f}'(x)$. We illustrate the methods by means of an example.

EXAMPLE 1 *The function* f *is given by* $\mathrm{f}(x)=x^3-5x^2-8x+532$. *Find the stationary points of the graph of* f *and deduce that the equation* $\mathrm{f}(x)=0$ *has one real root, which is negative. Given that this root is an integer, solve the equation* $\mathrm{f}(x)=0$.

The derivative $\mathrm{f}'(x)=3x^2-10x-8=(3x+2)(x-4)$, so that $\mathrm{f}'(x)=0$ for $x=-2/3$ and for $x=4$. Also $\mathrm{f}''(x)=6x-10$ and $\mathrm{f}''(-2/3)=-14$, $\mathrm{f}''(4)=14$.

When $x=-2/3$, $\mathrm{f}(x)=534{\cdot}81$ and **(−2/3, 534·81)** is a maximum.
When $x=4$, $\mathrm{f}(x)=484$ and **(4, 484) is a minimum.**
When $x=0$, $\mathrm{f}(x)=532$, and we use these results to sketch the graph of f in Fig. 8.1.

It is seen from the graph that f has one zero and that this is less than $-2/3$. If this zero is an integer, then it must be a factor of 532, by the Factor theorem. We test -2, -4 and -7 and find that $\mathrm{f}(-7)=0$. Thus $(x+7)$ is a factor of $\mathrm{f}(x)$ and the other factor will be of the form x^2+ax+b. Therefore

$$\mathrm{f}(x)=x^3-5x^2-8x+532=(x+7)(x^2+ax+b).$$

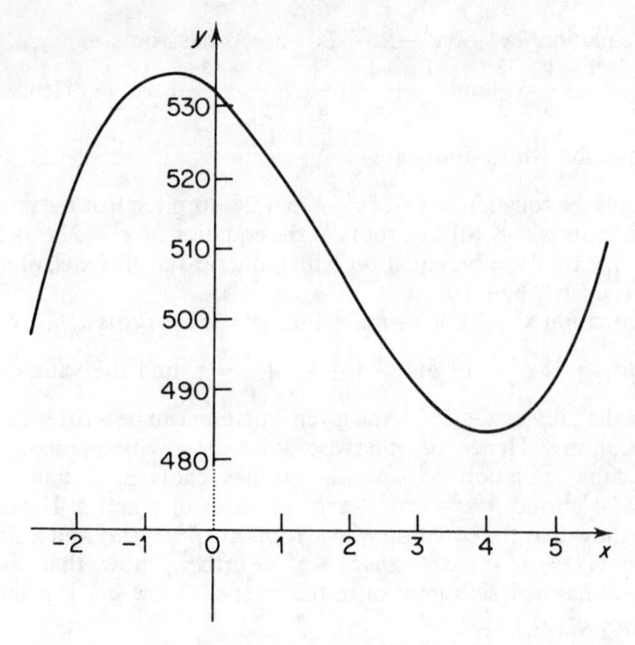

Fig. 8.1 Graph of f, $f(x) = x^3 - 5x^2 - 8x + 532$

On putting $x = 0$, $532 = 7b$, and so $b = 76$. Comparing the coefficients of x, $-8 = 7a + b = 7a + 76$, and so $a = -12$. Therefore

$$f(x) = (x+7)(x^2 - 12x + 76).$$

The quadratic factor, $x^2 - 12x + 76$ can be expressed as $(x-6)^2 + 40$, so it has zeros $6 + 2i\sqrt{10}$ and $6 - 2i\sqrt{10}$. Therefore the solution set of $f(x) = 0$ is $\{-7, 6+2i\sqrt{10}, 6-2i\sqrt{10}\}$.

EXAMPLE 2 *Show that the equation $x^9 + 2 = 0$ has only one real root and that this root is negative. Find an approximate value for this root, correct to two decimal places.*

Let $f(x) = x^9 + 2$. Then $f'(x) = 9x^8$, and this is never negative. Thus the graph of f is monotonic increasing and so it only crosses the x-axis in one point. Since $f(0) = 2$, the graph of f must cross the x-axis at a negative value of x. Therefore the given equation has one real root, which is negative. Suppose that this root is α then, since $f(-1) = 1$ and $f(-2) = -511$, $-2 < \alpha < -1$. We use Newton's method to find an approximate root, starting with the approximation $a_1 = -1$. A better approximation is a_2, where $a_2 = a_1 - f(a_1)/f'(a_1) = -1 - 1/9 = -1{\cdot}11$. The next two approximations are a_3 and a_4, where $a_3 = -1{\cdot}083$ and $a_4 = -1{\cdot}080$. The solution, to two decimal places, is $\mathbf{-1{\cdot}08}$.

EXERCISE 8.4

1 One root of the equation $x^3 - 3x^2 - 3x + a = 0$ is -2. Find the value of a and prove that the equation has no other real root. (*L*)

2 Find the stationary points on the curve given by the equation $y = x^3 - x^2 - 5x + 4$ and deduce that the equation $x^3 - x^2 - 5x + 4 = 0$ has three real roots. Prove that the largest root of this equation lies between 2 and 3 and, by using linear interpolation, find an approximation to the value of this root, correct to one decimal place.

3 Find the stationary points of the graph of $f(x)$ and deduce the number of real roots of the equation $f(x) = 0$. Use Newton's method to find an approximate value, correct to two decimal places, of the largest root of this equation: (i) $f(x) = x^7 + 2$, (ii) $f(x) = x^8 - 3$, (iii) $f(x) = x^3 - 3x^2 + 6x - 1$, (iv) $f(x) = x^4 - 4x^3 + 4x^2 - 1$.

4 The function f is given by $f(x) = 2x^3 - 3ax^2 + b$, where $a > 0$, $b > 0$. Prove that $f(x) = 0$ has: (i) one real root if $b > a^3$, (ii) three real roots if $b < a^3$. Solve the equation $f(x) = 0$, given that $b = a^3$.

5 Show that the equation $x^4 - 5x + 2 = 0$ has exactly two real roots, both of which are positive. Obtain, graphically or otherwise, these real roots, giving your results to one decimal place. Use these approximate results to obtain approximate values for the sum and product of the two complex roots. (*L*)

6 Show that, for all real values of k, the function f defined by $f(x) = x^3 - 6x^2 + 9x - k$, where x is real, has a maximum when $x = 1$ and a minimum when $x = -3$. Illustrate, by means of two separate sketches, the cases when two of the roots of the equation $f(x) = 0$ are real and equal. Find the range of values of k for which the equation has three real distinct roots. Find the value of k such that the product of two real and distinct roots of the equation $f(x) = 0$ is equal to 1. When k takes this value, verify that the three roots of the equation are in arithmetic progression. (*JMB*)

7 Sketch the curve $y = x^4 + 3x^3 + x^2$, giving the coordinates and nature of its turning points. For the equation $x^4 + 3x^3 + x^2 - k = 0$, (i) find the complex roots in the form $a + ib$, a and b being real, when $k = -4$, (ii) find the integer root when $k = 9$, and find two consecutive integers between which the other real root must lie in this case, (iii) calculate the sum of the squares of the roots, showing that it is independent of k. (*JMB*)

8 Sketch the curve $y = 3x^4 - 4x^3 - 12x^2$, giving the coordinates of its stationary points (you need not calculate the coordinates of any other point on the curve). Find the ranges of values of k for which the equation $3x^4 - 4x^3 - 12x^2 - k = 0$ has (i) no real roots, (ii) just two distinct real roots, (iii) four distinct real roots. Find the value of k for which 1 is a root of the equation, and, for this value of k, find two consecutive integers between which the other real root, β, must lie. Express the modulus and the real part of the complex roots in terms of β, and hence determine ranges in which these quantities must lie. (*JMB*)

8.5 Rational functions with quadratic denominators

The simplest of these functions is of the form $1/q$, where q is a real quadratic function. The graph of such a function depends upon whether $q(x)$ has distinct real roots, equal real roots or a pair of conjugate complex roots. These functions were considered in *Pure Mathematics* §27.1. We illustrate the different cases by means of examples.

EXAMPLE 1 *Sketch the graph of the function* f, *where:* (*i*) $f(x) = 1/(x^2 + x - 2)$, (*ii*) $f(x) = 1/(x-1)^2$, (*iii*) $f(x) = 1/(x^2 + x + 1)$.

(i) Let $g(x) = x^2 + x - 2 = (x-1)(x+2)$, so that $f(x) = 1/g(x)$. The graph of g is a parabola, with vertex downwards, crossing the x-axis at $(-2, 0)$ and $(1, 0)$. By symmetry, its axis is equidistant from these two points and is the line $x = -1/2$. Its vertex is the point $(-1/2, -9/4)$ and the graph of g is shown in Fig. 8.2. To obtain the graph of f, we note that the domain of f must exclude the points where $g(x)$ is zero so that it is $\mathbb{R}\backslash\{-2, 1\}$. Then for $x < -2$ and for $x > 1$, $g(x) > 0$ and so $f(x) > 0$. Also for $-2 < x < 1$, $-9/4 < g(x) < 0$ and so $f(x) < -4/9$. As x approaches the values -2 and $+1$, $f(x)$ tends to infinity so we have asymptotes $x = -2$ and $x = 1$ to the graph of f. $f(x)$ becomes large and positive for values of x which are just less than -2 and for values of x which are just greater than 1. On the other hand, for values of x which are just greater than -2 and for values just less than 1, $f(x)$ becomes large and negative. From this information, and from the fact that the graph crosses the y-axis at $(0, -2)$ we are able to sketch the graph, which is shown in Fig. 8.3.

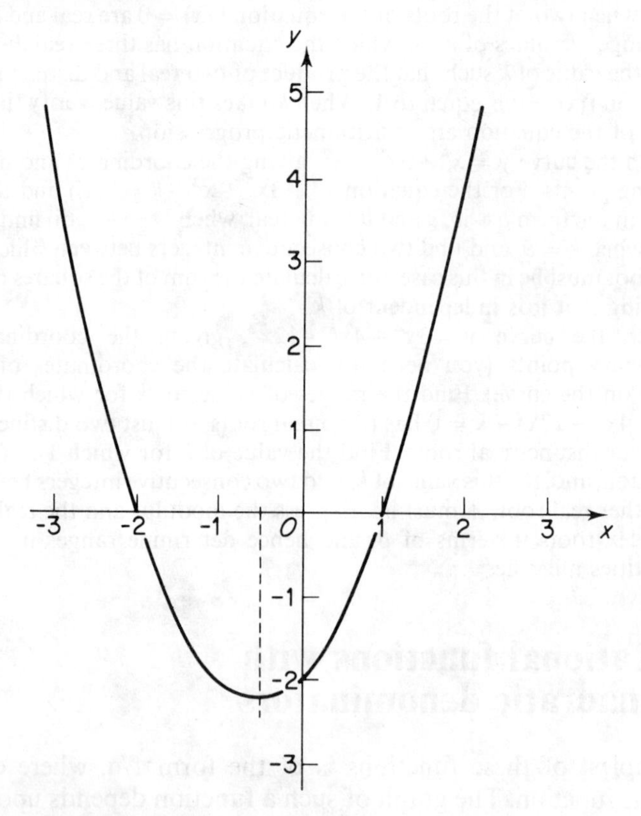

Fig. 8.2 Graph of g, $g(x) = (x-1)(x+2)$

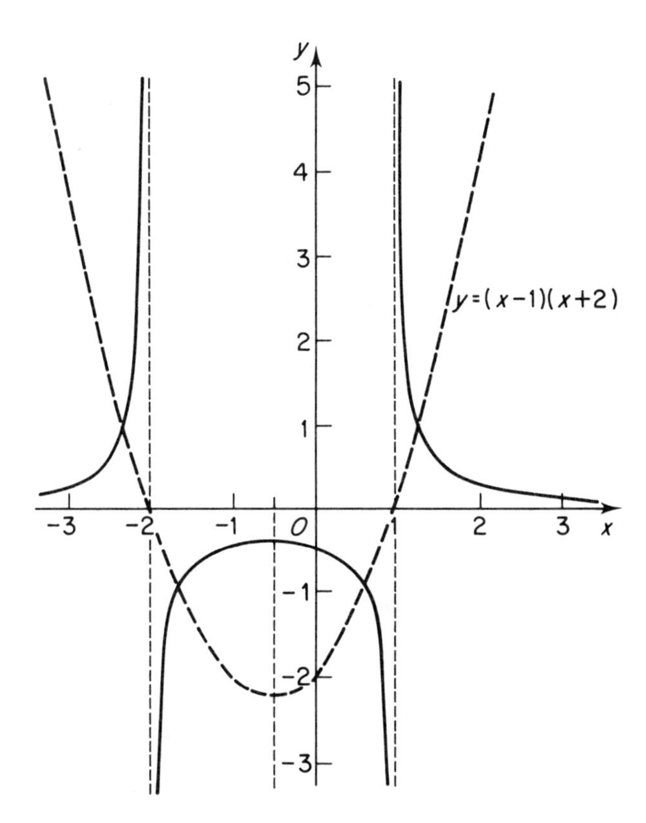

Fig. 8.3 Graph of f, $f(x) = 1/(x^2 + x - 2) = 1/\{(x-1)(x+2)\}$

(ii) Again, let $g(x) = (x-1)^2$ so that $f = 1/g$. The graph g is a parabola, with its vertex (1, 0) downwards, shown dotted in Fig. 8.4. The domain of f is $\mathbb{R}\backslash\{1\}$, $f(x)$ is always positive and tends to infinity as x tends to 1 from below and from above. Since $f'(x) = -2/(x-1)^3$, the gradient of f is positive and $f(x)$ is increasing for $x < 1$, while the gradient is negative and $f(x)$ is decreasing for $x > 1$. As x tends to infinity (positive or negative) $f(x)$ tends to zero. From these facts, we sketch the graph of f.
(iii) Let $g(x) = x^2 + x + 1 = (x + 1/2)^2 + 3/4$. Then, for all x, $3/4 \leqslant g(x)$ and so $0 < f(x) \leqslant 4/3$. There is a minimum of $g(x)$ at $(-1/2, 3/4)$ and so $f(x)$ is always positive and has a maximum at $(-1/2, 4/3)$. The graph is symmetrical about the line $x = -1/2$ and approaches the x-axis for large values of x, both positive and negative. The curve $y = x^2 + x + 1$ is shown dotted in Fig. 8.5 (on page 230) and, from all this information, the graph of f can now be sketched, as is shown in Fig. 8.5.

In other cases of a rational function f, with a quadratic denominator, the graph can be sketched using the following ideas.
(i) The graph crosses the y-axis at $(0, f(0))$,

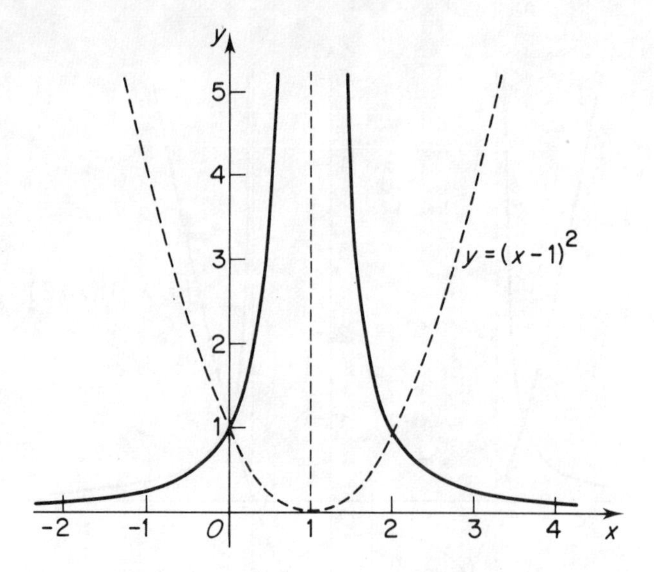

Fig. 8.4 Graph of f, $f(x) = 1/(x-1)^2$

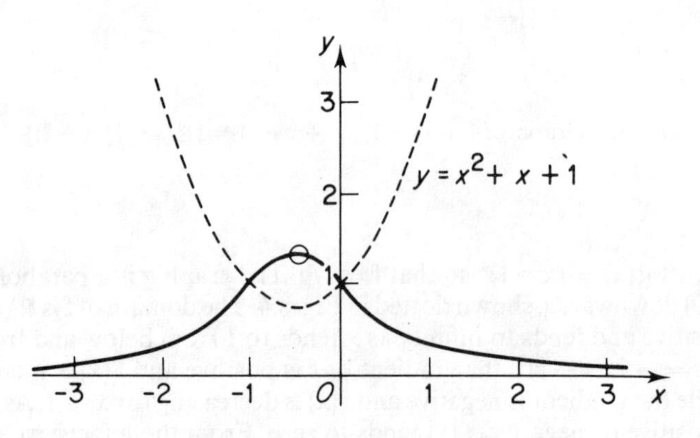

Fig. 8.5 Graph of f, $f(x) = 1/(x^2 + x + 1)$

(ii) the graph crosses the x-axis at points where the numerator of $f(x)$ vanishes,
(iii) any asymptotes parallel to the y-axis are given by the zeros of the denominator of $f(x)$,
(iv) for large values of x, the behaviour of $f(x)$ may be studied by rearranging it in terms of $1/x$ and then letting $1/x$ tend to zero.
(v) by writing $y = f(x)$ and looking for conditions on x for y to be real, it may be possible to determine regions of the plane in which no points of the graph lie.

EXAMPLE 2 *Sketch the graph of the function* f, *given by*

$$f(x) = (x-1)/(x^2-x+1).$$

Let $y = f(x)$ so that $y(x^2-x+1) = x-1$. Rearranging this as a quadratic equation in x, $yx^2-(y+1)x+(y+1) = 0$, and, for real values of x, the discriminant must be non-negative. Therefore

$$(y+1)^2 \geqslant 4y(y+1), \text{ or } (y+1)(y+1-4y) = (y+1)(1-3y) \geqslant 0.$$

Thus $(y+1)$ and $(1-3y)$ must have the same sign and so $-1 \leqslant y \leqslant 1/3$. Therefore the range of f is $\{x: -1 \leqslant x \leqslant 1/3\}$. As x tends to infinity, $f(x)$ tends to zero and so the x-axis is the only asymptote to the graph. The graph crosses the y-axis at $(0, -1)$ and the last piece of information required is the nature and position of any stationary point.

$$f'(x) = \frac{(x^2-x+1)-(x-1)(2x-1)}{(x^2-x+1)^2} = \frac{-x^2+2x}{(x^2-x+1)^2},$$

so that $f'(x) = 0$ when $x = 0$ or $x = 2$.
For $x < 0$ and for $x > 2$, $f'(x) < 0$ and for $0 < x < 2$, $f'(x) > 0$. Thus f had a minimum at $(0, -1)$ and a maximum at $(2, 1/3)$. The graph of f is shown in Fig. 8.6.

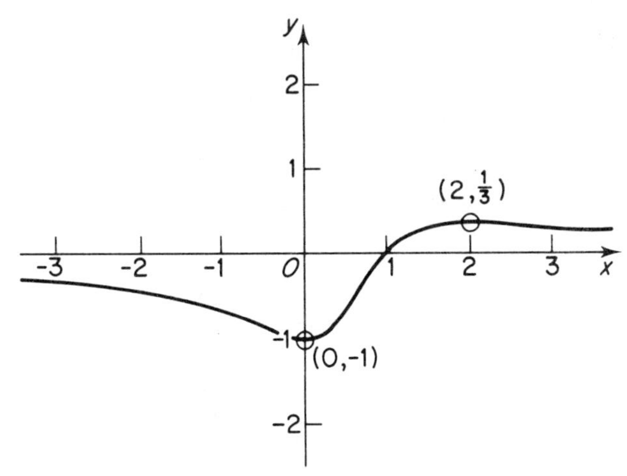

Fig. 8.6 Graph of f, $f(x) = (x-1)/(x^2-x+1)$

EXAMPLE 3 *Sketch the graph of the function* f, *where:*
(*i*) $f(x) = x^2/(x+1)^2$, (*ii*) $f(x) = (x+1)^2/(x^2+x-2)$,
(*iii*) $f(x) = (x^2+2x)/(x^2-1)$.

In each case, we summarise the relevant information and show the graph in a diagram.

(i) $f(x) \geqslant 0$ for all x and, since $f(x) = 1/(1+z)^2$, where $z = 1/x$, the graph has asymptotes $x = -1$ and $y = 1$.

$$f'(x) = \{2x(x+1)-2x^2\}/(x+1)^3 = 2x/(x+1)^3,$$

which is negative for $-1 < x < 0$ and is positive for $x < -1$ and for $x > 0$. Thus the graph has a minimum at $(0, 0)$ and is shown in Fig. 8.7.

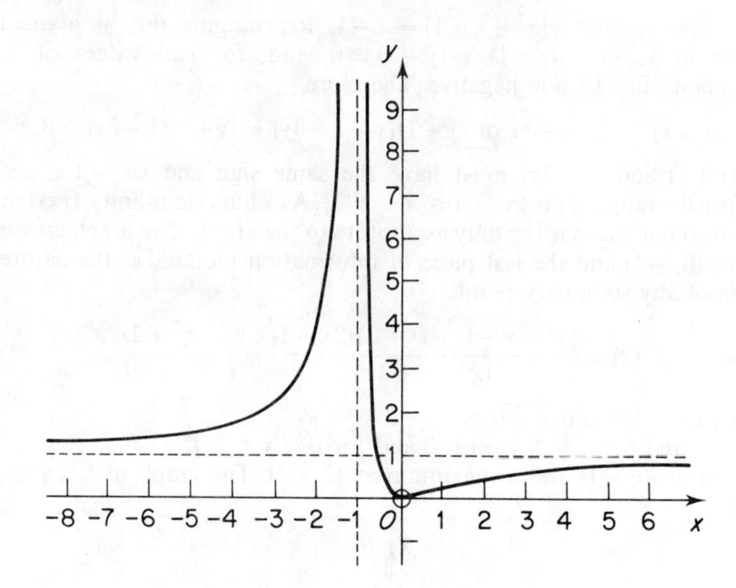

Fig. 8.7 Graph of f, $f(x) = x^2/(x+1)^2$

(ii) $$f(x) = \frac{(x+1)^2}{(x-1)(x+2)} = \frac{(1+z)^2}{(1-z)(1+2z)}, \text{ where } z = \frac{1}{x}.$$

Therefore the graph crosses the x-axis at $(-1, 0)$ and the y-axis at $(0, -1/2)$, and has three asymptotes $y = 1$, $x = 1$ and $x = -2$.

$$f'(x) = \frac{2(x+1)(x^2+x-2)-(x+1)^2(2x+1)}{(x^2+x-2)^2}$$

$$= \frac{(x+1)(2x^2+2x-4-2x^2-3x-1)}{(x^2+x-2)^2}$$

$$= -\frac{(x+1)(x+5)}{(x^2+x-2)^2}$$

Thus $f'(x) = 0$ for $x = -1$ and for $x = -5$, and $f(-1) = 0$, $f(-5) = 8/9$. For $x < -5$ and for $x > -1$, $f'(x) < 0$, and for $-5 < x < -1$, $f'(x) > 0$. So the graph has a minimum at $(-5, 8/9)$ and a maximum at $(-1, 0)$.
If $y = f(x)$, then $y(x^2+x-2) = (x+1)^2$, so that

$$(y-1)x^2 + (y-2)x - (2y+1) = 0.$$

For real values of x,

$$0 \leqslant (y-2)^2 + 4(y-1)(2y+1) = 9y^2 - 8y = y(9y-8),$$

and so either $y \leqslant 0$ or $y \geqslant 8/9$. Therefore there are no points on the graph in the strip $0 < y < 8/9$. The graph is shown in Fig. 8.8.

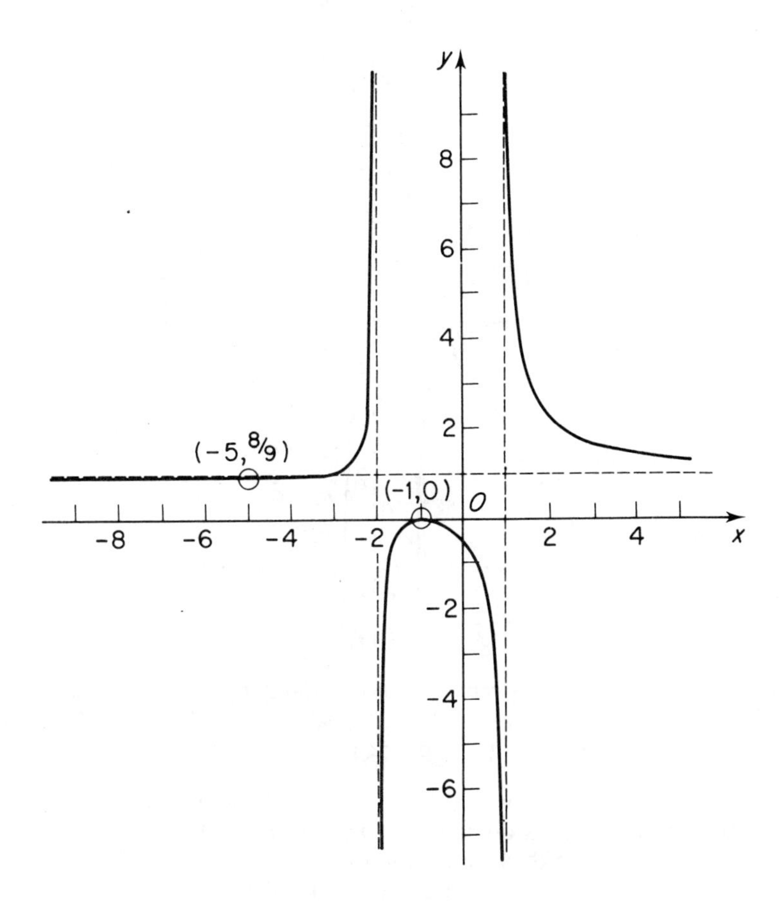

Fig. 8.8 Graph of f, $f(x) = (x+1)^2/(x^2+x-2)$

(iii)
$$f(x) = \frac{x^2+2x}{x^2-1} = \frac{1+2z}{1-z^2}, \text{ where } z = \frac{1}{x}.$$

The graph has asymptotes $y = 1, x = 1$ and $x = -1$ and crosses the axes at $(0, 0)$ and $(-2, 0)$. The derivative $f'(x)$ is always negative, since

$$\begin{aligned} f'(x) &= \{(2x+2)(x^2-1)-(x^2+2x)2x\}/(x^2-1)^2 \\ &= (-2x^2-2x-2)/(x^2-1)^2 = -\{(2x+1)^2+3\}/2(x^2-1)^2, \end{aligned}$$

so the graph has no turning points and its slope is always negative. The graph is shown in Fig. 8.9.

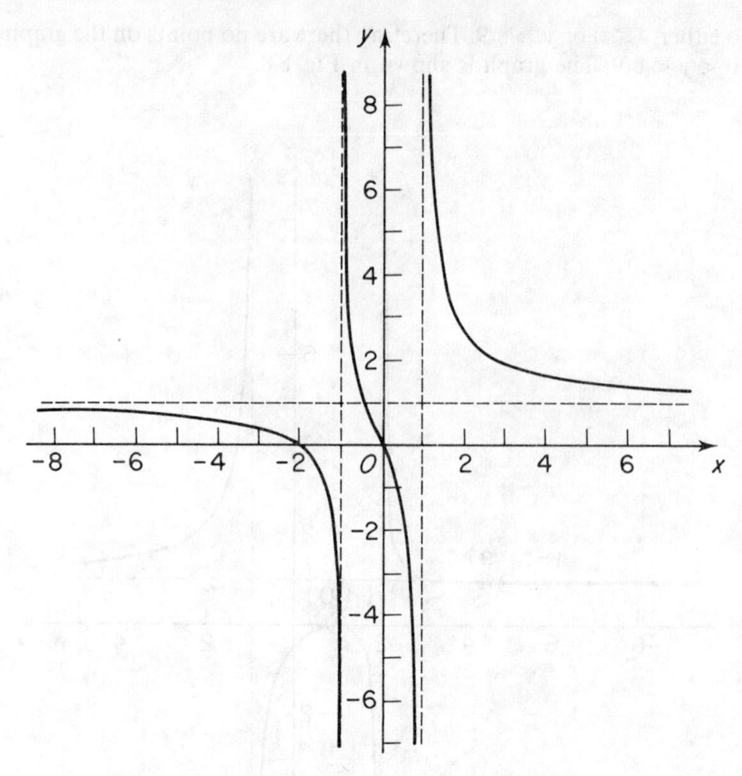

Fig. 8.9 Graph of f, $f(x) = (x^2 + 2x)/(x^2 - 1)$

EXERCISE 8.5

1 The function f is given by $f(x) = 2x/(x^2 - 4)$. Show that f is an increasing function throughout its domain, and that its graph has asymptotes $x = 2$, $x = -2$ and $y = 0$. Sketch the graph of f.

2 Find the turning points on the graph $y = (x-1)^2/(x+1)(x+3)$ and prove that there are no points on the graph for which $-8 < y < 0$. Sketch the graph.

3 Prove that the range of the function f, given by $f(x) = 3x/(x^2 + x + 1)$ is $\{x: -3 \leqslant x \leqslant 1\}$. Find the turning points on the graph of f and sketch this graph.

4 Find those values of y for which there are no points on the curve $y = (x+1)^2/(x-1)^2$. Sketch this curve.

5 Sketch the graph of the function f where:

(i) $f(x) = (x^2 - 3x)/(x^2 + 5x + 4)$, (ii) $f(x) = (2x^2 - 9x + 4)/(x-1)^2$,
(iii) $f(x) = (1 - x^2)/(1 + x^2)$, (iv) $f(x) = (x-1)^2/(x+1)^2$,
(v) $f(x) = (4x^2 - 3x)/(x^2 + 1)$, (vi) $f(x) = (1 + 2/x)^2$,
(vii) $f(x) = (4x+1)/(x^2 - 1)$, (viii) $f(x) = (x^2 + 1)/(x^2 - 9)$.

MISCELLANEOUS EXERCISE 8

1 Given that α, β, γ are the roots of the equation $x^3 + px^2 + qx + r = 0$, where the coefficients p, q, r are real, show that $\alpha^2 + \beta^2 + \gamma^2 = p^2 - 2q$. Find

$(\alpha-\beta)^2+(\beta-\gamma)^2+(\gamma-\alpha)^2$ in terms of p and q and show that α, β, γ cannot all be real if $p^2 < 3q$. *(L)*

2 Given that $y = (x^2+3x+1)/(x^2+3x+2)$, find the range or ranges of values that y can take for real values of x. Find also the equations of the asymptotes of the graph of this function. *(JMB)*

3 The equation $x^3+6x^2+\lambda x+\mu = 0$, where λ and μ are constants, has roots equal to 1 and -5. Find the third root and the value of μ. *(L)*

4 Three numbers p, q and r are such that $p+q+r = 6$ and $p^2+q^2+r^2 = 38$. Show that $pq+qr+rp = -1$. If also $pqr = -30$, form a cubic equation whose roots are p, q and r. Hence, or otherwise, find the three numbers. *(L)*

5 (i) The three roots of the equation $ax^3+bx^2+cx-1 = 0$ are α, β, γ. Form a cubic equation with roots $\alpha\beta, \beta\gamma, \gamma\alpha$ having its coefficients in terms of a, b, c.
(ii) Find integers m and n such that $(x+1)^2$ is a factor of x^5+2x^2+mx+n. *(L)*

6 Find the set of values of y for which x is real when $y = \dfrac{15+10x}{4+x^2}$. Find also the gradient of this curve when $x = 0$. *Sketch* the curve. *(L)*

7 Given that α, β, γ are the three roots of the equation $x^3-x^2+3x+10 = 0$ show that $\alpha^2+\beta^2+\gamma^2 = -5$. Hence, or otherwise, find the value of $\alpha^3+\beta^3+\gamma^3$. Find also a cubic equation whose roots are $\alpha^{-1}, \beta^{-1}, \gamma^{-1}$. *(L)*

8 The polynomial P is defined by $P(x) = (x-2)^3(rx+s)$. Show that $(x-2)$ is a repeated factor of $P'(x)$. When $P(x)$ is divided by $(x-1)$, the remainder is -2. When $P'(x)$ is divided by $(x-1)$, the remainder is 2. Find the values of r and s. *(JMB)*

9 The polynomial $P(x)$ is divided by (x^2-4). Show that the remainder is $\frac{1}{4}x\{P(2)-P(-2)\}+\frac{1}{2}\{P(2)+P(-2)\}$. Given further that (i) $P(x)$ is of degree 4, (ii) $P(x)$ has coefficient of x^4 equal to unity, (iii) $P(x) = P(-x)$, (iv) $P(2) = -16$, (v) $P(x)$ has a stationary value at $x = 1$, find the quotient when $P(x)$ is divided by x^2-4. State the zeros of $P(x)$ (a) in the field of real numbers (b) in the field of complex numbers. *(JMB)*

10 Find, in terms of the constant k, the maximum and minimum values of the function $f(x) = x^4-4x^3-2x^2+12x+k$. Sketch the graph of the function when k has the value 9. State the ranges of values of k for which the equation $f(x) = 0$ has (i) four real distinct roots, (ii) two real distinct roots and two non-real roots. Given that $k = -40$ and that one of the roots of the equation $f(x) = 0$ is $1+i\sqrt{3}$, express $f(x)$ as the product of two quadratic factors. Hence find all the roots of the equation $f(x) = 0$ when $k = -40$. *(JMB)*

11 Prove that the function $(x^2+1)/(x-1)^2$ is a minimum when $x = -1$. Sketch the graph of this function. With the aid of the sketch, or otherwise, determine the range of values of λ such that the equation $x^2+1 = \lambda x(x-1)^2$ has a real root between $x = 1$ and $x = 2$. *(JMB)*

12 Given that $P(x) = (x-\alpha)^2 Q(x)$, where $Q(x)$ is a polynomial, show that $P'(\alpha) = 0$. Given that the polynomial $2x^3+11x^2+12x-9$ has a factor $(x-\alpha)^2$, find the value of α. Express the polynomial as a product of linear factors. *(JMB)*

13 By considering the equation $yx^2+(y-1)x+(y-1) = 0$, where $y \neq 0$, as a quadratic equation in x, find the range of values of y for which this equation has real roots. State the value of x which satisfies this equation when $y = 0$.

Using these results, or otherwise, determine the range of the function f, where $f(x) = (x+1)/(x^2+x+1)$ and x is real. Sketch the graph of $f(x)$ giving the coordinates of any turning points. Explain why f has no inverse function. (*JMB*)

14 Given that the roots of the equation $x^3+bx^2+cx+d=0$, where b, c, d, are non-zero constants, are in geometric progression, find a relation between b, c, and d. Solve the equation $x^3-2x^2-4x+8=0$, and verify that your solutions are in geometric progression. (ii) Solve the equation $4x^3-24x^2+23x+18=0$, given that the roots are in arithmetic progression. (*L*)

15 The equation $px^3+qx^2+rx+s=0$ has roots α, $1/\alpha$ and β. Prove that $p^2-s^2=pr-qs$. Solve the equation $6x^3+11x^2-24x-9=0$. (*L*)

16 Sketch the curve $y=x/(1+x^2)$, finding its turning points, showing that the origin is a point of inflexion, and indicating the behaviour of y when $|x|$ is large. Calculate the area of the region defined by the inequalities $0 \leqslant x \leqslant a$, $0 \leqslant y \leqslant x/(1+x^2)$, and show that this area has the value 1 when $a=\sqrt{(e^2-1)}$. (*L*)

17 P is the statement: $(x-\alpha)^2$ is a factor of the polynomial $f(x)$. Q is the statement: $(x-\alpha)$ is a factor of the polynomial $f'(x)$. (a) Prove that $P \Rightarrow Q$. (b) Show by a counter-example that $Q \not\Rightarrow P$. Given that $f(x)=4x^4+4x^3-11x^2-6x+9$, write down $f'(x)$. Given also that $f(x)$ has a repeated linear factor, factorise $f'(x)$ and $f(x)$. (*L*)

18 (i) Show that $(a+b)$ is a root of the equation $x^3-3abx-(a^3+b^3)=0$. Express the equation $x^3-6x-6=0$ in the above form, giving your values for a^3 and b^3. Hence, find a real root of the equation $x^3-6x-6=0$, expressing your answer in the form $\sqrt[3]{m}+\sqrt[3]{n}$, where m and n are positive integers. (ii) Given that the equation $x^3+px^2+qx+r=0$ has three roots α, β, γ where $\alpha+\beta=\gamma$, show that $p^3+8r=4pq$. (*L*)

19 Sketch the graph of $y=(x-2)^2/(x-6)(x+1)$, giving the coordinates of the points of intersection with the coordinate axes and the coordinates of the stationary points. Give the equations of the asymptotes of the curve. (*L*)

20 Find the coordinates of the stationary point and the points of inflexion of the curve $y=x^2/(1+x^2)$. Show that the curve lies entirely in the region $0 \leqslant y < 1$, and that $\dfrac{dy}{dx} > 0$ for $x > 0$. Sketch the curve. Find the ranges of values of k for which the equation $x^2/(1+x^2)=kx$ has three real and distinct roots. (*JMB*)

21 The equation of a curve is $y=2(x+1)^2/(x^2+4)$. Show that, for all real values of x, $0 \leqslant y \leqslant 5/2$. Sketch the curve, showing its asymptote. Find the area of the region bounded by the x-axis, the y-axis, the curve and the line $x=2$. (*JMB*)

22 Sketch the graph of the function f, where:
(i) $f(x)=(x^2-2x-2)/(x-1)^2$, (ii) $f(x)=(x^2+1)/(x^2-x-2)$.

23 Show that the graph of $y=(x^2-1)/(x^2+4x)$ has no real stationary points and sketch this graph. (*JMB*)

24 (i) By considering the stationary values of the expression $f(x)$ where $f(x)=x^3+px+q$, $p>0$, $q>0$, or otherwise, show that the equation $f(x)=0$ has precisely one negative root and no positive root. (ii) Show that the equation $x^4+4x+3=0$ has precisely two non-real roots and that these non-real roots have positive real parts. (*L*)

9 Integration

9.1 Integration techniques (revision)

A number of techniques for integration were developed in *Pure Mathematics* Chapter 24 and in Chapter 7 of this book. In this section, we summarise the techniques, provide a few (more difficult) examples and conclude with a revision exercise. In the summary below, we omit the arbitrary constants of integration.

Standard Integrals

$\mathbf{f(x)}$	$\int \mathbf{f(x)dx}$
$x^n,\ n \neq -1,$	$x^{n+1}/(n+1)$
x^{-1}	$\ln\lvert x\rvert$
e^x	e^x
$\cos x$	$\sin x$
$\sin x$	$-\cos x$
$\tan x$	$\ln\lvert\sec x + \tan x\rvert$
$\sec^2 x$	$\tan x$
$\dfrac{1}{1+x^2}$	$\arctan x$
$\dfrac{1}{1-x^2}$	$\operatorname{artanh} x$ or $\ln\sqrt{\left\lvert\dfrac{1+x}{1-x}\right\rvert}$
$\dfrac{1}{\sqrt{(1-x^2)}}$	$\arcsin x$
$\dfrac{1}{\sqrt{(1+x^2)}}$	$\operatorname{arsinh} x$
$\dfrac{1}{\sqrt{(x^2-1)}}$	$\operatorname{arcosh} x$

Integration by substitution If $u = u(x)$, $f(x) = g(u)\dfrac{du}{dx}$, then

$$\int f(x)dx = \int g(u)\frac{du}{dx}\,dx = \int g(u)du.$$

Integration of rational functions

$$\text{Special case: } \int \frac{q'(x)}{q(x)}\,dx = \ln|q(x)|$$

In order to integrate $\dfrac{p(x)}{q(x)}$, a rational function,

(i) split the function into partial fractions,

(ii) for a term with a linear denominator, use

$$\int \frac{1}{ax+b}\,dx = \frac{1}{a}\ln|ax+b|,$$

(iii) for a term with a linear power denominator, with $n > 1$, use

$$\int \frac{1}{(ax+b)^n}\,dx = \frac{-1}{a(n-1)(ax+b)^{n-1}},$$

(iv) for a term with a quadratic denominator which does not factorise, complete the square of the terms in x so that the denominator has the form $(x+a)^2+b^2$, substitute $x+a=u$ and use

$$\int \frac{cu+d}{u^2+b^2}\,du = \frac{c}{2}\ln(u^2+b^2) + \frac{d}{b}\arctan\frac{u}{b}$$

and then substitute back again, $u = x+a$.

Integration by parts

$$\int u\frac{dv}{dx}\,dx = uv - \int v\frac{du}{dx}\,dx.$$

Integrand with a denominator which is the square root of a quadratic function. Complete the square in the quadratic function and substitute so that it is of the form $\sqrt{(\pm u^2 \pm b^2)}$, then use

$$\int \frac{1}{\sqrt{(b^2-u^2)}}\,du = \arcsin\frac{u}{b},$$

$$\int \frac{1}{\sqrt{(u^2-b^2)}}\,du = \operatorname{arcosh}\frac{u}{b},$$

$$\int \frac{1}{\sqrt{(u^2+b^2)}}\,du = \operatorname{arcsinh}\frac{u}{b},$$

and then substitute back again.

EXAMPLE 1 *Find the indefinite integral of* $\dfrac{1}{(1+x)^2\sqrt{(3+5x+2x^2)}}$, using the substitution $u = 1/(1+x)$.

Let $I(x) = \int \frac{1}{(1+x)^2\sqrt{(3+5x+2x^2)}}\,dx$, and substitute $u = \frac{1}{1+x}$ so that $1+x = \frac{1}{u}$, $\quad \frac{dx}{du} = \frac{-1}{u^2}$, $\quad 3+5x+2x^2 = (1+x)+2(1+x)^2 = \frac{1}{u}+2\frac{1}{u^2}$, $\sqrt{(3+5x+2x^2)} = u^{-1}\sqrt{(u+2)}$ and $I(x) = \int u^2 \frac{u}{\sqrt{(u+2)}}\left(\frac{-1}{u^2}\right) du$. Thus

$$
\begin{aligned}
I(x) &= \int \frac{-u}{\sqrt{(u+2)}}\,du = \int \frac{2-(u+2)}{\sqrt{(u+2)}}\,du = \int \frac{2}{\sqrt{(u+2)}}\,du - \int \sqrt{(u+2)}\,du \\
&= 4\sqrt{(u+2)} - \frac{2}{3}(u+2)^{3/2} + c = 4\sqrt{\left(\frac{3+2x}{1+x}\right)} - \frac{2}{3}\left(\frac{3+2x}{1+x}\right)^{3/2} + c. \\
&= \left\{4 - \frac{2}{3}\left(\frac{3+2x}{1+x}\right)\right\}\sqrt{\left(\frac{3+2x}{1+x}\right)} + c = \mathbf{\frac{2}{3}\left(\frac{3+4x}{1+x}\right)\sqrt{\left(\frac{3+2x}{1+x}\right)} + c.}
\end{aligned}
$$

EXAMPLE 2 *Evaluate* $\int_0^1 \frac{2x^3+x^2-1}{(x+1)^2(x^2+1)}\,dx$.

Split the integrand into partial fractions

$$\frac{2x^3+x^2-1}{(x+1)^2(x^2+1)} = \frac{A}{x+1} + \frac{B}{(x+1)^2} + \frac{Cx+D}{x^2+1}.$$

On putting $x = -1$, $-2 = 2B$ and so $B = -1$.
Comparing the coefficients of x^3 and x and the constant terms,

$$A+C = 2,\ A+C+2D = 0 \text{ and } A+B+D = -1.$$

Therefore $\qquad 2+2D = 0$ and $D = -1$, $A-1-1 = -1$

and $\qquad A = 1$, $1+C = 2$ and $C = 1$.

As a check, compare the coefficients of x,

$$A+C+2D = 1+1-2 = 0.$$

Therefore:

$$
\begin{aligned}
\int_0^1 \frac{2x^3+x^2-1}{(x+1)^2(x^2+1)}\,dx &= \int_0^1 \left\{\frac{1}{x+1} - \frac{1}{(x+1)^2} + \frac{x-1}{x^2+1}\right\} dx \\
&= \left[\ln|x+1| + \frac{1}{x+1} + \frac{1}{2}\ln(x^2+1) - \arctan x\right]_0^1 \\
&= \mathbf{\ln(2\sqrt{2}) - \frac{1}{2} - \frac{\pi}{4}}.
\end{aligned}
$$

EXAMPLE 3 *Evaluate* $\int_1^5 \frac{1}{\sqrt{(x^2-2x+17)}}\,dx$.

Completing the square, $x^2-2x+17 = (x-1)^2+16 = u^2+16$, where $u = x-1$.

Therefore, $\int_1^5 \frac{1}{\sqrt{(x^2-2x+17)}}\,dx = \int_0^4 \frac{1}{\sqrt{(u^2+16)}}\,du = \left[\operatorname{arsinh}\frac{u}{4}\right]_0^4$

$$= \operatorname{arsinh}(1) = \mathbf{\ln(1+\sqrt{2})}.$$

When an integrand has some properties of symmetry in the range of integration, this can sometimes be used to transform the integral into one that can be more easily handled. This is shown in the next example where the periodic symmetry of the trigonometrical functions is used.

EXAMPLE 4 *Evaluate* $\int_0^{2\pi} x\sin^2 x\,dx$.

Let $I = \int_0^{2\pi} x\sin^2 x\,dx$, and make the substitution $x = 2\pi - u$.

Then $I = \int_{2\pi}^{0} (2\pi - u)\sin^2(2\pi - u)(-du) = \int_0^{2\pi} (2\pi - u)\sin^2 u\,du$

$= \int_0^{2\pi} (2\pi - x)\sin^2 x\,dx = 2\pi\int_0^{2\pi} \sin^2 x\,dx - I$, on substituting $u = x$.

Thus, $2I = 2\pi\int_0^{2\pi} \sin^2 x\,dx = \pi\int_0^{2\pi} (1 - \cos 2x)\,dx = \pi\left[x - \frac{1}{2}\sin 2x\right]_0^{2\pi} = 2\pi^2$,

so that $I = \pi^2$.

EXERCISE 9.1

1 Find the indefinite integral of: (i) $\dfrac{1}{2x-1}$, (ii) $\dfrac{3}{(1-x)^2}$, (iii) $\dfrac{2}{x^2-1}$, (iv) $\dfrac{2}{x^2+4}$, (v) $\dfrac{x^2}{x^2+4}$, (vi) $\dfrac{x+1}{x^2+2x+2}$.

2 Evaluate: (i) $\int_0^2 \dfrac{6x^2}{2+x^3}\,dx$, (ii) $\int_1^2 x^2\sqrt{(x-1)}\,dx$, (iii) $\int_0^{1/2} \dfrac{x^2}{\sqrt{(1-x^2)}}\,dx$,

(iv) $\int_0^1 \sqrt{(4-3x)}\,dx$, (v) $\int_0^1 x\sqrt{(4-3x)}\,dx$, (vi) $\int_0^1 \dfrac{x^2}{1+x}\,dx$,

(vii) $\int_0^1 \dfrac{\arccos x}{\sqrt{(1-x^2)}}\,dx$.

3 Find: (i) $\int x^2 e^{-3x}\,dx$, (ii) $\int (1+x)\ln x\,dx$, (iii) $\int e^{ax}\sinh bx\,dx$,

(iv) $\int \operatorname{arsinh} x\,dx$, (v) $\int \ln(4+x^2)\,dx$, (vi) $\int \dfrac{x^5}{\sqrt{(x^2-1)}}\,dx$,

(vii) $\int (1-x^2)^{-3/2}\,dx$, (viii) $\int \dfrac{3}{x^3-1}\,dx$, (ix) $\int \dfrac{1}{\sqrt{(9x^2-12x-5)}}\,dx$.

4 Evaluate (i) $\int_0^1 \dfrac{x^3+4x+4}{(x-2)^2(x^2+1)}\,dx$, (ii) $\int_0^1 \dfrac{1}{(1+x)^2(1+x^2)}\,dx$.

5 Evaluate $\int_{-1/2}^{5/2} \dfrac{1}{\sqrt{(8+2x-x^2)}}\,dx$. (*L*)

6 (i) Evaluate $\int_{1/2}^{1} x \ln(2x)\,dx$.

(ii) Using the substitution $x = \sin^2\theta$, or otherwise, evaluate

$$\int_{1/2}^{3/4} \frac{1}{\sqrt{(x-x^2)}}\,dx.$$

(iii) Using integration by parts, or otherwise, evaluate

$$\int_0^1 x(1-x)^{3/2}\,dx.$$ (*L*)

7 Given that $f(x) = \dfrac{2(2x-1)}{x(x+1)(x+2)}$, express $f(x)$ in partial fractions. Hence (a) obtain $\sum_{r=1}^{n} \dfrac{2r-1}{r(r+1)(r+2)}$, (b) evaluate $\int_1^2 f(x)dx$, leaving your answer in terms of natural logarithms. (*L*)

8 Express $\dfrac{56(2x+3)}{(x+5)^2(x^2+3)}$ in partial fractions. Hence evaluate

$$\int_0^1 \frac{56(2x+3)}{(x+5)^2(x^2+3)}\,dx.$$ (*L*)

9 Express $f(x)$ in partial fractions, where $f(x) = \dfrac{8}{(x+1)^2(x^2+3)}$. Hence show that $\int_0^1 f(x)\,dx = 1 + \dfrac{1}{2}\ln 3 - \dfrac{\pi}{6\sqrt{3}}$. (*L*)

10 Given that $f(x) = \dfrac{x^3 - x^2 + 1}{(x-1)^2(x^2+1)}$, express $f(x)$ in partial fractions. Hence find (a) $\int f(x)\,dx$, (b) the first three terms when $f(x)$ is expanded in a series of ascending powers of x, stating the set of values of x for which the expansion is valid. (*L*)

11 (i) Find $\int \dfrac{1}{\sqrt{(x^2-2x+10)}}\,dx$. (ii) Find $\int \dfrac{1}{x^2-2x+10}\,dx$. (iii) By using the substitution $x = \sin\theta$, show that

$$\int_0^{1/2} \frac{x^4}{\sqrt{(1-x^2)}}\,dx = (4\pi - 7\sqrt{3})/64.$$ (*L*)

12 By means of the substitution $\theta = \pi/2 - \phi$, or otherwise, show that $\int_0^{\pi/2} \dfrac{\sin^3\theta}{\sin\theta + \cos\theta}\,d\theta = \int_0^{\pi/2} \dfrac{\cos^3\theta}{\sin\theta+\cos\theta}\,d\theta$ and hence, by adding these integrals, evaluate $\int_0^{\pi/2} \dfrac{\sin^3\theta}{\sin\theta+\cos\theta}\,d\theta$. (*JMB*)

13 Show that $\int_0^a f(x)\,dx = \int_0^a f(a-x)\,dx$. Hence evaluate $\int_0^{\pi} \dfrac{x\sin x}{1+\cos^2 x}\,dx$.

14 If f is an even function, prove that $\int_{-a}^{a} f(x)\,dx = 2\int_{0}^{a} f(x)\,dx$. If g is an odd function, prove that $\int_{-a}^{a} g(x)\,dx = 0$.

Evaluate $\int_{-1}^{1} (x^3+2x^2-x+7)\sqrt{(1-x^2)}\,dx$.

9.2 Reduction formulae

The use of the method of integration by parts sometimes leads to a formula which expresses an integral I_n in terms of I_{n-1} and/or I_{n-2}, where I_n involves some natural number n. Such a formula is called a *reduction formula* and is useful in the evaluation of certain integrals.

EXAMPLE 1 *Given that* $I_n = \int_0^{\pi/2} \sin^n x\,dx$, *prove that, for* $n > 1$, $nI_n = (n-1)I_{n-2}$. *Evaluate* $\int_0^{\pi/2} \sin^8 x\,dx$.

$$I_n = \int_0^{\pi/2} \sin^n x\,dx = -\int_0^{\pi/2} \sin^{n-1} x \frac{d}{dx}\cos x\,dx$$

$$= \Big[-\sin^{n-1} x \cos x\Big]_0^{\pi/2} + (n-1)\int_0^{\pi/2} \cos^2 x \sin^{n-2} x\,dx$$

$$= (n-1)\int_0^{\pi/2} (1-\sin^2 x)\sin^{n-2} x\,dx$$

$$= (n-1)\int_0^{\pi/2} \sin^{n-2} x\,dx - (n-1)\int_0^{\pi/2} \sin^n x\,dx = (n-1)(I_{n-2}-I_n).$$

Thus, for $n > 1$, $nI_n = (n-1)I_{n-2}$. Also $I_0 = \int_0^{\pi/2} dx = \pi/2$ and so

$$I_8 = \frac{7}{8}I_6 = \frac{7}{8}\cdot\frac{5}{6}I_4 = \frac{7}{8}\cdot\frac{5}{6}\cdot\frac{3}{4}I_2 = \frac{7}{8}\cdot\frac{5}{6}\cdot\frac{3}{4}\cdot\frac{1}{2}I_0 = \mathbf{\frac{105}{768}\pi}.$$

EXAMPLE 2 *Given that* $I_n = \int_0^{\pi} x^n \cos x\,dx$, *prove that, for* $n > 1$,

$$I_n = -n\pi^{n-1} - n(n-1)I_{n-2}.$$

Deduce that $\int_0^{\pi} x^4 \cos x\,dx = 24\pi - 4\pi^3$.

$$I_n = \int_0^{\pi} x^n \cos x\,dx = \left[x^n \sin x \right]_0^{\pi} - n\int_0^{\pi} x^{n-1} \sin x\,dx = n\int_0^{\pi} x^{n-1} \frac{d}{dx}\cos x\,dx$$

$$= n\left[x^{n-1}\cos x \right]_0^{\pi} - n(n-1)\int_0^{\pi} x^{n-2}\cos x\,dx = -n\pi^{n-1} - n(n-1)I_{n-2}.$$

$$I_0 = \int_0^{\pi} \cos x\,dx = \left[\sin x \right]_0^{\pi} = 0, \text{ so } I_4 = -4\pi^3 - 12I_2 = \mathbf{-4\pi^3 + 24\pi}.$$

EXAMPLE 3 *Given that* $I_n = \int_0^{\pi/4} \sec^n x\,dx$, *find a reduction formula relating* I_n *and* I_{n-2} *and hence evaluate* I_5.

$$I_n = \int_0^{\pi/4} \sec^n x\,dx = \int_0^{\pi/4} \sec^{n-2} x \frac{d}{dx}\tan x\,dx$$

$$= \left[\sec^{n-2} x \tan x \right]_0^{\pi/4} - \int_0^{\pi/4} (n-2)\sec^{n-2} x \tan^2 x\,dx$$

$$= \left[\sec^{n-2} x \tan x \right]_0^{\pi/4} - \int_0^{\pi/4} (n-2)\sec^{n-2} x(\sec^2 x - 1)\,dx$$

$$= (\sqrt{2})^{n-2} - (n-2)(I_n - I_{n-2}).$$

Thus $\mathbf{(n-1)I_n = (\sqrt{2})^{n-2} + (n-2)I_{n-2}}$.

Now $I_1 = \int_0^{\pi/4} \sec x\,dx = \left[\ln(\sec x + \tan x) \right]_0^{\pi/4} = \ln(1+\sqrt{2})$, and so

$$4I_5 = 2^{3/2} + 3I_3 = 2^{3/2} + \frac{3}{2}(2^{1/2} + I_1) \text{ and } I_5 = \mathbf{\frac{7}{8}\sqrt{2} + \frac{3}{8}\ln(1+\sqrt{2})}.$$

EXAMPLE 4 *Given that* $I_n = \int_0^1 x^n e^{2x} dx$, *show that, for* $n > 0$, $2I_n = e^2 - nI_{n-1}$. *Prove that* $\int_0^1 x^5 e^{2x}\,dx = \frac{1}{8}(15 - e^2)$.

$$2I_n = \int_0^1 x^n \frac{d}{dx} e^{2x}\,dx = \left[x^n e^{2x} \right]_0^1 - n\int_0^1 x^{n-1} e^{2x}\,dx = e^2 - nI_{n-1}.$$

Also $I_0 = \int_0^1 e^{2x}\,dx = \left[\frac{1}{2} e^{2x} \right]_0^1 = \frac{1}{2}(e^2 - 1)$, and so

$$\int_0^1 x^5 e^{2x}\,dx = I_5 = \frac{1}{2}(e^2 - 5I_4) = \frac{1}{2}e^2 - \frac{5}{4}(e^2 - 4I_3)$$

$$= -\frac{3}{4}e^2 + \frac{5}{2}(e^2 - 3I_2) = \frac{7}{4}e^2 - \frac{15}{4}(e^2 - 2I_1)$$

$$= -2e^2 + \frac{15}{4}(e^2 - I_0) = \frac{1}{8}(15 - e^2).$$

EXERCISE 9.2

1 Given that, for $n \geqslant 0$, $I_n = \int_0^1 x^n e^x \, dx$, show that, for $n \geqslant 1$, $I_n = e - nI_{n-1}$. Evaluate I_0 and I_3. (*L*)

2 If $I_n = \int_0^1 x^n e^{-x} \, dx$, where $n \geqslant 0$, find the relation between I_n and I_{n-1} where $n \geqslant 1$. Express $\int_0^1 x^4 e^{-x} \, dx$ in terms of e. (*L*)

3 Given that, for $n \geqslant 0$, $I_n = \int_0^\pi x^n \sin x \, dx$, prove that, for $n \geqslant 2$, $I_n = \pi^n - n(n-1)I_{n-2}$. Hence, evaluate $\int_0^\pi x^5 \sin x \, dx$. (*L*)

4 Given that $I_n = \int_1^2 x(\ln x)^n \, dx$, where n is a non-negative integer, prove that, for $n \geqslant 1$, $2I_n + nI_{n-1} = 4(\ln 2)^n$. Evaluate $\int_1^2 x(\ln x)^3 \, dx$. (*L*)

5 Given that $I_n = \int_0^1 x^n e^{-x^2/2} \, dx$, $n \geqslant 0$, prove that, for $n \geqslant 2$, $I_n = (n-1)I_{n-2} - e^{-1/2}$. Hence, evaluate I_3 in terms of e. (*JMB*)

6 Calculate $\int_0^1 x^4 e^{-2x} \, dx$.

7 Given that $I_n = \int \cosh^n x \, dx$, prove that, for $n > 1$, $nI_n = (n-1)I_{n-2} + \cosh^{n-1} x \sinh x$. Find $\int \cosh^5 x \, dx$.

8 Given that $I_n(x, k) = \int_0^x t^n e^{kt} \, dt$, $(k \neq 0,\ n \geqslant 0)$, show that, for $n \geqslant 1$, $I_n(x, k) = \{x^n e^{kx} - nI_{n-1}(x, k)\}/k$. (i) Find $I_3(-1, 1)$ in terms of e. (ii) Find the solution of the differential equation $2\dfrac{dy}{dx} + y = x^2$ such that $y = 0$ when $x = 0$. (*JMB*)

9 Given that $I_n = \int \operatorname{sech}^n x \, dx$, prove that

$$(n-1)I_n = \operatorname{sech}^{n-2} x \tanh x + (n-1)I_{n-2}.$$

By first expressing $\operatorname{sech} x$ in terms of exponential functions and then integrating, prove that $I_1 = 2\tan^{-1}(e^x) + \text{constant}$. Find I_4 and I_5. (*JMB*)

10 If $I_n = \int \dfrac{1}{(x^2+1)^n} \, dx$ show that $2nI_{n+1} = (2n-1)I_n + \dfrac{x}{(x^2+1)^n}$. Hence, find $\int \dfrac{1}{(x^2+1)^2} \, dx$. (*AEB*)

9.3 The arc length of a curve

The length of an arc of a curve can be found by using an integral as the limit of a sum, (see *Pure Mathematics* Chapter 20). In Fig. 9.1, let A and B be points on a curve C. Let P and Q be neighbouring points on C such that the arc lengths AP and PQ are respectively s and δs. Then δs is approximately equal to the chord distance $|PQ|$ between P and Q. The whole arc AB can be subdivided into n such elementary arcs PQ and the sum of the lengths of all these chords PQ is $\sum |PQ|$.

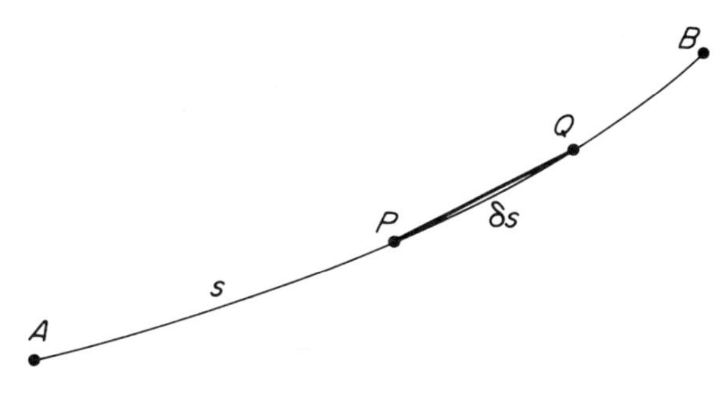

Fig. 9.1

If n tends to infinity as the lengths of all these chords PQ tend to zero, so that the sum of the chord lengths tends to a finite limit, then this sum becomes an integral, with value $s(B)$, the length of the arc AB

$$\lim \sum |PQ| = \lim \sum \delta s = \int_0^{s(B)} \mathrm{d}s = s(B).$$

The evaluation of the integral will depend upon the way in which the curve is described. This may be in terms of Cartesian coordinates, in terms of a parameter, or in terms of polar coordinates.

Arc length in Cartesian coordinates

Let the Cartesian equation of the curve C be $y = \mathrm{f}(x)$ so that A, B are the points $(a, \mathrm{f}(a))$, $(b, \mathrm{f}(b))$ and P, Q the neighbouring points (x, y), $(x+\delta x, y+\delta y)$ on C, as shown in Fig. 9.2. Then the length of the chord PQ is $\sqrt{(\delta x^2 + \delta y^2)}$ and the small element of arc PQ is of length δs, given by $\delta s \approx \sqrt{(\delta x^2 + \delta y^2)}$. As long as $P \neq Q$, $\delta x \neq 0$, and then

$$\frac{\delta s}{\delta x} \approx \sqrt{(\delta x^2 + \delta y^2)}/\delta x = \sqrt{\left\{1 + \left(\frac{\delta y^2}{\delta x^2}\right)\right\}}.$$

In the limit, as $\delta x \to 0$, we obtain the equation

$$\frac{\mathrm{d}s}{\mathrm{d}x} = \sqrt{\left\{1 + \left(\frac{\mathrm{d}y}{\mathrm{d}x}\right)^2\right\}}.$$

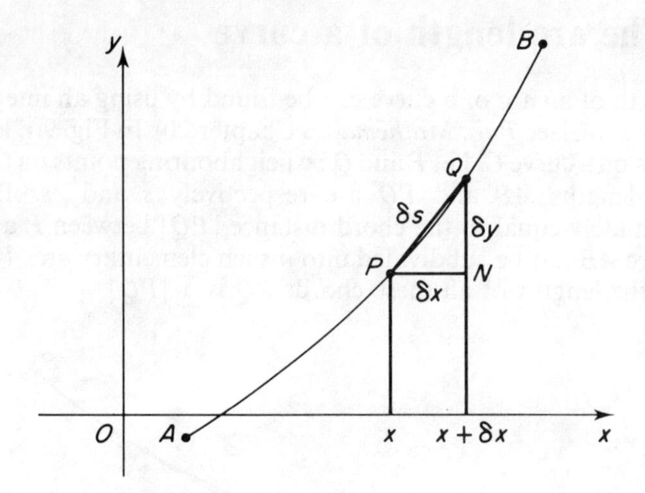

Fig. 9.2

The arc length AB on C is given by

$$AB = s = \int_a^b \frac{ds}{dx}\,dx = \int_a^b \sqrt{\left\{1 + \left(\frac{dy}{dx}\right)^2\right\}}\,dx.$$

EXAMPLE 1 *Find the arc length of the parabola $y = x^2$ between the origin* (0, 0) *and the point B with coordinates* (1, 1).

The arc length OB, denoted by s, is given by

$$s = \int_0^1 \sqrt{\left\{1 + \left(\frac{dy}{dx}\right)^2\right\}}\,dx = \int_0^1 \sqrt{(1+4x^2)}\,dx,$$

since $\dfrac{dy}{dx} = 2x$. In order to evaluate this integral, let $2x = \sinh z$ so that $2\dfrac{dx}{dz} = \cosh z$, and let $a = \operatorname{arsinh} 2$. Then

$$s = \int_0^a \sqrt{(1+\sinh^2 z)}\frac{\cosh z}{2}\,dz = \int_0^a \frac{\cosh^2 z}{2}\,dz$$

$$= \frac{1}{4}\int_0^a (1+\cosh 2z)\,dz = \frac{1}{4}\left[z + \frac{1}{2}\sinh 2z\right]_0^a$$

$$= \frac{1}{4}a + \frac{1}{8}\sinh 2a = \frac{1}{4}a + \frac{1}{4}\sinh a \cosh a$$

$$= \frac{1}{4}\operatorname{arsinh} 2 + \frac{2}{4}\sqrt{(1+4)} = \mathbf{\frac{1}{4}\,arsinh\,2 + \frac{1}{2}\sqrt{5}}.$$

Arc length in terms of a parameter

The coordinates (x, y) of the point P on the curve C may be given in terms of a parameter t, so that

$$x = x(t),\ y = y(t) \text{ and } s = s(t).$$

If the parameter takes the value $t + \delta t$ at a neighbouring point Q on C, with coordinates $(x + \delta x, y + \delta y)$, then

$$\delta x = x(t+\delta t) - x(t) \approx \frac{dx}{dt}\delta t \text{ and } \delta y = y(t+\delta t) - y(t) \approx \frac{dy}{dt}\delta t$$

and then $\delta s^2 \approx \delta x^2 + \delta y^2 \approx \left\{\left(\frac{dx}{dt}\right)^2 + \left(\frac{dy}{dt}\right)^2\right\}\delta t^2.$

In the limit, as $Q \to P$, $\delta t \to 0$ and $\frac{ds}{dt} = \sqrt{\left\{\left(\frac{dx}{dt}\right)^2 + \left(\frac{dy}{dt}\right)^2\right\}}.$

If the points A and B on C are given respectively by $t = \alpha$ and $t = \beta$, then the arc length s along C from A to B is given by

$$s = \int_\alpha^\beta \frac{ds}{dt}\,dt = \int_\alpha^\beta \sqrt{\left\{\left(\frac{dx}{dt}\right)^2 + \left(\frac{dy}{dt}\right)^2\right\}}\,dt.$$

EXAMPLE 2 *Find the total length of the curve given by*

$$x = a\cos^2 t,\ y = a\sin^2 t,\ 0 \leqslant t \leqslant 2\pi.$$

In terms of the parameter t, the arc length s is given by

$$s = \int_0^{2\pi} \sqrt{\left\{\left(\frac{dx}{dt}\right)^2 + \left(\frac{dy}{dt}\right)^2\right\}}\,dt.$$

Now $\frac{dx}{dt} = -2a\cos t\sin t$ and $\frac{dy}{dt} = 2a\sin t\cos t$, so that

$$\left(\frac{dx}{dt}\right)^2 + \left(\frac{dy}{dt}\right)^2 = 8a^2\cos^2 t\sin^2 t = 2a^2\sin^2 2t.$$

Therefore $\frac{ds}{dt} = \sqrt{2}a\,|\sin 2t\,|$ and so

$$s = \sqrt{2}a\int_0^{2\pi} |\sin 2t\,|\,dt = 4\sqrt{2}a\int_0^{\pi/2} \sin 2t\,dt$$

$$= \frac{4\sqrt{2}}{2}a\Big[-\cos 2t\Big]_0^{\pi/2} = \mathbf{4\sqrt{2}a}.$$

Note that we have used the periodic symmetry of the sine function in equating the integral of $|\sin 2t\,|$ from 0 to 2π to four times the integral of $\sin 2t$ from 0 to $\pi/2$.

Arc length in polar coordinates

Referring to Fig. 9.3, the arc length PQ, denoted by δs, is approximately equal to the chord length PQ. Using the right-angle triangle OMQ, where PM is the perpendicular from P on to OQ, $OM = r\cos\delta\theta \approx r$, since, for small θ, $\cos\theta \approx 1$, so $MQ \approx r + \delta r - r = \delta r$.

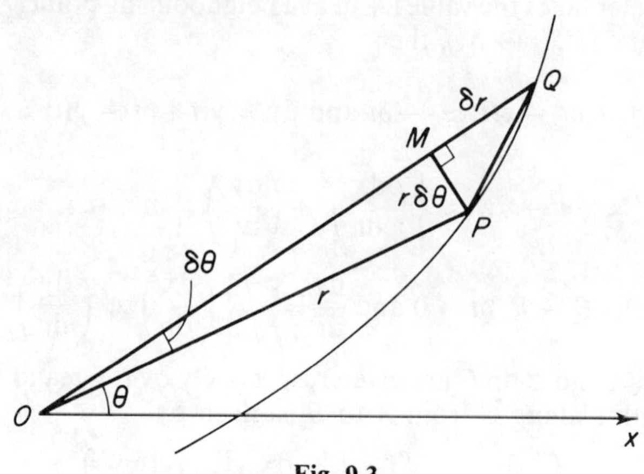

Fig. 9.3

Hence $$PQ^2 = PM^2 + MQ^2,$$

so $$\delta s^2 \approx r^2\delta\theta^2 + \delta r^2$$

and $$\delta s \approx \sqrt{\left\{r^2 + \left(\frac{\delta r}{\delta\theta}\right)^2\right\}}\,\delta\theta.$$

In the limit, as $Q \to P$, $\delta\theta \to 0$, and

$$\frac{\mathrm{d}s}{\mathrm{d}\theta} = \sqrt{\left\{r^2 + \left(\frac{\mathrm{d}r}{\mathrm{d}\theta}\right)^2\right\}},$$

and the length of the arc of the curve between $\theta = \alpha$ and $\theta = \beta$ is s, given by

$$s = \int_\alpha^\beta \sqrt{\left\{r^2 + \left(\frac{\mathrm{d}r}{\mathrm{d}\theta}\right)^2\right\}}\,\mathrm{d}\theta.$$

EXAMPLE 3 *Find the length of the cardioid $r = 2a(1 - \cos\theta)$.*

The length s is given by $s = \displaystyle\int_0^{2\pi} \sqrt{\left\{r^2 + \left(\frac{\mathrm{d}r}{\mathrm{d}\theta}\right)^2\right\}}\,\mathrm{d}\theta.$

Now $\dfrac{\mathrm{d}r}{\mathrm{d}\theta} = 2a\sin\theta$ and so

$$\begin{aligned} r^2 + \left(\frac{\mathrm{d}r}{\mathrm{d}\theta}\right)^2 &= 4a^2(1 + \cos^2\theta - 2\cos\theta + \sin^2\theta) \\ &= 4a^2(2 - 2\cos\theta) = 16a^2\sin^2(\theta/2). \end{aligned}$$

For $0 < \theta < 2\pi$, $\sin(\theta/2) > 0$ so that $\sqrt{\left\{r^2 + \left(\frac{dr}{d\theta}\right)^2\right\}} = 4a\sin(\theta/2)$.

Therefore

$$s = \int_0^{2\pi} 4a\sin(\theta/2)d\theta = \Big[-8a\cos(\theta/2)\Big]_0^{2\pi} = \mathbf{16a}.$$

The gradient of a curve

Let the tangent PT to the curve C meet the axis Ox at T and, as before, let P, Q be neighbouring points (x, y), $(x + \delta x, y + \delta y)$ on C, as shown in Fig. 9.4. Let N be the point $(x + \delta x, y)$ and let the angles PTx and QPN be denoted by ψ and ψ', respectively. Then the gradient of the curve at P is $\tan\psi$ and the gradient of its chord PQ is $\tan\psi'$.

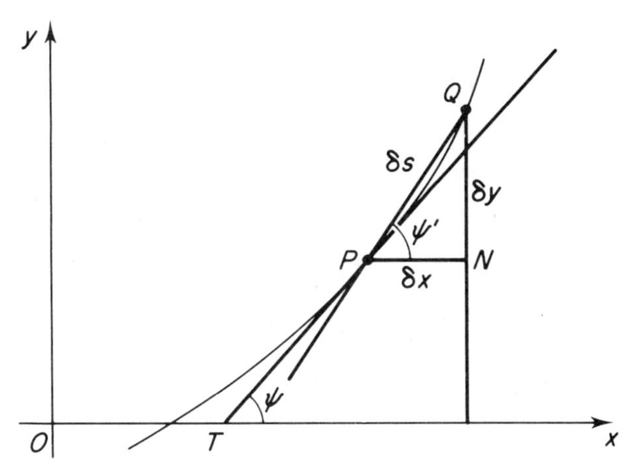

Fig. 9.4

Now

$$\cos\psi' = \frac{\delta x}{|PQ|} \approx \frac{\delta x}{\delta s}, \ \sin\psi' = \frac{\delta y}{|PQ|} \approx \frac{\delta y}{\delta s}, \ \tan\psi' = \frac{\delta y}{\delta x}.$$

In the limit, as $Q \to P$, $\psi' \to \psi$ and

$$\cos\psi = \frac{dx}{ds}, \ \sin\psi = \frac{dy}{ds}, \ \tan\psi = \frac{dy}{dx}.$$

These equations may sometimes be useful in changing the independent variable in an integral from s to x or y, or vice versa.

Angle between tangent and radius vector

In the case of polar coordinates, let ϕ be the angle OPT between the radius vector and the tangent to the curve then, from triangle OPT, $\theta + \phi = \psi$, as is seen in Fig. 9.5. Let PM be the perpendicular from P on to OQ and let

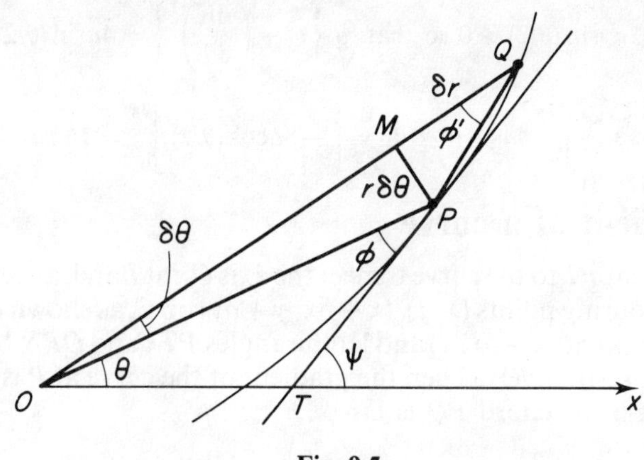

Fig. 9.5

angle MQP be denoted by ϕ'. Then

$$\cos\phi' = \frac{MQ}{PQ} \approx \frac{\delta r}{\delta s}, \quad \sin\phi' = \frac{MP}{PQ} \approx r\frac{\delta\theta}{\delta s} \text{ and}$$

$$\tan\phi' = \frac{MP}{MQ} \approx r\frac{\delta\theta}{\delta r}$$

In the limit, as $Q \to P$, $\phi' \to \phi$ and

$$\cos\phi = \frac{\mathrm{d}r}{\mathrm{d}s}, \quad \sin\phi = r\frac{\mathrm{d}\theta}{\mathrm{d}s}, \quad \tan\phi = r\frac{\mathrm{d}\theta}{\mathrm{d}r}.$$

EXAMPLE 4 *Show that, for the exponential spiral, $r = ae^{b\theta}$, the tangent to the curve at any point P makes a constant angle with the radius OP.*

Calculate $\frac{\mathrm{d}\theta}{\mathrm{d}r}$ as $1\Big/\left(\frac{\mathrm{d}r}{\mathrm{d}\theta}\right)$, and use the above results.

$$\frac{\mathrm{d}r}{\mathrm{d}\theta} = abe^{b\theta} = br \text{ so that } \tan\phi = r\frac{\mathrm{d}\theta}{\mathrm{d}r} = 1/b, \text{ a constant.}$$

EXAMPLE 5 *Show that the cardioid $r = 2a(1 - \cos\theta)$ touches the initial line Ox at the origin in a cusp.*

Since $r = 2a(1-\cos\theta)$, $\frac{\mathrm{d}r}{\mathrm{d}\theta} = 2a\sin\theta$,

$$r\frac{\mathrm{d}\theta}{\mathrm{d}r} = \frac{1-\cos\theta}{\sin\theta} = \frac{2\sin^2(\theta/2)}{2\sin(\theta/2)\cos(\theta/2)} = \tan(\theta/2).$$

Therefore, $\phi = \theta/2$ and, when $\theta = 0$, $\psi = \phi = 0$ and the tangent at the origin is the initial line. Since the curve approaches the origin from the right, both as θ tends to zero and as θ tends to 2π, and touches Ox at O, it has a cusp at O.

EXERCISE 9.3

1 Find the length of the curve given by: (i) $2y = \cosh 2x$, $0 \leqslant x \leqslant 3$, (ii) $3y = 2x^{3/2}$, $3 \leqslant x \leqslant 8$, (iii) $y = \ln(\sec x)$, $0 \leqslant x \leqslant \pi/4$, (iv) $y = \ln(x^2 - 1)$, $2 \leqslant x \leqslant 3$, (v) $y = \ln x$, $1 \leqslant x \leqslant 4$.

2 The curve C is given by parametric equations $x = a\cos^3 t$, $y = a\sin^3 t$, $0 \leqslant t \leqslant \pi/3$. Find the gradient of the curve at the point P, given by $t = p$. Prove that the angle ψ, made by the tangent at P with the axis Ox, is given by $\psi = \pi - p$. Calculate the length of the curve C.

3 Find the length of the curve given by parametric equations:
(i) $x = a\cos t$, $y = a\sin t$, $0 \leqslant t \leqslant \pi/3$,
(ii) $x = at^2$, $y = 4at$, $0 \leqslant t \leqslant 1$,
(iii) $x = t^3$, $y = at^2$, $0 \leqslant t \leqslant 2a$,
(iv) $x = a\cos^3 t$, $y = a\sin^3 t$, $0 \leqslant t \leqslant \pi/2$,
(v) $x = \tanh t$, $y = \operatorname{sech} t$, $0 \leqslant t \leqslant 1$,
(vi) $x = a(\theta - \sin\theta)$, $y = a(1 - \cos\theta)$, $0 \leqslant \theta \leqslant 2\pi$.

4 Find the length of the curve given, in polar coordinates, by:
(i) $r = a\theta$, $0 \leqslant \theta \leqslant 2\pi$,
(ii) $r = ae^{k\theta}$, $0 \leqslant \theta \leqslant 3$,
(iii) $r = a(1 + \cos\theta)$, $0 \leqslant \theta \leqslant \pi$,
(iv) $r = a\sin^3(\theta/3)$, $0 \leqslant \theta \leqslant 3\pi$.

9.4 The area of a surface of revolution

It is also possible to use an integral to define the area of a curved surface as a limit of a sum. In general, this is very difficult because a double integral is involved since a surface is two dimensional. This means that there are two limits of a double infinite sum to be considered and, unless great care is taken, different results can be obtained for the area by taking the limits in alternative ways.

However, the situation is straightforward in the case of a surface of revolution. Let C be the curve AB in the plane Oxy and let S be the area of the curved surface produced by rotating C through 2π about Ox. We shall assume that the whole of C lies in the region $y \geqslant 0$, see Fig. 9.6. Let $P(x, y)$ and $Q(x + \delta x, y + \delta y)$ be neighbouring points on C, with an arc length δs between P and Q. Let δS be the contribution to S from the rotation of the arc PQ about Ox. The arc PQ can be approximated by the chord PQ so that δS is approximately equal to the area of the frustum of a circular cone, as shown in Fig. 9.7.

Let the line PQ meet the axis Ox at V and make an angle γ there with Ox. Consider the two cones, with common vertex V and common axis Ox, formed by rotating the lines VP and VQ through 2π about Ox. The slant heights of these cones are VP and VQ respectively and their circular bases have radii y and $y + \delta y$, respectively. If these two cones are cut along a generator and flattened they form segments of circles of radii VP and VQ and with bounding arcs of lengths $2\pi y$ and $2\pi(y + \delta y)$ respectively. The

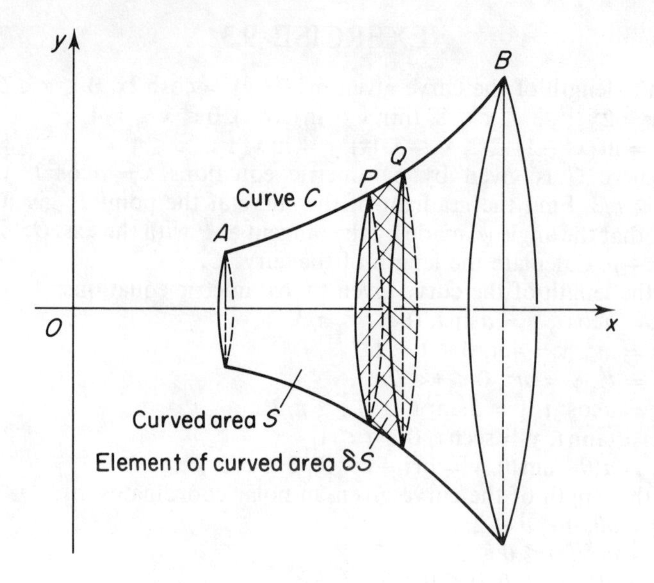

Fig. 9.6

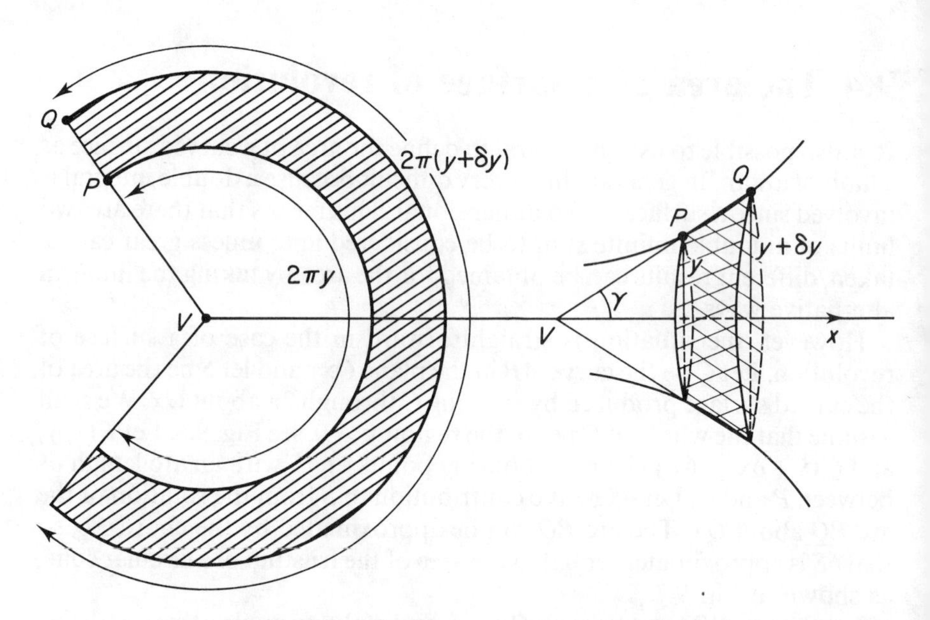

Fig. 9.7 Area of elementary curved surface δS

area of the smaller cone is therefore $\dfrac{2\pi y}{2\pi VP}\pi VP^2$, which can be written $\pi VP^2 \sin\gamma$. Similarly, the area of the larger cone is $\pi VQ^2 \sin\gamma$. The difference between these areas is the area of the required frustum and is approximately equal to δS, so

$$\begin{aligned}\delta S &\approx \pi\sin\gamma(VQ^2 - VP^2) = \pi\sin\gamma(VQ+VP)(VQ-VP)\\ &= \pi\sin\gamma(2VP+PQ)(PQ) = 2\pi yPQ + \pi\sin\gamma PQ^2\\ &= 2\pi y\delta s + \pi\sin\gamma\delta s^2.\end{aligned}$$

Therefore $$\frac{\delta S}{\delta s} \approx 2\pi y + \pi\sin\gamma\delta s$$

and, in the limit as $\delta s \to 0$, $\dfrac{\mathrm{d}S}{\mathrm{d}s} = 2\pi y$.

The area S of the curved surface of revolution is therefore given by

$$\int_{s(A)}^{s(B)} 2\pi y \mathrm{d}s.$$

In Cartesian coordinates, if $x = a$ at A and $x = b$ at B,

$$\frac{\mathrm{d}s}{\mathrm{d}x} = \sqrt{\left\{1+\left(\frac{\mathrm{d}y}{\mathrm{d}x}\right)^2\right\}} \text{ and so } S = \int_a^b 2\pi y\sqrt{\left\{1+\left(\frac{\mathrm{d}y}{\mathrm{d}x}\right)^2\right\}}\mathrm{d}x.$$

In parametric form, if $x = x(t)$, $y = y(t)$, and if A and B are given by $t = \alpha$ and $t = \beta$, then

$$\frac{\mathrm{d}s}{\mathrm{d}t} = \sqrt{\left\{\left(\frac{\mathrm{d}x}{\mathrm{d}t}\right)^2 + \left(\frac{\mathrm{d}y}{\mathrm{d}t}\right)^2\right\}}$$

and so $$S = \int_\alpha^\beta 2\pi y\sqrt{\left\{\left(\frac{\mathrm{d}x}{\mathrm{d}t}\right)^2 + \left(\frac{\mathrm{d}y}{\mathrm{d}t}\right)^2\right\}}\mathrm{d}t.$$

In terms of polar coordinates (r, θ), $x = r\cos\theta$, $y = r\sin\theta$. If A and B are given by $\theta = \alpha$ and $\theta = \beta$, respectively, then

$$\frac{\mathrm{d}s}{\mathrm{d}\theta} = \sqrt{\left\{r^2+\left(\frac{\mathrm{d}r}{\mathrm{d}\theta}\right)^2\right\}} \text{ and so } S = \int_\alpha^\beta 2\pi r\sin\theta\sqrt{\left\{r^2+\left(\frac{\mathrm{d}r}{\mathrm{d}\theta}\right)^2\right\}}\mathrm{d}\theta.$$

EXAMPLE 1 *The arc C, given by $y = \mathrm{e}^{-x}, 0 < x < \ln(2)$, is rotated through one complete revolution about the axis Ox. Find the area of the surface of revolution so generated.*

$\dfrac{\mathrm{d}y}{\mathrm{d}x} = -\mathrm{e}^{-x}$, so that the area S is given by $S = \displaystyle\int_0^{\ln 2} 2\pi\mathrm{e}^{-x}\sqrt{(1+\mathrm{e}^{-2x})}\,\mathrm{d}x$.

Let $u = \mathrm{e}^{-x}$, so that $\dfrac{\mathrm{d}u}{\mathrm{d}x} = -\mathrm{e}^{-x}$ and $S = -\displaystyle\int_1^{1/2} 2\pi\sqrt{(1+u^2)}\,\mathrm{d}u$.

Now let $u = \sinh\theta$, $\dfrac{\mathrm{d}u}{\mathrm{d}\theta} = \cosh\theta$ and, if $a = \operatorname{arsinh}(1/2)$, $b = \operatorname{arsinh}(1)$, then

$\cosh a = \sqrt{5}/2$, $\cosh b = \sqrt{2}$, $a = \ln\left(\dfrac{1+\sqrt{5}}{2}\right)$, $b = \ln(1+\sqrt{2})$ and

$$S = \int_a^b 2\pi\cosh^2\theta\,d\theta = \int_a^b \pi(1+\cosh 2\theta)d\theta = \left[\pi\theta + \frac{\pi}{2}\sinh 2\theta\right]_a^b$$
$$= \pi(b-a) + \pi(\sinh b\cosh b - \sinh a\cosh a)$$
$$= \boldsymbol{\pi\ln\left(\frac{2+2\sqrt{2}}{1+\sqrt{5}}\right) + \pi(\sqrt{2} - \sqrt{5}/4)}.$$

EXAMPLE 2 *Find the area of the surface of revolution formed by the rotation through 2π about the axis Ox of the curve given by the parametric equations $x = a\sin^3\theta$, $y = a\cos^3\theta$, $0 \leqslant \theta \leqslant \pi/2$.*

$$\frac{dx}{d\theta} = 3a\sin^2\theta\cos\theta, \quad \frac{dy}{d\theta} = -3a\cos^2\theta\sin\theta,$$

so

$$\frac{ds}{d\theta} = 3a\sin\theta\cos\theta\sqrt{(\sin^2\theta + \cos^2\theta)} = 3a\sin\theta\cos\theta,$$

and the area S of the surface of revolution is given by

$$S = \int_0^{\pi/2} 2\pi a\cos^3\theta\, 3a\sin\theta\cos\theta\,d\theta = 6\pi a^2 \int_0^{\pi/2} \cos^4\theta\sin\theta\,d\theta$$
$$= \frac{6}{5}\pi a^2\left[-\cos^5\theta\right]_0^{\pi/2} = \boldsymbol{\frac{6}{5}\pi a^2}.$$

EXAMPLE 3 *The cardioid, given by $r = a(1+\cos\theta)$, $0 < \theta < 2\pi$, is rotated through an angle π about the initial line. Find the area of the curved surface generated.*

That half of the curve, given by $0 < \theta < \pi$, is rotated about the initial line through an angle 2π to give a curved surface, of area S.

Since

$$\frac{dr}{d\theta} = -a\sin\theta, \quad \left(\frac{ds}{d\theta}\right)^2 = r^2 + \left(\frac{dr}{d\theta}\right)^2 = a^2(1 + 2\cos\theta + \cos^2\theta + \sin^2\theta)$$
$$= a^2(2 + 2\cos\theta) = 4a^2\cos^2(\theta/2),$$

$$S = \int_0^{\pi} 2\pi a(1+\cos\theta)\sin\theta\, 2a\cos(\theta/2)\,d\theta = \int_1^0 2\pi a^2(2c^2)4c^2(-2)\,dc,$$

using the substitution $c = \cos(\theta/2)$. Thus $S = 32\pi a^2\left[c^5/5\right]_0^1 = \boldsymbol{32\pi a^2/5}$.

Pappus Theorems

There is a useful connection between an area of revolution and the position of the centroid of its generating curve and, similarly, between a volume of revolution and the position of the centroid of its generating area. If the position of the centroid is known, then the area, or volume, of revolution can be found. Conversely, if the area (or volume) of revolution is known, then the distance of the centroid from the axis of revolution can be found.

Theorem 9.1 (Pappus)
(i) Let a surface of revolution, of area S, be generated by rotating through 2π a plane curve C about an axis in its plane, not meeting C. Then S is the product of the length of C and the length of the path of the centroid of C.
(ii) Let a volume of revolution, of volume V, be generated by rotating through 2π a plane area S about an axis in its plane, not meeting S. Then V is the product of the area of S and the length of the path of the centroid of S.

Proof Take the plane of the generating curve (or area) to be $O(x, y)$ and let Ox be the axis of revolution. We use the results on centroids and volumes of revolution from Chapter 20 of *Pure Mathematics* and the results on areas of revolution from earlier work in this chapter.
(i) Let P be a point (x, y) on the curve C, see Fig. 9.8 (i). Then, if the length of C is c and its centroid is at $(\bar{x}, \bar{y})$,

$$\bar{y}c = \int y\,ds, \text{ the integral being taken over } C.$$

Then the length of the path of the centroid is $2\pi\bar{y}$ and

$$2\pi\bar{y}c = \int 2\pi y\,ds = S,$$

where S is the area of the surface of revolution.

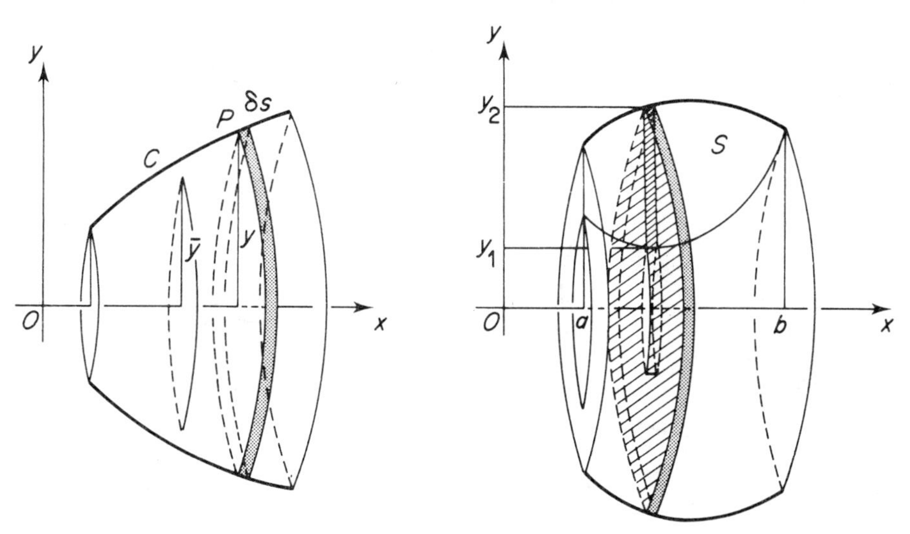

Fig. 9.8 (i) Surface of revolution. (ii) Volume of revolution

(ii) Let the generating area be S with centroid at $(\bar{x}, \bar{y})$. Using the notation indicated in Fig. 9.8 (ii),

$$S = \int_a^b (y_2 - y_1)\,dx.$$

The centroid of the strip, of length $(y_2 - y_1)$ and of area $(y_2 - y_1)\delta x$, is at the point $(x + \frac{1}{2}\delta x, \frac{1}{2}(y_2 + y_1))$, so that, on taking first moments about Ox,

$$\bar{y}S = \int_a^b \tfrac{1}{2}(y_2 + y_1)(y_2 - y_1)\,dx = \int_a^b \tfrac{1}{2}(y_2^2 - y_1^2)\,dx.$$

The length of the path of $(\bar{x}, \bar{y})$ is $2\pi\bar{y}$, so that

$$2\pi\bar{y}S = \int_a^b \pi(y_2^2 - y_1^2)\,dx = V,$$

the volume of revolution.

EXAMPLE 4 *A torus is formed by rotating a circle, of radius b, through an angle* 2π, *about an axis in its plane distance a from its centre, where* $a > b$. *Find (i) the surface area of the torus, (ii) the volume of the torus.*

The generating circle, shown in Fig. 9.9 (a), lies in the plane $O(x, y)$, with its centre C at a distance a from Ox, the axis of revolution. Then C is the centroid of both the circular disc and of the bounding circle C. The path of the centroid C is of length $2\pi a$. The area S of the surface of revolution (the torus) is thus given by

$$S = 2\pi a \,.\, 2\pi b = \mathbf{4\pi^2 ab}.$$

Similarly, the volume V of the torus is given by

$$V = 2\pi a \,.\, \pi b^2 = \mathbf{2\pi^2 ab^2}.$$

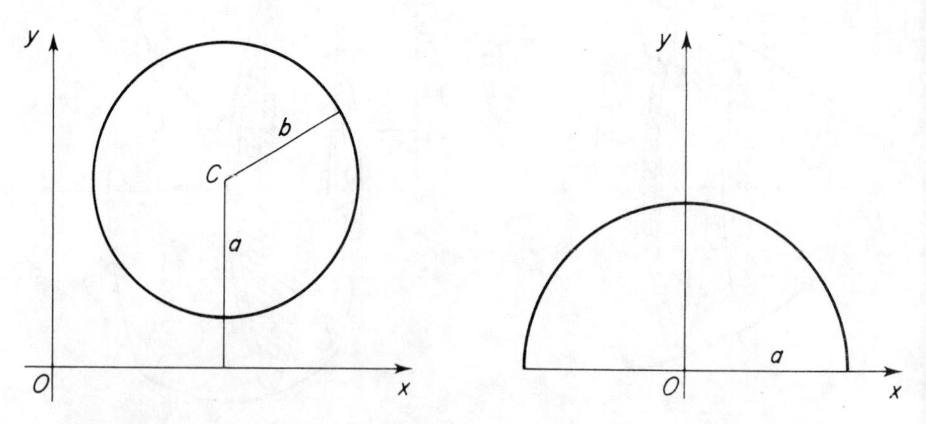

Fig. 9.9 (a) Generating circle of a torus. (b) Generating semicircle of a sphere

EXAMPLE 5 *Find the position of the centroid of: (i) a semicircular arc of radius a, (ii) a semicircular lamina of radius a.*

In Fig. 9.9 (b), let the semicircle lie in the region $y > 0$ with its centre at the origin of the x-y plane. The centroids of both the semicircular arc and the semicircular area both lie on the axis Oy and, on rotation through 2π about Ox, a spherical surface of area $4\pi a^2$ and a sphere of volume $4\pi a^3/3$ are generated.
(i) Let the centroid of the arc be at $(0, \bar{y})$. The length of the arc is πa so that $2\pi\bar{y}.\pi a = 4\pi a^2$ and $\bar{y} = \mathbf{2a/\pi}$.
(ii) Let the centroid of the semicircular area be at $(x, \bar{y})$. Its area is $\pi a^2/2$ so that $2\pi\bar{y}(\pi a^2/2) = 4\pi a^3/3$ and $\bar{y} = \mathbf{4a/3\pi}$.

EXERCISE 9.4

1 Find the area of the curved surface of revolution formed by rotating the given curve about the axis Ox through four right-angles:
(i) $y = e^x, 0 \leqslant x \leqslant 1$, (ii) $y = x^3, 0 \leqslant x \leqslant 1/2$.

2 A curve C is given parametrically, in terms of a parameter t. The curve is rotated through 2π about the axis Ox. Find the area of the curved surface of revolution so generated:

(i) $x = \operatorname{sech} t$,	$y = \tanh t$,	$0 \leqslant t \leqslant 1$,
(ii) $x = 2t^3$,	$y = 3t^2$,	$0 \leqslant t \leqslant 1$,
(iii) $x = t^3 - 3t$,	$y = 3t^2$,	$0 \leqslant t \leqslant 1$,
(iv) $x = 3\cos t - \cos 3t$,	$y = 3\sin t - \sin 3t$,	$0 \leqslant t \leqslant \pi/2$,
(v) $x = \tan t - t$,	$y = \ln(\sec t)$,	$0 \leqslant t \leqslant 1$,
(vi) $x = e^t \cos t$,	$y = e^t \sin t$,	$0 \leqslant t \leqslant \pi$.

3 For the curve with given polar equation, find the area of the curved surface generated by one complete revolution of the curve about the initial line:

(i) $r^2 = a^2\cos 2\theta$, $0 \leqslant \theta \leqslant \pi/4$,
(ii) $r = a\sec^2(\theta/2)$, $0 \leqslant \theta \leqslant \pi/2$,
(iii) $r = a(1 + \sin\theta)$, $0 \leqslant \theta \leqslant \pi/2$, $\{$hint, $(\sin\theta/2 + \cos\theta/2)^2 = 1 + \sin\theta\}$.

4 The area in the Cartesian plane bounded by the curves $x = 0$, $x = 1$ and $x = t^3, y = 3t^2, 0 \leqslant t \leqslant 1$, is rotated about the line $y = 0$ through the angle 2π to form a solid of revolution. Find the volume and the total surface area of the solid.

5 The arc of the curve $y^2 = 4ax$, for which $y > 0$ and $a \leqslant x \leqslant 4a$, is rotated through 2π about the x-axis. Find the area of the curved surface so generated. *(C)*

6 An arc of a circle of radius a subtends an angle 2α at the centre of the circle, where α is acute. The arc is rotated through the angle 2π about its bounding chord. Find the area of the surface which this generates.

7 The mirror of a telescope is a parabolic dish. Its shape is that surface formed by rotating a symmetrical arc of a parabola through π about its axis. The boundary of the mirror is a circle of radius a and the depth of the vertex below the plane of the bounding circle is h. Given that $5a = 24h$, show that the generating parabola can be given by the Cartesian equation $25y^2 = 576hx$. Find the area of the mirror.

8 Find the area of the ellipsoid of revolution formed by rotating the ellipse $x = a\cos\theta$, $y = b\sin\theta$, $0 \leqslant \theta \leqslant 2\pi$, through one half revolution about Ox.

9 The hyperbolic arc, given by $x^2 = y^2 + a^2$, $0 \leqslant y$, $a \leqslant x \leqslant \dfrac{a\cosh 2}{\sqrt{2}}$, is rotated through 2π about Ox. Find the area of the surface generated.

10 A square of side a is rotated through 2π about one side.
(i) Write down the volume of the cylinder so generated and the length of the path of the centroid of the square and verify Pappus Theorem in this case.
(ii) Write down the total surface area of the cylinder generated and use Pappus Theorem to find the distance of the centroid of the curve consisting of three sides of a square from the fourth side.

11 Use Pappus Theorem to find the position of the centroid of a quadrant of a circle of radius a, by rotating the quadrant through 2π about a bounding radius.

12 A ring is formed by rotating a plane circular region of radius a through 2π about a line in its plane distance $2a$ from its centre. Show that the volume of the ring is $4\pi^2 a^3$ and find its surface area.

13 Find the volume of a napkin ring, formed by rotating a semicircular area of radius a through 2π about a line in its plane, parallel to and distance a from the bounding diameter.

14 Find the surface area of the napkin ring of Question **13**.

9.5 Infinite and improper integrals

Infinite integrals

It is sometimes possible to attach a meaning to a definite integral in which one of the limits of integration is infinite. Thus, if

$$\int_a^N f(x)\,dx = F(N) \text{ and the limit } \lim_{N\to\infty} F(N) \text{ exists and is equal to } I,$$

then we say that the infinite integral $\displaystyle\int_a^\infty f(x)\,dx$ exists or, alternatively, that the integral *converges*. We then write

$$\int_a^\infty f(x)\,dx = \lim_{N\to\infty}\int_a^N f(x)\,dx = \lim_{N\to\infty} F(N) = I.$$

EXAMPLE 1 *Determine which of the integrals converge and evaluate those which do*:

(i) $\displaystyle\int_0^\infty \frac{1}{x^2+1}\,dx$, *(ii)* $\displaystyle\int_0^\infty \frac{2x-2}{(x+1)(x^2+1)}\,dx$, *(iii)* $\displaystyle\int_0^\infty \frac{1}{x+1}\,dx$.

(i) We write $F(t) = \displaystyle\int_0^t \frac{1}{x^2+1}\,dx$, then $F(t) = \Big[\arctan x\Big]_0^t = \arctan t$. Now $\displaystyle\lim_{t\to\infty} F(t) = \pi/2$ and so $\displaystyle\int_0^\infty \frac{1}{x^2+1}\,dx = \boldsymbol{\pi/2}$.

(ii) Let $F(t) = \int_0^t \frac{2x-2}{(x+1)(x^2+1)}\,dx$, and evaluate this integral by using partial fractions.

$$\frac{2x-2}{(x+1)(x^2+1)} = \frac{2x}{x^2+1} - \frac{2}{x+1},$$

so that $F(t) = \Big[\ln(x^2+1) - 2\ln(x+1)\Big]_0^t = \ln\frac{t^2+1}{(t+1)^2}$.

Now $F(t) = \ln\left[\left(1+\frac{1}{t^2}\right)\Big/\left(1+\frac{1}{t}\right)^2\right]$, and so $\lim_{t\to\infty} F(t) = \ln(1) = 0$.

Therefore $\int_0^\infty \frac{2x-2}{(x+1)(x^2+1)}\,dx = \lim_{t\to\infty}\int_0^t \frac{2x-2}{(x+1)(x^2+1)}\,dx = \mathbf{0}$.

(iii) Let $F(t) = \int_0^t \frac{1}{x+1}\,dx = \Big[\ln(x+1)\Big]_0^t = \ln(t+1)$.

As $t \to \infty$, $F(t) \to \infty$, so the infinite integral does not exist and we say that $\int_0^t \frac{1}{x+1}\,dx$ does **not converge** (or that it diverges).

Improper integrals

There are occasions where the integrand of a definite integral is undefined at some point, $x = c$, on the range of integration. Usually this happens because the integrand tends to infinity as x tends to c. The critical point, $x = c$, may be at one of the limits of integration or between the two limits. When c is the upper limit of integration, the integral may be regarded as a limit in the following way.

Let the integral be $\int_a^c f(x)\,dx$ and let $f(c)$ be undefined.

Let $F(d) = \int_a^d f(x)\,dx$, where $a < d < c$, and consider the limit of $F(d)$ as d tends to c. This limit may, or may not, exist. If the limit exists and $D = \lim_{d\to c} F(d)$, then we say that $D = \int_a^c f(x)\,dx$, or that $\int_a^c f(x)\,dx$ converges to D. If the limit does not exist, say if $\lim_{d\to c} F(d) = \infty$, then we say that $\int_a^c f(x)\,dx$ diverges.

The case when c is the lower limit of integration is dealt with similarly. When $a < c < b$, divide the integral into two integrals $\int_a^c f(x)\,dx$ and $\int_c^b f(x)dx$ so that the problem is reduced to one of the cases where c is a limit of integration.

We begin with some examples where the critical point is one of the limits of integration.

EXAMPLE 2 *Evaluate the integral: (i)* $\int_0^1 \frac{1}{\sqrt{(1-x)}}\,dx$, *(ii)* $\int_0^1 \frac{1}{1-x}\,dx$.

(i) Since the integrand does not exist at $x = 1$, consider $\lim_{d\to 1} F(d)$, where $F(d) = \int_0^d \frac{1}{\sqrt{(1-x)}}\,dx$. Now $F(d) = -2\Big[\sqrt{(1-x)}\Big]_0^d = 2-2\sqrt{(1-d)}$.

Clearly, $\lim_{d\to 1} F(d) = 2$ and so $\int_0^1 \frac{1}{\sqrt{(1-x)}}\,dx = \mathbf{2}$.

(ii) Let $G(d) = \int_0^d \frac{1}{1-x}\,dx$, so that $G(d) = -\Big[\ln(1-x)\Big]_0^d = -\ln(1-d)$. As d tends to 1, $\ln(1-d)$ tends to negative infinity so the integral does not exist and $\int_0^1 \frac{1}{1-x}\,dx$ **diverges.**

When the critical point occurs in the middle of the range of integration, the integral must be split into two parts.

EXAMPLE 3 *Evaluate the integral: (i)* $\int_0^2 \frac{1}{\sqrt{|1-x|}}\,dx$, *(ii)* $\int_0^2 \frac{1}{1-x}\,dx$.

(i) Consider the two integrals $\int_0^1 \frac{1}{\sqrt{|1-x|}}\,dx$, and $\int_1^2 \frac{1}{\sqrt{|1-x|}}\,dx$, and let $F(d) = \int_0^d \frac{1}{\sqrt{|1-x|}}\,dx$ and $G(e) = \int_e^2 \frac{1}{\sqrt{|1-x|}}\,dx$ then, as in example 2,

$$F(d) = 2-2\sqrt{(1-d)}, \qquad \text{whilst} \qquad G(e) = 2\Big[\sqrt{(x-1)}\Big]_e^2 = 2-2\sqrt{(e-1)}.$$

Then $\int_0^2 \frac{1}{\sqrt{|1-x|}}\,dx = \lim_{d\to 1-} F(d) + \lim_{e\to 1+} G(e) = \mathbf{4}$.

(ii) Let $F(d) = \int_0^d \frac{1}{1-x}\,dx$, $G(e) = \int_e^2 \frac{1}{1-x}\,dx$, so that $F(d) = -\Big[\ln(1-x)\Big]_0^d$ $= -\ln(1-d)$ and $G(e) = -\Big[\ln(x-1)\Big]_e^2 = \ln(e-1)$. Now $\lim_{d\to 1} F(d) = +\infty$ and $\lim_{e\to 1} G(e) = -\infty$, so the integral **diverges.**

EXERCISE 9.5

1 Prove that $\int_1^N \frac{1}{x(x+1)}\,dx = \ln\left(\frac{2N}{N+1}\right)$ and hence show that the integral $\int_0^\infty \frac{1}{x(x+1)}\,dx$ converges to $\ln(2)$.

2 Evaluate $\int_0^N xe^{-3x}\,dx$. Use the fact that xe^{-x} tends to zero as x tends to infinity to show that $\int_0^\infty xe^{-3x}\,dx = 1/9$.

3 Prove that $\int_0^\infty x^2e^{-x}\,dx = 2$.

4 Show that $\int_a^b \frac{1}{\sqrt{(x-x^2)}}\,dx = \sin^{-1}(2b-1)-\sin^{-1}(2a-1)$, for a and b satisfying the inequalities $0 < a < b < 1$. Deduce that $\int_0^1 \frac{1}{\sqrt{(x-x^2)}}\,dx = \pi$.

5 Given that $0 < a < 1 < b < 5$, evaluate the two integrals $\int_0^a \frac{1}{\sqrt{(1-x)}}\,dx$ and $\int_b^5 \frac{1}{\sqrt{(x-1)}}\,dx$. Deduce that the integral $\int_0^5 \frac{1}{\sqrt{(|1-x|)}}\,dx$ converges to the value 6.

6 Paying due attention to the question of convergence, evaluate:

(i) $\int_0^{32} x^{-1/5}\,dx$, (ii) $\int_3^\infty \frac{1}{x^2+2x+17}\,dx$,

(iii) $\int_0^1 \frac{1}{\sqrt{(5-4x-x^2)}}\,dx$, (iv) $\int_2^\infty \frac{4}{x^4-1}\,dx$,

(v) $\int_1^{1\cdot5} \frac{1}{\sqrt{(6x-4x^2)}}\,dx$.

MISCELLANEOUS EXERCISE 9

1 By means of the substitution $u = 1+\cosh x$, or otherwise, evaluate

$$\int_0^{\text{arcosh}\,2} \frac{\tanh x}{1+\cosh x}\,dx.$$ (*L*)

2 Given that $y = \frac{6x-10}{(x-3)^2(x+1)}$, express y in partial fractions. Determine the mean value of y with respect to x in the range $x = 4$ to $x = 6$. Find the expansion of y in ascending powers of x up to and including the term in x^2. (*AEB*)

3 Find $\int \frac{1}{\sqrt{(9x^2-12x-1)}}\,dx$.

4 Given that $I_n = \int_0^1 x^n\sqrt{(1-x^2)}\,dx$, $(n \geqslant 0)$, show that $(n+2)I_n = (n-1)I_{n-2}$, $(n \geqslant 2)$. Evaluate I_5. (*JMB*)

5 (i) If $I_n = \int_0^1 x^n e^x\,dx$, where n is a positive integer, express I_n in terms of I_{n-1}. Express I_5 in terms of e.
(ii) By using the substitution $x = 1/y$, or otherwise, show that $\int_{\sqrt{2}}^{2} \frac{1}{x\sqrt{(x^2-1)}}\,dx = \pi/12$. (*L*)

6 Prove that the area of the region between the curve whose equation is $y = a\cosh(x/a)$, the x-axis and any two ordinates is equal to as, where s is the length of the arc of the curve cut off by the two ordinates. Prove also that the volume of the solid swept out when this region is rotated once about the x-axis is equal to $aS/2$, where S is the curved surface area of the solid. Find in terms of a, e and π, the *total* surface area of the solid when the ordinates are $x = 0$ and $x = a$. (*JMB*)

7 (a) Prove that $\int_0^{\ln 2} x\sinh x\,dx = (5\ln 2 - 3)/4$.
(b) Find $\int \frac{1}{\sqrt{(x^2+4x-5)}}\,dx$, giving your answer in logarithmic form. (*C*)

8 Assuming the formula for the surface area of a sphere, deduce by a theorem of Pappus the distance of the centroid of a uniform semicircular arc of radius a from its bounding diameter. (*L*)

9 Find, in a simplified form, the derivative with respect to u of $\ln \frac{\sinh u}{1+\cosh u}$.
C is the curve with equation $y = \ln x$. Find the equation of the tangent to C at the point $(p, \ln p)$. Hence find the equation of the tangent to C which passes through the origin O. Sketch the curve C and its tangent from O. Show that the length of the arc of the curve C between the point where C cuts the x-axis and the point where the tangent from O meets C is given by $\int_1^{e} \frac{\sqrt{(1+x^2)}}{x}\,dx$.
Using the substitution $x = \sinh u$, or otherwise, evaluate this integral. (*JMB*)

10 (i) If $I_n = \int_0^x \frac{t^n}{\sqrt{(1+t^2)}}\,dt$, show that for $n > 1$,

$$nI_n + (n-1)I_{n-2} = x^{n-1}\sqrt{(1+x^2)}.$$

(ii) By means of the substitution $t = u^2$, find $\int_0^x \ln(t+\sqrt{t})\,dt$. (*L*)

11 Show that the area of the region enclosed by the two parabolas $y^2 = x$ and $x^2 = y$ is $1/3$. Find the first moment of this area about the y-axis. Show that the length of the boundary of this area is $\int_0^1 2\sqrt{(1+4x^2)}\,dx$. Use the substitution $x = (\sinh u)/2$ to show that the length of the boundary is $\sqrt{5} + \frac{1}{2}\ln(2+\sqrt{5})$. (*AEB*)

12 (a) For the curve whose equation is $y = x^2$ calculate the length of arc between the ordinates $x = 0$ and $x = \frac{1}{2}\sinh 2$, giving your answer in hyperbolic form. (b) Sketch the curve whose equation in polar coordinates is $r = a(1+\cos\theta)$, where a is a positive constant and $0 \leqslant \theta \leqslant 2\pi$. Calculate the total length of the curve. (*C*)

13 Given that $f(x) = \dfrac{x+1}{(x-1)(x^2+1)}$, (a) express $f(x)$ in partial fractions, (b) evaluate $\displaystyle\int_2^3 f(x)\,dx$ leaving your answer in terms of natural logarithms, (c) find an expression for $f(x)$ as a series of ascending powers of x up to and including the term in x^6. (*L*)

14 Use Pappus Theorem to prove that (i) the centroid of a circular arc, of radius a, subtending an angle 2θ at the centre, is distance $(a\sin\theta)/\theta$ from the centre, (ii) the centroid of a circular sector, of radius a, subtending an angle 2θ at the centre, is distance $(2a\sin\theta)/3\theta$ from the centre.

15 Given that $I_n = \displaystyle\int_0^1 x^n \cosh x\,dx$, prove that, for $n \geqslant 2$, $I_n = \sinh 1 - n\cosh 1 + n(n-1)I_{n-2}$. Evaluate (i) $\displaystyle\int_0^1 x^4 \cosh x\,dx$, (ii) $\displaystyle\int_0^1 x^3 \cosh x\,dx$, expressing each answer in terms of e. (*JMB*)

16 Given that $I_n = \displaystyle\int_0^{\pi/2} \sin^n\theta\,d\theta$, prove that, for $n \geqslant 2$, $nI_n = (n-1)I_{n-2}$. Hence, or otherwise, evaluate (i) $\displaystyle\int_0^{\pi/2} \sin^6\theta\,d\theta$, (ii) $\displaystyle\int_0^{\pi} \sin^5(\theta/2)\,d\theta$, (iii) $\displaystyle\int_{-\pi/2}^{\pi/2} \sin^7\theta\,d\theta$, (iv) $\displaystyle\int_0^{\pi/2} \sin^8\theta\,d\theta$. (*C*)

17 (i) Given that $I_n = \int \operatorname{cosec}^n x\,dx$, show that $(n-1)I_n = -\cot x \operatorname{cosec}^{n-2} x + (n-2)I_{n-2}$. (ii) Find, to three significant figures, the area of the region bounded by the curve $y = \operatorname{cosec}^3 x$, the x-axis and the lines $x = \pi/4$, $x = \pi/2$. (iii) The region described in part (ii) is rotated through 2π about the x-axis. Find the volume of revolution so obtained. (*C*)

18 Show that the length of the arc of the curve $y = x^2$ between $x = 0$ and $x = 1$ is given by $\displaystyle\int_0^1 \sqrt{(1+4x^2)}\,dx$. Hence, using the substitution $2x = \sinh\theta$, or otherwise, find this length. If this arc is rotated completely about the x-axis find the area of the surface of revolution. (*AEB*)

19 (i) Show that the mean value with respect to x of the function $2\sqrt{x}$ over the interval from $x = 1$ to $x = 3$ is $(6\sqrt{3}-2)/3$. A and B are the points $(1, 2)$ and $(3, 2\sqrt{3})$ respectively on the curve $y^2 = 4x$. Find the area of the surface generated when the arc AB of the curve is rotated through 2π radians about the x-axis, leaving your answer in terms of π.

(ii) Examine each of the integrals (a) $\displaystyle\int_0^{\infty} x^3 e^{-x^2}\,dx$, (b) $\displaystyle\int_0^{\pi/2} \frac{1}{1-\cos x}\,dx$, to decide whether it diverges and, where possible, evaluate the integral. (*L*)

20 (i) Given that $I_n = \int_1^e (\ln x)^n \, dx$, where n is a non-negative integer, prove that, for $n \geqslant 1$, $I_n + nI_{n-1} = e$. Evaluate I_5.
(ii) Sketch the curve whose equation in polar coordinates is $r = a\cos^2\theta$, $-\pi/2 \leqslant \theta \leqslant \pi/2$, where $a > 0$. Calculate the area of the region enclosed by the loop. (*L*)

21 (a) Differentiate $\frac{1}{4}x^4 (\ln x)^n$ with respect to x. Hence, or otherwise, show that if $I_n = \int_1^e x^3 (\ln x)^n \, dx$, then $I_n = \frac{1}{4}e^4 - \frac{1}{4}nI_{n-1}$. Evaluate I_2. (b) Sketch the curve whose equation is $y = \sinh^{-1} x$. Calculate the area of the finite region bounded by the curve, the x-axis and the line $x = 1$, giving your answer to three significant figures. (*C*)

22 The region bounded by the x-axis, the ordinates $x = 2$ and $x = 6$ and the arc of the parabola $y = \sqrt{x}$ between these ordinates, is rotated through 2π about the x-axis. Find (i) the coordinates of the centroid of the solid region so formed, (ii) the area of the curved surface of this solid.

23 By using the substitution $y = \sinh x$, or otherwise,

(i) find $\int \frac{y}{\sqrt{(y^2+1)}} \ln\{y + \sqrt{(y^2+1)}\} \, dy$,

(ii) show that $\int_0^{4/3} y \ln\{y + \sqrt{(y^2+1)}\} \, dy = \frac{41 \ln 3 - 20}{36}$. (*C*)

24 (i) Evaluate $\int_0^1 \arcsin x \, dx$. (ii) Express $\tanh x$ in terms of e^x, and hence evaluate $\int_0^1 e^x \tanh x \, dx$. (iii) Given that $I_n = \int_0^{\pi/2} \sin^n x \, dx$, $(n \geqslant 0)$, show that $nI_n = (n-1)I_{n-2}$, $(n \geqslant 2)$. Hence evaluate I_4. (*L*)

25 Sketch the curve $y = \cosh x$. Find the mean value of $\cosh x$ for $0 \leqslant x \leqslant \ln 2$. Find the perimeter of the region defined by the inequalities $0 \leqslant x \leqslant \ln 2$, $0 \leqslant y \leqslant \cosh x$. This region is rotated completely about the x-axis. Show that the *total* surface area of the solid of revolution formed is $\pi(7 + \ln 4)/2$. (*AEB*)

26 Given that $f(x) = \frac{x^3 + x + 1}{(x-1)^2(2x^2+1)}$, express $f(x)$ in partial fractions. Find (a) the value, to two significant figures, of $f'(x)$ when $x = 2$. (b) $\int_2^4 f(x) \, dx$, leaving natural logarithms in your answer. State the set of values of x for which $f(x)$ can be expanded as a series in ascending powers of x. (*L*)

27 Prove Pappus Theorem for the volume of a solid of revolution and *state* the corresponding theorem for the surface area. A ring is formed by revolving a plane semicircular area of radius a completely about a straight line in its plane, parallel to and distant $4a$ from the bounding diameter and on the same side of this diameter as the curve of the semicircle. Show that the surface area of the ring is $4\pi(2\pi + 3)a^2$ and find its volume. (*L*)

28 Differentiate $\dfrac{\sinh x}{\cosh^{n-1} x}$ with respect to x. Hence, or otherwise, given that

$I_n = \displaystyle\int_0^t \frac{1}{\cosh^n x}\,dx$, prove that $(n-1)I_n = \dfrac{\sinh t}{\cosh^{n-1} t} + (n-2)I_{n-2}$.
State the exact numerical values of sinh (ln 2) and cosh (ln 2) and show that $\displaystyle\int_0^{\ln 2} \frac{1}{\cosh^4 x}\,dx = \frac{66}{125}$. *(C)*

29 Prove that $\int 2\sec^3 x\,dx = \sec x \tan x + \int \sec x\,dx$. Sketch, for $-\pi/2 < x < \pi/2$, the curve whose equation is $y = 2\sec x$. Show that the length of the arc of this curve between $x = 0$ and $x = \pi/4$ is equal to $\displaystyle\int_0^{\pi/4} (2\sec^2 x - 1)\,dx$ and evaluate this integral. Find also the area of the surface generated when this arc is rotated once about the x-axis. *(JMB)*

30 (a) If $I_n = \displaystyle\int_0^1 x^n e^{-x}\,dx$, show that $I_n = -\dfrac{1}{e} + nI_{n-1}$ for $n > 0$. Deduce the value of I_2. (b) Prove that $\displaystyle\int_0^1 x^m(1-x)^n\,dx = \int_0^1 x^n(1-x)^m\,dx$, where $m > 0$ and $n > 0$. Hence, or otherwise, evaluate $\displaystyle\int_0^1 x^2(1-x)^{1/2}\,dx$. *(JMB)*

31 The parametric equations of a curve are $x = a\{\ln(\tan\theta) + \cos 2\theta\}$, $y = a\sin 2\theta$ $(0 < \theta < \pi/2, a > 0)$. Prove that $\dfrac{dx}{d\theta} = \dfrac{2a\cos^2 2\theta}{\sin 2\theta}$ and show that the angle between the tangent at the point whose parameter is θ and the x-axis is 2θ. Sketch the curve for $0 < \theta < \pi/2$. Show that the length of the arc of the curve joining the points whose parameters are $\pi/12$ and $\pi/4$ is $a\ln 2$. Find the area of the surface formed by rotating this arc through one revolution about the x-axis. *(JMB)*

32 Given that $I_n = \int x^n e^{ax}\,dx$, obtain a reduction formula in the form $I_n = f(x) - \dfrac{n}{a}I_{n-1}$. Hence, or otherwise, evaluate $\displaystyle\int_0^1 x^2 e^{3x}\,dx$. *(AEB)*

33 Sketch the curves C_1 and C_2 whose respective polar equations are $r = 2\sin\theta$, $(0 \leqslant \theta \leqslant \pi)$, and $r = 2\sin 2\theta$, $(0 \leqslant \theta \leqslant \pi/2)$. (i) Show that the area of the region enclosed by C_2 is half the area of the region enclosed by C_1. (ii) Show that the length of the curve C_2 is given by $2\displaystyle\int_0^{\pi/2} \sqrt{(1 + 3\cos^2 2\theta)}\,d\theta$. Use the Trapezium Rule with five ordinates to obtain an approximate value for the length of the curve C_2. *(C)*

34 A curve is given by the equations $x = 2\cos t - \cos 2t - 1$, $y = 2\sin t - \sin 2t$, $(0 \leqslant t < 2\pi)$. In any order, (i) show that the tangents to the curve at the points $t = \pi/3$, $t = 2\pi/3$, $t = \pi$, $t = 4\pi/3$ form a rectangle. (ii) sketch the curve, (iii) prove that the length of the curve is 16, (iv) show that the equation of the curve in polar coordinates may be expressed in the form $r = 2(1 - \cos\theta)$. *(C)*

10 Complex Numbers

10.1 Basic properties (revision)

The set, $\mathbb{R}$, of real numbers is extended to the set, $\mathbb{C}$, of complex numbers by adjoining the number i, with the property that $i^2 = -1$. Thus, $\mathbb{C} = \{z\colon\ z = x + iy,\ x, y \in \mathbb{R}\}$. The usual arithmetical operations are performed in $\mathbb{C}$, regarding i as a number and replacing i^2 by -1 wherever it occurs.

In what follows, let x, y, a and b be real numbers, with a and b not both zero. Then

$$(x+iy)+(a+ib) = (x+a)+i(y+b)$$
$$(x+iy)-(a+ib) = (x-a)+i(y-b)$$
$$(x+iy)(a+ib) = (xa-yb)+i(ya+xb)$$
$$(x+iy)/(a+ib) = \frac{xa+yb}{a^2+b^2}+i\frac{ya-xb}{a^2+b^2}.$$

Argand diagram

The number z, where $z = x+iy$, $x, y \in \mathbb{R}$, is represented on the Argand diagram by the point P, (x, y). The real part of z is x and the imaginary part of z is y and these are denoted by $\operatorname{Re}(z) = x$ and $\operatorname{Im}(z) = y$. If we write:

$$x = r\cos\theta,\ y = r\sin\theta,\ \text{with } r \geqslant 0 \text{ and } -\pi < \theta \leqslant \pi,$$

then
$r = |z|$ is the modulus of z,
$\theta = \arg z$ is the argument of z,
$z = x + iy$ is the Cartesian form of z,
$z = r(\cos\theta + i\sin\theta)$ is the trigonometric form of z,
$z = (r, \theta)$ is the polar form of z.

Of course, the addition, or subtraction, of any multiple of 2π to, or from, the argument of a complex number leaves that number unaltered. The argument of a complex number, z, can be regarded as a class of real numbers under the equivalence in which two real numbers are equivalent if they differ by a multiple of 2π. That is, the domain of the argument function on the complex numbers is the set of real numbers modulo 2π and the principal value of the argument of z is $\arg z$ and this is a representative of the class. We shall assume, without further reference, the equality of two complex numbers

$$r(\cos\theta + i\sin\theta) \text{ and } r\{\cos(\theta+2n\pi) + i\sin(\theta+2n\pi)\},\ n \in \mathbb{Z}.$$

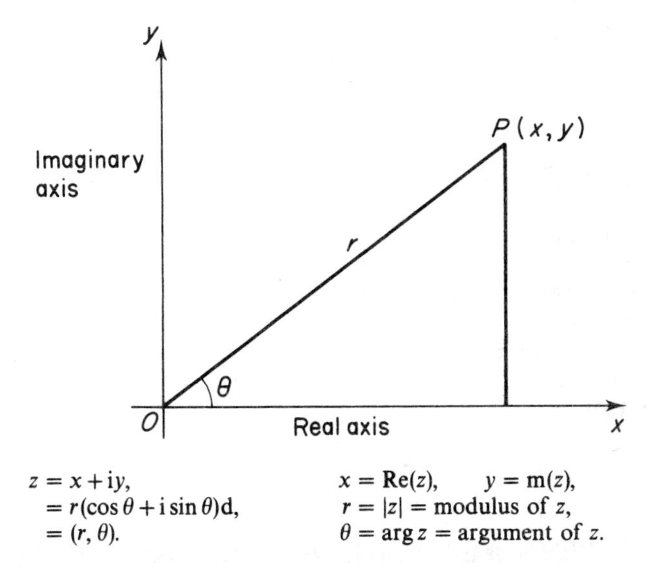

$z = x + iy,$ $x = \text{Re}(z),$ $y = \text{m}(z),$
$= r(\cos\theta + i\sin\theta)\text{d},$ $r = |z| = \text{modulus of } z,$
$= (r, \theta).$ $\theta = \arg z = \text{argument of } z.$

Fig. 10.1 The Argand diagram

These definitions are indicated on the Argand diagram in Fig. 10.1.

For products and quotients, assuming that $r_2 \neq 0$,

$$r_1(\cos\theta_1 + i\sin\theta_1)r_2(\cos\theta_2 + i\sin\theta_2)$$
$$= r_1r_2\{(\cos\theta_1\cos\theta_2 - \sin\theta_1\sin\theta_2) + i(\sin\theta_1\cos\theta_2 + \cos\theta_1\sin\theta_2)\}$$
$$= r_1r_2\{\cos(\theta_1+\theta_2) + i\sin(\theta_1+\theta_2)\}$$
$$r_1(\cos\theta_1 + i\sin\theta_1)/r_2(\cos\theta_2 + i\sin\theta_2)$$
$$= (r_1/r_2)(\cos\theta_1 + i\sin\theta_1)(\cos\theta_2 - i\sin\theta_2)$$
$$= (r_1/r_2)\{(\cos\theta_1\cos\theta_2 + \sin\theta_1\sin\theta_2) + i(\sin\theta_1\cos\theta_2 - \cos\theta_1\sin\theta_2)\}$$
$$= (r_1/r_2)\{\cos(\theta_1-\theta_2) + i\sin(\theta_1-\theta_2)\}.$$

Thus

$$|z_1z_2| = |z_1|\,|z_2|, \qquad \arg(z_1z_2) = \arg(z_1) + \arg(z_2),$$
$$|z_1/z_2| = |z_1|/|z_2|, \qquad \arg(z_1/z_2) = \arg(z_1) - \arg(z_2),$$

and, in polar form,

$$(r_1, \theta_1)(r_2, \theta_2) = (r_1r_2, \theta_1+\theta_2)$$
$$(r_1, \theta_1)/(r_2, \theta_2) = (r_1/r_2, \theta_1-\theta_2)$$

EXERCISE 10.1 (Revision)

1 Express in the form $a + ib$, $a, b \in \mathbb{R}$:

(i) $(3+2i)+(2-i)$, (ii) $2(1-i)+3(2+3i)$, (iii) $(4+2i)-(3+i)$,
(iv) $2(3-i)-5(2+3i)$, (v) $(2+i)(3-i)$, (vi) $(2-3i)(2+3i)$,
(vii) $i(3-4i)$, (viii) $(4+2i)/(3-4i)$, (ix) $(a+ib)/(x-iy)$.

2 Solve the equation: (i) $(2+3i)z = 1$, (ii) $1/z + 1/(1+i) = 2$, (iii) $z^2 - z + 1 = 0$, (iv) $z^2 - 12z + 40 = 0$.

3 Find the modulus and the argument of z_r and represent z_r on the Argand diagram:
(i) $z_1 = 3 - 4i$, (ii) $z_2 = 1 + i$, (iii) $z_3 = 12 + 5i$,
(iv) $z_4 = (\sqrt{3} - i)/2$, (v) $z_5 = i/(3 - 2i)$, (vi) $z_6 = (1 + 2i)(2 - \lambda)$,
(vii) $z_7 = (3 - 4i)/(4 + 3i)$.

4 Write down z^2, given that:
(i) $z = 2 + 3i$, (ii) $z = a + ib$, (iii) $z = 2(\cos \pi/3 + i \sin \pi/3)$,
(iv) $z = r(\cos\theta + i\sin\theta)$, (v) $z = (3, 2\pi/3)$.

5 Write down $z_1 z_2$ and z_1/z_2 in polar form, given that:
(i) $z_1 = (4, \pi/6)$, $z_2 = (2, \pi/3)$ (ii) $z_1 = (1, 0)$, $z_2 = (r, \theta)$
(iii) $z_1 = (3, \pi/2)$, $z_2 = (3, -\pi/2)$ (iv) $z_1 = 1 + i$, $z_2 = 1 - \sqrt{3}i$

6 Verify that:
(i) $(z^*)^* = z$ (ii) $(r, \theta)^* = (r, -\theta)$ (iii) $(z_1 + z_2)^* = z_1^* + z_2^*$
(iv) $(z_1 z_2)^* = z_1^* z_2^*$ (v) $(z_1/z_2)^* = z_1^*/z_2^*$ (vi) $|z^*| = |z| = \sqrt{(zz^*)}$.

10.2 Powers and roots of complex numbers

The powers of a complex number z are most easily found by using the polar form of z. Suppose that $z = r(\cos\theta + i\sin\theta) = (r, \theta)$, then

$$z^2 = zz = (r, \theta)(r, \theta) = (r^2, 2\theta),$$
$$z^3 = zz^2 = (r, \theta)(r^2, 2\theta) = (r^3, 3\theta).$$

This suggests a general result, which we prove by induction.

Theorem 10.1 (de Moivre) Given that $z = r(\cos\theta + i\sin\theta) = (r, \theta)$ and $n \in \mathbb{Z}$, then $z^n = r^n(\cos n\theta + i\sin n\theta) = (r^n, n\theta)$.

Proof The result has been proved for n equal to 2 and 3 and it is obviously true for n equal to 1 and 0, taking $z^0 = 1$. Assume that the result is true for $n = m$, where m is a positive integer. Then

$$z^{m+1} = zz^m = (r, \theta)(r^m, m\theta) = (r^{m+1}, (m+1)\theta),$$

and the result is true for $n = m + 1$. Therefore, by induction, the result is true for every natural number n.
Consider, for a positive integer n, the product

$$(r^n, n\theta)(r^{-n}, -n\theta) = (r^n r^{-n}, n\theta - n\theta) = (1, 0) = 1.$$

Therefore $z^{-n} = (z^n)^{-1} = (r^n, n\theta)^{-1} = (r^{-n}, -n\theta)$ and the result is true for every negative integer $-n$. Thus it is true for all integers.

Roots

For any positive real number, there are two square roots, one positive and one negative. A square root of a negative real number or of a complex number will be complex and therefore cannot be either positive or negative. However, there will be two square roots which will be negatives of each other. For example:

$$(2+3i)^2 = 4-9+6i+6i = -5+12i$$

and so $2+3i$ and $-2-3i$ are the two square roots of $-5+12i$. However, it is not so clear how we should calculate the square roots of another complex number.

EXAMPLE 1 *Solve the equation* $z^2 = 1+i$.

Method I Write $z = x+iy$, x and y real, and equate the real and imaginary parts of the equation. $x^2-y^2+2ixy = 1+i$, so $x^2-y^2 = 1$, $2xy = 1$. Therefore, $y = 1/2x$, $x^2-1/4x^2 = 1$ and so $4x^4-4x^2-1 = 0$, or $(2x^2-1)^2 = 2$. Since $(1-\sqrt{2})/2$ is negative, we reject this solution of the quadratic equation in x^2 and obtain $x^2 = (1+\sqrt{2})/2$ and $y^2 = 1/2(1+\sqrt{2}) = (\sqrt{2}-1)/2$. Since x and y are the same sign, the solution set of the equation is $\{z_1, z_2\}$ where $\mathbf{z_1 = \sqrt{\{(1+\sqrt{2})/2\}} + i\sqrt{\{(\sqrt{2}-1)/2\}}}$ and $\mathbf{z_2 = -z_1}$.
This method can not be generalised, that is, it will not be of use to find the cube root of a complex number. We show an alternative method which is capable of generalisation. It uses the polar form of the complex numbers.
Method II In polar form, $z^2 = 1+i = (\sqrt{2}, \pi/4) = (\sqrt{2}, 9\pi/4)$, since the addition of 2π to the argument does not change the complex number. Let $z = (r, \theta)$ in polar form, so that $z^2 = (r^2, 2\theta)$ and therefore $r^2 = \sqrt{2}$ and $r = \sqrt[4]{2}$ while $2\theta = \pi/4$ or $2\theta = 9\pi/4$, that $\theta = \pi/8$ or $9\pi/8$. There are thus two solutions, $\mathbf{z_1 = (\sqrt[4]{2}, \pi/8)}$ and $\mathbf{z_2 = (\sqrt[4]{2}, 9\pi/8)}$.
Note that,
since $\cos 9\pi/8 = \cos(\pi+\pi/8) = -\cos\pi/8$, $\sin 9\pi/8 = \sin(\pi+\pi/8) = -\sin\pi/8$,

$$z_1 = -z_2.$$

EXAMPLE 2 *Find the three roots of the equation* $z^3 = 1$.

Method I Use the polar form of z. If $z = (r, \theta)$, then $z^3 = (r^3, 3\theta)$, and, since $\cos 2m\pi = 1$ and $\sin 2m\pi = 0$ for any integer m,

$$1 = (1, 0) = (1, 2\pi) = (1, 4\pi) = \ldots = (1, -2\pi) = \ldots = (1, 2m\pi),\ m \in \mathbb{Z}.$$

Therefore $r^3 = 1$ and $3\theta = 2m\pi$, for any $m \in \mathbb{Z}$. So $r = 1$ and $\theta \in \{0, 2\pi/3, 4\pi/3\}$, the other values of m only giving repetitions of these three solutions. The solution set of the equation is $\{\mathbf{(1, 0), (1, 2\pi/3), (1, 4\pi/3)}\}$, that is $\{1, (-1+i\sqrt{3})/2, (-1-i\sqrt{3})/2\}$.
Method II In this particular case, it is possible to use a more elementary method to solve the equation by factorisation. Since

$$z^3-1 = (z-1)(z^2+z+1) = 0$$

either $z = 1$ or $z^2+z+1 = 0$, so the solution set is
$\mathbf{\{1, (-1+i\sqrt{3})/2, (-1-i\sqrt{3})/2\}}$.

The above examples lead to a method of calculating the nth root of a complex number, as we show in Example 3.

EXAMPLE 3 *Solve the equation $z^n = a$, where $a \in \mathbb{C}$.*

Write z and a in polar form, $z = (r, \theta)$, $a = (|a|, \alpha)$, where $\alpha = \arg(a)$. Then $z^n = (r^n, n\theta) = (|a|, \alpha)$, so $r^n = |a|$ and $r = \sqrt[n]{|a|}$. One solution for θ is given by $n\theta = \alpha$, that is $\theta = \alpha/n$, but there are $(n-1)$ other solutions. Since

$$a = (|a|, \alpha) = (|a|, \alpha + 2\pi) = (|a|, \alpha + 4\pi) = \ldots = (|a|, \alpha + 2(n-1)\pi),$$

the other solutions are given by $n\theta = \alpha + 2k\pi$, $1 \leqslant k \leqslant n-1$, that is $\theta = (\alpha + 2k\pi)/n$.
Thus the solution set of the equation $z^n = a$ is

$$\{z : z = (\sqrt[n]{|a|}, (\alpha + 2k\pi)/n), 0 \leqslant k \leqslant n-1\}.$$

EXERCISE 10.2A

1 By factorising $z^3 + 27$, solve the equation $z^3 + 27 = 0$. Verify that your solutions are correct by working out their cubes. Express -27 as a complex number in polar form and, hence, find the polar forms of the roots of the equation.

2 Find the four fourth roots of unity: (a) by factorising $z^4 - 1$, (b) by working with the complex numbers in polar form. Plot the four roots on the Argand diagram and check that they are symmetrically placed round the unit circle in the complex plane.

3 Find the polar form of the square roots of z in terms of π and/or α, where $\cos\alpha = 3/5$ and $\sin\alpha = 4/5$. Show the roots on the Argand diagram:
(i) $z = 4$ (ii) $z = -4$ (iii) $z = 4\text{i}$ (iv) $z = -4\text{i}$ (v) $z = 25\text{i}$
(vi) $z = 15 + 20\text{i}$ (vii) $z = 15 - 20\text{i}$ (viii) $z = 20 - 15\text{i}$ (ix) $z = -20 - 15\text{i}$.
Verify that these eighteen points on the Argand diagram lie on two circles and state why this is so.

4 Plot on the Argand Diagram all the points representing the solutions of the equation: (i) $z^3 = -1$ (ii) $z^6 = 1$ (iii) $z^2 = (-1 + \text{i}\sqrt{3})/2$
(iv) $z^2 = (-1 - \text{i}\sqrt{3})/2$.

The above results lead to a form of de Moivre's theorem which gives a value of z^q, where $z \in \mathbb{C}$ and $q \in \mathbb{Q}$.

Theorem 10.2 Let z be a complex number and let q be a rational number, given by $q = m/n$, $m \in \mathbb{Z}$, $n \in \mathbb{Z}^+$. Then,

if $\qquad z = r(\cos\theta + \text{i}\sin\theta) = (r, \theta)$,

one value of z^q is $\qquad r^q(\cos q\theta + \text{i}\sin q\theta)$, that is $(r^q, q\theta)$.

Proof By Theorem 10.1, for $m \in \mathbb{Z}$, $z^m = r^m(\cos m\theta + \text{i}\sin m\theta) = (r^m, m\theta)$. Since $n \in \mathbb{Z}$, we can use the same result to show that $(r^q, q\theta)^n = ((r^q)^n, nq\theta) = ((r^{m/n})^n, n(m/n)\theta) = (r^m, m\theta) = z^m$. Thus $(r^q, q\theta)$ is an nth root of z^m and so it is one value of $z^{m/n}$.

EXAMPLE 4 *Solve the equation $16z^8 + 15z^4 - 1 = 0$, giving the roots in polar form.*

Let $z^4 = x$ when the equation becomes $16x^2 + 15x - 1 = 0$, that is, $(16x-1)(x+1) = 0$, and so $x = 1/16$ or $x = -1$.
If $x = 1/16 = (1/16, 0) = (1/16, 2k\pi)$, $k \in \mathbb{Z}$, then, since $x = z^4$,

$$z \in \{(1/2, 0), (1/2, \pi/2), (1/2, \pi), (1/2, 3\pi/2)\}.$$

If $x = -1 = (1, \pi) = (1, 2k\pi + \pi)$, $k \in \mathbb{Z}$, then, since $x = z^4$,

$$z \in \{(1, \pi/4), (1, 3\pi/4), (1, 5\pi/4), (1, 7\pi/4)\}.$$

The final solution set for the equation is
$\mathbf{\{(1/2, 0),\ (1/2, \pi/2),\ (1/2, \pi),\ (1/2, 3\pi/2),\ (1, \pi/4),\ (1, 3\pi/4),\ (1, 5\pi/4), (1, 7\pi/4)\}}$.

EXAMPLE 5 *Prove that, for* $\sin\theta \neq 0$,

$$(\sin 5\theta)/(\sin\theta) = 16\cos^4\theta - 12\cos^2\theta + 1.$$

Show that $2\cos(\pi/5)$ *is one root of the equation* $x^4 - 3x^2 + 1 = 0$, *and state the values of the other three roots. Deduce that* $\cos\pi/5 = (1+\sqrt{5})/4$.

By de Moivre's Theorem

$$\begin{aligned}\cos 5\theta + \mathrm{i}\sin 5\theta &= (\cos\theta + \mathrm{i}\sin\theta)^5\\ &= \cos^5\theta + 5\mathrm{i}\cos^4\theta\sin\theta - 10\cos^3\theta\sin^2\theta - 10\mathrm{i}\cos^2\theta\sin^3\theta\\ &\quad + 5\cos\theta\sin^4\theta + \mathrm{i}\sin^5\theta\end{aligned}$$

and the imaginary part of this equation gives the result

$$\sin 5\theta = 5\cos^4\theta\sin\theta - 10\cos^2\theta\sin^3\theta + \sin^5\theta.$$

Hence,

$$\begin{aligned}(\sin 5\theta)/(\sin\theta) &= 5\cos^4\theta - 10\cos^2\theta(1-\cos^2\theta) + (1-\cos^2\theta)^2\\ &= 16\cos^4\theta - 12\cos^2\theta + 1\end{aligned}$$

Let $x = 2\cos\theta$, then

$$x^4 - 3x^2 + 1 = 16\cos^4\theta - 12\cos^2\theta + 1 = (\sin 5\theta)/(\sin\theta) = 0,$$

so that $\sin 5\theta = 0$ and $5\theta = k\pi$, $k \in \{1, 2, 3, 4\}$, since $\sin\theta \neq 0$.
Therefore, $\theta \in \{\pi/5, 2\pi/5, 3\pi/5, 4\pi/5\}$ and

$$x \in \{2\cos\pi/5, 2\cos 2\pi/5, 2\cos 3\pi/5, 2\cos 4\pi/5\}.$$

Of the four values of x, $2\cos\pi/5$ is the largest, since cosine is a decreasing function. Now

$$\begin{aligned}x^4 - 3x^2 + 1 &\Rightarrow x^2 = (3 \pm \sqrt{5})/2\\ &\Rightarrow x = \pm(1 \pm \sqrt{5})/2\end{aligned}$$

and, on choosing the largest value of x, $\cos\pi/5 = (1+\sqrt{5})/4$.

EXAMPLE 6 *Prove that, if* $\omega^3 = 1$, *then*

$$\begin{vmatrix} x & \omega & \omega^2 \\ \omega^2 & x & \omega \\ \omega & \omega^2 & x \end{vmatrix} = (x+2)(x-1)^2.$$

Using the symmetry of the determinant, we add the second and third rows to the first row and remove the common factor,

$$\Delta = \begin{vmatrix} x & \omega & \omega^2 \\ \omega^2 & x & \omega \\ \omega & \omega^2 & x \end{vmatrix} = \begin{vmatrix} x+\omega+\omega^2 & x+\omega+\omega^2 & x+\omega+\omega^2 \\ \omega^2 & x & \omega \\ \omega & \omega^2 & x \end{vmatrix}$$

$$= (x+\omega+\omega^2)\begin{vmatrix} 1 & 1 & 1 \\ \omega^2 & x & \omega \\ \omega & \omega^2 & x \end{vmatrix}$$

$$= (x+\omega+\omega^2)\begin{vmatrix} 1 & 0 & 0 \\ \omega^2 & x-\omega^2 & \omega-\omega^2 \\ \omega & \omega^2-\omega & x-\omega \end{vmatrix},$$

on subtracting the first column from the second and the third column. Expanding this last determinant along its first row,

$$\Delta = (x+\omega+\omega^2)\{x^2-(\omega+\omega^2)x+\omega^3+(\omega^2-\omega)^2\}$$
$$= (x+\omega+\omega^2)\ \{x^2-(\omega+\omega^2)x+\omega^3+\omega\omega^3-2\omega^3+\omega^2\}$$
$$= (x+\omega+\omega^2)\ \{x^2-(\omega+\omega^2)x+\omega+\omega^2-1\},$$

on using the given fact that $\omega^3 = 1$.
Now $\omega^3-1 = (\omega-1)(\omega^2+\omega+1) = 0$ so that either $\omega = 1$ or $\omega+\omega^2 = -1$.
If $\omega = 1$, then $\Delta = (x+2)(x^2-2x+1) = (x+2)(x-1)^2$.
If $\omega+\omega^2 = -1$, then $\Delta = (x-1)(x^2+x-2) = (x-1)(x-1)\ (x+2)$.
The required result follows in each case.

EXAMPLE 7 *Given that $\omega \neq 1$, $\omega^n = 1$, $n \in \mathbb{N}$, prove that*

$$1+\omega+\omega^2+\ \dots\ +\omega^{n-1} = 0.$$

The function $f: \mathbb{C} \to \mathbb{C}$ is given by $f(z) = mz+c$, where m and c are non-zero complex numbers. Write down $f(z)$, $f^2(z)$, $f^3(z)$, and, by considering the pattern of these expressions, conjecture the value of $f^r(z)$, where f^r is the function formed by the composition of r copies of the function f, and prove that your result is correct. Prove that, if $m^n = 1$ where $m \neq 1$ and $n \in \mathbb{N}$, then f^n is the identity function e. If $S = \{f^r\colon r \in \mathbb{Z}^+\}$, prove that $(S, \bar{\circ})$ is a cyclic group of order n if, and only if, $m^n = 1$ and $m \neq 1$.

Factorise ω^n-1,

$$\omega^n-1 = (\omega-1)\ (\omega^{n-1}+\omega^{n-2}+\ \dots\ +\omega+1)$$

so, if $\omega^n = 1$ and $\omega \neq 1$, then $\omega^{n-1}+\omega^{n-2}+\ \dots\ +\omega+1 = 0$.

Now $\qquad f(z) = mz+c$, so that

$$f^2(z) = ff(z) = f(mz+c) = m(mz+c)+c$$
$$= m^2z+(m+1)c,$$
$$f^3(z) = f^2f(z) = f^2(mz+c) = m^2(mz+c)+(m+1)c$$
$$= m^3z+(m^2+m+1)c.$$

From this pattern, we conjecture that

$$f^r(z) = m^rz+(m^{r-1}+m^{r-2}+\ \dots\ +m+1)c \qquad \text{P}(r)$$

and we have already shown that the statements P (1), P (2) and P (3) are all true. We now prove that P (r) is true for all positive integers r. Assume that P (r) is true, for some positive integer r. That is,

$$\mathrm{f}^r(z) = m^r z + (m^{r-1} + m^{r-2} + \ldots + m + 1)c$$

then $$\mathrm{f}^{r+1}(z) = \mathrm{f}^r\mathrm{f}(z) = m^r(mz + c) + (m^{r-1} + m^{r-2} + \ldots + m + 1)c$$
$$= m^{r+1}z + (m^r + m^{r-1} + m^{r-2} + \ldots + m + 1)c.$$

Therefore, P (r) $\Rightarrow$ P ($r+1$) and, since P (1) is true, the result follows by induction and P (r) is true, for all positive integers r. Now suppose that $n \in \mathbb{Z}^+$, $m^n = 1$, $m \neq 1$, then, on putting $\omega = m$,

$$m^{n-1} + m^{n-2} + \ldots + m + 1 = 0$$

so that $$\mathrm{f}^n(z) = m^n z + 0 . c = m^n z = z = \mathrm{e}(z),$$

and so $\mathrm{f}^n = \mathrm{e}$, the identity function. Therefore, if $m^n = 1$, $m \neq 1$,

$$S = \{\mathrm{f}, \mathrm{f}^2, \mathrm{f}^3, \ldots, \mathrm{f}^{n-1}, \mathrm{f}^n\},\ \mathrm{f}^n = \mathrm{e},$$

then S is the cyclic group of order n, generated by f. Conversely, if $(S, \circ)$ is a cyclic group of order n, then it is the cyclic group generated by f and therefore $\mathrm{f}^n = \mathrm{e}$. It follows that

$$z = \mathrm{e}(z) = \mathrm{f}^n(z) = m^n z + (m^{n-1} + m^{n-2} + \ldots + m + 1)c$$

so that $m^n = 1$ and, since $c \neq 0$, $m^{n-1} + m^{n-2} + \ldots + m + 1 = 0$, $m \neq 1$.

EXERCISE 10.2 B

1 Find one value of: (i) $\sqrt{(2 + \mathrm{i}2\sqrt{3})}$ (ii) $\sqrt[3]{(\mathrm{i} - 1)}$ (iii) $\sqrt[5]{(-1)}$ (iv) $\sqrt{(1+\mathrm{i})^3}$ (v) $\sqrt[4]{(\cos\theta - \mathrm{i}\sin\theta)}$ (vi) $(8 - 8\mathrm{i})^{2/3}$.

2 Given that $\omega = (-1 + \mathrm{i}\sqrt{3})/2$, prove that:
(i) the three cube roots of 1 are 1, ω, ω^2
(ii) $1 + \omega + \omega^2 = 0$
(iii) the three cube roots of -1 are -1, $-\omega$, $-\omega^2$
(iv) the six sixth roots of 1 are ± 1, $\pm\omega$, $\pm\omega^2$.

3 Express $(-1 + \mathrm{i}\sqrt{3})$ in polar form and evaluate $(-1 + \mathrm{i}\sqrt{3})^{10}$ in:
(i) polar form (ii) Cartesian form.

4 Write z in polar form and find its four fourth roots when:
(i) $z = -1$ (ii) $z = \mathrm{i}$ (iii) $z = -\mathrm{i}$ (iv) $z = 2\sqrt{3} + 2\mathrm{i}$.

5 Solve the equation and mark its solutions on an Argand diagram:
(i) $z^3 + \mathrm{i} = 1$ (ii) $z^4 = -16$ (iii) $z^4 + 5z^2 + 4 = 0$.

6 Show that the six sixth roots of 1 are the powers of $(1 + \mathrm{i}\sqrt{3})/2$. Factorise $z^6 - 1$ into a product of four real factors, (two linear and two quadratic).

7 Show that, under multiplication, the four fourth roots of 1 form a cyclic group. Show that the same is true of the five fifth roots of 1. In each case, list the subgroups of the group and state all the generators of the group.

8 Given that $z = \cos\theta + \mathrm{i}\sin\theta$, prove that $z + 1/z = 2\cos\theta$ and that $z^3 + 1/z^3 = 2\cos 3\theta$. By expanding $(z + 1/z)^3$ prove that $4\cos^3\theta = \cos 3\theta + 3\cos\theta$. Use $z - 1/z$ and $z^3 - 1/z^3$ to express $\sin^3\theta$ in terms of $\sin 3\theta$ and $\sin\theta$.

9 Use the methods of question 8 to express $\cos^4 \theta$ and $\sin^4 \theta$ in terms of the powers of $\cos \theta$.

10 Given that $z = 1 + i\sqrt{3}$, prove that $z^{14} = 2^{13}(-1 + i\sqrt{3})$. (*L*)

11 Represent the roots of the equation $z^5 = 243$ in an Argand diagram, clearly indicating the modulus and argument of each root. (*L*)

12 Use de Moivre's theorem to prove that $\cos 4\theta = 8\cos^4 \theta - 8\cos^2 \theta + 1$. (*L*)

13 Given that $z = 4(\cos \pi/3 + i \sin \pi/3)$ and $w = 2(\cos \pi/6 - i \sin \pi/6)$ write down the modulus and argument of each of the following:
(i) z^3, (ii) $1/w$, (iii) z^3/w. (*JMB*)

14 Solve the equation $(z+1)^5 = z^5$.

10.3 Loci in the complex plane

In *Pure Mathematics* §15.7, we considered loci in the complex plane, given by equations in the complex variable z and its conjugate z^*. We remind the reader of the methods used by means of an example and then we consider some more advanced examples.

EXAMPLE 1 *Draw the locus, given by the following equation, on an Argand diagram:* (*i*) $\arg(z) = \pi/4$, (*ii*) $|z| = 2$, (*iii*) $z + z^* = 1$, (*iv*) $zz^* = 4$, (*v*) $\arg\{(z-1)/(z+1)\} = \pi/2$.

The Cartesian coordinates (x, y) of a point on the Argand diagram are given by $z = x + iy$.
(i) $\text{Arg}(z) = \pi/4$ is the equation of the half line in the first quadrant, given by $y = x \geqslant 0$, Fig. 10.2(a). Note that the other half of the line $y = x$ is given by $\arg(z) = -3\pi/4$.
(ii) The locus $|z| = 2$ consists of all points at a distance 2 from the origin. This is the circle, centre O, radius 2, Fig. 10.2 (b).

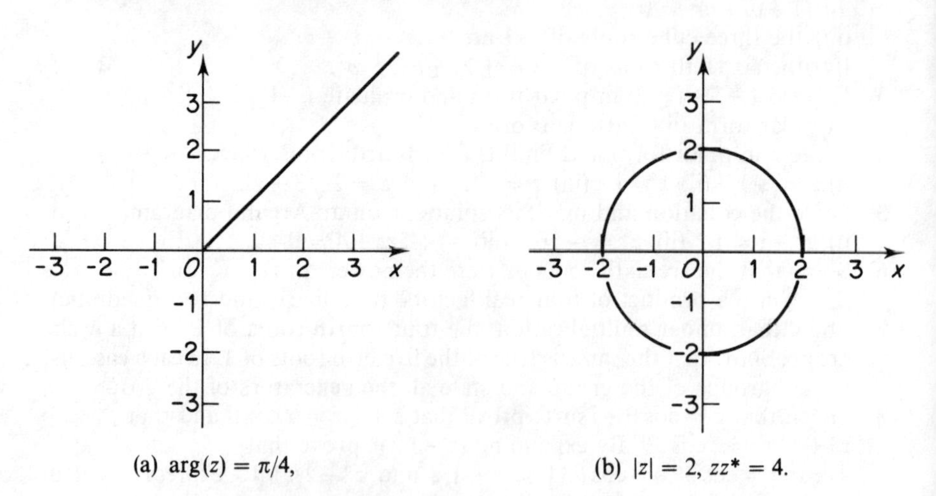

(a) $\arg(z) = \pi/4$, (b) $|z| = 2$, $zz^* = 4$.

Fig. 10.2 Loci in the complex plane

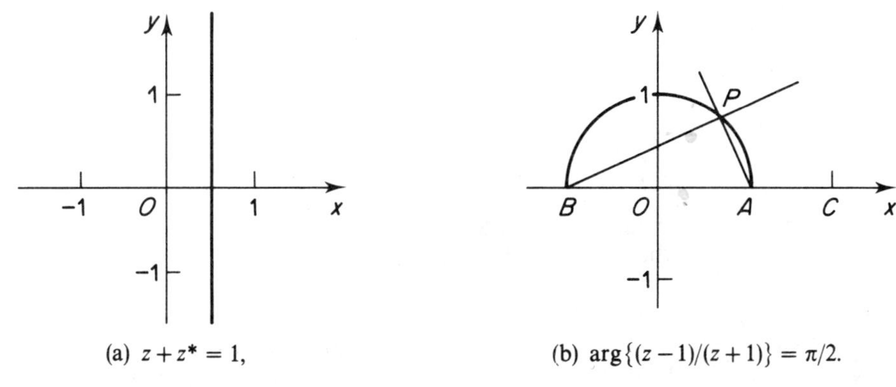

(a) $z+z^* = 1$,

(b) $\arg\{(z-1)/(z+1)\} = \pi/2$.

Fig. 10.3 Loci in the complex plane

(iii) $z+z^* = 2x$, so $z+z^* = 1$ gives the line $2x = 1$, Fig. 10.3 (a).
(iv) $zz^* = |z|^2$, so that $zz^* = 4 \Leftrightarrow |z| = 2$. The locus is the same circle as part (ii), shown in Fig. 10.2 (b).
(v) Let $P(x, y)$, $A(1, 0)$ and $B(-1, 0)$ represent the complex numbers z, 1 and -1 respectively on the Argand diagram. Then
$\arg\{(z-1)/(z+1)\} = \arg(z-1) - \arg(z+1)$, which is the angle APB. If this is a right-angle, then P lies on the upper half semicircle on the diameter AB, shown in Fig. 10.3 (b). Note that the corresponding semicircle in the lower half plane is given by $\arg\{(z+1)/(z-1)\} = \pi/2$.

EXAMPLE 2 *Shade that area of the Argand diagram occupied by points representing z where the following inequalities are both satisfied:* (i) $1 \leqslant |z| \leqslant 2$ *and* (ii) $\pi/3 \leqslant \arg(z) \leqslant 2\pi/3$.

If $1 \leqslant |z| \leqslant 2$, then the point P, representing z, lies in the annulus between the two circles, centre O, of radius 1 and 2.
If $\pi/3 \leqslant \arg(z) \leqslant 2\pi/3$, then P lies in the wedge in the upper half plane bounded by the lines $y = \pm x\sqrt{3}$. The intersection of the two regions is shown in Fig. 10.4.

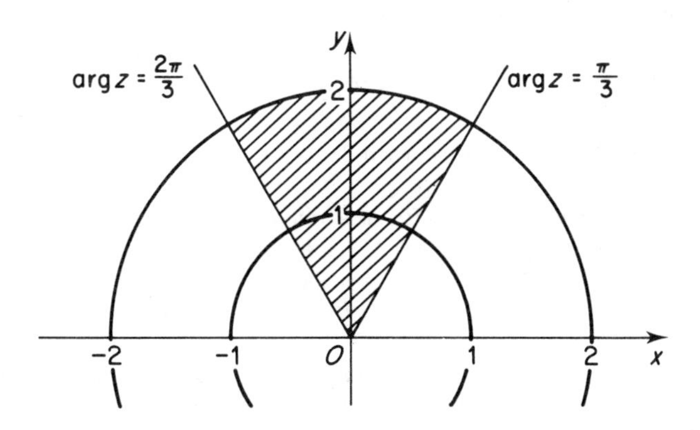

Fig. 10.4 Inequalities in the complex plane (i) $1 \leqslant |z| \leqslant 2$, and (ii) $\pi/3 \leqslant \arg z \leqslant 2\pi/3$

EXAMPLE 3 *Draw the locus in the Argand diagram given by the equation* $zz^* = \mathrm{i}z^* - \mathrm{i}z$.

We bring all the terms to one side of the equation and write the expression in terms of a product of a function of z and a function of z^*.

$$zz^* + \mathrm{i}z - \mathrm{i}z^* = (z-\mathrm{i})(z^*+\mathrm{i}) - 1 = 0.$$

Let $w = z - \mathrm{i}$, then, on taking complex conjugates, $w^* = z^* + \mathrm{i}$, and the equation becomes $ww^* = 1$, which is equivalent to $|w| = 1$. So the locus may be written $(z-\mathrm{i})(z-\mathrm{i})^* = 1$ or $|z-\mathrm{i}| = 1$. Now $|z-\mathrm{i}|$ is the distance of the point P, representing z, from the point Q, representing the number i. In Cartesian coordinates, P is the point (x, y) and Q is the point $(0, 1)$ and the locus is given by $|PQ| = 1$, that is, a circle of radius 1 and centre Q. The locus is shown in Fig. 10.5.

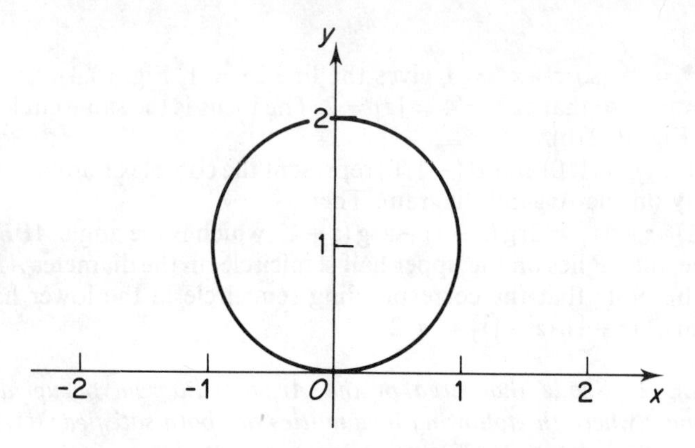

Fig. 10.5 The locus $zz^* = \mathrm{i}z^* - \mathrm{i}z$

EXAMPLE 4 *Given that* $w = (z+\mathrm{i})/(z-1)$, *for* $z \neq 1$, *and that the locus of* z *is a circle, centre* $z = 1$, *and radius 2, show that the locus of* w *is another circle with the same centre and find its radius.*

If z lies on the circle, centre 1, radius 2, then $|z-1| = 2$. Now

$$w - 1 = \frac{(z+\mathrm{i})-(z-1)}{(z-1)} = \frac{1+\mathrm{i}}{z-1}, \text{ so } |w-1| = \left|\frac{1+\mathrm{i}}{z-1}\right| = \frac{\sqrt{2}}{2} = \frac{1}{\sqrt{2}}$$

Hence, the locus of w is the circle with centre 1 and radius $\mathbf{1/\sqrt{2}}$.

EXAMPLE 5 *A transformation of the complex plane, from a z-plane to a w-plane, is given by* $w = (z-2)/(z-\mathrm{i})$, $z \neq \mathrm{i}$. *Find the locus in the w-plane, in terms of the real variables u and v, where* $w = u + \mathrm{i}v$, *if the locus of z is:*
(i) *the line given by* $z = z^* + 2\mathrm{i}$, (ii) *the line* $z + z^* = 1$, (iii) *the circle* $zz^* = 1$.

The equation expressing w in terms of z can be rearranged to make z the subject. Since $w = (z-2)/(z-\mathrm{i})$, $w(z-\mathrm{i}) = wz - \mathrm{i}w = z - 2$, and

$$z = (\mathrm{i}w - 2)/(w-1).$$

(i) Replacing z in the equation $z = z^* + 2\mathrm{i}$, for $z \neq \mathrm{i}$, $w \neq 1$,

$$\frac{\mathrm{i}w-2}{w-1} = \frac{-\mathrm{i}w^*-2}{w^*-1} + 2\mathrm{i}.$$

On multiplying by $(w-1)(w^*-1)$, this becomes

$$(\mathrm{i}w-2)(w^*-1) = -(w-1)(\mathrm{i}w^*+2) + 2\mathrm{i}(w-1)(w^*-1)$$

or $\mathrm{i}ww^* - 2w^* - \mathrm{i}w + 2 = -\mathrm{i}ww^* + \mathrm{i}w^* - 2w + 2 + 2\mathrm{i}ww^* - 2\mathrm{i}w - 2\mathrm{i}w^* + 2\mathrm{i}$ which simplifies to $2(w-w^*) + \mathrm{i}(w+w^*) = 2\mathrm{i}$, that is $\mathbf{2v + u = 1}$. So the locus of w is the line given by this equation.
(ii) In a similar manner, on substituting for z and multiplying the equation by $(w-1)(w^*-1)$, $z + z^* = 1$ becomes

$$(\mathrm{i}w-2)(w^*-1) - (w-1)(\mathrm{i}w^*+2) = (w-1)(w^*-1),$$

that is $$-(2+\mathrm{i})w - (2-\mathrm{i})w^* + 4 = ww^* - w - w^* + 1,$$

or $$ww^* + (1+\mathrm{i})w + (1-\mathrm{i})w^* = 3.$$

To interpret this equation, express the terms in w and w^* as a product of a function of w and function of w^*,

$$(w+1-\mathrm{i})(w^*+1+\mathrm{i}) = 3 + (1-\mathrm{i})(1+\mathrm{i}) = 3 + 2 = 5.$$

Therefore, $|w - (\mathrm{i}-1)| = \sqrt{5}$ and the locus of w is a **circle of radius $\sqrt{5}$ and centre at $w = i - 1$**.
(iii) Use the same methods as above.

$$zz^* = \frac{\mathrm{i}w-2}{w-1} \cdot \frac{-\mathrm{i}w^*-2}{w^*-1} = 1$$

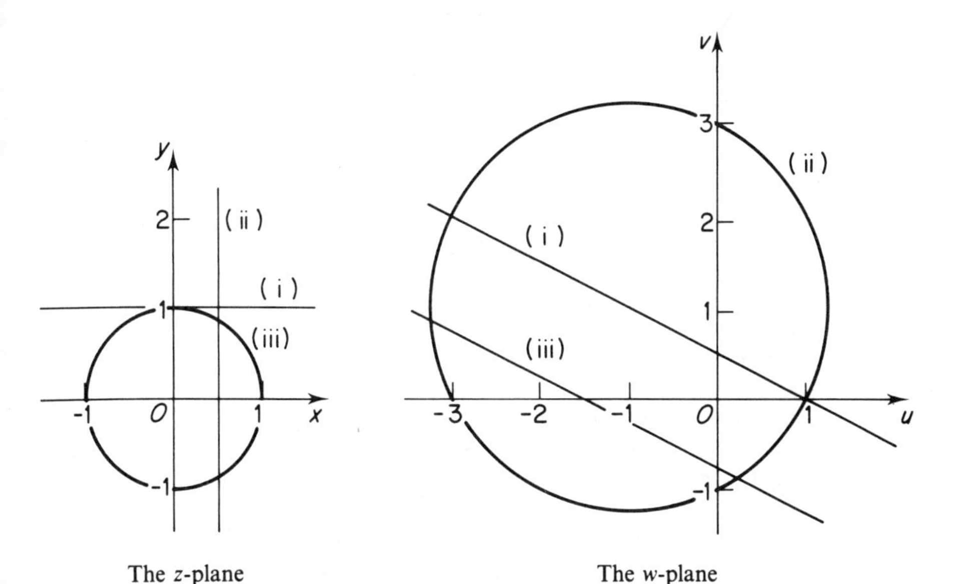

Fig. 10.6 The transformation $w = (z-2)/(z-\mathrm{i})$

so
$$-(\mathrm{i}w-2)(\mathrm{i}w^*+2) = (w-1)(w^*-1),$$
$$ww^*-2\mathrm{i}w+2\mathrm{i}w^*+4 = ww^*-w-w^*+1,$$
$$(w+w^*)-2\mathrm{i}(w-w^*)+3 = 0,$$

and the locus of w is the line $\mathbf{2u+4v+3=0}$.

The three loci in the z-plane and the three corresponding loci in the w-plane, for parts (i), (ii) and (iii) above are shown together in Fig. 10.6.

Summary of standard loci, where $a \in \mathbb{R}^+$, α, $\beta \in \mathbb{C}$.

Equation	Locus
$zz^* = a^2$ or $\lvert z\rvert = a$,	the circle, centre O, radius a
$(z-\alpha)(z^*-\alpha^*) = a^2$ or $\lvert z-\alpha\rvert = a$,	the circle, centre α, radius a
$\lvert z-\alpha\rvert = k\lvert z-\beta\rvert$	a circle if $0 < k \neq 1$
$\lvert z-\alpha\rvert = \lvert z-\beta\rvert$	the mediator of α and β

EXERCISE 10.3

1 Draw the locus on an Argand diagram given by:
(i) $\arg z = 3\pi/4$, (ii) $\arg z = -3/6$, (iii) $\arg\{(z+\mathrm{i})/(z-\mathrm{i})\} = \pi/2$,
(iv) $|z-1-\mathrm{i}| = \sqrt{2}$, (v) $|z^*| = 3$, (vi) $(z-1-\mathrm{i})(z^*-1+\mathrm{i}) = 2$.

2 Find the equation in $z\,(= x+\mathrm{i}y)$ and z^* which corresponds to the locus with the given Cartesian equation: (i) $x = -1$, (ii) $y = 3$, (iii) $x = y$, (iv) $x^2+2x+y^2=3$, (v) $x^2-2x+y^2-4y=4$, (vi) $x^2-2x+y^2-4y+5=0$.

3 Given that $w = (\sqrt{3}-\mathrm{i})/(\sqrt{3}+\mathrm{i})$, express w in (i) Cartesian form and (ii) polar form, and determine w^{12}. Sketch, on an Argand diagram, the locus given by the equation $|z-w| = 3$.

4 Determine the locus, given by the equation, and sketch it on an Argand diagram:
(i) $z+z^* = 3$, (ii) $z-z^* = 2$, (iii) $|z-1| = |z+1|$, (iv) $|z+1| = |z-\mathrm{i}|$,
(v) $zz^* = 9$, (vi) $(z-2)(z^*-2) = 1$, (vii) $(z+\mathrm{i})(z^*-\mathrm{i}) = 4$,
(viii) $(z-3-4\mathrm{i})(z^*-3+4\mathrm{i}) = 25$, (ix) $zz^*+2\mathrm{i}z-2\mathrm{i}z^* = 5$.

5 Shade the region in the complex plane satisfying the given inequalities and calculate its area:
(i) $0 \leqslant z+z^* \leqslant 6$ and $0 \leqslant \mathrm{i}(z-z^*) \leqslant 4$,
(ii) $2 \leqslant zz^* \leqslant 4$ and $0 \leqslant \arg z \leqslant \pi/2$.

6 Given that $z+w = 2$, find in terms of u and v, where $w = u+\mathrm{i}v$, the locus of the point w in the complex plane, given that: (i) $z+z^* = 2$, (ii) $|z-1| = 1$, (iii) $|z+2-\mathrm{i}| = 4$, (iv) $(1-\mathrm{i})z+(1+\mathrm{i})z^* = 6$.

7 Given that $w = z^2$, find the locus of w if the locus of z is the semicircle $|z| = 1$, $0 \leqslant \arg z \leqslant \pi$.

8 The variable non-zero complex number z satisfies the equation $|z-\mathrm{i}| = 1$. By writing z in the form $r(\cos\theta+\mathrm{i}\sin\theta)$ with $r > 0$, prove that the modulus of z is $2\sin\theta$. Obtain the modulus and argument of $1/z$ in terms of θ, and show that the imaginary part of $1/z$ is constant. Show in an Argand diagram the loci of the points which represent (i) z, (ii) $1/z$. (*JMB*)

9 Sketch, on separate Argand diagrams, the loci represented by (a) $|z| = 4$, (b) $|z-\mathrm{i}| = |z-3\mathrm{i}|$. The complex variables w and z are related by the transformation $w = 2z+5$. Sketch, also on separate Argand diagrams, the loci of w as z describes the loci (a) and (b) above. Write down a complex equation for each w-locus. *(L)*

10 Show that under the transformation given by $w = 1/(z+1)$, $(z \in \mathbb{C}, z \neq -1)$, the circle with centre -1 and radius r is transformed to the circle with centre O and radius $1/r$. Find, and illustrate in an Argand diagram, the image of the imaginary axis under this transformation. *(JMB)*

11 In an Argand diagram, P represents a complex number z such that $2|z-2| = |z-6i|$. Show that P lies on a circle, and find (a) the radius of this circle, (b) the complex number represented by its centre. If $\arg\{(z-2)/(z-6i)\} = \pi/2$, find the complex number represented by P. *(L)*

10.4 Exponential form of a complex number

The definition of a function by means of a convergent infinite series is well outside the scope of this book. However, if we assume certain results about the exponential series, then we can find some interesting connections between the polar form of a complex number and another form, called the exponential form. For our purposes, the exponential form may be regarded as a useful piece of notation and, if it is regarded in this way, it will not matter that we provide no proofs of the results on series.

The exponential series, §6.5, can be used to define the real exponential function e^x, that is

$$e^x = 1 + x + x^2/2 + x^3/3! + \dots + x^r/r! + \dots.$$

This infinite series converges for all real values of x and its sum defines the exponential function e^x which satisfies the rules of indices,

$$e^x e^y = e^{x+y}, \; e^{-x} = 1/e^x.$$

The corresponding series for complex values of x also converges and can be used to define the complex exponential function e^z for $z \in \mathbb{C}$. We shall state, without proof, some properties of e^z in the next theorem.

Definition $e^z = 1 + z + z^2/2 + z^3/3! + \dots + z^r/r! + \dots.$

Theorem 10.3 Let a, b, w, z be complex numbers, then

(i) the series e^z converges for all $z \in \mathbb{C}$,

(ii) the series with general term $aw^r/r! + bz^r/r!$ converges to the sum $ae^w + be^z$,

(iii) the series formed by multiplying the two series series e^w and e^z converges to the sum $e^w e^z = e^{w+z}$

The effect of this theorem is that we may perform arithmetical operations on the complex exponential series to obtain the corresponding arithmetical operations on the exponential functions. It should be remembered that these operations on series have not been justified since we give no proof of the Theorem 10.3.

Now let $z = x + iy$, x, $y \in \mathbb{R}$, then $e^z = e^{x+iy} = e^x e^{iy}$.
Since e^x is real, we look at the complex function e^{iy}.

$$\begin{aligned} e^{iy} &= 1 + iy + (iy)^2/2 + (iy)^3/3! + \ldots + (iy)^r/r! + \ldots \\ &= 1 - y^2/2 + y^4/4! + \ldots + (-1)^r y^{2r}/(2r)! + \ldots \\ &\quad + i\{y - y^3/3! + \ldots + (-1)^r y^{2r+1}/(2r+1)! + \ldots\}. \end{aligned}$$

Using the series for $\cos y$ and $\sin y$ from §6.5,

$$e^{iy} = \cos y + i \sin y = (1, y) \text{ in polar form.}$$

Changing the sign of i,

$$e^{-iy} = \cos y - i \sin y = (1, -y) \text{ in polar form,}$$

and hence

$$\cos y = (e^{iy} + e^{-iy})/2 \quad \text{and} \quad \sin y = (e^{iy} - e^{-iy})/2i.$$

Compare these equations with those for hyperbolic functions,

$$\cosh x = (e^x + e^{-x})/2 \quad \text{and} \quad \sinh x = (e^x - e^{-x})/2.$$

We can go further and define complex trigonometric and hyperbolic functions by means of series, since the series concerned all converge for all complex values of z.

Definition

$$\begin{aligned} \cos z &= 1 - z^2/2! + z^4/4! - z^6/6! + \ldots + (-1)^r z^{2r}/(2r)! + \ldots, \\ \sin z &= z - z^3/3! + z^5/5! - z^7/7! + \ldots + (-1)^r z^{2r+1}/(2r+1)! + \ldots, \\ \cosh z &= 1 + z^2/2! + z^4/4! + z^6/6! + \ldots + z^{2r}/(2r)! + \ldots, \\ \sinh z &= z + z^3/3! + z^5/5! + z^7/7! + \ldots + z^{2r+1}/(2r+1)! + \ldots. \end{aligned}$$

The same results are then obtained for complex functions as were previously obtained for functions of a real variable, namely:

$$\begin{aligned} e^{iz} &= \cos z + i \sin z, & e^z &= \cosh z + \sinh z, \\ e^{-iz} &= \cos z - i \sin z, & e^{-z} &= \cosh z - \sinh z, \\ \cos z &= (e^{iz} + e^{-iz})/2, & \cosh z &= (e^z + e^{-z})/2, \\ \sin z &= (e^{iz} - e^{-iz})/2i, & \sinh z &= (e^z - e^{-z})/2. \end{aligned}$$

It then follows that:

$$\begin{aligned} \cos iz &= \cosh z, & \cosh iz &= \cos z, \\ \sin iz &= i \sinh z, & \sinh iz &= i \sin z. \end{aligned}$$

With these definitions, all the sums and products formulae for trigonometric and hyperbolic functions are still true and may be verified by manipulation of the series or by using the exponential function. For example,

$$\cos 2z = (e^{2iz} + e^{-2iz})/2 = \{(e^{iz} + e^{-iz})^2 - 2\}/2 = 2\cos^2 z - 1,$$

$$\sin(w+z) = \{e^{i(w+z)} - e^{-i(w+z)}\}/2i$$
$$= \{(e^{iw} - e^{-iw})(e^{iz} + e^{-iz}) + (e^{iw} + e^{-iw})(e^{iz} - e^{-iz})\}/4i$$
$$= \sin w \cos z + \cos w \sin z.$$

Our use of the complex exponential is in the nature of a shorthand, using the fact that, for real θ,

$$e^{i\theta} = \cos\theta + i\sin\theta.$$

Then, if $|z| = r$ and $\arg z = \theta$, so that $z = (r, \theta)$ in polar form,

$$z = r(\cos\theta + i\sin\theta) = re^{i\theta},$$

which is the exponential form of a complex number.
In this exponential form, de Moivre's Theorem is very easy to remember since it becomes

$$z^n = (re^{i\theta})^n = r^n e^{in\theta}.$$

Note that
$$e^{2\pi i} = \cos 2\pi + i\sin 2\pi = 1,$$
$$e^{\pi i} = \cos\pi + i\sin\pi = -1.$$

EXERCISE 10.4

1 Using the definitions $\cos z = (e^{iz} + e^{-iz})/2$ and $\sin z = (e^{iz} - e^{-iz})/2i$, and properties of the exponential function, stated in Theorem 10.3, prove that:
(i) $\cos z$ is an even function and $\sin z$ is an odd function,
(ii) $\cos z$ and $\sin z$ have period 2π, that is, for any integer n,
$\cos(z + 2n\pi) = \cos z$ and $\sin(z + 2n\pi) = \sin z$,
(iii) $\cos(w+z) = \cos w \cos z - \sin w \sin z$,
(iv) $\sin(w-z) = \sin w \cos z - \cos w \sin z$,
(v) $\cos^2 z + \sin^2 z = 1$,
(vi) $\cos 2z = 1 - 2\sin^2 z$,
(vii) $\sin 2z = 2\sin z \cos z$.

2 Express, in the form $re^{i\theta}$, the complex number: (i) i, (ii) $-2i$, (iii) $1 + i\sqrt{3}$, (iv) $2 + 2i$, (v) $-2 + 2i$, (vi) $-3 - 4i$.

3 Verify that the transformation:
(i) $z \to z^*$ is a reflection of the complex plane in the real axis,
(ii) $z \to ze^{i\alpha}$ is a rotation of the complex plane about the origin through an angle α.
(iii) $z \to 2z$ is an enlargement of the complex plane by a scale factor 2,
(iv) $z \to wz$, where w is a constant complex number, is an enlargement of the complex plane by a scale factor $|w|$ together with a rotation of the plane about O through an angle $\arg w$.

4 Given that, in exponential form, $z = re^{i\theta}$, and that $w = z + 1/z$, prove that: (i) if the locus of z is the positive real axis, then the locus of w is the set $\{w: w \in \mathbb{R} \text{ and } w \geqslant 2\}$, (ii) if the locus of z is the unit circle $|z| = 1$, then the locus of w is that part of the real axis between the points -2 and 2.

5 By using the identity $\cos\theta = (e^{i\theta} + e^{-i\theta})/2$, or otherwise, show that $32\cos^6\theta = \cos 6\theta + 6\cos 4\theta + 15\cos 2\theta + 10$. Evaluate $\int_0^{\pi/2} \cos^6\theta \, d\theta$. (*JMB*)

6 Express $\sqrt{3}-\mathrm{i}$ in the form $r\mathrm{e}^{\mathrm{i}\theta}$, where $r > 0$ and $-\pi < \theta \leqslant \pi$. Hence show that, when n is a positive integer, $(\sqrt{3}-\mathrm{i})^n + (\sqrt{3}+\mathrm{i})^n = 2^{n+1}\cos(n\pi/6)$. (*L*)

7 Given that $2z = 1+\mathrm{e}^{\mathrm{i}\phi}$, $-\pi < \phi < \pi$, show that the real part of $1/z$ is independent of ϕ. Let C be the part of the circle $|2z-1| = 1$ in the z-plane for which the imaginary part of z is positive. Find the image of C in the w-plane under each of the transformations (i) $w = 1/z$, (ii) $w = z/(2z-1)$. (*JMB*)

MISCELLANEOUS EXERCISE 10

1 Represent on an Argand diagram the roots of the equation $z^3+8=0$. (*L*)

2 Show that $\mathrm{e}^{\mathrm{i}\pi/5}$ is a root of the equation $z^5+1=0$ and determine the other roots of this equation. Indicate, on an Argand diagram, the points which represent all these roots. Express the product of the non-real roots of the equation $z^5+1=0$ as a real number. (*JMB*)

3 Assuming de Moivre's theorem for an index which is a positive integer, prove that one value of $(\cos\theta+\mathrm{i}\sin\theta)^{p/q}$, where p and q are positive integers, is $\cos(p\theta/q)+\mathrm{i}\sin(p\theta/q)$. Determine, in the form $a+\mathrm{i}b$, the three cube roots of i. (*JMB*)

4 (i) Find z in the form $x+\mathrm{i}y$, where x and y are real, when $(z+1)(2-\mathrm{i}) = 3-4\mathrm{i}$ and show the position of z on an Argand diagram. (ii) By using de Moivre's theorem, or otherwise, express $\sin 4\theta$ and $\cos 4\theta$ in terms of powers of $\sin\theta$ and $\cos\theta$ and show that

$$\tan 4\theta = \frac{4\tan\theta(1-\tan^2\theta)}{1-6\tan^2\theta+\tan^4\theta}. \qquad (L)$$

5 (i) Show that the points in the Argand diagram corresponding to the values of z for which $(1-z)^n = z^n$ all lie on the line whose equation is $\mathrm{Re}(z) = 1/2$. (ii) Using de Moivre's theorem, or otherwise, expand $\cos 6\theta$ in powers of $\cos\theta$. (*L*)

6 Express $1+\mathrm{i}\sqrt{3}$ and $1-\mathrm{i}\sqrt{3}$ in the form $r\mathrm{e}^{\mathrm{i}\theta}$, where $r>0$ and $-\pi<\theta\leqslant\pi$. Hence evaluate $\left(\dfrac{1+\mathrm{i}\sqrt{3}}{1-\mathrm{i}\sqrt{3}}\right)^{20}$, expressing your result in the form $a+\mathrm{i}b$, where a and b are real numbers. (*L*)

7 The two complex numbers z and w are connected by the equation $z = \mathrm{e}^{\mathrm{i}w}$, where $w = t+\mathrm{i}k$, t and k are real and k is constant. Show that, as t varies, the point representing z in the Argand diagram describes a circle and give the centre and radius of this circle. (*L*)

8 (i) Shade on an Argand diagram the region R for which $|z| \leqslant |z-\mathrm{i}|$. If $w = z-3\mathrm{i}/2$, find the region for which $|w| \leqslant |w-\mathrm{i}|$. (ii) By using $2\cos n\theta = \mathrm{e}^{\mathrm{i}n\theta}+\mathrm{e}^{-\mathrm{i}n\theta}$ for suitable values of n, or otherwise, show that $2^5\cos^6\theta = \cos 6\theta + 6\cos 4\theta + 15\cos 2\theta + 10$. Hence, or otherwise, evaluate

$$\int_0^{\pi/4} 2^5\cos^6\theta\,\mathrm{d}\theta. \qquad (L)$$

9 Given that the points A, B, C represent the complex numbers a, b, c, respectively, in the Argand diagram and that $\dfrac{b-a}{c-a} = \mathrm{e}^{\mathrm{i}\pi/3}$, show that the triangle ABC is equilateral. Find the value of $\dfrac{c-b}{a-b}$ and deduce that $a^2+b^2+c^2 = bc+ca+ab$. (*JMB*)

10 (i) Simplify $(1+\omega)(1+\omega^2)$, where $\omega = \cos(2\pi/3)+\mathrm{i}\sin(2\pi/3)$. Express in the form $r(\cos\theta+\mathrm{i}\sin\theta)$ each of the three roots of the equation $(z-\omega)^3 = 1$. (ii) Express in terms of θ the roots α and β of the equation $z+z^{-1} = 2\cos\theta$. In the Argand diagram the points P and Q represent the numbers $(\alpha^n+\beta^n)$ and $(\alpha^n-\beta^n)$ respectively. Show that the length of PQ does not depend on the integer n. (*L*)

11 Using de Moivre's theorem, or otherwise, show that $\tan 5\theta = \dfrac{5t-10t^3+t^5}{1-10t^2+5t^4}$, where $t = \tan\theta$. Hence (i) show that $\tan\pi/5 = (5-2\sqrt{5})^{1/2}$, (ii) write down the value of the sum of the roots of the equation of the fifth degree in t obtained by putting $\tan 5\theta = 1$, and deduce that

$$\tan\frac{\pi}{20}+\tan\frac{9\pi}{20}+\tan\frac{13\pi}{20}+\tan\frac{17\pi}{20} = 4. \qquad (JMB)$$

12 (i) Solve the equation $z^5 = 1$ and represent the roots on an Argand diagram. If ω denotes any one of the non-real roots of $z^5 = 1$, show that $1+\omega+\omega^2+\omega^3+\omega^4 = 0$. (ii) By expressing $\sin\theta$ and $\cos\theta$ in terms of $\mathrm{e}^{\mathrm{i}\theta}$ and $\mathrm{e}^{-\mathrm{i}\theta}$, or otherwise, prove that

$$2^5\sin^4\theta\cos^2\theta = \cos 6\theta-2\cos 4\theta-\cos 2\theta+2. \qquad (L)$$

13 Prove de Moivre's theorem, that $(\cos\theta+\mathrm{i}\sin\theta)^n = \cos n\theta+\mathrm{i}\sin n\theta$, for $n\in\mathbb{Z}^+$. Putting $z = \cos\theta+\mathrm{i}\sin\theta$, write down expressions in terms of z and n for $\cos n\theta$ and $\sin n\theta$, and hence or otherwise prove that $\sin^4\theta\cos^2\theta = (\cos 6\theta-2\cos 4\theta-\cos 2\theta+2)/32$.

Evaluate $\displaystyle\int_0^{\pi/2}\sin^4\theta\cos^2\theta\,\mathrm{d}\theta$. (*C*)

14 (i) Given that $z = 1-\mathrm{i}$, find the values of r (>0) and θ, $-\pi<\theta\leqslant\pi$, such that $z = r(\cos\theta+\mathrm{i}\sin\theta)$. Hence, or otherwise, find $1/z$ and z^6, expressing your answers in the form $p+\mathrm{i}q$, where $p, q\in\mathbb{R}$. (ii) Sketch on an Argand diagram the set of points corresponding to the set A, where $A = \{z\colon z\in\mathbb{C}, \arg(z-\mathrm{i}) = \pi/4\}$. Show that the set of points corresponding to the set B, where $B = \{z\colon z\in\mathbb{C}, |z+7\mathrm{i}| = 2|z-1|\}$, forms a circle in the Argand diagram. If the centre of this circle represents the number z_1, show that z_1 is in A. (*L*)

15 The transformation $w = (z+1)^2+3$ maps the complex number $z = x+\mathrm{i}y$ to the complex number $w = u+\mathrm{i}v$. Show that as z moves along the y-axis from the origin to the point $(0, 2)$ in the z-plane, w moves from the point $(4, 0)$ to the point $(0, 4)$ along a curve in the w-plane. Write down the equation of this curve. (*JMB*)

16 If (r, θ) denotes, in polar form, the complex number with modulus r and argument θ, prove by induction that, for all $n\in\mathbb{N}$, $(r, \theta)^n = (r^n, n\theta)$, where $n\theta$ is reduced mod 2π. {You may assume that $(r, \theta)\times(s, \phi) = (rs, \theta+\phi)$, where the argument $\theta+\phi$ is reduced mod 2π.} By writing each side of the equation $z^5 = 1$ in polar form, and using the result above, find 5 *distinct* roots of the equation. Deduce the roots of the equation $z^5 = 32$. Show that, under multiplication, the set S of distinct roots of the equation $z^5 = 1$ forms an abelian group. Explain why the set of distinct roots of the equation $z^5 = 32$ is not a group under multiplication. Name a group isomorphic to S. (*SMP*)

17 Express z_1, where $z_1 = \dfrac{10 - \mathrm{i}2\sqrt{3}}{1 - \mathrm{i}3\sqrt{3}}$, in the form $p + \mathrm{i}q$, where p and q are real. Given that $z_1 = r(\cos\theta + \mathrm{i}\sin\theta)$, where $r > 0$ and $-\pi < \theta \leqslant \pi$, obtain values for r and θ. Hence determine ${z_1}^9$. Sketch on an Argand diagram the locus of the points representing the complex number z such that $|z - z_1| = \sqrt{3}$. (*L*)

18 Find, in trigonometric form, the roots of the equation $z^5 - 1 = 0$. Deduce that $z^4 + z^3 + z^2 + z + 1 = \left(z^2 - 2z\cos\dfrac{2\pi}{5} + 1\right)\left(z - 2z\cos\dfrac{4\pi}{5} + 1\right)$. Hence find a quadratic equation with integer coefficients whose roots are $\cos\dfrac{2\pi}{5}$ and $\cos\dfrac{4\pi}{5}$. (*JMB*)

19 (i) Find, in the form $r\mathrm{e}^{\mathrm{i}\theta}$, where $r > 0$ and $-\pi < \theta \leqslant \pi$, the four fourth roots of $8(-1 + \mathrm{i}\sqrt{3})$, and plot them on an Argand diagram. (ii) The points P and Q lie in the z-plane and the w-plane respectively. Given that $wz = 2$, find an equation of the path of Q when P describes the line $\operatorname{Re}(z) = c$, where c is a positive constant. Draw a sketch of the path of P and a sketch of the corresponding path of Q, giving the coordinates of any points where the paths cut the coordinate axes. (*L*)

20 Find the solutions in the range $0 \leqslant \theta < \pi$ of the equation $\sin 5\theta = \sin 4\theta$. By considering $(\cos\theta + \mathrm{i}\sin\theta)^5$, or otherwise, express $\sin 5\theta/\sin\theta$ in terms of c where $c = \cos\theta$ and $\sin\theta \neq 0$. Assuming, without proof, that $\sin 4\theta/\sin\theta = 8c^3 - 4c$ show that when $\sin\theta \neq 0$ $\dfrac{\sin 5\theta - \sin 4\theta}{\sin\theta} = 16c^4 - 8c^3 - 12c^2 + 4c + 1$. By writing $x = 2c$, show that the roots of the equation $x^4 - x^3 - 3x^2 + 2x + 1 = 0$ are $2\cos(\pi/9)$, $2\cos(5\pi/9)$, $2\cos(7\pi/9)$ and 1. (*JMB*)

21 Express the roots of the equation $x^7 - 1 = 0$ as powers of a, where $a = \mathrm{e}^{2\pi\mathrm{i}/7}$. Find the numerical value of $a + a^2 + a^3 + a^4 + a^5 + a^6$. Given that $p = a(1 + a + a^3)$ and $q = a^3(1 + a^2 + a^3)$, show that p and q are the roots of the equation $x^2 + x + 2 = 0$. By solving this quadratic equation, or otherwise, find the values of $\cos(2\pi/7) + \cos(4\pi/7) + \cos(8\pi/7)$ and $\sin(2\pi/7) + \sin(4\pi/7) + \sin(8\pi/7)$. (*JMB*)

22 Show that the transformation of the complex plane given by $w = (z - a)/(z - b)$, $z \neq b$, where a and b are real numbers, transforms the circle with centre b and radius r, where $r > 0$, into the circle with centre 1 and radius $|b - a|/r$. Let T be the set of all transformations of the form $z \to (z - a)/(z - b)$, $z \neq b$, where a and b are real numbers. A relation R is defined on T by $\mathrm{f\,R\,g} \Leftrightarrow \mathrm{f(i) - g(i)}$ has equal real and imaginary parts, where $\mathrm{f, g} \in T$. Prove that R is an equivalence relation. Find two members of T which transform the circle with centre 1 and radius r, where $r > 0$, into itself. Show that these two transformations are related by R. (*JMB*)

23 The transformation T: $z \to w$ in the complex plane is defined by $w = (az + b)/(z + c)$, $(a, b, c \in \mathbb{R})$. Given that $w = 3\mathrm{i}$ when $z = -3\mathrm{i}$, and $w = 1 - 4\mathrm{i}$ when $z = 1 + 4\mathrm{i}$, find the values of a, b and c. (i) Show that the points for which z is transformed to z^* lie on a circle and give the centre and radius of this circle. (ii) Show that the line through the point $z = 4$ and perpendicular to the real axis is invariant under T. (*C*)

24 (i) Prove that the set of complex numbers which are solutions of the equation $z^8 = 1$ form a group G_1 under the operation of multiplication of complex numbers (which may be assumed to be associative). Identify all proper subgroups of G_1.

(ii) Prove that the set of matrices of the form $\begin{pmatrix} \cos\theta & -\sin\theta \\ \sin\theta & \cos\theta \end{pmatrix}$, where $\theta \in \mathbb{R}$, form a group G_2 under the operation of matrix multiplication, (which may be assumed to be associative).

(iii) Write down a subgroup of G_2 which is isomorphic to G_1. (*C*)

25 Given that $z_1 = \dfrac{9+\mathrm{i}\sqrt{3}}{5-\mathrm{i}\sqrt{3}}$, $z_2 = \dfrac{6\mathrm{i}}{\sqrt{3}+\mathrm{i}}$, express z_1 and z_2 in the form $p+\mathrm{i}q$, where $p, q \in \mathbb{R}$. Show also that $z_1 = \sqrt{3}\{\cos(\pi/6) + \mathrm{i}\sin(\pi/6)\}$ and $z_2 = 3\{\cos(\pi/3) + \mathrm{i}\sin(\pi/3)\}$. Given that $z_3 = z_1^4 - z_2^2$ express z_3 in the form $a + \mathrm{i}b$, where $a, b \in \mathbb{R}$. Plot on an Argand diagram the points representing z_1, z_2 and z_3. (*L*)

26 (i) Show on an Argand diagram (a) the three roots of the equation $z^3 - 1 = 0$, (b) the four roots of the equation $z^4 - 16 = 0$.

(ii) If $z = \mathrm{e}^{\mathrm{i}\theta}$, show that (a) $z + z^{-1} = 2\cos\theta$, (b) $z^n + z^{-n} = 2\cos n\theta$. Show also that $\cos^6\theta = (\cos 6\theta + 6\cos 4\theta + 15\cos 2\theta + 10)/32$. (*L*)

27 (i) Sketch on an Argand diagram the locus represented by the equation $|z-1| = 1$. Shade on your diagram the region for which $|z-1| < 1$ and $\pi/6 < \arg z < \pi/3$. (ii) Show that the points in the Argand diagram representing the solutions of the equation $z^8 = 256$ lie at the vertices of a regular octagon. (*L*)

28 If $z = \cos\theta + \mathrm{i}\sin\theta$, show that $z - 1/z = 2\mathrm{i}\sin\theta$, $z^n - 1/z^n = 2\mathrm{i}\sin n\theta$. Hence, or otherwise, show that $16\sin^5\theta = \sin 5\theta - 5\sin 3\theta + 10\sin\theta$. (*L*)

29 By using de Moivre's theorem, or otherwise, show that, when $\cos\theta \neq 0$, $\cos 8\theta = \cos^8\theta\,(1 - 28\tan^2\theta + 70\tan^4\theta - 28\tan^6\theta + \tan^8\theta)$. (*L*)

30 Shade on an Argand diagram the region R for which $|z| < 1$. If z is any point in the region R, and z^* is the complex conjugate of z, find the corresponding regions for w when (a) $w = z + 3 + 4\mathrm{i}$, (b) $|wz| = 1$, (c) $w = zz^*$. (*L*)

31 Prove, by induction or otherwise, de Moivre's theorem that $(\cos\theta + \mathrm{i}\sin\theta)^n = \cos n\theta + \mathrm{i}\sin n\theta$, $(n \in \mathbb{Z}^+)$. If $z = \cos\theta + \mathrm{i}\sin\theta$, show that $z - 1/z = 2\mathrm{i}\sin\theta$ and that $z^n + 1/z^n = 2\cos n\theta$, $(n \in \mathbb{Z}^+)$. Hence, or otherwise, express $\sin^4\theta$ and $\sin^6\theta$ in terms of cosines of multiples of θ and prove that $\displaystyle\int_0^{\pi/6} \sin^4\theta\cos 2\theta\,\mathrm{d}\theta = \frac{5\sqrt{3}}{64} - \frac{\pi}{24}$. (*C*)

11 Differential Equations

11.1 Introduction and revision

This chapter extends the work on Differential Equations of Chapter 25 of *Pure Mathematics*. We begin with an example.

EXAMPLE 1 *Sketch the family of solutions of the differential equation* $\frac{dy}{dx} = (x+1)(y-2)$, *and find the particular solution passing through the origin.*

We first demonstrate a method of sketching the family of solutions, sometimes called the *compass needle* method. It provides a quick way of seeing the general shape of the various solutions. Also it may easily be programmed for a computer.

The equation $\frac{dy}{dx} = (x+1)(y-2)$ provides a formula for the gradient of a curve, given any particular pair of values (x, y). The compass needle method repeats this calculation for a representative set of (x, y) coordinates. A small line segment is plotted at (x, y) and this is a part of the tangent to the graph of the solution passing through that point. It is then relatively easy to imagine these line segments joined up to form the *family of solution curves*. The calculation is laid out in table form.

Table 11.1 Values of $\frac{dy}{dx}$ ***at (x, y)***

$x=$	-3	-2	-1	0	1	2
$y=$ 4	-4	-2	0	2	4	6
3	-2	-1	0	1	2	3
2	0	0	0	0	0	0
1	2	1	0	-1	-2	-3
0	4	2	0	-2	-4	-6
-1	6	3	0	-3	-6	-9

The line segments are then plotted, using this data and the results are shown in Fig. 11.1.

A greater number of coordinate plottings will give even more information but it can be seen that the solution curves appear to be symmetrical about two axes $x = -1$ and $y = 2$.

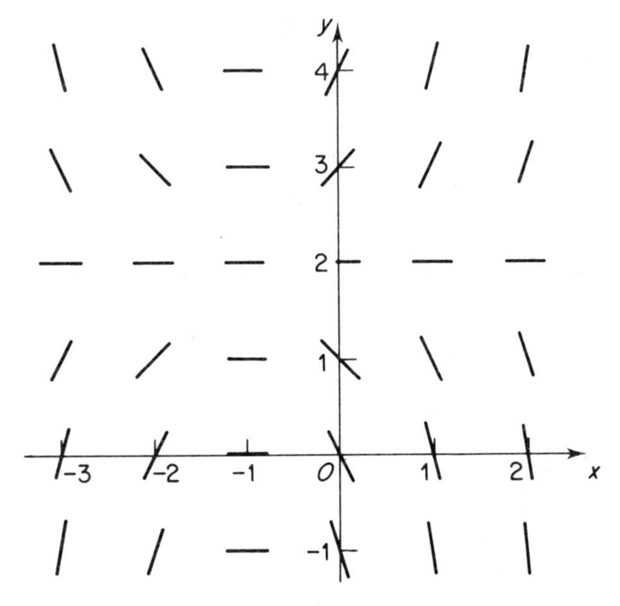

Fig. 11.1 Line segments showing $\frac{dy}{dx}$ at (x, y), $\frac{dy}{dx} = (x+1)(y-2)$

We now formally solve this differential equation by *separation of variables.*

$$\frac{dy}{dx} = (x+1)(y-2) \text{ gives } \frac{1}{y-2}\frac{dy}{dx} = x+1, \text{ if } y \neq 2.$$

Integrate both sides with respect to x

$$\int \frac{1}{y-2}\,dy = \int \frac{1}{y-2}\frac{dy}{dx}\,dx = \int (x+1)\,dx,$$

hence, $\ln|y-2| = x^2/2 + x + A$, where A is a constant of integration. On putting $B = e^A$, this may be rewritten

$$y = 2 + Be^{(x^2/2+x)}.$$

This is called the *general solution* of the differential equation and it involves one arbitrary (unknown) constant B.

The *particular solution* which passes through (0, 0) is found by substitution, $0 = 2 + Be^0$. Thus $B = -2$ and $\mathbf{y = 2\{1 - e^{(x^2/2+x)}\}}$. We sketch this and other members of the family of solutions in Fig. 11.2, labelling the curves with the values of the constant B.

The method of separation of variables is applicable to any first order differential equation $\frac{dy}{dx} = f(x, y)$, where the function f can be split into a product of two functions, one of x and one of y, $f(x, y) = g(x)h(y)$. Certain

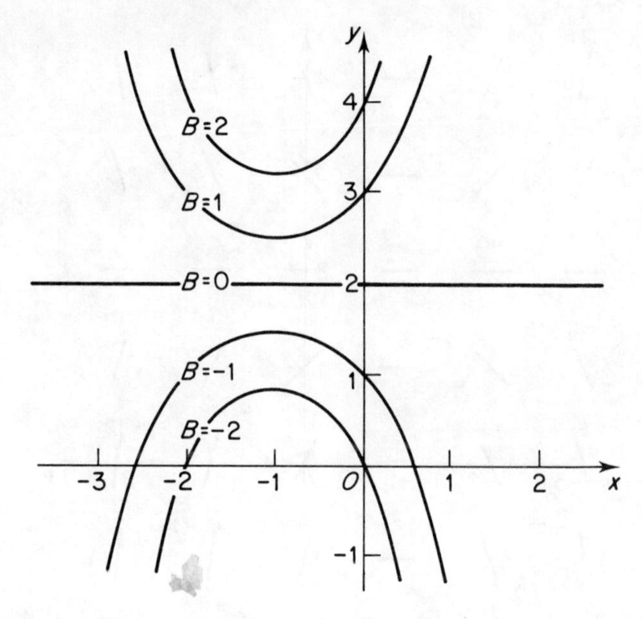

Fig. 11.2 Family of solutions of $\dfrac{dy}{dx} = (x+1)(y-2)$; $y = 2 + Be^{(x^2/2+x)}$

other first order equations, called *homogeneous equations*, can also be solved by this method after a suitable substitution, as is shown in Example 2.

EXAMPLE 2 *Find the general solution of the equation* $\dfrac{dy}{dx} = \dfrac{3x^2+y^2}{4x^2}$.

Rewrite $\dfrac{3x^2+y^2}{4x^2}$ as $\dfrac{1}{4}\left\{3+\left(\dfrac{y}{x}\right)^2\right\}$. A suitable substitution will then be $\dfrac{y}{x} = u$,

then $y = xu$ and $\dfrac{dy}{dx} = u + x\dfrac{du}{dx}$, and the differential equation becomes

$$u + x\frac{du}{dx} = \frac{1}{4}(3+u^2),$$

$$\frac{4}{u^2-4u+3}\frac{du}{dx} = \frac{1}{x},$$

$$2\int\left\{\frac{1}{u-3} - \frac{1}{u-1}\right\}du = \int\frac{1}{x}\,dx,$$

$$\ln\left|\frac{u-3}{u-1}\right| = \tfrac{1}{2}\ln|x| + A, \text{ where } A \text{ is a constant,}$$

or $$\frac{u-3}{u-1} = Bx^{\frac{1}{2}}, \text{ where } B = e^A.$$

Replacing u by y/x, the required general solution is $\dfrac{y-3x}{y-x} = Bx^{\frac{1}{2}}$.

EXERCISE 11.1

1 Use the compass needle method to sketch the family of solution curves of the differential equations:

(i) $\dfrac{dy}{dx} = x^2$, (ii) $\dfrac{dy}{dx} = -\dfrac{x}{y}$, (iii) $\dfrac{dy}{dx} = \dfrac{y}{x}$, (iv) $\dfrac{dy}{dx} = 4y$.

2 Find the general solution of the differential equations, given in Question 1. Calculate the particular solution which passes through the following points, respectively: (i) (3, 10), (ii) (3, 4), (iii) (1, 2), (iv) (0, 2).

3 Solve the differential equation by separating the variables: (i) $y\dfrac{dy}{dx} = x$,

(ii) $\dfrac{dy}{dx} = x(1-y)$, (iii) $x\dfrac{dy}{dx} = \sin y$, (iv) $x + 3y\dfrac{dy}{dx} = 1 - x^2$,

(v) $\dfrac{dy}{dx} = \dfrac{\cos x}{(1-y)^2}$, (vi) $\dfrac{dy}{dx} = y - yx^2$, (vii) $\dfrac{dx}{dt} + t = x^2 t$, (viii) $t + x\dfrac{dx}{dt} = 2$.

4 Use the substitution $y = ux$ to solve the differential equation:

(i) $\dfrac{dy}{dx} = \dfrac{2y-x}{2x-y}$, (ii) $2\dfrac{dy}{dx} = \dfrac{y^2}{x^2} + \dfrac{y}{x}$, (iii) $\dfrac{dy}{dx} = \dfrac{xy}{x^2-y^2}$,

(iv) $x\dfrac{dy}{dx} = y + \sqrt{(x^2+y^2)}$.

5 A body, falling under gravity through a liquid, experiences a resistive force proportional to v^2, where v is the velocity of the body at time t. The motion of the body is described by the equation $\dfrac{dv}{dt} = g - kv^2$, where g and k are positive constants. Assuming that $v = 0$ when $t = 0$, and that $k = g/4$, show that $v = 2\tanh(gt/2)$. Hence find the terminal (or limiting) velocity of the body.

6 The size P of a population at time t (years) approximately satisfies the differential equation $\dfrac{dP}{dt} = kP$, where k is a constant. Find P as a function of t. The population numbered 32 million in the year 1900 and had increased to 48 million by 1970. Prove that $k \approx 0{\cdot}0058$ and estimate the size of the population in the year 2000.

7 A flask contains 20 litres of a solution of a chemical in water. The concentration of the chemical is reduced by running in pure water at a rate of 2 litres per minute and allowing 3 litres per minute of the (well mixed) solution to run out of the flask. Suppose that there are x kg of chemical in the flask at time t minutes after the dilution begins. Show that $\dfrac{dx}{dt} = \dfrac{-3x}{20-t}$. Find the general solution of this differential equation. Assuming that there is 1 kg of chemical in the flask initially, find the mass of chemical in the flask after 5 minutes.

8 Which of the following differential equations can be solved by separation of variables, which are exact and can be integrated immediately and which cannot be solved by either of these methods:

(i) $\dfrac{dy}{dx} = yx^2$, (ii) $\dfrac{dy}{dx} = y + x^2$,

(iii) $x^2\dfrac{dy}{dx} + 2xy = e^x$, (iv) $3x\dfrac{dy}{dx} - x^2y = 4$,

(v) $e^{-x}\dfrac{dy}{dx} - e^{-x}y = \sin x$, (vi) $\dfrac{dy}{dx} = xy + 2x - y - 2$.

9 Express y in terms of x, given that $\dfrac{dy}{dx} = (y+1)(2x+1)$ and that $y = 2$ when $x = 0$. (*L*)

10 Differentiate $\ln(\sec x)$ with respect to x. Find y in terms of x, given that $\dfrac{dy}{dx} = y(\tan x + \sec^2 x)$ and that $y = 1$ when $x = 0$. (*L*)

11 A tank, with a rectangular base and having vertical sides of height h, is initially full of water. The water leaks out of a small hole in the horizontal base of the tank at a rate which, at any instant, is proportional to the square root of the depth of the water at that instant. If x is the depth of the water at time t after the leak started, write down a differential equation connecting x and t. If the tank is exactly half empty after one hour, find the further time that elapses before the tank becomes completely empty. (*C*)

12 A colony of bacteria is being grown in a shallow dish of area 100 cm^2. After t days the area of the colony is x cm^2. The growth of the colony is modelled by the differential equation $\dfrac{dx}{dt} = t(100 - x)/25$. Given that initially the colony occupies 1 cm^2, solve the differential equation, and calculate the number of days that will elapse before the colony has covered half the remaining area of the dish (i.e. $x = 50{\cdot}5$). Show that the rate of growth is at its maximum when $t = 5$, and that the area of the colony is then nearly 40 cm^2. Sketch the graph of x against t for $t \geqslant 0$. (*C*)

11.2 First order linear differential equations—integrating factors

The method of separation of variables only works for a limited number of differential equations, those for which $\dfrac{dy}{dx}$ can be written as a product of a function of x and a function of y. Consider the following three first order differential equations.

$$\text{(i)}\ \frac{dy}{dx} = y^2x, \quad \text{(ii)}\ \frac{dy}{dx} = y^2 + x, \quad \text{(iii)}\ \frac{dy}{dx} = y + x^2.$$

Equation (i) can be separated. Equation (ii) cannot be solved exactly and the best method of solution would be a numerical method, perhaps a refinement of the compass needle method. Equation (iii) can be solved because it is a member of a particular family of differential equations. It is a *linear* differential equation because both y and $\frac{dy}{dx}$ occur in linear form, with no product or square, as in (i) and (ii).

Definition A differential equation in x, y and derivatives of y is *linear* if y and its derivatives all appear in linear form, with no products or powers.

In a linear differential equation the coefficients of y and its derivatives are any functions of x, the independent variable. Thus the general first order linear differential equation is of the form

$$F(x)\frac{dy}{dx} + G(x)y = H(x).$$

Such an equation can often be solved by changing the left-hand side into an exact derivative by multiplication by a factor, called an *integrating factor*. The problem is to find the integrating factor.

EXAMPLE 1 *Solve* $x\frac{dy}{dx} + y = e^x$

In this case, the left-hand side is already exact being the derivative of xy, since $\frac{d}{dx}(xy) = x\frac{dy}{dx} + y$. Then $\frac{d}{dx}(xy) = e^x$ and so $xy = \int e^x\,dx = e^x + C$, where C is a constant, and the general solution is $\mathbf{y = (e^x + C)/x}$.

EXAMPLE 2 *Find the general solution of the differential equation* $x \ln x\frac{dy}{dx} + y = 3 \ln x$.

The left-hand side is not exact in this case but it becomes exact after division by x. $\ln x\frac{dy}{dx} + \frac{1}{x}y = \frac{d}{dx}(y \ln x)$, so we divide the equation throughout by x and integrate.

$$(\ln x)y = \int \frac{3 \ln x}{x}\,dx = \frac{3}{2}(\ln x)^2 + C,$$

where C is a constant, and so $\mathbf{y = \frac{3}{2} \ln x + \frac{C}{\ln x}}$.

We now consider the general case $F(x)\dfrac{dy}{dx} + G(x)y = H(x)$.

The first step is to reduce the coefficient of $\dfrac{dy}{dx}$ to unity by dividing by $F(x)$, thus

$$\frac{dy}{dx} + P(x)y = Q(x), \text{ where } P(x) = \frac{G(x)}{F(x)},\ Q(x) = \frac{H(x)}{F(x)}.$$

We now try to discover a function $R(x)$ with which to multiply the equation so that the left-hand side becomes the exact derivative $\dfrac{d}{dx}(R(x)y)$.

That is $R(x)\dfrac{dy}{dx} + R(x)P(x)y = \dfrac{d}{dx}(R(x)y) = R(x)\dfrac{dy}{dx} + R'(x)y$.

The required condition is that $R'(x) = R(x)P(x)$ or $\dfrac{1}{R(x)}R'(x) = P(x)$, and, on integrating, $\ln R(x) = \int P(x)dx$, so that $R(x) = e^{\int P(x)dx}$. $R(x)$ is called the integrating factor and the solution continues:

$$\frac{d}{dx}(R(x)y) = R(x)Q(x)$$

$$R(x)y = \int R(x)Q(x)\,dx$$

or

$$e^{\int P(x)dx}y = \int e^{\int P(x)dx}Q(x)\,dx.$$

We state this result as a theorem. Note that we use the form involving $P(x)$ and $Q(x)$ rather than that involving $F(x)$, $G(x)$ and $H(x)$.

Theorem 11.1 The first order linear differential equation

$$\frac{dy}{dx} + P(x)y = Q(x)$$

can be transformed into a form with exact left-hand side by the multiplication by the integrating factor $e^{\int P(x)dx}$.

EXAMPLE 3 *Solve the following equations, by use of the integrating factor method where possible:*

(*i*) $\sin x\dfrac{dy}{dx} + 3y\cos x = 2$, (*ii*) $\dfrac{dy}{dx} + 4y = x^2$, (*iii*) $x^2\dfrac{dy}{dx} + y^2 = 3e^x$.

(i) First, divide by $\sin x$ to make the coefficient of $\dfrac{dy}{dx}$ equal to 1.

$\dfrac{dy}{dx} + 3y\cot x = 2\operatorname{cosec} x$. This equation is linear with an integrating factor

$R(x)$, where $R(x) = e^{3\int \cot x\,dx} = e^{3\ln\sin x} = \sin^3 x$. On multiplying by $\sin^3 x$, $\frac{d}{dx}(\sin^3 x\, y) = 2\operatorname{cosec} x \sin^3 x = 2\sin^2 x$. On integrating, $\sin^3 x\, y = \int 2\sin^2 x\,dx = \int (1-\cos 2x)\,dx = x - \frac{1}{2}\sin 2x + C$. Therefore,

$$y = \frac{2x - \sin 2x + D}{2\sin^3 x},$$ where D is an arbitrary constant.

(ii) This equation is particularly simple, with $P(x) = 4$ and integrating factor $R(x)$ given by $R(x) = e^{\int 4dx} = e^{4x}$. Then $\frac{d}{dx}(e^{4x}y) = e^{4x}x^2$, and

$$e^{4x}y = \int x^2 e^{4x}\,dx = \frac{1}{4}x^2 e^{4x} - \int \frac{1}{2}xe^{4x}\,dx = \frac{1}{4}x^2 e^{4x} - \frac{1}{8}xe^{4x} + \int \frac{1}{8}e^{4x}\,dx$$

$$= \left(\frac{1}{4}x^2 - \frac{1}{8}x + \frac{1}{32}\right)e^{4x} + C,$$

and $$y = \frac{1}{4}x^2 - \frac{1}{8}x + \frac{1}{32} + Ce^{-4x}.$$

(iii) This differential equation is not linear so the integrating factor method does not work. This is demonstrated by using the method in an attempt to find an integrating factor.

$$R(x) = e^{\int (1/x^2)dx} = e^{(-1/x)}, \quad \text{but} \quad \frac{d}{dx}(e^{-1/x}y) = e^{-1/x}\frac{dy}{dx} + \frac{1}{x^2}e^{-1/x}y$$

$$= e^{-1/x}\left(\frac{dy}{dx} + \frac{1}{x^2}y\right),$$

and, for the method to work, we would need the derivative to be $e^{-1/x}\left(\frac{dy}{dx} + \frac{y^2}{x^2}\right)$. The linearity condition is a necessary one in the method.

EXERCISE 11.2

1 Calculate the integrating factor $R(x)$, $R(x) = e^{\int P(x)dx}$, given the function $P(x)$: (i) $P(x) = -6$, (ii) $P(x) = 3x^2 - 2$, (iii) $P(x) = \frac{1}{x}$, (iv) $P(x) = \tan x$, (v) $P(x) = -\frac{2}{x}$, (vi) $P(x) = \frac{1}{x\ln x}$.

2 Multiply by the integrating factors, calculated in question 1, to solve the differential equation: (i) $\frac{dy}{dx} - 6y = 0$, (ii) $\frac{dy}{dx} + (3x^2 - 2)y = 0$,

(iii) $x\dfrac{dy}{dx}+y=0$, (iv) $\cos x\dfrac{dy}{dx}+\sin x\, y=0$, (v) $x\dfrac{dy}{dx}-2y=0$,

(vi) $x\ln x\dfrac{dy}{dx}+y=0$.

3 Solve the differential equation using the integrating factor method:

(i) $\dfrac{dy}{dx}+2y=3x$, (ii) $x\dfrac{dy}{dx}-2y=x^3+1$, (iii) $\cos x\dfrac{dy}{dx}-\sin x\, y=2\sec^2 x$,

(iv) $\ln x\dfrac{dy}{dx}-\dfrac{y}{x}=\dfrac{2\ln x}{x}$, (v) $x\cos x\dfrac{dy}{dx}+(x\sin x+\cos x)y=2$,

(vi) $\dfrac{dy}{dx}=\dfrac{3y}{x-1}+(x-1)^4$.

4 Find the particular solution of the following equations, given that $y=1$ when $x=1$:

(i) $(1+x)\dfrac{dy}{dx}-y=2$, (ii) $x\dfrac{dy}{dx}-y=6x^3\,e^x$,

(iii) $(1+x^2)\dfrac{dy}{dx}+4xy=3$, (iv) $2\dfrac{dy}{dx}+5y=4e^{-3x}$.

5 Obtain the general solution of the differential equation $x\dfrac{dy}{dx}-y=x^2+1$ in the form $y=x^2+Cx-1$, where C is an arbitrary constant. Show that each solution curve of the differential equation has one minimum point. Find the equation of the curve on which all these minimum points lie. Sketch three members of the family of solution curves, one for some negative value of C, one for some positive value of C, and $C=0$. *(C)*

6 Use the substitution $z=1/y^2$ to prove that the differential equation

$$x^2\frac{dy}{dx}+xy=y^3 \text{ can be reduced to } \frac{dz}{dx}-\frac{2z}{x}=-\frac{2}{x^2}.$$

Hence, or otherwise, find the particular solution of the differential equation in y and x for which $y=1$ when $x=1$. *(C)*

7 Find the general solution of the differential equation

$$t\frac{dx}{dt}+3x=\sqrt{(3+4/t^2)}. \qquad (C)$$

8 Find the solution of the differential equation $\dfrac{dy}{dx}+y\cot x=\cos 3x$ for which $y=1$ when $x=\pi/6$. *(JMB)*

9 Find the solution of $(x-1)\dfrac{dy}{dx}-2y=x-1$ for which $y=1$ when $x=\dfrac{1}{2}$. *(JMB)*

10 Solve the differential equation $x^2\dfrac{dy}{dx}+y=0$ for $x>0$, given that $y=1$ when $x=1$. *(SMP)*

11 Solve the differential equation $\dfrac{dz}{d\theta} - z\tan\theta + \sec\theta = 0$. (*AEB*)

12 Solve the differential equation $x\dfrac{dy}{dx} + 2y = xe^x$, given that $y = 2e$ when $x = 1$. (*AEB*)

13 Find the general solution of the differential equation $\dfrac{dy}{dx} + \dfrac{y}{x} = \dfrac{1}{x+1}$, $(x > 0)$. Find also the particular solution for which $y = 1$ when $x = 1$. (*C*)

11.3 Linear differential equations with constant coefficients

We now restrict our attention to a particular class of linear differential equations in x and y, namely those in which the coefficients of y and its derivatives are constant. This type of differential equation occurs frequently in mechanics problems and examples were met in Questions 2(i) and 3(i) of Exercise 11.2. It should be noted that, in these equations, the integrating factor was particularly simple: thus the integrating factor was

$$e^{-6x} \text{ for } \frac{dy}{dx} - 6y = 0 \quad \text{and was} \quad e^{2x} \text{ for } \frac{dy}{dx} + 2y = 3x.$$

The general first and second order linear differential equations with constant coefficients can be written

$$\frac{dy}{dx} + ay = f(x) \quad \text{and} \quad \frac{d^2y}{dx^2} + a\frac{dy}{dx} + by = f(x),$$

with $a, b \in \mathbb{R}$ and with f some real function.

We begin by considering the homogeneous equations in which $f(x) = 0$.

In the first order case, $\dfrac{dy}{dx} + ay = 0$, $\dfrac{dy}{dx} = -ay$, $\displaystyle\int \frac{1}{y}\,dy = \int -a\,dx$, $\ln|y| = -ax + C$, for constant C, and the solution is $y = e^{-ax+C}$ or $y = Ae^{-ax}$, where $A = e^C$ is also constant. Notice that the coefficient of x in e^{-ax} is $-a$, which is the root of the equation $m + a = 0$, for some variable m. Just as $\dfrac{d}{dx}$ may be regarded as a differential operator, operating on y, $\dfrac{dy}{dx} + ay$ may be written $\left(\dfrac{d}{dx} + a\right)y$ and $\left(\dfrac{d}{dx} + a\right)$ may be regarded as a differential operator, operating on y. The left-hand side of the equation $m + a = 0$ is obtained by replacing $\dfrac{d}{dx}$ by m in the operator $\left(\dfrac{d}{dx} + a\right)$. This leads to the following definition.

Definition For a linear differential equation with constant coefficients the *auxiliary* (or characteristic) equation is obtained from the differential operator by replacing $\dfrac{d}{dx}$ by m and equating to 0.

It so happens that the solution of the general homogeneous second order linear differential equation can be similarly obtained. Write $\dfrac{d^2y}{dx^2}+a\dfrac{dy}{dx}+by=\left(\dfrac{d^2}{dx^2}+a\dfrac{d}{dx}+b\right)y$ and equate to zero the quadratic in m obtained by replacing $\dfrac{d}{dx}$ by m in the operator $\dfrac{d^2}{dx^2}+a\dfrac{d}{dx}+b$. This gives the auxiliary equation $m^2+am+b=0$. Three cases arise since the roots of the quadratic may be (i) two real and distinct roots, (ii) one repeated real root, (iii) two complex roots. We show how the solution is obtained in each case.

Case (i) Two real distinct roots α and β.

$$m^2+am+b=(m-\alpha)(m-\beta) \text{ so}$$

$$\left(\frac{d^2}{dx^2}+a\frac{d}{dx}+b\right)y=\left(\frac{d}{dx}-\alpha\right)\left(\frac{d}{dx}-\beta\right)y=0.$$

The factorisation of the operator is justified since the differential operator $\dfrac{d}{dx}$ commutes with a constant and the distributive law holds. This becomes a first order equation by the substitution $z=\left(\dfrac{d}{dx}-\beta\right)y$, namely $\left(\dfrac{d}{dx}-\alpha\right)z=0$ and its solution is $z=Ce^{\alpha x}$, C constant. Substituting back for z gives the linear equation $\left(\dfrac{d}{dx}-\beta\right)y=Ce^{\alpha x}$. This equation has integrating factor $e^{-\beta x}$, so that

$$e^{-\beta x}y=\int Ce^{\alpha x}e^{-\beta x}\,dx=\frac{C}{\alpha-\beta}e^{(\alpha-\beta)x}+B, \text{ where } B \text{ is constant.}$$

On multiplying by $e^{\beta x}$, $y=Ae^{\alpha x}+Be^{\beta x}$, putting $A=C/(\alpha-\beta)$, which is also constant. Notice that it is essential that α and β are distinct since we divide by $\alpha-\beta$.

Case (ii) 1 real repeated root α.

Put $\beta=\alpha$ in (i) and follow through to the second integration, where

$$e^{-\alpha x}y=\int Ce^{\alpha x}e^{-\alpha x}\,dx=Cx+B, \text{ where } B \text{ is constant.}$$

Then $\qquad y=(Ax+B)e^{\alpha x}$, where A and B are constants.

Case (iii) 2 complex roots $\alpha \pm i\beta$.

If a function f of a real variable x contains some coefficients which are complex, these may be split into their real and imaginary parts so that $f(x) = g(x) + ih(x)$, where g and h are real functions. We may then regard the differentiation and integration of f as the result of performing the corresponding operation on g and h and then adding i times the second to the first. In this case, we therefore follow the work of case (i) and obtain

$y = Ce^{(\alpha + i\beta)x} + De^{(\alpha - i\beta)x}$, where C and D are complex constants.

Then $y = e^{\alpha x}(Ce^{i\beta x} + De^{-i\beta x})$

$$= e^{\alpha x}\left\{\left(\frac{C+D}{2}\right)(e^{i\beta x} + e^{-i\beta x}) + \left(\frac{C-D}{2}\right)(e^{i\beta x} - e^{-i\beta x})\right\}$$

$$= e^{\alpha x}(A\cos\beta x + B\sin\beta x), \text{ where } A = \frac{C+D}{2} \text{ and}$$

$$B = i\left(\frac{C-D}{2}\right).$$

Since y is real for all values of x, A and B are real and so C and D are complex conjugates. Thus, the solution is

$y = e^{\alpha x}(A\cos\beta x + B\sin\beta x)$, where A and B are arbitrary constants.

These results are summarised in the following theorem:

Theorem 11.2 The general solution of the second order differential equation with constant coefficient $\frac{d^2y}{dx^2} + a\frac{dy}{dx} + by = 0$ depends on the roots of the auxiliary equation $m^2 + am + b = 0$. If there are:

(i) 2 real distinct roots α, β then $y = Ae^{\alpha x} + Be^{\beta x}$,
(ii) 1 real repeated root α then $y = (Ax + B)e^{\alpha x}$,
(iii) 2 complex roots $\alpha \pm i\beta$ then $y = (A\cos\beta x + B\sin\beta x)e^{\alpha x}$.

Each of these solutions contains two arbitrary constants, A and B, as is to be expected, since the solution of a second order differential equation involves two integrations. That they are indeed solutions is checked by substitution in the differential equation. We show this for the case of two complex roots, with $A = 1$ and $B = 0$, and leave the other cases as exercises for the reader.

Let $y = e^{\alpha x}\cos\beta x$, then

$$\frac{dy}{dx} = \alpha e^{\alpha x}\cos\beta x - \beta e^{\alpha x}\sin\beta x = e^{\alpha x}(\alpha\cos\beta x - \beta\sin\beta x) \text{ and}$$

$$\frac{d^2y}{dx^2} = \alpha e^{\alpha x}(\alpha\cos\beta x - \beta\sin\beta x) + e^{\alpha x}(-\alpha\beta\sin\beta x - \beta^2\cos\beta x)$$

$$= e^{\alpha x}\{(\alpha^2 - \beta^2)\cos\beta x - 2\alpha\beta\sin\beta x\}.$$

Then $\dfrac{d^2y}{dx^2}+a\dfrac{dy}{dx}+by = e^{\alpha x}\{(\alpha^2-\beta^2)\cos\beta x - 2\alpha\beta\sin\beta x\}$

$$+ae^{\alpha x}(\alpha\cos\beta x - \beta\sin\beta x)+be^{\alpha x}\cos\beta x$$
$$= e^{\alpha x}\{(\alpha^2-\beta^2+a\alpha+b)\cos\beta x-(2\alpha+a)\beta\sin\beta x\}.$$

This is zero because $\alpha\pm i\beta$ are the roots of $m^2+am+b=0$ so that $-a=2\alpha$ and $b=\alpha^2+\beta^2$, therefore

$$\alpha^2-\beta^2+a\alpha+b=\alpha^2-\beta^2-2\alpha^2+\alpha^2+\beta^2=0,\ 2\alpha+a=0.$$

The use of Theorem 11.2 is demonstrated in the next example.

EXAMPLE 1 *Solve the differential equations: (i)* $\dfrac{dy}{dx}+4y=0$,

(ii) $\dfrac{d^2y}{dx^2}-5\dfrac{dy}{dx}+4y=0$, *(iii)* $\dfrac{d^2y}{dx^2}-4\dfrac{dy}{dx}+4y=0$, *(iv)* $2\dfrac{d^2y}{dx^2}-4\dfrac{dy}{dx}+8y=0$.

(i) The auxiliary equation is $m+4=0$ so $m=-4$ and the general solution is $\boldsymbol{y=Ae^{-4x}}$.
(ii) The auxiliary equation is $m^2-5m+4=0$ so $m=4$ or 1 and the general solution is $\boldsymbol{y=Ae^{4x}+Be^{x}}$.
(iii) The auxiliary equation is $m^2-4m+4=0$ so $m=2$, twice, and the general solution is $\boldsymbol{y=(Ax+B)e^{2x}}$.
(iv) Divide the equation by 2, giving the auxiliary equation $m^2-2m+4=0$, with roots $\{2\pm\sqrt{(4-16)}\}/2=1\pm i\sqrt{3}$. That is, $\alpha=1$, $\beta=\sqrt{3}$ and the general solution is $\boldsymbol{y=(A\cos\sqrt{3}x+B\sin\sqrt{3}x)e^{x}}$.

EXAMPLE 2 *Find the general solution of the differential equation* $\dfrac{d^2x}{dt^2}+4x=0$. *Find the particular solution such that $x=2$ when $t=0$ and x has a stationary value when $t=\pi/2$.*

The auxiliary equation is $m^2+4=0$ with roots $m=\pm 2i$, so $\alpha=0$, $\beta=2$, and the general solution is $x=A\cos 2t+B\sin 2t$. For the particular solution, $x=2$ when $t=0$ so that $2=A$. Also $\dfrac{dx}{dt}=-4\sin 2t+2B\cos 2t$ and this must be zero when $t=\pi/2$, for the stationary value. Thus $0=-2B$ and $B=0$. The required particular solution is $\boldsymbol{x=2\cos 2t}$.

EXERCISE 11.3

1 Solve the differential equation:

(i) $3\dfrac{dy}{dx}-4y=0$, (ii) $\dfrac{dx}{dt}=-2x$, (iii) $\dfrac{d^2y}{dx^2}+2\dfrac{dy}{dx}-3y=0$,

(iv) $\dfrac{d^2x}{dt^2}-2\dfrac{dx}{dt}+x=0$, (v) $\dfrac{d^2y}{dx^2}-3\dfrac{dy}{dx}+3y=0$,

(vi) $\dfrac{d^2x}{dt^2}+9\dfrac{dx}{dt}=0$, (vii) $\dfrac{d^2y}{dx^2}=16y$, (viii) $\dfrac{d^2y}{dx^2}=-25y$.

2 Find the particular solution of the equations:

(i) $\frac{dy}{dx} - 8y = 0$, $y = 4$ when $x = 0$,

(ii) $\frac{d^2y}{dx^2} - 5\frac{dy}{dx} - 6y = 0$, $y = 3$ and $\frac{dy}{dx} = 4$ when $x = 0$,

(iii) $\frac{d^2y}{dx^2} + 6\frac{dy}{dx} + 9y = 0$, $y = 0$ and $\frac{dy}{dx} = 2$ when $x = 0$.

3 Find the general solution of the differential equation $\frac{d^2y}{dx^2} + \frac{dy}{dx} + y = 0$. (*L*)

4 Find y in terms of k and x, given that $\frac{d^2y}{dx^2} + k^2y = 0$, where k is a constant, and $y = 1$ and $\frac{dy}{dx} = 1$ when $x = 0$. (*L*)

5 Eliminate y between the differential equations $\frac{dx}{dt} = -ky$ and $\frac{dy}{dt} = -kx$ (where k is a constant) to obtain a second-order differential equation in x and t only. Solve this equation and obtain expressions for x and for y in terms of t, given that $x = X$ and $y = Y$ when $t = 0$. If $X = 2Y$, prove that, when $y = 0$, $x = \sqrt{3}X/2$. (*C*)

6 Solve the simultaneous differential equations $\frac{dx}{dt} + ky = 0$ and $\frac{dy}{dt} + kx = 0$ with the initial conditions that $x = 0$ and $y = 1$ when $t = 0$. Show that $y^2 - x^2 = 1$ for all values of t.

11.4 Particular integral functions

We now return to the general case where $f(x)$ is non-zero in the first order linear differential equation with constant coefficients, that is, $\frac{dy}{dx} + ay = f(x)$. We may still apply the integrating factor method, as follows.

$$e^{ax}\left(\frac{dy}{dx} + ay\right) = \frac{d}{dx}(e^{ax}y) = e^{ax}f(x),$$

$$\text{so } e^{ax}y = \int e^{ax}f(x)\,dx \text{ and } y = e^{-ax}\int e^{ax}f(x)\,dx.$$

In the next example, we apply this method to certain types of function f and look for patterns in the results.

EXAMPLE 1 *Solve the first order linear differential equations:*

(*i*) $\frac{dy}{dx} - 3y = 0$, (*ii*) $\frac{dy}{dx} + 2y = e^{5x}$, (*iii*) $\frac{dy}{dx} - y = 3x$, (*iv*) $\frac{dy}{dx} + 4y = \sin x$.

(i) $a = -3$, $f(x) = 0$, so $\int f(x)e^{ax}\,dx = \int 0\,dx = A$, for some constant A and the general solution is $\boldsymbol{y = Ae^{3x}}$, as expected.
(ii) $a = 2$, $f(x) = e^{5x}$, so $\int f(x)e^{ax}\,dx = \int e^{5x}e^{2x}\,dx = \frac{1}{7}e^{7x} + A$, for some constant A and the general solution is $\boldsymbol{y = Ae^{-2x} + \frac{1}{7}e^{5x}}$.
(iii) $a = -1$, $f(x) = 3x$, so

$$\int f(x)e^{ax}\,dx = \int 3xe^{-x}\,dx = [-3xe^{-x}] + \int 3e^{-x}\,dx$$

$$= -3xe^{-x} - 3e^{-x} + A,$$

using integration by parts. The general solution is $\boldsymbol{y = Ae^{x} - 3x - 3}$.
(iv) $a = 4$, $f(x) = \sin x$, so $\int f(x)e^{ax}\,dx = \int \sin xe^{4x}\,dx = I$, say.

Then $$I = \left[\frac{\sin x}{4}e^{4x}\right] - \int \frac{\cos x}{4}e^{4x}\,dx$$

$$= \frac{\sin x}{4}e^{4x} - \left[\frac{\cos x}{16}e^{4x} + \int \frac{\sin x}{16}e^{4x}\,dx\right]$$

$$= e^{4x}\left[\frac{\sin x}{4} - \frac{\cos x}{16}\right] - \frac{1}{16}I.$$

So $$I = \frac{16}{17}e^{4x}\left[\frac{\sin x}{4} - \frac{\cos x}{16}\right] + A$$ and the general solution is

$$\boldsymbol{y = Ae^{-4x} + \frac{4}{17}\sin x - \frac{1}{17}\cos x.}$$

Notice that the solution is in two distinct parts. The first part Ae^{-4x}, containing the arbitrary constant of integration, is the solution of the homogeneous equation $\frac{dy}{dx} + ay = 0$, and is known as the *Complementary Function*. The second part $\frac{4}{17}\sin x - \frac{1}{17}\cos x$ contains no arbitrary constants and depends upon the particular function f, so it is called the *Particular Integral.*

Definition Let $Dy = f(x)$ be a linear differential equation of order n, so that the differential operator D is a linear combination of

$\frac{d^n}{dx^n}, \frac{d^{n-1}}{dx^{n-1}}, \ldots, \frac{d}{dx}, 1$. Then
(i) a Particular Integral (P.I.) is a solution of the equation containing no arbitrary constants,
(ii) a Complementary Function, (C.F.), is a general solution, containing n arbitrary constants, of the corresponding homogeneous equation $Dy = 0$.

Theorem 11.3 The general solution of a linear differential equation is the sum of a Complementary Function and a Particular Integral.

Proof Let $F(x)$ be a P.I. of the equation $Dy = f(x)$, so that $y = F(x)$ is a solution of the equation and $DF(x) = f(x)$. Let $y = w + F(x)$, then $Dy = D\{w + F(x)\} = Dw + DF(x) = Dw + f(x)$, and so $Dw = 0$. Therefore, the general solution for w is a Complementary Function $G(x)$ and so the solution is given by $y = G(x) + F(x) = \text{C.F.} + \text{P.I.}$.

We now tabulate the results from Example 1 to show the various types of particular integrals that occurred.

$f(x)$	*Particular Integral*
0	0
e^{kx}	Be^{kx}, for some particular constant B,
kx	$Bx + C$, for some particular constants B, C,
$\sin kx$	$B\sin kx + C\cos x$, for some particular constants B, C.

The process of finding the particular integral may be speeded up by making a sensible guess for the required type of function and then evaluating the particular constants B, C involved.

EXAMPLE 2 *Solve the differential equation* $\dfrac{dy}{dx} - 3y = x^2 - 1$.

The complementary function is Ae^{3x}, as in Example 1(i). Since the right hand side is a quadratic function of x, suppose that the particular integral is of the form $Bx^2 + Cx + D$. Then $\dfrac{dy}{dx} = 2Bx + C$ and, by substitution in the differential equation, we find that $(2Bx + C) - 3(Bx^2 + Cx + D) = x^2 - 1$. By equating the coefficients, $-3B = 1, 2B - 3C = 0, C - 3D = -1$, so $B = -1/3, C = -2/9$ and $D = 7/27$. Thus, a particular solution is $y = (-9x^2 - 6x + 7)/27$ and the general solution is $\boldsymbol{y = Ae^{3x} - (9x^2 + 6x - 7)/27}$.

EXERCISE 11.4A

1 Use the suggested particular functions to solve the following differential equations:

(i) $\dfrac{dy}{dx} + 2y = 3e^x$, $y = Be^x$, (ii) $\dfrac{dy}{dx} - 4y = \cos 3x$, $y = B\cos 3x + C\sin 3x$,

(iii) $\dfrac{dy}{dx} - 5y = 7$, $y = B$, (iv) $\dfrac{dy}{dx} + y = 4 - x$, $y = Bx + C$,

(v) $2\dfrac{dy}{dx} - y = x^2$, $y = Bx^2 + Cx + D$,

(vi) $\dfrac{dy}{dx} + 7y = 3e^{-7x}$, $y = (Bx + C)e^{-7x}$.

2 Solve the following differential equations: (i) $3\dfrac{dy}{dx} + y = x$,

(ii) $\dfrac{dy}{dx}+2y = 7\sin 2x$, (iii) $\dfrac{dy}{dx} = e^{2x}-y$, (iv) $2\dfrac{dy}{dx}-x^2y = 0$.

3 Solve the differential equation $\dfrac{dx}{dt}+2x = t$, given that $\dfrac{dx}{dt} = 0$ when $t = 0$. *(L)*

4 Solve the differential equation $\dfrac{dy}{dx} = y+e^{-x}$, given that $y = 1$ when $x = 0$. *(L)*

5 A chemical substance X decays, at a rate equal to twice the quantity of X present, so that $\dfrac{dx}{dt} = -2x$, where x is the quantity of X present at time t. Given that initially $x = a$, find an expression for x in terms of a and t. The quantity, y, of another substance Y changes so that its rate of increase is equal to $2ae^{-2t}-(y/2)$. Given that initially $y = 0$, find an expression for y at time t and determine the time at which y is a minimum, leaving your answer in terms of natural logarithms. *(L)*

Theorem 11.3 may now be applied to solve the general second order linear differential equation with constants. This is the equation

$$\frac{d^2y}{dx^2}+a\frac{dy}{dx}+by = f(x),$$

and its solution is the sum of a C.F. and a P.I.. The form of the C.F. is given by Theorem 11.2 and depends upon the roots of the auxiliary equation $m^2+am+b = 0$. The P.I. is a function, $F(x)$, which is a solution of the equation. It may be found by using certain trial functions, chosen by studying the differential equation. The method is demonstrated in Example 3.

EXAMPLE 3 *Find Particular Integrals for the differential equations:*
(i) $\dfrac{d^2y}{dx^2}-25y = \sin 2x$, *(ii)* $\dfrac{d^2y}{dx^2}-8\dfrac{dy}{dx}+16y = x^2-1$,

(iii) $\dfrac{d^2y}{dx^2}+\dfrac{dy}{dx}+y = xe^{3x}$.

(i) $f(x) = \sin 2x$, so it seems sensible to start with a linear combination of $\sin 2x$ and $\cos 2x$ since the derivatives of these two functions are multiples of each other. Try $F(x) = C\sin 2x+D\cos 2x$. Then $F'(x) = 2C\cos 2x-2D\sin 2x$ and $F''(x) = -4C\sin 2x-4D\cos 2x$, and, on substitution in the equation, we need

$$-4C\sin 2x-4D\cos 2x-25(C\sin 2x+D\cos 2x) = \sin 2x, \text{ for all } x.$$

The coefficients of $\sin 2x$ and $\cos 2x$ must each vanish, so $D = 0$ and $C = -1/29$, and $\mathbf{F(x) = -\dfrac{1}{29}\sin 2x}$.

(ii) $f(x) = x^2-1$ so the sensible choice for the P.I. is a general quadratic, and we put $F(x) = Cx^2 + Dx + E$. Then $F'(x) = 2Cx + D$ and $F''(x) = 2C$. On substituting in the equation,

$$2C - 8(2Cx + D) + 16(Cx^2 + Dx + E) = x^2 - 1, \text{ for all } x.$$

Equating the coefficients, $16C = 1$, $-16C + 16D = 0$, $2C - 8D + 16E = -1$ and so $C = 1/16 = D$ and $E = -5/128$. The P.I. is the function F, given by $\mathbf{F(x) = (8x^2 + 8x - 5)/128}$.

(iii) $f(x) = xe^{3x}$ and its derivatives consist of linear functions multiplied by e^{3x}. A sensible choice of a P.I. will be of the same form so let $F(x) = (Cx + D)e^{3x}$. Then $F'(x) = (C + 3Cx + 3D)e^{3x}$ and $F''(x) = (3C + 3C + 9Cx + 9D)e^{3x}$. Substituting in the equation, $e^{3x}(6C + 9Cx + 9D) + e^{3x}(C + 3Cx + 3D) + e^{3x}(Cx + D) = xe^{3x}$, giving $6C + 9D + C + 3D + D = 0$ and $9C + 3C + C = 1$. Thus $C = 1/13$, $D = -7/169$ and the P.I. is given by $\boldsymbol{y = e^{3x}(13x - 7)/169}$.

The following table may be of use in selecting a suitable trial function for the P.I.

$f(x)$	Trial $F(x)$
e^{ax}	Ce^{ax}
$\cos bx$ or $\sin bx$	$C \sin bx + D \cos bx$
k	C
$kx + 1$	$Cx + D$
polynomial	general polynomial of the same degree
polynomial times e^{ax}	general polynomial of the same degree times e^{ax}
polynomial times $\sin bx$ or $\cos bx$	general polynomial of the same degree times $\sin bx$ + another times $\cos bx$

Problems can arise when the trial P.I. is a function of a class similar to a term in the C.F.. This is dealt with by choosing a trial function formed by multiplying the term in the C.F. by a polynomial, as is illustrated in Examples 4 and 5.

EXAMPLE 4 *Show that $y = Ce^{2x}$ is not a P.I. for the differential equation $\dfrac{d^2y}{dx^2} - 7\dfrac{dy}{dx} + 10y = 3e^{2x}$. Solve the equation by using Cxe^{2x} as a trial function instead.*

Let $G(x) = Ce^{2x}$, then $\dfrac{dG}{dx} = 2Ce^{2x}$ and $\dfrac{d^2G}{dx^2} = 4Ce^{2x}$. Then, on putting $y = G(x)$, the left hand side of the differential equation becomes $4Ce^{2x} - 7(2Ce^{2x}) + 10(Ce^{2x}) = 0$. $G(x)$ is part of the C.F. and so it cannot be a P.I. satisfying the equation. The auxiliary equation is $m^2 - 7m + 10 = 0$, with roots 2 and 5. The C.F. is $Ae^{5x} + Be^{2x}$, which contains e^{2x} as was expected. Now let $y = Cxe^{2x}$, then $\dfrac{dy}{dx} = (C + 2Cx)e^{2x}$, $\dfrac{d^2y}{dx^2} = (2C + 2C + 4Cx)e^{2x}$ giving $(4C + 4Cx)e^{2x} - 7(C + 2Cx)e^{2x} + 10(Cxe^{2x}) = 3e^{2x}$. On equating the coefficients of e^{2x} and xe^{2x}, $4C - 14C + 10C = 0$ and $4C - 7C = 3$, so $C = -1$. Therefore, the P.I. is $-xe^{2x}$ and the general solution is $\boldsymbol{y = Ae^{5x} + Be^{2x} - xe^{2x}}$.

The same type of difficulty can also occur when the auxiliary equation has complex roots and $f(x)$ is a trigonometric function occurring in the C.F.. As before, the appropriate trial function for the P.I. is the C.F. multiplied by x, as shown in Example 5.

EXAMPLE 5 *Find the value of C, given that $Cx \sin 3x$ is a particular solution of the differential equation $\frac{d^2y}{dx^2} + 9y = 6\cos 3x$, and hence find the general solution of this equation. Given that, when $x = 0$, $y = 1/3$ and $\frac{dy}{dx} = -3$, solve the equation. Find the values of x for which the solution has stationary values.*

Let $y = Cx \sin 3x$, so $\frac{dy}{dx} = C\sin 3x + 3Cx\cos 3x$, $\frac{d^2y}{dx^2} = 6C\cos 3x - 9Cx\sin 3x$ and, on substitution in the differential equation, $6C\cos 3x - 9Cx\sin 3x + 9Cx\sin 3x = 6\cos 3x$. Thus $\mathbf{C=1}$ and the required P.I. is $x\sin 3x$. The auxiliary equation is $m^2+9=0$, with roots 3i and -3i, so the C.F. is $A\cos 3x + B\sin 3x$, giving a general solution $\boldsymbol{y = A\cos 3x + B\sin 3x + x\sin 3x}$.

When $x=0$, $y=A$ and, since $\frac{dy}{dx} = -3A\sin 3x + 3B\cos 3x + \sin 3x + 3x\cos 3x$, $\frac{dy}{dx} = 3B$. Hence $A = 1/3$, $B = -1$ and the solution is

$\boldsymbol{y = \frac{1}{3}\cos 3x - \sin 3x + x\sin 3x}$. Then $\frac{dy}{dx} = (3x-3)\cos 3x$ and so, for a stationary value of y, either $x = 1$ or $\cos 3x = 0$. The required set of values of x is $\{1\} \cup \{(2n+1)\pi/6 : n \in \mathbb{Z}\}$.

EXERCISE 11.4B

1 Find suitable values for the constants C, D, E, so that the given functions form particular integrals for the given differential equations:

(i) Ce^x, $\frac{d^2y}{dx^2} + 3\frac{dy}{dx} - 10y = 3e^x$, (ii) $Cx + D$, $\frac{d^2y}{dx^2} - 2\frac{dy}{dx} + 4y = 2x - 8$,

(iii) $C\sin x + D\cos x$, $\frac{d^2y}{dx^2} + 9y = 4\cos x$,

(iv) $Cx^2 + Dx + E$, $\frac{d^2y}{dx^2} - 3\frac{dy}{dx} + 4y = 1 - x^2$.

2 Show that $y = e^x$ is a solution of the differential equation $\frac{d^2y}{dx^2} + 4\frac{dy}{dx} + 4y = 9e^x$, and find the general solution of this differential equation. (*L*)

3 Find the value of the constant p for which pe^{2x} is a solution of the differential equation $\frac{d^2y}{dx^2} - 6\frac{dy}{dx} + 13y = 10e^{2x}$. Solve this differential equation com-

pletely given that $y = 0$ and $\frac{dy}{dx} = 0$ when $x = 0$. (*L*)

4 Solve the differential equation $\frac{d^2x}{dt^2} + 6\frac{dx}{dt} + 5x = e^{-t}$, given that $x = 0$ and $\frac{dx}{dt} = 1$ when $t = 0$.

5 Find the values of the constants a and b for which $ax + b$ is a solution of the differential equation $\frac{d^2y}{dx^2} + 2\frac{dy}{dx} + 10y = 100x + 50$. Hence solve this differential equation completely, given that $y = 3$ and $\frac{dy}{dx} = 0$ when $x = 0$. (*L*)

6 Solve the differential equation $\frac{d^2x}{dt^2} + 2\frac{dx}{dt} + cx = 0$ in each of the cases $c = -3$ and $c = 2$. Find the solution of the equation $\frac{d^2x}{dt^2} + 2\frac{dx}{dt} + 2x = t$ for which $x = 0$ and $\frac{dx}{dt} = 0$ when $t = 0$. (*C*)

7 Find the general solution of the differential equation

$$\frac{d^2y}{dx^2} - 2\frac{dy}{dx} + 2y = \sin 2x. \qquad (JMB)$$

8 (i) Find the constant c such that $y = c$ is a solution of the differential equation $\frac{d^2y}{dx^2} - 9y = 4$, and give the general solution of this differential equation.
(ii) Find the values of the constants a and b for which $a \sin x + b \cos x$ is a solution of the differential equation $\frac{d^2y}{dx^2} + 4\frac{dy}{dx} + 3y = \sin x$. (*L*)

9 Solve the differential equation $\frac{d^2x}{dt^2} + 4\frac{dx}{dt} + 13x = 80 \sin 3t$, given when $t = 0$, $x = 0$ and $\frac{dx}{dt} = -6$. (*C*)

10 Find the value of the constant a for which $y = axe^{-x}$ is a solution of the differential equation $\frac{d^2y}{dx^2} + 3\frac{dy}{dx} + 2y = 2e^{-x}$. Find the solution of this differential equation for which $y = 1$ and $\frac{dy}{dx} = 3$ when $x = 0$. (*L*)

11 (i) Obtain the general solution of the differential equation

$$\frac{d^2y}{dx^2} + k^2y = 5 \sin 2x, \text{ where } k \text{ is real and } |k| \neq 2.$$

(ii) Find the value of p for which $px \cos 2x$ is a particular integral of the differential equation $\frac{d^2y}{dx^2} + 4y = 5 \sin 2x$. Hence find the solution of this equation which satisfies the conditions $y = 2$ when $x = 0$ and $y = 0$ when $x = \pi/4$. (*L*)

11.5 Summary

In order to use the techniques, described in this chapter, the differential equation must first be classified in order to select the appropriate technique. The following flow diagrams, shown in Figs. 11.3 and 11.4, can assist in this problem of classification.

In A-level examination questions, substitutions are usually suggested in the question and the solution is obtained by substituting, solving the new equation and then substituting back again.

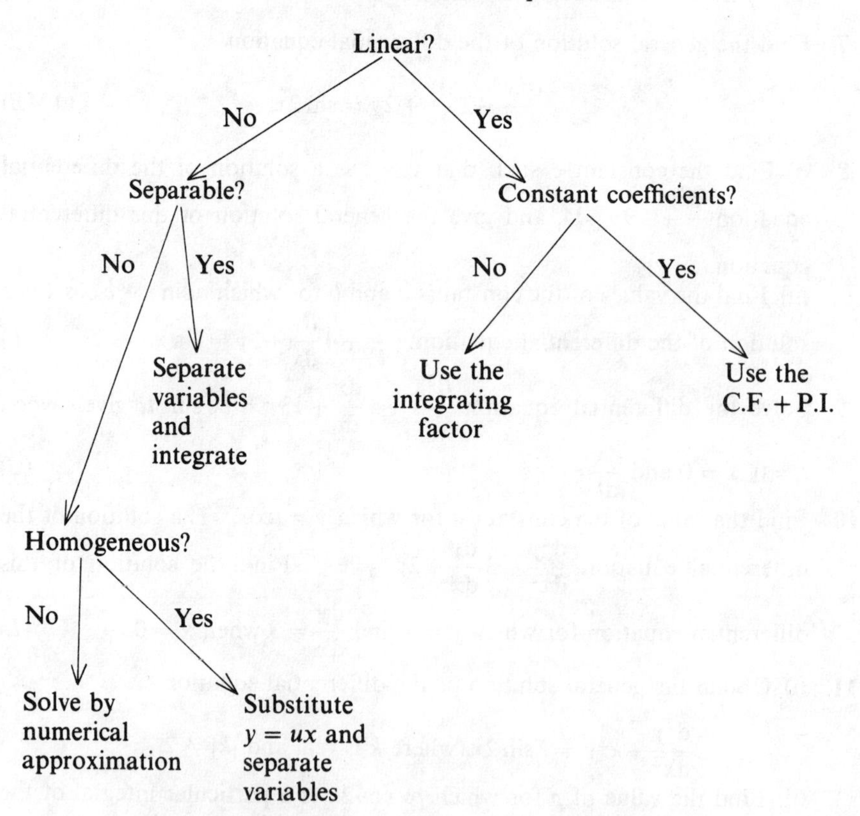

Fig. 11.3 Classification of 1st Order Differential Equations

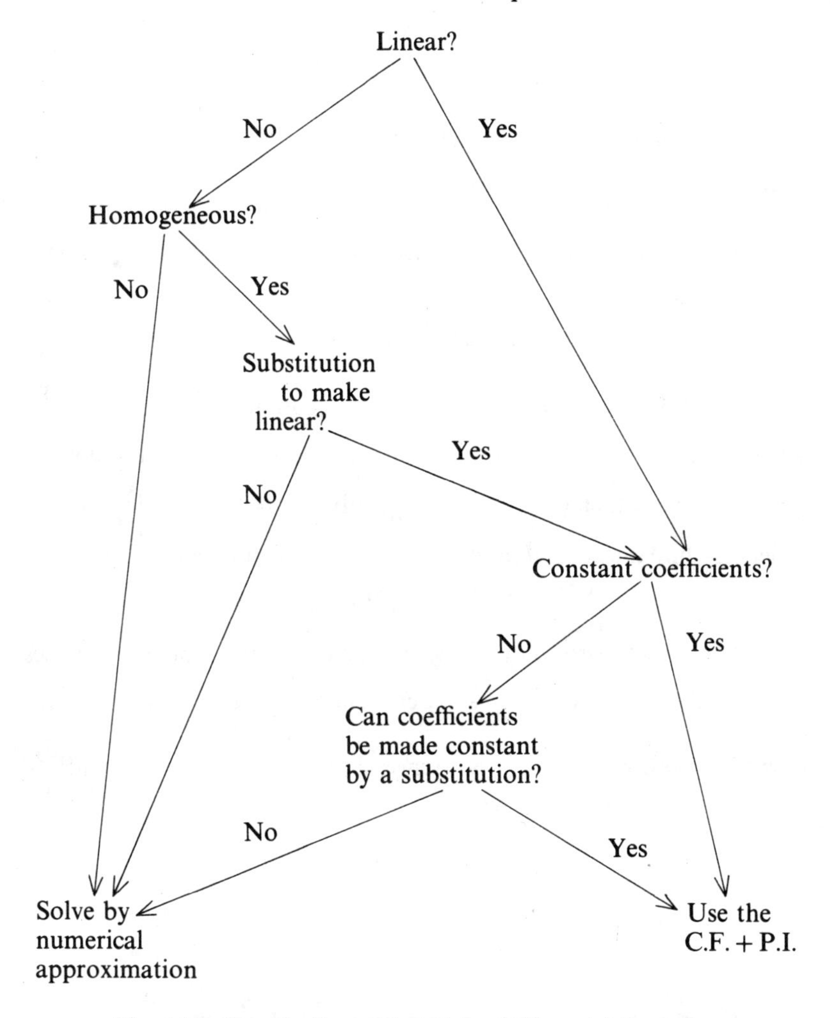

Fig. 11.4 Classification of 2nd Order Differential Equations

EXAMPLE 1 *By making the substitution* $y = vx$, *where* v *is a function of* x, *show that the differential equation* $x^2\dfrac{d^2y}{dx^2} - 2x\dfrac{dy}{dx} + 2y = x^3$ *reduces to* $\dfrac{d^2v}{dx^2} = 1$. *Hence find the general solution.*

If $y = vx$, then $\dfrac{dy}{dx} = x\dfrac{dv}{dx} + v$ and $\dfrac{d^2y}{dx^2} = 2\dfrac{dv}{dx} + x\dfrac{d^2v}{dx^2}$. On substituting, $x^2\left(2\dfrac{dv}{dx} + x\dfrac{d^2v}{dx^2}\right) - 2x\left(x\dfrac{dv}{dx} + v\right) + 2vx = x^3$, that is, $x^3\dfrac{d^2v}{dx^2} = x^3$ and $\dfrac{d^2v}{dx^2} = 1$, as required. This differential equation can be integrated twice directly giving $\dfrac{dv}{dx} = x + A$, $v = x^2/2 + Ax + B$. Since $y = vx$, the general solution of the original equation is $\boldsymbol{y = x^3/2 + Ax^2 + Bx}$, with arbitrary constants A and B.

EXERCISE 11.5

1 Use the substitution $z = y/x$ to solve the equation $\dfrac{dy}{dx} = \dfrac{y}{x} + x\sec\left(\dfrac{y}{x}\right)$.

2 Given that y is a function of x, where $x > 0$, show that, if the substitution $x = \sqrt{t}$ is made, then (i) $\dfrac{dy}{dx} = 2\sqrt{t}\dfrac{dy}{dt}$, (ii) $\dfrac{d^2y}{dx^2} = 4t\dfrac{d^2y}{dt^2} + 2\dfrac{dy}{dt}$. Hence, or otherwise, find the general solution of the differential equation

$$\frac{d^2y}{dx^2} - \frac{1}{x}\left(\frac{dy}{dx}\right) + 4x^2(9y+6) = 0. \qquad (C)$$

3 Show that the substitution $y = z\cos x$, where z is a function of x, reduces the differential equation $\cos^2 x\dfrac{d^2y}{dx^2} + 2\cos x\sin x\dfrac{dy}{dx} + 2y = x\cos^3 x$ to the differential equation $\dfrac{d^2z}{dx^2} + z = x$. Hence, find y given that $y = 0$ and $\dfrac{dy}{dx} = 1$ when $x = 0$. (*L*)

11.6 Numerical methods for solving differential equations

There are many different methods of solving differential equations numerically and most of them rely on the type of approximation techniques, described in Chapter 6 on Power Series Expansions. The simplest methods are of the type where a linear or quadratic approximation is used to calculate the next value of the function from values previously calculated. The calculation is done in steps and the process is called a *step-by-step* method of solution. The differential equation is used to find the derivative of the solution at any point on the solution curve and this is then employed as the gradient of the approximation polynomial. We begin with some methods called Euler type.

Euler's method

Suppose that the differential equation to be solved is written in the form $\frac{dy}{dx} = f(x, y)$, with a starting point (x_0, y_0). The argument of the function f is the pair of values (x, y) and f will have a domain consisting of such pairs. The linear approximation method of §6.3 is used to calculate values $y(x_r)$ at a sequence (x_r) of values of x. This is done in steps, with step length h, and

$$x_r = x + rh \text{ so that } x_{r+1} = x_r + h.$$

h is assumed to be small and to a first approximation

$$y(x_1) \approx y(x_0) + hy'(x_0) = y(x_0) + hf(x_0, y(x_0))$$

To make the work easier to follow, we introduce the notation

$$y_r = y(x_r) \text{ and } y_r' = y'(x_r) = \frac{dy}{dx} \text{ evaluated at } x = x_r$$

and Euler's method can then be written iteratively as follows:

$$y_{r+1} = y_r + hf(x_r, y_r), \text{ given starting values } (x_0, y_0).$$

The solution curve which starts at the point (x_0, y_0) is being approximated to by a sequence of short straight lines, as shown in Fig. 11.5. It is important to note that at each step the gradient of one member of the solution family is used to continue to the next step but that a different member of the family may be used at each step.

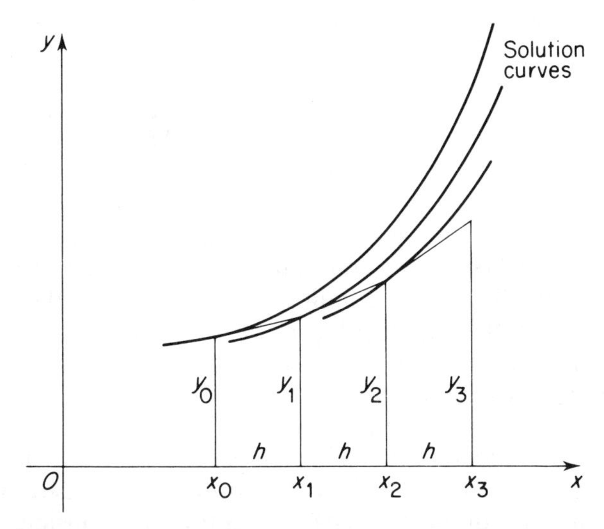

Fig. 11.5 Euler's method $y_{r+1} = y_r + hy_r'$, $r = 0, 1, 2, \ldots$

EXAMPLE 1 *Use Euler's method, with a step length of* 0·1, *to solve the differential equation* $\frac{dy}{dx} = 1 + y^2$, *where* $y = 0$ *when* $x = 0$. *Compare the computed values for* y *for* $x \in \{0·1, 0·2, 0·3, 0·4\}$ *with the exact values.*

$y' = 1 + y^2$ and so the iteration equation is $y_{r+1} = y_r + h(1 + y_r^2)$. We lay out the work in Table 11.2, rounding numbers to four decimal places as necessary. The final column is computed from the exact solution $y = \tan x$.

Table 11.2 Euler's method

r	x_r	y_r	y'_r	exact $y(x_r)$
0	0	0	1	0
1	0·1	0·1	1·01	0·1003
2	0·2	0·201	1·0404	0·2027
3	0·3	0·305	1·0930	0·3093
4	0·4	0·4143		0·4228

The accuracy of the method may be improved by reducing the step length h, since this gives a better approximation, but there is a penalty of increased computation. The next table shows the values obtained for y in the above example by using different step lengths. The calculation for the last column requires 100 steps of length 0·001 in order to find each value of y and this is feasible only with the use of a computer.

Table 11.3 Values of y_r for different h

x	$h = 0·1$	$h = 0·05$	$h = 0·01$	$h = 0·001$
0·1	0·1	0·1001	0·1003	0·1003
0·2	0·201	0·2018	0·2025	0·2027
0·3	0·305	0·3070	0·3088	0·3093
0·4	0·4143	0·4182	0·4218	0·4226

Unfortunately, such a large number of calculations can lead to a build-up of rounding errors, since a computer only works to a fixed number of significant figures. This gives a limit on the accuracy of the approximate solution. Indeed, if the step length continues to be reduced, a point will be reached where the solution begins to become less accurate due to the accumulation of rounding errors.

An alternative way of improving the accuracy is to use a better approximation in the iteration. The above method is one sided in that it uses y_{r-1} and y_r to estimate y'_r. A better approximation is found by using points on either side of the original point, as follows. Consider x_{-1} and x_1, so that $x_1 = x_0 + h$ and $x_{-1} = x_0 - h$. Then the slope of the chord $P_{-1}P_1$

is $(y_1 - y_{-1})/2h$ and this is used as the approximation for y'_0. The resulting approximation is shown in Fig. 11.6 and this leads to the following iteration.

$$y_{r+1} = y_{r-1} + 2hy'_r = y_{r-1} + 2hf(x_r, y_r).$$

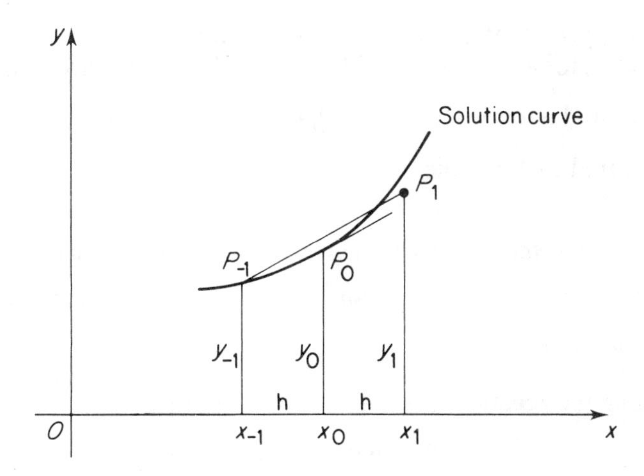

Fig. 11.6 Modified Euler's method $y_{r+1} = y_{r-1} + 2hy'_r$, $r = 0, 1, 2, \ldots$

Notice that this needs the two initial points (x_0, y_0) and (x_{-1}, y_{-1}) in order to begin the iteration.

EXAMPLE 2 *Solve the equation of Example 1 using the modified Euler's method described above.*

The iteration becomes $y_{r+1} = y_{r-1} + 2h(1 + y_r^2)$. We shall need to use the original Euler's method to find $y(-h)$ $(= y_{-1})$ in order to start the iteration and the results are shown in Table 11.4.

If this method is repeated for $h = 0{\cdot}05$, the same accuracy can be achieved as with the original method with $h = 0{\cdot}005$.

Table 11.4 Modified Euler's method

r	x_r	y_r	y'_r	exact $y(x_r)$
−1	−0·1	−0·1		
0	0	0	1	0
1	0·1	0·1	1·01	0·1003
2	0·2	0·202	1·0408	0·2027
3	0·3	0·3082	1·0950	0·3093
4	0·4	0·4210		0·4228

As might be expected, similar, or better, accuracy can be achieved by using the second derivative in a second order Taylor series expansion,

$$y_1 \approx y_0 + hy'_0 + \left(\frac{h^2}{2}\right)y''_0.$$

To use this approximation, it is necessary to find the value of the second derivative $y''(x)$ for any value of x. This may be found by differentiating the given differential equation $y' = \mathrm{f}(x, y)$, that is $y'' = \dfrac{\mathrm{d}}{\mathrm{d}x}\mathrm{f}(x, y)$. The process is demonstrated in Example 3.

EXAMPLE 3 *Use the second order Taylor series approximation to find values of* $y(0{\cdot}2)$, $y(0{\cdot}4)$ *and* $y(0{\cdot}6)$, *where* y *satisfies the differential equation* $\dfrac{\mathrm{d}y}{\mathrm{d}x} + xy = 1$, *and* $y = 1$ *when* $x = 0$.

Differentiating the equation, $y'' + y + xy' = 0$. Therefore

$$y' = 1 - xy \quad \text{and} \quad y'' = y(x^2 - 1) - x$$

and the iteration can be written

$$y_{r+1} = y_r + h(1 - x_r y_r) + h^2\{y_r(x_r^2 - 1) - x_r\}/2.$$

As before, the work is laid out in Table 11.5, working to three decimal places.

Table 11.5 Second order Taylor approximation

r	x_r	y_r	y'_r	y''_r
0	0	1	1	-1
1	0·2	1·18	0·764	$-1{\cdot}333$
2	0·4	1·31	0·476	$-1{\cdot}550$
3	0·6	1·37		

As before, more accurate results may be found by reducing the step width, or by applying the idea of the modified Euler's method to this second order case. This is done as follows:

$$y_1 = y_0 + hy'_0 + \left(\frac{h^2}{2}\right)y''_0,$$

$$y_{-1} = y_0 - hy'_0 + \left(\frac{h^2}{2}\right)y''_0.$$

On adding these equations, the first step becomes

$$y_1 + y_{-1} = 2y_0 + h^2 y''_0 \quad \text{or} \quad y_1 = 2y_0 - y_{-1} + h^2 y''_0.$$

The general iteration therefore becomes

$$y_{r+1} = 2y_r - y_{r-1} + h^2 y_r''.$$

The value of the second derivative is calculated in the same way as in Example 3 and the iterative formula is applied to find a step by step solution, as before.

EXAMPLE 4 *Solve the differential equation of Example 3, using the modified Euler's method.*

The iteration formula is $y_{r+1} = 2y_r - y_{r-1} + h^2\{y_r(x_r^2 - 1) - x_r\}$, and the work is again set out in table form. Notice that, in order to start, y_{-1} is needed and an estimate of this is found by the Euler's method.

Table 11.6 Modified Euler's method

r	x_r	y_r	y_r'	y_r''
−1	−0·2	0·8		
0	0	1	1	−1
1	0·2	1·16	0·768	−1·314
2	0·4	1·267	0·493	−1·464
3	0·6	1·315		

All the methods considered so far rely on a step by step approach, using a simple approximating polynomial. The final method shown here uses the higher derivatives, which can be found by differentiating the original equation to form a higher degree polynomial approximation at the starting point. If this polynomial approximation is good enough, it may be used to find the values of the solution directly. In effect, the method is to find a Taylor series expansion at the starting point and the accuracy of the solution will depend upon the rate of convergence of this series and the distance away from the starting point.

If the solution of the differential equation is supposed to have a polynomial approximation

$$y = a_0 + a_1 x + a_2 x^2 + \ldots + a_r x^r + \ldots + a_k x^k,$$

then, on differentiating this polynomial r times and putting $x = 0$,

$$y^{(r)}(0) = r!\,a_r \quad \text{and so} \quad a_r = y^{(r)}(0)/r!.$$

The process is demonstrated in Example 5.

EXAMPLE 5 *Find an approximate value for y when $x = 1$, given that $y'' + xy' + y = 0$ and that $y = 1$, $y' = 0$ at $x = 0$.*

Using the above Taylor series method, $a_0 = y(0) = 1$, $a_1 = y'(0) = 0$,

$$\begin{aligned}
y'' &= -y \quad -xy' &&\text{so}\quad a_2 = y''(0)/2 &&= -1/2,\\
y^{(3)} &= -2y' \quad -xy^{(2)} &&\text{so}\quad a_3 = y^{(3)}(0)/3! &&= 0,\\
y^{(4)} &= -3y^{(2)} - xy^{(3)} &&\text{so}\quad a_4 = y^{(4)}(0)/4! &&= 3/24,\\
y^{(5)} &= -4y^{(3)} - xy^{(4)} &&\text{so}\quad a_5 = y^{(5)}(0)/5! &&= 0,\\
y^{(6)} &= -5y^{(4)} - xy^{(5)} &&\text{so}\quad a_6 = y^{(6)}(0)/6! &&= -15/720.
\end{aligned}$$

So $y = 1 - x^2/2 + x^4/8 - x^6/48$ and, when $x = 1$, $y = 0{\cdot}60417$. The next non-zero term in the series will be the coefficient of x^8. $a_8 = 105/8! = 0{\cdot}003$ and so we may assume that $y = 0{\cdot}61$ to 2 D.P..

EXERCISE 11.6

1 Solve the following differential equations (by Euler's Method), using the step lengths as indicated, to find estimates of the values stated:

(i) $\dfrac{dy}{dx} = y^2 + x$, $y(0) = 1$, $h = 0{\cdot}1$, $y(0{\cdot}1)$, $y(0{\cdot}2)$, $y(0{\cdot}3)$,

(ii) $\dfrac{dy}{dx} = xy$, $y(0) = -2$, $h = 0{\cdot}2$, $y(0{\cdot}2)$, $y(0{\cdot}4)$, $y(0{\cdot}6)$,

(iii) $\dfrac{dy}{dx} = \dfrac{1 - y^2}{x}$, $y(2) = 0$, $h = 0{\cdot}5$, $y(2{\cdot}5)$, $y(3)$, $y(4)$,

(iv) $\dfrac{dy}{dx} + 3x^2 y = \dfrac{1}{x}$, $y(-4) = 3$, $h = 0{\cdot}01$, $y(-3{\cdot}99)$, $y(-3{\cdot}98)$, $y(-3{\cdot}95)$.

2 Follow the method given in example 2 to solve the 4 parts of question **1**, using a modified Euler's method.

3 Use the modified Euler's method to find y when $x = 1$, using a step length of $0{\cdot}2$, given that $(1 + x^2)\dfrac{dy}{dx} = yx - 2x^2$ and that $y(0) = 1$, $y(-0{\cdot}2) = 0{\cdot}9$.

4 Use the second order Taylor series approximation $y_1 \approx y_0 + hy'_0 + \dfrac{h^2}{2}y''_0$ to find $y(0{\cdot}1)$, $y(0{\cdot}2)$ and $y(0{\cdot}3)$ where y satisfies the differential equation $y' - yx^2 = 2$, and $y(0) = 1$.

5 Use the method given in example 4 to solve question **4**, using the modified second order iteration $y_{r+1} \approx 2y_r - y_{r-1} + h^2 y''_r$.

6 Find a fourth degree polynomial approximation to the solution of the following differential equation, by using a Taylor series expansion at $x = 0$. Find approximations to the value of y when $x = 1$ and when $x = 2$.

(i) $y'' - x^2 y' - y = 0$, $y = 2$, $y' = 0$ when $x = 0$,
(ii) $(1 + x)y'' + y = 0$, $y = 0$, $y' = 1$ when $x = 0$,
(iii) $y'' - 3xy' + x^2 y = 0$, $y = 1$, $y' = -4$ when $x = 0$.

7 Show that Euler's method for solving numerically the differential equation $\frac{dy}{dx} = x+y$ leads to the relation $y_{n+1} = (1+h)y_n + hx_n$. Evaluate y_2 taking $x_0 = 1$, $y_0 = 0{\cdot}5$ and $h = 0{\cdot}1$. *(L)*

8 The differential equation $\frac{dy}{dx} = 2xy - y^2$ is to be solved numerically using the second order Taylor series $y(x_0+h) \approx y(x_0) + hy'(x_0) + \frac{h^2}{2}y''(x_0)$. Show that, for the given differential equation

$$y(x_0+h) \approx y_0\{1 + (2x_0 - y_0)h + ((x_0 - y_0)(2x_0 - y_0) + 1)h^2\},$$

where y_0 denotes $y(x_0)$. Given that $y(0) = 1$ and $h = 0{\cdot}2$ calculate successively $y(0{\cdot}2)$, $y(0{\cdot}4)$ and $y(0{\cdot}6)$, giving your answer to three decimal places. *(C)*

9 The differential equation $\frac{dy}{dx} = y + 10x^2$, with $y = 0{\cdot}2$ when $x = 0$, is to be solved approximately by a numerical method. The values of y when $x = 0$, $0{\cdot}1$, $0{\cdot}2$ are y_0, y_1, y_2 respectively. Use the Euler's formula $y_1 \approx y_0 + hy'_0$ as a predictor and the modified Euler's formula $y_1 \approx y_0 + h(y'_0 + y'_1)/2$ as a corrector to find estimates for y_1, and then for y_2, giving three decimal places in your answers. *(C)*

MISCELLANEOUS EXERCISE 11

1 Solve the differential equation $\frac{dy}{dx} + 2y\cot x = \cos x$, given that $y = 2$ when $x = \pi/2$. *(AEB)*

2 (a) Solve the differential equation $\frac{d^2x}{dt^2} + 5\frac{dx}{dt} + 4x = 5 + 4t$, given that, when $t = 0$, $x = 0$ and $\frac{dx}{dt} = 1$.

(b) Find the general solution of the differential equation

$$(x^2+1)\frac{dy}{dx} + y = \tan^{-1} x.$$ *(C)*

3 (a) Solve the differential equation $\frac{d^2y}{dx^2} + 4\frac{dy}{dx} + 4y = 0$, given that $y = 1$ when $x = 0$, and y has a stationary value when $x = 1$. Sketch the graph of the solution for $x \geqslant 0$.

(b) Obtain a particular integral of the differential equation

$$\frac{d^2y}{dx^2} + 4\frac{dy}{dx} + 4y = 2e^{-2x}.$$ *(C)*

4 Given that $(l \cos 4x + m \sin 4x)$ is a particular integral of the differential equation $\frac{d^2y}{dx^2} + 4y = 12 \cos 4x$, find the values of the constants l and m, and hence obtain the general solution of this differential equation. Find y in terms of x given that $y = 1$ when $x = \pi/4$, and $y = -3$ when $x = \pi/2$. Find the x-coordinates of the stationary points of the curve given by this solution. *(L)*

5 (a) Find the solution of the differential equation $\frac{d^2y}{dx^2} - 6\frac{dy}{dx} + 13y = 1$ for which both y and $\frac{dy}{dx}$ are zero when $x = 0$.

(b) Find the solution of the differential equation $\frac{d^2z}{dt^2} - z = e^{-2t}$ for which $z = 0$ when $t = 0$ and $z \to 0$ as $t \to \infty$. *(C)*

6 Show that an integrating factor for the differential equation $x\frac{dy}{dx} + (x-1)y = x^2e^{2x}$, is $\frac{e^x}{x}$. Solve this differential equation given that $y = 2e^2$ when $x = 1$. *(AEB)*

7 The differential equation $\frac{dy}{dx} = x - y^2$ is to be solved numerically using a second order Taylor series method. Given that $y(1) = 1$, calculate successively $y(1{\cdot}2)$, $y(1{\cdot}4)$, $y(1{\cdot}6)$, giving your answer to three decimal places. *(C)*

8 Find the general solution of the differential equation $\frac{dy}{dx} + 2xy = 2x(x^2+1)$. Show that, if a particular solution curve of the differential equation passes through the point $(0, k)$, then that solution curve will have a maximum or a minimum at $(0, k)$ according as k is greater than or less than 1. On one diagram sketch the three solution curves passing respectively through the points $(0, 0)$, $(0, 1)$, $(0, 2)$. *(C)*

9 (i) Solve the differential equation $x\frac{dy}{dx} + 2y = x^4$ given that $y = 1$ when $x = -2$. (ii) Given that $p \sin 3x + q \cos 3x$, where p and q are constants, is a particular integral of the differential equation $\frac{d^2y}{dx^2} + 4\frac{dy}{dx} + 3y = -30 \sin 3x$, find the values of p and q, and obtain the general solution of the differential equation. *(L)*

10 Find the particular solution of each of the differential equations

(i) $\frac{d^2x}{dt^2} + 9x = 18t$, (ii) $\frac{d^2x}{dt^2} + 9\frac{dx}{dt} = 0$, for each of which, when $t = 0$, $x = 0$ and $\frac{dx}{dt} = 6$. Sketch the solution curve for the second equation. *(C)*

11 (a) Obtain the general solution of the differential equation

$$\frac{dy}{dx} = y \tan x + 2x \sec x.$$

(b) A particle is propelled from rest at the point O by a force acting along Ox. The magnitude of the force is such that the speed $v\ \text{m s}^{-1}$ and the displacement x m of the particle are related by $v\dfrac{dv}{dx} = 50(1-10x)$, $(0 \leqslant x \leqslant 0{\cdot}1)$. Calculate the speed of the particle when it has travelled 0·1 m. *(C)*

12 Show that the substitution $u = 1/y^3$ transforms the differential equation $3\dfrac{dy}{dx} + yf(x) = y^4g(x)$ into the differential equation $\dfrac{du}{dx} - uf(x) = -g(x)$. Find y in terms of x when (i) $f(x) = -g(x) = \dfrac{3x^2}{1+x^3}$, (ii) $f(x) = \dfrac{1}{x}$, $g(x) = \ln x$, given that, in each case, $y = 1$ when $x = 1$. *(JMB)*

13 Euler's Method for obtaining a numerical solution of the differential equation $\dfrac{dy}{dx} = f(x, y)$, with $y = y_0$ when $x = x_0$, is to evaluate $y_1 = y_0 + hf'(x_0, y_0)$, where $h = x_1 - x_0$ and to use y_1 as an approximation to $y(x_1)$. Copy the diagram in Fig. 11.7 and mark clearly on it y_1 and $y(x_1)$. Use Euler's Method with $h = 0{\cdot}1$ to estimate in succession $y(0{\cdot}1)$, $y(0{\cdot}2)$, $y(0{\cdot}3)$ when $\dfrac{dy}{dx} + xy = 1 + x^2$ and $y(0) = 1$. Verify that $y = x + e^{-x^2/2}$ is the exact solution of this differential equation and hence find the percentage error in the approximate value of $y(0{\cdot}3)$ obtained by Euler's method. *(C)*

14 (a) Solve the differential equation $\dfrac{dy}{dx} = y \cot x$, given that $y = 1$ when $x = \pi/6$. (b) Solve the differential equation $\dfrac{d^2x}{dt^2} - \dfrac{dx}{dt} - 2x = 3e^{-t}$, given

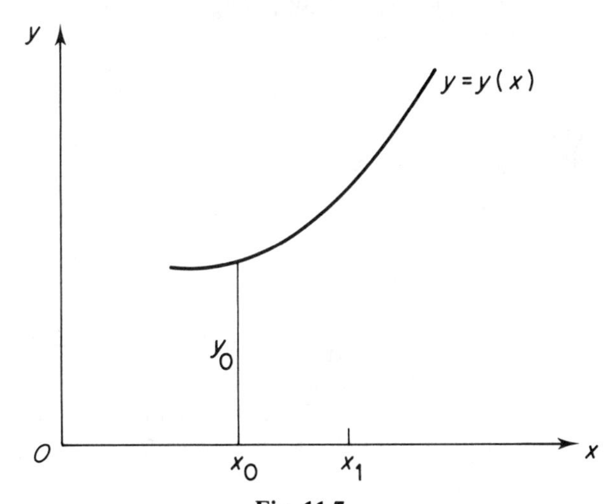

Fig. 11.7

that $x = 1$ when $t = 0$, and that $\lim\limits_{t \to \infty} x = 0$. (You may assume that $\lim\limits_{z \to \infty} ze^{-z} = 0$.) *(C)*

15 A particular integral of the differential equation $\frac{d^2y}{dx^2}+n^2y=\sin nx$ has the form $x(a\sin nx+b\cos nx)$, where a, b and n are constants. Determine a and b in terms of n. Hence find y in terms of x given that $y=0$ and $\frac{dy}{dx}=0$ when $x=0$. *(L)*

16 (a) Given that $y=\tan(3\tan^{-1}x)$, show that $(1+x^2)\frac{dy}{dx}=3(1+y^2)$.

(b) Solve the differential equation $x\frac{dy}{dx}=2y+x^3\ln x$, given that $y=2$ when $x=1$. *(C)*

17 (a) Solve the differential equation $x\frac{dy}{dx}-2y=x^4\cos 2x$.

(b) By means of the substitution $y=xu$ transform the differential equation $(x^2+y^2)\frac{dy}{dx}=xy$ into a differential equation containing only u, x and $\frac{du}{dx}$.

Hence, or otherwise, solve the differential equation $(x^2+y^2)\frac{dy}{dx}=xy$. Show that the solution for which $y=1$ when $x=1$ is $x=y\sqrt{(1+2\ln y)}$. *(C)*

18 Integrate the differential equation $\frac{dy}{dx}+y=2xy$, given that $y=e^2$ when $x=1$. *(L)*

19 In a chemical reaction, a mass $3m$ of a compound A can decompose into a mass m of an element B and a mass $2m$ of an element C, the rate of decomposition being equal to k times the mass of A present, where k is a constant. At the same time a mass m of B and a mass $2m$ of C can recombine into a mass $3m$ of A, the rate of combination being equal to $2k$ times the product of the masses of B and C present. Show that, if the masses of A, B, C present at time t are $1-3y$, y, $2y$ respectively, then $3\frac{dy}{dt}=k(1-4y)(1+y)$. Given that initially only compound A is present, find y in terms of k and t and deduce that y is always less than 1/4. *(L)*

20 Justify graphically the approximation $y_{n+1}-y_n\approx h\left(\frac{dy}{dx}\right)_{x=x_n}$, where y_n and y_{n+1} are the ordinates at $x=x_n$ and $x=x_n+h$ respectively. Use this step-by-step method, with $h=0{\cdot}5$, to find approximations to the values of y at $x=0{\cdot}5$ and $x=1{\cdot}0$ given that $\frac{dy}{dx}=y^2$ and $y(0)=0{\cdot}2$. Obtain the exact solution of this differential equation and find the relative errors in your approximate values. Explain why the relative errors in the values found by the step-by-step method will increase as x increases. *(L)*

21 It is given that $y=f(x)$ satisfies the differential equation $x^2\frac{dy}{dx}=x^2+y-1$, and that $f(1)=0{\cdot}72$, $f(1{\cdot}2)=0{\cdot}96$. Use a suitable approximation for $f'(x)$ to estimate $f(1{\cdot}4)$ and $f(1{\cdot}6)$, working to two decimal places. *(JMB)*

22 (i) Solve the equation $\frac{dy}{dx}=xy$ given that $y=5$ when $x=0$. (ii) Given that

$\dfrac{dy}{dx} = x + y^2$ and that $y(0{\cdot}1) = 0{\cdot}005$, use the modified Euler's method to estimate the values of $y(0{\cdot}2)$ and $y(0{\cdot}3)$, working to three decimal places. (*L*)

23 Obtain the general solution of the differential equation $\dfrac{d^2y}{dx^2} + 4\dfrac{dy}{dx} + 8y = 0$. (*L*)

24 Given a function u, the function y is defined by $y(x) = \{u(x)\}^2$. Write down expressions for $\dfrac{dy}{dx}$ and $\dfrac{d^2y}{dx^2}$ in terms of u, $\dfrac{du}{dx}$ and $\dfrac{d^2u}{dx^2}$. Hence rewrite the differential equation $u\dfrac{d^2u}{dx^2} + \left(\dfrac{du}{dx}\right)^2 + u\dfrac{du}{dx} = 0$ as a differential equation in y. Hence find the solution of the given differential equation which satisfies the initial conditions $u = 2$ and $\dfrac{du}{dx} = -1$ when $x = 0$. (*JMB*)

25 (i) A particular integral of the differential equation $\dfrac{d^2y}{dx^2} + 2\dfrac{dy}{dx} + 5y = 26 + 15x$ is given by $y = a + bx$, where a, b are constants. Find the values of a, b and determine the general solution of the differential equation.
(ii) Using the substitution $y = xv$, where v is a function of x, transform the differential equation $x^2\dfrac{dy}{dx} = x^2 + xy + y^2$, where $x > 0$, into a differential equation relating v and x. Hence find y in terms of x, given that $y = 0$ when $x = 1$. (*L*)

26 The motion of a particle, which moves along the x-axis, is governed by the differential equation $\dfrac{d^2x}{dt^2} + \dfrac{dx}{dt} = k \sin \omega t$ where k and ω are positive constants. The particle starts from rest at the point $x = x_0$. Determine x in terms of t subject to the given initial conditions. Find the values between which x oscillates if t is so large that e^{-t} can be neglected. (*JMB*)

27 The differential equation $\dfrac{dy}{dx} = x^2 - y^2 + 2$ for which $y = 1$ when $x = 1$ is to be solved numerically. Use a Taylor series method to find, correct to three decimal places, the values of y for $x = 1{\cdot}1$ and $x = 1{\cdot}2$. (*C*)

28 The gradient of a curve is given by the differential equation

$\dfrac{dy}{dx} + \dfrac{2xy}{x^2 - 1} = \dfrac{2x - 4}{x^2 - 1}$ and the curve passes through the point (2, 0). By solving this differential equation, show that the equation of the curve is $y = \dfrac{(x-2)^2}{x^2 - 1}$. Find the equations of the asymptotes and the coordinates of the stationary points of the curve. Sketch the curve. (*JMB*)

29 (a) Indicate on a sketch the family of solution curves for the differential equation $\dfrac{dy}{dx} = x - y$. (b) Solve the differential equation

$x\dfrac{dy}{dx} + (x+1)y = 2xe^{-x}$, given that $y = 0$ when $x = 1$. (*C*)

12 Vector Products

12.1 Definitions

The scalar product **a** . **b**, of two vectors **a** and **b**, was introduced in *Pure Mathematics* §12.6.

If, in $\mathbb{R}^3$, $\mathbf{a} = \begin{pmatrix} a_1 \\ a_2 \\ a_3 \end{pmatrix}$, $\mathbf{b} = \begin{pmatrix} b_1 \\ b_2 \\ b_3 \end{pmatrix}$, then $\mathbf{a} \cdot \mathbf{b} = a_1 b_1 + a_2 b_2 + a_3 b_3$.

A geometric interpretation of the scalar product is that

$$\mathbf{a} \cdot \mathbf{b} = |\mathbf{a}|\,|\mathbf{b}| \cos\theta = ab\cos\theta,$$

where θ is the angle between **a** and **b**.

We now define another type of product of two vectors, **a** and **b**, which is itself a vector. It is called the *vector product* of **a** and **b** and is denoted by $\mathbf{a} \times \mathbf{b}$. Another notation, used by some authors, is $\mathbf{a} \wedge \mathbf{b}$. In words, the vector product may be expressed as 'a cross b' or as 'a vec b', just as the scalar product is expressed as 'a dot b'. We define the vector product in $\mathbb{R}^3$, using the components of **a** and **b**.

Definition Let $\mathbf{a} = \begin{pmatrix} a_1 \\ a_2 \\ a_3 \end{pmatrix}$, $\mathbf{b} = \begin{pmatrix} b_1 \\ b_2 \\ b_3 \end{pmatrix}$, then $\mathbf{a} \times \mathbf{b} = \begin{pmatrix} a_2 b_3 - a_3 b_2 \\ a_3 b_1 - a_1 b_3 \\ a_1 b_2 - a_2 b_1 \end{pmatrix}$.

This definition may appear difficult to remember but, if it is rewritten in terms of the Cartesian unit vectors, **i**, **j**, **k**, then $\mathbf{a} \times \mathbf{b}$ can be expressed in the form of a determinant. Thus:

If $\mathbf{a} = a_1\mathbf{i} + a_2\mathbf{j} + a_3\mathbf{k}$ and $\mathbf{b} = b_1\mathbf{i} + b_2\mathbf{j} + b_3\mathbf{k}$ then

$$\mathbf{a} \times \mathbf{b} = (a_2 b_3 - a_3 b_2)\mathbf{i} + (a_3 b_1 - a_1 b_3)\mathbf{j} + (a_1 b_2 - a_2 b_1)\mathbf{k}$$

$$= \begin{vmatrix} \mathbf{i} & \mathbf{j} & \mathbf{k} \\ a_1 & a_2 & a_3 \\ b_1 & b_2 & b_3 \end{vmatrix}.$$

Thus $\mathbf{a} \times \mathbf{b}$ is the value of the determinant whose top row consists of the three unit vectors (**i** **j** **k**) and whose second and third rows consist respectively of the components $(a_1\ a_2\ a_3)$ of **a** and the components $(b_1\ b_2\ b_3)$ of **b**. Using the properties of determinants, it is immediately seen that, for all vectors **a** and **b**,

$$\mathbf{a} \times \mathbf{a} = \mathbf{0},\ \mathbf{a} \times (k\mathbf{a}) = \mathbf{0},\ \mathbf{b} \times \mathbf{a} = -\mathbf{a} \times \mathbf{b}.$$

Thus the vector product is quite different from any previous product we have met. Although the set $\mathbb{R}^3$ is closed under the binary operation of vector product, the vector product is non-commutative and the vector product of two parallel vectors is zero. It is also non-associative, as is seen from the products:

$$\mathbf{i} \times (\mathbf{i} \times \mathbf{j}) = \mathbf{i} \times \mathbf{k} = -\mathbf{j}, \ (\mathbf{i} \times \mathbf{i}) \times \mathbf{j} = \mathbf{0} \times \mathbf{j} = \mathbf{0}.$$

Cartesian unit vectors

From the definition of the vector product, it is found that:

$$\begin{array}{lll} \mathbf{i} \times \mathbf{i} = \mathbf{0}, & \mathbf{j} \times \mathbf{j} = \mathbf{0}, & \mathbf{k} \times \mathbf{k} = \mathbf{0}, \\ \mathbf{i} \times \mathbf{j} = \mathbf{k}, & \mathbf{j} \times \mathbf{k} = \mathbf{i}, & \mathbf{k} \times \mathbf{i} = \mathbf{j}, \\ \mathbf{j} \times \mathbf{i} = -\mathbf{k}, & \mathbf{k} \times \mathbf{j} = -\mathbf{i}, & \mathbf{i} \times \mathbf{k} = -\mathbf{j}. \end{array}$$

For two different Cartesian unit vectors in the cyclic order **i**, **j**, **k**, **i**, their vector product is the third, and for two in the reversed cyclic order **i**, **k**, **j**, **i**, their vector product is the negative of the third.

Note The above definition of the vector product $\mathbf{a} \times \mathbf{b}$ in terms of the Cartesian components of **a** and **b** depends upon the choice of Cartesian origin and axes. If a new set of Cartesian axes is chosen, with the same scales, still mutually at right-angles and with the unit vectors still right-handedly related, then the above definition still gives the same vector as the vector product $\mathbf{a} \times \mathbf{b}$. This property is summarised by saying that the vector product $\mathbf{a} \times \mathbf{b}$ is *invariant* with respect to an orthogonal change of axes. We cannot prove this result in this book so the reader is asked to accept it without proof. The following geometrical interpretation of the vector product is dependent on it.

Geometrical interpretation

In order to interpret $\mathbf{a} \times \mathbf{b}$ geometrically, choose the Cartesian axes in a convenient way. Let the axis Ox be chosen in the direction of **a** so that $\mathbf{a} = |\mathbf{a}|\mathbf{i} = a\mathbf{i}$. If **b** is parallel to **a**, then $\mathbf{a} \times \mathbf{b} = \mathbf{0}$. Suppose that **b** is not parallel to **a** so that **a** and **b** define a plane. Choose this as the coordinate plane Oxy in such a way that **b** makes an acute angle with Oy. Then **b** lies in the top half of the Oxy plane. Let θ be the angle between **a** and **b**. Then the picture in the Oxy plane is as shown in Fig. 12.1. Then $\mathbf{b} = \begin{pmatrix} b_1 \\ b_2 \\ 0 \end{pmatrix}$, and the two cases, (i) θ acute, $b_1 \geqslant 0$, (ii) θ obtuse, $b_1 < 0$, are shown in Fig. 12.1 (i) and (ii).

The corresponding situation is shown as a three dimensional picture in Fig. 12.2. Let the origin be O, let $\mathbf{a} = \overrightarrow{OA} = a\mathbf{i}$ and let $\mathbf{b} = \overrightarrow{OB} = b_1\mathbf{i} + b_2\mathbf{j}$. Again the case when θ is acute is shown in (i) and the case when θ is obtuse

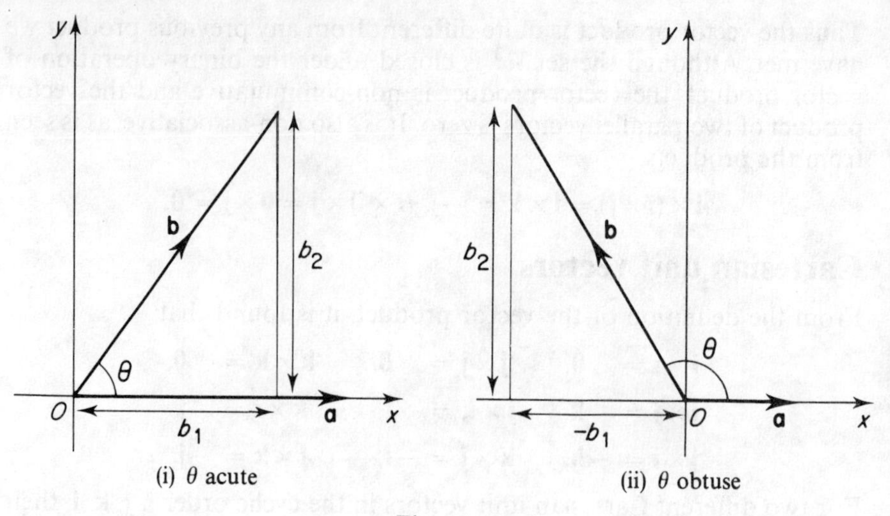

Fig. 12.1

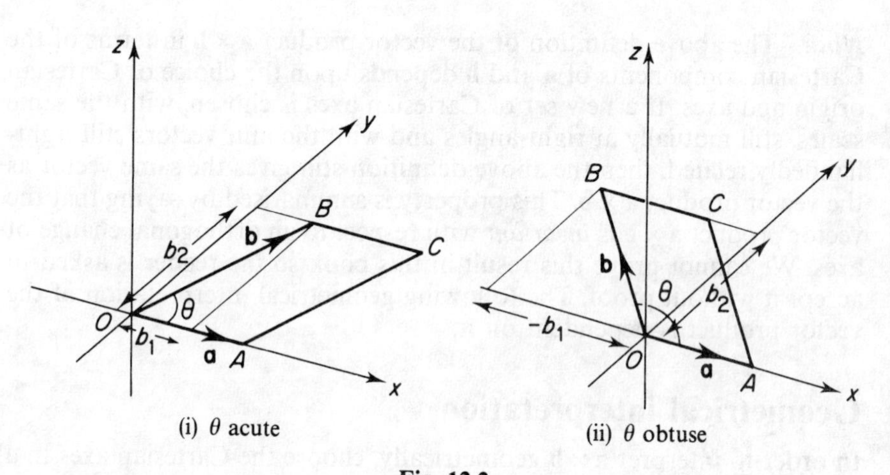

Fig. 12.2

is shown in (ii). Let C be the point $(a+b_1, b_2, 0)$, so that $OACB$ is a parallelogram. Then

$$\mathbf{a}\times\mathbf{b} = (a\mathbf{i})\times(b_1\mathbf{i}+b_2\mathbf{j}) = ab_1\mathbf{i}\times\mathbf{i}+ab_2\mathbf{i}\times\mathbf{j} = ab_2\mathbf{k}.$$

Now, if $|\mathbf{b}| = b$, $b_1 = b\cos\theta$ and $b_2 = b\sin\theta$. Hence

$$\mathbf{a}\times\mathbf{b} = ab\sin\theta\,\mathbf{k} = |\mathbf{a}|\,|\mathbf{b}|\sin\theta\,\mathbf{k} = S\mathbf{k},$$

where S is the area of the parallelogram $OACB$.

This is used to give an alternative definition of $\mathbf{a}\times\mathbf{b}$, equivalent to the original definition.

Definition Let $|\mathbf{a}| = a$, $|\mathbf{b}| = b$, let θ be the angle between $\mathbf{a}$ and $\mathbf{b}$ and let $\mathbf{n}$ be the unit vector perpendicular to the plane of $\mathbf{a}$ and $\mathbf{b}$, such that $(\mathbf{a}, \mathbf{b}, \mathbf{n})$ is a right-handed set, then $\mathbf{a} \times \mathbf{b} = ab \sin\theta \mathbf{n}$.

Remember that the statement that $(\mathbf{a}, \mathbf{b}, \mathbf{n})$ is right-handed means that a rotation from $\mathbf{a}$ to $\mathbf{b}$ through the angle θ about $\mathbf{n}$ is in the clockwise direction about $\mathbf{n}$. The situation is shown in Fig. 12.3.

The important properties of the vector product are summarised in Theorem 12.1. They follow in a straightforward manner from the definition and the proofs are left as an exercise for the reader.

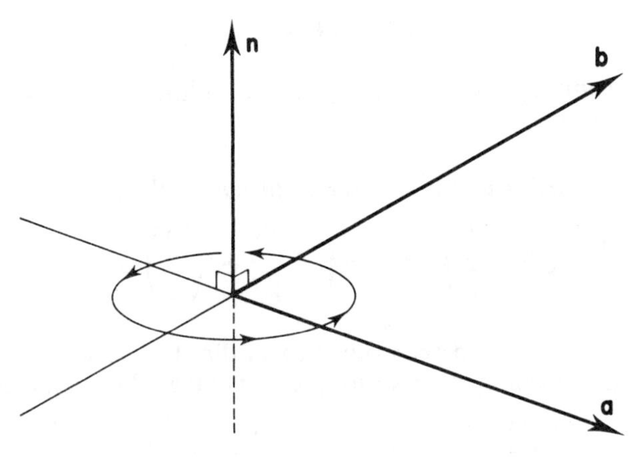

Fig. 12.3 $(\mathbf{a}, \mathbf{b}, \mathbf{n})$ right-handed

Theorem 12.1 For any vectors $\mathbf{a}$, $\mathbf{b}$, $\mathbf{c}$ in $\mathbb{R}^3$, and any scalar k,

(i) $\mathbf{a} \times \mathbf{b} = -\mathbf{b} \times \mathbf{a}$,
(ii) $k(\mathbf{a} \times \mathbf{b}) = \mathbf{a} \times (k\mathbf{b}) = (k\mathbf{a}) \times \mathbf{b}$,
(iii) $\mathbf{a} \times (\mathbf{b} + \mathbf{c}) = \mathbf{a} \times \mathbf{b} + \mathbf{a} \times \mathbf{c}$,
(iv) if $\mathbf{a} \times \mathbf{b} = \mathbf{0}$ then either $\mathbf{a}$ is parallel to $\mathbf{b}$ or $\mathbf{a} = \mathbf{0}$ or $\mathbf{b} = \mathbf{0}$.

The fact that a vector product vanishes for parallel vectors means that cancellation cannot be performed in an equation of vector products. If $\mathbf{a} \times \mathbf{b} = \mathbf{a} \times \mathbf{c}$ then $\mathbf{a} \times (\mathbf{b} - \mathbf{c}) = \mathbf{0}$ and all that can be deduced is that either $\mathbf{a} = \mathbf{0}$ or $\mathbf{b} = \mathbf{c}$ or $\mathbf{a}$ is parallel to $(\mathbf{b} - \mathbf{c})$.

Note It must be remembered that the vector product $\mathbf{a} \times \mathbf{b}$ is a vector, whereas the scalar product $\mathbf{a} \cdot \mathbf{b}$ is a scalar. If θ is the angle between $\mathbf{a}$ and $\mathbf{b}$, then

$$\mathbf{a} \times \mathbf{b} = |\mathbf{a}|\ |\mathbf{b}| \sin\theta\ \mathbf{n},$$

$$\mathbf{a} \cdot \mathbf{b} = |\mathbf{a}|\ |\mathbf{b}| \cos\theta.$$

If neither $\mathbf{a}$ nor $\mathbf{b}$ is the zero vector, then

$$\mathbf{a} \times \mathbf{b} = \mathbf{0} \Leftrightarrow \mathbf{a} \text{ and } \mathbf{b} \text{ are parallel},$$

$$\mathbf{a} \cdot \mathbf{b} = 0 \Leftrightarrow \mathbf{a} \text{ and } \mathbf{b} \text{ are perpendicular}.$$

EXAMPLE 1 *Evaluate:* (i) $\begin{pmatrix}1\\2\\3\end{pmatrix}\times\begin{pmatrix}2\\-1\\1\end{pmatrix}$, (*ii*) $(\mathbf{i}-\mathbf{j}+2\mathbf{k})\times(2\mathbf{j}+3\mathbf{k})$.

$$\text{(i)}\ \begin{pmatrix}1\\2\\3\end{pmatrix}\times\begin{pmatrix}2\\-1\\1\end{pmatrix}=\begin{vmatrix}\mathbf{i}&\mathbf{j}&\mathbf{k}\\1&2&3\\2&-1&1\end{vmatrix}=\begin{pmatrix}2.1+1.3\\3.2-1.1\\-1.1-2.2\end{pmatrix}=\begin{pmatrix}2+3\\6-1\\-1-4\end{pmatrix}=\begin{pmatrix}\mathbf{5}\\\mathbf{5}\\\mathbf{-5}\end{pmatrix}.$$

$$\begin{aligned}\text{(ii)}\ (\mathbf{i}-\mathbf{j}+2\mathbf{k})\times(2\mathbf{j}+3\mathbf{k}) &= 2\mathbf{i}\times\mathbf{j}+3\mathbf{i}\times\mathbf{k}-2\mathbf{j}\times\mathbf{j}-3\mathbf{j}\times\mathbf{k}+4\mathbf{k}\times\mathbf{j}+6\mathbf{k}\times\mathbf{k}\\ &= 2\mathbf{k}-3\mathbf{j}-2\mathbf{0}-3\mathbf{i}-4\mathbf{i}+6\mathbf{0}\\ &= \mathbf{-7i-3j+2k}.\end{aligned}$$

The vector product provides an easy way of finding a vector perpendicular to two other vectors.

EXAMPLE 2 *Find a unit vector perpendicular to the two lines given by the equations* $\mathbf{r}=\begin{pmatrix}1\\2\\3\end{pmatrix}+s\begin{pmatrix}1\\2\\0\end{pmatrix}$ *and* $\mathbf{r}=\begin{pmatrix}2\\0\\-1\end{pmatrix}+t\begin{pmatrix}0\\2\\1\end{pmatrix}$.

The required unit vector is perpendicular to the direction vectors of the two lines and so it is in the direction of the vector product of their direction vectors. This is given by

$$\begin{pmatrix}1\\2\\0\end{pmatrix}\times\begin{pmatrix}0\\2\\1\end{pmatrix}=\begin{vmatrix}\mathbf{i}&\mathbf{j}&\mathbf{k}\\1&2&0\\0&2&1\end{vmatrix}=\begin{pmatrix}2\\-1\\2\end{pmatrix}.$$

This vector has length $\sqrt{(2^2+(-1)^2+2^2)}=3$, and so the required unit vector is $\begin{pmatrix}\mathbf{2/3}\\\mathbf{-1/3}\\\mathbf{2/3}\end{pmatrix}$.

EXAMPLE 3 *Using the scalar and vector products of* $\begin{pmatrix}a\\b\\0\end{pmatrix}$ *and* $\begin{pmatrix}c\\d\\0\end{pmatrix}$, *prove that* $(a^2+b^2)(c^2+d^2)=(ac+bd)^2+(ad-bc)^2$.

Let θ be the angle between the vectors $\begin{pmatrix}a\\b\\0\end{pmatrix}$ and $\begin{pmatrix}c\\d\\0\end{pmatrix}$, then

$$\begin{pmatrix}a\\b\\0\end{pmatrix}.\begin{pmatrix}c\\d\\0\end{pmatrix}=ac+bd=\sqrt{(a^2+b^2)}\sqrt{(c^2+d^2)}\cos\theta,\text{ and}$$

$$\begin{pmatrix}a\\b\\0\end{pmatrix}\times\begin{pmatrix}c\\d\\0\end{pmatrix}=(ad-bc)\mathbf{k}=\sqrt{(a^2+b^2)}\sqrt{(c^2+d^2)}\sin\theta\,\mathbf{k},$$

and so $ad-bc=\sqrt{(a^2+b^2)}\sqrt{(c^2+d^2)}\sin\theta$.

Therefore $(a^2+b^2)(c^2+d^2) = (a^2+b^2)(c^2+d^2)(\cos^2\theta+\sin^2\theta)$
$$= (ac+bd)^2+(ad-bc)^2.$$

EXERCISE 12.1

1 Prove Theorem 12.1, with $\mathbf{a} = \begin{pmatrix} a_1 \\ a_2 \\ a_3 \end{pmatrix}$, $\mathbf{b} = \begin{pmatrix} b_1 \\ b_2 \\ b_3 \end{pmatrix}$, $\mathbf{c} = \begin{pmatrix} c_1 \\ c_2 \\ c_3 \end{pmatrix}$.

2 Write down $\mathbf{a}\times\mathbf{b}$, given that:

(i) $\mathbf{a} = \begin{pmatrix} 1 \\ 2 \\ 0 \end{pmatrix}$, $\mathbf{b} = \begin{pmatrix} -1 \\ 3 \\ 0 \end{pmatrix}$ (ii) $\mathbf{a} = \begin{pmatrix} 1 \\ 0 \\ 1 \end{pmatrix}$, $\mathbf{b} = \begin{pmatrix} 1 \\ 0 \\ -1 \end{pmatrix}$

(iii) $\mathbf{a} = \begin{pmatrix} 1 \\ -2 \\ 3 \end{pmatrix}$, $\mathbf{b} = \begin{pmatrix} 2 \\ 0 \\ 1 \end{pmatrix}$ (iv) $\mathbf{a} = \begin{pmatrix} 1 \\ 1 \\ 1 \end{pmatrix}$, $\mathbf{b} = \begin{pmatrix} 1 \\ -2 \\ 1 \end{pmatrix}$

(v) $\mathbf{a} = \begin{pmatrix} 3 \\ -1 \\ 4 \end{pmatrix}$, $\mathbf{b} = \begin{pmatrix} 2 \\ 1 \\ -3 \end{pmatrix}$ (vi) $\mathbf{a} = \begin{pmatrix} u \\ v \\ w \end{pmatrix}$, $\mathbf{b} = \begin{pmatrix} x \\ y \\ z \end{pmatrix}$.

3 Find the vector product $\mathbf{a}\times\mathbf{b}$, given that: (i) $\mathbf{a} = \mathbf{i}$, $\mathbf{b} = \mathbf{i}+2\mathbf{j}$, (ii) $\mathbf{a} = \mathbf{i}-\mathbf{j}$, $\mathbf{b} = \mathbf{j}-\mathbf{k}$ (iii) $\mathbf{a} = \mathbf{i}+\mathbf{j}+\mathbf{k}$, $\mathbf{b} = \mathbf{i}-\mathbf{k}$ (iv) $\mathbf{a}=\mathbf{i}+2\mathbf{j}$, $\mathbf{b}=\mathbf{i}-3\mathbf{k}$ (v) $\mathbf{a} = \mathbf{i}+2\mathbf{j}-\mathbf{k}$, $\mathbf{b} = 2\mathbf{i}-\mathbf{j}+\mathbf{k}$ (vi) $\mathbf{a} = -\mathbf{i}+\mathbf{j}+\mathbf{k}$, $\mathbf{b} = 2\mathbf{i}-2\mathbf{j}-2\mathbf{k}$.

4 Given that $\mathbf{a}\times(\mathbf{b}+\mathbf{c}) = \mathbf{a}\times\mathbf{b}+\mathbf{a}\times\mathbf{c}$, prove that $(\mathbf{b}+\mathbf{c})\times\mathbf{a} = \mathbf{b}\times\mathbf{a}+\mathbf{c}\times\mathbf{a}$.

5 Assuming: (i) the geometrical definition of a vector product: namely that if $|\mathbf{a}| = a$, $|\mathbf{b}| = b$, θ is the angle between $\mathbf{a}$ and $\mathbf{b}$ and $\mathbf{n}$ is the unit vector perpendicular to the plane of $\mathbf{a}$ and $\mathbf{b}$, such that $(\mathbf{a}, \mathbf{b}, \mathbf{n})$ is a right-handed set, then $\mathbf{a}\times\mathbf{b} = ab\sin\theta\,\mathbf{n}$, (ii) the distributive law for vector products: namely that $\mathbf{a}\times(\mathbf{b}+\mathbf{c}) = \mathbf{a}\times\mathbf{b}+\mathbf{a}\times\mathbf{c}$, evaluate the nine vector products of the Cartesian unit vectors and hence prove that

$$\text{if}\quad \mathbf{a} = \begin{pmatrix} a_1 \\ a_2 \\ a_3 \end{pmatrix},\ \mathbf{b} = \begin{pmatrix} b_1 \\ b_2 \\ b_3 \end{pmatrix},\ \text{then } \mathbf{a}\times\mathbf{b} = \begin{pmatrix} a_2b_3-a_3b_2 \\ a_3b_1-a_1b_3 \\ a_1b_2-a_2b_1 \end{pmatrix}.$$

6 Find the magnitude of the vector $(\mathbf{i}+\mathbf{j}-\mathbf{k})\times(\mathbf{i}-\mathbf{j}+\mathbf{k})$. (*L*)

7 If $\mathbf{a} = -\mathbf{i}+2\mathbf{j}-5\mathbf{k}$ and $\mathbf{b} = 5\mathbf{i}-2\mathbf{j}+\mathbf{k}$ find (i) $\mathbf{a}.\mathbf{b}$, (ii) $\mathbf{a}\times\mathbf{b}$, (iii) the unit vector in the direction of $\mathbf{a}\times\mathbf{b}$. (*L*)

8 Given that $\mathbf{u} = 2\mathbf{i}-\mathbf{j}+2\mathbf{k}$, $\mathbf{v} = a\mathbf{i}+b\mathbf{k}$ and $\mathbf{u}\times\mathbf{v} = \mathbf{i}+c\mathbf{k}$, find a, b and c. Find, in surd form, the cosine of the angle between the vectors $\mathbf{u}$ and $\mathbf{v}$. (*JMB*)

9 (i) Define the scalar and vector products of two vectors $\mathbf{a}$ and $\mathbf{b}$. Deduce that $|\mathbf{a}\times\mathbf{b}|^2 = |\mathbf{a}|^2|\mathbf{b}|^2-(\mathbf{a}.\mathbf{b})^2$. (ii) Simplify $(\mathbf{a}-\mathbf{b})\times(\mathbf{a}+\mathbf{b})$ and show that, if $\mathbf{a}$ is perpendicular to $\mathbf{b}$, then $|(\mathbf{a}-\mathbf{b})\times(\mathbf{a}+\mathbf{b})| = 2|\mathbf{a}|\,|\mathbf{b}|$. (*JMB*)

10 Given that $\mathbf{r} = a\mathbf{i}+b\mathbf{j}+c\mathbf{k}$, $\mathbf{k}\times\mathbf{r} = \mathbf{p}$, $\mathbf{r}\times\mathbf{p} = \mathbf{k}$, where $\mathbf{i}, \mathbf{j}$ and $\mathbf{k}$ are a triad of mutually orthogonal unit vectors and a, b, c are constants, show that $a^2+b^2 = 1$, $c = 0$. (*L*)

11 The vectors $\mathbf{u}$, $\mathbf{v}$ and $\mathbf{n}$, where $\mathbf{n}$ is a unit vector, are such that $\mathbf{n}\times\mathbf{u} = 3\mathbf{n}\times\mathbf{v}$. Deduce that $\mathbf{u}-3\mathbf{v} = \lambda\mathbf{n}$, where λ is a scalar. Given that $\mathbf{u}$ and $\mathbf{v}$ are

perpendicular, show that $\lambda = \pm\sqrt{(u^2+9v^2)}$, where $u = |\mathbf{u}|$ and $v = |\mathbf{v}|$. For the case $\lambda > 0$, find the cosine of the angle between **u** and **n** in terms of u and v.
(JMB)

12.2 Geometrical applications

Solution of vector equations

It is useful to be able to solve simple vector equations, so we give the solutions of two equations in a theorem.

Theorem 12.2 Let **n** be a non-zero vector. Then:

(i) $\mathbf{x}.\mathbf{n} = 0 \Leftrightarrow \mathbf{x} = \mathbf{n}\times\mathbf{w}$, (ii) $\mathbf{x}\times\mathbf{n} = \mathbf{0} \Leftrightarrow \mathbf{x} = t\mathbf{n}$,

where **w** is a vector parameter and t is a scalar parameter.

Proof (i) If $\mathbf{x}.\mathbf{n} = 0$ and $\mathbf{x} \neq \mathbf{0}$, then **x** and **n** are perpendicular and $\mathbf{x}\times\mathbf{n} = xn\mathbf{u}$, where **u** is a unit vector and (**x**, **n**, **u**) is a mutually perpendicular right-handed set. Then $\mathbf{n}\times\mathbf{u} = n.1\,(\mathbf{x}/x)$ so $\mathbf{x} = x(\mathbf{n}\times\mathbf{u})/n$. That is, $\mathbf{x} = \mathbf{n}\times\mathbf{w}$, where $\mathbf{w} = x\mathbf{u}/n = (\mathbf{x}\times\mathbf{n})/n^2$.
Conversely, let **w** be any vector and let $\mathbf{x} = \mathbf{n}\times\mathbf{w}$. Then either $\mathbf{x} = \mathbf{0}$ or **x** is perpendicular to **n** and hence $\mathbf{x}.\mathbf{n} = 0$.
(ii) If $\mathbf{x}\times\mathbf{n} = \mathbf{0}$, either $\mathbf{x} = \mathbf{0}$ or **x** is parallel to **n**, hence $\mathbf{x} = t\mathbf{n}$ for some scalar t. Conversely, if $\mathbf{x} = t\mathbf{n}$, then $\mathbf{x}\times\mathbf{n} = t\mathbf{n}\times\mathbf{n} = \mathbf{0}$.

EXAMPLE 1 *Given perpendicular vectors* **a** *and* **b**, *prove that* $a^2\mathbf{b} = (\mathbf{a}\times\mathbf{b})\times\mathbf{a}$. *Solve the equation* $\mathbf{x}\times\mathbf{a} = \mathbf{b}$.

Since **a** and **b** are perpendicular, (**a** × **b**, **a**, **b**) forms a right-handed set of mutually perpendicular vectors. Hence $|\mathbf{a}\times\mathbf{b}| = ab$, $|(\mathbf{a}\times\mathbf{b})\times\mathbf{a}| = a^2b$ and the unit vector in the direction of $(\mathbf{a}\times\mathbf{b})\times\mathbf{a}$ is $\mathbf{b}/b$. Therefore,

$$(\mathbf{a}\times\mathbf{b})\times\mathbf{a} = a^2b(\mathbf{b}/b) = a^2\mathbf{b}.$$

Now $\mathbf{x}\times\mathbf{a} = \mathbf{b} \Leftrightarrow a^2\mathbf{x}\times\mathbf{a} = a^2\mathbf{b} = (\mathbf{a}\times\mathbf{b})\times\mathbf{a} \Leftrightarrow (a^2\mathbf{x}-\mathbf{a}\times\mathbf{b})\times\mathbf{a} = \mathbf{0}$

$$\Leftrightarrow a^2\mathbf{x}-\mathbf{a}\times\mathbf{b} = t\mathbf{a} \qquad \text{(by Theorem 12.2)}$$

$$\Leftrightarrow \mathbf{x} = \{\mathbf{a}\times\mathbf{b}+t\mathbf{a}\}/a^2.$$

So, in terms of a parameter t, the solution of the equation is $\mathbf{x} = \{\mathbf{a}\times\mathbf{b}+t\mathbf{a}\}/a^2$.

EXAMPLE 2 *The points A, B and C have position vectors* **a**, **b** *and* **c**, *respectively. Prove that the area S of the triangle ABC is given by* $2S = |\mathbf{b}\times\mathbf{c}+\mathbf{c}\times\mathbf{a}+\mathbf{a}\times\mathbf{b}|$.

The area S is given by $2S = AB.AC\sin A = |\overrightarrow{AB}\times\overrightarrow{AC}|$

$$= |(\mathbf{b}-\mathbf{a})\times(\mathbf{c}-\mathbf{a})|$$

$$= |\mathbf{b}\times\mathbf{c}-\mathbf{a}\times\mathbf{c}-\mathbf{b}\times\mathbf{a}+\mathbf{a}\times\mathbf{a}|$$

$$= |\mathbf{b}\times\mathbf{c}+\mathbf{c}\times\mathbf{a}+\mathbf{a}\times\mathbf{b}|.$$

Vector equation of a line

Let L be the line in the direction of the non-zero vector $\mathbf{u}$ passing through the point A, of position vector $\mathbf{a}$.
Then a point P, of position vector $\mathbf{r}$, lies on L if, and only if, $\overrightarrow{AP}$ is parallel to $\mathbf{u}$. Hence the equation of L is $\mathbf{r}-\mathbf{a} = t\mathbf{u}$, or $\mathbf{r} = \mathbf{a}+t\mathbf{u}$, in terms of a scalar parameter t. (See *Pure Mathematics* §5.6.) The situation is shown in Fig. 12.4. Now, by Theorem 12.2 (ii),

$$\mathbf{r}-\mathbf{a} = t\mathbf{u} \Leftrightarrow (\mathbf{r}-\mathbf{a}) \times \mathbf{u} = \mathbf{0} \Leftrightarrow \mathbf{r} \times \mathbf{u} = \mathbf{a} \times \mathbf{u}.$$

Therefore, the line L is also given by the equation $\mathbf{r} \times \mathbf{u} = \mathbf{a} \times \mathbf{u}$. We state these results as a theorem.

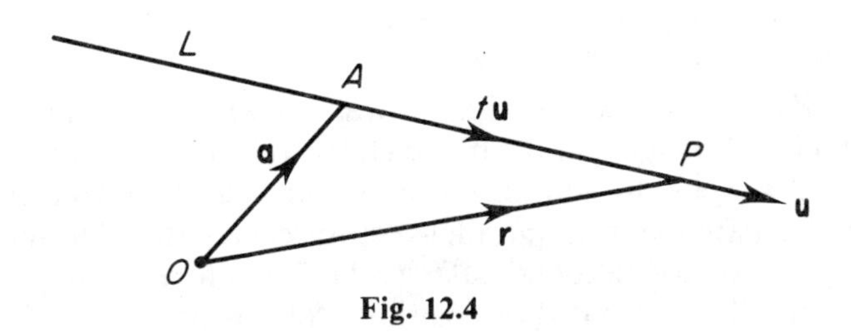

Fig. 12.4

Theorem 12.3 The line in the direction of the vector $\mathbf{u}$ and passing through the point A, with position vector $\mathbf{a}$, is given by either of the equations:
(i) $\mathbf{r} \times \mathbf{u} = \mathbf{a} \times \mathbf{u}$,
and (ii) $\mathbf{r} = \mathbf{a}+t\mathbf{u}$, in terms of a parameter t.

EXAMPLE 3 *Relative to the origin O, three points A, B, C, have position vectors* $\mathbf{a}$, $\mathbf{b}$, $\mathbf{c}$, *respectively, and no two of these three vectors are parallel. Given that* $\mathbf{a} \times \mathbf{b} = \mathbf{b} \times \mathbf{c}$, *prove that:* (*i*) *O, A, B and C are coplanar,* (*ii*) *OB bisects AC.*

Let $\mathbf{n}$ be the unit vector in the direction of $\mathbf{a} \times \mathbf{b}$ so that $\mathbf{n}$ is perpendicular to the plane OAB. Since $\mathbf{a} \times \mathbf{b} = \mathbf{b} \times \mathbf{c}$, $\mathbf{n}$ is also perpendicular to the plane OBC, but both planes contain O so they are the same plane and $OABC$ are coplanar. This is shown in Fig. 12.5.

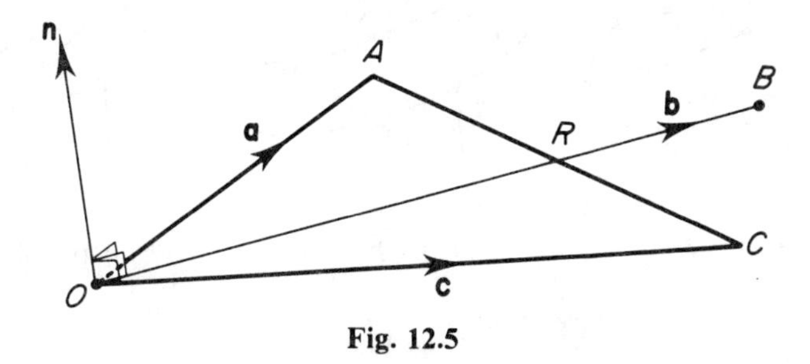

Fig. 12.5

Let R be the mid-point of AC, with position vector $\mathbf{r}$. Then $2\mathbf{r}=\mathbf{a}+\mathbf{c}$, so that $2\mathbf{r}\times\mathbf{b}=\mathbf{a}\times\mathbf{b}+\mathbf{c}\times\mathbf{b}=\mathbf{a}\times\mathbf{b}-\mathbf{b}\times\mathbf{c}=\mathbf{0}$. Therefore, $\mathbf{r}$ is parallel to $\mathbf{b}$, that is, the mid-point of AC lies on OB, and hence OB bisects AC.

Vector equation of a plane

In *Pure Mathematics* §12.8, it was shown that the equation of the plane Π through the point A, with position vector $\mathbf{a}$, and perpendicular to the vector $\mathbf{n}$ is $(\mathbf{r}-\mathbf{a}).\mathbf{n}=0$. This is because, for a point P in Π, with position vector $\mathbf{r}$, $\overrightarrow{AP}=\mathbf{r}-\mathbf{a}$ and $\overrightarrow{AP}$ is perpendicular to $\mathbf{n}$, see Fig. 12.6 (i). By Theorem 12.2 (i),

$$(\mathbf{r}-\mathbf{a}).\mathbf{n}=0 \Leftrightarrow \mathbf{r}-\mathbf{a}=\mathbf{n}\times\mathbf{w} \Leftrightarrow \mathbf{r}=\mathbf{a}+\mathbf{n}\times\mathbf{w},$$

so the plane Π is also given by the equation $\mathbf{r}=\mathbf{a}+\mathbf{n}\times\mathbf{w}$, in terms of a vector parameter $\mathbf{w}$, see Fig. 12.6 (ii).

Also, in *Pure Mathematics* §12.9, it was shown that if $\mathbf{p}$ and $\mathbf{q}$ are two non-parallel vectors lying in the plane Π, then the plane Π is also given by the equation $\mathbf{r}=\mathbf{a}+s\mathbf{p}+t\mathbf{q}$, in terms of two scalar parameters s and t. Now if $\mathbf{p}$ and $\mathbf{q}$ are two non-parallel vectors, each parallel to the plane Π, then $\mathbf{p}\times\mathbf{q}$ is a vector perpendicular to Π. Therefore, the equation of a plane Π, through the point A of position vector $\mathbf{a}$, and parallel to the two non-parallel vectors $\mathbf{p}$ and $\mathbf{q}$, is $(\mathbf{r}-\mathbf{a}).(\mathbf{p}\times\mathbf{q})=0$, see Fig. 12.6 (iii). These results are summarised in a theorem.

Theorem 12.4 Let the point A have position vector $\mathbf{a}$ and let $\mathbf{n}$ be any non-zero vector and let $\mathbf{p}$ and $\mathbf{q}$ be non-parallel vectors. Then
(a) the plane through A perpendicular to $\mathbf{n}$ is given by either
 (i) $(\mathbf{r}-\mathbf{a}).\mathbf{n}=0$
or (ii) $\mathbf{r}=\mathbf{a}+\mathbf{n}\times\mathbf{w}$, in terms of a vector parameter $\mathbf{w}$.
(b) the plane through A parallel to $\mathbf{p}$ and to $\mathbf{q}$ is given by either
 (i) $(\mathbf{r}-\mathbf{a}).(\mathbf{p}\times\mathbf{q})=0$,
or (ii) $\mathbf{r}=\mathbf{a}+s\mathbf{p}+t\mathbf{q}$, in terms of two scalar parameters s and t.

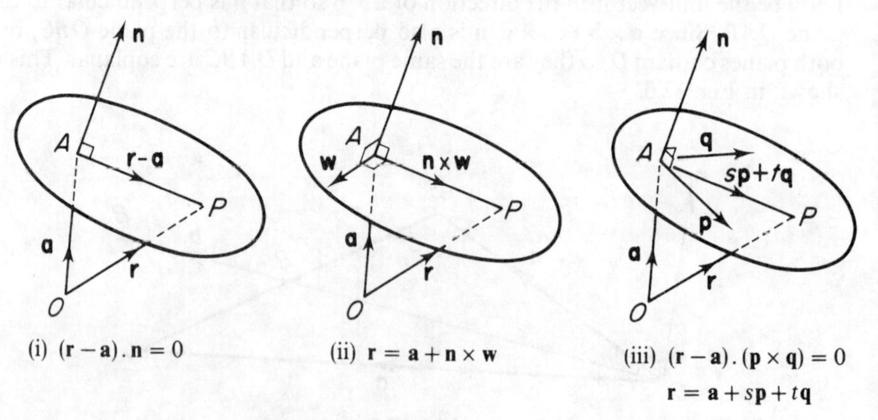

(i) $(\mathbf{r}-\mathbf{a}).\mathbf{n}=0$ (ii) $\mathbf{r}=\mathbf{a}+\mathbf{n}\times\mathbf{w}$ (iii) $(\mathbf{r}-\mathbf{a}).(\mathbf{p}\times\mathbf{q})=0$, $\mathbf{r}=\mathbf{a}+s\mathbf{p}+t\mathbf{q}$

Fig. 12.6

EXAMPLE 4 *The position vectors* **a**, **b** *and* **c**, *of A, B and C respectively, are given by* $\mathbf{a} = \mathbf{i}+2\mathbf{j}+3\mathbf{k}$, $\mathbf{b} = 3\mathbf{i}+\mathbf{j}+\mathbf{k}$, $\mathbf{c} = 3\mathbf{i}-\mathbf{j}-3\mathbf{k}$. *Find (i) the cosine of the angle BAC, (ii) the area of the triangle ABC, (iii) a vector equation of the plane ABC.*

(i) $\overrightarrow{AB} = \mathbf{b}-\mathbf{a} = 2\mathbf{i}-\mathbf{j}-2\mathbf{k}$, $\overrightarrow{AC} = \mathbf{c}-\mathbf{a} = 2\mathbf{i}-3\mathbf{j}-6\mathbf{k}$, so $\overrightarrow{AB}.\overrightarrow{AC} = 4+3+12 = 19$, $|\overrightarrow{AB}| = \sqrt{(4+1+4)} = 3$, $|\overrightarrow{AC}| = \sqrt{(4+9+36)} = 7$, so $\cos A = \mathbf{19/21}$.
(ii) $\overrightarrow{AB}\times\overrightarrow{AC} = 0\mathbf{i}+8\mathbf{j}-4\mathbf{k}$ which has length $\sqrt{(80)}$, therefore, the area of the triangle is $\mathbf{2\sqrt{5}}$.
(iii) The plane ABC has a normal vector $2\mathbf{j}-\mathbf{k}$, so it has an equation $(\mathbf{r}-\mathbf{a}).(2\mathbf{j}-\mathbf{k}) = 0$, that is, $\mathbf{r}.(2\mathbf{j}-\mathbf{k}) = (\mathbf{i}+2\mathbf{j}+3\mathbf{k}).(2\mathbf{j}-\mathbf{k})$, or $\mathbf{r.(2j-k)=1}$.

EXAMPLE 5 *The points A, B and C have coordinates* (2, 1, 3), (1, 4, 5) *and* (1, 3, 3), *respectively, and O is the origin. Find (i) the area of triangle ABC, (ii) a Cartesian equation for the plane ABC, (iii) the volume of the tetrahedron OABC.*

(i) $\overrightarrow{AB} = \begin{pmatrix}-1\\3\\2\end{pmatrix}$, $\overrightarrow{AC} = \begin{pmatrix}-1\\2\\0\end{pmatrix}$, $\overrightarrow{AB}\times\overrightarrow{AC} = \begin{pmatrix}-4\\-2\\1\end{pmatrix}$ and $|\overrightarrow{AB}\times\overrightarrow{AC}| = \sqrt{(21)}$.
Therefore the area of triangle ABC is $\mathbf{\sqrt{(21)}/2}$.
(ii) The equation of the plane ABC is $(\mathbf{r}-\mathbf{a}).(\overrightarrow{AB}\times\overrightarrow{AC}) = 0$, that is,

$$\mathbf{r}.\begin{pmatrix}-4\\-2\\1\end{pmatrix} = \begin{pmatrix}2\\1\\3\end{pmatrix}.\begin{pmatrix}-4\\-2\\1\end{pmatrix} = -7, \text{ or } \mathbf{4x+2y-z=7}.$$

(iii) Regarding ABC as the base of the tetrahedron, the height, h, of the tetrahedron is the projection of $\overrightarrow{OA}$ on the unit vector normal to ABC, so $h = \left|\begin{pmatrix}2\\1\\3\end{pmatrix}.\begin{pmatrix}4\\2\\-1\end{pmatrix}\right|/\sqrt{(21)} = 7/\sqrt{(21)}$. Therefore the volume of the tetrahedron is V, given by $V = \frac{1}{3}.\frac{7}{\sqrt{(21)}}.\frac{\sqrt{(21)}}{2} = \mathbf{\frac{7}{6}}$.

Shortest distance between two skew lines

Let L and M be skew lines, with direction vectors $\mathbf{u}$ and $\mathbf{v}$, respectively. Then $\mathbf{u}\times\mathbf{v}$ is perpendicular to both lines. Let Π be the plane containing the line L and with normal vector $\mathbf{u}\times\mathbf{v}$. Then M is parallel to Π and, if M' is the projection of M on to Π in the direction of $\mathbf{u}\times\mathbf{v}$, let M' meet L at a point P, see Fig. 12.7 (i). The line through P perpendicular to Π, that is in the direction $\mathbf{u}\times\mathbf{v}$, will meet M, so let this point of intersection be Q. Then $\overrightarrow{PQ}$ is perpendicular to both lines L and M and it is called their *common perpendicular*. If S is any point on L and T is any point on M, then $|\overrightarrow{PQ}| = |\overrightarrow{ST}.(\mathbf{u}\times\mathbf{v})|/|\mathbf{u}\times\mathbf{v}|$ and clearly the distance ST is least when $S = P$ and $T = Q$. Thus the length PQ of the common perpendicular of L and M is the shortest distance between the lines L and M.

Let the lines L and M be given by the equations $\mathbf{r} = \mathbf{a}+s\mathbf{u}$ and $\mathbf{r} = \mathbf{b}+t\mathbf{v}$, respectively, where $\mathbf{a}$ and $\mathbf{b}$ are the position vectors of fixed

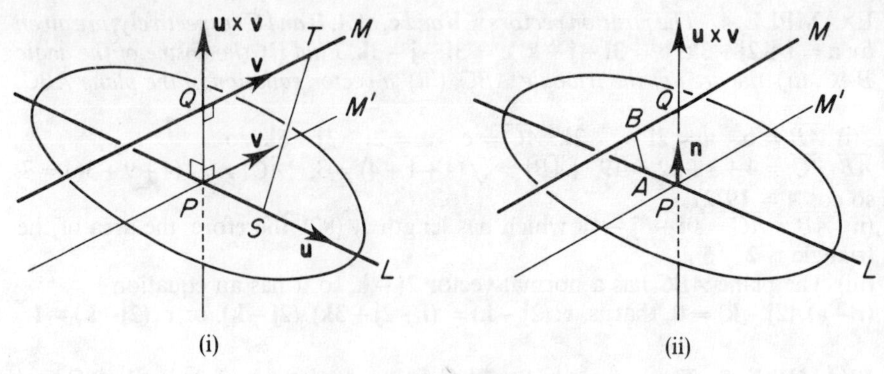

Fig. 12.7

points A on L and B on M, respectively, see Fig. 12.7 (ii). The unit vector $\mathbf{n}$ along $\overrightarrow{PQ}$ is $(\mathbf{u}\times\mathbf{v})/|\mathbf{u}\times\mathbf{v}|$ and the shortest distance between the two lines is the projection of $\overrightarrow{AB}$ on $\mathbf{n}$, that is $|(\mathbf{b}-\mathbf{a}).(\mathbf{u}\times\mathbf{v})|/|\mathbf{u}\times\mathbf{v}|$.

EXAMPLE 6 *Find the shortest distance between the two lines given by*

$$\mathbf{r}=\begin{pmatrix}1\\2\\3\end{pmatrix}+s\begin{pmatrix}-2\\0\\2\end{pmatrix} \text{ and } \mathbf{r}=\begin{pmatrix}0\\3\\1\end{pmatrix}+t\begin{pmatrix}4\\7\\3\end{pmatrix}.$$

Using the above notation, $\mathbf{u}=\begin{pmatrix}-2\\0\\2\end{pmatrix}$, $\mathbf{v}=\begin{pmatrix}4\\7\\3\end{pmatrix}$, $\mathbf{u}\times\mathbf{v}=\begin{pmatrix}-14\\14\\-14\end{pmatrix}$, so choose $\mathbf{n}=\frac{1}{\sqrt{3}}\begin{pmatrix}1\\-1\\1\end{pmatrix}$. Also $\mathbf{a}=\begin{pmatrix}1\\2\\3\end{pmatrix}$, $\mathbf{b}=\begin{pmatrix}0\\3\\1\end{pmatrix}$, $\mathbf{b}-\mathbf{a}=\begin{pmatrix}-1\\1\\-2\end{pmatrix}$, so the shortest distance d between the two lines is given by $d=|(\mathbf{b}-\mathbf{a}).\mathbf{n}|=4/\sqrt{3}$.

EXERCISE 12.2

1 Given that $\mathbf{a}$ is a non-zero vector and that $\mathbf{a}\times\mathbf{x}=\mathbf{x}\times\mathbf{a}$, state what conclusions may be drawn about the vector $\mathbf{x}$. (*JMB*)

2 The points A, B and R have positions vectors $\mathbf{a}$, $\mathbf{b}$ and $\mathbf{r}$ respectively, A and B are fixed points and R varies. Describe geometrically the locus of R in each of the cases (i) $(\mathbf{r}-\mathbf{a}).\mathbf{b}=0$, (ii) $(\mathbf{r}-\mathbf{a})\times\mathbf{b}=\mathbf{0}$. (*JMB*)

3 For the given vectors $\mathbf{u}$ and $\mathbf{v}$, solve the equation $\mathbf{r}\times\mathbf{u}=\mathbf{v}$, using the following steps: (a) check that $\mathbf{u}.\mathbf{v}=0$, (b) put $\mathbf{r}=x\mathbf{i}+y\mathbf{j}+z\mathbf{k}$ and use the fact that $\mathbf{r}.\mathbf{v}=0$ to express $\mathbf{r}$ in terms of just x and y, (c) use the equation $\mathbf{r}\times\mathbf{u}=\mathbf{v}$ to express $\mathbf{r}$, in terms of a single parameter t, in the form $\mathbf{r}=\mathbf{w}+t\mathbf{u}$, where $\mathbf{w}$ is to be found.
(i) $\mathbf{u}=\mathbf{i}+\mathbf{k}$, $\mathbf{v}=\mathbf{i}-\mathbf{k}$ (ii) $\mathbf{u}=\mathbf{i}-\mathbf{j}+\mathbf{k}$, $\mathbf{v}=\mathbf{i}+2\mathbf{j}+\mathbf{k}$.

4 Solve the equation and interpret the solution geometrically:

(i) $\mathbf{r}\times\mathbf{i}=\mathbf{j}$ (ii) $\mathbf{r}\times(\mathbf{j}+\mathbf{k})=\mathbf{i}+\mathbf{j}-\mathbf{k}$ (iii) $\mathbf{r}\times\begin{pmatrix}1\\2\\3\end{pmatrix}=\begin{pmatrix}1\\-2\\1\end{pmatrix}$.

5 The position vectors of two fixed points A and B are $\mathbf{a}$ and $\mathbf{b}$, respectively, and P is a variable point with position vector $\mathbf{r}$. Describe geometrically the locus of P in *each* of the following cases: (i) when $(\mathbf{r}-\mathbf{a}) \times (\mathbf{r}-\mathbf{b}) = \mathbf{0}$, (ii) when $(\mathbf{r}-\mathbf{a}).(\mathbf{r}-\mathbf{b}) = 0$. Find the values of $\mathbf{r}$ that satisfy both these equations simultaneously. (*JMB*)

6 Three non-collinear points A, B, C have position vectors $\mathbf{a}, \mathbf{b}, \mathbf{c}$ relative to an origin O, which does not lie in the plane ABC. Given that P is a variable point with position vector $\mathbf{r}$ relative to O, show that the equation $(\mathbf{r}-\mathbf{a}) \times \mathbf{b} = \mathbf{0}$ represents a line through A in the direction of the vector $\mathbf{b}$. Find a vector equation for each of the planes (i) through C and perpendicular to the line $(\mathbf{r}-\mathbf{a}) \times \mathbf{b} = \mathbf{0}$, (ii) through C and containing the line $(\mathbf{r}-\mathbf{a}) \times \mathbf{b} = \mathbf{0}$. (*JMB*)

7 The points $A\,(1, 0, -2)$, $B\,(2, -2, 1)$ and $C\,(5, -4, 0)$ have position vectors $\mathbf{a}$, $\mathbf{b}$ and $\mathbf{c}$ respectively. Find the vector product of $\overrightarrow{AC}$ and $\overrightarrow{AB}$ and hence, or otherwise, find (i) an equation for the plane ABC, (ii) the area of the triangle ABC. Find the equation of the plane which passes through A and which is perpendicular to the plane ABC and to the plane $(\mathbf{r}-\mathbf{a}).\mathbf{b} = 0$. (*JMB*)

8 For the points A, B, C, with the given coordinates, determine: (a) $\overrightarrow{AB} \times \overrightarrow{AC}$, (b) the area of the triangle ABC, (c) a vector equation of the plane ABC, (d) a Cartesian equation of the plane ABC, (e) the volume of the tetrahedron $ABCD$, where D is the point $(2, 3, 4)$:

(i) $A\,(1, 2, 3)$, $B\,(1, 1, 1)$, $C\,(3, 2, -1)$,
(ii) $A\,(1, 0, 2)$, $B\,(2, 1, 3)$, $C\,(-1, 1, 3)$,
(iii) $A\,(0, 1, 2)$, $B\,(0, 0, 0)$, $C\,(4, 2, 3)$,
(iv) $A\,(2, 3, 5)$, $B\,(1, 3, 3)$, $C\,(1, 2, 4)$.

9 Find the shortest distance between the lines AB and CD, given the points: (i) $A\,(0, 0, 0)$, $B\,(1, 2, 1)$, $C\,(1, 1, -1)$, $D\,(2, 3, 2)$ (ii) $A\,(0, 2, 1)$, $B\,(1, 3, 2)$, $C\,(1, 2, 1)$, $D\,(2, 0, 2)$ (iii) $A\,(2, 1, -1)$, $B\,(-2, 0, 3)$, $C\,(1, 2, 3)$, $D\,(3, 0, 4)$.

10 Prove that $|\overrightarrow{PR} \times \overrightarrow{QR}|$ is twice the area of the triangle PQR. The points P, Q and R in 3-dimensional Euclidean space are $(1, 3, 2)$, $(4, 5, 1)$ and $(3, 3, 1)$ respectively. (i) Find the Cartesian equation of the plane PQR. (ii) Find the coordinates of the foot N of the perpendicular from the origin O to the plane PQR. (iii) Find the volume of the tetrahedron $OPQR$. (*JMB*)

11 (a) Given $\mathbf{a} = 4\mathbf{i}+2\mathbf{j}-2\mathbf{k}$, $\mathbf{b} = \mathbf{i}+6\mathbf{j}-4\mathbf{k}$, $\mathbf{c} = \mathbf{i}+2\mathbf{j}+3\mathbf{k}$, verify that $(\mathbf{a} \times \mathbf{b}) \times \mathbf{c} = (\mathbf{a}.\mathbf{c})\mathbf{b} - (\mathbf{b}.\mathbf{c})\mathbf{a}$. Hence show that the points A, B, K with position vectors $\mathbf{a}$, $\mathbf{b}$, $(\mathbf{a} \times \mathbf{b}) \times \mathbf{c}$ respectively are collinear, and obtain the ratio $AK:KB$.

(b) Given the vectors $\mathbf{p} = p\mathbf{i}$, $\mathbf{q} = q\mathbf{j}$, where $p \neq 0$, show that the general solution of the equation $\mathbf{x} \times \mathbf{p} = \mathbf{q}$ is $\mathbf{x} = (\mathbf{p} \times \mathbf{q})/p^2 + \lambda\mathbf{p}$, where λ is an arbitrary scalar. (*C*)

12 (a) The vectors $\mathbf{a}$, $\mathbf{b}$, $\mathbf{c}$ are distinct and each of unit length and are such that $\lambda\mathbf{a} + \mu\mathbf{b} + v\mathbf{c} = \mathbf{0}$, where λ, μ, v are scalars, not all zero. Show that $\mu(\mathbf{a} \times \mathbf{b}) = v(\mathbf{c} \times \mathbf{a})$. Hence, or otherwise, prove that $|\lambda|$, $|\mu|$, $|v|$ are proportional to the sines of the angles between $\mathbf{b}$ and $\mathbf{c}$, $\mathbf{c}$ and $\mathbf{a}$, $\mathbf{a}$ and $\mathbf{b}$ respectively.

(b) The vertices O, P, Q, R, S, T, U, V of a cuboid have position vectors $\mathbf{0}$, $a\mathbf{i}$, $a\mathbf{i}+b\mathbf{j}$, $b\mathbf{j}$, $c\mathbf{k}$, $a\mathbf{i}+c\mathbf{k}$, $a\mathbf{i}+b\mathbf{j}+c\mathbf{k}$, $b\mathbf{j}+c\mathbf{k}$ respectively. Prove that the distance between PT and OU is $ab/\sqrt{(a^2+b^2)}$, and give the distances between RQ and OV, and between SV and OU. (*C*)

12.3 Vector products in mechanics

Vector products play an important part in the vectorial treatment of mechanics. We cannot go into this very deeply but we give two applications: (i) the kinematics of a particle, described in terms of the derivatives of the position vector, when that vector is a given function of the time, (ii) the moment of a force about a point.

Derivatives of vector and scalar products

Let the vectors **a** and **b** be differentiable functions of the time t so that, in Cartesian form,

$\mathbf{a}(t) = a_1(t)\mathbf{i} + a_2(t)\mathbf{j} + a_3(t)\mathbf{k}, \quad \mathbf{b}(t) = b_1(t)\mathbf{i} + b_2(t)\mathbf{j} + b_3(t)\mathbf{k}.$

Then

$$\frac{d\mathbf{a}}{dt} = \frac{da_1}{dt}\mathbf{i} + \frac{da_2}{dt}\mathbf{j} + \frac{da_3}{dt}\mathbf{k}, \quad \frac{d\mathbf{b}}{dt} = \frac{db_1}{dt}\mathbf{i} + \frac{db_2}{dt}\mathbf{j} + \frac{db_3}{dt}\mathbf{k},$$

$$\begin{aligned}
\frac{d}{dt}(\mathbf{a}\,.\,\mathbf{b}) &= \frac{d}{dt}(a_1b_1 + a_2b_2 + a_3b_3) \\
&= a_1\frac{db_1}{dt} + \frac{da_1}{dt}b_1 + a_2\frac{db_2}{dt} + \frac{da_2}{dt}b_2 + a_3\frac{db_3}{dt} + \frac{da_3}{dt}b_3 \\
&= a_1\frac{db_1}{dt} + a_2\frac{db_2}{dt} + a_3\frac{db_3}{dt} + \frac{da_1}{dt}b_1 + \frac{da_2}{dt}b_2 + \frac{da_3}{dt}b_3 \\
&= \mathbf{a}\,.\,\frac{d\mathbf{b}}{dt} + \frac{d\mathbf{a}}{dt}\,.\,\mathbf{b}.
\end{aligned}$$

Similarly

$$\begin{aligned}
\frac{d}{dt}(\mathbf{a}\times\mathbf{b}) &= \frac{d}{dt}\{(a_2b_3 - a_3b_2)\mathbf{i} + (a_3b_1 - a_1b_3)\mathbf{j} + (a_1b_2 - a_2b_1)\mathbf{k}\} \\
&= \left(a_2\frac{db_3}{dt} + \frac{da_2}{dt}b_3 - a_3\frac{db_2}{dt} - \frac{da_3}{dt}b_2\right)\mathbf{i} \\
&\quad + \left(a_3\frac{db_1}{dt} + \frac{da_3}{dt}b_1 - a_1\frac{db_3}{dt} - \frac{da_1}{dt}b_3\right)\mathbf{j} \\
&\quad + \left(a_1\frac{db_2}{dt} + \frac{da_1}{dt}b_2 - a_2\frac{db_1}{dt} - \frac{da_2}{dt}b_1\right)\mathbf{k} \\
&= \left(a_2\frac{db_3}{dt} - a_3\frac{db_2}{dt}\right)\mathbf{i} + \left(a_3\frac{db_1}{dt} - a_1\frac{db_3}{dt}\right)\mathbf{j} \\
&\quad + \left(a_1\frac{db_2}{dt} - a_2\frac{db_1}{dt}\right)\mathbf{k} + \left(\frac{da_2}{dt}b_3 - \frac{da_3}{dt}b_2\right)\mathbf{i} \\
&\quad + \left(\frac{da_3}{dt}b_1 - \frac{da_1}{dt}b_3\right)\mathbf{j} + \left(\frac{da_1}{dt}b_2 - \frac{da_2}{dt}b_1\right)\mathbf{k} \\
&= \mathbf{a}\times\frac{d\mathbf{b}}{dt} + \frac{d\mathbf{a}}{dt}\times\mathbf{b}.
\end{aligned}$$

Thus the differentiation of a scalar product or a vector product of two vectors follows the same product rule for the differentiation of the product of two scalar functions. The only special point which must be remembered is that the order of the vector product must be preserved. We have proved the following theorem.

Theorem 12.5 Let $\mathbf{a}(t)$ and $\mathbf{b}(t)$ be differentiable vector functions of a scalar variable t. Then

$$\frac{\mathrm{d}}{\mathrm{d}t}(\mathbf{a}\,.\,\mathbf{b}) = \mathbf{a}\,.\frac{\mathrm{d}\mathbf{b}}{\mathrm{d}t} + \frac{\mathrm{d}\mathbf{a}}{\mathrm{d}t}\,.\,\mathbf{b}, \quad \frac{\mathrm{d}}{\mathrm{d}t}(\mathbf{a}\times\mathbf{b}) = \mathbf{a}\times\frac{\mathrm{d}\mathbf{b}}{\mathrm{d}t} + \frac{\mathrm{d}\mathbf{a}}{\mathrm{d}t}\times\mathbf{b}.$$

EXAMPLE 1 *A particle P, with position vector* $\mathbf{r}$, *has velocity* $\dfrac{\mathrm{d}\mathbf{r}}{\mathrm{d}t}$ *which is perpendicular to* $\mathbf{r}$ *at all times. Prove that P moves on a sphere.*

$\dfrac{\mathrm{d}}{\mathrm{d}t}r^2 = \dfrac{\mathrm{d}}{\mathrm{d}t}(\mathbf{r}\,.\,\mathbf{r}) = \mathbf{r}\,.\dfrac{\mathrm{d}\mathbf{r}}{\mathrm{d}t} + \dfrac{\mathrm{d}\mathbf{r}}{\mathrm{d}t}\cdot\mathbf{r} = 2\mathbf{r}\cdot\dfrac{\mathrm{d}\mathbf{r}}{\mathrm{d}t} = 0$, since $\mathbf{r}$ and $\dfrac{\mathrm{d}\mathbf{r}}{\mathrm{d}t}$ are perpendicular.
Hence, r^2 is a constant and so r is a constant and P moves on a sphere, centre at the origin.

EXAMPLE 2 *A particle P, with position vector* $\mathbf{r}$ *relative to the origin O, moves such that* $\mathbf{r}\times\dfrac{\mathrm{d}\mathbf{r}}{\mathrm{d}t} = \mathbf{a}$, *where* $\mathbf{a}$ *is a constant vector. Prove that:* (*i*) *P moves in a plane,* (*ii*) *the acceleration of P is always parallel to* $\overrightarrow{OP}$.

Since the constant vector $\mathbf{a}$ is perpendicular to $\mathbf{r}$, $\mathbf{r}\,.\,\mathbf{a} = 0$ and this means that P always lies in the plane through O perpendicular to $\mathbf{a}$. Also, since $\mathbf{a}$ is constant

$$\mathbf{0} = \frac{\mathrm{d}\mathbf{a}}{\mathrm{d}t} = \frac{\mathrm{d}}{\mathrm{d}t}\left(\mathbf{r}\times\frac{\mathrm{d}\mathbf{r}}{\mathrm{d}t}\right) = \mathbf{r}\times\frac{\mathrm{d}^2\mathbf{r}}{\mathrm{d}t^2} + \frac{\mathrm{d}\mathbf{r}}{\mathrm{d}t}\times\frac{\mathrm{d}\mathbf{r}}{\mathrm{d}t} = \mathbf{r}\times\frac{\mathrm{d}^2\mathbf{r}}{\mathrm{d}t^2}$$

and therefore the acceleration $\dfrac{\mathrm{d}^2\mathbf{r}}{\mathrm{d}t^2}$ is parallel to $\mathbf{r}$.

Vector moments

The moment about an origin O of a force $\mathbf{F}$, acting on a body at a point A, is the turning effect of $\mathbf{F}$ about O. As shown in Fig. 12.8, this turning effect is about an axis through O in the direction of a unit vector $\mathbf{u}$, perpendicular to $\overrightarrow{OA}$ and to $\mathbf{F}$. The magnitude of the turning effect is proportional to $F\ (= |\mathbf{F}|)$ and to the distance of the line of action of $\mathbf{F}$ from O. Let $\overrightarrow{OA} = \mathbf{a}$, $a = |\mathbf{a}|$ and let the angle between $\overrightarrow{OA}$ and $\mathbf{F}$ be θ, then the distance d of the line of action of $\mathbf{F}$ from O is given by $d = a\sin\theta$. These results lead to the definition of the moment of $\mathbf{F}$ about O as the vector $\mathbf{M}$, given by $\mathbf{M} = Fa\sin\theta\mathbf{u} = \mathbf{a}\times\mathbf{F} = \overrightarrow{OA}\times\mathbf{F}$. This can be generalised to the moment about any point, as follows:

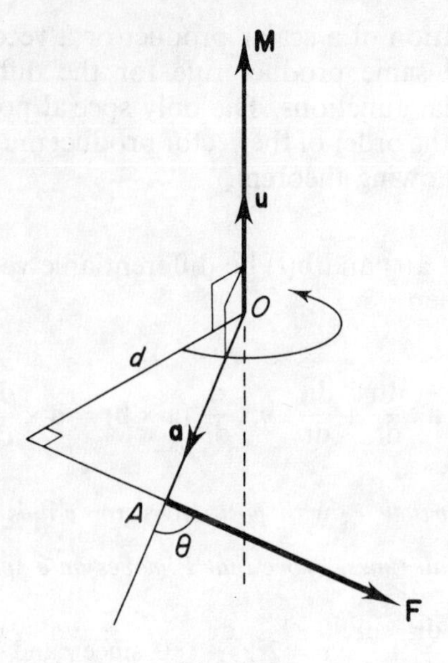

Fig. 12.8

Definition The moment about B of the force $\mathbf{F}$ acting at the point A is the vector $\mathbf{M}$ given by $\mathbf{M} = \overrightarrow{BA} \times \mathbf{F}$.

EXAMPLE 2 *Find the moment of the force* $\mathbf{F}$, *of magnitude* $\sqrt{7}$, *acting along the line* $\mathbf{r} = \begin{pmatrix} 1 \\ 2 \\ 3 \end{pmatrix} + t\begin{pmatrix} 2 \\ -1 \\ 3 \end{pmatrix}$, *about the origin.*

The point A (1, 2, 3) lies on the line of action of the force $\mathbf{F}$ and the direction of $\mathbf{F}$ is $\begin{pmatrix} 2 \\ -1 \\ 3 \end{pmatrix}$. So $\mathbf{F} = \dfrac{1}{\sqrt{2}}\begin{pmatrix} 2 \\ -1 \\ 3 \end{pmatrix}$ and the moment of $\mathbf{F}$ about O is

$$\overrightarrow{OA} \times \mathbf{F} = \begin{pmatrix} 1 \\ 2 \\ 3 \end{pmatrix} \times \frac{1}{\sqrt{2}}\begin{pmatrix} 2 \\ -1 \\ 3 \end{pmatrix} = \frac{1}{\sqrt{2}}\begin{pmatrix} 9 \\ 3 \\ -5 \end{pmatrix}.$$

EXERCISE 12.3

1 For the given function $\mathbf{r}$ of t, prove that:
(a) $\ddot{\mathbf{r}} = k\mathbf{r}$, for some fixed scalar k, (b) $\mathbf{r} \times \dot{\mathbf{r}}$ is a constant vector, (c) $\dot{\mathbf{r}}$ and $\mathbf{r}$ have equal magnitude: (i) $\mathbf{r} = \cos t\mathbf{i} + \sin t\mathbf{j}$, (ii) $\mathbf{r} = \sinh 2t\mathbf{i} + \cosh 2t\mathbf{k}$.

2 Given that $\mathbf{r}$ is a function of t and that $\dfrac{d\mathbf{r}}{dt} = \dot{\mathbf{r}}$, write down $\dfrac{d}{dt}(\mathbf{r} \cdot \dot{\mathbf{r}})$, $\dfrac{d}{dt}(\mathbf{r} \times \dot{\mathbf{r}})$ and $\dfrac{d}{dt}(\mathbf{r} \cdot \mathbf{r})$ in terms of $\mathbf{r}$, $\dot{\mathbf{r}}$ and $\ddot{\mathbf{r}}$. Given that $\ddot{\mathbf{r}} = k\mathbf{r}$, for some constant k, prove that $\dot{r}^2 - kr^2$ is a constant scalar and that $\dot{\mathbf{r}} \times \mathbf{r}$ is a constant vector.

3 A particle P moves so that its position vector $\mathbf{r}$ at time t is given by $\mathbf{r} = \mathbf{a}\cos\omega t + \mathbf{b}\sin\omega t$, where $\mathbf{a}$, $\mathbf{b}$ are constant vectors and ω is a constant. Show that $\mathbf{r} \times \frac{d\mathbf{r}}{dt}$ is independent of t. (*L*)

4 A vector $\mathbf{u}$ is given by $\mathbf{u} = x\mathbf{a} + y\mathbf{b}$ where $\mathbf{a}$ and $\mathbf{b}$ are constant vectors with $\mathbf{a} \times \mathbf{b} \neq \mathbf{0}$ and x and y are functions of the time t such that at all times $\mathbf{u} \times \frac{d\mathbf{u}}{dt} = \mathbf{0}$. Show that y/x is constant. (*JMB*)

5 Find the vector moment of the force $3\mathbf{i} + 2\mathbf{j} - \mathbf{k}$ about the point $(-1, 0, 2)$, given that the line of action of the force passes through the point $(1, -2, 1)$. (*L*)

6 The moment of a non-zero force $\mathbf{F}$ about a point A is the same as its moment about another point B. Show that the line of action of the force $\mathbf{F}$ is parallel to AB. (*JMB*)

7 A force $\mathbf{F}$ acts at the point with position vector $\mathbf{r}$. Express as a vector product the moment of this force about the point with position vector $\mathbf{a}$. A force of unit magnitude has equal moments about the points $(0, 1, 0)$ and $(1, 2, -1)$. Find the possible values of the components of the force. (*JMB*)

MISCELLANEOUS EXERCISE 12

1 Given that $\mathbf{a} = -2\mathbf{i} + 4\mathbf{k}$, $\mathbf{b} = 3\mathbf{j} - 2\mathbf{k}$, find $\mathbf{a} \cdot \mathbf{b}$ and $\mathbf{a} \times \mathbf{b}$. (*L*)

2 (a) By considering the vector product $\mathbf{a} \times \mathbf{b}$, where $\mathbf{a} = a_1\mathbf{i} + a_2\mathbf{j} + a_3\mathbf{k}$ and $\mathbf{b} = b_1\mathbf{i} + b_2\mathbf{j} + b_3\mathbf{k}$, prove that
$(a_1^2 + a_2^2 + a_3^2)(b_1^2 + b_2^2 + b_3^2) \geqslant (a_2b_3 - a_3b_2)^2 + (a_3b_1 - a_1b_3)^2 + (a_1b_2 - a_2b_1)^2$.
(b) For each of the following assertions, state whether it is true or false and justify your answer: (i) if $\mathbf{a} \times \mathbf{b} = \mathbf{b} \times \mathbf{a}$ then $\mathbf{a} \times \mathbf{b} = \mathbf{0}$,
(ii) $(\mathbf{a} \times \mathbf{b}) \times \mathbf{c} = \mathbf{a} \times (\mathbf{b} \times \mathbf{c})$, where $\mathbf{a}$, $\mathbf{b}$, $\mathbf{c}$ are any vectors,
(iii) $(\mathbf{a} \times \mathbf{b}) \times \mathbf{a} = \mathbf{a} \times (\mathbf{b} \times \mathbf{a})$, where $\mathbf{a}$, $\mathbf{b}$ are any vectors.
{No credit will be given for answers unsupported by valid reasons.} (*C*)

3 The point A has position vector $\mathbf{a}$ referred to a point O and $\overrightarrow{OU}$ $(= \mathbf{u})$ is a unit vector. The point B is the reflection of A in the line OU. Show that the position vector $\mathbf{b}$ of B is given by $\mathbf{b} = 2(\mathbf{a} \cdot \mathbf{u})\mathbf{u} - \mathbf{a}$. Show further that (i) $\mathbf{b} \times \mathbf{a} = 2(\mathbf{a} \cdot \mathbf{u})(\mathbf{b} \times \mathbf{u})$, (ii) $\mathbf{b} \times \mathbf{u} = \mathbf{u} \times \mathbf{a}$. (*JMB*)

4 Find the moment about the point $(-1, 1, 2)$ of the force $-\mathbf{i} + \mathbf{j} + 3\mathbf{k}$ given that its line of action passes through the point $(-1, 3, 7)$. Explain why this force has the same moment about the point $(-3, 3, 8)$. (*JMB*)

5 For each of the following assertions, state whether it is true or false and justify your answer. (i) Given that the position vectors, relative to O, of the points A and B are $\mathbf{a}$ and $\mathbf{b}$ respectively, then the area of the triangle OAB is $|\mathbf{a} \times \mathbf{b}|/2$. (ii) Given that $\mathbf{a}$, $\mathbf{b}$, $\mathbf{c}$ are coplanar position vectors, then $\mathbf{a} \times (\mathbf{b} \times \mathbf{c}) = \mathbf{0}$. (iii) Given that $\mathbf{a} = \mathbf{i} + 2\mathbf{j} + \mathbf{k}$, $\mathbf{b} = 2\mathbf{i} + \mathbf{j} - 2\mathbf{k}$, $\mathbf{c} = \mathbf{i} - \mathbf{j} - \mathbf{k}$, then the four points with position vectors $\mathbf{a}$, $\mathbf{b}$, $\mathbf{c}$, $\mathbf{a} + \mathbf{b} + \mathbf{c}$ are coplanar. (iv) The line with equation $\mathbf{r} = (2 + 3\lambda)\mathbf{i} + (1 + \lambda)\mathbf{j} + (-1 - 3\lambda)\mathbf{k}$ meets the plane with equation $\mathbf{r} \cdot (2\mathbf{i} - 3\mathbf{j} + \mathbf{k}) = 0$ in a single point. (*C*)

6 (i) Given that $\mathbf{r} = (\cos 2t)\mathbf{a} + (\sin 2t)\mathbf{b}$, where $\mathbf{a}$ and $\mathbf{b}$ are constant vectors, show that (a) $\frac{d^2\mathbf{r}}{dt^2} + 4\mathbf{r} = \mathbf{0}$, (b) $\mathbf{r} \times \frac{d\mathbf{r}}{dt} = 2\mathbf{a} \times \mathbf{b}$. (ii) If $\frac{d^2\mathbf{r}}{dt^2} = \mu\mathbf{r}$, where μ is

a constant scalar, verify that (a) $\left(\frac{d\mathbf{r}}{dt}\right)^2 = \mu \mathbf{r}^2 + c$, where c is a constant scalar, (b) $\mathbf{r} \times \frac{d\mathbf{r}}{dt}$ is a constant vector. (*L*)

7 Find a unit vector which is perpendicular to the vector $(4\mathbf{i}+4\mathbf{j}-7\mathbf{k})$ and to the vector $(2\mathbf{i}+2\mathbf{j}+\mathbf{k})$. (*L*)

8 The points A, B, C and D have coordinates $(4, 3, -2)$, $(3, 1, 0)$, $(-1, 1, 1)$ and $(3, 4, 2)$ respectively referred to a system of mutually perpendicular axes with O as the origin. The vectors $\mathbf{a}$, $\mathbf{b}$, $\mathbf{c}$ and $\mathbf{d}$ are defined by $\mathbf{a} = \overrightarrow{OA}$, $\mathbf{b} = \overrightarrow{OB}$, $\mathbf{c} = \overrightarrow{OC}$ and $\mathbf{d} = \overrightarrow{OD}$.
(i) Verify, in the case of these four vectors, the identity
$(\mathbf{a} \times \mathbf{b}) . (\mathbf{c} \times \mathbf{d}) = (\mathbf{b} . \mathbf{d})(\mathbf{c} . \mathbf{a}) - (\mathbf{b} . \mathbf{c})(\mathbf{d} . \mathbf{a})$.
(ii) Show that, if P is a point on OA and Q is a point on BC such that $\overrightarrow{PQ}$ is perpendicular to both $\overrightarrow{OA}$ and $\overrightarrow{BC}$, then the length l of $\overrightarrow{PQ}$ is given by

$$l = \frac{|\{\mathbf{a} \times (\mathbf{c}-\mathbf{b})\} . \mathbf{b}|}{|\mathbf{a} \times (\mathbf{c}-\mathbf{b})|}.$$

Hence, or otherwise, find the numerical value of l. (*L*)

9 The position vectors of three non-collinear points A, B and C are $\mathbf{a}$, $\mathbf{b}$ and $\mathbf{c}$ respectively relative to the origin O. (i) Show that the length of the perpendicular from C to the line AB is $|\mathbf{b} \times \mathbf{c} + \mathbf{c} \times \mathbf{a} + \mathbf{a} \times \mathbf{b}|/|\mathbf{b}-\mathbf{a}|$. (ii) Given that $\mathbf{a} = 3\mathbf{i}+2\mathbf{j}+\mathbf{k}$, $\mathbf{b} = -4\mathbf{j}+4\mathbf{k}$, $\mathbf{c} = -\mathbf{i}+5\mathbf{k}$, find the position vector of the reflection of C in the line AB. (*C*)

10 The points A, B, C and D are $(-2, 0, 1)$, $(6, 6, 2)$, $(6, 6, 1)$ and $(1, -4, 2)$, respectively. Calculate the vector product $\overrightarrow{AB} \times \overrightarrow{AC}$. Hence, or otherwise, find (i) a Cartesian equation of the plane containing A, B and C, (ii) the area of the triangle ABC, (iii) the volume of the tetrahedron $ABCD$. (Volume of a tetrahedron is one third of the area of its base multiplied by its perpendicular height.) Find the equations of two planes such that for any point P in either of these planes the volume of the tetrahedron $ABCP$ equals that of $ABCD$. Find the coordinates of the point D' having the three properties: D and D' are on opposite sides of the plane ABC, AD' is perpendicular to the plane ABC, the volume of $ABCD'$ equals the volume of $ABCD$. (*JMB*)

11 If $\mathbf{r}$ is a function of t, show that $\frac{d}{dt}(\mathbf{r} \times \dot{\mathbf{r}}) = \mathbf{r} \times \ddot{\mathbf{r}}$. A particle, whose position vector referred to the origin at time t is $\mathbf{r}$, is moving so that $\ddot{\mathbf{r}} = -n^2\mathbf{r}$, where n is a real constant. Prove that, at any instant in the motion, (a) $\mathbf{r} \times \ddot{\mathbf{r}} = \mathbf{0}$, (b) $\mathbf{r} \times \dot{\mathbf{r}}$ is a constant vector. Show also that the path of the particle lies in a plane. (*L*)

12 The points A, B, C are $(1, 0, -4)$, $(4, 3, 0)$, $(4, 12, 0)$, respectively. Find (i) the vector product of $\overrightarrow{AB}$ and $\overrightarrow{AC}$, (ii) a Cartesian equation of the plane ABC, (iii) the area of the triangle ABC. (*JMB*)

13 (a) Show that if it is possible to find a vector $\mathbf{r}$ such that $\mathbf{r} \times \mathbf{a} = \mathbf{b}$, where $\mathbf{a}$ and $\mathbf{b}$ are given vectors, then $\mathbf{a} . \mathbf{b} = 0$. Find the set of vectors $\mathbf{r}$ which satisfy $\mathbf{r} \times \mathbf{a} = \mathbf{b}$ in the following cases:

(i) $\mathbf{a} = \begin{pmatrix} 2 \\ 1 \\ -3 \end{pmatrix}$, $\mathbf{b} = \begin{pmatrix} 3 \\ -3 \\ 1 \end{pmatrix}$, (ii) $\mathbf{a} = \begin{pmatrix} 4 \\ -1 \\ 5 \end{pmatrix}$, $\mathbf{b} = \begin{pmatrix} 8 \\ 0 \\ 7 \end{pmatrix}$.

(b) Given that $2\mathbf{x}+(\mathbf{x}\,.\,\mathbf{b})\mathbf{a}=\mathbf{c}$, where $\mathbf{a}\,.\,\mathbf{b}\neq -2$, show that $\mathbf{x}\,.\,\mathbf{b}=\mathbf{c}\,.\,\mathbf{b}/(2+\mathbf{a}\,.\,\mathbf{b})$. Deduce an expression for $\mathbf{x}$ in terms of $\mathbf{a}$, $\mathbf{b}$ and $\mathbf{c}$. (*C*)

14 The position vectors $\mathbf{a}, \mathbf{b}, \mathbf{c}, \mathbf{d}$ of the points A, B, C, D respectively are given by $\mathbf{a}=2\mathbf{i}+\mathbf{j}+2\mathbf{k}$, $\mathbf{b}=-7\mathbf{i}-\mathbf{j}+4\mathbf{k}$, $\mathbf{c}=7\mathbf{i}+2\mathbf{j}-2\mathbf{k}$, $\mathbf{d}=-8\mathbf{i}-2\mathbf{j}$. Find, in the form $\mathbf{r}\,.\,\mathbf{n}=p$, the equation of the plane containing A and B and parallel to the line CD. Show that the shortest distance between the lines AB and CD is $8/7$ units. Find the cosine of the acute angle between the directions of the lines AC and BD. (*C*)

15 The three non-collinear points A, B, C have non-zero position vectors $\mathbf{a}, \mathbf{b}, \mathbf{c}$. The variable point P in three dimensions has position vector $\mathbf{r}$. Describe each of the following loci, λ and μ being scalar parameters and k a positive constant, (a) $\mathbf{r}=\mathbf{a}+\lambda(\mathbf{c}-\mathbf{a})$, (b) $|\mathbf{r}-\mathbf{a}|=k$, (c) $|\mathbf{r}-\mathbf{a}|=|\mathbf{r}-\mathbf{b}|$, (d) $\mathbf{r}=\lambda(\mathbf{b}-\mathbf{a})\times(\mathbf{c}-\mathbf{a})$, (e) $\mathbf{r}=\mathbf{a}+\lambda(\mathbf{c}-\mathbf{a})+\mu(\mathbf{b}-\mathbf{a})$. Write down one other equation for $\mathbf{r}$ which, together with two of the equations (a) to (e) that are to be specified, determines the centre of the circle passing through A, B and C. (*L*)

16 The position vectors of the points A, B, C are respectively $\mathbf{a}=\mathbf{i}+2\mathbf{j}+\mathbf{k}$, $\mathbf{b}=2\mathbf{i}+4\mathbf{j}+3\mathbf{k}$, $\mathbf{c}=6\mathbf{i}+6\mathbf{j}+6\mathbf{k}$. Find (i) $\overrightarrow{AB}\,.\,\overrightarrow{AC}$, (ii) $\overrightarrow{AB}\times\overrightarrow{AC}$, (iii) the angle BAC, to the nearest degree, (iv) the area of the triangle ABC, (v) *either* the vector *or* the Cartesian equation of the plane ABC. (*L*)

17 Given that $\overrightarrow{OA}=\boldsymbol{\alpha}=(-\mathbf{i}+\mathbf{j}+\mathbf{k})$, $\overrightarrow{OB}=\boldsymbol{\beta}=(\mathbf{i}-\mathbf{j}+\mathbf{k})$, $\overrightarrow{OC}=\boldsymbol{\gamma}=(\mathbf{i}+\mathbf{j}-\mathbf{k})$, find (a) the angle between the straight lines BC and CA, (b) a unit vector normal to the plane ABC, (c) the perpendicular distance from the origin O to the plane ABC. Further, show that the points P and Q with position vectors $\boldsymbol{\alpha}+\boldsymbol{\beta}+\boldsymbol{\gamma}$ and $\boldsymbol{\beta}\times\boldsymbol{\gamma}+\boldsymbol{\gamma}\times\boldsymbol{\alpha}+\boldsymbol{\alpha}\times\boldsymbol{\beta}$ are collinear with the origin. (*L*)

18 A force of magnitude 6N acts in the direction of the vector $\begin{pmatrix}-2\\1\\2\end{pmatrix}$ through the point with position vector $\begin{pmatrix}6\\4\\5\end{pmatrix}$ m.

(i) Write down the equation of the line of action of the force in the form $\dfrac{x-x_1}{l}=\dfrac{y-y_1}{m}=\dfrac{z-z_1}{n}$.

(ii) Find the moment of the force about the point A whose position vector is $\begin{pmatrix}-1\\1\\-1\end{pmatrix}$ m. Hence, or otherwise, find the length of the perpendicular from A to the line in part (i). (*JMB*)

19 (a) The vertices A, B, C, D of a plane convex quadrilateral have position vectors $\mathbf{a}, \mathbf{b}, \mathbf{c}, \mathbf{d}$ respectively. Prove that the area of the triangle ABC is given by $|\mathbf{b}\times\mathbf{c}+\mathbf{c}\times\mathbf{a}+\mathbf{a}\times\mathbf{b}|/2$. Prove also that the area of the quadrilateral $ABCD$ is given by $|\mathbf{a}\times\mathbf{b}+\mathbf{b}\times\mathbf{c}+\mathbf{c}\times\mathbf{d}+\mathbf{d}\times\mathbf{a}|/2$.
(b) The points P, Q, R have position vectors $6\mathbf{i}+2\mathbf{k}$, $3\mathbf{i}+2\mathbf{j}+4\mathbf{k}$, $-2\mathbf{j}+3\mathbf{k}$ respectively relative to the origin O. Find the equation of the plane PQR in the form $\mathbf{r}\,.\,\mathbf{n}=p$. Obtain the volume of the tetrahedron $OPQR$. (*C*)

20 At time t, the position vector $\mathbf{r}(t)$ of a moving particle satisfies the relation $\ddot{\mathbf{r}}=\mathbf{E}+\dot{\mathbf{r}}\times\mathbf{B}$, where $\mathbf{E}$ and $\mathbf{B}$ are constant vectors. Show that $\mathbf{r}=\cos t\,\mathbf{i}+(\sin t+3t)\mathbf{k}$ satisfies this relation provided $\mathbf{E}$ and $\mathbf{B}$ take certain values which are to be determined. (*JMB*)

13 Limits and Convergence

13.1 Continuity and differentiability

We repeat the definition of a continuous function, given in *Pure Mathematics* §18.2.

Definition The function f is *continuous at* $x = a \Leftrightarrow \lim_{x \to a} f(x) = f(a)$. The function f is *continuous* $\Leftrightarrow$ f is continuous at all points of its domain.

Note that, if f has domain D, then for f to be continuous at $x = a$ it is necessary for both of the following conditions to hold:
(i) $f(a)$ exists, that is $a \in D$,
(ii) $\lim_{x \to a} f(x)$ exists and is equal to $f(a)$.

If $a \notin D$ but the limit $\lim_{x \to a} f(x)$ exists, then it is possible to extend the domain of f to form a new function f_1, with domain $D \cup \{a\}$, by defining: $f_1(x) = f(x)$ for $x \in D$, $f_1(a) = \lim_{x \to a} f(x)$, and the function f_1 will be continuous at $x = a$.

We shall need to consider the limit of $f(x)$ as x tends to a through values that are all greater than a and, similarly through values that are all less than a. The following notation is easily understood and will be very useful.

Notation $\lim_{x \to a+} f(x)$ is the limit of $f(x)$ as x tends to a from above, that is, the limit as x tends to a through values all greater than a.

$\lim_{x \to a-} f(x)$ is the limit of $f(x)$ as x tends to a from below, that is, the limit as x tends to a through values all less than a.

EXAMPLE 1 *Show that the real function* f, *defined by:*

$$f(x) = |x|/x \text{ for } x > 0, \quad f(x) = \cos x \text{ for } x \leqslant 0,$$

is continuous at $x = 0$.

$f(0) = 1$, so we consider the limit of $f(x)$ as x tends to 0. Since the equations defining $f(x)$ are different for positive and for negative values of x, we must look at the limits from above and from below 0.

$$\text{For } x > 0, \lim_{x \to 0+} f(x) = \lim \frac{|x|}{x} = \lim_{x \to 0+} 1 = 1, \text{ and}$$

$$\text{for } x < 0, \lim_{x \to 0-} f(x) = \lim_{x \to 0-} \cos x = 1.$$

Since these two limits are the same $f(x)$ tends to the limit 1 as x tends to 0 and hence f is continuous at $x = 0$. The graph of f is shown in Fig. 13.1.

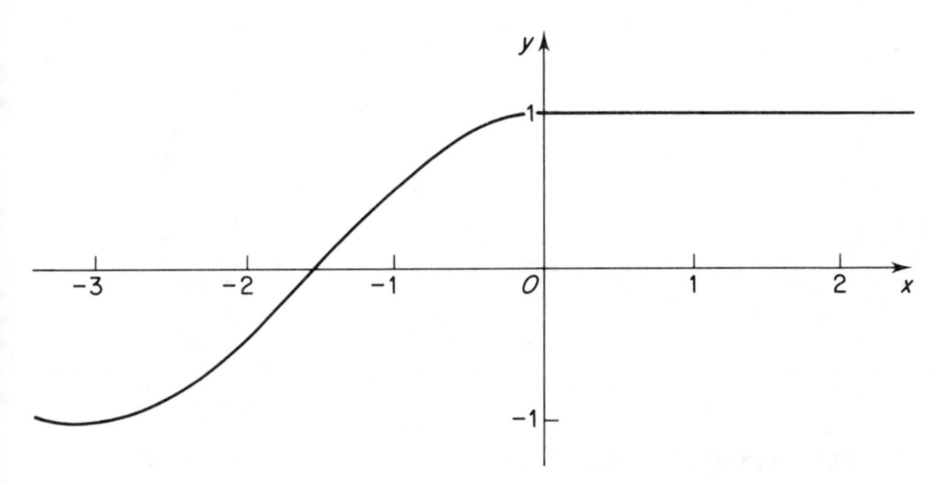

Fig. 13.1 Graph of f where $f(x) = |x|/x$ for $x > 0$, and $f(x) = \cos x$ for $x \leqslant 0$

EXAMPLE 2 *The function* f *has domain D, given by* $D = \{x: 0 \leqslant x \leqslant 3\}$. *For* $0 \leqslant x \leqslant 2$, $f(x) = a/(2x-5)$, *and for* $2 < x \leqslant 3$, $f(x) = x^2$. *Given that* f *is a continuous function, find the value of a.*

Clearly f is continuous at all points of its domain except possibly at $x = 2$. $\lim_{x \to 2-} f(x) = \lim_{x \to 2} a/(2x-5) = -a$ and $\lim_{x \to 2+} f(x) = \lim_{x \to 2} x^2 = 4$. Since f is continuous at $x = 2$, $-a = 4$ and so $\boldsymbol{a = -4}$. The graph of f is shown in Fig. 13.2.

We repeat a definition of a differentiable function, and of its derived function, which was given in *Pure Mathematics* §18.6.

Definition The function f is differentiable, and has derived function f′, if, at each point x in the domain of f,

$$f'(x) = \lim_{h \to 0} \frac{f(x+h) - f(x)}{h},$$

The condition involves the existence of a limit at $h = 0$ and, as in the case of continuity, it may be necessary to consider the limit for positive values

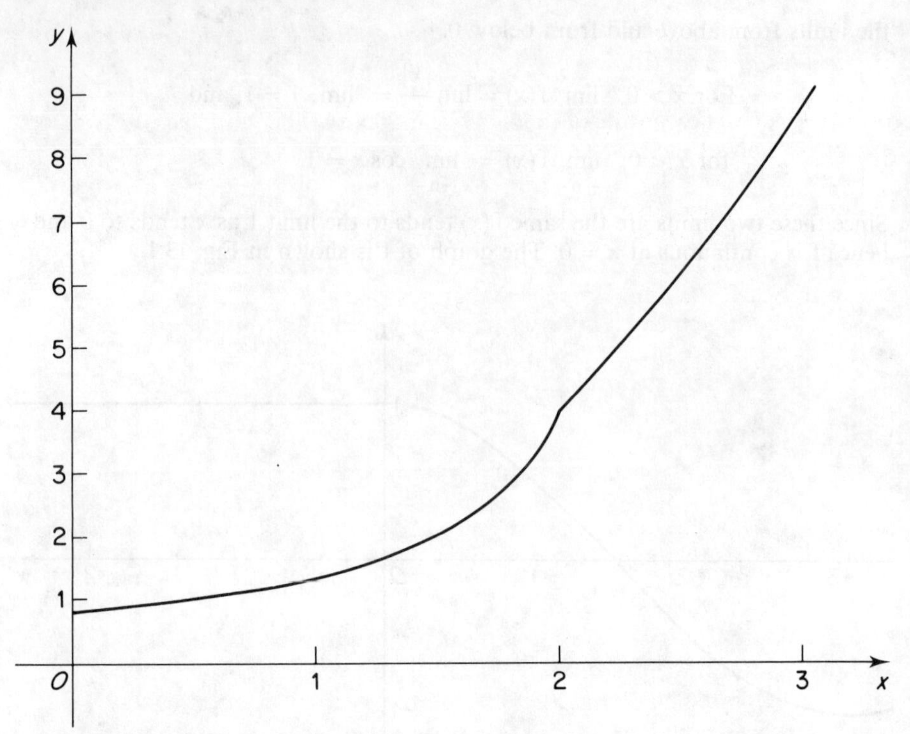

Fig. 13.2 Graph of f where $f(x) = -4/(2x-5)$ for $0 \leqslant x \leqslant 2$ and $f(x) = x^2$ for $2 < x \leqslant 3$

of h and the limit for negative values of h separately. It should be remembered that it is necessary (but not sufficient) that a function be continuous in order that it be differentiable.

When the functional form for $f(x)$ on one side of the critical point, $x = a$, is differentiable, then it may be possible to use the derivative to calculate the limit on that side. Suppose that f is continuous at $x = a$ and is given by the equations

$$f(x) = g(x) \quad \text{for } x \geqslant a,$$
$$f(x) = k(x) \quad \text{for } x \leqslant a,$$

with $g(a) = k(a)$, where g and k are differentiable functions. Use a notation for right (and left) derivatives of f at $x = a$, corresponding to that introduced above for limits.

Notation

$$f'(a+) = \lim_{h \to 0+} \frac{f(a+h) - f(a)}{h} = \lim_{h \to 0+} \frac{g(a+h) - g(a)}{h} = g'(a+),$$

$$f'(a-) = \lim_{h \to 0-} \frac{f(a+h) - f(a)}{h} = \lim_{h \to 0-} \frac{k(a+h) - k(a)}{h} = k'(a-),$$

The function f is then differentiable at $x = a$ if $f'(a+) = f'(a-)$.

EXAMPLE 3 *Determine whether or not the function* f *is differentiable where* f *is the function given in:* (*i*) *Example* 1, (*ii*) *Example* 2.

(i) The function f is continuous and it is clear that it is differentiable everywhere, except possibly at the origin.

For $x \geqslant 0$, $f(x) = 1$, $f'(x) = 0$, so $f'(0+) = 0$.

For $x \leqslant 0$, $f(x) = \cos x$, $f'(x) = -\sin x$, so $f'(0-) = -\sin 0 = 0$.

The right and left derivatives are the same and so f is **differentiable** and $f'(0) = 0$. In Fig. 13.1, it will be seen that the graph is smooth with no kink in it at $x = 0$.

(ii) In this case, f is differentiable, except possible at $x = 2$.

Case 1: if $a \neq -4$, then f is not continuous at $x = 2$ and so it is not differentiable either.

Case 2: if $a = -4$, we consider the derivatives on each side of the critical point $x = 2$.

For $x > 2$, $f(x) = x^2$, $f'(x) = 2x$, so $f'(2+) = 4$.

For $x < 2$, $f(x) = -4/(2x-5)$, $f'(x) = 8/(2x-5)^2$, so $f'(2-) = 8$.

Therefore $f'(2+) \neq f'(2-)$ and f is **not differentiable** at $x = 2$. There is a sharp change in the tangent to the graph of f at $x = 2$, the gradient is 8 on the left of the point and is 4 on the right. This kink in the graph at $x = 2$ is shown in Fig. 13.2.

The calculation of a derivative involves finding the limits on the right and on the left of a function which tends to the form 0/0. Limits of this type may be found by using power series expansions. The power series $f(x)$ of a function f can give information about the behaviour of the function for small values of the argument. When a function is constructed from other functions, these may separately be expanded to as high a power of x as is necessary to obtain the first few non-vanishing terms of the required power series. The process is valid for $|x| < R$, where R is the smallest radius of convergence of all the powers series concerned, see §6.6. This is illustrated in the next two examples.

EXAMPLE 4 *Find the limit as* $x \to 0$ *of* $\dfrac{\sin(a+x) - \sin(a-x)}{x}$.

$$\frac{\sin(a+x) - \sin(a-x)}{x} = \frac{2\cos a \sin x}{x} \to \mathbf{2\cos a}, \text{ using the fact that}$$

$$\lim_{x \to 0} \frac{\sin x}{x} = 1.$$

Alternatively, this limit is found by expanding $\sin x$ in a power series. For small values of x,

$$\frac{\sin x}{x} \approx \frac{x - x^3/3!}{x} = 1 - x^2/6 \to 1.$$

EXAMPLE 5 *Evaluate* $\lim\limits_{\theta \to 0} \dfrac{2 - 2\cos\theta - \theta\sin\theta}{\theta^4}$.

Since the denominator consists of θ^4, the numerator must be expanded in a power series at least as far as θ^4.
$2-2\cos\theta-\theta\sin\theta = 2-2(1-\theta^2/2+\theta^4/24)-\theta(\theta-\theta^3/6)+\ldots,$
so, omitting powers of θ greater than the fourth,
$2-2\cos\theta-\theta\sin\theta \approx 2-2+\theta^2-\theta^4/12-\theta^2+\theta^4/6=\theta^4/12.$

Therefore $\lim\limits_{\theta\to 0}\dfrac{2-2\cos\theta-\theta\sin\theta}{\theta^4}=\lim\limits_{\theta\to 0}\dfrac{\theta^4}{12\theta^4}=\mathbf{1/12}.$

The behaviour of a function f for large values of the argument x will be needed in a later section. The limit of $f(x)$ as x tends to infinity may be found by considering the limit as $1/x$ tends to zero.

EXAMPLE 6 *Show that, for any given real number a,* $\lim\limits_{x\to\infty} x^a e^{-x}=0.$

Since the limit as x tends to infinity is to be found, it may be assumed that $x>1$ throughout the work. For $a<0$ and $x>1$, $0<x^a e^{-x}<e^{-x}$ and $e^{-x}\to 0$ as $x\to\infty$, so $x^a e^{-x}\to 0$ as $x\to\infty$. For $a>0$, let b be an integer greater than a. Then, for $x>1$, $e^x>x^{b+1}/(b+1)!$ so that $0<x^a e^{-x}<x^b e^{-x}<(b+1)!/x$. Since $(b+1)!/x\to 0$, $x^a e^{-x}\to 0$ as $x\to\infty$.

EXERCISE 13.1

1 State whether or not the function f is continuous and/or differentiable at the origin, when f is given by:
(i) for all real x, $f(x)=|x|$,
(ii) for $x<0$, $f(x)=\sin x$, and for $x\geqslant 0$, $f(x)=x$,
(iii) for $x<0$, $f(x)=x^2+1$, and for $x\geqslant 0$, $f(x)=x^2$,
(iv) for $x\leqslant 0$, $f(x)=\cosh x$, and for $x>0$, $f(x)=\cos x$,
(v) for $x<0$, $f(x)=\sin 3x$, and for $x\geqslant 0$, $f(x)=2x+\tan x$,
(vi) for $x\leqslant 0$, $f(x)=xe^x$, and for $x>0$, $f(x)=\sinh x$.

2 Find the values of a and b, given that the function f is differentiable, when f is given by:
(i) $f(x)=ax+b$ for $x<0$, $f(x)=\sin x$ for $x\geqslant 0$,
(ii) $f(x)=x^3+a$ for $x<0$, $f(x)=2+\sin bx$ for $x\geqslant 0$,
(iii) $f(x)=a\sinh x+b$ for $x<0$, $f(x)=(1-x)^2$ for $x\geqslant 0$,
(iv) $f(x)=a\ln(1-x)$ for $x<0$, $f(x)=\sin 3x+b$ for $x\geqslant 0$.

3 The function f is continuous and differentiable and its domain is the interval $\{x: 0\leqslant x\leqslant 1\}$. Write down conditions which make it possible to extend f to a periodic, differentiable function g, with period 1 and with domain $\mathbb{R}$.

4 Evaluate the limit as x tends to 0, of the function f with domain $\mathbb{R}\setminus\{0\}$, given by:
(i) $f(x)=\{x\sin x-\cos x+1\}/x^2$, (ii) $f(x)=\{x\cosh x-\sinh x\}/x^3$,
(iii) $f(x)=\{e^x-\sin x-\cos x\}/\{e^x-\tan x-1\}$,
(iv) $f(x)=\{x\cos 2x-\sin x\}/x^3$.

5 Find: (i) $\lim\limits_{x\to 0}(x\tan x)/3x^2$, (ii) $\lim\limits_{x\to\infty}(\cos x)/x$,

(iii) $\lim\limits_{x\to 0}\dfrac{x^4}{2\cos x+x^2-2}$, (iv) $\lim\limits_{x\to\infty}x^2e^{-x}$, (v) $\lim\limits_{x\to\infty}xe^{-x^2}$.

13.2 Convergence of sequences

An infinite sequence (s_n) is an ordered set consisting of an unending list of real numbers. We shall use the notation

$$(s_n) = s_1, s_2, s_3, \ldots, s_{n-1}, s_n, s_{n+1}, \ldots.$$

Note that we use round brackets for a sequence, and not the curly brackets of set notation, since a sequence is an *ordered* set.

Sequences can behave in various ways as is seen in the following examples. If s_n is given by:

(i) $s_n = \dfrac{1}{n+1}$, s_n tends to zero as n tends to infinity,
(ii) $s_n = n$, s_n tends to infinity as n tends to infinity,
(iii) $s_n = (-1)^n$, s_n oscillates finitely,
(iv) $s_n = n^2 \sin(n\pi)$, s_n oscillates infinitely,
(v) $s_n = a + e^{-n}$, s_n tends to a as n tends to infinity.

In examples (i) and (v), the sequence is said to *converge* and in all the other cases the sequence diverges. We assume that the concept of a sequence *converging to a limit* is intuitively obvious and we will deal with the matter informally. For those who wish for a formal definition, it may be given in the form that the difference between s_n and the limit can be made as small as you like as long as n is large enough.

Definition The sequence (s_n) *converges* to zero if, given any positive number e, however small, there exists a number N such that, for all n greater than N, $|s_n|$ is less than e. The sequence (s_n) converges to a if $(s_n - a)$ converges to zero. A sequence which does not converge is said to *diverge*.

Notation The statement that the sequence (s_n) converges to s is written $s_n \to s$, or $\lim_{n\to\infty} s_n = s$, or $\lim(s_n) = s$.

s_n tends to infinity is written $s_n \to \infty$,
s_n tends to negative infinity is written $s_n \to -\infty$.

A sequence is a special kind of real function with domain equal to $\mathbb{N}$ and there are rules for the limits of sequences which are the same as the rules for limits, given in *Pure Mathematics* §18.1. These rules can be proved from the formal definition, given above, but we shall assume them without proof. We restate those we need as a theorem.

Theorem 13.1 Let $s_n \to s$, $t_n \to t$ and let $k \in \mathbb{R}$, then

Sum of sequences: $(s_n + t_n) \to s + t$,
Product by constant: $(ks_n) \to ks$,

A sequence can be given in an iterative form. That is, the first one or two terms of the sequence are given and then the nth term is given in terms of earlier terms of the sequence by some formula, called a recurrence relation. Such a sequence arises as the successive approximations when the solution of an equation is evaluated by an iterative process. Before giving an example of such a sequence, we state a theorem concerning the convergence of a monotonic sequence, that is a sequence which is either always increasing or always decreasing.

Definition

The sequence (s_n) is *monotonic increasing* if, for $m < n$, $s_m \leqslant s_n$,
the sequence (s_n) is *monotonic decreasing* if, for $m < n$, $s_m \geqslant s_n$.
The sequence (s_n) is *bounded above* by a if, for all n, $s_n \leqslant a$,
the sequence (s_n) is *bounded below* by b if, for all n, $s_n \geqslant b$.
A sequence which is *either* monotonic increasing *or* monotonic decreasing is **monotonic**. A sequence which is bounded above *and* bounded below is **bounded**.

Theorem 13.2 Let (s_n) be a monotonic increasing sequence. Then either (i) (s_n) is bounded, in which case (s_n) is convergent, so, for some s, $s_n \to s$, or (ii) (s_n) is unbounded, in which case (s_n) diverges, so $s_n \to \infty$. Corresponding results hold for a monotonic decreasing sequence.

The proof of this theorem is beyond the scope of this book. It uses the idea that, if a set of real numbers is bounded above then there must be a smallest of all the upper bounds. If a sequence (s_n) is monotonic increasing and is bounded above, then let s be its least upper bound. Then it can be shown that, as might be expected, s_n tends to s as n tends to infinity. The result for a decreasing sequence is obtained by changing the sign of s_n.

EXAMPLE 1 *Determine for what values of a the sequence (a^n) converges and, for these values, state the limit of the sequence.*

Consider different values of a, in turn.

(i) $a > 1$. Then $a^{n+1} = aa^n > 1 . a^n = a^n$, so (a^n) is monotonic increasing. Suppose that (a^n) is bounded and tends to a limit s. Then also $(a^{n+1}) \to s$. But $a^{n+1} = aa^n$ and, by Theorem 13.1, $aa^n \to as$ which is strictly greater than s. This is false so (a^n) is unbounded and $a^n \to \infty$.
(ii) $a = 1$. Then $a^n = 1$, for all n, so $a^n \to \mathbf{1}$.
(iii) $0 < a < 1$. Then $1/a > 1$ so $1/a^n \to \infty$ and $a^n \to \mathbf{0}$, on using (i).
(iv) $a = 0$. Then $a^n \to \mathbf{0}$.
(v) $-1 < a < 0$. Then $0 < |a| < 1$, $|a^n| \to 0$ and so $a^n \to \mathbf{0}$.
(vi) $a = -1$. Then a^n oscillates finitely.
(vii) $a < -1$. Then $|a| > 1$, so $|a^n| \to \infty$ and a^n oscillates infinitely.
In conclusion: for $a = 1$, (a^n) converges to 1,
for $|a| < 1$, (a^n) converges to 0,
otherwise (a^n) diverges.

EXAMPLE 2 *The sequence (x_n) is defined by:*

$$x_1 = 1 \text{ and, for all } n \in \mathbb{Z}^+, x_{n+1} = 2 - 1/(x_n + 2).$$

Prove that, for all $n \in \mathbb{Z}^+$, (i) $0 < x_n < \sqrt{3}$, (ii) $x_n < x_{n+1}$. Show that (x_n) converges to $\sqrt{3}$. Examine the convergence of the sequence if, initially, $x_1 = 3$.

When a sequence is given by a recurrence relation, it is helpful to find out what the limit must be *if* the sequence converges. Therefore, suppose that the sequence converges to the real number a. Then both $x_n \to a$ and $x_{n+1} \to a$ and, in the limit, the recurrence relation becomes $a = 2 - 1/(a+2)$. Then $a = (2a+3)/(a+2)$, $a^2 + 2a = 2a + 3$, $a^2 = 3$ and so either $a = \sqrt{3}$ or $a = -\sqrt{3}$. Now prove the inequalities asked for in the question.
(i) $x_1 = 1$. Then $0 < x_1 < \sqrt{3}$. Suppose that, $0 < x_n < \sqrt{3}$ for some $n \in \mathbb{Z}^+$. Then $x_{n+1} = (2x_n + 3)/(x_n + 2)$ and so $x_{n+1} > 0$. Also

$$x_{n+1} - \sqrt{3} = \frac{2x_n + 3 - \sqrt{3}(x_n + 2)}{x_n + 2} = \frac{(2 - \sqrt{3})(x_n - \sqrt{3})}{x_n + 2} < 0 \text{ since}$$

$x_n < \sqrt{3} < 2$. Hence $0 < x_{n+1} < \sqrt{3}$ and, by induction, for all $n \in \mathbb{Z}^+$, $0 < x_n < \sqrt{3}$. Also $x_{n+1} - x_n = \dfrac{2x_n + 3}{x_n + 2} - x_n = \dfrac{3 - x_n^2}{x_n + 2} > 0$, so $x_n < x_{n+1}$. Therefore (x_n) is a monotonic increasing sequence, bounded above by $\sqrt{3}$ so it converges to some limit which must be either $\sqrt{3}$ or $-\sqrt{3}$. Since $x_1 = 1, x_n > 1$, so the limit must be greater than 1 and is therefore $\sqrt{3}$.
(b) If $x_1 = 3$, then $x_1 > \sqrt{3}$ and, if $x_n > \sqrt{3}$, then $2 + \sqrt{3} < x_n + 2$, so

$$\sqrt{3} = \frac{3 + 2\sqrt{3}}{2 + \sqrt{3}} = 2 - \frac{1}{2 + \sqrt{3}} < 2 - \frac{1}{x_n + 2} = x_{n+1}.$$

By induction, $\sqrt{3} < x_n$, for all $n \in \mathbb{Z}^+$. Also $x_{n+1} - x_n = \dfrac{3 - x_n^2}{x_n + 2} < 0$, so (x_n) is monotonic decreasing and bounded below. Hence (x_n) converges and, as in case (a), its limit is $\sqrt{3}$.

EXERCISE 13.2

1. Describe the behaviour of the sequence (s_n) as $n \to \infty$, stating whether the sequence (a) converges, (b) oscillates finitely, (c) diverges to $+\infty$, (d) diverges to $-\infty$, (e) diverges in a manner different from that of (b), (c) or (d).
(i) $s_n = 2^n$, (ii) $s_n = (-2)^n$, (iii) $s_n = n^2$, (iv) $s_n = n^{-2}$, (v) $s_n = (-n)^2$, (vi) $s_n = (-n)^{-2}$, (vii) $s_n = \cos n\pi$, (viii) $s_n = n/(n^2+1)$, (ix) $s_n = (n^2+2)/n^2$.
2. State which of the sequences, if any, given in question 1, are (a) monotonic increasing, (b) monotonic decreasing, (c) not monotonic.
3. State which of the sequences, if any, given in question 1, are (a) bounded, (b) unbounded below and bounded above, (c) unbounded above and bounded below, (d) unbounded above and below.
4. Show that the given sequence is monotonic and bounded. Deduce that it converges and find its limit.
(i) $s_n = n^{-2}$, (ii) $s_n = (n+1)/n^2$, (iii) $s_n = (n-1)/(n+1)$, (iv) $s_n = (n - 3n^2)/(n^2 + 5n + 1)$, (v) $s_n = (1 - 2n^2)/(n^2 + n + 3)$.
5. The sequence (x_n) is given by $x_{n+1} = (x_n^2 + 3)/2x_n$, $x_1 = 2$. Prove that, for all $n \in \mathbb{Z}^+$, $\sqrt{3} < x_{n+1} < x_n$. Deduce that (x_n) converges to $\sqrt{3}$. Calculate x_4 to give an approximation to $\sqrt{3}$, correct to 7 significant figures.

6 The sequence (s_n) is defined by: $s_1 = 2$, $s_{n+1} = 2 + 1/(s_n + 2)$. Prove that, for $n \in \mathbb{Z}^+$, (i) if $s_n < \sqrt{5}$ then $s_{n+1} > \sqrt{5}$, and if $s_n > \sqrt{5}$ then $s_{n+1} < \sqrt{5}$, (ii) $s_{2n+1} < s_{2n+3} < s_{2n+2} < s_{2n}$. Deduce that (s_n) converges and express its limit as a surd.

7 The real number a is greater than 1 and the sequence (s_n) is defined by: $s_1 > a$ and, for all $n \in \mathbb{Z}^+$, $s_{n+1} = \dfrac{2s_n^3 + a^3}{3s_n^2}$. Factorise $s_{n+1} - a$ and prove that, for all $n \in \mathbb{Z}^+$, (i) $s_n > a$, (ii) $s_{n+1} < s_n$. Deduce that (s_n) converges to a. Put $s_1 = 2$ and $a^3 = 5$, and calculate s_4 to find an approximation to $\sqrt[3]{5}$, correct to six decimal places.

13.3 Convergence of series

As was stated in §6.2, the series $\sum a_r$ converges to a limit s if its sequence (s_n) of partial sums converges to s, where

$$s_n = a_1 + a_2 + a_3 + \ldots + a_n = \sum_{r=1}^{n} a_r.$$

If (s_n) diverges then $\sum a_r$ is said to diverge.

Notation $\sum a_r = s$ means that $\sum a_r$ is a convergent series with sum s.

The results of Theorem 13.1 for sequences immediately translate into corresponding results for series and some are stated in Theorem 13.3.

Theorem 13.3 Let two convergent series have sums given by $\sum a_r = s$ and $\sum b_r = t$ and let $k \in \mathbb{R}$. Then

Sum $\qquad \sum (a_r + b_r) = s + t,$

Product by a constant $\qquad \sum (ka_r) = ks.$

The usual method of determining whether or not a series converges is to compare the series in some way with a known series. In order to show how this may be done, it is necessary to use some simple results, given in the next theorems.

Theorem 13.4 If $\sum a_r$ converges then $a_r \to 0$ as $r \to \infty$.

Proof Let $\sum a_r = s = \lim s_n$, where s_n is the nth partial sum of $\sum a_r$. Then $a_n = s_n - s_{n-1}$. Hence

$$\lim_{n\to\infty} a_n = \lim_{n\to\infty} s_n - \lim_{n\to\infty} s_{n-1} = s - s = 0.$$

The converse of Theorem 13.4 is false. This is illustrated by the series $\sum 1/r$ which is shown to be divergent in Example 1.

Theorem 13.5 (Comparison theorem) Let $0 < a_r \leqslant b_r$, for all r.
Then (i) if $\sum b_r$ converges $\sum a_r$ converges,
(ii) if $\sum a_r$ diverges $\sum b_r$ diverges.

Proof The two statements are logically equivalent so we only have to prove the first. Let (s_n) and (t_n) be the sequences of partial sums of the series $\sum a_r$ and $\sum b_r$. Then $0 < s_n = \sum_{r=1}^{n} a_r \leqslant \sum_{r=1}^{n} b_r = t_n$. If $\sum b_r$ converges to t, then (t_n) converges to t. Since $t_{n+1} = t_n + b_{n+1} > t_n$, (t_n) is monotonic increasing and $t_n < t$. Therefore (s_n) is monotonic increasing and is bounded above by t. Hence (s_n) converges to some real number s and $\sum a_r = s$.

Theorem 13.5 refers to series which have all terms non-negative. It can be useful to consider the series formed by using the moduli of the terms of a given series, and this leads to the definition of absolute convergence.

Definition The series $\sum a_r$ is *absolutely convergent* if $\sum |a_r|$ is convergent.

Subseries A *subseries* of a given series $\sum a_r$ is a series obtained by deleting any number of terms of $\sum a_r$.

Theorem 13.6 Let $\sum a_r$ be a convergent series of non-negative terms. Then any subseries of $\sum a_r$ also converges.

Proof The sequence (s_n) of partial sums of $\sum a_r$ is monotonic increasing and is bounded above by $\sum a_r$. If $\sum d_r$ is any subseries of $\sum a_r$ then, for all r, $0 \leqslant d_r \leqslant a_m$, where d_r was the mth term in $\sum a_r$. Thus the sequence (t_n) of partial sums of the subseries $\sum d_r$ is monotonic increasing and bounded above by $\sum a_r$. Hence (t_n) converges to some real number d and $\sum d_r = d$.

The terms of a series may vary in sign and the series may be split into two subseries, one containing all the positive terms and one containing all the negative terms. Given the series $\sum a_r$,

define $\quad b_r = (|a_r| + a_r)/2, \quad c_r = (|a_r| - a_r)/2,$

then $\quad |a_r| = b_r + c_r, \quad a_r = b_r - c_r$, with $b_r \geqslant 0$, $c_r \geqslant 0$.

This is used in the next theorem.

Theorem 13.7 If $\sum a_r$ is absolutely convergent then $\sum a_r$ is convergent.

Proof Since $\sum a_r$ converges absolutely, let $\sum |a_r| = d$. Let

$$b_r = (|a_r| + a_r)/2 \quad \text{and} \quad c_r = (|a_r| - a_r)/2,$$

so that

$$a_r = b_r \text{ if } a_r \geqslant 0, \qquad a_r = -c_r \text{ if } a_r < 0,$$

$$|a_r| = b_r + c_r, \qquad a_r = b_r - c_r.$$

Now $\sum b_r$ and $\sum c_r$ are subseries of $\sum |a_r|$, a convergent series of non-negative terms. Hence they both converge so let $\quad \sum b_r = b$ and $\sum c_r = c$.

Then $\quad \sum a_r = \sum (b_r - c_r) = \sum b_r - \sum c_r = b - c.$

The converse of this theorem is not true. A convergent series may not be absolutely convergent as is shown in the next example.

EXAMPLE 1 *Show that (i) $\sum(-1)^r/r$ converges, (ii) $\sum 1/r$ diverges.*

(i) Let s_n be the nth partial sum of the series $\sum(-1)^r/r$.

Then for $n = 2k$, $s_{2k+1} = s_{2k} - 1/(2k+1) \quad < s_{2k}$,

and
$$s_{2k+2} = s_{2k} - 1/(2k+1) + 1/(2k+2)$$
$$= s_{2k} - 1/(2k+1)(2k+2) \quad < s_{2k}.$$

Also
$$s_{2k+3} = s_{2k+1} + 1/(2k+2) - 1/(2k+3)$$
$$= s_{2k+1} + 1/(2k+2)(2k+3) \quad > s_{2k+1}.$$

Hence
$$s_{2k+1} < s_{2k+3} < s_{2k+2} < s_{2k},$$

(s_{2n}) is a monotonic decreasing sequence, bounded below by s_1, (s_{2n+1}) is a monotonic increasing sequence, bounded above by s_2. There exist limits s and t such that $s_{2n} \to s$ and $s_{2n+1} \to t$. Then $(s_{2n} - s_{2n+1}) \to s - t$. But $s_{2n} - s_{2n+1} = 1/2n+1$ and this tends to zero, so $s - t = 0$ and $s = t = \sum(-1)^r/r$.
(ii) To show that $\sum 1/r$ diverges, we divide the series up into a set of finite series, each twice the length of the one before and each having a sum greater than 1/2. Let t_n be the nth partial sum of $\sum 1/r$, and consider $n = 2^k$, for $k \in \mathbb{Z}^+$.

$k = 1, \quad t_2 = 1 + \frac{1}{2},$

$k = 2, \quad t_4 = t_2 + \frac{1}{3} + \frac{1}{4} > t_2 + \frac{1}{4} + \frac{1}{4} > 1 + \frac{1}{2} + \frac{1}{2},$

$k = 3, \quad t_8 = t_4 + \frac{1}{5} + \frac{1}{6} + \frac{1}{7} + \frac{1}{8} > t_4 + \frac{1}{8} + \frac{1}{8} + \frac{1}{8} + \frac{1}{8}$
$\quad = t_4 + \frac{1}{2} > 1 + \frac{1}{2} + \frac{1}{2} + \frac{1}{2}.$

Generally, $t_{2^k} = t_{2^{k-1}} + \dfrac{1}{2^{k-1}+1} + \ldots + \dfrac{1}{2^k}$

$$> t_{2^{k-1}} + \frac{1}{2^k} + \ldots \frac{1}{2^k} = t_{2^{k-1}} + 2^{k-1}\left(\frac{1}{2^k}\right)$$

$$= t_{2^{k-1}} + \frac{1}{2}.$$

This means that, by induction, $t_{2^k} > 1 + \dfrac{k}{2}$, and so $(t_{2^n}) \to \infty$. Therefore, since (t_n) is monotonic increasing, $\sum 1/r$ diverges.

In order to use the comparison theorem, some test series are needed with which a given series may be compared. An obvious test series is the geometrical series which converges when its common ratio is less than 1 in modulus and diverges otherwise. Theorem 13.8 gives another very useful test series, obtained from a generalisation of Example 1.

Theorem 13.8 $\sum 1/r^p$ converges if $p > 1$ and diverges if $p \leqslant 1$.

Proof The terms of the series are all positive and convergence is determined by comparison with a geometric series. Let s_n be the nth partial sum of $\sum 1/r^p$ and consider $In = 2^k - 1$, $IIn = 2^k$, for $k \in \mathbb{Z}^+$.

Case I $p > 1$. If $p > 1$, $p - 1 > 0$, $2^{p-1} > 1$, so, if $a = 1/2^{p-1}$, then $0 < a < 1$.

$k = 1$, $s_1 = 1$,

$$k = 2,\ s_3 = s_1 + \frac{1}{2^p} + \frac{1}{3^p} < s_1 + \frac{1}{2^p} + \frac{1}{2^p} = 1 + \frac{2}{2^p} = 1 + a,$$

$$k = 3,\ s_7 = s_3 + \frac{1}{4^p} + \frac{1}{5^p} + \frac{1}{6^p} + \frac{1}{7^p} < s_3 + \frac{4}{4^p} = 1 + a + a^2,$$

$$k=4,\ s_{15} = s_7 + \frac{1}{8^p} + \frac{1}{9^p} + \ldots + \frac{1}{15^p} < s_7 + \frac{8}{8^p}$$
$$= 1 + a + a^2 + a^3.$$

Generally, $$s_{2^k-1} = s_{2^{k-1}-1} + \frac{1}{(2^{k-1})^p} + \ldots + \frac{1}{(2^k-1)^p} < s_{2^{k-1}-1} + \frac{2^{k-1}}{(2^{k-1})^p}$$
$$< 1 + a + a^2 + \ldots + a^{k-1} = (1 - a^k)/(1 - a) < 1/(1 - a).$$

The series $\sum 1/r^p$ is monotonic increasing and is bounded above by $1/(1 - a)$ and hence it converges.

Case II $p \leqslant 1$. If $p \leqslant 1$ then $2^p \leqslant 2$ and $1/2^p \geqslant 1/2$.

$$k = 1,\ s_2 = 1 + \frac{1}{2^p} \geqslant 1 + \frac{1}{2},$$

$$k = 2,\ s_4 = s_2 + \frac{1}{3^p} + \frac{1}{4^p} > s_2 + \frac{2}{4^p} > s_2 + \frac{2}{4} \geqslant 1 + \frac{1}{2} + \frac{1}{2},$$

$$k = 3,\ s_8 = s_4 + \frac{1}{5^p} + \frac{1}{6^p} + \frac{1}{7^p} + \frac{1}{8^p} > s_4 + \frac{4}{8^p} > 1 + \frac{1}{2} + \frac{1}{2} + \frac{1}{2},$$

Generally, $$s_{2^k} = s_{2^{k-1}} + \frac{1}{(2^k-1)^p} + \ldots + \frac{1}{(2^k)^p} > s_{2^{k-1}} + \frac{2^{k-1}}{(2^k)^p} > 1 + k/2.$$

So $s_{2^k} \to \infty$ and therefore $\sum 1/r^p$ diverges.

EXAMPLE 2 *Test for convergence the series:* (*i*) $\sum (r+1)/(2r^2 - 1)$, (*ii*) $\sum (r-1)/(r^2+3)$, (*iii*) $\sum (r+1)/r(r+2)(r+3)$.

The method is to look at the form, as n becomes large, of the general term a_r of the series and attempt to find some test series with which to compare it.
(i) $a_r = (r+1)/(2r^2-1) = (1+1/r)/r(2-1/r) \approx 1/2r$, for large r. We expect the series to diverge, on comparison with $\sum 1/r$. To show this divergence, since $r+1 > r$ and $2r^2-1 < 2r^2$, $a_r > r/2r^2 = 1/2r$. Since $\sum 1/r$ diverges, $\sum 1/2r$, and hence $\sum (r+1)/(2r^2-1)$, **diverges.**
(ii) $a_r = (r-1)/(r^2+3) = (1-1/r)/r(1+3/r^2)$ and again, for large r, $a_r \approx 1/r$, and we expect divergence. However, we cannot just omit the constants -1 and 3, as was done in (i), since this gives $a_r < 1/r$. We use the idea that $(1-1/r)/(1+3/r^2)$ tends to 1 as r tends to infinity and so this fraction will be larger than its value for some small value of r, say for $r = 2$.
Thus for $r > 2$, $1-1/r > 1-1/2 = 1/2$, $1+3/r^2 < 1+3/4 = 7/4$, and hence $(1-1/r)/(1+3/r^2) > 2/7$, and $a_r > 2/7r$. Again, by comparison with the divergent series $\sum 2/7r$, $\sum a_r$ **diverges.**
(iii) $a_r = (r+1)/r(r+2)(r+3) = (1+1/r)/r^2(1+2/r)(1+3/r) \approx 1/r^2$, for large r. Therefore, in comparison with $\sum 1/r^2$, we expect convergence. To show that a_r is less than some multiple of $1/r^2$, $r+2 > r$, $r+3 > r$ and, for $r > 1$, $1+1/r < 2$, so that $a_r = (1+1/r)/(r+2)(r+3) < 2/r^2$. Therefore, in comparison with the convergent series $\sum 2/r^2 = 2\sum 1/r^2$, the series $\sum a_r$ **converges.**

EXERCISE 13.3

1 Determine whether or not the series $\sum a_r$ converges, where:
(i) $a_r = (3/4)^r$, (ii) $a_r = (3^r-1)/4^r$, (iii) $a_r = (3^r+1)/4^r$, (iv) $a_r = (r+1)/r^3$, (v) $a_r = (\sqrt{r}-2)/r^2$, (vi) $a_r = (\sqrt{r}+2)/(2r^2-1)$, (vii) $a_r = r/\sqrt{(1+r^3)}$, (viii) $a_r = (-1)^r/r\sqrt{r}$, (ix) $a_r = r/r!$.

2 For what values of x does the series converge? (i) $\sum e^{rx}$, (ii) $\sum rx^r$, (iii) $\sum (2/x)^r$, (iv) $\sum 1/(1-x)^r$.

3 Test for convergence: (i) $\sum (\sqrt{n}-1)/(n^2+4)$, (ii) $\sum (2n-1)/(n^2+2)$, (iii) $\sum \{\sin(n\pi+\pi/2)\}/n^2$.

4 Find the values of x for which the series converges: (i) $\sum nx^{n-1}$, (ii) $\sum (\sinh nx)/2^n$, (iii) $\sum nx^{-n}$, (iv) $\sum (\sin nx)/n^2$, (v) $\sum \cos nx/2^n$, (vi) $\sum (n+1)(3x)^n$.

5 Prove that, for $x > 0$, $e^x > x^2/2$ and deduce that the series $\sum ne^{-n}$ converges.

13.4 Convergence of power series

In §6.5, we stated the power series expansions of certain functions and gave their radii of convergence. For completeness, we now indicate the proofs of their convergence. These proofs are quite difficult and are best omitted on the first reading.

Exponential series

We prove the result, stated in §6.5, that the exponential series, $\sum x^r/r!$, converges for all x. Let s_n be the nth partial sum of the series so that

$$s_n = \sum_{r=0}^{n-1} x^r/r!.$$

Case I $x \geqslant 0$. If x is large, the first few terms of the power series for e^x increase rapidly in size. It is only when r becomes large that $x^r/r!$ becomes small. We therefore look at the 'infinite tail' of the series, that is the sum of all the terms beyond the Nth term, where N is chosen to be larger than x. Choose any integer N, such that $N > x$ and let $\rho = x/(N+1)$ so that $\rho < 1$. Then, for $n > N$,

$$
\begin{aligned}
s_n &= s_{N-1} + \sum_{r=N}^{n-1} x^r/r! \\
&= s_{N-1} + \frac{x^N}{N!} + \frac{x^{N+1}}{(N+1)!} + \frac{x^{N+2}}{(N+2)!} + \ldots + \frac{x^{n-1}}{(n-1)!} \\
&= s_{N-1} + \frac{x^N}{N!}\left\{1 + \frac{x}{N+1} + \frac{x^2}{(N+1)(N+2)} + \ldots \right. \\
&\qquad \left. + \frac{x^{n-N-1}}{(N+1)\ldots(n-1)}\right\} \\
&< s_{N-1} + \frac{x^N}{N!}\left\{1 + \frac{x}{N+1} + \left(\frac{x}{N+1}\right)\left(\frac{x}{N+1}\right) + \ldots \right. \\
&\qquad \left. + \left(\frac{x}{N+1}\right)\cdots\left(\frac{x}{N+1}\right)\right\} \\
&< s_{N-1} + \frac{x^N}{N!}\{1 + \rho + \rho^2 + \ldots + \rho^{n-N-1}\}
\end{aligned}
$$

Since $0 \leqslant \rho < 1$, the geometric series $\sum \rho^r$ converges and s_n is bounded above. Since s_n is also monotonic increasing, it converges to a limit and this limit we define to be e^x,

$$\lim_{n\to\infty} s_n = \sum_{r=0}^{\infty} x^r/r! = \sum x^r/r! = e^x.$$

Case II $x < 0$. By case I, $\sum x^r/r!$ is absolutely convergent since $\sum |x|^r/r!$ converges and so $\sum x^r/r!$ converges, its limit again being defined as e^x. Therefore the exponential series e^x converges for all real x.

The series for $\cosh x$ and $\sinh x$ are each subseries of e^x and the series for $\cos x$ and $\sin x$ are subseries of e^x with alternate terms having their signs changed. Since e^x is absolutely convergent, the other four series are also absolutely convergent and hence convergent for all real x.

Logarithmic series

Since $\sum 1/r$ diverges, the series $\sum x^r/r$ diverges for all x such that $x \geqslant 1$. For $|x| < 1$, $0 < |x|^r/r < |x|^r$ and so $\sum |x|^r/r$ converges, in comparison

with the convergent geometric series $\sum |x|^r$. Hence $\sum x^r/r$ is absolutely convergent and therefore convergent.
For $x = -1$, the series $\sum x^r/r$ converges as shown in §13.3 Example 1.
For $x < -1$, let $x = -1 - y$, where $y > 0$. Then

$$|x|^r = (1+y)^r > 1 + ry + r(r-1)y^2/2 > r(r-1)y^2/2,$$

and hence $|x|^r/r \to \infty$ as $r \to \infty$. Therefore by theorem 13.4, $\sum x^r/r$ diverges.
We conclude that the logarithmic series $\sum -(-x)^r/r$ converges for $-1 < x \leqslant 1$ and its sum is $\ln(1+x)$.

EXERCISE 13.4

1 Find the sum of the series and state the values of x for which it converges: (i) $\sum x^n/(2n!)$, (ii) $\sum (2x)^n/n$, (iii) $\sum (x-1)^n/n!$, (iv) $\sum (x-1)^n/n$.

2 Express the sum of the infinite series in terms of hyperbolic and/or trigonometric functions: (i) $\sum_{n=1}^{\infty} x^n/(2n+1)!$, (ii) $\sum_{n=0}^{\infty} (-x)^n/(2n+1)!$,

(iii) $\sum_{r=1}^{\infty} x^{r+1}/r!$, (iv) $\sum_{r=1}^{\infty} (r-1)x^r/r!$.

3 Prove that the radius of convergence of the binomial series $\sum a_r$ for $(1+x)^n$, $n \notin \mathbb{N}$, is 1, by the following steps.
(i) Prove that, for $r > n$, $|a_{r+1}/a_r| = (r-n)|x|/(r+1)$. (ii) If $d = |1-|x||$ and $r > |x|(|n|+1)/d$, then if $|x| > 1$, $(r-n)|x|/(r+1) > |x| - d > 1$, whilst if $|x| < 1$, $(r-n)|x|/(r+1) < |x| + d < 1$. (iii) Compare with the geometric series with common ratio $(1+|x|)/2$.

MISCELLANEOUS EXERCISE 13

1 Evaluate (i) $\lim_{\theta \to 0} \left(\dfrac{\sin 5\theta + \sin 7\theta}{6\theta} \right)$, (ii) $\lim_{\theta \to 0} \left(\dfrac{1 + \theta - \cos\theta}{\sin\theta} \right)$. (*L*)

2 Find the limit as $x \to 0$ of $\dfrac{\sin x - x\cos x}{x^3}$. (*L*)

3 The function f is defined by $f(x) = \sin x$ for $x \leqslant 0$, $f(x) = x$ for $x > 0$. Sketch the graphs of $f(x)$ and its derivative $f'(x)$ for $-\pi/2 \leqslant x \leqslant \pi/2$ and decide whether the functions f and f′ arc continuous at $x = 0$ or not. (*L*)

4 The sequence (a_n) of positive real numbers is given by $a_1 = 3$, $a_{n+1} = \frac{1}{2}\sqrt{(12 + a_n^2)}$, $n \geqslant 1$. Prove by induction that $a_n \geqslant 2$ for each n. Show that $a_{n+1}^2 \leqslant a_n^2$ for each n. Show also that the sequence converges and find its limit. (*JMB*)

5 Test for convergence the series: (i) $\sum_{r=2}^{\infty} \dfrac{2r+1}{\sqrt{(r^3-2)}}$, (ii) $\sum_{n=1}^{\infty} \dfrac{\cos n\pi/6}{n}$,

(iii) $\sum_{r=0}^{\infty} \dfrac{2r+1}{r^3-2}$, (iv) $\sum n e^{-n^2}$

6 The continuous function f is given by $f(x) = \sinh x, x < 0, f(x) = c + \cosh x$, $x \geqslant 0$, where c is a constant. Find the value of c. Sketch the graph of f. State whether or not the function is differentiable at $x = 0$, giving brief reasons. (*JMB*)

7 Sequences (a_n) and (b_n) are defined by $a_1 = 7, b_1 = 1, a_{n+1} = (a_n + b_n)/2$ and $b_{n+1} = 2a_nb_n/(a_n + b_n)$ $(n \geqslant 1)$. Prove that $a_{n+1} \geqslant b_{n+1}$ for each $n \geqslant 0$. Deduce that $a_{n+1} - a_n \leqslant 0$ for each $n \geqslant 1$. Prove that $a_nb_n = 7$ and that $b_{n+1} \geqslant b_n$ for each $n \geqslant 1$. Show that the sequences (a_n) and (b_n) converge to the same limit and find this limit. (*JMB*)

8 Find the limit as $\theta \to 0$ of $(1 - \cos 4\theta + \theta \sin 3\theta)/\theta^2$. (*L*)

9 A function f is defined in the interval $0 \leqslant x \leqslant 2$ by the relations $f(x) = 2 + 4x - 3x^2$ when $0 \leqslant x \leqslant 1$, $f(x) = a + b/(2x + 1)$ when $1 < x \leqslant 2$, where the constants a and b are to be chosen so that the graph of $y = f(x)$ is continuous at $x = 1$ and has a (unique) tangent at $x = 1$. Find a and b.

10 Find the values of x for which the series $\sum a_n(x)$ converges, given that: (i) $a_n(x) = x^n/n^2$, (ii) $a_n(x) = x^n/n(n+2)$, (iii) $a_n(x) = (n+1)x^n/n(n+2)$.

11 The periodic function f is given by $f(x) = 1,\ 0 \leqslant x < 1,\ f(x) = -1$, $1 \leqslant x < 2$, $f(x) = f(x + 2)$. Sketch the graph of f for $-4 \leqslant x \leqslant 4$. Explain why you consider this function to be continuous or discontinuous at $x = 3$, and write down the value of $\displaystyle\int_{-\pi}^{\pi} f(x)\,dx$. (*L*)

12 Find the limit as θ tends to 0 of $\dfrac{(\sin\theta + \cos\theta)^2 - 1}{\theta}$. (*L*)

13 (i) Test each of the following series to decide whether or not it converges:

$$\text{(a)}\ \sum_{n=1}^{\infty} \frac{(n-1)}{2n^2(n+1)}, \quad \text{(b)}\ \sum_{n=1}^{\infty} \frac{1}{\sqrt{(n^2+n)}}, \quad \text{(c)}\ \sum_{n=1}^{\infty} \frac{1}{n}\left(\frac{1}{2}\right)^n.$$

(ii) Expand $f(x)$, where $f(x) = e^x \cos x + \sin x$, in powers of x as far as the term in x^3. Draw a sketch of the curve $y = f(x)$ for small values of x. Show on the same diagram the tangent to this curve at $x = 0$. (*L*)

14 The function f has domain $\mathbb{R}$ and is continuous and differentiable at $x = 0$. The function g, defined by $g(x) = \{f(x) + 1\}^3$ for $x < 0$ and $g(x) = -f(x) - 1$ for $x \geqslant 0$, is also continuous and differentiable at $x = 0$. Find $f(0)$ and $f'(0)$. (*JMB*)

15 Find the sum of the series, stating the values of x for which the result is valid:

$$\text{(i)}\ \sum_{n=1}^{\infty} \frac{(-x)^n}{(2n)!}, \quad \text{(ii)}\ \sum_{r=0}^{\infty} \frac{(x+1)^r}{r}, \quad \text{(iii)}\ \sum n(2x)^{n-1}, \quad \text{(iv)}\ \sum_{n=0}^{\infty} \frac{nx^n}{(n+1)!}.$$

16 If $y = \tan x$, show that $\dfrac{dy}{dx} = 1 + y^2$. By repeated differentiation of this equation show that $y_5 = 6y_2^2 + 8y_1y_3 + 2yy_4$ where $y_r = \dfrac{d^ry}{dx^r}$. Use Maclaurin's Theorem to find the first three non-zero terms of the expansion of $\tan x$ in ascending powers of x. Hence, or otherwise, evaluate

$$\lim_{x \to 0} \left\{ \frac{\tan x - \sin x}{x(\cosh x - \cos x)} \right\}.$$

(*AEB*)

14 Vector Spaces

14.1 Definitions

The set of all real column vectors in two dimensions, that is, with two rows, has the following properties:
(i) under addition it is an abelian group with zero element **0**,
(ii) under multiplication by real numbers (scalars) it is closed and the associativity and distributivity laws hold, namely that for all vectors **a** and **b** and all real numbers x and y,

$$(xy)\mathbf{a} = x(y\mathbf{a}), \quad (x+y)\mathbf{a} = x\mathbf{a} + y\mathbf{a}, \quad x(\mathbf{a}+\mathbf{b}) = x\mathbf{a} + x\mathbf{b} \quad \text{and} \quad 1\mathbf{a} = \mathbf{a}.$$

This is an example of an algebraic system called a *vector space* (or a *linear space*). Vector spaces occur frequently in mathematics and their properties, which are due to the fact that they are closed under the taking of linear combinations, are worth investigation. A vector space is a set of vectors which is an abelian group under addition and is closed under multiplication by scalars, taken from a set called a field. A field is an algebraic system with two operations, addition and multiplication, which is an abelian group under addition and, when the additive zero is removed, is an abelian group under multiplication, such that multiplication distributes over addition.

Definition A field F is an algebraic system $F(+, \times)$, with binary operations of addition and multiplication, such that:

$$(F, +) \text{ is an abelian group with zero } 0,$$
$$(F\backslash\{0\}, \times) \text{ is an abelian group with identity } 1,$$
$$\text{for all } x, y, z \in F, \; x(y+z) = xy + xz, \text{ and } 1x = x.$$

In a field, the inverse of x under addition is the negative $-x$, the inverse of x ($\neq 0$) under multiplication is x^{-1}. A field is closed under the operations of addition, subtraction, changing the sign, multiplication, division by a non-zero and taking the inverse of a non-zero. It contains no zero-divisors. The set $\mathbb{Q}$ of rational numbers, the set $\mathbb{R}$ of real numbers and the set $\mathbb{C}$ of complex numbers are familiar examples of fields. The set $\mathbb{Z}$ of integers is not a field since it is not closed under division. The set of all 2×2 real matrices is not a field since it is not commutative under multiplication and it contains singular matrices which have no inverses and are zero-divisors. We shall be concerned mostly with the field $\mathbb{R}$, but sometimes we need the field $\mathbb{C}$ in order that we can be sure that a polynomial equation has a solution.

Definition A vector space V is a set of vectors, $V = \{\mathbf{0}, \mathbf{a}, \mathbf{b}, \mathbf{c}, \ldots\}$, which is an abelian group under addition, with zero $\mathbf{0}$, and which is closed under multiplication by the elements of a field F, $F = \{0, 1, x, y, \ldots\}$, such that for all $\mathbf{a}, \mathbf{b} \in V$ and $x, y \in F$,

$$(xy)\mathbf{a} = x(y\mathbf{a}), \; (x+y)\mathbf{a} = x\mathbf{a} + y\mathbf{a},$$
$$x(\mathbf{a}+\mathbf{b}) = x\mathbf{a} + x\mathbf{b}, \qquad 1\mathbf{a} = \mathbf{a}.$$

It can be shown that, in a vector space, the usual rules of algebra hold. For example, $0\mathbf{a} = \mathbf{0}$, $-(x\mathbf{a}) = (-x)\mathbf{a}$. We shall not be concerned with proving results from the definitions but will investigate the consequences of the closure of a vector space under linear combinations. We shall mostly be concerned with vector spaces of real column vectors of small dimension and use the notation $\mathbb{R}^n$ of §1.3, giving a more formal definition.

Definition $\mathbb{R}^n$ is the vector space of real column vectors with n rows, over the field of real numbers.

Thus, $\mathbb{R}^2 = \left\{\begin{pmatrix} x \\ y \end{pmatrix} : x, y \in \mathbb{R}\right\}$ and $\mathbb{R}^3 = \left\{\begin{pmatrix} x \\ y \\ z \end{pmatrix} : x, y, z \in \mathbb{R}\right\}$ define the familiar vector spaces in 2 and 3 dimensions. Of course, $\mathbb{R}^1$ is isomorphic to $\mathbb{R}$, as a real vector space.

EXAMPLE 1 *Prove that the following are vector spaces over* $\mathbb{R}$:
(*i*) *The set of all solutions of the simultaneous equations*

$$2x + 3y + 4z = 0, \quad 3x - y - 5z = 0, \quad x + 4y - 2z = 0.$$

(*ii*) *The set of all solutions of the differential equation*

$$\frac{\mathrm{d}^3 y}{\mathrm{d}x^3} + 3\frac{\mathrm{d}^2 y}{\mathrm{d}x^2} - 4\frac{\mathrm{d}y}{\mathrm{d}x} + 4y = 0.$$

(i) A solution of the set of three equations is a triplet of values of x, y and z, which we may write as $\mathbf{u} = (x, y, z)$. Let V be the set of all solutions, so that $V = \{(a, b, c)\colon x = a,\ y = b,\ z = c \text{ is a solution}\}$. Suppose that $\mathbf{u}_1, \mathbf{u}_2 \in V$, with $\mathbf{u}_1 = (a_1, b_1, c_1)$, $\mathbf{u}_2 = (a_2, b_2, c_2)$, and let $k \in \mathbb{R}$, then it is easily verified, by substituting into the three equations, that

$$x = a_1 + a_2, \; y = b_1 + b_2, \; z = c_1 + c_2 \text{ and}$$
$$x = ka_1, \qquad y = kb_1, \qquad z = kc_1 \text{ are also solutions.}$$

Therefore $\mathbf{u}_1 + \mathbf{u}_2 \in V$ and $k\mathbf{u}_1 \in V$, which means that V is closed under the taking of linear combinations. It follows that V is closed under the associative binary operation of addition, with a zero element $\mathbf{0}$, given by $\mathbf{0} = (0, 0, 0)$, which is a solution, and with inverse given by negatives. Hence $(V, +)$ is an abelian group. The distributive rules also hold and so V is a vector space over $\mathbb{R}$.

(ii) Suppose that $y = \mathrm{f}(x)$ and $y = \mathrm{g}(x)$ are solutions of the differential equation and that $k \in \mathbb{R}$. Let $w = \mathrm{f}(x) + \mathrm{g}(x)$, $z = k\mathrm{f}(x)$,

$$\begin{aligned}\frac{\mathrm{d}^3 w}{\mathrm{d}x^3} + 3\frac{\mathrm{d}^2 w}{\mathrm{d}x^2} - 4\frac{\mathrm{d}w}{\mathrm{d}x} + 4w &= \frac{\mathrm{d}^3}{\mathrm{d}x^3}\{\mathrm{f}(x)+\mathrm{g}(x)\} + 3\frac{\mathrm{d}^2}{\mathrm{d}x^2}\{\mathrm{f}(x)+\mathrm{g}(x)\} \\ &\quad - 4\frac{\mathrm{d}}{\mathrm{d}x}\{\mathrm{f}(x)+\mathrm{g}(x)\} + 4\{\mathrm{f}(x)+\mathrm{g}(x)\} \\ &= \frac{\mathrm{d}^3}{\mathrm{d}x^3}\mathrm{f}(x) + 3\frac{\mathrm{d}^2}{\mathrm{d}x^2}\mathrm{f}(x) - 4\frac{\mathrm{d}}{\mathrm{d}x}\mathrm{f}(x) + 4\mathrm{f}(x) \\ &\quad + \frac{\mathrm{d}^3}{\mathrm{d}x^3}\mathrm{g}(x) + 3\frac{\mathrm{d}^2}{\mathrm{d}x^2}\mathrm{g}(x) \\ &\quad - 4\frac{\mathrm{d}}{\mathrm{d}x}\mathrm{g}(x) + 4\mathrm{g}(x) \\ &= 0,\end{aligned}$$

$$\begin{aligned}\frac{\mathrm{d}^3 z}{\mathrm{d}x^3} + 3\frac{\mathrm{d}^2 z}{\mathrm{d}x^2} - 4\frac{\mathrm{d}z}{\mathrm{d}x} + 4z &= \frac{\mathrm{d}^3}{\mathrm{d}x^3}k\mathrm{f}(x) + 3\frac{\mathrm{d}^2}{\mathrm{d}x^2}k\mathrm{f}(x) - 4\frac{\mathrm{d}}{\mathrm{d}x}k\mathrm{f}(x) + 4k\mathrm{f}(x) \\ &= k\left\{\frac{\mathrm{d}^3}{\mathrm{d}x^3}\mathrm{f}(x) + 3\frac{\mathrm{d}^2}{\mathrm{d}x^2}\mathrm{f}(x) - 4\frac{\mathrm{d}}{\mathrm{d}x}\mathrm{f}(x) + 4\mathrm{f}(x)\right\} \\ &= 0.\end{aligned}$$

Therefore, w and z are also solutions of the differential equation and on using the same argument as was used in (i) it is seen that the set of all solutions is a vector space over $\mathbb{R}$.

In Example 1, the sets of solutions form vector spaces because the equations concerned are linear and this ensures that solutions can be added together and multiplied by a constant to give other solutions. This would not be so in the case of non-linear equations. Note that we do not have to solve the equations but just use their linearity.

Subspaces If a subset of vectors in a vector space is also a vector space over the same field then it is a *subspace* of the vector space.

Theorem 14.1 Let V be a vector space over the field F and let U be a non-empty subset of V. Then U is a subspace of V if U is closed under taking linear combinations with scalar multipliers in F.

Proof Suppose that $U \subseteq V$ and $U \neq \phi$, and that U is closed under taking linear combinations. Then U contains at least one vector, say $\mathbf{a}$, and $\mathbf{0} = \mathbf{a} - \mathbf{a} \in U$. For any $\mathbf{a}, \mathbf{b} \in U$ and any $k \in F$, $\mathbf{a} + \mathbf{b}$, $-\mathbf{a}$, $k\mathbf{a} \in U$ and, since the axioms of associativity, commutativity and distributivity are satisfied in V, they are also satisfied in U. Therefore U is a vector space and hence a subspace of V.

EXAMPLE 2 *Prove that the set of all vectors in* $\mathbb{R}^3$ *which are linear combinations of the two vectors* $\begin{pmatrix} 1 \\ -1 \\ 0 \end{pmatrix}$ *and* $\begin{pmatrix} 1 \\ 0 \\ -1 \end{pmatrix}$ *is a subspace and is the set* $\left\{\begin{pmatrix} x \\ y \\ z \end{pmatrix} : x+y+z=0\right\}$.

Let $\mathbf{a} = \begin{pmatrix} 1 \\ -1 \\ 0 \end{pmatrix}$, $\mathbf{b} = \begin{pmatrix} 1 \\ 0 \\ -1 \end{pmatrix}$, then $\mathbf{u} = p\mathbf{a} + q\mathbf{b} = \begin{pmatrix} p+q \\ -p \\ -q \end{pmatrix}$ and so $\begin{pmatrix} x \\ y \\ z \end{pmatrix}$ is a linear combination of $\mathbf{a}$ and $\mathbf{b}$ if, and only if, $x+y+z=0$. Let $U = \left\{\begin{pmatrix} x \\ y \\ z \end{pmatrix} : x+y+z=0\right\}$, then U is closed under linear combinations and, since $\mathbf{a} \in U$, $U \neq \phi$ and hence U is a subspace of $\mathbb{R}^3$. Note that U can be thought of as the set of all solutions of the linear equation $x+y+z=0$, c.f. Example 1.

It follows from Theorem 14.1 that the set of all linear combinations of the vectors in any set S, of vectors in a vector space, forms a subspace.

Linear equations A set of linear equations in variables $(x, y, \ldots, z)$ is said to be *homogeneous linear* if each equation is linear in each of the variables. This means that every term in each equation is just a multiple of one of the variables. There is no constant term and no products, or powers, of the variables. In Example 1 (i), it was shown that the set of all solutions of a set of homogeneous linear equations in (x, y, z) forms a vector space. This result generalises to any set of linear equations and the space is called the *solution space.*

Linear transformations In chapter 4, linear transformations of $\mathbb{R}^n$ into $\mathbb{R}^m$ were considered, together with their associated matrices. Let $\mathrm{T}: \mathbb{R}^n \to \mathbb{R}^m$ be a linear transformation. Then the range of T is the set of vectors in $\mathbb{R}^m$ given by

$$\begin{aligned} \text{range T} &= \{\mathbf{v} \in \mathbb{R}^m : \text{for some } \mathbf{u} \text{ in } \mathbb{R}^n, \mathrm{T}(\mathbf{u}) = \mathbf{v}\} \\ &= \{\mathrm{T}(\mathbf{u}) : \text{for all } \mathbf{u} \in \mathbb{R}^n\}. \end{aligned}$$

Some vectors in $\mathbb{R}^n$ may be transformed by T into the zero vector $\mathbf{0}_m$ in $\mathbb{R}^m$. The set of all such vectors is called the *null space* of T, or the *kernel* of T.

Definition The null space, or kernel, of a linear transformation T is the inverse image N of the zero vector in range T, that is, N = null space of T = Kernel T = $\{\mathbf{u}: \mathrm{T}(\mathbf{u}) = \mathbf{0}\}$.

We show that the range and the null space of T are both vector spaces.

Theorem 14.2 Let $\mathrm{T}: \mathbb{R}^n \to \mathbb{R}^m$ be a linear transformation. Then
(i) the range R of T is a subspace of $\mathbb{R}^m$,
(ii) the null space N of T is a subspace of $\mathbb{R}^n$.

Proof (i) $R = \{\mathrm{T}(\mathbf{u})\colon \mathbf{u} \in \mathbb{R}^n\}$ and R is a non-empty subset of $\mathbb{R}^m$ since $\mathbf{0}_m = \mathrm{T}\mathbf{0}_n \in R$. If $\mathbf{u}, \mathbf{v} \in R$, then for some $\mathbf{a}, \mathbf{b} \in \mathbb{R}^n$, $\mathrm{T}(\mathbf{a}) = \mathbf{u}$ and $\mathrm{T}(\mathbf{b}) = \mathbf{v}$. Then for any scalars, h and k, $h\mathbf{u} + k\mathbf{v} = h\mathrm{T}(\mathbf{a}) + k\mathrm{T}(\mathbf{b}) = \mathrm{T}(h\mathbf{a} + k\mathbf{b}) \in R$. Hence R is a subspace of $\mathbb{R}^m$.
(ii) $\mathrm{T}\mathbf{0}_n = \mathbf{0}_m$ and so $\mathbf{0}_n \in N = \{\mathbf{a} \in \mathbb{R}^n\colon \mathrm{T}\mathbf{a} = \mathbf{0}_m\}$, so N is a non-empty subset of $\mathbb{R}^n$. Let $\mathbf{a}$, $\mathbf{b} \in N$ and let h, k be any scalars. Then

$$\mathrm{T}(h\mathbf{a} + k\mathbf{b}) = h(\mathrm{T}\mathbf{a}) + k(\mathrm{T}\mathbf{b}) = h\mathbf{0}_m + k\mathbf{0}_m = \mathbf{0}_m,$$

so that $h\mathbf{a} + k\mathbf{b} \in N$. Therefore N is a subspace of $\mathbb{R}^n$, the domain of T.

EXAMPLE 3 *The linear transformation* $\mathrm{T}\colon \mathbb{R}^3 \to \mathbb{R}^2$ *is given by the matrix* $\mathbf{M}$, *where* $\mathbf{M} = \begin{pmatrix} 1 & 0 & 2 \\ -1 & 0 & -2 \end{pmatrix}$. *Find the range and the null space of T.*

For $\begin{pmatrix} x \\ y \\ z \end{pmatrix} \in \mathbb{R}^3$, $\mathrm{T}\begin{pmatrix} x \\ y \\ z \end{pmatrix} = \begin{pmatrix} x + 2z \\ -x - 2z \end{pmatrix}$, and so $\begin{pmatrix} a \\ b \end{pmatrix} \in$ range T if $b = -a$.

The range of T is the set $\left\{\mathbf{u} \in \mathbb{R}^2\colon \mathbf{u} = k\begin{pmatrix} 1 \\ -1 \end{pmatrix}\right\}$, which is the line $x + y = 0$, in $\mathbb{R}^2$.

Let N be the null space of T. Then $\begin{pmatrix} x \\ y \\ z \end{pmatrix} \in N$ if $\mathrm{T}\begin{pmatrix} x \\ y \\ z \end{pmatrix} = \begin{pmatrix} 0 \\ 0 \end{pmatrix}$, that is, if $x + 2z = 0$, or $x = -2z$. Hence $N = \left\{\begin{pmatrix} -2z \\ y \\ z \end{pmatrix}\colon \text{for all } y, z \in \mathbb{R}\right\}$ and this is the plane $x + 2z = 0$, in $\mathbb{R}^3$.

EXERCISE 14.1

1 Show that: (i) $\mathbb{R}^2$ is a vector space over the field $\mathbb{R}$,
(ii) $\mathbb{R}^2$ is a vector space over $\mathbb{Q}$,
(iii) $\mathbb{R}^2$ is not a vector space over $\mathbb{C}$,
(iv) $\mathbb{C}^2$ is a vector space over $\mathbb{R}$,
(v) $\mathbb{C}^2$ is a vector space over $\mathbb{C}$.

2 Prove that every vector space V contains the two (special) subspaces V and $\{\mathbf{0}\}$.

3 Describe the subspace of $\mathbb{R}^2$ consisting of all linear combinations of the vectors $\mathbf{a}$ and $\mathbf{b}$, where:

(i) $\mathbf{a} = \begin{pmatrix} 1 \\ 1 \end{pmatrix}$, $\mathbf{b} = \begin{pmatrix} 1 \\ -1 \end{pmatrix}$, (ii) $\mathbf{a} = \mathbf{b} = \begin{pmatrix} 1 \\ 1 \end{pmatrix}$, (iii) $\mathbf{a} = \mathbf{b} = \begin{pmatrix} 0 \\ 0 \end{pmatrix}$.

4 Show that the set of all 2×2 real matrices forms a vector space over $\mathbb{R}$.

5 Show that the set of all non-singular 2×2 real matrices is not a vector spacc.

6 Find the range and the null space of the linear transformation given by the matrix: (i) $\begin{pmatrix} 2 & 3 \\ 3 & 5 \end{pmatrix}$, (ii) $\begin{pmatrix} 2 & 3 \\ -4 & -6 \end{pmatrix}$, (iii) $\begin{pmatrix} 1 & 0 & 3 \\ 2 & 1 & 4 \end{pmatrix}$,

(iv) $\begin{pmatrix} 2 & 1 & 3 \\ 0 & 1 & 0 \\ 1 & 0 & 3 \end{pmatrix}$, (v) $\begin{pmatrix} 1 & 0 & 2 \\ 1 & 2 & -3 \\ 0 & 2 & -5 \end{pmatrix}$, (vi) $\begin{pmatrix} 1 & -1 & 0 & -2 \\ 3 & 2 & 5 & -1 \\ 2 & 3 & 5 & 1 \end{pmatrix}$.

14.2 Bases

Linear dependence and spanning sets

We recall the definitions of §1.3 concerning a non-empty, finite set S of vectors. The set of all linear combinations of the vectors in S is a vector space V and we say that

$$S \text{ is a spanning set of } V \quad \text{or} \quad S \text{ spans } V$$

$$\text{or} \quad V \text{ is the span of } S \qquad \text{or} \quad V = \text{span}\,(S).$$

S is linearly independent if the only linear combination of vectors in S equal to the zero vector is that with all zero coefficients. S is linearly dependent if some linear combination of vectors in S, with not all zero coefficients, is the zero vector.

We shall prove that the 'largest' linearly independent subset S of a vector space V is also the 'smallest' spanning set of V. Such a set will be a basis of V, see §1.3, and we shall show that the number of elements in any basis of V is an invariant, called the dimension of V. This work involves a long list of theorems and various counting arguments and it may be helpful to the reader to first read the summary at the end of the section.

Minimal and maximal sets

A set S with a given property P such that the set obtained by removing any one vector from S no longer has the property P is called *minimal* with respect to P. A maximal set is defined similarly.

Definition A set S is *minimal* with the property P if S has the property P but no proper subset of S has the property P.

A set S is *maximal* with the property P if S has the property P but the set obtained by adding one vector to S no longer has the property P.

Theorem 14.3 Let S be a spanning set of the vector space V. Then S contains some subset T which is a minimal spanning set of V.

Proof Either S is a minimal spanning set of V or S contains a vector $\mathbf{a}_1$ such that $S\backslash\{\mathbf{a}_1\}$ spans V. In this case let $S_1 = S\backslash\{\mathbf{a}_1\}$.
Repetition of this argument will produce a sequence of spanning sets of V, each containing one less vector than its predecessor, until a minimal spanning set T is obtained.

Theorem 14.4 In a vector space V, let $S = \{\mathbf{a}_1, \mathbf{a}_2, \ldots, \mathbf{a}_k\}$, with $\mathbf{a}_i \in V$ and $\mathbf{a}_i \neq 0$ for $1 \leqslant i \leqslant k$. Then S is linearly dependent if and only if one vector in S is a linear combination of its predecessors.

Note: In this theorem, S is regarded as an ordered set, the vectors being ordered by their suffices.

Proof Assume first that, for some j, $\mathbf{a}_j$ is a linear combination of its predecessors, say $\mathbf{a}_j = x_1\mathbf{a}_1 + x_2\mathbf{a}_2 + \ldots + x_{j-1}\mathbf{a}_{j-1}$. Put $x_j = -1$ and $x_i = 0$ for $j+1 \leqslant i \leqslant k$, then $\sum_{i=1}^{j} x_i\mathbf{a}_i = \mathbf{0}$, $x_j \neq 0$, so S is linearly dependent.

Conversely, assume that S is linearly dependent. Then $\sum_{i=1}^{k} x_i\mathbf{a}_i = \mathbf{0}$, with not all x_i equal to zero. Let j be the largest integer such that $x_j \neq 0$, so that $\sum_{i=1}^{j} x_i\mathbf{a}_i = \mathbf{0}$, $x_j \neq 0$. Then $j > 1$, since, if $j = 1$, $x_1\mathbf{a}_1 = \mathbf{0}$, $x_1 \neq 0$, so $\mathbf{a}_1 = \mathbf{0}$, which is false. Hence $\mathbf{a}_j = -\sum_{i=1}^{j-1} \frac{x_i}{x_j}\mathbf{a}_i$, as was required.

Note: The fact that a vector $\mathbf{a}$ is a linear combination of the vectors in some set S, that is that $\mathbf{a} \in \text{span}(S)$, may be described by saying that $\mathbf{a}$ is linearly dependent on the vectors in S, or that $\mathbf{a}$ is linearly dependent on S. This must not be confused with the statement that a set of vectors is linearly dependent. Clearly, if $\mathbf{a}$ is linearly dependent on S then $S \cup \{\mathbf{a}\}$ is linearly dependent. The converse is not true as may be seen by choosing S to be any linearly dependent set. Then $S \cup \{\mathbf{a}\}$ is linearly dependent but $\mathbf{a}$ need not lie in span (S).

Theorem 14.5 Let S be a minimal spanning set of the vector space V. Then S is linearly independent.

Proof Suppose that S is linearly dependent. Then one vector, say $\mathbf{a}$, in S is a linear combination of its predecessors. This means that $S\backslash\{\mathbf{a}\}$ spans V and then S is not a minimal spanning set. Therefore, if S is a minimal spanning set of V, it must be linearly independent.

We remind the reader of the definition of a basis, §1.3.

Definition A basis of a vector space V is a linearly independent spanning set of V.

Theorem 14.6 The following are equivalent statements:

(i) S is a basis of the vector space V,
(ii) S is a minimal spanning set of V,
(iii) S is a maximal linearly independent set in V.

Proof We prove the equivalence cyclically by showing that (i) → (ii) → (iii) → (i).

(i) → (ii). Let S be a basis of V. Then S spans V. Suppose that, $\mathbf{a} \in S$ and $S\backslash\{\mathbf{a}\}$ spans V. Then $\mathbf{a}$ can be written as a linear combination of the vectors in $S\backslash\{\mathbf{a}\}$ so that S is linearly dependent. Since this is false, S is a minimal spanning set of V.

(ii) → (iii). Let S be a minimal spanning set of V. Then, by theorem 14.5, S is linearly independent. Let $\mathbf{a} \in V$ and let $T = S \cup \{\mathbf{a}\}$. Then $\mathbf{a}$ can be

written as a linear combination of the vectors in S and so T is linearly dependent. Therefore S is a maximal linearly independent set.
(iii) → (i). Let S be a maximal linearly independent set in V. Let $\mathbf{a} \in V$, then $S \cup \{\mathbf{a}\}$ is linearly dependent. By theorem 14.4 one vector, say $\mathbf{b}$, in $S \cup \{\mathbf{a}\}$ is a linear combination of its predecessors. Since S is linearly independent, $\mathbf{b} \notin S$ and hence $\mathbf{a} \in \operatorname{span}(S)$. Therefore S spans V and is a basis of V.

Any spanning set of a vector space V must contain a subset which is a minimal spanning set (a basis). Also any linearly independent set of vectors in V must be contained in a basis, that is, must be able to be extended to a basis of V. These results are illustrated in Example 1 and then proved in general in Theorems 14.7 and 14.8.

EXAMPLE 1 *The four vectors* $\mathbf{a}$, $\mathbf{b}$, $\mathbf{c}$, $\mathbf{d}$ *in* $\mathbb{R}^3$ *are given by*

$$\mathbf{a} = \begin{pmatrix} 1 \\ -1 \\ 1 \end{pmatrix}, \mathbf{b} = \begin{pmatrix} 2 \\ 0 \\ 3 \end{pmatrix}, \mathbf{c} = \begin{pmatrix} 0 \\ 4 \\ 2 \end{pmatrix}, \mathbf{d} = \begin{pmatrix} -1 \\ 2 \\ 3 \end{pmatrix}.$$

Prove that $\{\mathbf{a}, \mathbf{b}, \mathbf{c}, \mathbf{d}\}$ *spans* $\mathbb{R}^3$ *and find a basis of* $\mathbb{R}^3$ *from the vectors of this set.*

The procedure is to test for linear dependence the sets $\{\mathbf{a}\}$, $\{\mathbf{a}, \mathbf{b}\}$, $\{\mathbf{a}, \mathbf{b}, \mathbf{c}\}$, $\{\mathbf{a}, \mathbf{b}, \mathbf{c}, \mathbf{d}\}$, in turn. On finding a linearly dependent set the last vector is omitted since it must lie in the span of the (linearly independent) set consisting of the previous vectors which have not been omitted so far. In order to simplify the calculation, try to solve the equation $w\mathbf{a} + x\mathbf{b} + y\mathbf{c} + z\mathbf{d} = \mathbf{0}$ and, in turn, take $x = 0 = y = z$, $y = 0 = z$, $z = 0$. This can be done by performing row operations on the matrix $\mathbf{M}$, whose columns are the column vectors $\mathbf{a}$, $\mathbf{b}$, $\mathbf{c}$, $\mathbf{d}$. Then the equation to be solved is $\mathbf{M}\mathbf{r} = \mathbf{0}$, where $\mathbf{r} = \begin{pmatrix} w \\ x \\ y \\ z \end{pmatrix}$.

$$\mathbf{M} = \begin{pmatrix} 1 & 2 & 0 & -1 \\ -1 & 0 & 4 & 2 \\ 1 & 3 & 2 & 3 \end{pmatrix}, \quad \text{form } \mathbf{M}_1 \text{ by } \mathbf{r}_2 \leftarrow \mathbf{r}_2 + \mathbf{r}_1,\ \mathbf{r}_3 \leftarrow \mathbf{r}_3 - \mathbf{r}_1,$$

$$\mathbf{M}_1 = \begin{pmatrix} 1 & 2 & 0 & -1 \\ 0 & 2 & 4 & 1 \\ 0 & 1 & 2 & 4 \end{pmatrix}, \quad \text{form } \mathbf{M}_2 \text{ by } \mathbf{r}_3 \leftarrow 2\mathbf{r}_3 - \mathbf{r}_2,$$

$$\mathbf{M}_2 = \begin{pmatrix} 1 & 2 & 0 & -1 \\ 0 & 2 & 4 & 1 \\ 0 & 0 & 0 & 7 \end{pmatrix}.$$

Now, if $x = 0 = y = z$, then $w = 0$, by row 1, and so $\{\mathbf{a}\}$ is linearly independent. (This step could be omitted).
If $y = 0 = z$, then $x = 0$, by row 2, and so $\{\mathbf{a}, \mathbf{b}\}$ is linearly independent.
If $z = 0$, then, by row 3, $\{\mathbf{a}, \mathbf{b}, \mathbf{c}\}$ is linearly dependent and, by rows 1 and 2, $w + 2x = 0$, $2x + 4y = 0$, so that $x = -2y$ and $w = -2x = 4y$ and then $4\mathbf{a} - 2\mathbf{b} + \mathbf{c} = \mathbf{0}$. Therefore $\mathbf{c} = -2\mathbf{b} - 4\mathbf{a}$ and $\mathbf{c}$ can be omitted from the spanning set.

This leaves the spanning set $\{\mathbf{a}, \mathbf{b}, \mathbf{d}\}$ which is linearly independent by consideration of the three rows (omitting the third column), as follows.
If $w\mathbf{a}+x\mathbf{b}+z\mathbf{d}=0$ then $z=0$, by row 3, then $y=0$, by row 2, and finally $x=0$, by row 1.

Let $\mathbf{N}$ be the matrix $\mathbf{M}$ with the third column omitted, $\mathbf{N}=\begin{pmatrix}1 & 2 & -1\\ 0 & 2 & 1\\ 0 & 0 & 7\end{pmatrix}$.

Then $\mathbf{N}$ can be used to demonstrate that $\{\mathbf{a}, \mathbf{b}, \mathbf{d}\}$ spans $\mathbb{R}^3$, because $\mathbf{N}\begin{pmatrix}x\\ y\\ z\end{pmatrix}=\begin{pmatrix}p\\ q\\ r\end{pmatrix}$ if $7z=r$, $2y+z=q$, $x+2y-z=p$, that is, $z=r/7$, $y=(7q-r)/14$, $x=(7p-7q+2r)/7$. Hence $\{\mathbf{a}, \mathbf{b}, \mathbf{d}\}$ is a basis of $\mathbb{R}^3$.

EXAMPLE 2 *Find a basis of* $\mathbb{R}^3$ *containing the vectors* $\begin{pmatrix}1\\ 1\\ 1\end{pmatrix}$ *and* $\begin{pmatrix}1\\ -1\\ -1\end{pmatrix}$.

Use the standard basis $\left\{\begin{pmatrix}1\\ 0\\ 0\end{pmatrix}, \begin{pmatrix}0\\ 1\\ 0\end{pmatrix}, \begin{pmatrix}0\\ 0\\ 1\end{pmatrix}\right\}$ of $\mathbb{R}^3$ and the method of Example 3 on the spanning set $\left\{\begin{pmatrix}1\\ 1\\ 1\end{pmatrix}, \begin{pmatrix}1\\ -1\\ -1\end{pmatrix}, \begin{pmatrix}1\\ 0\\ 0\end{pmatrix}, \begin{pmatrix}0\\ 1\\ 0\end{pmatrix}, \begin{pmatrix}0\\ 0\\ 1\end{pmatrix}\right\}$.

Let $\mathbf{M}=\begin{pmatrix}1 & 1 & 1 & 0 & 0\\ 1 & -1 & 0 & 1 & 0\\ 1 & -1 & 0 & 0 & 1\end{pmatrix}$, form $\mathbf{M}_1$ by $\mathbf{r}_2 \leftarrow \mathbf{r}_2-\mathbf{r}_1$, $\mathbf{r}_3 \leftarrow \mathbf{r}_3-\mathbf{r}_1$,

$\mathbf{M}_1=\begin{pmatrix}1 & 1 & 1 & 0 & 0\\ 0 & -2 & -1 & 1 & 0\\ 0 & -2 & -1 & 0 & 1\end{pmatrix}$, form $\mathbf{M}_2$ by $\mathbf{r}_3 \leftarrow \mathbf{r}_3-\mathbf{r}_2$,

$\mathbf{M}_2=\begin{pmatrix}1 & 1 & 1 & 0 & 0\\ 0 & -2 & -1 & 1 & 0\\ 0 & 0 & 0 & -1 & 1\end{pmatrix}$. This shows that the first two vectors are linearly independent, but the first three are not. So the third vector is omitted. Then the first two and the fourth vector are linearly independent whilst the first, second, fourth and fifth are linearly dependent. We therefore find a basis

$$\left\{\begin{pmatrix}1\\ 1\\ 1\end{pmatrix}, \begin{pmatrix}1\\ -1\\ -1\end{pmatrix}, \begin{pmatrix}0\\ 1\\ 0\end{pmatrix}\right\}.$$

Theorem 14.7 Any spanning set of a vector space contains a subset which is a basis.

Proof Let S span the vector space V. If S is linearly independent then it is a basis and $S \subseteq S$. If S is linearly dependent, then one vector $\mathbf{a}$ in S is a linear combination of its predecessors. Remove this vector from S leaving a spanning set $S\backslash\{\mathbf{a}\}$ and $S\backslash\{\mathbf{a}\} \subset S$. Repeat this argument, using $S\backslash\{\mathbf{a}\}$ instead of S, to obtain a sequence of subsets of S which span V, until a spanning set is obtained which is linearly independent and hence is a basis of V.

Theorem 14.8 A linearly independent set of vectors in a vector space V can be extended to a basis of V.

Proof Let T be linearly independent and let S be any spanning set of V. Then $T \cup S$ is linearly dependent and so some of its vectors are linearly dependent of their predecessors. Remove one such vector at a time leaving a spanning set containing T as a subset. No vector in T will be removed since T is linearly independent. Continue removing vectors, which are linearly dependent upon their (current) predecessors, until a linearly independent spanning set is obtained. This set is a basis which contains T as a subset.

Theorem 14.9 Let T be a linearly independent set and let S be a spanning set of a vector space V. Then

$$\text{if } S = \{\mathbf{a}_1, \mathbf{a}_2, \ldots, \mathbf{a}_m\} \text{ and } T = \{\mathbf{b}_1, \mathbf{b}_2, \ldots, \mathbf{b}_r\}, \text{ then } r \leqslant m.$$

Proof Since T is linearly independent, any non-empty subset of T is also linearly independent. Since S spans V, $\{\mathbf{b}_1, \mathbf{a}_1, \mathbf{a}_2, \ldots, \mathbf{a}_m\}$ is linearly dependent and, since $\{\mathbf{b}_1\}$ is linearly independent, one of the vectors, $\mathbf{a}_i$, in this set is linearly dependent on its predecessors. Re-label the vectors in S so that this vector $\mathbf{a}_i$ becomes $\mathbf{a}_1$. Then this vector may be removed leaving a spanning set

$$\{\mathbf{b}_1, \mathbf{a}_2, \mathbf{a}_3, \ldots, \mathbf{a}_m\}.$$

Repeat this argument. $\{\mathbf{b}_1, \mathbf{b}_2, \mathbf{a}_2, \mathbf{a}_3 \ldots, \mathbf{a}_m\}$ is linearly dependent and $\{\mathbf{b}_1, \mathbf{b}_2\}$ is linearly independent, so that some vector $\mathbf{a}_i$, where $2 \leqslant i \leqslant m$, is linearly dependent on its predecessors in the set $\{\mathbf{b}_1, \mathbf{b}_2, \mathbf{a}_2, \mathbf{a}_3, \ldots, \mathbf{a}_m\}$. Re-label the vectors so that this vector is $\mathbf{a}_2$, remove it from the set leaving a spanning set $\{\mathbf{b}_1, \mathbf{b}_2, \mathbf{a}_3, \ldots, \mathbf{a}_m\}$. The process can be repeated $(r-2)$ times because, since T is linearly independent, no proper subset of T can span V and so the elements of S cannot all be used up before the $(r-2)$nd repetition. This leaves a spanning set

$$\{\mathbf{b}_1, \mathbf{b}_2, \ldots, \mathbf{b}_r, \ldots, \mathbf{a}_m\}, \text{ with } r \leqslant m.$$

Theorem 14.10 Any two bases of a vector space V contain the same number of elements.

Proof Let S and T be bases of V, then use Theorem 14.9. S spans V and T is linearly independent so $\mathrm{n}(T) \leqslant \mathrm{n}(S)$. T spans V and S is linearly independent so $\mathrm{n}(S) \leqslant \mathrm{n}(T)$. Hence $\mathrm{n}(S) = \mathrm{n}(T)$.

Theorem 14.10 confirms what may have been deduced from all the previous examples in $\mathbb{R}^2$ and $\mathbb{R}^3$, namely that any basis of $\mathbb{R}^2$ contains 2 vectors and any basis of $\mathbb{R}^3$ contains 3 vectors. This is described by saying that the spaces have dimensions 2 and 3 respectively.

Definition The dimension of a vector space (with a finite spanning set) is the number of vectors in any basis.

EXAMPLE 3 *Prove that, in* $\mathbb{R}^3$,
(i) a linearly independent set of 3 *vectors is a basis,*
(ii) a spanning set of 3 *vectors is a basis.*

Let $S = \{\mathbf{a}_1, \mathbf{a}_2, \mathbf{a}_3\}$. (i) If S is linearly independent, then S can be extended to a basis which will contain 3 vectors and is therefore equal to S. (ii) If S spans $\mathbb{R}^3$, then S contains a subset (of 3 vectors) which is a basis and so this basis is S.

EXAMPLE 4 *The linear transformation* T *of* $\mathbb{R}^3$ *into* $\mathbb{R}^3$ *is given by the matrix* **M**, *so that, for any vector* $\mathbf{r} \in \mathbb{R}^3$, $\mathrm{T}(\mathbf{r}) = \mathbf{M}\mathbf{r}$, *where* $\mathbf{M} = \begin{pmatrix} 1 & 3 & -5 \\ 0 & -1 & 2 \\ -1 & 4 & -9 \end{pmatrix}$.
Find a basis for (i) range T, *(ii) the null space of* T.

The columns of **M** are the images under T of the basis vectors

$$\begin{pmatrix} 1 \\ 0 \\ 0 \end{pmatrix}, \begin{pmatrix} 0 \\ 1 \\ 0 \end{pmatrix}, \begin{pmatrix} 0 \\ 0 \\ 1 \end{pmatrix} \text{ and } \mathrm{T}\begin{pmatrix} x \\ y \\ z \end{pmatrix} = x\mathrm{T}\begin{pmatrix} 1 \\ 0 \\ 0 \end{pmatrix} + y\mathrm{T}\begin{pmatrix} 0 \\ 1 \\ 0 \end{pmatrix} + z\mathrm{T}\begin{pmatrix} 0 \\ 0 \\ 1 \end{pmatrix}.$$

Therefore the range of T is spanned by the columns of **M** and we look for a basis which is a subset of this spanning set. We use column operations on **M** in order to obtain a lower triangular matrix whose non-zero columns give the required basis vectors

$$\mathbf{M} = \begin{pmatrix} 1 & 3 & -5 \\ 0 & -1 & 2 \\ -1 & 4 & -9 \end{pmatrix}, \text{ form } \mathbf{M}_1 \text{ by the operations: } \begin{cases} \mathbf{c}_2 \leftarrow \mathbf{c}_2 - 3\mathbf{c}_1, \\ \mathbf{c}_3 \leftarrow \mathbf{c}_3 + 5\mathbf{c}_1, \end{cases}$$

$$\mathbf{M}_1 = \begin{pmatrix} 1 & 0 & 0 \\ 0 & -1 & 2 \\ -1 & 7 & -14 \end{pmatrix}, \text{ form } \mathbf{M}_2 \text{ by the operation: } \mathbf{c}_3 \leftarrow \mathbf{c}_3 + 2\mathbf{c}_2,$$

$\mathbf{M}_2 = \begin{pmatrix} 1 & 0 & 0 \\ 0 & -1 & 0 \\ -1 & 7 & 0 \end{pmatrix}$. This shows that the 3 columns of **M** are linearly dependent, but the first two are linearly independent. In fact, $\mathbf{c}_3 + 5\mathbf{c}_1 + 2(\mathbf{c}_2 - 3\mathbf{c}_1) = \mathbf{c}_3 - \mathbf{c}_1 + 2\mathbf{c}_2 = \mathbf{0}$. A basis of the range of T is

$$\left\{\begin{pmatrix} 1 \\ 0 \\ -1 \end{pmatrix}, \begin{pmatrix} 3 \\ -1 \\ 4 \end{pmatrix}\right\}, \text{ and so is } \left\{\begin{pmatrix} \mathbf{1} \\ \mathbf{0} \\ \mathbf{-1} \end{pmatrix}, \begin{pmatrix} \mathbf{0} \\ \mathbf{-1} \\ \mathbf{7} \end{pmatrix}\right\}.$$

The null space of T is found by solving the equation $\mathbf{M}\mathbf{r} = \mathbf{0}$, so we use row operations to reduce **M** to an upper triangular matrix.

$$\mathbf{M} = \begin{pmatrix} 1 & 3 & -5 \\ 0 & -1 & 2 \\ -1 & 4 & -9 \end{pmatrix}, \text{ form } \mathbf{M}_3 \text{ by the operations: } \mathbf{r}_3 \leftarrow \mathbf{r}_3 + \mathbf{r}_1,$$

$$\mathbf{M}_3 = \begin{pmatrix} 1 & 3 & -5 \\ 0 & -1 & 2 \\ 0 & 7 & -14 \end{pmatrix}, \text{ form } \mathbf{M}_4 \text{ by the operation: } \mathbf{r}_3 \leftarrow \mathbf{r}_3 + 7\mathbf{r}_2,$$

$\mathbf{M}_4 = \begin{pmatrix} 1 & 3 & -5 \\ 0 & -1 & 2 \\ 0 & 0 & 0 \end{pmatrix}$. This shows that $\mathbf{M}\begin{pmatrix} x \\ y \\ z \end{pmatrix} = \mathbf{0}$ if $y = 2z$ and $x = 5z - 3y = -z$. The null space of T therefore has dimension 1 and has a basis $\left\{\begin{pmatrix} -1 \\ 2 \\ 1 \end{pmatrix}\right\}$.

Summary In the vector space V:
Every spanning set contains a minimal spanning set.
Every L.I. set is contained in a maximal L.I. set.
An ordered set S is linearly dependent $\Leftrightarrow$ one vector in S is a linear combination of its predecessors.
A minimal spanning set is linearly independent.
A maximal L.I. set is a spanning set.
A basis is a minimal spanning set and is a maximal L.I. set.
Any spanning set contains a basis as a subset.
Any L.I. set can be extended to a basis.
Let L be any L.I. set and let S be any spanning set, then $n(L) \leqslant n(S)$.
Let B be a basis, then $n(B)$ is an invariant and is called the dimension of V.

EXERCISE 14.2

1 Prove that a non-empty subset of a linearly independent set of vectors is linearly independent.

2 Prove that a set of vectors, which contains, as a subset, a spanning set of a vector space V, is a spanning set of V.

3 Prove that the set of all solutions of the differential equation $\dfrac{d^2y}{dx^2} + \omega^2 y = 0$ is a vector space of dimension 2 and that $\{\cos \omega x, \sin \omega x\}$ is a basis of this solution space.

4 Prove that any set of two dimensional vectors which contains three or more vectors is linearly dependent.

5 Prove that any set of three dimensional vectors which contains four or more vectors is linearly dependent.

6 V is the vector space of all triples (x, y, z), such that $x + 2y = 3z$. Find a basis for V and state the dimension of V.

7 The vector space V is the set of all solutions of the equations $x + 2y + 3z = 0$ and $2x - y + z = 0$. Find a basis for V.

8 The set of all real cubic polynomials in x forms a vector space V. Show that $\{1, x, x^2, x^3\}$ is a basis of V. Find a basis of the set of all polynomials in V which: (i) vanish at $x = 0$, (ii) vanish at $x = 2$, (iii) have a factor $(x^2 - 1)$.

9 Given that U and W are subspaces of a vector space V, define $U + W$ by $\mathbf{v} \in U + W$ if there exist vectors $\mathbf{u} \in U$ and $\mathbf{w} \in W$ such that $\mathbf{v} = \mathbf{u} + \mathbf{w}$. Prove that $U + W$ and $U \cap W$ are both subspaces of V.

10 The linear transformation T of $\mathbb{R}^3$ into $\mathbb{R}^3$ transforms the unit vectors $\begin{pmatrix} 1 \\ 0 \\ 0 \end{pmatrix}, \begin{pmatrix} 0 \\ 1 \\ 0 \end{pmatrix}, \begin{pmatrix} 0 \\ 0 \\ 1 \end{pmatrix}$ into $\begin{pmatrix} 2 \\ 1 \\ 0 \end{pmatrix}, \begin{pmatrix} 1 \\ 2 \\ 1 \end{pmatrix}, \begin{pmatrix} 4 \\ -1 \\ -2 \end{pmatrix}$, respectively. Write down the matrix $\mathbf{M}$ such that, for $\mathbf{v} \in \mathbb{R}^3$, $\mathrm{T}(\mathbf{v}) = \mathbf{M}\mathbf{v}$. Find a basis for (i) the range of T, (ii) the null space of T.

11 For each of the following sets, either extend it to a basis of $\mathbb{R}^3$ or give a reason why it cannot be extended to a basis:
(i) $\{\mathbf{i}, \mathbf{i}+\mathbf{j}\}$, (ii) $\{\mathbf{i}+\mathbf{j}, \mathbf{i}-\mathbf{k}\}$, (iii) $\{\mathbf{i}+\mathbf{j}, \mathbf{j}+\mathbf{k}, \mathbf{i}-\mathbf{k}\}$,
(iv) $\{\mathbf{i}+\mathbf{j}+\mathbf{k}, \mathbf{i}-\mathbf{j}+\mathbf{k}\}$.

12 In the space V of all real polynomials in x of degree less than or equal to 3, state the dimension of V and write down a basis for V. Let U and W be subspaces of V, defined by:

$$\mathrm{p}(x) \in U \Leftrightarrow \mathrm{p}(2) = 0, \quad \mathrm{q}(x) \in W \Leftrightarrow \mathrm{q}'(1) = 0.$$

Find bases for the subspaces: (i) U, (ii) W, (iii) $U \cap W$.
Prove that the set $T = \{\mathbf{v}: \mathbf{v} = \mathbf{u} + \mathbf{w}$, for all $\mathbf{u}$ in U and all $\mathbf{w}$ in $W\}$, which is denoted by $U + W$, is a subspace of V and find a basis for T. Show that $U \cup W$ is not a vector space.

13 Show that the vectors $\mathbf{e}_1 = (3\ 7\ 5\ 1)$ and $\mathbf{e}_2 = (-1\ \ 3\ -3\ \ 1)$ generate the same space as $\mathbf{f}_1 = (0\ \ 4\ -1\ \ 1)$ and $\mathbf{f}_2 = (5\ \ 1\ \ 11\ -1)$. Determine whether $\mathbf{g} = (5\ 5\ 10\ 2)$ is a vector in this space. *(JMB)*

14 Show that, if $\mathbf{a} = (3\ \ 1\ -2)$, $\mathbf{b} = (-2\ -1\ \ 1)$, $\mathbf{c} = (1\ 5\ 5)$, $\mathbf{d} = (1\ -1\ -2)$, then $\{\mathbf{a}, \mathbf{b}, \mathbf{c}\}$ forms a basis for $\mathbb{R}^3$, and $\{\mathbf{a}, \mathbf{b}, \mathbf{d}\}$ does not. Express the vector $(3\ 5\ 3)$ in the form $\lambda\mathbf{a} + \mu\mathbf{b} + \eta\mathbf{c}$, where λ, μ, η are constants to be determined. Show that, if S is the subspace spanned by $\{\mathbf{a}, \mathbf{b}, \mathbf{d}\}$, then the vector $(-1\ 1\ 1) \notin S$. *(L)*

15 Let $\mathbf{A} = \begin{pmatrix} 1 & 0 & 1 & 2 \\ 0 & 1 & 3 & -1 \\ 1 & -1 & -1 & 4 \end{pmatrix}$. V and W are subsets of $\mathbb{R}^4$ defined by

$V = \{\mathbf{X} \in \mathbb{R}^4: \mathbf{AX} = \mathbf{0}\}$, $W = \left\{\mathbf{X} \in \mathbb{R}^4: \mathbf{AX} = \begin{pmatrix} 4 \\ 3 \\ 3 \end{pmatrix}\right\}$. Prove that V is a subspace of $\mathbb{R}^4$ and show that W is not a subspace of $\mathbb{R}^4$. Find a basis for V, and hence, or otherwise, solve the equation $\mathbf{AX} = \begin{pmatrix} 4 \\ 3 \\ 3 \end{pmatrix}$. *(C)*

16 A linear transformation $\mathrm{L}: \mathbb{R}^3 \to \mathbb{R}^3$ is defined by $\mathrm{L}\begin{pmatrix} x \\ y \\ z \end{pmatrix} = \begin{pmatrix} x-y \\ y+z \\ x+z \end{pmatrix}$. Find bases for the null space of L and for the range of L. Find also the images under L of

(i) the line with equation $\mathbf{r} = \begin{pmatrix} 2 \\ -1 \\ 1 \end{pmatrix} + \lambda \begin{pmatrix} 0 \\ 1 \\ 1 \end{pmatrix}$,

(ii) the line with equation $\mathbf{r} = \begin{pmatrix} 2 \\ -1 \\ 1 \end{pmatrix} + \lambda \begin{pmatrix} 1 \\ 1 \\ -1 \end{pmatrix}$,

(iii) the plane with equation $\mathbf{r}.\begin{pmatrix} 2 \\ -5 \\ -3 \end{pmatrix} = 0$. *(C)*

17 (a) Show that the matrix $\mathbf{X}$, where $\mathbf{X} = \begin{pmatrix} -2 & 3 \\ 5 & 4 \end{pmatrix}$, belongs to the linear space (over $\mathbb{R}$) spanned by the matrices $\mathbf{A}$, $\mathbf{B}$, $\mathbf{C}$, where $\mathbf{A} = \begin{pmatrix} 1 & 0 \\ 2 & 1 \end{pmatrix}$, $\mathbf{B} = \begin{pmatrix} 1 & -1 \\ 0 & 0 \end{pmatrix}$, $\mathbf{C} = -\begin{pmatrix} -1 & 0 \\ 1 & 2 \end{pmatrix}$. Find the dimension of this space.
(b) Find a relationship between x, y, z equivalent to the statement that $(x \;\; y \;\; z)$ belongs to the subspace of $\mathbb{R}^3$ spanned by the vectors $\mathbf{a} = (1 \quad 2 \; -1)$, $\mathbf{b} = (3 \; -1 \quad 3)$, $\mathbf{c} = (1 \; -5 \quad 5)$. Find the dimension of this subspace.
(c) Determine the value of a given that the vectors $\begin{pmatrix} 1 \\ 2 \\ 3 \end{pmatrix}$, $\begin{pmatrix} -1 \\ 3 \\ 0 \end{pmatrix}$, $\begin{pmatrix} a \\ 1 \\ 2 \end{pmatrix}$ do not form a spanning set for $\mathbb{R}^3$. (*C*)

18 The linear transformation σ: $\mathbb{R}^3 \to \mathbb{R}^3$ is represented by the matrix $\begin{pmatrix} 2 & 1 & 3 \\ 0 & 3 & -3 \\ -1 & 2 & -4 \end{pmatrix}$ with respect to the standard basis of $\mathbb{R}^3$. (i) Show that the range of σ has dimension 2. (ii) Show that all points of the line $\mathbf{r} = \begin{pmatrix} 3 \\ 0 \\ -1 \end{pmatrix} + \lambda \begin{pmatrix} -2 \\ 1 \\ 1 \end{pmatrix}$ are mapped to the same point by σ, and find the position vector of this point. (iii) Find the subset of $\mathbb{R}^3$ whose image under σ is the point with position vector $\begin{pmatrix} 5 \\ -3 \\ -5 \end{pmatrix}$. (*C*)

14.3 Orthogonality in $\mathbb{R}^n$

Orthogonal vectors

In a vector space V, in which a scalar product is defined, two vectors with a zero scalar product are called *orthogonal*, rather than perpendicular, although both words may be used in $\mathbb{R}^2$ and $\mathbb{R}^3$.

The results proved in this chapter concern real vector spaces, $\mathbb{R}^m$, $\mathbb{R}^n$ of any dimension. We shall often work in $\mathbb{R}^3$, when all vector spaces referred to will be subspaces of $\mathbb{R}^3$. The theory generalises to $\mathbb{R}^n$, and we give general proofs when possible, but it is often easier to see what is happening in $\mathbb{R}^3$. The scalar product $\mathbf{a}\,.\,\mathbf{b}$ of the two vectors $\mathbf{a}$ and $\mathbf{b}$ in $\mathbb{R}^3$, where

$$\mathbf{a} = \begin{pmatrix} a_1 \\ a_2 \\ a_3 \end{pmatrix}, \; \mathbf{b} = \begin{pmatrix} b_1 \\ b_2 \\ b_3 \end{pmatrix}, \text{ is given by } \mathbf{a}\,.\,\mathbf{b} = a_1b_1 + a_2b_2 + a_3b_3.$$

If the column vector $\mathbf{a}$ is regarded as a (3×1) matrix its transpose $\mathbf{a}^\mathrm{T}$ is a (1×3) matrix, that is a row vector, and the scalar product $\mathbf{a}\,.\,\mathbf{b}$ can be written as a matrix product

$$\mathbf{a}\,.\,\mathbf{b} = (a_1 \;\; a_2 \;\; a_3)\begin{pmatrix} b_1 \\ b_2 \\ b_3 \end{pmatrix} = \mathbf{a}^\mathrm{T}\mathbf{b}.$$

Similarly, in $\mathbb{R}^n$, $\mathbf{a}\,.\,\mathbf{b} = \mathbf{a}^{\mathrm{T}}\mathbf{b}$, the product of a row vector, $\mathbf{a}$ $(1 \times n)$ matrix, and a column vector, $\mathbf{b}$ $(n \times 1)$ matrix.

Definition Let V be a vector space, let $\mathbf{a}, \mathbf{b} \in V$ and let $S \subseteq V$. Then:
(i) $\mathbf{a}$ and $\mathbf{b}$ are orthogonal if $\mathbf{a}\,.\,\mathbf{b} = \mathbf{a}^{\mathrm{T}}\mathbf{b} = 0$,
(ii) S is orthogonal if, for all $\mathbf{a}, \mathbf{b} \in S$, $\mathbf{a}$ and $\mathbf{b}$ are orthogonal.

Theorem 14.11 An orthogonal set of non-zero vectors is linearly independent.

Proof Let S be an orthogonal set, $S = \{\mathbf{a}_1, \mathbf{a}_2, \ldots, \mathbf{a}_k\}$, and suppose that $\sum_{i=1}^{k} x_i\mathbf{a}_i = \mathbf{0}$. Then, for each j, $1 \leqslant j \leqslant k$,

$$x_j a_j^2 = x_j\mathbf{a}_j\,.\,\mathbf{a}_j = \sum_{i=1}^{k} x_i\mathbf{a}_i\,.\,\mathbf{a}_j = 0, \text{ so } x_j = 0,$$

because, for $i \neq j$, $\mathbf{a}_i\,.\,\mathbf{a}_j = 0$. Hence S is linearly independent.

Orthogonal complements

Let U be a subspace of a vector space V. Then the set of all vectors in V which are orthogonal to every vector in U, including the zero vector, is called the *orthogonal complement* of U.

Definition Let U be a subspace of V. Then the orthogonal complement W of U is given by:

$$W = \{\mathbf{v} \in V\text{: for every } \mathbf{u} \in U,\ \mathbf{v}\,.\,\mathbf{u} = 0\}.$$

Theorem 14.12 The orthogonal complement of a subspace U of a vector space V is also a subspace of V.

Proof Let U be a subspace of V and let

$$W = \{\mathbf{v} \in V\text{: for every } \mathbf{u} \in U,\ \mathbf{v}\,.\,\mathbf{u} = 0\}.$$

Then $\mathbf{0} \in W$ since $\mathbf{0}\,.\,\mathbf{u} = 0$, for every $\mathbf{u}$ in U. Suppose that $\mathbf{v}, \mathbf{w} \in W$ and let a, b be scalars. Then for every $\mathbf{u}$ in U,

$$(a\mathbf{v} + b\mathbf{w})\,.\,\mathbf{u} = a\mathbf{v}\,.\,\mathbf{u} + b\mathbf{w}\,.\,\mathbf{u} = a\,.\,0 + b\,.\,0 = 0, \text{ so } a\mathbf{v} + b\mathbf{w} \in W.$$

Hence W is a subspace of V.

A set of vectors which are all orthogonal to one another and, at the same time, are all unit vectors, is called **orthonormal**. The set $\{\mathbf{i}, \mathbf{j}, \mathbf{k}\}$, of Cartesian unit vectors, is an orthonormal set.

Definition An orthonormal set of vectors is a set of mutually orthogonal unit vectors.

Theorem 14.13 Any subspace V of $\mathbb{R}^n$ has an orthonormal basis.

Proof As a basis for an induction, let $\mathbf{e}_1$ be any unit vector in V. Let $S_1 = \{\mathbf{e}_1\}$ and let $V_1 = \text{span}(S_1)$, with an orthonormal basis S_1.

Inductive hypothesis: Suppose that $S_t = \{\mathbf{e}_1, \ldots, \mathbf{e}_t\}$ and that S_t is an orthonormal basis of a subspace V_t of V. If $V_t = V$, then V has an orthonormal basis S_t. If $V_t \neq V$, then let $\mathbf{v}$ be a vector in V and not in V_t. Define $\mathbf{w} = \mathbf{v} - \sum_{i=1}^{t} (\mathbf{e}_i \cdot \mathbf{v})\mathbf{e}_i$. Then, if $1 \leqslant j \leqslant t$,

$\mathbf{w} \cdot \mathbf{e}_j = \mathbf{v} \cdot \mathbf{e}_j - \sum_{i=1}^{t} (\mathbf{e}_i \cdot \mathbf{v})(\mathbf{e}_i \cdot \mathbf{e}_j) = \mathbf{v} \cdot \mathbf{e}_j - \mathbf{e}_j \cdot \mathbf{v} = 0$, so $\mathbf{w}$ is orthogonal to all vectors in S_t and, therefore, to V_t. Let $\mathbf{e}_{t+1} = \mathbf{w}/|\mathbf{w}|$, let $S_{t+1} = S_t \cup \{\mathbf{e}_{t+1}\}$ and let $V_{t+1} = \text{span}(S_{t+1})$. Then S_{t+1} is an orthonormal basis of V_{t+1}. By induction, we obtain a sequence (V_t) of subspaces, of increasing dimension and, since the dimension of V is finite, for some t, $V_t = V$ and so V has an orthonormal basis S_t.

Theorem 14.14 Let U and V be orthogonal complements in $\mathbb{R}^n$, with orthonormal bases S_1 and S_2, respectively and let
$S_1 = \{\mathbf{u}_1, \ldots, \mathbf{u}_p\}$ and $S_2 = \{\mathbf{v}_1, \ldots, \mathbf{v}_q\}$. Then:

(i) $U \cap V = \{\mathbf{0}\}$,
(ii) $S_1 \cup S_2$ is an orthonormal basis of $\mathbb{R}^n$,
(iii) $p + q = n$.

Proof (i) Let $\mathbf{x} \in U \cap V$, then $\mathbf{x} \in U$ and $\mathbf{x} \in V$ so $x^2 = \mathbf{x} \cdot \mathbf{x} = 0$ so $\mathbf{x} = \mathbf{0}$.
(ii) Let $\mathbf{x} \in \mathbb{R}^n$ and define $\mathbf{y} = \sum_{i=1}^{p} (\mathbf{x} \cdot \mathbf{u}_i)\mathbf{u}_i$, then $\mathbf{y} \in U$ and, for all j such that $1 \leqslant j \leqslant p$, $\mathbf{y} \cdot \mathbf{u}_j = \mathbf{x} \cdot \mathbf{u}_j$, as in Theorem 14.13. Also, for all j, such that $1 \leqslant j \leqslant q$, $\mathbf{y} \cdot \mathbf{v}_j = 0$. Now let $\mathbf{z} = \mathbf{x} - \mathbf{y}$, then $\mathbf{z} \cdot \mathbf{u}_i = \mathbf{x} \cdot \mathbf{u}_i - \mathbf{y} \cdot \mathbf{u}_i = 0$ for all i, $1 \leqslant i \leqslant p$. So $\mathbf{z}$ lies in V, the orthogonal complement of U. Thus $\mathbf{x} = \mathbf{y} + \mathbf{z}$, $\mathbf{y} \in U = \text{span}(S_1)$ and $\mathbf{z} \in V = \text{span}(S_2)$, so $S_1 \cup S_2$ spans $\mathbb{R}^n$. But $S_1 \cup S_2$ is orthonormal and, therefore, linearly independent and so it is an orthonormal basis of $\mathbb{R}^n$. (iii) From (ii) $n = p + q$.

Orthogonal matrices

Let $\mathbf{A}$ be a (3×3) real matrix and let $\mathbf{c}_1$, $\mathbf{c}_2$, $\mathbf{c}_3$, be the columns of $\mathbf{A}$, regarded as vectors in $\mathbb{R}^3$. Then their transposes $\mathbf{c}_1^{\mathrm{T}}, \mathbf{c}_2^{\mathrm{T}}, \mathbf{c}_3^{\mathrm{T}}$, are row vectors, namely the rows of $\mathbf{A}^{\mathrm{T}}$. Suppose that the columns of $\mathbf{A}$ form an orthonormal set of vectors. Then, for $i, j = 1, 2, 3$, with $i \neq j$, $\mathbf{c}_i \cdot \mathbf{c}_j = 0$ and $\mathbf{c}_i \cdot \mathbf{c}_i = 1$. Now, for any i, j, $\mathbf{c}_i \cdot \mathbf{c}_j = \mathbf{c}_i^{\mathrm{T}}\mathbf{c}_j$ and this is the (i, j) entry in $\mathbf{A}^{\mathrm{T}}\mathbf{A}$. Hence $\mathbf{A}^{\mathrm{T}}\mathbf{A} = \mathbf{I}_3$ and so $\mathbf{A}^{\mathrm{T}} = \mathbf{A}^{-1}$. But then $\mathbf{A}\mathbf{A}^{\mathrm{T}} = \mathbf{I}_3$ and so, if $\mathbf{r}_1, \mathbf{r}_2, \mathbf{r}_3$, are the row vectors of $\mathbf{A}$, then $\mathbf{r}_i\mathbf{r}_j^{\mathrm{T}} = 0$ if $i \neq j$ and $\mathbf{r}_i\mathbf{r}_i^{\mathrm{T}} = 1$. Therefore the rows of $\mathbf{A}$ are also mutually orthogonal unit vectors. Now suppose, conversely, that the (3×3) matrix $\mathbf{A}$ has an inverse which is equal to its transpose, that is $\mathbf{AA}^T = \mathbf{A}^T\mathbf{A} = \mathbf{I}_3$. Then, by the same argument as before, the columns of **A** are mutually orthogonal unit vectors and so are the rows of **A**. Such a matrix **A** is called orthogonal. Obviously an orthogonal matrix must be non-singular.

Definition The square matrix **A** is orthogonal if $\mathbf{A}^T = \mathbf{A}^{-1}$.

We summarise the results which we have just proved in a theorem.

Theorem 14.15 For an $(n \times n)$ real matrix **A**, the following are equivalent:

(i) **A** is orthogonal,

(ii) $\mathbf{AA}^T = \mathbf{A}^T\mathbf{A} = \mathbf{I}_n$,

(iii) the columns of **A** form a mutually orthogonal set of unit column vectors,

(iv) the rows of **A** form a mutually orthogonal set of unit row vectors.

Orthogonal linear transformations

Some linear transformations are isometries, that is, they preserve the distance between two points representing vectors. This means that they also preserve angle, length, area and volume and hence scalar product.

Let T be a linear transformation of $\mathbb{R}^3$, given by the matrix **A** so that for any vector $\mathbf{u} \in \mathbb{R}^3$, $\mathrm{T}(\mathbf{u}) = \mathbf{Au}$. Consider the effect of T upon a scalar product. Let $\mathbf{u}, \mathbf{v} \in \mathbb{R}^3$, then $\mathrm{T}(\mathbf{u}.\mathbf{v}) = \mathbf{Au}.\mathbf{Av} = (\mathbf{Au})^T\mathbf{Av} = \mathbf{u}^T\mathbf{A}^T\mathbf{Av}$. If **A** is orthogonal, $\mathbf{A}^T\mathbf{A} = \mathbf{I}_3$, so $\mathrm{T}(\mathbf{u}.\mathbf{v}) = \mathbf{u}^T\mathbf{I}_3\mathbf{v} = \mathbf{u}^T\mathbf{v} = \mathbf{u}.\mathbf{v}$ and so T is an isometry.

Conversely, if T is an isometry, then, for all vectors **u** and **v**, $\mathrm{T}(\mathbf{u}.\mathbf{v}) = \mathbf{u}.\mathbf{v}$. Choose $\mathbf{u}_i$ and $\mathbf{v}_j$ to be the unit vectors having all entries zero except for the ith and jth places respectively, where the entry is 1. Then $\mathbf{Au}_i = \mathbf{c}_i$, the ith column of **A** and $\mathbf{Av}_j = \mathbf{c}_j$, the jth column of **A**, so that

$$(\mathbf{Au}_i)^T(\mathbf{Av}_j) = \mathbf{c}_i^T\mathbf{c}_j = (\mathbf{A}^T\mathbf{A})_{ij}, \text{ the } (i, j) \text{ entry of } \mathbf{A}^T\mathbf{A}.$$

If $i \neq j$, then $\mathbf{u}_i.\mathbf{v}_j = 0$ and so $\mathrm{T}(\mathbf{u}_i.\mathbf{v}_j) = (\mathbf{Au}_i)^T(\mathbf{Av}_j) = 0$, whilst, since $\mathbf{u}_i.\mathbf{v}_i = 1$, $\mathrm{T}(\mathbf{u}_i.\mathbf{v}_i) = (\mathbf{Au}_i)^T(\mathbf{Av}_i) = 1$. This means that $\mathbf{A}^T\mathbf{A} = \mathbf{I}_3$ and so **A** is orthogonal. We have therefore proved that a linear transformation T is an isometry if and only if its matrix is orthogonal.

Since an isometry preserves length and angle, a linear transformation which is an isometry transforms a basis of $\mathbb{R}^3$, consisting of a mutually perpendicular set of unit vectors, into a similar basis. Such a transformation is called *orthogonal* and we have proved Theorem 14.16.

Definition A linear transformation which is an isometry is called orthogonal.

Theorem 14.16 Let T be a linear transformation of $\mathbb{R}^n$ with a corresponding matrix **A**. Then the following are equivalent:

(i) T is an isometry,
(ii) T is orthogonal,
(iii) **A** is orthogonal.

EXERCISE 14.3A

1 Prove that the following pairs of vectors are orthogonal:

(i) $\begin{pmatrix}1\\2\\0\end{pmatrix}, \begin{pmatrix}2\\-1\\0\end{pmatrix}$ (ii) $\begin{pmatrix}0\\-1\\1\end{pmatrix}, \begin{pmatrix}0\\1\\1\end{pmatrix}$ (iii) $\begin{pmatrix}1\\1\\1\end{pmatrix}, \begin{pmatrix}1\\0\\-1\end{pmatrix}$

(v) $\begin{pmatrix}1\\2\\3\end{pmatrix}, \begin{pmatrix}-2\\-2\\2\end{pmatrix}$ (vi) $\begin{pmatrix}\cos\theta\\ \sin\theta\\0\end{pmatrix}, \begin{pmatrix}-\sin\theta\\ \cos\theta\\0\end{pmatrix}$.

2 Which of the following pairs of vectors are orthogonal?

(i) $\begin{pmatrix}0\\3\\4\end{pmatrix}, \begin{pmatrix}0\\8\\-6\end{pmatrix}$ (ii) $\begin{pmatrix}1\\-1\\1\end{pmatrix}, \begin{pmatrix}1\\1\\1\end{pmatrix}$ (iii) $\begin{pmatrix}1\\-1\\1\end{pmatrix}, \begin{pmatrix}2\\1\\1\end{pmatrix}$

(iv) $\begin{pmatrix}a\\b\\c\end{pmatrix}, \begin{pmatrix}bc\\ac\\-2ab\end{pmatrix}$ (v) $\begin{pmatrix}\cos\theta\\ \sin\theta\cos\phi\\ \sin\theta\sin\phi\end{pmatrix}, \begin{pmatrix}-\sin\theta\\ \cos\theta\cos\phi\\ \cos\theta\sin\phi\end{pmatrix}$.

3 Prove that the given matrix is orthogonal:

(i) $\begin{pmatrix}1/\sqrt{2} & 1/\sqrt{2} & 0\\ -1/\sqrt{2} & 1/\sqrt{2} & 0\\ 0 & 0 & 1\end{pmatrix}$ (ii) $\begin{pmatrix}3/5 & 0 & -4/5\\ 0 & 1 & 0\\ 4/5 & 0 & 3/5\end{pmatrix}$

(iii) $\begin{pmatrix}1/3 & 2/3 & 2/3\\ 2/3 & 1/3 & -2/3\\ 2/3 & -2/3 & 1/3\end{pmatrix}$.

4 Determine whether or not the matrix is orthogonal:

(i) $\begin{pmatrix}-1 & -1\\ 1 & -1\end{pmatrix}$ (ii) $\begin{pmatrix}-1/\sqrt{2} & -1/\sqrt{2}\\ 1/\sqrt{2} & -1/\sqrt{2}\end{pmatrix}$ (iii) $\begin{pmatrix}\cos\theta & -\sin\theta\\ \sin\theta & \cos\theta\end{pmatrix}$

(iv) $\begin{pmatrix}-2 & 1 & 2\\ -1 & 2 & -2\\ 2 & 2 & 1\end{pmatrix}$ (v) $\begin{pmatrix}3 & -4 & 6\\ 2 & 6 & 3\\ -6 & 3 & 4\end{pmatrix}$ (vi) $\begin{pmatrix}3 & 2 & 22\\ 4 & -3 & 9\\ 6 & 1 & -17\end{pmatrix}$

(vii) $\begin{pmatrix}3/\sqrt{(61)} & 2/\sqrt{(14)} & 22/\sqrt{(854)}\\ 4/\sqrt{(61)} & -3/\sqrt{(14)} & 9/\sqrt{(854)}\\ 6/\sqrt{(61)} & 1/\sqrt{(14)} & -17/\sqrt{(854)}\end{pmatrix}$ (viii) $\begin{pmatrix}3 & -6 & 2\\ 2 & 3 & 6\\ 6 & 2 & -3\end{pmatrix}$.

5 Of the matrices in question **4** which are not orthogonal, determine which can be made orthogonal by multiplying rows and/or columns by certain constants and write down the orthogonal matrices so formed.

Rank

The results on orthogonality are now used to show that a matrix has the same number of linearly independent rows as columns, the number being called the rank of the matrix.

Definition The *column space* C and the *row space* R of an $m \times n$ matrix $\mathbf{A}$ are subspaces of $\mathbb{R}^m$ and $\mathbb{R}^n$ respectively, given by

$$C = \text{span}\{\mathbf{c}_1, \mathbf{c}_2, \ldots, \mathbf{c}_n\}, \; R = \text{span}\{\mathbf{r}_1^T, \mathbf{r}_2^T, \ldots, \mathbf{r}_m^T\},$$

where $\mathbf{c}_j$, $1 \leqslant j \leqslant n$, are the column vectors of $\mathbf{A}$, $\mathbf{c}_j \in \mathbb{R}^m$,
and $\mathbf{r}_i$, $1 \leqslant i \leqslant m$, are the row vectors of $\mathbf{A}$, $\mathbf{r}_i^T \in \mathbb{R}^n$.
The *row rank* of $\mathbf{A}$ is the dimension of R.
The *column rank* of $\mathbf{A}$ is the dimension of C.

Theorem 14.17 Let T be a linear transformation of $\mathbb{R}^n$ to $\mathbb{R}^m$, with an $m \times n$ matrix $\mathbf{A}$. Let C and R be the column and row spaces of $\mathbf{A}$. Then:
(i) the range of T is C,
(ii) the null space (kernel) of T is the orthogonal complement of R.
(iii) the row rank of $\mathbf{A}$ is equal to the column rank of $\mathbf{A}$.

Proof (i) Range(T) $= \{\mathbf{Ax}: \mathbf{x} \in \mathbb{R}^n\} = \text{span}\{\mathbf{Ae}_1, \mathbf{Ae}_2, \ldots, \mathbf{Ae}_n\}$, where $\{\mathbf{e}_1, \mathbf{e}_2, \ldots, \mathbf{e}_n\}$ is the Cartesian basis of $\mathbb{R}^n$. But $\mathbf{Ae}_j$ is the jth column of $\mathbf{A}$ and so range(T) $= C$. (ii) Let N be the null space (or kernel) of T so that

$$\mathbf{x} \in N \Leftrightarrow \mathbf{Ax} = \mathbf{0} \Leftrightarrow \mathbf{r}_i^T . \mathbf{x} = \mathbf{r}_i \mathbf{x} = 0, \text{ for } 1 \leqslant i \leqslant m.$$

Thus $\mathbf{x}$ lies in N if and only if $\mathbf{x}$ lies in the orthogonal complement of the row space R of $\mathbf{A}$, so N is the orthogonal complement of R in $\mathbb{R}^n$. (iii) Using (ii) and theorem 14.14, if the row rank of $\mathbf{A}$ is r, then the dimension of N is $n-r$ and we can choose an orthogonal basis $(\mathbf{f}_1, \mathbf{f}_2, \ldots, \mathbf{f}_n\}$ of $\mathbb{R}^n$, such that $\{\mathbf{f}_1, \mathbf{f}_2, \ldots, \mathbf{f}_r\}$ is a basis of R and $\{\mathbf{f}_{r+1}, \ldots, \mathbf{f}_n\}$ is a basis of N. Then, for all $\mathbf{x} \in \mathbb{R}^n$, $\mathbf{x} = \sum_{i=1}^{n} c_i \mathbf{f}_i$ and

$\mathrm{T}(\mathbf{x}) = \mathbf{Ax} = \mathbf{A} \sum_{i=1}^{n} c_i \mathbf{f}_i = \sum_{i=1}^{n} c_i \mathbf{Af}_i = \sum_{i=1}^{r} c_i \mathbf{Af}_i$, since $\mathbf{f}_i \in N$ for $r+1 \leqslant i \leqslant n$, so the range of T is spanned by $\{\mathbf{Af}_i: 1 \leqslant i \leqslant r\}$. Suppose that $\sum_{i=1}^{r} d_i \mathbf{Af}_i = \mathbf{0}$, then $\mathbf{A} \sum_{i=1}^{r} d_i \mathbf{f}_i = 0$ and $\sum_{i=1}^{r} d_i \mathbf{f}_i \in N$. Therefore, for some d_i, $r+1 \leqslant i \leqslant n$, $\sum_{i=1}^{r} d_i \mathbf{f}_i = \sum_{i=r+1}^{n} -d_i \mathbf{f}_i$, since $\{\mathbf{f}_{r+1}, \ldots, \mathbf{f}_n\}$ is a basis of N. But then $\sum_{i=1}^{n} d_i \mathbf{f}_i = \mathbf{0}$ and, since $\{\mathbf{f}_1, \mathbf{f}_2, \ldots, \mathbf{f}_n\}$ is a basis of $\mathbb{R}^n$, $d_i = 0$, for $1 \leqslant i \leqslant n$, and, in particular, $d_i = 0$, for $1 \leqslant i \leqslant r$. This shows that $\{\mathbf{Af}_i\colon 1 \leqslant i \leqslant r\}$ is linearly independent and therefore a basis of the range of T. Hence the range of T, which is the column space C of **A**, has dimension r, that is, the column rank of **A** is equal to its row rank.

The common dimension of the row space and the column space of a matrix is called its *rank*.

Definition Let T: $\mathbb{R}^n \to \mathbb{R}^m$ be a linear transformation, with matrix **A**. Then the *rank* of T is equal to the *rank* of **A** and is defined as the row rank of **A** (equal to the column rank of **A**).

We are now in a position to prove Theorem 1.8(ii) and (iii), using the row space and the column space of a matrix. We restate parts (ii) and (iii) of Theorem 1.8 as Theorem 14.18, and then prove them using Theorem 14.17 with $m = n$.

Theorem 14.18 Let **A** be an $(n \times n)$ matrix with row space R and column space C. Then

(i) **A** is non-singular $\Leftrightarrow$ the rows of **A** are linearly independent
$\Leftrightarrow$ the columns of **A** are linearly independent.

(ii) **A** is singular $\Leftrightarrow$ the rows of **A** are linearly dependent
$\Leftrightarrow$ the columns of **A** are linearly dependent.

Proof Let T: $\mathbb{R}^n \to \mathbb{R}^n$ be the linear transformation with matrix **A**. Then by Theorem 4.2

A is non-singular $\Leftrightarrow$ T is non-singular $\Leftrightarrow$ T is one-one
$\Leftrightarrow$ the range of T is $\mathbb{R}^n$ and the null space of T is $\{\mathbf{0}\}$.

Using Theorem 14.17, with $n = m$,
the range of T is $\mathbb{R}^n \Leftrightarrow C = \mathbb{R}^n$
$\Leftrightarrow$ the n columns of **A** span $\mathbb{R}^n$
$\Leftrightarrow$ the columns of **A** are linearly independent.

Similarly,
the null space of T is $\{\mathbf{0}\} \Leftrightarrow R = \mathbb{R}^n$
$\Leftrightarrow$ the transposes of the rows of **A** span $\mathbb{R}^n$
$\Leftrightarrow$ the rows of **A** are linearly independent.

This completes the proof of part (i) and also part (ii) since the two parts are logically equivalent.

EXAMPLE 1 *The linear transformation* T *of* $\mathbb{R}^3$ *has matrix* **A**, *given by* $\mathbf{A} = \begin{pmatrix} 1 & -2 & 4 \\ 2 & 1 & 3 \\ 3 & 2 & 4 \end{pmatrix}$. *Show that the range C of* T *is a plane and find its equation.*

Show that C has an orthogonal basis $\{\mathbf{e}, \mathbf{f}\}$ *where* $\mathbf{e} = \begin{pmatrix} -2 \\ 1 \\ 2 \end{pmatrix}$ *and* $\mathbf{f} = \begin{pmatrix} 7 \\ 4 \\ 5 \end{pmatrix}$. *Show* *that the line* $\dfrac{x+5}{2}=\dfrac{y-5}{-1}=\dfrac{z-1}{-1}$ *is mapped by* T *on to a single point P and express the position vector of P in terms of* **e** *and* **f**.

The range C of T is the column space of **A**. To study C, reduce **A** to lower triangular form by column operations.

$$\begin{pmatrix}1&-2&4\\2&1&3\\3&2&4\end{pmatrix}\begin{matrix}\mathbf{c}_2\leftarrow\mathbf{c}_2+2\mathbf{c}_1\\\mathbf{c}_3\leftarrow\mathbf{c}_3-4\mathbf{c}_1\end{matrix}\begin{pmatrix}1&0&0\\2&5&-5\\3&8&-8\end{pmatrix}\ \mathbf{c}_3\leftarrow\mathbf{c}_3+\mathbf{c}_2\ \begin{pmatrix}1&0&0\\2&5&0\\3&8&0\end{pmatrix}\text{ and}$$

$C=\text{span}\left\{\begin{pmatrix}1\\2\\3\end{pmatrix},\begin{pmatrix}0\\5\\8\end{pmatrix}\right\}$. Hence C has equation $\mathbf{r}=\begin{pmatrix}s\\2s+5t\\3s+8t\end{pmatrix}$.

To see if **e** and **f** lie in C, try to match their top rows by using a multiple of $\begin{pmatrix}1\\2\\3\end{pmatrix}$ and then add an appropriate multiple of $\begin{pmatrix}0\\5\\8\end{pmatrix}$.

$$-2\begin{pmatrix}1\\2\\3\end{pmatrix}=\begin{pmatrix}-2\\-4\\-6\end{pmatrix}\quad\text{and}\quad -2\begin{pmatrix}1\\2\\3\end{pmatrix}+\begin{pmatrix}0\\5\\8\end{pmatrix}=\begin{pmatrix}-2\\1\\2\end{pmatrix}=\mathbf{e},$$

$$7\begin{pmatrix}1\\2\\3\end{pmatrix}=\begin{pmatrix}7\\14\\21\end{pmatrix}\quad\text{and}\quad 7\begin{pmatrix}1\\2\\3\end{pmatrix}-2\begin{pmatrix}0\\5\\8\end{pmatrix}=\begin{pmatrix}7\\4\\5\end{pmatrix}=\mathbf{f}.$$

Also $\mathbf{e}.\mathbf{f}=-14+4+10=0$, so **e** and **f** are orthogonal vectors in C and, since C has dimension 2, $\{\mathbf{e},\mathbf{f}\}$ is an orthogonal basis of C. A general point on the line $\dfrac{x+5}{2}=\dfrac{y-5}{-1}=\dfrac{z-1}{-1}$ has position vector $\begin{pmatrix}-5+2t\\5-t\\1-t\end{pmatrix}$ and is mapped by T to the point P with position vector **p**, given by $\mathbf{p}=\begin{pmatrix}1&-2&4\\2&1&3\\3&2&4\end{pmatrix}\begin{pmatrix}-5+2t\\5-t\\1-t\end{pmatrix}=\begin{pmatrix}-11\\-2\\-1\end{pmatrix}$, so the whole line is mapped on to the one point P. Finally, if $\mathbf{p}=a\mathbf{e}+b\mathbf{f}$, then $-2a+7b=-11$, $a+4b=-2$, $2a+5b=-1$, and $a=2$, $b=-1$, $\mathbf{p}=2\mathbf{e}-\mathbf{f}$.

Solution of simultaneous linear equations

The above theory can be used to study the solution of linear equations. A set of m simultaneous linear equations in n unknowns can be written $\mathbf{Ax}=\mathbf{b}$, where **A** is an $n\times m$ matrix, $\mathbf{x}\in\mathbb{R}^n$ and $\mathbf{b}\in\mathbb{R}^m$. Let $\mathrm{T}:\mathbb{R}^n\to\mathbb{R}^m$ be the linear transformation with matrix **A**, let C and R be the respective column and row spaces of **A** and let r be the rank of **A**. Then C is the range of T and so $\mathbf{Ax}=\mathbf{b}$ can have a solution only if $\mathbf{b}\in C$. Suppose, therefore, that $\mathbf{b}\in C$ and that $\mathbf{x}=\mathbf{z}$ is a particular solution of $\mathbf{Ax}=\mathbf{b}$. Then $\mathbf{Az}=\mathbf{b}$ and, if N is the null space of T,

$$\mathbf{Ax}=\mathbf{b}\Leftrightarrow\mathbf{Ax}=\mathbf{Az}\Leftrightarrow\mathbf{A}(\mathbf{x}-\mathbf{z})=\mathbf{0}\Leftrightarrow\mathbf{x}-\mathbf{z}\in N.$$

Hence the general solution of $\mathbf{Ax} = \mathbf{b}$ is the set $\{\mathbf{x} = \mathbf{z} + \mathbf{y} : \mathbf{y} \in N\}$. Since N has dimension $(n-r)$, the general solution of $\mathbf{Ax} = \mathbf{b}$ contains $(n-r)$ parameters.

EXAMPLE 2 *Let* $\mathbf{A} = \begin{pmatrix} 1 & 0 & 2 & 3 \\ -1 & 1 & -3 & -5 \\ 2 & 3 & 1 & 0 \end{pmatrix}$ *and let* $\mathrm{T}: \mathbb{R}^4 \to \mathbb{R}^3$ *be the linear transformation with matrix* $\mathbf{A}$. *Find (i) the rank of* T *(ii) a basis for the null space of* T *(iii) a basis for the range of* T. *Solve the equations* $\begin{matrix} w & & +2y+3z = a \\ -w & +x & -3y-5z = b \\ 2w & +3x & +y \quad = c \end{matrix}$, *in the cases (a)* $a = 0 = b = c$, *(b)* $a = 1, b = 2, c = 3$, (c) $a = 2, b = -1, c = 7$.

Reduce the augmented matrix $(\mathbf{A}\,|\,\mathbf{b})$ to upper triangular form by row operations,

$$\left(\begin{array}{cccc|c} 1 & 0 & 2 & 3 & a \\ -1 & 1 & -3 & -5 & b \\ 2 & 3 & 1 & 0 & c \end{array}\right) \begin{matrix} \mathbf{r}_2 \leftarrow \mathbf{r}_2 + \mathbf{r}_1 \\ \mathbf{r}_3 \leftarrow \mathbf{r}_3 - 2\mathbf{r}_1 \end{matrix} \left(\begin{array}{cccc|c} 1 & 0 & 2 & 3 & a \\ 0 & 1 & -1 & -2 & a+b \\ 0 & 3 & -3 & -6 & c-2a \end{array}\right) \mathbf{r}_3 \leftarrow \mathbf{r}_3 - 3\mathbf{r}_2$$

$$\left(\begin{array}{cccc|c} 1 & 0 & 2 & 3 & a \\ 0 & 1 & -1 & -2 & a+b \\ 0 & 0 & 0 & 0 & c-5a-3b \end{array}\right).$$ (i) The rank of T is the rank of A which is **2**.

(ii) The null space N of T is the set $\left\{\begin{pmatrix} w \\ x \\ y \\ z \end{pmatrix} : \begin{matrix} w + 2y + 3z = 0 \\ x - \; y - 2z = 0 \end{matrix}\right\}$ which has a basis given by $y=1$, $z=0$ and $y=0$, $z=1$, that is $\left\{\begin{pmatrix} \mathbf{-2} \\ \mathbf{1} \\ \mathbf{1} \\ \mathbf{0} \end{pmatrix}, \begin{pmatrix} \mathbf{-3} \\ \mathbf{2} \\ \mathbf{0} \\ \mathbf{1} \end{pmatrix}\right\}$.

(iii) The range of T is C, the column space of $\mathbf{A}$, with a basis $\left\{\begin{pmatrix} \mathbf{1} \\ \mathbf{-1} \\ \mathbf{2} \end{pmatrix}, \begin{pmatrix} \mathbf{0} \\ \mathbf{1} \\ \mathbf{3} \end{pmatrix}\right\}$ since these are the first two columns of $\mathbf{A}$ and C has dimension $4 - r = 4 - 2 = 2$.

For there to be a solution of $\mathbf{Ax} = \mathbf{b} = \begin{pmatrix} a \\ b \\ c \end{pmatrix}$, the necessary condition is that $c = 5a + 3b$. (a) If $a = 0 = b = c$, the condition is satisfied and the solution set is $N = \left\{\begin{pmatrix} \mathbf{-2y-3z} \\ \mathbf{y+2z} \\ \mathbf{y} \\ \mathbf{z} \end{pmatrix} : \mathbf{y, z \in R}\right\}$.

(b) If $a = 1, b = 2, c = 3$, then $c \neq 5a + 3b$ and the solution set is $\boldsymbol{\phi}$.
(c) If $a = 2, b = -1, c = 7$, then $c = 5a + 3b$ and one solution is found by taking

$y = 0 = z$, that is $x = a+b = 1$ and $w = a = 2$. Therefore, the general solution is the set $\left\{ \begin{pmatrix} 2-2y-3z \\ 1+y+2z \\ y \\ z \end{pmatrix} : y, z \in \mathbf{R} \right\}$.

In the special case of a set of n homogeneous linear equations in n unknowns, $\mathbf{b} = \mathbf{0}$ and $m = n$. The equations in matrix form are $\mathbf{Ax} = \mathbf{0}$, where $\mathbf{A}$ is an $n \times n$ matrix and $\mathbf{x} \in \mathbb{R}^n$. These equations always have the solution $\mathbf{x} = \mathbf{0}$ and the question is whether there is a non-zero solution.

Let T be the linear transformation of $\mathbb{R}^n$ with matrix $\mathbf{A}$. Then, by Theorem 14.18,

$$\begin{aligned} \mathbf{Ax} = \mathbf{0}, \mathbf{x} \neq \mathbf{0}, &\Leftrightarrow \text{the columns of } \mathbf{A} \text{ are linearly dependent} \\ &\Leftrightarrow \mathbf{A} \text{ is singular, that is, } \det(\mathbf{A}) = 0, \end{aligned}$$

on using Theorem 1.7. This has proved the Theorem:

Theorem 14.19 Let $\mathbf{A}$ be an $n \times n$ matrix and let $\mathbf{x} \in \mathbb{R}^n$, then

$$\mathbf{Ax} = \mathbf{0} \text{ has a non-zero solution} \Leftrightarrow \det(\mathbf{A}) = 0.$$

EXAMPLE 3 *Find those values of k for which the simultaneous equations*

$$\begin{aligned} 6x + (k-2)y + \quad 3z &= 0, \\ (k-5)x + 3y - \quad 3z &= 0, \\ 12x - 6y + (k+7)z &= 0, \end{aligned}$$

have a non-zero solution. Solve the equations for each of these values of k and, in each case, give a geometrical interpretation of the result.

Write the equations in matrix form $\mathbf{Ax} = \mathbf{0}$ then, for a non-zero solution, the necessary and sufficient condition is that $\det(\mathbf{A}) = 0$.

$$\det(\mathbf{A}) = \begin{vmatrix} 6 & k-2 & 3 \\ k-5 & 3 & -3 \\ 12 & -6 & k+7 \end{vmatrix} = \begin{vmatrix} 6 & k-2 & 3 \\ k+1 & k+1 & 0 \\ 12 & -6 & k+7 \end{vmatrix} \quad \text{(on adding } \mathbf{r}_1 \text{ to } \mathbf{r}_2)$$

$$= (k+1)\begin{vmatrix} 6 & k-2 & 3 \\ 1 & 1 & 0 \\ 12 & -6 & k+7 \end{vmatrix} \quad \text{(on taking out a factor } k+1 \text{ from } \mathbf{r}_2)$$

$$= (k+1)\begin{vmatrix} 6 & k-8 & 3 \\ 1 & 0 & 0 \\ 12 & -18 & k+7 \end{vmatrix} \quad \text{(on subtracting } \mathbf{c}_1 \text{ from } \mathbf{c}_2)$$

$$\begin{aligned} &= (k+1)\{6(0-0)\} - (k-8)(k+7-0) + 3(-18-0)\} \\ &= (k+1)(-k^2+k+56-54) = -(k+1)(k^2-k+2) \\ &= -(k+1)(k+1)(k-2). \end{aligned}$$

Thus $|\mathbf{A}| = 0$ if and only if $k = -1$ or $k = 2$ and so these are the values of k for which the equations have a non-zero solution.

If $k = -1$, then $\mathbf{A} = \begin{pmatrix} 6 & -3 & 3 \\ -6 & 3 & -3 \\ 12 & -6 & 6 \end{pmatrix}$ and, after the row operations

$\mathbf{r}_2 \leftarrow \mathbf{r}_2 + \mathbf{r}_1$, $\mathbf{r}_3 \leftarrow \mathbf{r}_3 - 2\mathbf{r}_1$, this matrix becomes $\begin{pmatrix} 6 & -3 & 3 \\ 0 & 0 & 0 \\ 0 & 0 & 0 \end{pmatrix}$. The three equations are therefore equivalent to the one equation $2x - y + z = 0$. The solution may be given in terms of two parameters, s, t, in the form $x = t$, $y = s$, $z = s - 2t$. Geometrically, the planes represented by the three equations are identical.

If $k = 2$, then $\mathbf{A} = \begin{pmatrix} 6 & 0 & 3 \\ -3 & 3 & -3 \\ 12 & -6 & 9 \end{pmatrix}$ and we use row reduction. Divide each row by 3, $\begin{pmatrix} 2 & 0 & 1 \\ -1 & 1 & -1 \\ 4 & -2 & 3 \end{pmatrix}$ $\mathbf{r}_1 \leftarrow \mathbf{r}_1 + \mathbf{r}_2$ $\begin{pmatrix} 1 & 1 & 0 \\ -1 & 1 & -1 \\ 4 & -2 & 3 \end{pmatrix}$

$\begin{matrix} \mathbf{r}_2 \leftarrow \mathbf{r}_2 + \mathbf{r}_1 \\ \mathbf{r}_3 \leftarrow \mathbf{r}_3 - 4\mathbf{r}_1 \end{matrix} \begin{pmatrix} 1 & 1 & 0 \\ 0 & 2 & -1 \\ 0 & -6 & 3 \end{pmatrix}$ $\mathbf{r}_3 \leftarrow \mathbf{r}_3 + 3\mathbf{r}_2$ $\begin{pmatrix} 1 & 1 & 0 \\ 0 & 2 & -1 \\ 0 & 0 & 0 \end{pmatrix}$.

Only two of the three equations are linearly independent and, in terms of one parameter t, the solution is $x = -t$, $y = t$, $z = 2t$. Geometrically, the three equations represent three planes through O having one common line. That is, a sheaf of planes through the line $-2x = 2y = z$, which is the line of solutions.

Note When a row reduction is made on a singular matrix **A** the resultant echelon matrix must have at least one zero row. If not, then some mistake has been made in the calculations.

EXERCISE 14.3 B

1 Find a basis of (i) the row space, (ii) the column space of the given matrix:

(a) $\begin{pmatrix} 1 & 2 \\ 3 & 4 \end{pmatrix}$ (b) $\begin{pmatrix} 1 & 2 \\ 2 & 4 \end{pmatrix}$ (c) $\begin{pmatrix} 1 & 2 & 3 \\ 3 & -1 & 2 \\ -1 & 1 & 0 \end{pmatrix}$ (d) $\begin{pmatrix} 1 & 2 & 3 \\ 2 & 4 & 6 \\ -1 & -2 & -3 \end{pmatrix}$

(e) $\begin{pmatrix} 1 & -1 & 0 \\ 2 & 1 & 3 \end{pmatrix}$ (f) $\begin{pmatrix} 2 & -3 \\ 1 & 4 \\ 3 & 1 \\ 5 & -2 \end{pmatrix}$ (g) $\begin{pmatrix} 1 & 2 & 5 & 3 \\ 2 & 1 & 4 & 2 \\ 4 & -1 & 2 & 1 \end{pmatrix}$.

2 Given that T is a linear transformation of $\mathbb{R}^n$ to $\mathbb{R}^m$, with matrix **A**, find a basis for (i) the null space of T, (ii) the range of T, where **A** is the matrix given in question 1 in cases (a) to (g).

3 Let $\mathbf{A} = \begin{pmatrix} 1 & -2 & -3 \\ 2 & 1 & 4 \end{pmatrix}$ and let T: $\mathbb{R}^3 \to \mathbb{R}^2$ be the linear transformation with matrix **A**. Find (i) the null space of T (ii) the range of T (iii) the rank of T.

4 The *column rank* of a matrix is the number of linearly independent columns (each column being considered as a vector), the *row rank* of the matrix is defined in a similar way. Determine the column and the row rank of each of the matrices $\begin{pmatrix} 1 & -2 & 7 \\ 2 & 5 & -4 \\ -2 & -3 & 0 \end{pmatrix}$ and $\begin{pmatrix} 1 & -2 & 1 \\ 2 & 5 & 1 \\ -2 & -3 & 1 \end{pmatrix}$. (*JMB*)

5 Find a linear relation connecting the rows of the matrix $\mathbf{M}$, where $\mathbf{M} = \begin{pmatrix} 1 & 2 & 0 & 3 \\ -2 & 1 & -5 & -1 \\ 0 & -1 & 1 & -1 \end{pmatrix}$. State the rank of $\mathbf{M}$. The linear transformation T is defined by $\mathrm{T}\begin{pmatrix} x \\ y \\ z \\ t \end{pmatrix} = \mathbf{M}\begin{pmatrix} x \\ y \\ z \\ t \end{pmatrix}$. State the dimension of the kernel of T, and find a basis for the range of T. *(SMP)*

6 The set of vectors $\{\mathbf{e}_1, \mathbf{e}_2, \mathbf{e}_3\}$ is a linearly independent set in $\mathbb{R}^3$, and $\{\mathbf{f}_1, \mathbf{f}_2\}$ is a linearly independent set in $\mathbb{R}^2$. The linear transformation L: $\mathbb{R}^3 \to \mathbb{R}^2$ is defined by $\mathrm{L}(\mathbf{e}_1) = \mathbf{f}_1 + \mathbf{f}_2$, $\mathrm{L}(\mathbf{e}_2) = \mathbf{f}_1 - \mathbf{f}_2$, $\mathrm{L}(\mathbf{e}_3) = 2\mathbf{f}_2$. Find (i) a vector in $\mathbb{R}^3$ which is mapped to $\mathbf{f}_1$ by L, (ii) a basis for the null space of L, (iii) the set of vectors in $\mathbb{R}^3$ which are mapped to $\mathbf{f}_2$ by L. *(C)*

7 The linear transformation σ: $\mathbb{R}^4 \to \mathbb{R}^3$ is represented by the matrix $\mathbf{A} = \begin{pmatrix} 2 & 1 & 3 & 3 \\ 0 & -3 & 1 & -2 \\ 4 & 5 & 5 & 8 \end{pmatrix}$ with respect to the standard bases in $\mathbb{R}^4$ and $\mathbb{R}^3$. Find a basis for the null space of σ, and deduce that the range of σ has dimension 2. Find a basis for the range. Solve the equation $\mathbf{AX} = \begin{pmatrix} 5 \\ 3 \\ 7 \end{pmatrix}$, and state, with a reason, whether the set of solutions forms a subspace of $\mathbb{R}^4$. *(C)*

8 Find the dimension of the row space of the matrix $\mathbf{A}$, where $\mathbf{A} = \begin{pmatrix} 3 & -1 & 2 \\ 1 & 1 & -2 \\ 7 & -5 & 10 \end{pmatrix}$. Find the solution, if any, of the equation $\mathbf{AX} = \mathbf{K}$ in the cases: (i) $\mathbf{K} = \begin{pmatrix} 0 \\ 0 \\ 0 \end{pmatrix}$, (ii) $\mathbf{K} = \begin{pmatrix} 2 \\ 1 \\ 2 \end{pmatrix}$, (iii) $\mathbf{K} = \begin{pmatrix} 1 \\ 3 \\ -3 \end{pmatrix}$. *(C)*

9 (a) $\mathbf{A}$ is a real 3×3 matrix, $\mathbf{B}$ and $\mathbf{X}$ are vectors in $\mathbb{R}^3$. Prove that the equation $\mathbf{AX} = \mathbf{B}$ has a solution if and only if $\mathbf{B}$ is in the subspace spanned by the columns of the matrix $\mathbf{A}$.

(b) Find the value of c for which the equations $\begin{aligned} x + y + z &= c \\ x - 3y + 2z &= -1 \\ 2x + 6y + z &= 4 \end{aligned}$ have a solution, and find the solution in this case. *(C)*

10 The matrix $\mathbf{A}$ is given by $\mathbf{A} = \begin{pmatrix} 2 & 0 & -1 & 1 \\ 0 & 1 & 3 & 1 \\ 4 & -1 & -5 & 1 \end{pmatrix}$. Find a basis of the subspace V of $\mathbb{R}^4$ where $V = \{\mathbf{X}: \mathbf{X} \in \mathbb{R}^4, \mathbf{AX} = \mathbf{0}\}$. Hence, or otherwise, find the solutions, if any, of the equation $\mathbf{AX} = \mathbf{K}$ in each of the following cases:

(i) $\mathbf{K} = \begin{pmatrix} 0 \\ 0 \\ 0 \end{pmatrix}$, (ii) $\mathbf{K} = \begin{pmatrix} 2 \\ 1 \\ 3 \end{pmatrix}$, (iii) $\mathbf{K} = \begin{pmatrix} 1 \\ 0 \\ 0 \end{pmatrix}$.

14.4 Eigenvalues and eigenvectors

The standard 2×2 matrix transformations were classified in §4.3 in terms of their invariants:

rotations	—	leave the origin, O, fixed,
shears	—	leave one line fixed,
reflections and one way stretches	—	leave one line fixed and every perpendicular line invariant,
enlargements	—	leave lines through O invariant.

If T is a transformation of $\mathbb{R}^n$ with matrix $\mathbf{M}$, and $\mathbf{x} \in \mathbb{R}^n$, then:
the fixed points under T are found by solving the equation $\mathbf{Mx} = \mathbf{x}$,
the invariant lines under T are found by solving the equation $\mathbf{Mx} = \lambda\mathbf{x}$.
If $\mathbf{Mx} = \mathbf{x}$, then $\mathbf{M}(k\mathbf{x}) = k(\mathbf{Mx}) = k\mathbf{x}$, giving a fixed line. We therefore consider the equation $\mathbf{Mx} = \lambda\mathbf{x}$ and note that, for solutions with $\lambda = 1$ we obtain a fixed line.

Now let $\mathbf{I}$ be the unit $n \times n$ matrix, then

$$\mathbf{Mx} = \lambda\mathbf{x} \Leftrightarrow \mathbf{Mx} = \lambda\mathbf{Ix} \Leftrightarrow \mathbf{Mx} - \lambda\mathbf{Ix} = 0$$
$$\Leftrightarrow (\mathbf{M} - \lambda\mathbf{I})\mathbf{x} = \mathbf{0}.$$

$(\mathbf{M} - \lambda\mathbf{I})\mathbf{x} = \mathbf{0}$ is a set of n homogeneous linear equations in n unknowns which always has the solution $\mathbf{x} = \mathbf{0}$. This just means that the origin is a fixed point of any transformation. By Theorem 14.19, there will be non-zero solutions if and only if the coefficient matrix is singular, that is, if and only if $\det(\mathbf{M} - \lambda\mathbf{I}) = \mathbf{0}$. This result is stated in the next theorem.

Theorem 14.20
There is a non-zero vector $\mathbf{x}$ with $\mathbf{Mx} = \lambda\mathbf{x} \Leftrightarrow \det(\mathbf{M} - \lambda\mathbf{I}) = 0$.

The equation $\det(\mathbf{M} - \lambda\mathbf{I}) = 0$ is a polynomial equation of degree n in λ and is called the *characteristic equation* of $\mathbf{M}$. $\det(\mathbf{M} - \lambda\mathbf{I})$ is called the *characteristic polynomial* of $\mathbf{M}$. In the complex field, $\mathbb{C}$, the characteristic equation has n roots λ_i, $1 \leqslant i \leqslant n$ which may be distinct, or repeated, and corresponding to each root λ_i there will be a solution $\mathbf{x}_i$ of the equation $\mathbf{Mx} = \lambda\mathbf{x}$.

Definition The roots λ_i, $1 \leqslant i \leqslant n$ of the equation $\det(\mathbf{M} - \lambda\mathbf{I}) = 0$ are called the *eigenvalues* of $\mathbf{M}$ (and of the transformation T with matrix $\mathbf{M}$). To each eigenvalue λ_i, there corresponds an *eigenvector* $\mathbf{x}_i$ satisfying $\mathbf{Mx}_i = \lambda_i\mathbf{x}_i$ and the line $\mathbf{r} = t\mathbf{x}_i$ is an invariant line for T.

If an eigenvalue is complex, the associated eigenvector is also complex and we shall not be concerned with this complex case. However, it can be shown that, when $\mathbf{M}$ is a real symmetric matrix the eigenvalues are all real and n mutually perpendicular eigenvectors can be found. We shall not prove this result but state it without proof in a theorem.

Theorem 14.21 Let **M** be a real symmetric $n \times n$ matrix. Then **M** has n real eigenvalues (counting repeated roots) and a corresponding set of n mutually perpendicular eigenvectors may be chosen.

Note Eigenvalues and eigenvectors are also called characteristic values and characteristic vectors.

These ideas are illustrated in two and three dimensions in the next examples and we restrict our work to real symmetric matrices.

EXAMPLE 1 *Find the eigenvalues and the corresponding eigenvectors for the transformation* T, *with matrix* **M** *given by* $\mathbf{M} = \begin{pmatrix} 1/2 & \sqrt{3}/2 \\ \sqrt{3}/2 & -1/2 \end{pmatrix}$ *and hence describe the transformation.*

$$\mathbf{M} - \lambda\mathbf{I} = \begin{pmatrix} 1/2 & \sqrt{3}/2 \\ \sqrt{3}/2 & -1/2 \end{pmatrix} - \lambda\begin{pmatrix} 1 & 0 \\ 0 & 1 \end{pmatrix} = \begin{pmatrix} 1/2-\lambda & \sqrt{3}/2 \\ \sqrt{3}/2 & -1/2-\lambda \end{pmatrix}$$

so the characteristic equation becomes

$$\det(\mathbf{M} - \lambda\mathbf{I}) = (1/2-\lambda)(-1/2-\lambda) - (\sqrt{3}/2)(\sqrt{3}/2) = 0,$$

that is $-1/4 + \lambda^2 - 3/4 = 0$ or $\lambda^2 = 1$, so $\lambda = 1$ or $\lambda = -1$.
There are two distinct eigenvalues **1** and $-$**1** and we now consider each in turn.
$\lambda = 1$: $\mathbf{Mx} = \mathbf{x}$, so $\begin{pmatrix} 1/2 & \sqrt{3}/2 \\ \sqrt{3}/2 & -1/2 \end{pmatrix}\begin{pmatrix} x \\ y \end{pmatrix} = \begin{pmatrix} x \\ y \end{pmatrix}$, which gives two equations

$$\begin{aligned} x/2 + \sqrt{3}y/2 &= x \\ \sqrt{3}x/2 - y/2 &= y \end{aligned}, \quad \text{both of which give } x = \sqrt{3}y.$$

(If, at this point, the two equations do not give the same result, a mistake must have been made, so the pair of equations provide a check on the work.)

Any multiple of $\begin{pmatrix} \sqrt{3} \\ 1 \end{pmatrix}$ is an eigenvector and the line $x = \sqrt{3}y$ is invariant. Since $\lambda = 1$ it is, in fact, a fixed line.

$\lambda = -1$: $\mathbf{Mx} = -\mathbf{x}$, so $\begin{pmatrix} 1/2 & \sqrt{3}/2 \\ \sqrt{3}/2 & -1/2 \end{pmatrix}\begin{pmatrix} x \\ y \end{pmatrix} = -\begin{pmatrix} x \\ y \end{pmatrix}$, which gives two equations

$$\begin{aligned} x/2 + \sqrt{3}y/2 &= -x \\ \sqrt{3}x/2 - y/2 &= -y \end{aligned}, \quad \text{both of which give } y = -\sqrt{3}x.$$

Any multiple of $\begin{pmatrix} -1 \\ \sqrt{3} \end{pmatrix}$ is an eigenvector and the line $y = -\sqrt{3}x$ is invariant. Since $\lambda = -1$, this line is reflected in the origin under T. From these two results, T has a fixed line $x = \sqrt{3}y$ and an invariant line $y = -\sqrt{3}x$, perpendicular to the fixed line and so T is a reflection in the line $x = \sqrt{3}y$. Note that the two eigenvectors are perpendicular.

In the two dimensional case, the description of T is obtained much more quickly from a diagram showing the transformation of the unit square. The real value of the eigenvector technique lies in its use in three dimensions where the geometrical methods are less obvious.

EXAMPLE 2 *Find the characteristic equation and the eigenvalues of the transformation* T *of three dimensional space represented by the matrix* **M**, *given by* $\mathbf{M} = \begin{pmatrix} 3 & 2 & 4 \\ 2 & 0 & 2 \\ 4 & 2 & 3 \end{pmatrix}$. *Substitute to find the corresponding eigenvectors and hence determine the equations of an invariant line and an invariant plane for* T.

Expand det $(\mathbf{M} - \lambda\mathbf{I})$ to obtain the characteristic polynomial.

$$\begin{vmatrix} 3-\lambda & 2 & 4 \\ 2 & -\lambda & 2 \\ 4 & 2 & 3-\lambda \end{vmatrix} = (3-\lambda)\{\lambda(\lambda-3)-2.2\}+2\{2.4-2(3-\lambda)\} +4\{2.2-4(-\lambda)\}$$

$$= -\lambda^3+6\lambda^2+15\lambda+8.$$

To solve the characteristic equation, we look for factors. By inspection, $\lambda = -1$ is a root so factorise out the factor $(\lambda+1)$ and then factorise the other (quadratic) factor.

$$-\lambda^3+6\lambda^2+15\lambda+8 = (\lambda+1)\,(-\lambda^2+7\lambda+8) = (\lambda+1)\,(8-\lambda)\,(\lambda+1) = 0.$$

The roots are -1, -1 and 8 so there is a repeated eigenvalue $\mathbf{-1}$ and a second eigenvalue **8**. Consider the latter first.

$\lambda = 8$: The corresponding eigenvector satisfies $\mathbf{Mx} = \mathbf{8x}$, which gives three equations

$$\begin{aligned} 3x+2y+4z &= 8x \\ 2x \quad\; +2z &= 8y, \\ 4x+2y+3z &= 8z \end{aligned} \quad \text{that is,} \quad \begin{aligned} 2y+4z &= 5x \\ x+\; z &= 4y. \\ 4x+2y &= 5z \end{aligned}$$

Eliminating z gives $2y = x$ and eliminating x gives $z = 2y$ and all equations are satisfied by $x = 2y = z$. These are the equations of the invariant line and the corresponding eigenvector is $\begin{pmatrix} 2 \\ 1 \\ 2 \end{pmatrix}$.

$\lambda = -1$: For this repeated eigenvalue, the equation $\mathbf{Mx} = -\mathbf{x}$ gives

$$\begin{aligned} 3x+2y+4z &= -x \\ 2x \quad\; +2z &= -y, \\ 4x+2y+3z &= -z \end{aligned} \quad \text{that is,} \quad \begin{aligned} 4x+2y+4z &= 0 \\ 2x+\; y+2z &= 0. \\ 4x+2y+4z &= 0 \end{aligned}$$

All three equations give the same plane $2x+y+2z = 0$ so there is a plane of solutions. Any line through the origin in this plane is invariant and its direction is an eigenvector. For the repeated eigenvalue -1, we can choose a pair of perpendicular eigenvectors $\begin{pmatrix} \mathbf{1} \\ \mathbf{0} \\ \mathbf{-1} \end{pmatrix}$ and $\begin{pmatrix} \mathbf{1} \\ \mathbf{-4} \\ \mathbf{1} \end{pmatrix}$. Then the three eigenvectors given above are mutually perpendicular. T has an invariant plane $\mathbf{2x+y+2z=0}$, corresponding to the eigenvalue -1, and an invariant line $\mathbf{x=2y=}z$ perpendicular to the invariant plane, corresponding to the eigenvalue 8. Therefore T represents a rotation through π about the invariant line together with a one-way stretch along this line with scale factor 8.

Note In the work so far, an eigenvalue which gave one distinct eigenvector gave rise to an invariant line or, in the case $\lambda = 1$, a fixed line.

However, a repeated eigenvalue gave rise to an invariant plane so that it was possible to choose two linearly independent eigenvectors in this plane. Also, so far, the eigenvalues have all been real and there were 2 in the 2×2 case and 3 in the 3×3 case. This is not always so but when the $n \times n$ matrix is real and symmetric, the eigenvalues are real and a set of n mutually orthogonal eigenvectors can be found, as was stated in Theorem 14.21. For non-symmetric matrices, the eigenvalues may be complex and n orthogonal eigenvectors cannot be found.

Diagonalisation

Suppose that a 3×3 symmetric matrix $\mathbf{M}$ has three real eigenvalues, λ_1, λ_2 and λ_3, possibly some repeated, and a set $\{\mathbf{e}_1, \mathbf{e}_2, \mathbf{e}_3\}$ of corresponding mutually orthogonal unit eigenvectors. Then $\mathbf{M}\mathbf{e}_i = \lambda_i \mathbf{e}_i$, for $1 \leqslant i \leqslant 3$. Form the orthogonal matrix $\mathbf{E}$, with columns equal to these eigenvectors so that $\mathbf{E} = (\mathbf{e}_1\ \mathbf{e}_2\ \mathbf{e}_3)$. Then $\mathbf{ME} = (\lambda_1\mathbf{e}_1\ \ \lambda_2\mathbf{e}_2\ \ \lambda_3\mathbf{e}_3) = \mathbf{ED}$, where $\mathbf{D}$ is the diagonal matrix, given by $\mathbf{D} = \begin{pmatrix} \lambda_1 & 0 & 0 \\ 0 & \lambda_2 & 0 \\ 0 & 0 & \lambda_3 \end{pmatrix}$. Now, since $\mathbf{E}$ is orthogonal, it is non-singular with an inverse $\mathbf{E}^{\mathrm{T}}$, so that

$$\mathbf{E}^{\mathrm{T}}\mathbf{ME} = \mathbf{E}^{-1}\mathbf{ME} = \mathbf{E}^{-1}\mathbf{ED} = \mathbf{D}$$

and we say that $\mathbf{M}$ has been reduced to *diagonal form* $\mathbf{D}$, or that $\mathbf{M}$ has been *diagonalised*. Conversely, $\mathbf{M}$ is constructed from the diagonal matrix $\mathbf{D}$ by $\mathbf{M} = \mathbf{EDE}^{-1} = \mathbf{EDE}^{\mathrm{T}}$. This is now applied to the previous examples.

EXAMPLE 3 *Diagonalise the matrices of* (i) *Example* 1 (ii) *Example* 2.

(i) The unit orthogonal eigenvectors $\mathbf{e}_1$ and $\mathbf{e}_2$ are given by $\mathbf{e}_1 = \begin{pmatrix} \sqrt{3}/2 \\ 1/2 \end{pmatrix}$ and $\mathbf{e}_2 = \begin{pmatrix} -1/2 \\ \sqrt{3}/2 \end{pmatrix}$ so let $\mathbf{E} = \begin{pmatrix} \sqrt{3}/2 & -1/2 \\ 1/2 & \sqrt{3}/2 \end{pmatrix}$. Then, since the eigenvalues are 1 and -1,

$$\mathbf{M}\begin{pmatrix} \sqrt{3}/2 \\ 1/2 \end{pmatrix} = \begin{pmatrix} 1/2 & \sqrt{3}/2 \\ \sqrt{3}/2 & -1/2 \end{pmatrix}\begin{pmatrix} \sqrt{3}/2 \\ 1/2 \end{pmatrix} = \begin{pmatrix} \sqrt{3}/2 \\ 1/2 \end{pmatrix}$$

and
$$\mathbf{M}\begin{pmatrix} -1/2 \\ \sqrt{3}/2 \end{pmatrix} = \begin{pmatrix} 1/2 & \sqrt{3}/2 \\ \sqrt{3}/2 & -1/2 \end{pmatrix}\begin{pmatrix} -1/2 \\ \sqrt{3}/2 \end{pmatrix} = -\begin{pmatrix} -1/2 \\ \sqrt{3}/2 \end{pmatrix}.$$

Then $\mathbf{ME} = \mathbf{ED}$ and $\mathbf{M} = \mathbf{EDE}^{\mathrm{T}}$ since, as is verified by matrix products,

$$\begin{pmatrix} 1/2 & \sqrt{3}/2 \\ \sqrt{3}/2 & -1/2 \end{pmatrix}\begin{pmatrix} \sqrt{3}/2 & -1/2 \\ 1/2 & \sqrt{3}/2 \end{pmatrix} = \begin{pmatrix} \sqrt{3}/2 & -1/2 \\ 1/2 & \sqrt{3}/2 \end{pmatrix}\begin{pmatrix} 1 & 0 \\ 0 & -1 \end{pmatrix}$$

and
$$\begin{pmatrix} 1/2 & \sqrt{3}/2 \\ \sqrt{3}/2 & -1/2 \end{pmatrix} = \begin{pmatrix} \sqrt{3}/2 & -1/2 \\ 1/2 & \sqrt{3}/2 \end{pmatrix}\begin{pmatrix} 1 & 0 \\ 0 & -1 \end{pmatrix}\begin{pmatrix} \sqrt{3}/2 & 1/2 \\ -1/2 & \sqrt{3}/2 \end{pmatrix}.$$

Now the matrix $\mathbf{E}$ is a rotation of $\pi/6$ about the origin so $\mathbf{E}^{\mathrm{T}}(=\mathbf{E}^{-1})$ is the opposite rotation. The matrix $\begin{pmatrix} 1 & 0 \\ 0 & -1 \end{pmatrix}$ represents a reflection in the x-axis and T can be seen as a rotation of the axes of $-\pi/6$ followed by a reflection in the x-axis followed by the opposite rotation of $\pi/6$. T represents a reflection in the line $y = x\tan(\pi/6)$ but, relative to new axes $O(X, Y)$ in the direction of the eigenvectors $\mathbf{e}_1$ and $\mathbf{e}_2$ of $\mathbf{M}$, this is a reflection in the axis OX. This is shown in Fig. 14.1.

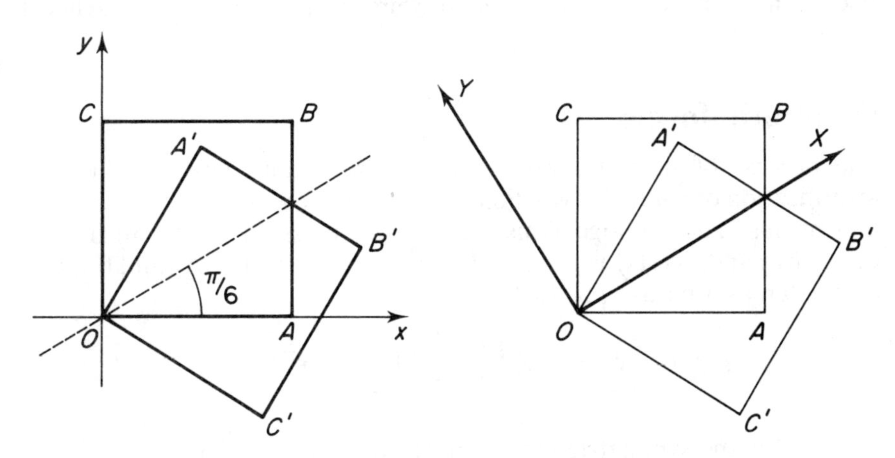

(i) Reflection in $y = x \tan \pi/6$ (ii) Reflection in $Y=0$

Fig. 14.1 Transformation T relative to axes $O(x, y)$ and $O(X, Y)$.

(ii) Corresponding to the eigenvalues 8, -1, -1, the three eigenvectors found were $\begin{pmatrix} 2 \\ 1 \\ 2 \end{pmatrix}, \begin{pmatrix} 1 \\ 0 \\ -1 \end{pmatrix}$ and $\begin{pmatrix} 1 \\ -4 \\ 1 \end{pmatrix}$ and we choose the orthogonal unit eigenvectors $\begin{pmatrix} 2/3 \\ 1/3 \\ 2/3 \end{pmatrix}, \begin{pmatrix} 1/\sqrt{2} \\ 0 \\ -1/\sqrt{2} \end{pmatrix}, \begin{pmatrix} 1/3\sqrt{2} \\ -4/3\sqrt{2} \\ 1/3\sqrt{2} \end{pmatrix}$.

Therefore $\mathbf{E} = \begin{pmatrix} 2/3 & 1/\sqrt{2} & 1/3\sqrt{2} \\ 1/3 & 0 & -4/3\sqrt{2} \\ 2/3 & -1/\sqrt{2} & 1/3\sqrt{2} \end{pmatrix}, \mathbf{M} = \begin{pmatrix} 3 & 2 & 4 \\ 2 & 0 & 2 \\ 4 & 2 & 3 \end{pmatrix}$

Then $\mathbf{ME} = \mathbf{ED}$ and $\mathbf{M} = \mathbf{EDE}^{\mathrm{T}}$, as is verified by matrix products,

$$\begin{pmatrix} 3 & 2 & 4 \\ 2 & 0 & 2 \\ 4 & 2 & 3 \end{pmatrix}\begin{pmatrix} 2/3 & 1/\sqrt{2} & 1/3\sqrt{2} \\ 1/3 & 0 & -4/3\sqrt{2} \\ 2/3 & -1/\sqrt{2} & 1/3\sqrt{2} \end{pmatrix} = \begin{pmatrix} 2/3 & 1/\sqrt{2} & 1/3\sqrt{2} \\ 1/3 & 0 & -4/3\sqrt{2} \\ 2/3 & -1/\sqrt{2} & 1/3\sqrt{2} \end{pmatrix}\begin{pmatrix} 8 & 0 & 0 \\ 0 & -1 & 0 \\ 0 & 0 & -1 \end{pmatrix}$$

$$\begin{pmatrix} 3 & 2 & 4 \\ 2 & 0 & 2 \\ 4 & 2 & 3 \end{pmatrix} = \begin{pmatrix} 2/3 & 1/\sqrt{2} & 1/3\sqrt{2} \\ 1/3 & 0 & -4/3\sqrt{2} \\ 2/3 & -1/\sqrt{2} & 1/3\sqrt{2} \end{pmatrix}\begin{pmatrix} 8 & 0 & 0 \\ 0 & -1 & 0 \\ 0 & 0 & -1 \end{pmatrix}\begin{pmatrix} 2/3 & 1/3 & 2/3 \\ 1/\sqrt{2} & 0 & -1/\sqrt{2} \\ 1/3\sqrt{2} & -4/3\sqrt{2} & 1/\sqrt{2} \end{pmatrix}.$$

$\mathbf{E}$ and $\mathbf{E}^{\mathrm{T}}$ are matrices corresponding to inverse isometries, so if a new set of axes $O(X, Y, Z)$ are chosen along the directions of the eigenvectors $\mathbf{e}_1$, $\mathbf{e}_2$, $\mathbf{e}_3$, then T has matrix **M** corresponding to axes $O(x, y, z)$ and matrix $\begin{pmatrix} 8 & 0 & 0 \\ 0 & -1 & 0 \\ 0 & 0 & -1 \end{pmatrix}$ corresponding to axes $O(X, Y, Z)$. This diagonal form of **M** shows that T consists of a stretch with S.F. 8 and two reflections. Thus diagonalisation gives a method of selecting a new set of axes relative to which the transformation has a particularly simple form, namely the diagonal form corresponding to stretches. Since we have restricted consideration to symmetric matrices, we have excluded shears.

Quadratic forms

The process of diagonalisation has an important application in the identification of conics in two dimensions and of quadric surfaces in three dimensions. The homogeneous quadratic expression (or form) in two variables x and y is $\mathrm{f}(x, y)$ given by $\mathrm{f}(x, y) = ax^2 + 2hxy + by^2$ and this can be written as a matrix product

$$\mathrm{f}(x, y) = (x\ y)\begin{pmatrix} a & h \\ h & b \end{pmatrix}\begin{pmatrix} x \\ y \end{pmatrix} = \mathbf{x}^{\mathrm{T}}\mathbf{M}\mathbf{x},$$

where **M** is the symmetric matrix given by $\mathbf{M} = \begin{pmatrix} a & h \\ h & b \end{pmatrix}$.

Similarly, in three dimensions the general quadratic form is given by $\mathrm{f}(x, y, z) = ax^2 + by^2 + cz^2 + 2fyz + 2gzx + 2hxy$ and this can be written in matrix form, $\mathrm{f}(x, y, z) = (x\ y\ z)\begin{pmatrix} a & h & g \\ h & b & f \\ g & f & c \end{pmatrix}\begin{pmatrix} x \\ y \\ z \end{pmatrix} = \mathbf{x}^{\mathrm{T}}\mathbf{M}\mathbf{x}$, where

$$\mathbf{M} = \begin{pmatrix} a & h & g \\ h & b & f \\ g & f & c \end{pmatrix}.$$

In two dimensions, the equation $\mathrm{f}(x, y) = 1$ represents a conic and in three dimensions the equation $\mathrm{f}(x, y, z) = 1$ represents a quadric surface. In each case the origin of coordinates is the centre of the conic or quadric. The quadric (conic) can be identified by a transformation of the Cartesian axes in order to remove the product terms in the quadratic form, leaving just the square terms. This corresponds to diagonalising **M**. Let **E** be the orthogonal matrix whose columns are eigenvectors of **M** corresponding to eigenvalues $\lambda_1, \lambda_2, \lambda_3$. The required transformation of coordinates from (x, y, z) to (X, Y, Z) is given by the matrix $\mathbf{E}^{\mathrm{T}}$. Since **E** is orthogonal, the transformation is an isometry so distances and angles and, hence, the shape of curves or surfaces is unaltered. If $\mathbf{r} = \begin{pmatrix} x \\ y \\ z \end{pmatrix}$, $\mathbf{R} = \begin{pmatrix} X \\ Y \\ Z \end{pmatrix}$, the coordinate change is given by $\mathbf{R} = \mathbf{E}^{\mathrm{T}}\mathbf{r} = \mathbf{E}^{-1}\mathbf{r}$ or $\mathbf{r} = \mathbf{E}\mathbf{R}$. Then the quadratic form $f(x, y, z)$ is transformed to $F(X, Y, Z)$ thus,

$$f(x, y, z) = \mathbf{r}^T\mathbf{M}\mathbf{r} = \mathbf{R}^T\mathbf{E}^T\mathbf{M}\mathbf{E}\mathbf{R} = \mathbf{R}^T\mathbf{D}\mathbf{R} = F(X, Y, Z)$$
$$= \lambda_1 X^2 + \lambda_2 Y^2 + \lambda_3 Z^2.$$

When the equation is in this diagonal form, the corresponding axes are called the principal axes of the quadric surface (conic). If all the eigenvalues are positive then the surface is an ellipsoid but if any eigenvalues are negative it is a hyperboloid. In two dimensions $F(X, Y) = \lambda_1 X^2 + \lambda_2 Y^2$. If the eigenvalues are both positive, the curve is an ellipse, if one is negative, the curve is a hyperbola.

EXAMPLE 4 *Find the eigenvalues of the symmetric matrix* $\mathbf{M}$ *given by*

$$(x\ y\ z)\,\mathbf{M}\begin{pmatrix} x \\ y \\ z \end{pmatrix} = x^2 + 4y^2 + z^2 + 4xz.$$

Using the eigenvectors of $\mathbf{M}$, *or otherwise, obtain an orthogonal matrix* $\mathbf{E}$ *such that* $\mathbf{E}^{-1}\mathbf{M}\mathbf{E} = \mathbf{D}$, *a diagonal matrix. Verify that the diagonal elements of* $\mathbf{D}$ *are the eigenvalues of* $\mathbf{M}$. *Show that the transformation* T, *given by*

$\begin{pmatrix} X \\ Y \\ Z \end{pmatrix} = T\begin{pmatrix} x \\ y \\ z \end{pmatrix} = E^{-1}\begin{pmatrix} x \\ y \\ z \end{pmatrix}$, *reduces* $x^2 + 4y^2 + z^2 + 4xz$ *to the form* $aX^2 + bY^2 + cZ^2$.

$x^2 + 4y^2 + z^2 + 4xz = 1.x^2 + 4.y^2 + 1.z^2 + 0.yz + 2.2xz + 0.xy$ so let

$\mathbf{M} = \begin{pmatrix} 1 & 0 & 2 \\ 0 & 4 & 0 \\ 2 & 0 & 1 \end{pmatrix}$, then

$$x^2 + 4y^2 + z^2 + 4xz = (x\ y\ z)\begin{pmatrix} 1 & 0 & 2 \\ 0 & 4 & 0 \\ 2 & 0 & 1 \end{pmatrix}\begin{pmatrix} x \\ y \\ z \end{pmatrix} = \mathbf{r}^T\mathbf{M}\mathbf{r}.$$

Find the eigenvalues and eigenvectors of $\mathbf{M}$.

$\text{Det}\,(\mathbf{M} - \lambda\mathbf{I}) = \begin{vmatrix} 1-\lambda & 0 & 2 \\ 0 & 4-\lambda & 0 \\ 2 & 0 & 1-\lambda \end{vmatrix} = 0$ if λ is an eigenvalue. Expanding the determinant

$$(1-\lambda)\{(4-\lambda)(1-\lambda)-0\}+0+2\{0-2(4-\lambda)\} = -(\lambda-1)(\lambda-4)(\lambda-1)+4(\lambda-4)$$
$$= -(\lambda-4)\{(\lambda-1)^2-4\} = -(\lambda-4)(\lambda-1+2)(\lambda-1-2)$$
$$= -(\lambda-4)(\lambda+1)(\lambda-3),$$

so $\det(\mathbf{M} - \lambda\mathbf{I}) = -(\lambda-4)(\lambda+1)(\lambda-3) = 0$ if $\lambda \in \{-1, 3, 4\}$. There are three eigenvalues $\lambda_1 = -1$, $\lambda_2 = 3$, $\lambda_3 = 4$, and we calculate the corresponding unit eigenvectors $\mathbf{e}_1$, $\mathbf{e}_2$, $\mathbf{e}_3$.

Expanding $\mathbf{M}\mathbf{r} = \lambda\mathbf{r}$, $(1-\lambda)x + 2z = 0$, $(4-\lambda)y = 0$, $2x + (1-\lambda)z = 0$.
(i) $\lambda = -1$ gives $x + z = 0$, $y = 0$, (ii) $\lambda = 3$ gives $-x + z = 0$, $y = 0$, (iii) $\lambda = 4$

gives $3x=2z$, $2x=-3z$ so $x=z=0$. Hence $\mathbf{e}_1 = \begin{pmatrix} 1/\sqrt{2} \\ 0 \\ -1/\sqrt{2} \end{pmatrix}$, $\mathbf{e}_2 = \begin{pmatrix} 1/\sqrt{2} \\ 0 \\ 1/\sqrt{2} \end{pmatrix}$,

$\mathbf{e}_3 = \begin{pmatrix} 0 \\ 1 \\ 0 \end{pmatrix}$, $\mathbf{E} = \begin{pmatrix} 1/\sqrt{2} & 1/\sqrt{2} & 0 \\ 0 & 0 & 1 \\ -1/\sqrt{2} & 1/\sqrt{2} & 0 \end{pmatrix}$ and $\mathbf{E}^{-1} = \begin{pmatrix} 1/\sqrt{2} & 0 & -1/\sqrt{2} \\ 1/\sqrt{2} & 0 & 1/\sqrt{2} \\ 0 & 1 & 0 \end{pmatrix}$.
The new coordinates are given by $X = (x-z)/\sqrt{2}$, $Y = (x+z)/\sqrt{2}$, $Z = y$ and, referred to these coordinates, $\mathbf{M}$ becomes $\mathbf{E}^{-1}\mathbf{ME} = \mathbf{D} = \begin{pmatrix} -1 & 0 & 0 \\ 0 & 3 & 0 \\ 0 & 0 & 4 \end{pmatrix}$ so the quadratic form is $-X^2+3Y^2+4Z^2$. We verify this as follows. $\mathbf{r} = \mathbf{ER}$ so $x = (X+Y)/\sqrt{2}$, $y = Z$, $z = (-X+Y)/\sqrt{2}$, and so

$$\begin{aligned} x^2+4y^2+z^2+4xz &= (X+Y)^2/2+4Z^2+(-X+Y)^2/2+4(X+Y)(-X+Y)/2 \\ &= X^2/2+XY+Y^2/2+4Z^2+X^2/2-XY+Y^2/2-4X^2/2+4Y^2/2 \\ &= -X^2+3Y^2+4Z^2. \end{aligned}$$

EXERCISE 14.4

1 Find the eigenvalues and the eigenvectors of the following matrices:
(a) $\begin{pmatrix} 4 & 3 \\ 1 & 2 \end{pmatrix}$ (b) $\begin{pmatrix} 2 & 0 \\ 3 & -1 \end{pmatrix}$ (c) $\begin{pmatrix} -2 & 1 \\ -2 & -5 \end{pmatrix}$ (d) $\begin{pmatrix} 2 & 0 & 0 \\ 0 & 2 & -3 \\ 0 & 3 & -4 \end{pmatrix}$
(e) $\begin{pmatrix} 3 & -1 & 2 \\ 1 & 1 & 2 \\ 2 & 1 & 1 \end{pmatrix}$.

2 Find the eigenvalues and corresponding eigenvectors for the matrix $\begin{pmatrix} 1 & 1 & -2 \\ -1 & 2 & 1 \\ 0 & 1 & -1 \end{pmatrix}$.

3 Show that the eigenvalues of the matrix $\mathbf{P}$, given by $\mathbf{P} = \begin{pmatrix} 1/4 & 1/2 \\ 3/4 & 1/2 \end{pmatrix}$ are $-1/4$ and 1, and find the corresponding eigenvectors Deduce a matrix $\mathbf{A}$ such that $\mathbf{PA} = \mathbf{AD}$ where $\mathbf{D} = \begin{pmatrix} -1/4 & 0 \\ 0 & 1 \end{pmatrix}$. Hence show that $\mathbf{P}^n \to \begin{pmatrix} 2/5 & 2/5 \\ 3/5 & 3/5 \end{pmatrix}$ as $n \to \infty$. *(SMP)*

4 Reflections in the x- and y-axes are denoted by $\mathbf{X}$ and $\mathbf{Y}$ respectively, while reflection in the line $y = x/\sqrt{3}$ is denoted by $\mathbf{M}$. It is given that $\mathbf{M}$ may be represented by the matrix $\begin{pmatrix} 1/2 & \sqrt{3}/2 \\ \sqrt{3}/2 & -1/2 \end{pmatrix}$. Find the matrix that represents $\mathbf{YMX}$. Show that its eigenvalues are ± 1, and find the corresponding eigenvectors. Hence describe geometrically a single transformation equivalent to $\mathbf{YMX}$. *(SMP)*

5 Find the values of k for which the equations $x-6y-z=kx$, $x-7y-z=ky$, $2x+3y-2z=kz$ have solutions other than $x=y=z=0$.

For each of the above values of k determine the ratios $x:y:z$, and interpret the results geometrically. Show that when $k = -4$ the solutions of the above equations must also satisfy the equation $4x-3y=0$. (*L*)

6 Show that 4 is one of the roots of the characteristic equation of the matrix $\begin{pmatrix} 5 & 2 & -3 \\ 3 & 3 & -2 \\ -5 & 4 & 5 \end{pmatrix}$. Give an interpretation of such a root when the matrix is used to define a linear transformation of geometrical space. (*L*)

7 Find the principal axis of the ellipsoid $3x^2+6y^2+2z^2-2xy-2yz-2zx=1$ associated with the integer eigenvalue of the corresponding matrix. (*SMP*)

8 Verify that the matrices **M** and **L** given by $\mathbf{M} = \begin{pmatrix} 2 & 1 & 2 \\ 1 & 2 & -2 \\ -2 & 2 & 1 \end{pmatrix}$, $\mathbf{L} = \frac{1}{9}\begin{pmatrix} 2 & 1 & -2 \\ 1 & 2 & 2 \\ 2 & -2 & 1 \end{pmatrix}$ are inverses of each other. Prove that the lines OA, OB, OC drawn from the origin and represented by the vectors $\begin{pmatrix} 2 \\ 1 \\ -2 \end{pmatrix}$, $\begin{pmatrix} 1 \\ 2 \\ 2 \end{pmatrix}$ and $\begin{pmatrix} 2 \\ -2 \\ 1 \end{pmatrix}$ respectively form three edges of a cube. A rotation about the diagonal OD, represented by $\begin{pmatrix} x \\ y \\ z \end{pmatrix} \mapsto \mathbf{P}\begin{pmatrix} x \\ y \\ z \end{pmatrix}$, maps A to B, B to C and C to A. Write down the matrix **PM**, and deduce the matrix **P**. If $\mathbf{T} = \begin{pmatrix} 0 & 0 & 1 \\ 1 & 0 & 0 \\ 0 & 1 & 0 \end{pmatrix}$, verify that $\mathbf{P} = \mathbf{MTM}^{-1}$, and give a geometric interpretation of this relationship. Give a geometric reason why **P** and **T** have the same set of eigenvalues. Calculate the eigenvalues of **T**, and verify that one of them is equal to $\cos(2\pi/3) + i\sin(2\pi/3)$. Find corresponding eigenvectors, and deduce a set of eigenvectors for **P**. (*SMP*)

9 (a) If **A**, **P**, **D** are non-singular 2×2 matrices such that $\mathbf{AP} = \mathbf{PD}$, prove that $\mathbf{A}^2 = \mathbf{PD}^2\mathbf{P}^{-1}$ and that $\mathbf{A}^3 = \mathbf{PD}^3\mathbf{P}^{-1}$.

(b) If $\begin{pmatrix} a & b \\ c & d \end{pmatrix}\begin{pmatrix} x \\ y \end{pmatrix} = k\begin{pmatrix} x \\ y \end{pmatrix}$, prove that k must satisfy the equation $k^2-(a+d)k+(ad-bc)=0$. If the roots of this quadratic equation are α and β, write down, in terms of a, b, c, d, the values of $\alpha+\beta$ and $\alpha\beta$. Hence or otherwise prove that $\begin{pmatrix} a & b \\ c & d \end{pmatrix}\begin{pmatrix} b & b \\ \alpha-a & \beta-a \end{pmatrix} = \begin{pmatrix} b & b \\ \alpha-a & \beta-a \end{pmatrix}\begin{pmatrix} \alpha & 0 \\ 0 & \beta \end{pmatrix}$.

(*C*)

10 The plane isometry $\mathbf{x} \mapsto \mathbf{Mx}$ is such that the matrix **M** has eigenvalues 1 and -1, with corresponding eigenvectors **u** and **v**. It is given that for such an isometry $\mathbf{M}^{\mathrm{T}}\mathbf{M} = \mathbf{I}$. By considering $(\mathbf{Mu})^{\mathrm{T}}(\mathbf{Mv})$ in two different ways, show that **u** is perpendicular to **v**. Deduce the nature of the isometry by writing the position vector of a point of the plane in the form $\alpha\mathbf{u}+\beta\mathbf{v}$. Hence, or otherwise, find the image of the point (10, 10) under reflection in the line $y = 3x$. *(SMP)*

11 Write down the equation which express the fact that a matrix **A** has an eigenvector **v** with eigenvalue λ. If a 2×2 matrix **A** has independent eigenvectors $\mathbf{v}_1$, $\mathbf{v}_2$ with corresponding non-zero eigenvalues λ_1, λ_2, prove that $\alpha\mathbf{v}_1 + \beta\mathbf{v}_2$ (where $\alpha, \beta \neq 0$) can only be an eigenvector if $\lambda_1 = \lambda_2$. Given that such an eigenvector $\alpha\mathbf{v}_1 + \beta\mathbf{v}_2$ does exist, what can you say about the matrix **A**, its eigenvectors and the transformation which it represents? A 2×2 matrix **B** has *only one* independent eigenvector $\mathbf{v}_1$, with corresponding non-zero eigenvalue λ_1. If **B** transforms a certain vector $\mathbf{v}_2$, independent of $\mathbf{v}_1$, into $\kappa\mathbf{v}_1 + \mu\mathbf{v}_2$ (where $\kappa, \mu \neq 0$), find the image under the transformation of a general vector **v**, where $\mathbf{v} = \alpha\mathbf{v}_1 + \beta\mathbf{v}_2$ in the plane. Write down the condition for **v** to be an eigenvector, and show that this condition is satisfied if $\alpha = \kappa$ and $\beta = \mu - \lambda_1$. State why this is impossible unless $\mu = \lambda_1$. Hence show that the image of a *general* vector **v**, where $\mathbf{v} = \alpha\mathbf{v}_1 + \beta\mathbf{v}_2$ is $\lambda_2\mathbf{v} + \beta\kappa\mathbf{v}_1$. Deduce that the transformation is the combination of a shear and an enlargement (which is the identity if $\lambda_1 = 1$). *(SMP)*

12 Show that the characteristic equation of the matrix **A** given by

$$\mathbf{A} = \begin{pmatrix} 1 & -4 & 1 \\ -4 & 1 & 1 \\ 4 & 4 & 4 \end{pmatrix}$$

has two equal roots. In a linear transformation the point with position vector $\begin{pmatrix} x \\ y \\ z \end{pmatrix}$ becomes the point with position vector $\mathbf{A}\begin{pmatrix} x \\ y \\ z \end{pmatrix}$. Show that the straight line $x = y = -z$ is transformed into itself. Find the equation of the plane which is transformed into itself, but which does not contain this line. *(L)*

MISCELLANEOUS EXERCISE 14

1 Let S be the vector space of ordered triples of real numbers defined by $S = \{(x, y, z): x - 2y - 2z = 0\}$. Find a basis for S and state the dimension of S. *(JMB)*

2 The vectors $\mathbf{a}_1, \mathbf{a}_2, \mathbf{a}_3, \mathbf{a}_4$ form a basis for the vector space V over the field $\mathbb{R}$ of real numbers. The vector $\mathbf{y} = x_1\mathbf{a}_1 + x_2\mathbf{a}_2 + x_3\mathbf{a}_3 + x_4\mathbf{a}_4$, where $x_1, x_2, x_3, x_4 \in \mathbb{R}$. State for each of the following subsets of V whether or not it constitutes a subspace, giving brief but sufficient reasons for your answer: (i) all **y** with $x_1 = 0$, (ii) all **y** with $2x_1 + x_2 = 3$, (iii) all **y** with $x_1 \in \mathbb{Z}$. When the subset is a subspace, give a basis for the subspace in terms of $\mathbf{a}_1$, $\mathbf{a}_2$, $\mathbf{a}_3$ and $\mathbf{a}_4$. *(SMP)*

3 (a) Show that if $\mathbf{u}_1$, $\mathbf{u}_2$, $\mathbf{u}_3$ are linearly independent vectors, then $\mathbf{u}_1 + \mathbf{u}_2$, $\mathbf{u}_2 + \mathbf{u}_3$, $\mathbf{u}_3 + \mathbf{u}_1$ are also linearly independent. (b) The vector space V over $\mathbb{R}$ consists of $\{(z, t): z \in \mathbb{C} \text{ and } t \in \mathbb{R}\}$, with addition defined by $(z_1, t_1) + (z_2, t_2) = (z_1 + z_2, t_1 + t_2)$ and scalar multiplication defined by $k(z, t) = (kz, kt)$, where $k \in \mathbb{R}$.

(i) Verify that scalar multiplication is distributive over addition for these operations.

(ii) Show that (1, 0), (i, 0) and (0, 1) are linearly independent.
(iii) Prove that V is of dimension 3.
(iv) Using the result proved in (a), or otherwise, show that $\{(1+2\mathrm{i}, 1), (1+\mathrm{i}, 2), (2+\mathrm{i}, 1)\}$ is a basis for V.
(v) Given that $\{(z, (z+z^*)/2): z \in \mathbb{C}\}$ is a subspace of V, find a basis for this subspace. *(JMB)*

4 Show that the vectors $\mathbf{X}_1$, $\mathbf{X}_2$ and $\mathbf{X}_3$, where $\mathbf{X}_1 = \begin{pmatrix} 1 \\ 2 \\ 3 \end{pmatrix}$, $\mathbf{X}_2 = \begin{pmatrix} 3 \\ 2 \\ 2 \end{pmatrix}$, $\mathbf{X}_3 = \begin{pmatrix} 3 \\ 2 \\ 1 \end{pmatrix}$, are linearly independent. Express $\mathbf{X}_4$, where $\mathbf{X}_4 = \begin{pmatrix} 4 \\ 4 \\ 3 \end{pmatrix}$, as a linear combination of $\mathbf{X}_1, \mathbf{X}_2$ and $\mathbf{X}_3$. Find the value of a such that the vector $\begin{pmatrix} a \\ 3 \\ 4 \end{pmatrix}$ is orthogonal to $\mathbf{X}_4$. *(L)*

5 Show that the vectors $\mathbf{p}$, $\mathbf{q}$, $\mathbf{r}$, where $\mathbf{p} = \begin{pmatrix} 1 \\ 1 \\ 0 \end{pmatrix}$, $\mathbf{q} = \begin{pmatrix} 0 \\ 1 \\ 1 \end{pmatrix}$, $\mathbf{r} = \begin{pmatrix} 1 \\ 1 \\ 1 \end{pmatrix}$, form a basis for the linear space $\mathbb{R}^3$. Given that $\begin{pmatrix} x \\ y \\ z \end{pmatrix} = \alpha\mathbf{p} + \beta\mathbf{q} + \gamma\mathbf{r}$, find a matrix $\mathbf{Q}$ such that $\begin{pmatrix} \alpha \\ \beta \\ \gamma \end{pmatrix} = \mathbf{Q}\begin{pmatrix} x \\ y \\ z \end{pmatrix}$. Let L be the linear transformation from $\mathbb{R}^3$ to $\mathbb{R}^3$ defined by $\mathrm{L}\begin{pmatrix} x \\ y \\ z \end{pmatrix} = \mathbf{A}\begin{pmatrix} x \\ y \\ z \end{pmatrix}$, where $\mathbf{A}$ is the matrix $\begin{pmatrix} 2 & 0 & 1 \\ 0 & -4 & 1 \\ 3 & 1 & 0 \end{pmatrix}$. Show that the image of $\mathbf{p}$ under this transformation is $-8\mathbf{p} - 6\mathbf{q} + 10\mathbf{r}$, and find (also in terms of $\mathbf{p}$, $\mathbf{q}$ and $\mathbf{r}$) the images of $\mathbf{q}$ and $\mathbf{r}$ under the transformation. Given that the vector $\mathbf{u}$ is such that the image of $\mathbf{u}$ under the transformation L is $\mathbf{p}$, express $\mathbf{u}$ as a linear combination of $\mathbf{p}$, $\mathbf{q}$ and $\mathbf{r}$. *(C)*

6 In the four-dimensional vector space $\mathbb{R}^4$, let L be the subspace spanned by (1 1 0 0), (0 1 1 0) and (0 0 1 1), and let M be the subspace consisting of all vectors of the form $(\alpha \;\; 3\alpha \;\; \beta \;\; 2\beta)$. Write down an expression for a general vector in L. Show that the subspace $\mathrm{L} \cap \mathrm{M}$ *is one-dimensional.* *(SMP)*

7 (a) Show that the set S, given by $S = \left\{\begin{pmatrix} 1 \\ 0 \\ 1 \end{pmatrix}, \begin{pmatrix} 0 \\ -1 \\ 0 \end{pmatrix}, \begin{pmatrix} 0 \\ 1 \\ 2 \end{pmatrix}\right\}$ forms a basis for the linear (vector) space $\mathbb{R}^3$. (b) L is a linear transformation from $\mathbb{R}^3$ to $\mathbb{R}^3$ defined by $\mathrm{L}\begin{pmatrix} x \\ y \\ z \end{pmatrix} = \begin{pmatrix} 2x - y \\ x + z \\ 3x - 2y - z \end{pmatrix}$. (i) Find the null space of L, and state its dimension. (ii) Show that $\mathrm{L}\begin{pmatrix} 1 \\ 0 \\ 1 \end{pmatrix} = 2\begin{pmatrix} 1 \\ 0 \\ 1 \end{pmatrix} - 2\begin{pmatrix} 0 \\ -1 \\ 0 \end{pmatrix}$, and express $\mathrm{L}\begin{pmatrix} 0 \\ -1 \\ 0 \end{pmatrix}$ and $\mathrm{L}\begin{pmatrix} 0 \\ 1 \\ 2 \end{pmatrix}$ as linear combinations of the vectors in S. *(C)*

8 The set Q consists of the functions f defined for all real x by

$f(x) = e^x (a \cos 2x + b \sin 2x)$, where a and b are real. Thus Q is a vector space over the real numbers, with the functions f_1 and f_2 as a basis, where $f_1(x) = e^x \cos 2x$ and $f_2(x) = e^x \sin 2x$. With this basis for Q the element f may be expressed by the vector $\begin{pmatrix} a \\ b \end{pmatrix}$. Find a 2×2 matrix which premultiplies this vector to give the vector representing the derived function f' of f. Use this matrix to find in vector form (i) the second derivative function f'' of f, (ii) the element g of Q such that $g'(x) = e^x (a \cos 2x + b \sin 2x)$. *(JMB)*

9 Let V and W be subspaces of $\mathbb{R}^n$. Show that $V \cap W$ is also a subspace of $\mathbb{R}^n$. Find a basis for $V \cap W$ in the following cases:

(i) $V = \{(x_1, x_2, x_3, x_4) \in \mathbb{R}^4 : x_1 = x_3 = 0\}$,
$W = \{(x_1, x_2, x_3, x_4) \in \mathbb{R}^4 : x_4 = 2x_2\}$,

(ii) $V = \{(x_1, x_2, x_3, x_4) \in \mathbb{R}^4 : x_1 + 2x_2 = 0\}$,
$W = \{(x_1, x_2, x_3, x_4) \in \mathbb{R}^4 : 7x_4 = 3x_2 = 0\}$,

(iii) $V = \left\{\mathbf{X} \in \mathbb{R}^3 : \mathbf{AX} = \mathbf{0}, \text{ where } \mathbf{A} = \begin{pmatrix} 1 & 2 & 1 \\ 0 & 1 & 3 \end{pmatrix}\right\}$,

$W = \left\{\mathbf{X} \in \mathbb{R}^3 : \mathbf{BX} = \mathbf{0}, \text{ where } \mathbf{B} = \begin{pmatrix} 2 & 1 & -7 \\ -4 & -2 & 14 \end{pmatrix}\right\}$. *(C)*

10 Find a necessary and sufficient condition on α which ensures that the vectors $\begin{pmatrix} -1 \\ 4 \\ 7 \end{pmatrix}, \begin{pmatrix} 3 \\ -2 \\ 1 \end{pmatrix}, \begin{pmatrix} 4 \\ -1 \\ \alpha \end{pmatrix}$ are linearly independent in $\mathbb{R}^3$. Investigate the solution of the equations
$$\begin{aligned} -x + 4y + 7z &= 2, \\ 3x - 2y + z &= 4, \\ 4x - y + \alpha z &= \alpha + 2 \end{aligned}$$
for all values of the constant α. *(JMB)*

11 The vector space V consists of all 2×2 matrices with real elements. Show that **I**, **J**, **M**, **N**, given by $\mathbf{I} = \begin{pmatrix} 1 & 0 \\ 0 & 1 \end{pmatrix}$, $\mathbf{J} = \begin{pmatrix} 0 & -1 \\ 1 & 0 \end{pmatrix}$, $\mathbf{M} = \begin{pmatrix} 0 & 1 \\ 1 & 0 \end{pmatrix}$, $\mathbf{N} = \begin{pmatrix} 1 & 0 \\ 0 & -1 \end{pmatrix}$ are linearly independent vectors in V. Show that the matrix **A**, given by $\mathbf{A} = \begin{pmatrix} a & b \\ c & d \end{pmatrix}$, can be written as the sum of two matrices **P** and **R**, given by $\mathbf{P} = \begin{pmatrix} p & q \\ q & p \end{pmatrix}$ and $\mathbf{R} = \begin{pmatrix} r & -s \\ s & -r \end{pmatrix}$. Hence show that $\{\mathbf{I}, \mathbf{J}, \mathbf{M}, \mathbf{N}\}$ forms a basis of V. *(JMB)*

12 Explain what is meant by the statement that the m column vectors $\mathbf{x}_1, \mathbf{x}_2, \ldots, \mathbf{x}_m$, where each vector consists of n elements, are linearly independent. Find the matrix **A**, where $\mathbf{y} = \mathbf{Ax}$ represents a linear transformation for which $\begin{pmatrix} 3 \\ 1 \\ 2 \end{pmatrix} = \mathbf{A}\begin{pmatrix} 1 \\ 0 \\ 0 \end{pmatrix}, \begin{pmatrix} 2 \\ 3 \\ 1 \end{pmatrix} = \mathbf{A}\begin{pmatrix} 0 \\ 1 \\ 0 \end{pmatrix}, \begin{pmatrix} 3 \\ 2 \\ 1 \end{pmatrix} = \mathbf{A}\begin{pmatrix} 0 \\ 0 \\ 1 \end{pmatrix}$.
Find the images of $\mathbf{x}_1$, $\mathbf{x}_2$ and $\mathbf{x}_3$ under this transformation, where $\mathbf{x}_1 = \begin{pmatrix} 1 \\ 1 \\ 1 \end{pmatrix}$, $\mathbf{x}_2 = \begin{pmatrix} 4 \\ 3 \\ -1 \end{pmatrix}$ and $\mathbf{x}_3 = \begin{pmatrix} 5 \\ 4 \\ 0 \end{pmatrix}$. Show that (a) $\mathbf{x}_1$ and $\mathbf{x}_2$ are

linearly independent, (b) $\mathbf{Ax}_1$ and $\mathbf{Ax}_2$ are linearly independent, (c) $\mathbf{x}_1, \mathbf{x}_2$ and $\mathbf{x}_3$ are linearly dependent. *(L)*

13 In each of the following cases, state whether, with the usual operations, the given set forms a linear space over the field $\mathbb{R}$. For each of those which you consider to be a linear space, give a basis for the space, and for each of those which you consider not to be a linear space, justify your answer.

(i) $S_1 = \{\text{solutions of } \frac{d^2y}{dx^2} - 3\frac{dy}{dx} + 2y = 0\}$.

(It may be assumed that the general solution of the differential equation is $y = Ae^{2x} + Be^x$.)

(ii) $S_2 = \left\{\text{solutions of } \frac{d^2y}{dx^2} - 3\frac{dy}{dx} + 2y = 2x - 3\right\}$.

(It may be assumed that the general solution of the differential equation is $y = Ae^{2x} + Be^x + x$.)

(iii) $S_3 = \{(x, y): y = 2x + 1\}$.

(iv) $S_4 = \{(x, y, z): x + y + z = 0\}$.

(v) $S_5 = \{z: z \in \mathbb{C}\}$. *(C)*

14 (i) Let V and W be subspaces of the linear (vector) space $\mathbb{R}^4$. The set of vectors of the form $\mathbf{v} + \mathbf{w}$, where $\mathbf{v} \in V$ and $\mathbf{w} \in W$ is denoted by $V + W$. Show that $V + W$ is also a subspace of $\mathbb{R}^4$.

(ii) S is the subspace of $\mathbb{R}^4$ spanned by the set of vectors $\{(1\ 0\ 1\ 1), (5\ 0\ 2\ 2), (-2\ 0\ 1\ 1), (3\ 0\ 0\ 0)\}$ and T is the subspace of $\mathbb{R}^4$ spanned by the set of vectors $\{(1\ -1\ 0\ 1), (0\ 0\ 0\ 0), (0\ 1\ 1\ 0), (0\ 3\ 3\ 0)\}$. Give bases for the subspaces $S, T, S \cap T, S + T$. Give also a basis for a subspace U of $\mathbb{R}^4$, which is such that $S \cap U = \{(0\ 0\ 0\ 0)\}$ and $S + U = \mathbb{R}^4$. *(C)*

15 The following statements concern vectors in a vector space. Two of the statements are true and one is false. Give a proof for each true statement and provide a counter-example for the false statement. (i) The vectors $\mathbf{a}$ and $\mathbf{b}$ span the same subspace as $\mathbf{a} + \mathbf{b}$ and $\mathbf{a} - \mathbf{b}$. (ii) If $\mathbf{a}$, $\mathbf{b}$, $\mathbf{c}$ are linearly independent vectors and $\mathbf{d}$ is any other vector, then $\mathbf{a} + \mathbf{d}$, $\mathbf{b} + \mathbf{d}$, $\mathbf{c} + \mathbf{d}$ are always linearly independent vectors. (iii) If $\mathbf{a}$, $\mathbf{b}$, $\mathbf{c}$ are linearly independent vectors and k is a non-zero scalar, then $k\mathbf{a}$, $k\mathbf{b}$, $k\mathbf{c}$ are linearly independent vectors. *(JMB)*

16 Given that $\mathbf{p} = \begin{pmatrix} 1 \\ 1 \\ 1 \end{pmatrix}$, $\mathbf{A} = \begin{pmatrix} 1 & -1 & 1 \\ -1 & 2 & -1 \\ 2 & -1 & 1 \end{pmatrix}$, $\mathbf{B} = \begin{pmatrix} -1 & 0 & 1 \\ 1 & 1 & 0 \\ 3 & 1 & -1 \end{pmatrix}$, show that $\mathbf{AB} = \mathbf{BA}$, and that the column vectors of $\mathbf{A}$ are linearly independent. Express $\mathbf{p}$ as a linear combination of the column vectors of $\mathbf{A}$. Find the point which maps on to the point (1, 1, 1) under the transformation with matrix $\mathbf{A}$, and the point on to which the point (1, 1, 1) is mapped under the transformation with matrix $\mathbf{B}$. *(L)*

17 Let $\mathbf{c}_1 = \begin{pmatrix} 1 \\ 1 \end{pmatrix}$, $\mathbf{c}_2 = \begin{pmatrix} -1 \\ 1 \end{pmatrix}$, $\mathbf{c}_3 = \begin{pmatrix} 2 \\ 1 \end{pmatrix}$, $\mathbf{c}_4 = \begin{pmatrix} 1 \\ 3 \end{pmatrix}$. Find linear relations between $\mathbf{c}_1, \mathbf{c}_2, \mathbf{c}_3$ and between $\mathbf{c}_2, \mathbf{c}_3, \mathbf{c}_4$. Hence, or otherwise, find the kernel of the transformation T given by $T\begin{pmatrix} x \\ y \\ z \\ t \end{pmatrix} = \begin{pmatrix} 1 & -1 & 2 & 1 \\ 1 & 1 & 1 & 3 \end{pmatrix}\begin{pmatrix} x \\ y \\ z \\ t \end{pmatrix}$, and solve the equations $x - y + 2z + t = 1$, $x + y + z + 3t = 5$. *(SMP)*

18 The linear transformation σ: $\mathbb{R}^4 \to \mathbb{R}^3$ is represented by the matrix $\begin{pmatrix} 1 & -4 & 5 & -2 \\ 0 & 9 & -3 & 3 \\ 4 & 5 & 13 & -1 \end{pmatrix}$ with respect to the standard bases of $\mathbb{R}^4$ and $\mathbb{R}^3$. Find bases for the null space of σ and for the range (image space) of σ. Find the set of vectors in $\mathbb{R}^4$ which are mapped to the vector $\begin{pmatrix} 0 \\ 9 \\ 21 \end{pmatrix}$ by σ. (*C*)

19 (i) Find the values of k for which the equations $\begin{matrix} -x-6y+4z=kz \\ -x-4y+z=ky \\ x+3y+2z=kx \end{matrix}$ have a solution other than $x=y=z=0$. (ii) A mapping of three-dimensional space is defined by the matrix $\mathbf{M}$ where $\mathbf{M}=\begin{pmatrix} 0 & 2 & 4 \\ 1 & 0 & 2 \\ -1 & 1 & 0 \end{pmatrix}$. Interpret in geometric terms the mappings defined by $\mathbf{M}^2$ and $\mathbf{M}^3$. (*L*)

20 Find the inverse $\mathbf{A}^{-1}$ and the characteristic equation of the matrix $\mathbf{A}$ given by $\mathbf{A}=\begin{pmatrix} 3 & -2 & 3 \\ 1 & 2 & 1 \\ 1 & 3 & 0 \end{pmatrix}$. Find a constant λ and a non-zero vector $\mathbf{x}$ such that $\mathbf{A}\mathbf{x}=\mathbf{A}^{-1}\mathbf{x}=\lambda\mathbf{x}$. (*L*)

21 Let V be the subspace of $\mathbb{R}^3$ (three dimensional space) spanned by $\mathbf{v}_1=\begin{pmatrix} 1 \\ -1 \\ -1 \end{pmatrix}$, $\mathbf{v}_2=\begin{pmatrix} 2 \\ 1 \\ -1 \end{pmatrix}$ and $\mathbf{v}_3=\begin{pmatrix} -1 \\ -5 \\ -1 \end{pmatrix}$. Determine the dimension of V. The linear transformation T is given by

$$\mathbf{w}=\mathrm{T}(\mathbf{x})=\begin{pmatrix} 1 & -1 & 2 \\ 1 & 0 & 1 \\ -2 & 1 & -3 \end{pmatrix}\mathbf{x}.$$

(i) Show that all points in $\mathbb{R}^3$ which are mapped by T on to the origin O are points in V. (ii) Show that all points in V are mapped on to a line, and give its equations. (iii) Let A, B be any two points in $\mathbb{R}^3$. Show that AB is perpendicular to the plane $x-y-z=0$ if A and B are mapped by T on to the same point. (*JMB*)

22 Show that $\left\{\begin{pmatrix} 1 \\ 2 \end{pmatrix}, \begin{pmatrix} 3 \\ 7 \end{pmatrix}\right\}$ is a basis for $\mathbb{R}^2$, and express $\begin{pmatrix} x \\ y \end{pmatrix}$ in the form $\alpha\begin{pmatrix} 1 \\ 2 \end{pmatrix}+\beta\begin{pmatrix} 3 \\ 7 \end{pmatrix}$, where α, β are to be found in terms of x, y. Let L: $\mathbb{R}^3 \to \mathbb{R}^2$ be the linear mapping defined by $\mathrm{L}\begin{pmatrix} x \\ y \\ z \end{pmatrix}=\begin{pmatrix} x+2y-z \\ 2x-y+z \end{pmatrix}$. Find the images under L of the vectors $\begin{pmatrix} 0 \\ 0 \\ 1 \end{pmatrix}, \begin{pmatrix} 0 \\ 1 \\ 1 \end{pmatrix}, \begin{pmatrix} 1 \\ 1 \\ 1 \end{pmatrix}$. Hence, or otherwise, find the matrix of L with respect to the bases $\left\{\begin{pmatrix} 0 \\ 0 \\ 1 \end{pmatrix}, \begin{pmatrix} 0 \\ 1 \\ 1 \end{pmatrix}, \begin{pmatrix} 1 \\ 1 \\ 1 \end{pmatrix}\right\}$ and $\left\{\begin{pmatrix} 1 \\ 2 \end{pmatrix}, \begin{pmatrix} 3 \\ 7 \end{pmatrix}\right\}$ of $\mathbb{R}^3$ and $\mathbb{R}^2$ respectively. (*C*)

15 Conics

The parabola

In the Cartesian plane, let S be the point $(a, 0)$ and let D be the line $x = -a$. The parabola with focus S and directrix D is defined to be the locus of points P equidistance from S and D, Fig. 15.1. Let PN be the perpendicular from P (x, y) on to D so that N is the point $(-a, y)$. Then $PN = PS$ so $x + a = \sqrt{((x-a)^2 + y^2)}$, that is

$$x^2 + 2ax + a^2 = x^2 - 2ax + a^2 + y^2 \text{ or } y^2 = 4ax.$$

This is the equation of the parabola. For the corresponding polar form, take the pole at S and SO as the initial line, so that $SP = r$ and angle $PSO = \theta$. Then $PN = 2a - r\cos\theta$ and the parabola is the locus

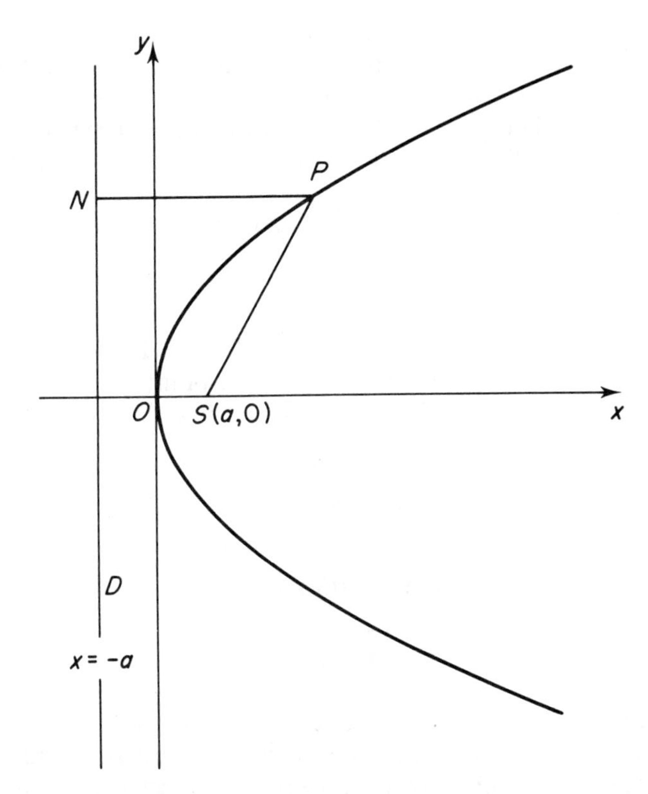

Fig. 15.1 The parabola $y^2 = 4ax$

$r(1+\cos\theta) = 2a$. (This is the form of the general conic, with $e = 1$ and $ep = 2a$, see §7.2.)

EXAMPLE 1 *Find the equation of the parabola with focus at* (3, 2) *and directrix* $y = 0$.

The point (x, y) is equidistant from (3, 2) and from $y = 0$ if $(x-3)^2+(y-2)^2 = y^2$ so the equation of the parabola is

$$\mathbf{x^2 - 6x - 4y + 13 = 0.}$$

EXAMPLE 2 *Show that the equation* $x^2+4x = 4y$ *represents a parabola. Find its focus and directrix.*

Complete the square of the terms in x,

$$x^2+4x+4 = 4y+4, \text{ or } (x+2)^2 = 4(y+1).$$

Changing the origin to $(-2, -1)$, with new coordinates (X, Y) given by $X = x+2$, $Y = y+1$, the equation becomes $X^2 = 4Y$, which represents a parabola with focus at $X = 0$, $Y = 1$, and directrix $Y = -1$. Referred to the original axes, the parabola has focus $\mathbf{(-2, 0)}$ and directrix $\mathbf{y = -2}$.

The ellipse

The ellipse with focus S, directrix D and eccentricity e is defined as the locus of a point P whose distance from the point S is e times its distance from the line D, where $0 < e < 1$. By choosing the Cartesian axes carefully, it is possible to obtain a simple form for the Cartesian equation of the ellipse.

Let the distance of S from D be p and define a by the equation

$$p = a/e - ea, \text{ so that } a = ep/(1-e^2).$$

Then choose the Cartesian axes so that S is the point $(ea, 0)$ and the perpendicular from S on to D is the axis Ox, see Fig. 15.2.
Then D is the line $x = a/e$ and, if P is the point (x, y), let PN be the perpendicular from P on to D. Then $SP = e\,PN$ if $SP^2 = e^2PN^2$, that is,

$$\begin{aligned}(x-ea)^2+y^2 &= e^2(a/e-x)^2\\ x^2-2aex+e^2a^2+y^2 &= a^2-2aex+e^2x^2,\\ x^2(1-e^2)+y^2 &= a^2(1-e^2).\end{aligned}$$

Let $b^2 = a^2(1-e^2)$, then the equation becomes

$$\frac{x^2}{a^2}+\frac{y^2}{b^2} = 1.$$

For all points on the ellipse, $-a \leqslant x \leqslant a$ and $-b \leqslant y \leqslant b$. By symmetry of the equation in x and y, the ellipse is symmetrical about both axes and has a second focus S' at $(-ea, 0)$ and a second directrix D', $x = -a/e$. The ellipse crosses the axes at $A(a, 0)$, $A'(-a, 0)$, $B(0, b)$, $B'(0, -b)$. A and A' are

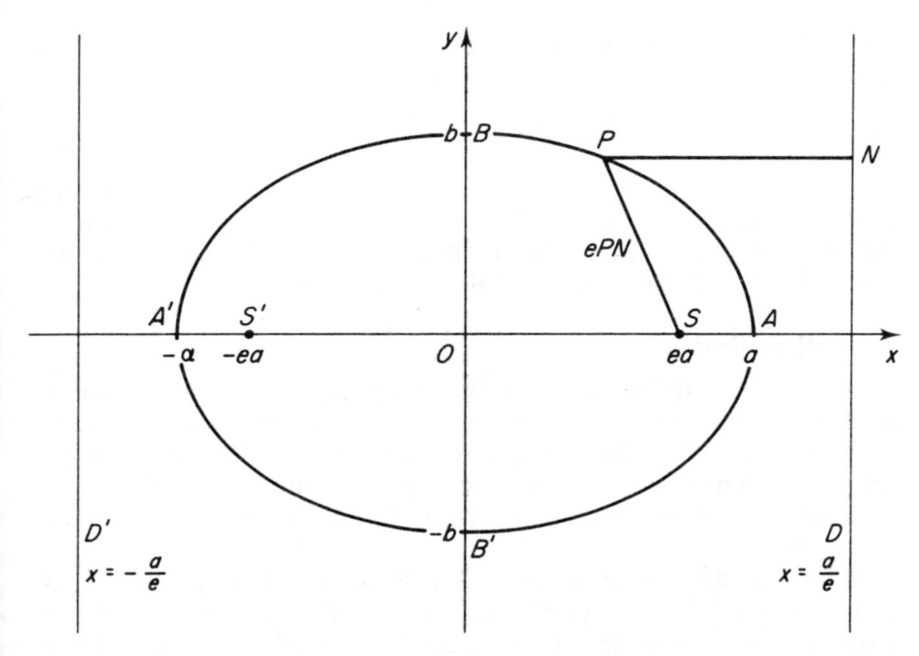

Fig. 15.2 The ellipse $x^2/a^2 + y^2/b^2 = 1$

the points on the ellipse furthest from O, B and B' are the points nearest to O. The line AA' is called the major axis of the ellipse and BB' is the minor axis.

For the polar equation, take the pole at S and the initial line SA, so that $SP = r$ and angle $PSA = \theta$. Then $PN = a/e - ae - r\cos\theta$ and $r = e(a/e - ae - r\cos\theta)$, giving the polar equation

$$r(1 + e\cos\theta) = a(1 - e^2) = ep, \text{ as found in §7.2.}$$

EXAMPLE 3 *Let S, S' be the foci of the ellipse with the equation*

$$\frac{x^2}{a^2} + \frac{y^2}{b^2} = 1, 0 < b < a.$$

Prove that, for a point P on the ellipse, $SP + S'P = 2a$.

Using the figure 15.2, let PN and PN' be the perpendiculars from P on to D and D', then $SP + S'P = ePN + ePN' = e(PN + PN') = e(2a/e) = 2a$.

EXAMPLE 4 *Find the equation of the ellipse with eccentricity $1/2$, focus at the point $(4, 0)$ and directrix $x = 16$.*

The ellipse is the locus of a point whose distance from $(4, 0)$ is one half of its distance from the line $x = 16$. This is given by

$$(x - 4)^2 + y^2 = (x - 16)^2/4, \quad \text{or} \quad x^2 - 8x + 16 + y^2 = x^2/4 - 8x + 64,$$

$$\mathbf{3x^2/4 + y^2 = 48, \quad or \quad x^2/64 + y^2/48 = 1.}$$

EXAMPLE 5 *Find the coordinates of the foci and the equations of the directrices of the ellipse with equation* $25x^2+4y^2 = 100$.

The equation can be written $x^2/4+y^2/25 = 1$, and this is an ellipse with major axis $x = 0$. Writing the equation
$y^2/a^2+x^2/b^2 = 1, a = 5, b = 2$, so the eccentricity e is given by $(1-e^2)a^2 = b^2$ or $(1-e^2)25 = 4$. Thus $e^2 = (25-4)/25, e = \sqrt{(21)}/5$. Then $ae = \sqrt{(21)}$ so that the foci are at $\mathbf{(0, \sqrt{(21)})}$, $\mathbf{(0, -\sqrt{(21)})}$. The directrices are the lines $ey = a$ and $ey = -a$, that is the lines $\mathbf{y\sqrt{(21)} = 25}$ and $\mathbf{y\sqrt{(21)} = -25}$.

The hyperbola

The hyperbola with focus S, directrix D and eccentricity $e, e > 1$, is defined as the locus of a point P whose distance from S is e times its distance from D. As with the ellipse, by appropriate choice of Cartesian axes, a simple form of the equation of the hyperbola can be found. Again, let p be the distance of S from D and define $a = ep/(e^2-1)$, so that $p = ea-a/e$ and D is the line $x = a/e$, see Fig. 15.3.
Let PN be the perpendicular from P on to D, then $SP = e\,PN$ and $(x-ea)^2+y^2 = e^2(a/e-x)^2$, so $x^2(e^2-1)-y^2 = a^2(e^2-1)$. Let $b^2 = a^2(e^2-1)$ and then the equation becomes $x^2/a^2-y^2/b^2 = 1$. For all points on the hyperbola, $|x| \geqslant a$ and the points $A\,(a, 0)$ and $A'\,(-a, 0)$ are the nearest points to the origin. Again, by symmetry, the hyperbola is symmetric about both axes. Its major axis is AA', it has a second focus at S' $(-ea, 0)$ and a second directrix D', $x = -a/e$. In polar form, with pole at S and initial line SO, so that $PS = r$ and angle $PSO = \theta$,
$r = ePN = e(ae-a/e-r\cos\theta)$. This gives the polar equation
$r(1+e\cos\theta) = a(e^2-1) = ep$, as in §7.2.

EXAMPLE 6 *Given the foci S and S′ of the hyperbola* $x^2/a^2-y^2/b^2 = 1$, *prove that, for a point P on the hyperbola,* $|SP-S'P| = 2a$.

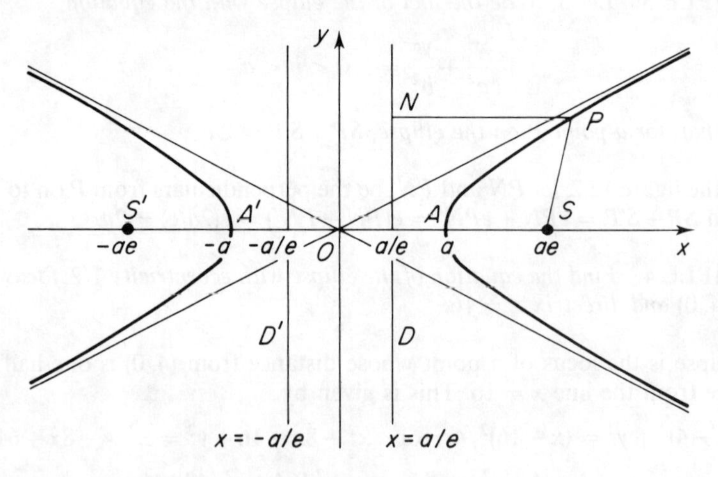

Fig. 15.3 The hyperbola $x^2/a^2-y^2/b^2 = 1$

From Fig. 15.3, if N, N' are the feet of the perpendiculars from P on to D and D', $|SP - S'P| = e|PN - PN'| = eNN' = e(2a/e) = 2a$.

EXAMPLE 7 *Identify the conic* $x^2 + 2\sqrt{3}xy - y^2 = 2$.

Write the equation $f(x, y) = 1$, where $f(x, y) = x^2/2 + \sqrt{3}xy - y^2/2$. Then $f(x, y)$ is the quadratic form given by $f(x, y) = \mathbf{r}^T\mathbf{M}\mathbf{r}$, where $\mathbf{r} = \begin{pmatrix} x \\ y \end{pmatrix}$ and $\mathbf{M} = \begin{pmatrix} 1/2 & \sqrt{3}/2 \\ \sqrt{3}/2 & -1/2 \end{pmatrix}$.

By example 3 of §14.4, the orthogonal transformation of coordinates $\begin{pmatrix} x \\ y \end{pmatrix} = \begin{pmatrix} \sqrt{3}/2 & -1/2 \\ -1/2 & \sqrt{3}/2 \end{pmatrix} \begin{pmatrix} x \\ y \end{pmatrix}$ diagonalises the quadratic form and transforms the equation to $X^2 - Y^2 = 1$. Since the transformation is an isometry, the shape of the conic is unaltered, so it is a hyperbola with eccentricity e, given by $1 = e^2 - 1$, that is $e = \sqrt{2}$.

Asymptotes

Using the same arguments as were used in *Pure Mathematics*, chapter 27, the asymptotes to the hyperbola are obtained by considering the terms of degree 2 in the equation, since these will be the dominant terms for large values of x and y. The hyperbola $x^2/a^2 - y^2/b^2 = 1$ has asymptotes given by $x^2/a^2 - y^2/b^2 = 0$. This equation factorises, $(x/a - y/b)(x/a + y/b) = 0$ so it represents the pair of lines $x/a = y/b$ and $x/a = -y/b$, referred to as a line pair.

EXAMPLE 8 *Determine the shape of the curve with equation*

$$x^2 + 4xy + y^2 - 2x = 1.$$

The curve is a conic and we consider the possibility of the existence of real asymptotes. The equation of the asymptotes (if any) is $x^2 + 4xy + y^2 = 0$. Completing the square of the terms involving x, this becomes $(x + 2y)^2 - 3y^2 = 0$, which factorises into a line pair $(x + (2 + \sqrt{3})y)(x + (2 - \sqrt{3})y) = 0$. Thus the conic has real asymptotes, so it is a **hyperbola**.

EXERCISE 15

1 Find the focus and the directrix of the parabola with equation:
(i) $y^2 = -4x$, (ii) $x^2 = y$, (iii) $x^2 = -9y$, (iv) $(y-2)^2 = 4(x-1)$,
(v) $y^2 + 2y + x = 0$, (vi) $x^2 + x + y = 1$.

2 Find the Cartesian equation of the parabola:
(i) with focus $(-1, 0)$ and directrix $x = 1$,
(ii) with focus $(-1, 0)$ and directrix $x = 0$,
(iii) with focus $(0, 1)$ and directrix $y = 3$,
(iv) with focus $(1, 2)$ and directrix $y = -1$,
(v) with focus $(2, 2)$ and directrix $x + y = 0$.

3 Find the eccentricity, foci and directrices of the ellipse with equation: (i) $4x^2+9y^2=36$, (ii) $x^2+2y^2=4$, (iii) $x^2/16+y^2/25=1$.

4 Find the equation of the ellipse:
(i) with eccentricity 1/2, focus $(1, 0)$ and directrix $x=4$,
(ii) with eccentricity 1/2, focus $(-1, 0)$ and directrix $x=-4$,
(iii) with eccentricity 1/3, focus $(0, 2)$ and directrix $y=18$,
(iv) with eccentricity e, focus $(0, ek)$ and directrix $y=k/e$.

5 Obtain the equation of the ellipse having major axis of length 10 units and foci at the points $(3, 0)$ and $(-3, 0)$. *(L)*

6 Find the coordinates of the foci of the ellipse whose equation is $16x^2+25y^2=100$. *(L)*

7 Find the equation of the hyperbola:
(i) with eccentricity 2, focus $(2, 0)$ and directrix $x=1/2$,
(ii) with eccentricity 4, focus $(-5, 0)$ and directrix $x+1=0$,
(iii) with focus $(0, 3)$ and vertices at $(0, 1)$ and $(0, -1)$,
(iv) with focus $(4, 0)$ and with asymptotes $4y=3x$, $4y=-3x$.

8 Find the set of values of h for which the curve

$$3x^2+2hxy+5y^2+7x+13y+19=0$$

has distinct real asymptotes. *(L)*

9 A curve has equation $x^2-y^2+10x-2y=0$.
Write down the equation of the tangent to the curve at the origin, and find the angle between the asymptotes of the curve. *(L)*

10 Find the coordinates of the foci of the ellipse whose equation is $64x^2+25y^2=100$.

11 Use the focus-directrix property to obtain, in any form, the equation of the parabola which has the point $(4, 1)$ as focus and the line $y=2$ as directrix. *(L)*

12 Find the equation of the tangent to the parabola $y^2=4ax$ at the point $T\ (at^2, 2at)$. If S is the focus, find the equation of the chord QSR which is parallel to the tangent at T. Prove that $QR=4TS$. *(L)*

13 Find the equation of the parabola whose directrix is the y-axis and whose focus is the point $(2, 0)$. Find also the equations of two tangents to the parabola, one parallel to and the other perpendicular to the line $y=mx$, where $m\neq 0$. Write down the coordinates of P, the point of intersection of these two tangents and determine the locus of P as m varies. *(L)*

14 The distance between the foci of an ellipse is 8 units, and the distance between the directrices is 18 units. Find the equation of the ellipse referred to its centre as origin and to the major axis as axis of x. *(L)*

15 Assuming the focus-directrix property for an ellipse, prove that the sum of the focal distances of any point P on the ellipse is equal to the length of the major axis of the ellipse. *(L)*

16 The foci of the ellipse $x^2/a^2+y^2/b^2=1$ are S, S'. Show that the normal at a point P on the ellipse is equally inclined to the lines SP, $S'P$. The perpendicular from the centre O of the ellipse to the tangent at P meets SP, produced if necessary, in G. Show that the locus of G is a circle of radius a whose centre is S. *(L)*

17 A point P moves so that its distance from the line $x=12$ is twice its distance from the point $Q(3, 0)$. Find the equation of the locus of P, and sketch this locus, showing the coordinates of the points where it crosses the axes of coordinates. (*L*)

18 State the equations of the asymptotes of the hyperbola

$$x^2/a^2 - y^2/b^2 = 1$$

The point P on the curve lies in the first quadrant. The line through P parallel to Oy meets an asymptote of the curve at the point Q which also lies in the first quadrant. The normal at P meets the x-axis at G. Prove that QG is perpendicular to the asymptote. (*L*)

19 A hyperbola of the form $x^2/a^2 - y^2/b^2 = 1$ has asymptotes $y^2 = m^2x^2$ and passes through the point $(c, 0)$. Find the equation of the hyperbola in terms of x, y, c and m. A point P on this hyperbola is equidistant from one of its asymptotes and the x-axis. Prove that, for all values of m, P lies on the curve $(x^2 - y^2)^2 = 4x^2(x^2 - c^2)$. (*L*)

20 Write down the equation of the tangent at the origin to the curve with equation $x^2 + 2xy + 3y^2 + 2x - y = 0$. By considering only the second degree terms in the equation, establish whether the curve has real asymptotes, and hence determine whether it is an ellipse or a hyperbola. (*L*)

Formulae

Algebraic structures

1. A *group* G is a set of elements $a, b, c, \ldots$ with a binary operation $*$ such that:
   (i) $a*b$ is in G for all a, b in G,
   (ii) $a*(b*c) = (a*b)*c$ for all a, b, c in G,
   (iii) G contains an element e, called the identity element, such that $e*a = a = a*e$ for all a in G,
   (iv) given any a in G, there exists in G an element a^{-1}, called the element inverse to a, such that $a^{-1}*a = e = a*a^{-1}$.

A *commutative* (or *Abelian*) *group* is one for which $a*b = b*a$ for all a, b in G.

2. A *field* F is a set of elements $a, b, c, \ldots$ with two binary operations $+$ and $.$ such that:
   (i) F is a commutative group with respect to $+$ with identity 0,
   (ii) the non-zero elements of F form a commutative group with respect to $.$ with identity 1,
   (iii) $a.(b+c) = a.b + a.c$ for all a, b, c in F.

3. A *vector space* V over a field F is a set of elements $\mathbf{a}, \mathbf{b}, \mathbf{c}, \ldots$ with a binary operation $+$ such that:
   (i) they form a commutative group under $+$; and, for all λ, μ in F and all $\mathbf{a}, \mathbf{b}$ in V,
   (ii) $\lambda\mathbf{a}$ is defined and is in V,
   (iii) $\lambda(\mathbf{a}+\mathbf{b}) = \lambda\mathbf{a} + \lambda\mathbf{b}$,
   (iv) $(\lambda+\mu)\mathbf{a} = \lambda\mathbf{a} + \mu\mathbf{a}$,
   (v) $(\lambda.\mu)\mathbf{a} = \lambda(\mu\mathbf{a})$,
   (vi) if 1 is an element of F such that $1.\lambda = \lambda$ for all λ in F, then $1\mathbf{a} = \mathbf{a}$.

4. An *equivalence relation* R between the elements $a, b, c, \ldots$ of a set C is a relation such that, for all a, b, c in C:
   (i) aRa (R is reflexive),
   (ii) $aRb \Rightarrow bRa$ (R is symmetric),
   (iii) $(aRb$ and $bRc) \Rightarrow aRc$ (R is transitive).

Finite series

1. $\sum_{r=1}^{n} r = \frac{1}{2}n(n+1)$

2. $\sum_{r=1}^{n} r^2 = \frac{1}{6}n(n+1)(2n+1)$

3. $\sum_{r=1}^{n} r^3 = \frac{1}{4}n^2(n+1)^2$

Arithmetic series

nth term is $a+(n-1)d$

$S_n = \frac{1}{2}n(a+l) = \frac{1}{2}n\{2a+(n-1)d\}$

Geometric series

nth term is ar^{n-1}

$$S_n = \frac{a(1-r^n)}{1-r}$$

$$S_\infty = \frac{a}{1-r} \text{ for } |r| < 1$$

Power series

1. $$(1+x)^n = 1 + nx + \frac{n(n-1)}{2!}x^2 + \ldots + \binom{n}{r}x^r + \ldots,$$

where $\binom{n}{r} = \frac{n(n-1)(n-2)\ldots(n-r+1)}{r!}$

If n is a positive integer, the series terminates and is convergent for all x. If n is not a positive integer, the series is infinite and converges for $|x| < 1$.

2. $e^x = 1 + x + \frac{x^2}{2!} + \ldots + \frac{x^r}{r!} + \ldots$ (all x)

3. $\ln(1+x) = x - \frac{x^2}{2} + \frac{x^3}{3} - \ldots + (-1)^{r+1}\frac{x^r}{r} + \ldots \; (-1 < x \leqslant 1)$

4. $\cos x = \frac{e^{ix}+e^{-ix}}{2} = 1 - \frac{x^2}{2!} + \frac{x^4}{4!} - \ldots + \frac{(-1)^r x^{2r}}{(2r)!} + \ldots$ (all x)

$\sin x = \frac{e^{ix}-e^{-ix}}{2i} = x - \frac{x^3}{3!} + \frac{x^5}{5!} - \ldots + \frac{(-1)^r x^{2r+1}}{(2r+1)!} + \ldots$ (all x)

5. $\cosh x = \dfrac{e^x + e^{-x}}{2} = 1 + \dfrac{x^2}{2!} + \dfrac{x^4}{4!} + \ldots + \dfrac{x^{2r}}{(2r)!} + \ldots$ (all x)

$\sinh x = \dfrac{e^x - e^{-x}}{2} = x + \dfrac{x^3}{3!} + \dfrac{x^5}{5!} + \ldots + \dfrac{x^{2r+1}}{(2r+1)!} + \ldots$ (all x)

6. Taylor's series

$$f(x+h) = f(x) + hf'(x) + \frac{h^2}{2!}f''(x) + \ldots + \frac{h^{n-1}}{(n-1)!}f^{(n-1)}(x) + \ldots$$

Inverse hyperbolic functions

$\sinh^{-1} x = \ln[x + \sqrt{(x^2+1)}]$

$\cosh^{-1} x = \ln[x \pm \sqrt{(x^2-1)}]$ $(x \geqslant 1)$

$\tanh^{-1} x = \frac{1}{2}\ln\left[\dfrac{1+x}{1-x}\right]$ $(|x| < 1)$

Trigonometry

$$\cos^2 A + \sin^2 A = 1$$

$$\sec^2 A = 1 + \tan^2 A$$

$$\operatorname{cosec}^2 A = 1 + \cot^2 A$$

$$\cos(A \pm B) = \cos A \cos B \mp \sin A \sin B$$

$$\sin(A \pm B) = \sin A \cos B \pm \cos A \sin B$$

$$\tan(A \pm B) = \frac{\tan A \pm \tan B}{1 \mp \tan A \tan B}$$

$$\cos 2A = \cos^2 A - \sin^2 A$$

$$\sin 2A = 2 \sin A \cos A$$

$$\tan 2A = \frac{2 \tan A}{1 - \tan^2 A}$$

$$\text{If } t = \tan\tfrac{1}{2}A,\ \sin A = \frac{2t}{1+t^2},\ \cos A = \frac{1-t^2}{1+t^2}$$

$$2 \sin A \cos B = \sin(A+B) + \sin(A-B)$$

$$2 \cos A \cos B = \cos(A+B) + \cos(A-B)$$

$$2 \sin A \sin B = \cos(A-B) - \cos(A+B)$$

$$\sin A + \sin B = 2 \sin\frac{A+B}{2} \cos\frac{A-B}{2}$$

$$\sin A - \sin B = 2\cos\frac{A+B}{2}\sin\frac{A-B}{2}$$

$$\cos A + \cos B = 2\cos\frac{A+B}{2}\cos\frac{A-B}{2}$$

$$\cos A - \cos B = -2\sin\frac{A+B}{2}\sin\frac{A-B}{2}$$

$$a\cos\theta + b\sin\theta = R\cos(\theta - \alpha), \text{ where } R = \sqrt{(a^2+b^2)}$$
$$\text{and } \cos\alpha = a/R,\ \sin\alpha = b/R.$$

In the triangle ABC $\quad \dfrac{a}{\sin A} = \dfrac{b}{\sin B} = \dfrac{c}{\sin C} = 2R$

$$a^2 = b^2 + c^2 - 2bc\cos A$$

$$\text{area} = \tfrac{1}{2}ab\sin C$$

Ranges of the inverse trigonometric functions

$$-\tfrac{1}{2}\pi \leqslant \sin^{-1} x \leqslant \tfrac{1}{2}\pi$$
$$0 \leqslant \cos^{-1} x \leqslant \pi$$
$$-\tfrac{1}{2}\pi < \tan^{-1} x < \tfrac{1}{2}\pi$$

Differentiation

function	*derived function*
uv	$u'v + uv'$
u/v	$(u'v - uv')/v^2$
composite	$\dfrac{dz}{dx} = \dfrac{dz}{dy}\dfrac{dy}{dx}$
x^n	nx^{n-1}
e^x	e^x
$a^x\ (a > 0)$	$(\ln a)a^x$
$\ln x$	$1/x$
$\sin x$	$\cos x$
$\cos x$	$-\sin x$
$\tan x$	$\sec^2 x$
$\operatorname{cosec} x$	$-\operatorname{cosec} x \cot x$
$\sec x$	$\sec x \tan x$
$\cot x$	$-\operatorname{cosec}^2 x$
$\sinh x$	$\cosh x$
$\cosh x$	$\sinh x$

Leibnitz's theorem

$$\frac{d^n}{dx^n}(f.g) = f^{(n)}.g + nf^{(n-1)}.g' + \frac{n(n-1)}{2!}f^{(n-2)}.g'' + \ldots + \binom{n}{r}f^{(n-r)}.g^{(r)} + \ldots + f.g^{(n)}$$

Integration

In the following table the constants of integration have been omitted.

function	*integral*
$u\dfrac{dv}{dx}$	$uv - \displaystyle\int \frac{du}{dx} v\,dx$
$x^n (n \neq -1)$	$\dfrac{x^{n+1}}{n+1}$
$1/x$	$\ln\|x\|$
$\cos x$	$\sin x$
$\sin x$	$-\cos x$
$\tan x$	$\ln\|\sec x\|$
$\operatorname{cosec} x$	$\ln\|\tan \frac{1}{2}x\|$
$\sec x$	$\ln\|\sec x + \tan x\| = \ln\|\tan(\frac{1}{4}\pi + \frac{1}{2}x)\|$
$\cot x$	$\ln\|\sin x\|$
$\dfrac{1}{a^2+x^2}$	$\dfrac{1}{a}\tan^{-1}\dfrac{x}{a}$
$\dfrac{1}{a^2-x^2}$	$\dfrac{1}{2a}\ln\left\|\dfrac{a+x}{a-x}\right\|$
$\dfrac{1}{\sqrt{(a^2-x^2)}}$	$\sin^{-1}\dfrac{x}{a}$
$\dfrac{1}{\sqrt{(a^2+x^2)}}$	$\sinh^{-1}\dfrac{x}{a}$ or $\ln\|x+\sqrt{(a^2+x^2)}\|$
$\dfrac{1}{\sqrt{(x^2-a^2)}}$	$\cosh^{-1}\dfrac{x}{a}$ or $\ln\|x+\sqrt{(x^2-a^2)}\|$
$\sinh x$	$\cosh x$
$\cosh x$	$\sinh x$
$\dfrac{f'(x)}{f(x)}$	$\ln\|f(x)\|$

Area of sector is $\displaystyle\int_{\alpha}^{\beta} \frac{1}{2}r^2\,d\theta$.

Length of arc

$$s=\int_{x_1}^{x_2}\sqrt{\left[1+\left(\frac{\mathrm{d}y}{\mathrm{d}x}\right)^2\right]}\mathrm{d}x=\int_{t_1}^{t_2}\sqrt{\left[\left(\frac{\mathrm{d}x}{\mathrm{d}t}\right)^2+\left(\frac{\mathrm{d}y}{\mathrm{d}t}\right)^2\right]}\mathrm{d}t$$

Area of surface of rotation

$$S=2\pi\int_{x_1}^{x_2}y\sqrt{\left[1+\left(\frac{\mathrm{d}y}{\mathrm{d}x}\right)^2\right]}\mathrm{d}x=2\pi\int_{t_1}^{t_2}y\sqrt{\left[\left(\frac{\mathrm{d}x}{\mathrm{d}t}\right)^2+\left(\frac{\mathrm{d}y}{\mathrm{d}t}\right)^2\right]}\mathrm{d}t$$

Newton's approximation

If a is an approximation to a root of $\mathrm{f}(x)=0$ then

$$a-\mathrm{f}(a)/\mathrm{f}'(a)$$

is usually a better approximation.

Approximate differentiation

$$\mathrm{f}'(x)\approx\frac{\mathrm{f}(x+h)-\mathrm{f}(x-h)}{2h}$$

$$\mathrm{f}''(x)\approx\frac{\mathrm{f}(x+h)-2\mathrm{f}(x)+\mathrm{f}(x-h)}{h^2}$$

Approximate integration

In the following approximations, $f_r=\mathrm{f}(x_r)$, where $x_r=x_0+rh$.

1. Trapezium rule

$$\int_{x_0}^{x_n}\mathrm{f}(x)\,\mathrm{d}x\approx\tfrac{1}{2}h\left[f_0+2(f_1+f_2+\ldots+f_{n-1})+f_n\right]$$

2. Simpson's rule (in which n must be even, giving an odd number of ordinates)

$$\int_{x_0}^{x_n}\mathrm{f}(x)\,\mathrm{d}x\approx\tfrac{1}{3}h\left[f_0+f_n+4(f_1+f_3+\ldots+f_{n-1})+2(f_2+f_4+\ldots+f_{n-2})\right]$$

Vectors

Line through point, position vector **a**, parallel to **b**

$$\mathbf{r} = \mathbf{a} + t\mathbf{b}$$

Plane through point, position vector **a**, perpendicular to **n**

$$(\mathbf{r} - \mathbf{a}) \, . \, \mathbf{n} = 0$$

Scalar product $= \mathbf{a}_1 \, . \, \mathbf{a}_2 = a_1 a_2 \cos\theta = x_1 x_2 + y_1 y_2 + z_1 z_2$.

Vector product $= \mathbf{a}_1 \times \mathbf{a}_2 = \mathbf{a}_1 \wedge \mathbf{a}_2 = a_1 a_2 \sin\theta \, \mathbf{n} = \begin{vmatrix} \mathbf{i} & \mathbf{j} & \mathbf{k} \\ x_1 & y_1 & z_1 \\ x_2 & y_2 & z_2 \end{vmatrix}$

Geometry

The distance of the point (h, k) from the straight line $ax + by + c = 0$ is

$$\frac{|ah + bk + c|}{\sqrt{(a^2 + b^2)}}$$

Parabola $y^2 = 4ax$, focus $(a, 0)$, directrix $x = -a$, $e = 1$

Ellipse $\dfrac{x^2}{a^2} + \dfrac{y^2}{b^2} = 1$, foci $(\pm ae, 0)$, directrices $x = \pm\dfrac{a}{e}$,

$$b^2 = a^2(1 - e^2), \; e < 1$$

Hyperbola $\dfrac{x^2}{a^2} - \dfrac{y^2}{b^2} = 1$, foci $(\pm ae, 0)$, directrices $x = \pm\dfrac{a}{e}$,

$$b^2 = a^2(e^2 - 1), \; e > 1$$

Answers

Exercise 1.1 (p. 12)

1 (a) 2 (b) 21 (c) $\begin{pmatrix} 5 & 7 & 2 & 4 & 6 \\ 0 & 3 & 0 & 2 & 1 \\ 0 & 0 & 2 & 0 & 0 \end{pmatrix}$ (d) $\begin{pmatrix} 10 & 15 & 4 & 4 & 12 \\ 0 & 5 & 0 & 2 & 2 \\ 0 & 1 & 3 & 0 & 0 \end{pmatrix}$

(e) (105 251 79 0 149) pence. **2** (a) 9 (b) Everton (c) $\begin{pmatrix} 16 \\ 12 \\ 8 \\ 14 \\ 6 \end{pmatrix}$

(d) (8×6) (e) Yes $\begin{matrix} \text{Ev.} \\ \text{Ar.} \end{matrix}\begin{pmatrix} 3-8+4{\cdot}5-8{\cdot}4 = -8{\cdot}9 \\ 1-8+3{\cdot}5-6{\cdot}3 = -9{\cdot}8 \end{pmatrix}$.

3 (a) $\begin{matrix} & A & B & C \\ A & 2 & 0 & 1 \\ B & 0 & 0 & 2 \\ C & 1 & 2 & 0 \end{matrix}$ (b) $\begin{matrix} & D & E & F & G & H \\ D & 2 & 1 & 0 & 0 & 1 \\ E & 1 & 0 & 0 & 0 & 2 \\ F & 0 & 0 & 0 & 1 & 1 \\ G & 0 & 0 & 1 & 0 & 2 \\ H & 1 & 2 & 1 & 2 & 0 \end{matrix}$ (c) $\begin{matrix} & J & K & L & M \\ J & 0 & 1 & 0 & 0 \\ K & 0 & 0 & 0 & 1 \\ L & 0 & 1 & 0 & 0 \\ M & 2 & 0 & 1 & 1 \end{matrix}$

(d) $\begin{matrix} & P & Q & R & S & T & U \\ P & 0 & 1 & 0 & 0 & 0 & 0 \\ Q & 0 & 0 & 1 & 0 & 0 & 1 \\ R & 0 & 0 & 0 & 1 & 0 & 0 \\ S & 0 & 1 & 0 & 0 & 1 & 0 \\ T & 0 & 0 & 0 & 0 & 0 & 1 \\ U & 1 & 0 & 0 & 1 & 0 & 0 \end{matrix}$ Sums are 8, 18, 7, 9, which are for (a), (b) (number of routes) × 2 for (c), (d) number of routes.

(e) $\begin{matrix} & X & Y & Z \\ X & 2 & 0 & 1 \\ Y & 0 & 0 & 2 \\ Z & 1 & 2 & 0 \end{matrix}$ which is the same matrix as in (a). **4** (a) (3×1)

(b) (i) $\begin{pmatrix} 5 \\ 2 \\ 4 \end{pmatrix}$ (ii) $\begin{pmatrix} -3 \\ 12 \\ 6 \end{pmatrix}$ (iii) $\begin{pmatrix} 4 \\ -2 \\ -13 \end{pmatrix}$ (c) $\begin{pmatrix} 1 & 9 \\ 8 & 1 \\ 1 & 6 \end{pmatrix}$ (d) $\begin{pmatrix} 8 & -3 \\ 4 & -12 \\ -7 & 13 \end{pmatrix}$.

5 (a) (i) t_{32} (ii) t_{yx} (b) $\sum_{i=1}^{m} t_{i4}$ (c) $\sum_{j=1}^{n} t_{rj}$ (d) $\sum_{j=1}^{n} \sum_{i=1}^{m} t_{ij}$

(e) $(n \times m)$ (f) $t_{ij} = t_{ji}$ for $1 \leqslant i \leqslant n$, $1 \leqslant j \leqslant n$, $i \neq j$. Sum of leading diagonal entries.

6 (a) $\begin{array}{c|ccccc} & 2 & 3 & 4 & 6 & 8 \\ \hline 2 & 1 & 0 & 1 & 1 & 1 \\ 3 & 0 & 1 & 0 & 1 & 0 \\ 4 & 0 & 0 & 1 & 0 & 1 \\ 6 & 0 & 0 & 0 & 1 & 0 \\ 8 & 0 & 0 & 0 & 0 & 1 \end{array}$ (b) $\begin{array}{c|ccccccc} & -2 & -1 & 0 & 1 & 2 & 3 & 4 \\ \hline -2 & 0 & 0 & 0 & 0 & 0 & 0 & 0 \\ -1 & 0 & 0 & 0 & 0 & 0 & 0 & 0 \\ 0 & 0 & 0 & 1 & 0 & 0 & 0 & 0 \\ 1 & 0 & 1 & 0 & 1 & 0 & 0 & 0 \\ 2 & 0 & 0 & 0 & 0 & 0 & 0 & 0 \\ 3 & 0 & 0 & 0 & 0 & 0 & 0 & 0 \\ 4 & 1 & 0 & 0 & 0 & 1 & 0 & 0 \end{array}$ (c) $\begin{array}{c|cccc} & k & l & m & n \\ \hline k & 0 & 1 & 0 & 0 \\ l & 1 & 0 & 1 & 1 \\ m & 0 & 1 & 0 & 0 \\ n & 0 & 1 & 0 & 0 \end{array}$

The matrices for the inverse relations are the transposes of these matrices, that is, (a)^T, (b)^T, (c)^T as defined in question 5 (f).

Exercise 1.2 (p. 20)

1 (i) $\begin{pmatrix} 6 & -5 & 1 \\ 4 & -3 & 1 \end{pmatrix}$ (ii) $\begin{pmatrix} 2 & -7 & 7 \\ 12 & 0 & 12 \end{pmatrix}$ (iii) $\begin{pmatrix} 16 & -4 & -9 \\ -8 & -9 & -15 \end{pmatrix}$

(iv) $\begin{pmatrix} 0 & 0 & 0 \\ 0 & -3 & -11 \end{pmatrix}$ (v) $\begin{pmatrix} 22 & -13 & -3 \\ 4 & -6 & 14 \end{pmatrix}$.

3 (a) (i) $\begin{pmatrix} 2 & 3 \\ -4 & -1 \end{pmatrix}$ (ii) $\begin{pmatrix} 0 & 3 \\ 0 & 2 \end{pmatrix}$ (iii) $\begin{pmatrix} a & 0 \\ 0 & -b \end{pmatrix}$

(b) (i) $2\mathbf{X}+5\mathbf{Y}-\mathbf{Z}$ (ii) $3\mathbf{W}+5\mathbf{Y}+2\mathbf{Z}$ (iii) $4(\mathbf{W}+\mathbf{X}+\mathbf{Y}+\mathbf{Z})$
(iv) $a\mathbf{W}+b\mathbf{X}+c\mathbf{Y}+d\mathbf{Z}$.

4 (i) $\begin{pmatrix} 26 \\ 4 \end{pmatrix}$ (ii) $\begin{pmatrix} 12 & -6 & -3 \\ 16 & -8 & -4 \end{pmatrix}$ (iii) $\begin{pmatrix} 32 & 21 & -16 \\ -10 & -7 & 5 \\ 6 & 3 & -3 \end{pmatrix}$

(iv) $\begin{pmatrix} 15 & 6 & -5 \\ -11 & 2 & 1 \\ -15 & 6 & 0 \\ 0 & 12 & -5 \end{pmatrix}$ (v) $\begin{pmatrix} 13 & 3 \\ 20 & -26 \end{pmatrix}$ (vi) $\begin{pmatrix} 12 & 1 & 5 \\ 4 & 1 & 5 \\ -4 & 0 & -2 \end{pmatrix}$.

5 (i) $\begin{pmatrix} 8 & 0 & 4 \\ 3 & 9 & 18 \end{pmatrix}$ (ii) not possible (iii) $\begin{pmatrix} 16 & 0 & 4 \\ 26 & 20 & 1 \end{pmatrix}$

(iv) $\begin{pmatrix} 10 & 16 \\ 3 & 7 \end{pmatrix}$ (v) $\begin{pmatrix} 16 & 12 & 30 \\ 9 & 3 & 10 \\ -4 & 0 & -2 \end{pmatrix}$ (vi) not possible

(vii) $\begin{pmatrix} 3 & 7 \\ 6 & 3 \\ 16 & 18 \end{pmatrix}$ (viii) $\begin{pmatrix} 6 & 12 \\ 4 & 3 \\ -2 & 0 \end{pmatrix}$ (ix) $\begin{pmatrix} 32 & 0 & 8 \\ 78 & 60 & 3 \end{pmatrix}$

(x) $\begin{pmatrix} 32 & 0 & 8 \\ 78 & 60 & 3 \end{pmatrix}$.

6 (i) $m = p$, $n = q$ (ii) $q = m$ (iii) $n = p = q$
(iv) $p = n$, $q = m$ (v) $m = n = p = q$ (vi) $m = p$ (only!).

7 $\mathbf{J}^T = \begin{pmatrix} 3 & 2 \\ 1 & -1 \\ 0 & 4 \end{pmatrix}$, $\mathbf{K}^T = \begin{pmatrix} 4 & -2 & 1 \\ 0 & 3 & -1 \end{pmatrix}$, $\mathbf{J}^T\mathbf{K}^T = \begin{pmatrix} 12 & 0 & 1 \\ 4 & -5 & 2 \\ 0 & 12 & -4 \end{pmatrix}$,

$\mathbf{JK} = \begin{pmatrix} 10 & 3 \\ 14 & -7 \end{pmatrix}$, $\mathbf{K}^T\mathbf{J}^T = \begin{pmatrix} 10 & 14 \\ 3 & -7 \end{pmatrix}$, $\mathbf{KJ} = \begin{pmatrix} 12 & 4 & 0 \\ 0 & -5 & 12 \\ 1 & 2 & -4 \end{pmatrix}$, **JK**.

8 (a) (i) $\begin{array}{c} \\ I_1 \\ I_2 \end{array}\begin{array}{c} \begin{array}{ccc} P_1 & P_2 & P_3 \end{array} \\ \begin{pmatrix} 1 & 1 & 0 \\ 1 & 1 & 1 \end{pmatrix} \end{array}$ (ii) $\begin{array}{c} \\ P_1 \\ P_2 \\ P_3 \end{array}\begin{array}{c} \begin{array}{cc} T_1 & T_2 \end{array} \\ \begin{pmatrix} 1 & 1 \\ 1 & 0 \\ 1 & 1 \end{pmatrix} \end{array}$, $\begin{pmatrix} 2 & 1 \\ 3 & 2 \end{pmatrix}$, routes from I_1, I_2 to T_1, T_2 (b) (i) 2 (ii) 3 (iii) 2.

9 $\mathbf{M} = \begin{array}{c} \\ A \\ B \\ C \end{array}\begin{array}{c} \begin{array}{ccc} A & B & C \end{array} \\ \begin{pmatrix} 1 & 1 & 0 \\ 0 & 0 & 1 \\ 1 & 1 & 0 \end{pmatrix} \end{array}$, $\mathbf{M}^2 = \begin{array}{c} \\ A \\ B \\ C \end{array}\begin{array}{c} \begin{array}{ccc} A & B & C \end{array} \\ \begin{pmatrix} 1 & 1 & 1 \\ 1 & 1 & 0 \\ 1 & 1 & 1 \end{pmatrix} \end{array}$, (i) $A \to A \to A$ (ii) $A \to A \to B$ (iii) $C \to B \to C$,

$\mathbf{M}^3 = \begin{array}{c} \\ A \\ B \\ C \end{array}\begin{array}{c} \begin{array}{ccc} A & B & C \end{array} \\ \begin{pmatrix} 2 & 2 & 1 \\ 1 & 1 & 1 \\ 2 & 2 & 1 \end{pmatrix} \end{array}$, (i) $A \to A \to A \to A$, $A \to B \to C \to A$

(ii) $C \to B \to C \to B$, $C \to A \to A \to B$ (iii) $C \to B \to C \to A$, $C \to A \to A \to A$ 3, number of n stage routes from A to A.

10 $x = 1$, $y = -2$, $\mathbf{P} = \begin{pmatrix} 1 \\ -2 \end{pmatrix} = \begin{pmatrix} x \\ y \end{pmatrix}$, $\begin{pmatrix} 4 & -1 & 1 \\ 3 & 2 & -1 \\ 1 & -1 & 2 \end{pmatrix}\begin{pmatrix} x \\ y \\ z \end{pmatrix} = \begin{pmatrix} 2 \\ 1 \\ 0 \end{pmatrix}$.

11 $\begin{pmatrix} 0 & 1 \\ 2 & 1 \end{pmatrix}$, $\begin{pmatrix} 0 & -1 \\ -2 & -1 \end{pmatrix}$, $\begin{pmatrix} 4/9 & 1/9 \\ 2/9 & 5/9 \end{pmatrix}$, $\begin{pmatrix} -4/9 & -1/9 \\ -2/9 & -5/9 \end{pmatrix}$.

Exercise 1.3 (p. 26)

1 (i) L.I. (ii) L.I. (iii) L.D. (iv) L.D. **2** (i) L.I. (ii) L.D. (iii) L.I. (iv) L.D. (v) L.I. (vi) L.I. **5** (i) L.I. (ii) L.I. (iii) L.D. (iv) L.D. (v) L.D. (vi) L.D. (vii) L.D. (viii) L.I. (ix) L.D. (x) L.I. **6** (i) 2 (ii) 3 (iii) 2 (iv) 2 (v) 2 (vi) 2 (vii) 3 (viii) 3 (ix) 2 (x) 3.
8 (i) L.I. (ii) L.D. (iii) L.I. (iv) L.I. (v) L.I. (vi) L.I. **9** (i) $2\mathbf{a} + 3\mathbf{b} - \mathbf{c}$ (ii) $(2x - y - z)\mathbf{a} + (x - z)\mathbf{b} + (y + z - x)\mathbf{c}$. **10** $\mathbf{e} = -\mathbf{a} + \mathbf{b} + 2\mathbf{d} = -\mathbf{c} + 2\mathbf{d}$.
11 (i) $\{\mathbf{i}, \mathbf{j}, \mathbf{k}, \mathbf{i} + \mathbf{j}\}$ (ii) $\{\mathbf{i}, \mathbf{j}\}$ (iii) $\{\mathbf{i}, \mathbf{j}, \mathbf{k}\}$ (iv) $\{\mathbf{i}, \mathbf{j}, \mathbf{i} + \mathbf{j}\}$ (v) $\{\mathbf{i}, \mathbf{j}\}$.

Exercise 1.4A (p. 31)

2 Suppose $\mathbf{I}_n\mathbf{A} = \mathbf{X} = (x_{ij})$ then $x_{ij} = \sum_{k=1}^{n} i_{ik}a_{kj} = a_{ij}$ since $i_{ik} = 0$ for $k \neq i$ and $i_{ii} = 1$, hence $\mathbf{X} = \mathbf{A}$.

5 (i) $\begin{pmatrix} 3 & -3 \\ -1 & 1 \end{pmatrix}$ (ii) $\begin{pmatrix} 1 & 2 \\ 0 & 0 \end{pmatrix}$ (iii) $\begin{pmatrix} 2 & 0 \\ -1 & 0 \end{pmatrix}$ (iv) $\begin{pmatrix} 1 & 2 \\ 1 & 2 \end{pmatrix}$.

Exercise 1.4B (p. 38)

2 $\begin{pmatrix} 0 & 1 \\ -1 & 0 \end{pmatrix}$, $\begin{pmatrix} 1 & 0 \\ 0 & 1 \end{pmatrix}^{-1} = \begin{pmatrix} 1 & 0 \\ 0 & 1 \end{pmatrix}$, $\begin{pmatrix} 0 & -1 \\ 1 & 0 \end{pmatrix}^{-1} = \begin{pmatrix} 0 & 1 \\ -1 & 0 \end{pmatrix}$,
$\begin{pmatrix} 0 & 1 \\ -1 & 0 \end{pmatrix}^{-1} = \begin{pmatrix} 0 & -1 \\ 1 & 0 \end{pmatrix}$, $\begin{pmatrix} -1 & 0 \\ 0 & -1 \end{pmatrix}^{-1} = \begin{pmatrix} -1 & 0 \\ 0 & -1 \end{pmatrix}$.

3 (i) $\begin{pmatrix} 4 & -5 \\ 3 & -4 \end{pmatrix}$ (ii) $\dfrac{1}{2}\begin{pmatrix} 10 & -7 \\ 14 & -10 \end{pmatrix}$ (iii) $\dfrac{1}{4}\begin{pmatrix} -2 & 0 \\ -4 & 2 \end{pmatrix}$ (iv) $\dfrac{1}{a^2 + bc}\begin{pmatrix} a & b \\ c & -a \end{pmatrix}$

4 (i) $x = -2, y = 11/2$ (ii) $x = 11/3, y = 4/3, z = -7/3$
(iii) $x = 1, y = 2, z = 3$ (iv) $x = -3, y = -8, z = 3$.

5 (i) $\begin{pmatrix} w & x \\ y & z \end{pmatrix} = \begin{pmatrix} 8 & -5 \\ -3 & 2 \end{pmatrix}\begin{pmatrix} 3 & 5 \\ -2 & 2 \end{pmatrix} = \begin{pmatrix} 34 & 30 \\ -13 & -11 \end{pmatrix}$

(ii) $\begin{pmatrix} x \\ y \end{pmatrix} = \frac{1}{5}\begin{pmatrix} 2 & 3 \\ -1 & 1 \end{pmatrix}\begin{pmatrix} -1 \\ 3 \end{pmatrix} = \begin{pmatrix} 7/5 \\ 4/5 \end{pmatrix}$

(iii) no inverse matrix, no solution (iv) no inverse matrix, solution $\begin{pmatrix} k \\ -(4+k)/3 \end{pmatrix}$ for all k in $\mathbb{R}$.

6 $\begin{pmatrix} 1/p & 0 \\ 0 & 1 \end{pmatrix}\begin{pmatrix} 1 & 0 \\ 0 & 1/p \end{pmatrix}\begin{pmatrix} 0 & 1 \\ 1 & 0 \end{pmatrix}\begin{pmatrix} 1 & -p \\ 0 & 1 \end{pmatrix}\begin{pmatrix} 1 & 0 \\ -p & 1 \end{pmatrix}$.

7 $\mathbf{A} = \mathbf{P}^{-1}\mathbf{Q}^{-1}\mathbf{S}^{-1}\mathbf{R}^{-1}$, $\mathbf{A}^{-1} = \begin{pmatrix} 2 & -5/2 \\ -1 & 3/2 \end{pmatrix}$.

8 (i) $\begin{pmatrix} 1/2 & 0 \\ 0 & 1 \end{pmatrix}, \begin{pmatrix} 1 & 0 \\ -3 & 1 \end{pmatrix}, \begin{pmatrix} 1 & -1/2 \\ 0 & 1 \end{pmatrix}, \begin{pmatrix} 1 & 0 \\ 0 & 2/5 \end{pmatrix}$

(ii) $\begin{pmatrix} -1/2 & 0 \\ 0 & 1 \end{pmatrix}, \begin{pmatrix} 1 & 0 \\ -3 & 1 \end{pmatrix}, \begin{pmatrix} 1 & 1/2 \\ 0 & 1 \end{pmatrix}, \begin{pmatrix} 1 & 0 \\ 0 & 2/7 \end{pmatrix}$

(iii) $\begin{pmatrix} 1/6 & 0 \\ 0 & 1 \end{pmatrix}, \begin{pmatrix} 1 & 0 \\ 1 & 1 \end{pmatrix}, \begin{pmatrix} 1 & -1/3 \\ 0 & 1 \end{pmatrix}, \begin{pmatrix} 1 & 0 \\ 0 & 3/13 \end{pmatrix}$

(iv) $\begin{pmatrix} 1/a & 0 \\ 0 & 1 \end{pmatrix}, \begin{pmatrix} 1 & 0 \\ -1 & 1 \end{pmatrix}, \begin{pmatrix} 1 & -1/a \\ 0 & 1 \end{pmatrix}, \begin{pmatrix} 1 & 0 \\ 0 & a \end{pmatrix}$.

9 (i) $p = a + d$, $q = ad - bc$ (ii) $\begin{pmatrix} 2 & 1 \\ 1 & 1 \end{pmatrix}, \begin{pmatrix} 1 & 1 \\ 1 & 2 \end{pmatrix}$.

10 (i) F (ii) T (iii) F (iv) F (v) T.
11 (ii) $\mathbf{P} = 3\mathbf{C}$ or $\mathbf{P} = -2\mathbf{C}$ (iii) $\alpha = a - b$, $\beta = b$.
12 (i) $x = \lambda c$, $y = \mu + \lambda d$ (ii) $\alpha = \lambda$, $\beta = \mu$ (iv) $\mathbf{A} = \begin{pmatrix} 5 & 5 \\ 2 & 2 \end{pmatrix}$ or $\begin{pmatrix} -2 & 5 \\ 2 & -5 \end{pmatrix}$.

Exercise 1.5 (p. 48)

1 (i) 10 (ii) 11 (iii) -42 (iv) 17. **2** (i) $\mathbf{r}_3 = 2\mathbf{r}_1$ (ii) $\mathbf{c}_3 = \mathbf{c}_1 + \mathbf{c}_2$
(iii) $\mathbf{r}_4 = (0 \ \ 0 \ \ 0 \ \ 0)$. **4** 0. **5** $(x-1)(y-1)(y-x)(y+x+1)$. **6** $3\lambda, -21$.
7 $\alpha\beta + \beta\gamma + \gamma\alpha - \alpha^2 - \beta^2 - \gamma^2, -4k^3$. **8** $-(a-1)(b-1)(c-1)(a-b)(b-c)(c-a)$.
9 $(x+a+y)(x-a)(y-a)$, $\{0, \pi/2, 2\pi/3, \pi\}$.

Miscellaneous Exercise 1 (p. 49)

1 (a) $\begin{pmatrix} a & b & a \\ b & a+b & b \\ a & b & a \end{pmatrix}$ (b) $\begin{pmatrix} 0 & 2ab & 0 \\ ab & ab & ab \\ 0 & 2ab & 0 \end{pmatrix}$. **2** $p = 2, q = -1$.

4 $\mathbf{AB} = \begin{pmatrix} 1 & 0 & 0 \\ 0 & 1 & 0 \\ 0 & 0 & 1 \end{pmatrix} = \mathbf{BA}$, (a) $\begin{pmatrix} x \\ y \\ z \end{pmatrix} = \begin{pmatrix} 5 \\ 1 \\ 0 \end{pmatrix}$ (b) $\begin{pmatrix} x \\ y \\ z \end{pmatrix} = \begin{pmatrix} -5 \\ -7 \\ -16 \end{pmatrix}$.

6 (i) T (ii) T (iii) F, $\begin{pmatrix} 1 & 2 \\ 2 & -1 \end{pmatrix}\begin{pmatrix} 0 & 1 \\ 1 & 0 \end{pmatrix} = \begin{pmatrix} 2 & 1 \\ -1 & 2 \end{pmatrix}$, (iv) T.

7 (i) L.D. (ii) L.I. (iii) L.D. **9** $\mathbf{A}^2 = \begin{pmatrix} 1 & 3 \\ 0 & 4 \end{pmatrix}$, $\mathbf{A}^3 = \begin{pmatrix} 1 & 7 \\ 0 & 8 \end{pmatrix}$, $\mathbf{A}^n = \begin{pmatrix} 1 & 2^n - 1 \\ 0 & 2^n \end{pmatrix}$.

10 $\begin{pmatrix} b^2 - c^2 & (a-d)(c-b) \\ (a-d)(c-b) & c^2 - b^2 \end{pmatrix}$. **11** $\mathbf{d} = 2\mathbf{a} - 3\mathbf{b} + \mathbf{c}$.

12 $\mathbf{A}^{-1} = \begin{pmatrix} 0 & -1 & 0 \\ -1 & -3 & -1 \\ 0 & -2 & -1 \end{pmatrix}$. **13** $\mathbf{A}^{-1} = \frac{1}{4}\mathbf{B}$, $\mathbf{B}^{-1}\mathbf{A}^{-1} = \frac{1}{4}\mathbf{I}_3$,

$\begin{pmatrix} x \\ y \\ z \end{pmatrix} = \begin{pmatrix} 2 \\ 3 \\ 4 \end{pmatrix}$, $\begin{pmatrix} -8 \\ 17 \\ 22 \end{pmatrix} = 2\begin{pmatrix} -1 \\ 2 \\ 3 \end{pmatrix} + 3\begin{pmatrix} 2 \\ -1 \\ 4 \end{pmatrix} + 4\begin{pmatrix} -3 \\ 4 \\ 1 \end{pmatrix}$.

14 1. **16** $(a^2 + b^2 + c^2 - ab - ac - bc)$ (i) $x = (1 \pm i\sqrt{3})/2$ or -1, (ii) $x = \pm 1/\sqrt{3}$.

17 (a) $q - r$. **18** (b) (i) L.D. (ii) $\begin{pmatrix} 0 \\ 1 \\ 0 \end{pmatrix}$ (iii) L.D.

(c) (i) $\left\{ \begin{pmatrix} 1 \\ 0 \\ 0 \end{pmatrix}, \begin{pmatrix} 0 \\ 1 \\ 0 \end{pmatrix}, \begin{pmatrix} 1 \\ 1 \\ 1 \end{pmatrix} \right\}$ (ii) $\begin{pmatrix} 0 \\ 1 \\ 0 \end{pmatrix}$ not in span (iii) $\begin{pmatrix} 0 \\ 0 \\ 1 \end{pmatrix}$ not in span.

19 $(a-b)(b-c)(a-c)$, $\begin{pmatrix} 0 & 0 & 0 \\ 0 & 0 & 0 \\ 1 & 0 & 0 \end{pmatrix}$, $\begin{pmatrix} 0 & 0 & 0 \\ 0 & 0 & 0 \\ 0 & 0 & 0 \end{pmatrix}$.

Exercise 2.1 (p. 54)

1 (i) $-1 < x < 2$ or $x > 3$ (ii) $x < -4$ (iii) $x < -1$ or $0 < x < 1$ or $x > 2$
(iv) $-3 < x < 1$ or $2 < x < 4$. **2** (i) $x < 2$ or $2 < x < 3$ or $x > 4$
(ii) $1/2 < x < 1, 2 < x$ (iii) $x > 1$ (iv) $-3 < x < -2$ or $2 < x < 3$.
3 $5/3 < x < 5$. **4** (i) $x < -1$ or $x > 0$ (ii) $x < 1/2$ or $x > 1$ (iii) $x < 5/3$ or $x > 7$
(iv) $-2 < x < 2$. **5** $-\sqrt{(5/2)} < x < \sqrt{(5/2)}$. **6** $2 < x < 3$ or $4 < x < 6$.
7 (a) $-4/5 < x < 2$ (b) $-3 \leqslant x \leqslant 1/3$ or $x \geqslant 2$.
8 (a) $-1 < x < 4$ (b) $-4 < x < -1$ or $2 < x < 4$ (c) $4/5 < x < 4$.

Exercise 2.2 (p. 55)

4 $a = -1, b = 0$.

Exercise 2.3 (p. 58)

6 (a) (0, 1) (b) (2/3, 2/3) (c) (9/11, 8/11). **7** 18/5. **8** 20. **9** 20, 120.
10 2 of X and 3 of Y.

Miscellaneous Exercise 2 (p. 59).

5 (0, 4). **7** (i) $-2 < x < 0$ or $2 < x < 4$. **8** $x < 0$ or $x > 2$.
10 (i) $x < -6$ or $-2 \leqslant x < 2/3$ (ii) $1 \leqslant x \leqslant 3$. **11** (i) 1 (ii) $-\sqrt{2}$.
12 (ii) (1, 2). **13** 7, 29. **14** -12, 8.

Exercise 3.1 (p. 61)

1 (i), (iii), (iv), (v), (vi), (viii), (ix), (xi), (xiii), (xiv) are all closed, and (ii), (vii), (x), (xii) are not closed. **2** For all x in S, define $f^2(x) = ff(x) = f(f(x))$.

Exercise 3.2 (p. 63)

1 (ii) $x = 2$, $y = 3$, (iii) $x = \mathbf{i}$, $y = \mathbf{i}$.
2 (i), (iii), (iv), (v), (vii) are closed, and (ii), (vi) are not closed.

3 (i)

	2	4	6	8	10
2	4	6	8	10	12
4	6	8	10	12	14
6	8	10	12	14	16
8	10	12	14	16	18
10	12	14	16	18	20

(ii)

	15	18	21	22
15	15	3	3	1
18	3	18	3	2
21	3	3	21	1
22	1	2	1	22

(iii)

	1	3	5	7	9	11
1	1	3	5	7	9	11
3	3	3	5	7	9	11
5	5	5	5	7	9	11
7	7	7	7	7	9	11
9	9	9	9	9	9	11
11	11	11	11	11	11	11

(iv)

	0	1	2	3	4	5	6
0	0	1	2	3	4	5	6
1	1	0	1	2	3	4	5
2	2	1	0	1	2	3	4
3	3	2	1	0	1	2	3
4	4	3	2	1	0	1	2
5	5	4	3	2	1	0	1
6	6	5	4	3	2	1	0

4

$\cap$	A	ϕ
A	A	ϕ
ϕ	ϕ	ϕ

5

	A	B	C	D
A	A	B	C	D
B	B	B	D	D
C	C	D	C	D
D	D	D	D	D

6 (ii) is unary and (i), (iii), (iv) are binary.

Exercise 3.3 (p. 66)

1 (i)

+	0	1	2	3
0	0	1	2	3
1	1	2	3	0
2	2	3	0	1
3	3	0	1	2

(ii)

×	0	1	2	3
0	0	0	0	0
1	0	1	2	3
2	0	2	0	2
3	0	3	2	1

(iii)

−	0	1	2	3
0	0	3	2	1
1	1	0	3	2
2	2	1	0	3
3	3	2	1	0

(iv)

−	0	1	2
0	0	2	1
1	1	0	2
2	2	1	0

2

∘	F	A
F	F	A
A	A	F

3

∘	F	A	R	L
F	F	A	R	L
A	A	F	L	R
R	R	L	A	F
L	L	R	F	A

4 (i)

+	0	1
0	0	1
1	1	2

(ii)

×	0	1
0	0	0
1	0	1

(iii)

+	1	−1
1	2	0
−1	0	−2

(iv)

×	1	−1
1	1	−1
−1	−1	1

(v)

×	1	i	−1	−i
1	1	i	−1	−i
i	i	−1	−i	1
−1	−1	−i	1	i
−i	−i	1	i	−1

(vi)

+	0	1
0	0	1
1	1	0

(vii)

×	0	1
0	0	0
1	0	1

The systems with the same structures are: (a) (ii), (vii) and Qu **4** Ex. 3.2. (b) (iv), (vi) and Qu **2**, (c) (v), Qu **1** (i) and Qu **3**.

Exercise 3.4 (p. 68)

2 (ii), (iii), (iv) and (v) are symmetric, (ii) and (v) are reflexive, (i), (ii) and (v) are transitive.

Exercise 3.5 (p. 70)

1 (i) $\{1, 3, 5\}, \{2, 4\}$, (ii) $\{5n+k: n \in \mathbb{N}\}, k = 0, 1, 2, 3, 4$, (iii) $\{-4, 4\}, \{-2, 2\}, \{-1, 1\}$. **2** (i) ref., tran., $1 \geqslant 0, 0 \ngeqslant 1$, (ii) sym., $3+3$ is even so not ref., $3+6$ and $6+3$ are odd but $3+3$ is even so not tran., (iii) sym., not ref. e.g. 1, 1, not trans. e.g. 2R6, 6R9, not 2R9.
4 (i) $\{0, 1, 2\}, \{0, 1\}$, (ii) $\{-1, 0, 1, 2\}, \{-1, 0, 1\}$, (iii) $\mathbb{Z}, E$, (iv) E, E, (v) O, E, (vi) E, O.

5

+	*E*	*O*
E	*E*	*O*
O	*O*	*E*

×	*E*	*O*
E	*E*	*E*
O	*E*	*O*

6

AND	0	1
0	0	0
1	0	1

OR	0	1
0	0	1
1	1	0

7

+	0	1	2	3	4
0	0	1	2	3	4
1	1	2	3	4	0
2	2	3	4	0	1
3	3	4	0	1	2
4	4	0	1	2	3

−	0	1	2	3	4
0	0	4	3	2	1
1	1	0	4	3	2
2	2	1	0	4	3
3	3	2	1	0	4
4	4	3	2	1	0

Solutions: (i) $\{4\}$, (ii) $\{4\}$, (iii) $\{1, 4\}$, (iv) $\{2, 3\}$, (v) ϕ, (vi) ϕ.

8

+	**0**	**1**	**2**
0	**0**	**1**	**2**
1	**1**	**2**	**0**
2	**2**	**0**	**1**

×	**0**	**1**	**2**
0	**0**	**0**	**0**
1	**0**	**1**	**2**
2	**0**	**2**	**1**

$(\mathbb{Z}_3, +)$	0	1	2
0	0	1	2
1	1	2	0
2	2	0	1

$(\mathbb{Z}_3, \times)$	0	1	2
0	0	0	0
1	0	1	2
2	0	2	1

11 The equivalence classes consist of sets of pairs (a, b) of non-zero integers in which a/b is a given rational number. Thus the set of all equivalence classes corresponds one-one to the set of non-zero rational numbers. **12** R is an equivalence, P is not sym. $3 \leqslant 4$ but not $4 \leqslant 3$, Q is not tran. 3Q4 and 4Q5 but not 3Q5. **13** $S = \mathbb{Z}$, $a\mathrm{R}b \Leftrightarrow ab \neq 0$.

Exercise 3.6 (p. 75)

1 (i)

∘	E	N
E	E	N
N	N	E

(ii)

∘	E	R
E	E	R
R	R	E

(iii)

∘	E
E	E

(iv)

∘	E	R	M	N
E	E	R	M	N
R	R	E	N	M
M	M	N	E	R
N	N	M	R	E

(v)

∘	E	M
E	E	M
M	M	E

2

∘	E	R	P
E	E	R	P
R	R	P	E
P	P	E	R

3

∘	E	R	P	L	M	N
E	E	R	P	L	M	N
R	R	P	E	M	N	L
P	P	E	R	N	L	M
L	L	N	M	E	P	R
M	M	L	N	R	E	P
N	N	M	L	P	R	E

4

∘	E	P	R	Q
E	E	P	R	Q
P	P	R	Q	E
R	R	Q	E	P
Q	Q	E	P	R

Exercise 3.7 (p. 77)

1 Isomorphism f, values of f(0), f(1): (ii) A, ϕ; (iii) ϕ, A; (iv) 4, 2.
2 Isomorphism f values of f(0), f(1), f(2): (ii) 8, 2, 4; (iii) 0, 4, 8; (iv) E, R, R^2.
3 Isomorphism f, values of f(0), f(1), f(2), f(3): (ii) F, R, A, L; (iii) 1, 2, 4, 8; (iv) 1, i, −1, −i; (v) 1, 3, 9, 7.
4 Isomorphism f, values of f(0), f(1), f(2): (ii) 0, 4, 2; (iii) 4, 1, 7.
5 Isomorphism f, values of f(0), f(1): (ii) 1, −1; (iii) 1, 2; (iv) E, N.

Exercise 3.8A (p. 80)

1 (i) (a) yes, (b) yes, b, (c) b and c; (ii) (a) no, $bc \neq cb$, (b) no; (iii) (a) yes, (b) yes, a, (c) all elements. **2** $-x/(1+x)$.
3 (i) neutral 1, $\mathbb{R}\backslash\{0\}$; (ii) neutral 0, $\mathbb{R}$; (iii) no neutral; (iv) no neutral; (v) neutral 1, $\{1, 3\}$.

Exercise 3.8B (p. 82)

1 (i) closed, non-commutative, associative, no neutral; (ii) closed, commutative, associative, neutral c, only c has an inverse; (iii) closed, commutative, neutral a, a, c and d have inverses, non-associative since $d(bb) \neq (db)\,b$.
2 (i) closed, commutative, associative, no neutral; (ii) closed, commutative, associative, no neutral; (iii) closed, commutative, non-associative, neutral 0, every element is a self inverse; (iv) closed, commutative, non-associative, no neutral; (v) closed, commutative, associative, no neutral; (vi) closed, commutative, associative, neutral B, A has no inverse: (vii) closed, commutative, associative, neutral ϕ, only ϕ has an inverse; (viii) closed, commutative, associative, neutral $\begin{pmatrix} 0 & 0 \\ 0 & 0 \end{pmatrix}$, every matrix A has an inverse $-A$: (ix) closed, commutative, associative, neutral $\begin{pmatrix} 1 & 0 & 0 \\ 0 & 1 & 0 \\ 0 & 0 & 1 \end{pmatrix}$, every matrix A has an inverse A^{-1};
(x) closed, commutative, associative, neutral $\begin{pmatrix} 1 & 0 \\ 0 & 1 \end{pmatrix}$, every non-singular matrix has an inverse.
3 (i) Y, (ii) N, (i) N, (ii) Y, (iii) N.

Exercise 3.9 (p. 89)

2 *a.* **4** (i) G4, (ii) G2, G3, G4, (iii) G1, G4, (iv) G2, G3, G4, (v) G3, G4, (vi) G2, G3, G4, (vii) G4, (viii) G2, G3, G4, (ix) G2. **5** All are groups.

8 $\begin{pmatrix}0\\0\\0\end{pmatrix}, \begin{pmatrix}-1\\2\\-3\end{pmatrix}$. **10** The group with the given table is isomorphic to: (i) $(\mathbb{Z}_3, +)$, (ii) $(S, \circ)$ of example 4, (iii) $(\mathbb{Z}_4, +)$.

11 (i)

	0	1	2	3
0	0	0	0	0
1	0	1	2	3
2	0	2	0	2
3	0	3	2	1

(ii)

	1	2	3
1	1	2	3
2	2	0	2
3	3	2	1

(iii)

	1	2	3	4
1	1	2	3	4
2	2	4	1	3
3	3	1	4	2
4	4	3	2	1

(iv)

	1	2	3	4	5
1	1	2	3	4	5
2	2	4	0	2	4
3	3	0	3	0	3
4	4	2	0	4	2
5	5	4	3	2	1

(i) Not a group, 0 and 2 have no inverses, (ii) not a group, not closed, (iii) a group isomorphic to $(\mathbb{Z}_4, +)$, (iv) not a group, not closed. **15** inverse pairs (2, 4), (3, 5), 1 and 6 are self inverse, 1 has order 1, 2 and 4 have order 3, 3 and 5 have order 6, 6 has order 2, $5 = 2 \times 6$. **16** e, e, b, d, closure, identity, inverse, $\{e, a\}$, $\{e, b\}$.

Exercise 3.10 (p. 96)

2 Rotations of the plane through 0, $2\pi/3$ and $4\pi/3$. **3** Translation (-1) parallel to Ox, translation (2) parallel to Ox. **4** R^4, R^3. **5** (i) (a) 1 (b) $n = 1$ (c) $\{0\}$, (ii) (a) 1 (b) $n = 1$ (c) $\{1\}$, (iii) (a) 2 (b) $n = 2$ (c) $\{-1\}$, (iv) (a) infinite (b) $(\mathbb{Z}, +)$ (c) $\{a, a^{-1}\}$. **11** (iii) $ab = ba$ so G is commutative, (iv) G is cyclic.

Exercise 3.11 (p. 102)

1 (i) 1, (ii) 0, (iii) 2, (iv) 2, (v) 3, (vi) 4.

5

∘	F	R	A	L
F	F	R	A	L
R	R	A	L	F
A	A	L	F	R
L	L	F	R	A

, period of F is 1, of A is 2, of R and L is 4.

6 Period of E is 1, of R, M and N is 2. **8** $\{E, R, M\} = \{E, R\} \cup \{E, M\}$, and this is not a subgroup, since it is not closed.

Miscellaneous Exercise 3 (p. 103)

4

	e	a	b	p	q	r
e	e	a	b	p	q	r
a	a	b	e	r	p	q
b	b	e	a	q	r	p
p	p	q	r	e	a	b
q	q	r	p	b	e	a
r	r	p	q	a	b	e

7 (i) 2, (ii) 2, (iii) 4, (iv) ∞, (v) 2, (vi) 3, (vii) 8, (viii) 4.
10 Axioms G1, G2, G3, G4. For all a, b in S, $a * b = b * a$.

$\cup$	R	T	V	W
R	R	R	R	R
T	R	T	R	T
V	R	R	V	V
W	R	T	V	W

$\cap$	R	T	V	W
R	R	T	V	W
T	T	T	W	W
V	V	W	V	W
W	W	W	W	W

(i) $(S, \cup)$ is not a group, since W is the neutral but R, T and V have no inverses. (ii) $(S, \cap)$ is not a group, since R is the neutral, but T, V and W have no inverses.
12 (i) sym., (ii) equivalence, classes odds, evens, (iii) sym.,
(iv) equivalence, all fractions with the same denominator when in lowest terms,
(v) equivalence, classes $\{a, c\}$, $\{b, d, e\}$, $\{f\}$. **13** Not ass. e.g. $a = 5$, $b = 3$, $c = 1$.
14 $(G, +) = P$ is a group, $(G, \times)$ is not a group since 0 has no inverse, $H \cong P$, $(G, *)$ is not a group (no identity).
15 $\{1, z^3\}$, $\{1, z^2, z^4\}$. **16** (i) and (iii), $b = a^2 = a^{-1}$, Fig. 3.5 and an equilateral triangle. **17** (i) $\{\{l\}\}$, (ii) $\{\{l\}\}$, $\{\{k, l\}\}$, (iii) ϕ, $r \in \{2, 4\}$.

18 (i)

S	1	3	5	7
1	1	3	5	7
3	3	1	7	5
5	5	7	1	3
7	7	5	3	1

T	2	4	6	8
2	4	8	2	6
4	8	6	4	2
6	2	4	6	8
8	6	2	8	4

U	1	−1	j	−j
1	1	−1	j	−j
−1	−1	1	−j	j
j	j	−j	−1	1
−j	−j	j	1	−1

$T \cong U$ under isomorphisms f and g, $f(6) = g(6) = 1$, $f(4) = g(4) = -1$, $f(2) = g(8) = j$, $f(8) = g(2) = -j$. (i) 3, (ii) 1, (iii) 1.
19 1 has order (period) 1, 6 has order 2, 2 and 4 have order 3, 3 and 5 have order 6. The generators are 3 and 5. Proper subgroups $\{1, 6\}$, $\{1, 2, 4\}$. Second generator of S is $p^{-1} = (a\ f\ e\ d\ c\ b)$.

20

	I	R^2	L	M
I	I	R^2	L	M
R^2	R^2	I	M	L
L	L	M	I	R^2
M	M	L	R^2	I

, $\{I, R, R^2, R^3\}$.

21 (i) (a) Identity symmetry (b) all symmetries are self inverses, (ii) (a) identity A (b) C is self inverse, B and D are mutual inverses. (i) and (ii) are not isomorphic, having different numbers of elements of period 2.
23 T is non-abelian (non-commutative), U is abelian (commutative), W is not a group, V, X, Y are groups, $V \not< T$, $V < U$, $X < T$, $X \not< U$, $Y < T$, $Y < U$.

24 $I = \begin{pmatrix} 1 & 0 \\ 0 & 1 \end{pmatrix}, J = \begin{pmatrix} 0 & 1 \\ -1 & 0 \end{pmatrix}, \{z : z \in \mathbb{C}, |z| = 1\}$, 3 elements $\theta = 0, \theta = 2\pi/3, \theta = 4\pi/3$.

25 p_1 is the identity, p_2 and p_3 have order (period) 3, p_4, p_5 and p_6 have order 2,

	p_1	p_2	p_3	p_4	p_5	p_6
p_1	p_1	p_2	p_3	p_4	p_5	p_6
p_2	p_2	p_3	p_1	p_6	p_4	p_5
p_3	p_3	p_1	p_2	p_5	p_6	p_4
p_4	p_4	p_5	p_6	p_1	p_2	p_3
p_5	p_5	p_6	p_4	p_3	p_1	p_2
p_6	p_6	p_4	p_5	p_2	p_3	p_1

$\{p_1, p_2, p_3\}$, generator p_2, $\{p_1, p_4\}$, $\{p_1, p_5\}$, $\{p_1, p_6\}$.

26 2 and 3, $g^2 : x \to 1/(1-x)$, $fg : x \to (1-x)$, $gf : x \to 1/x$, $\{e, f\}$, $\{e, g, g^2\}$, $\{e, fg\}$, $\{e, gf\}$, **27** $f_2(x) = -1/x$, $f_3(x) = (1+x)/(1-x)$.

29

	A	B	C	D	E	F
A	A	B	C	D	E	F
B	B	A	E	F	C	D
C	C	F	A	E	D	B
D	D	E	F	A	B	C
E	E	D	B	C	F	A
F	F	C	D	B	A	E

$\{A, E, F\}$, $\{A, B\}$, $\{A, C\}$, $\{A, D\}$.

30 ρ_1 not refl. e.g. $\begin{pmatrix} 1 & 1 \\ 0 & 1 \end{pmatrix}$, ρ_2 not tran. e.g. $\begin{pmatrix} 1 & 0 \\ 1 & 1 \end{pmatrix}$ commutes with $\mathbf{I}_2$, $\mathbf{I}_2$ commutes with $\begin{pmatrix} 1 & 1 \\ 0 & 1 \end{pmatrix}$, but $\begin{pmatrix} 1 & 0 \\ 1 & 1 \end{pmatrix}$ does not commute with $\begin{pmatrix} 1 & 1 \\ 0 & 1 \end{pmatrix}$, ρ_3 not refl. e.g. $\begin{pmatrix} 1 & 1 \\ 0 & 1 \end{pmatrix}$ is not self inverse.

31

	e	a	a^2	a^3	b	ab	a^2b	a^3b
e	e	a	a^2	a^3	b	ab	a^2b	a^3b
a	a	a^2	a^3	e	ab	a^2b	a^3b	b
a^2	a^2	a^3	e	a	a^2b	a^3b	b	ab
a^3	a^3	e	a	a^2	a^3b	b	ab	a^2b
b	b	a^3b	a^2b	ab	e	a^3	a^2	a
ab	ab	b	a^3b	a^2b	a	e	a^3	a^2
a^2b	a^2b	ab	b	a^3b	a^2	a	e	a^3
a^3b	a^3b	a^2b	ab	b	a^3	a^2	a	e

e has order 1, a and a^3 have order 4, rest have order 2, 5, (i) $\{e, a^2, a, a^3\}$, (ii) $\{e, a^2, b, a^2b\}$.

Exercise 4.1 (p. 121)

1 $\mathbf{r}' = \mathbf{Ar}$, where $\mathbf{A} =$ (i) $\begin{pmatrix} 1 & 0 \\ 0 & -1 \end{pmatrix}$ (ii) $\begin{pmatrix} 2 & 0 \\ 0 & -3 \end{pmatrix}$ (iii) $\begin{pmatrix} 0 & 1 \\ 1 & 0 \end{pmatrix}$ (iv) $\begin{pmatrix} 0 & 1 \\ -1 & 0 \end{pmatrix}$ (v) $\begin{pmatrix} 1 & 1 \\ 0 & 1 \end{pmatrix}$ (vi) $\begin{pmatrix} 0 & 0 & 1 \\ 0 & 1 & 0 \\ 1 & 0 & 0 \end{pmatrix}$ (vii) $\begin{pmatrix} 2 & 0 & 0 \\ 0 & 2 & 0 \\ 0 & 0 & 2 \end{pmatrix}$ (viii) $\begin{pmatrix} 1 & 0 & 0 \\ 0 & 1 & 0 \\ 0 & 2 & 1 \end{pmatrix}$,

the geometrical interpretations are:
(i) Refl. in $x = 0$ (ii) 2 way Str. S.F. 2 in x, -3 in y (iii) Refl. in $y = x$ (iv) Rot. $\pi/2$
(v) Shear, inv. line $y = 0$, $\begin{pmatrix}0\\1\end{pmatrix} \to \begin{pmatrix}1\\1\end{pmatrix}$ (vi) Refl. in plane $x = z$ (vii) Enlgmt. S.F. 2
(viii) Shear fixing plane $y = 0$.

2 (i) $\begin{pmatrix}3 & 0\\0 & 1\end{pmatrix}$ (ii) $\begin{pmatrix}-3/5 & 4/5\\4/5 & 3/5\end{pmatrix}$ (iii) $\begin{pmatrix}-1/\sqrt{2} & -1/\sqrt{2}\\1/\sqrt{2} & -1/\sqrt{2}\end{pmatrix}$ (iv) $\begin{pmatrix}1 & -2\\0 & 1\end{pmatrix}$

(v) $\begin{pmatrix}2 & 0 & 0\\0 & 1 & 0\\0 & 0 & 5\end{pmatrix}$ (vi) $\begin{pmatrix}1 & 0 & 0\\0 & 0 & -1\\0 & -1 & 0\end{pmatrix}$ (vii) $\begin{pmatrix}0 & -1 & 0\\1 & 0 & 0\\0 & 0 & 1\end{pmatrix}$ (viii) $\begin{pmatrix}0 & 0 & 1\\1 & 0 & 0\\0 & 1 & 0\end{pmatrix}$.

4 $y = x$, $y = 6x$. **5** (i) (a) $(2, -1)$ (b) $y = 3x$ (ii) (a) $y = 3x/2$ (b) $(1, 2)$.
6 $x'y' = 1$. **7** $4x + 5y = 0$, $4y = (1 \pm \sqrt{17})x$. **8** $3y' = 8x' - 5$.

9 (i) $\begin{pmatrix}-1 & 0\\0 & 1\end{pmatrix}\begin{pmatrix}\cos\theta & -\sin\theta\\\sin\theta & \cos\theta\end{pmatrix}$. **10** $OP'^2 = (ax + by)^2 + (cx + dy)^2$.

11 (i) $\begin{pmatrix}0 & 1\\1 & 2\end{pmatrix}\begin{pmatrix}x\\y\end{pmatrix} = \begin{pmatrix}x+4\\y-2\end{pmatrix}$, $x = -3$, $y - 1$ (ii) $\begin{pmatrix}1 & 2\\0 & 1\end{pmatrix}\begin{pmatrix}x\\y\end{pmatrix} = \begin{pmatrix}x+4\\y-2\end{pmatrix}$,
no solution, the shear does not change the y-coordinate.
12 $(9, 18, 18)$, $1:9$, $(2, -2, 1)$, -3, $(2, 1, -2)$.
13 one way stretch S.F. k in the y-direction, πab, $2a$, $2b$, 56000, 8·4.

14 $\begin{pmatrix}0 & 1 & 0\\0 & 0 & 1\\-1 & 0 & 0\end{pmatrix}$, opposite.

Exercise 4.2 (p. 131)

1 (a) $\cos\alpha$, $-\sin\alpha$, $\sin\alpha$, $\cos\alpha$ (b) $\cos\beta$, $\sin\beta$, $\sin\beta$, $-\cos\beta$.
2 $\{0, 1, 1, 0\}$, $\{0, -1, -1, 0\}$, $\sqrt{(x_0^2 + y_0^2)}$.

3 $\begin{pmatrix}\cos\theta + 5\sin\theta\\-\sin\theta + 5\cos\theta\end{pmatrix}$, $\begin{pmatrix}4\cos\theta + \sin\theta\\-4\sin\theta + \cos\theta\end{pmatrix}$, $\begin{pmatrix}19/5\\0\end{pmatrix}$, $\begin{pmatrix}76/25\\57/25\end{pmatrix}$.

4 (a) (i) $\begin{pmatrix}-5 & -3\\2 & 1\end{pmatrix}$ (ii) $\begin{pmatrix}3/2 & -2\\-1/2 & 1\end{pmatrix}$ (iii) $\begin{pmatrix}5/2 & -1\\1/2 & 0\end{pmatrix}$

(b) (i) $\begin{pmatrix}1 & 0\\-2 & 1\end{pmatrix}\begin{pmatrix}1 & 3\\0 & 1\end{pmatrix}$ (ii) $\begin{pmatrix}2 & 1\\0 & 1\end{pmatrix}\begin{pmatrix}1 & 0\\1 & 1\end{pmatrix}\begin{pmatrix}1 & 2\\0 & 1\end{pmatrix}$

(iii) $\begin{pmatrix}5/2 & -1\\1/2 & 0\end{pmatrix}\begin{pmatrix}1 & 1\\0 & 1\end{pmatrix}\begin{pmatrix}1 & 0\\-1 & 1\end{pmatrix}\begin{pmatrix}1 & 0\\0 & 2\end{pmatrix}\begin{pmatrix}1 & -3\\0 & 1\end{pmatrix}$.

7 D is a one way stretch S.F. 2 in the x-direction, R is a rotation of $-53{\cdot}1°$ about O,
$3y = 4x$. **8** Refl. in $y = 2x$. **9** (a) $y + x = 1$ (b) $\begin{pmatrix}\cos\alpha & -\sin\alpha\\\sin\alpha & \cos\alpha\end{pmatrix}$,

$\begin{pmatrix}\cos 2\beta & \sin 2\beta\\\sin 2\beta & -\cos 2\beta\end{pmatrix}$, $\begin{pmatrix}\cos(\alpha + 2\beta) & \sin(\alpha + 2\beta)\\\sin(\alpha + 2\beta) & -\cos(\alpha + 2\beta)\end{pmatrix}$, refl. in line $y = x\tan(\alpha/2 + \beta)$.

10 $\begin{pmatrix}\cos\alpha & \sin\alpha\\-\sin\alpha & \cos\alpha\end{pmatrix}$, $\begin{pmatrix}\cos(\alpha + \beta) & -\sin(\alpha + \beta)\\\sin(\alpha + \beta) & \cos(\alpha + \beta)\end{pmatrix}$.

11 $\begin{pmatrix}3 & 1 & 2\\7 & 2 & 5\\-5 & -2 & -4\end{pmatrix}$. **12** $\begin{pmatrix}1 & -2 & -3\\3 & -5 & -9\\-2 & 4 & 7\end{pmatrix}$. **13** $\begin{pmatrix}1 & 0 & 0\\-9 & 1 & 0\\ac - b & -c & 1\end{pmatrix}$.

14 $\begin{pmatrix} 11 & -12 & -7 \\ -8 & 9 & 5 \\ 7 & -8 & -4 \end{pmatrix}$, (i) $\begin{pmatrix} 11 & -8 & 7 \\ -12 & 9 & -8 \\ -7 & 5 & -4 \end{pmatrix}$, (ii) $\begin{pmatrix} 11/2 & -12 & -7 \\ -4 & 9 & 5 \\ 7/2 & -8 & -4 \end{pmatrix}$,

(iii) $\begin{pmatrix} -2 & -4 & -3/2 \\ -3/2 & -5/2 & -1/2 \\ -1/2 & -2 & -3/2 \end{pmatrix}$. **15** $\frac{1}{9}\begin{pmatrix} 10 & -4 & 1 \\ 7 & -1 & -2 \\ -8 & 5 & 1 \end{pmatrix}$, $\begin{pmatrix} x \\ y \\ z \end{pmatrix} = \begin{pmatrix} 6 \\ 2 \\ -5 \end{pmatrix}$.

16 $\begin{pmatrix} 0 & 1 \\ 1 & 0 \end{pmatrix}, \begin{pmatrix} 1 & 0 \\ 0 & 1/6 \end{pmatrix}, \begin{pmatrix} 1 & -3 \\ 0 & 1 \end{pmatrix}$, refl. in $y = x$, one way stretch S.F. 1/6 in y-direction, shear x-axis fixed $\begin{pmatrix} 0 \\ 1 \end{pmatrix} \to \begin{pmatrix} -3 \\ 1 \end{pmatrix}$.

Exercise 4.3 (p. 138)

1 Determinant 1 and sense preserved for (ii), (iii), (vii), (viii), determinant -1 and sense reversed for (i), (iv), (v), (vi).

5 $\begin{pmatrix} 3 & 4 \\ -4 & 3 \end{pmatrix}$ a rotation and an enlargement S.F. 5, $\begin{pmatrix} 3 & 4 \\ 4 & -3 \end{pmatrix}$ a reflection and an enlargement S.F. 5.

Exercise 4.4 (p. 149)

1 (i) $x = 2$, $y = 1$, $z = 10$ (ii) $x = -1$, $y = 3$, $z = 5$.
2 $x = t$, $y = -1 - 2t$, $z = t$, for all $t \in \mathbb{R}$ (i) a line (ii) a plane (iii) a sheaf of three planes with a line in common. **3** 1, $2x - 4 = y + 1 = 2z$.

4 $\begin{pmatrix} x' \\ y' \\ z' \end{pmatrix} = \begin{pmatrix} 7 & 5 & 6 \\ 4 & 3 & 3 \\ 10 & 7 & \lambda \end{pmatrix}\begin{pmatrix} x \\ y \\ z \end{pmatrix}$, $\begin{matrix} x = -3x' - 2y' + 3z' \\ y = 2x' + 4y' - 3z', \\ z = 2x' - y' - z' \end{matrix}$ (5, −7, 1), $\lambda = 9$.

5 (a) $x = 1, y = z = 0$, three planes meet in a point (b) $x = 1, y = z$, three planes form a sheaf meeting in a line. **6** $a = -3$, (i) $\begin{pmatrix} 9t-4 \\ 2-5t \\ 11t-6 \end{pmatrix}$ (ii) no solution.

7 $\begin{pmatrix} x \\ y \\ z \end{pmatrix} = \begin{pmatrix} 2+2t \\ -1-3t \\ t \end{pmatrix}$, this is the solution for $\lambda = 4$ and the three planes form a sheaf, the solution for $\lambda \neq 4$ is $x = 4/3$, $y = 0$, $z = -1/3$, the three planes meeting at one point. **8** $\begin{pmatrix} 1 & 2 & -1 & 1 \\ 0 & 1 & a+2 & 1 \\ 0 & 0 & (7-a)(a+3) & 7-a \end{pmatrix}$, (i) $a = -3$

(ii) $a = 7$ with the solution $x = 19z - 1$, $y = 1 - 9z$, $z \in \mathbb{R}$ (iii) $a \notin \{-3, 7\}$.
9 $(-2t, 5t/2, t)$. **10** (a) $3x = 2y$ (b) $15x = 5y - 3z$. **11** 3.

12 $a = 2$ with solution $\frac{x-2}{8} = \frac{y+2}{-13} = \frac{z-1}{6}$, $a = 3$ with no solution.

13 $\begin{pmatrix} 1 & 2 & -1 & \beta \\ 0 & 3 & -(\alpha+2) & 2\beta \\ 0 & 0 & -2(\alpha^2-4) & \beta(\alpha-2) \end{pmatrix}$ (i) $\left\{\begin{pmatrix} -1/2 \\ 1/2 \\ -1/2 \end{pmatrix}\right\}$ (ii) $\left\{\begin{pmatrix} t \\ 0 \\ t \end{pmatrix} : t \in \mathbf{R}\right\}$.

14 $\begin{pmatrix} x \\ y \\ z \end{pmatrix} = \begin{pmatrix} 5-5t \\ -1+2t \\ t \end{pmatrix}$, $5x + 6y + 13z = 19$, $\begin{pmatrix} 5/\sqrt{(230)} \\ 6/\sqrt{(230)} \\ 13/\sqrt{(230)} \end{pmatrix}$, $31/\sqrt{(230)}$.

15 $\{-2, -1, 3\}$, $1/\sqrt{2}$.

Miscellaneous Exercise 4 (p. 151)

1 $\begin{pmatrix} 24 & -23 & -21 \\ -8 & 8 & 7 \\ -1 & 1 & 1 \end{pmatrix}$ (i) $\begin{pmatrix} 0 \\ 0 \\ 0 \end{pmatrix}$ (ii) $\begin{pmatrix} -47 \\ 16 \\ 2 \end{pmatrix}$ (iii) $\begin{pmatrix} -85 \\ 29 \\ 4 \end{pmatrix}$.

2 $\begin{pmatrix} -1 & 0 & 0 \\ 0 & -1 & 0 \\ 0 & 0 & 1 \end{pmatrix}$, rotations about Oz, **M**, ($\mathbf{M}^2$), through $\pi/2$, (π).

3 det A $\neq 0$, $\mathbf{A}^{-1} = \dfrac{1}{17}\begin{pmatrix} 7 & 2 & -3 \\ -5 & 1 & 7 \\ -6 & 8 & 5 \end{pmatrix}$, $\begin{pmatrix} x \\ y \\ z \end{pmatrix} = \begin{pmatrix} 1 \\ -2 \\ 4 \end{pmatrix}$.

4 (i) $\begin{pmatrix} 1/\sqrt{2} & -1/\sqrt{2} \\ 1/\sqrt{2} & 1/\sqrt{2} \end{pmatrix}\begin{pmatrix} 1 & 1 \\ 0 & 1 \end{pmatrix}$ (ii) $\begin{pmatrix} 2/\sqrt{2} & 0 \\ 1/\sqrt{2} & 1/\sqrt{2} \end{pmatrix}$, a shear.

5 (i) $5y_1 = x_1^2 - 10x_1 + 25$, refl. in Oy and enlgmt S.F. 5

(ii) $x = (-a+b+c)/3$, $y = (2a-b)/3$, $z = (2a-c)/3$, $\begin{pmatrix} -1/3 & 1/3 & 1/3 \\ 2/3 & -1/3 & 0 \\ 2/3 & 0 & -1/3 \end{pmatrix}$.

6 $\mathbf{P} = \begin{pmatrix} 0 & 1 & 0 \\ 0 & 1 & 1 \\ 2 & 3 & -1 \end{pmatrix}$, $\mathbf{Q} = \begin{pmatrix} 1 & 0 & 0 \\ -3 & 1 & 0 \\ 11 & -2 & 5 \end{pmatrix}$, $\mathbf{B}^{-1} = \begin{pmatrix} -20 & 4 & -9 \\ -5 & 1 & -2 \\ 11 & -2 & 5 \end{pmatrix}$, **A** has no inverse, $\alpha = 2$, $\beta = 3$, $\gamma = -1$. **7** Rotation $\pi/2$ about Ox, $\mathbf{T}_2 = \begin{pmatrix} 1 & 0 & 0 \\ 0 & -1 & 0 \\ 0 & 0 & 1 \end{pmatrix}$,

$\mathbf{T}_3 = \begin{pmatrix} 1 & 0 & 0 \\ 0 & 0 & 1 \\ 0 & 1 & 0 \end{pmatrix}$, $\mathbf{T}_3\mathbf{T}_1 = \begin{pmatrix} 1 & 0 & 0 \\ 0 & 1 & 0 \\ 0 & 0 & -1 \end{pmatrix}$, a reflctn in plane Oyz,

$\mathbf{T}_1\mathbf{T}_3 = \begin{pmatrix} 1 & 0 & 0 \\ 0 & -1 & 0 \\ 0 & 0 & 1 \end{pmatrix} = \mathbf{T}_2$. **8** $x = y\sqrt{3} = z\sqrt{3}$.

9 $\begin{pmatrix} x \\ y \\ z \end{pmatrix} = \begin{pmatrix} 2s+k \\ 2t+k \\ 2k-s-t \end{pmatrix}$, $\begin{pmatrix} x \\ y \\ z \end{pmatrix} = \begin{pmatrix} k \\ -5k \\ 2k \end{pmatrix}$.

10 $\begin{pmatrix} 1 & 1 & -4 \\ -2 & -1 & 7 \\ 1 & -1 & -1 \end{pmatrix}$, $\begin{pmatrix} -8 \\ 13 \\ 0 \end{pmatrix}$, $(5 \;\; 0 \;\; -13)$. **11** $2x = z+2$, $\begin{pmatrix} 2/\sqrt{5} \\ 0 \\ -1/\sqrt{5} \end{pmatrix}$.

12 **A** is a squash on to the line L, $3y = x$, **AB** is an enlgmnt S.F. 5 followed by the squash on to L, points move further along L under **AB**.

13 $\lambda = 3$, (1, 2), $\lambda = 6$, $(-1/5, 1/5)$. **14** $\begin{pmatrix} 1 & p \\ 0 & 1 \end{pmatrix}$, $\begin{pmatrix} -4/5 & -3/5 \\ 3/5 & -4/5 \end{pmatrix}$, $p = 1$.

15 Enlargement S.F. a. **16** $a = 5c - 6b$, 0, $6x' - 5y' + z' = 0$.

17 $(a-b)(b-c)(c-a)(a+b+c)$, $a+b+c$.

18 (i) line $x/5 = -y/2 = -z/9$, $7\lambda = 9\mu$ (ii) plane $x + z = 2y$ (iii) line $y = 0$, $z + x = 0$, $\lambda = -\mu$.

Exercise 5.1 (p. 159)

2 (i) 3/4 and 3/5 or $-3/4$ and $-3/5$, (ii) 5/12 and 5/13. **3** (i) $\ln(5/3)$, (ii) 0 or $\ln 2$. **4** (i) 0 or $\ln(5/3)$, (ii) $\ln 3$, (iii) ϕ, (iv) $\ln 2$ or $-\ln 4$, (v) $\ln\{(5+\sqrt{31})/2\}$. **5** $\ln(2/3)$. **7** (i) $4\sinh^3 x + 3\sinh x$, (ii) $\cosh(x+y) - \cosh(x-y)$, (iii) $\sinh(x+y) - \sinh(x-y)$, (iv) $2\cosh\{(x+y)/2\}\cosh\{(x-y)/2\}$, (v) $2\sinh\{(x+y)/2\}\cosh\{(x-y)/2\}$.

Exercise 5.2 (p. 163)

3 (i) $c-\frac{1}{2}\coth 2x$, (ii) $c-\frac{1}{3}\operatorname{sech} 3x$. **4** (i) $2\sinh 2x$, (ii) $3\cosh 2x$, (iii) $3\cosh(3x-2)$, (iv) $\cosh x + x\sinh x$, (v) e^{2x}, (vi) $(x\operatorname{sech}^2 x - \tanh x)/x^2$, (vii) $\tanh x$, (viii) $-\sinh x \sin(\cosh x)$, (ix) $\operatorname{sech} x$, (x) $\{\cosh(\ln x)\}/x$
5 (i) $(\sinh x\cosh x - x)/2 + c$, (ii) $c - \operatorname{sech} x$, (iii) $\frac{1}{3}\cosh^3 x - \cosh x + c$, (iv) $x - \tanh x + c$, (v) $c - \frac{1}{3}\operatorname{sech}^3 x$, (vi) $\frac{1}{4}e^{2x} - \frac{1}{2}x + c$, (vii) $x\sinh x - \cosh x + c$, (viii) $\frac{1}{12}\cosh 6x - \frac{1}{4}\cosh 2x + c$. **6** $y = \cosh 2 + 3(x-1)\sinh 2$, $y = \cosh 2 - (x-1)/(3\sinh 2)$. **7** $ay\sinh p = bx\cosh p - ab$, $by\cosh p + ax\sinh p = (a^2+b^2)\sinh p\cosh p$, $ab(\sinh p\cosh p - p)$.
8 $2\sinh 1 + 3\cosh 3$, $2\cosh 1 + 9\sinh 3$. **10** (i) $c + \ln(\sinh x)$, (ii) $\frac{1}{4}\tanh^4 x + c$, (iii) $x - \tanh x - \frac{1}{3}\tanh^3 x + c$, (iv) $\frac{1}{4}\cosh^4 x + c$, (v) $\frac{1}{5}\cosh^5 x - \frac{1}{3}\cosh^3 x + c$, (vi) $\frac{1}{3}x^2\cosh 3x - \frac{2}{9}x\sinh 3x + \frac{2}{27}\cosh 3x + c$.
12 $\pi(\sinh 4 - 4)/4$.

Exercise 5.3 (p. 166)

2 (i) $\ln(2+\sqrt{3})$, (ii) $\ln(\sqrt{10}-3)$, (iii) $\ln\sqrt{3}$. **3** (i) $3\ln(2+\sqrt{5}) + 4\ln(2+\sqrt{3})$, (ii) $3\ln(1+\sqrt{2})$, (iii) does not exist. **4** (i) $\ln\{(2+\sqrt{5})/(1+\sqrt{2})\}$, (ii) $\ln\{(3+\sqrt{5})/2\}$. **6** (i) $\{0\}$, (ii) $\{\ln(2+\sqrt{3}),\ \ln(3+\sqrt{8})\}$, (iii) $\{\ln\sqrt{(7/3)}\}$.
7 (i) $2\ln(1+\sqrt{2}) + 2\sqrt{2}$, (iii) $\sqrt{3} - \ln\sqrt{(2+\sqrt{3})}$.

Exercise 5.4 (p. 170)

2 (i) $3/\sqrt{(9x^2-24x+17)}$, (ii) $1/2(x-x^2)$, (iii) $2x/\sqrt{(x^4-1)}$, (iv) $-1/\sqrt{(x^2-x^4)}$, (v) $2x/\sqrt{(x^4+4x^2+5)}$, (vi) $\sinh^{-1} x + x/\sqrt{(x^2+1)}$, (vii) $(\sinh x)/\sqrt{(1+\cosh^2 x)}$. **3** (i) $\operatorname{arcosh}(x/2) + c$, (ii) $\operatorname{arsinh}(x/3) + c$, (iii) $c - (\operatorname{artanh}(x/3))/3$, (iv) $(\arctan(x/3))/3 + c$, (v) $c - \arccos(x/3)$.
4 (i) $\operatorname{arsinh}(x-2) + c$, (ii) $\operatorname{arcosh}(x-2) + c$, (iii) $(\operatorname{arcosh}((2x+3)/3))/2 + c$. **5** (i) $\operatorname{arsinh}(x/2) + c$, (ii) $(\operatorname{arcosh} 3x)/3 + c$, (iii) $\operatorname{arsinh}(x+1) + c$, (iv) $\operatorname{arcosh}((x+2)/2) + c$, (v) $(\ln\{(2x-3)/(2x+3)\})/12 + c$, (vi) $\frac{1}{2}\operatorname{arsinh}(2x/3) + c$. **6** (i) $\ln(1+\sqrt{2})$, (ii) $\ln\{(3+\sqrt{8})/(2+\sqrt{3})\}$, (iii) $\ln\{(b+\sqrt{(b^2+1)})/(a+\sqrt{(a^2+1)})\}$, (iv) $\ln\{(2+\sqrt{5})/(1+\sqrt{2})\}$, (v) $\ln\{(3+\sqrt{10})/(2+\sqrt{5})\}$, (vi) $\ln\{(3+\sqrt{8})/(2+\sqrt{3})\}$, (vii) $2\sqrt{8} - 2\sqrt{3}$, (viii) $\sqrt{8} - \sqrt{3} - 2\ln\{(3+\sqrt{8})/(2+\sqrt{3})\}$. **7** $\ln(3+2\sqrt{2})$.

Miscellaneous Exercise 5 (p. 171)

1 $\{0, \ln 4\}$. **2** $\{1/\sqrt{2}, -1/\sqrt{2}\}$. **3** (a) $2\operatorname{cosech} 2x$, (b) $2\cos 2x \sinh(\sin 2x)$.
4 $\ln 2$, $\ln(1/4)$. **5** $9/8$, $\pi(19 - 22\ln 2)/8$. **6** (a) $(\operatorname{arsinh} 3x)/3 + c$, (b) $(\sinh 6x)/12 - x/2 + c$. **7** $\ln(4/3)$. **8** (i) $\{\ln(5/3), \ln 3\}$. **9** 3, 3, (i) 3 (ii) $\ln 9$.
10 2, 1, -1. **11** $y + x + 1 = 0$. **12** $\tan^{-1}(\tanh x) + c$.
13 $\pi/4$, $3\pi(\ln(1+\sqrt{2}))/2 - \pi/\sqrt{2}$. **15** (i) $\sqrt{(y^2+1)}\ln\{y+\sqrt{(y^2+1)}\} - y$.
16 $\ln(7+5\sqrt{2})$. **17** e. **18** $(2^{1/3} - 2^{-1/3})/2$. **19** $1/12$.

Exercise 6.1 (p. 176)

1 (i) 0·099834, 1 (ii) 0·841469, 4 (iii) $-0·544$, 16. **2** 7, 11.

Exercise 6.2 (p. 178)

1 (a) 1, 1, 1, 1; 1, 1·5, 1·75, 1·875; 1, 2, 3, 4; 1, -4, 21, -104; 1, 11, 111, 1111; no
(b) 1, 1, 1, 1; 1, 1·5, 1·625, 1·667; 1, 2, 2·5, 2·833; 1, -1, 1, $-1·667$; 1, 9, 41, 211·6; no

(c) 1, 1, 1, 1; 1, 0·75, 0·812, 0·797; 1, 0·5, 0·75, 0·625; 1, 2, 3, 4; 1, −3, 13, −51; no
(d) 1, 1, 1, 1; 1, 1·125, 1·1276, 1·1276; 1, 1·5, 1·542, 1·543; 1, 9, 19·67, 25·356; yes
(e) 0, 0, 0, 0; −0·5, −0·479, −0·4794, −0·4794; 1, 0·83, 0·8416, 0·8415;
−2, −0·66, −0·933, −0·9079; yes. **2** 1·175201.
3 (i) yes (ii) yes (iii) yes (iv) no (v) yes (vi) no.
4 $f'(x) = 4 - 10x$, $f''(x) = -10$, 3, 4, −10. **5** $a = 1$, $b = 1$, $c = 1/2$, $d = 1/6$.

Exercise 6.3 (p. 181)

1 (i) $-4x$ (ii) $2x$ (iii) $1 + x$ (iv) x. **2** (i) 1·1, 1·5, 2 (ii) 0·1, 0·5, 1
(iii) 0·1, 0·5, 1 (iv) 1·105, 1·22, 1·48 (v) 1·1103, 1·2427, 3·33 (vi) 1, 1, 1 (vii) 0·9, 0·5, 0.
3 0·8778. **5** (i) 8·0625 (ii) 7·75. **6** (i) 0·8469 (ii) 0·5319 (iii) 0·78747.
7

x	1·02	1·03	1·04	1·05
$\sin x$	0·8522	0·8575	0·8626	0·8677
$\cos x$	0·5234	0·5149	0·5063	0·4977

Exercise 6.4 (p. 185)

1 (i) 0·00001 (ii) 0·00033 (iii) 0·01 (iv) 0·07 (v) 2·03. **2** (i) 4×10^{-8}%
(ii) $3{\cdot}5 \times 10^{-5}$% (iii) $2{\cdot}5 \times 10^{-3}$% (iv) 5000%.
3 (i) $1 - x/2 - x^2/8$, 0·94875, 1·095, 0·7187, 0·375 (ii) $2 - 2x + 2x^2$, 1·82, 2·48, 1·5, 2
(iii) $1 - x^2/2$, 0·995, 0·98, 0·875, 0·5. **4** e^x (i) 2 (ii) 3 (iii) 5 (iv) 6 (v) 9.
5 $\cos x \approx 1 - x^2/2! + x^4/4! - x^6/6!$, $\sin x \approx x - x^3/3! + x^5/5! - x^7/7!$,
$e^x \approx 1 + x + x^2/2! + x^3/3!$, $\ln(1 + x) \approx x - x^2/2 + x^3/3 - x^4/4$.
6 (i) x, $x + x^3/3$, $x + x^3/3$ (ii) x^2, x^2, $x^2 - x^4/6$ (iii) $1 - x^2$, $1 - x^2$, $1 - x^2 + x^4/2$.
7 $(\pi - x) - (\pi - x)^3/6$, 1·2%. **8** $x^4 - 5x^3 + 10x^2 - 10x + 5$, 0·5, 10·67, 0·0625.

Exercise 6.5 (p. 191)

2 $x - x^2/2 + x^3/3$, $(-1)^{r+1}x^r/r$, −0·030459. **3** $a = 2$, $b = -7/6$.
5 $-x - 13x^2/2 - 19x^3/3$. **6** $a = -1/2$, $b = 1/2$. **8** (i) $1 - x/2 - x^2/8 - x^3/16$
(ii) $-ax - a^2x^2/2 - a^3x^3/3$, $a = 2$, $p = 3/8$, $q = 125/48$.
9 $x - x^3/6 + x^5/120$, $(-1)^{n+1}x^{2n-1}/(2n-1)!$.
10 $\ln(1 - \sin x) = -x - x^2/2 - x^3/6 - x^4/12 + \ldots$. **12** 0·080043.
13 $(-1)^n(1 + n - n^2)/n!$. **14** $x + x^2/2 - 5x^3/6$. **15** 0·005. **16** $2x^3/3$.
17 1, 0, $x + x^3/3 + x^5/5 + \ldots + x^{2n+1}/(2n+1)$. **18** $x - x^5/30$.
19 (i) 0, $-n^2/2$, $-n^3/3$ (ii) $x + x^3/3 + \ldots + x^{2r+1}/(2r+1) + \ldots$, $|x| < 1$.
20 $2x + 2x^3/3 + 2x^5/5$, 0, $2/(2n+1)$, $|x| < 1$,
$x + x^3/6 + 3x^5/40 + 5x^7/112 + \ldots$, $(2n)!/4^n(n!)^2(2n+1)$.

Exercise 6.6 (p. 197)

2 $1 + x + x^2/2 + x^3/3$. **3** (i) $1 + x^2/2! + x^4/4!$ (ii) $1 - x^2 + x^4/3$
(iii) $e - ex^2/2 + ex^4/6$, $a = -2e/3$, $b = -e/3$.
4 (i) $f(x) = (e^x + e^{-x})/2$, $g(x) = (e^x - e^{-x})/2$, $1 + (2x)^2/2! + \ldots + (2x)^{2n}/(2n)! + \ldots$,
(ii) $3x - 3x^2/2 + 3x^3 - 15x^4/4$, $|x| < 1/2$. **5** $(\ln p)^2 + (2\ln p/p)\alpha + ((1 - \ln p)/p^2)\alpha^2$.
6 $\cosh\sqrt{x} - 1$. **7** $(e^x - e^{-x} + 2\sin x)/4$, $x \in \mathbb{R}$.

Miscellaneous Exercise 6 (p. 198)

1 $x + x^3/3$, 0·0014%, 14·4%. **2** 2·3979. **3** −0·0513. **4** $x - 3x^2/2$.
5 $x + x^3/3 + 2x^5/15$. **6** $a = 2$, $b = -7/3$. **7** $a = 1/2$, $b = -1/2$.

8 (a) $a = -3, b = -1, 470/49$ (b) 0·2231. **9** $1+2x^2-8x^3/3, |x| < 1/2$.
10 $1/n! - 2^{2n}/2n, |x| < 1/2$. **12** $-2/3, 2^{n-2}(n-4)(n-1)/n!$.
13 $(n^2c + n(b-c) + a)/n!, a = 2, b = 4, c = 1$. **14** $(-2)^n/n(n-1)$.
15 $p^2-2, p^3-3p, -px-(p^2-2)x^2-(p^3-3p)x^3$. **16** $1+x^2/2!+x^4/4!, x^n/(2n)!$.
17 $-3x^2-2x^3$. **18** $2x^3/27, (-1)^{n+1}(3^n-3(2^n-1))/3^n n$.
19 $3x/2+x^2/4+x^3/2, a = -1/6, k = 1/2$.
20 $a = 1/2, b = 1/4$. **21** $1+y/2-y^2/8, \quad a = -1/2, \quad b = -9/8$.
22 $1+3x+9x^2/2+4x^3+15x^4/8, 0$. **23** $a = 1, b = 2/45$.
24 $a = 3/4, b = 0, c = -1/2, d = 0$.
25 $y+2x, a_0 = 0, a_1 = 1, a_2 = 0, a_3 = 1/3, a_4 = 0, a_5 = -2/15, a_6 = 0, a_7 = 8/105$.
26 (a) $1+2x-2x^2+4x^3$, 2·4495 (b) (i) $x-x^2/2+x^3/3-x^4/4$
(ii) $1+x+x^2/2!+x^3/3!+x^4/4!, y = -\ln(1-x), x = y-y^2/2+y^3/6-y^4/24$.
27 $2-(x-1)+3(x-1)^2/2+(x-1)^3/2-(x-1)^4/8$, 1·915488.
28 $1, 0, 1+x-x^3/3-x^4/6-x^5/30$, 1·099641. **30** 2, 1·937, 1·939.

Exercise 7.1 (p. 206)

2 (2·07, 2·07). **3** (i) 1, (1, 0) (ii) 1·5, $(1{\cdot}5, \pi/2)$ (iii) 1/4, $(1/4, \pi)$
(iv) $1/\sqrt{2}, (1/\sqrt{2}, \pi/4)$. **4** (i) $r\cos\theta = 2$ (ii) $r\sin\theta = 1$ (iii) $r\sin\theta = -1$.
5 $(1, 3\pi/2)$. **6** $(a, \pi/3), (a, -\pi/3), 2r\cos\theta = a$. **7** $(\sqrt{2}a, \pi/4)$.
8 $(2\sqrt{2}, \pi/4), (2\sqrt{2}, -\pi/4)$. **9** $4a/\pi$.

Exercise 7.2 (p. 211)

1 $(2a, \pi)$. **2** $(\sqrt{5}-1)/2$. **5** Rectangular hyperbola $x^2-y^2 = a^2$.

Exercise 7.3 (p. 214)

1 $\pi^3a^2/6$. **2** $(\sqrt{2}, \pi/4), r\cos\theta = 1, r\sin\theta = 1, \theta = \pi/4, \pi/2-1$. **3** $5\pi a^2$.
4 $9\pi/8$. **5** $\sqrt{3}a^2/8$.

Miscellaneous Exercise 7 (p. 215)

1 $3\pi^3/16, 3\pi^3/32$. **2** $(a, 2\pi/3), (a, 4\pi/3)$. **3** $(6, \pi/3), (6, -\pi/3), 9\pi+3\sqrt{3}/2$.
4 $n\pi \pm \pi/6, \sqrt{3}a^2/2 - a^2\{\ln(2+\sqrt{3})\}/2$. **5** 2π.
6 $r = 4\cos\theta, (4, 0), (2\sqrt{2}, \pi/4), \theta = \pi/4, \quad \theta = \pi/4 - 2\arctan(1/2)$.
7 3, 4, 6, 2, 90°, 270°, 0°, 180°, 360°.
8 $(3a/2, \pi/3), (3a/2, -\pi/3), \theta = \pi/3, \theta = -\pi/3, 4r\cos\theta = 3a$.
9 $5\pi/4+3\sqrt{3}/8$. **10** $r\cos\theta = \pm a, r\sin(\theta-\alpha) = \pm a, 4a^2\sec\theta$.
11 $\pi a^2/8, (2a/3, \pm\arcsin(1/\sqrt{6}))$.

Exercise 8.1 (p. 218)

1 (i) $3/(x^2+3x+1)$, (ii) $(2x^2-4x+3)/(x^2-9)$, (iii) $4/(4-x^2)$,
(iv) $(3x^2+5x+1)/(x+1)$. **2** (i) $x-3, 2$, (ii) $x^2, 0$, (iii) $x^2+3, -2$,
(iv) $(16x^4+8x^3-44x^2-22x-11)/32, 53/32$. **3** (i) 4, (ii) 14, (iii) $-21/2$, (iv) 1,
(v) 44, (vi) 15/4. **4** (i) $(x-1)(x+1)(x-\mathrm{i})(x+\mathrm{i})$, (ii) $(x+3)(x-4)(x+5)$,
(iii) $(x-2)(x-2+\mathrm{i}\sqrt{2})(x-2-\mathrm{i}\sqrt{2})$, (iv) $(x-1)(x+(1+\mathrm{i}\sqrt{3})/2)(x+(1-\mathrm{i}\sqrt{3})/2)$,
(v) $(t-2)(2t-1)(2t+3)$, (vi) $(x+\mathrm{i}y)^2(x-\mathrm{i}y)^2$.
5 (i) $x-3+2/(x+1), 1-2/(x+1)^2$, (ii) $2x+4-3/(x-2), 2+3/(x-2)^2$,
(iii) $(8x^3-12x^2+10x-15+45/(2x+3))/16, (12x^2-12x+5-45/(2x+3)^2)/8$.
6 (i) $\ln 2 - 1/2$, (ii) $16/3+5\ln 2$, (iii) $(-60+505\ln(7/5))/32$.

Exercise 8.2 (p. 221)

1 (i) x, x, (ii) $2x+4, 8x+16$, (iii) $x+1, -7x$, (iv) $x^2+3x+10, 30x+2$, (v) $3x^2+10x+15, -50$, (vi) $(4x^2+6x+7)/8, (15x-7)/8$. **2** (i) $x, x-2$, (ii) $2x+1, -3x-7$, (iii) $x^2+3x+4, 7x-9$, (iv) $x^2-1, 2$, (v) $x^2-2x+4, -11x+2$, (vi) $(3x^2+6x-7)/9, (49-69x)/9$. **3** (i) $x-2$ (ii) $-4x-6$ (iii) $(13x-17)/4$ (iv) $(25x+14)/8$ (v) $(246-475x)/16$ (vi) $54x+4$. **4** (i) $3x$ (ii) $12x-20$ (iii) $(22x-13)/4$ (iv) $7x-4$ (v) $(8x+21)/16$ (vi) $(527-372x)/16$. **5** (i) $14x+20$ (ii) $5x-1$ (iii) $9x-5$ (iv) $918x-1512$ (v) $171x-170$ (vi) $63(x+1)$.
6 (i) $(x+1)^2(x-2)$ (ii) $(2x-3)^2(x+9)$ (iii) $(2x+1)^2(x-3)(x+2)$ (iv) $(2x+1+i\sqrt{3})^2(2x+1-i\sqrt{3})^2/16$ (v) $(2x-1)^2(x-2)^2$ (vi) $(x-1-i)^2(x-1+i)^2$. **7** 13. **8** 0. **9** $\{-1/2, 9/4\}$.

Exercise 8.3 (p. 224)

1 (i) $-2, 1, 4$ (ii) $-5/3, -1/3, 1$. **2** (i) $-1, 1/2, 2$ (ii) $-3/2, 9/8, 2$. **3** 10, 31.
5 -216. **7** (i) $\sqrt{3}, -\sqrt{3}, (-1+i\sqrt{3})/2, (-1-i\sqrt{3})/2$, (ii) $i\sqrt{3}, -i\sqrt{3}, (-1+\sqrt{5})/2, (-1-\sqrt{5})/2$. **8** $3x^4-2x^3+x^2-3x+2=0$.
9 (a) $-6, 1/2, 4/3$, (b) $x^3+9x^2+2x-25=0$.
10 $a=-4, b=-4$, solution $2\pm\sqrt{2}, 1\pm i$. **11** $-2p, p^2$.

Exercise 8.4 (p. 226)

1 14. **2** Max. $(-1, 7)$, min. $(5/3, -67/27)$, 2·4. **3** (i) Inflex. $(0, 2)$, $-1{\cdot}10$ (ii) min. $(0, -3)$, 1·15 (iii) none, 0·18 (iv) min. $(0, -1)$, min. $(2, -1)$, max. $(1, 0)$, 2·41.
4 $\{-a/2, a\}$. **5** 0·4, 1·5. **6** $0<k<4, k=2$.

Exercise 8.5 (p. 234)

2 Min. $(1, 0)$, max. $(-5/3, 8)$. **3** Max. $(1, 1)$, min. $(-1, -3)$. **4** $y<0$.

Miscellaneous Exercise 8 (p. 234)

1 $2p^2-6q$. **2** $y<1$ or $y>5$, $y=1, x=-1, x=-2$. **3** $-2, -10$. **4** $-2, 3, 5$.
5 (i) $a^2x^3-acx^2-bx-1=0$ (ii) $m=-1$, $n=-2$. **6** $-5/4<y<5$, $5/2$.
7 22, $10x^3+3x^2-x+1=0$. **8** $r=4$, $s=-2$. **9** x^2+2, (a) $\{\sqrt{6}, -\sqrt{6}\}$ (b) $\{\sqrt{6}, -\sqrt{6}, 2i, -2i\}$. **10** Max. $k+7$, min. $k-9$, (i) $-7<k<9$, (ii) $k<-7$, $(x^2-2x+4)(x^2-2x-10)$, $1\pm i\sqrt{3}, 1\pm\sqrt{11}$. **11** $\lambda>5/2$.
12 -3, $(x+3)(x+3)(2x-1)$.
13 $-1/3\leqslant y\leqslant 1$, $x=-1$, $\{x: -1/3\leqslant x\leqslant 1\}$, $(0, 1)$, $(-2, -1/3)$, not single valued.
14 (i) $c^3=b^3d$, $\{2, -2\}$ (ii) $\{-1/2, 2, 9/2\}$. **15** $-1/3, 3/2, -3$.
16 Max. $(1, 1/2)$, min. $(-1, -1/2)$, area $\ln\sqrt{(1+a^2)}$.
17 $f(x)=1+(x-a)^2$, $16x^3+12x^2-22x-6$, $2(x-1)(2x+3)(4x+1)$, $(x-1)(x-1)(2x+3)(2x+3)$. **18** (i) 2, 4, $\sqrt[3]{2}+\sqrt[3]{4}$.
19 $(0, -2/3)$, $(2, 0)$, $(2, 0)$, $(-22, 0{\cdot}98)$, $x=-1$, $x=6$, $y=1$.
20 Min. $(0, 0)$, inflexions $(1/\sqrt{3}, 1/4)$, $(-1/\sqrt{3}, 1/4)$, $0<|k|<1$. **21** $4+\ln 4-3\pi/4$.

Exercise 9.1 (p. 240)

The arbitrary constants in indefinite integrals are omitted:
1 (i) $\{\ln|2x-1|\}/2$, (ii) $3/(1-x)$, (iii) $\ln|(x-1)/(x+1)|$, (iv) $\arctan(x/2)$, (v) $x-2\arctan(x/2)$, (vi) $\ln\sqrt{(x^2+2x+2)}$, **2** (i) $\ln 25$, (ii) 184/105, (iii) $\pi/12-\sqrt{3}/8$, (iv) 14/9, (v) 94/135, (vi) $\ln 2-1/2$, (vii) $\pi^2/8$.
3 (i) $-e^{-3x}(9x^2+6x+2)/27$, (ii) $\{(4x+2x^2)\ln x-4x-x^2\}/4$, (iii) $e^{ax}(a\sinh bx-b\cosh bx)/(a^2-b^2)$ if $a^2\neq b^2$, $(e^{2ax}-2ax)/4a$ if $a=b\neq 0$, $(2ax-e^{2ax})/4a$ if $a=-b\neq 0$, (iv) $x\operatorname{arsinh}x-\sqrt{(1+x^2)}$, (v) $x\ln(4+x^2)-2x+4\arctan(x/2)$, (vi) $\sqrt{(x^2-1)}(3x^4+4x^2+8)/15$,

(vii) $x/\sqrt{(1-x^2)}$, (viii) $\frac{1}{2}\ln\{(x-1)^2/(x^2+x+1)\}-\sqrt{3}\arctan\{(2x+1)/\sqrt{3}\}$, (ix) $\frac{1}{3}\operatorname{arcosh}(x-2/3)$. **4** (i) $2+\ln\sqrt{2}$, (ii) $(1+\ln 2)/4$. **5** $\pi/3$.
6 (i) $(8\ln 2-3)/16$, (ii) $\pi/6$, (iii) $4/35$. **7** $-1/x+6/(x+1)-5/(x+2)$, (a) $3/4+1/(2n+2)-5/(2n+4)$, (b) $11\ln 3-17\ln 2$.
8 $(x+9)/(x^2+3)-1/(x+5)-14/(x+5)^2$, $\ln(5/3\sqrt{3})+\sqrt{3}\pi/2-7/15$.
9 $1/(x+1)+2/(x+1)^2-(x+1)/(x^2+3)$. **10** $1/2(x-1)^2+(x+\frac{1}{2})/(x^2+1)$, (a) $-1/2(x-1)+\frac{1}{2}\ln(x^2+1)+\frac{1}{2}\arctan x$, (b) $1+2x+2x^2$, for $|x|<1$.
11 (i) $\operatorname{arsinh}\{(x-1)/3\}$, (ii) $\frac{1}{3}\arctan\{(x-1)/3\}$. **12** $(\pi+1)/4$.
13 $\pi/4$. **14** $29\pi/8$.

Exercise 9.2 (p. 244)

1 $e-1, 6-2e$, **2** $I_n=nI_{n-1}-e^{-1}$, $24-65e^{-1}$. **3** $\pi^5-20\pi^3+120\pi$.
4 $2(\ln 2)^3-3(\ln 2)^2+3(\ln 2)-9/8$. **5** $2-3e^{-1/2}$. **6** $(3-21e^{-2})/4$.
7 $\sinh x(5\cosh^4 x+4\cosh^2 x+8)/15$.
8 (i) $6-16e^{-1}$, (ii) $y=x^2-4x+8-8e^{-x/2}$.
9 $(\operatorname{sech}^2 x\tanh x+2\tanh x)/3$, $(2\operatorname{sech}^3 x\tanh x+3\operatorname{sech} x\tanh x+6\arctan e^x)/8$.
10 $x/(x^2+1)^2+\frac{1}{2}\tan^{-1}x$.

Exercise 9.3 (p. 251)

1 (i) $(\sinh 6)/2$, (ii) $38/3$, (iii) $\ln(\sqrt{2}+1)$, (iv) $1+\ln(3/2)$, (v) $\sqrt{(17)}-\sqrt{2}+\frac{1}{2}\ln\{(9-\sqrt{17})(3+2\sqrt{2})/16\}$. **2** $9a/8$. **3** (i) $2\pi a$, (ii) $a(2\pi+3\sqrt{3})/3$, (iii) $56a^4/9$, (iv) $3a/2$, (v) $\ln\{(e^2+1)/(e^2+2e-1)\}$, (vi) $8a$.
4 (i) $\frac{1}{2}a\ln\{2\pi+\sqrt{(1+4\pi^2)}\}+\pi a\sqrt{(1+4\pi^2)}$, (ii) $a\sqrt{(1+k^2)}(e^{6k}-1)/2k$, (iii) $4a$, (iv) $3\pi a/2$.

Exercise 9.4 (p. 257)

1 (i) $\pi\left\{\ln\dfrac{e+\sqrt{(1+e^2)}}{1+\sqrt{2}}+e\sqrt{(1+e^2)}-\sqrt{3}\right\}$, (ii) $61\pi/5184$. **2** (i) $2\pi(1-\operatorname{sech}1)$, (ii) $12\pi(2+\sqrt{2})/5$, (iii) $12\pi/5$, (iv) $9\pi^2$, (v) $2\pi\{(\sec 1)(\ln\sec 1+1)\}$.
3 (i) $\pi a^2(2-\sqrt{2})$, (ii) $8\pi a^2(2\sqrt{2}-1)/3$, (iii) $20\sqrt{2}\pi a^2/3$.
4 $V=27\pi/7$, $S=3\pi(128-50\sqrt{5})/5$. **5** $8\pi a^2(5\sqrt{5}-2\sqrt{2})/3$.
6 $4\pi a^2(\sin\alpha-\alpha\cos\alpha)$. **7** $15008\pi h^2/625$.
8 $2\pi b^2+2\pi a^2 b\arccos(b/a)/\sqrt{(a^2-b^2)}$. **9** $\pi a^2\left\{\ln\dfrac{1+\sqrt{2}}{2+\sqrt{3}}+\sqrt{3}\cosh 2-\sqrt{2}\right\}/\sqrt{2}$.
10 (i) $4\pi a^2$ (ii) πa^3 (iii) πa (iv) $2a/3$. **11** $4a/3\pi$ from each bounding radius.
12 $8\pi a^2$.

Exercise 9.5 (p. 260)

2 $\{1-(3N+1)e^{-3N}\}/9$. **5** $2-2\sqrt{(1-a)}$, $4-2\sqrt{(b-1)}$. **6** (i) 20, (ii) $\pi/16$, (iii) $\pi/2-\arcsin(2/3)$, (iv) $2\arctan 2+\ln 3-\pi$, (v) $\pi/2-\arcsin(1/3)$.

Miscellaneous Exercise 9 (p. 261)

1 $\ln(4/3)$. **2** $1/(x-3)+2/(x-3)^2-1/(x+1)$, $\{4+3\ln(15/7)\}/6$, $(-26x^2+27x-30)/27$. **3** $\dfrac{1}{3}\operatorname{arcosh}\dfrac{3x-2}{\sqrt{5}}$. **4** $8/105$. **5** $I_n=e-nI_{n-1}$,

$120-44e$. **6** $\pi a^2(5+e^2)/2$. **7** $\ln\{(x+2)/3+\sqrt{(x^2+2x-5)}\}$. **8** $2a/\pi$.
9 $\operatorname{cosech} u$, $py - p\ln p = x - p$, $ey = x$, $\sqrt{(1+x^2)}-\sqrt{2}+\ln\dfrac{e(\sqrt{2}+1)}{1+\sqrt{(1+e^2)}}$.
10 (ii) $x\ln(x+\sqrt{x})-\ln(\sqrt{x}+1)-x+\sqrt{x}$. **11** $3/20$. **12** (a) $(4+\sinh 4)/8$, (b) $8a$. **13** (a) $1/(x-1)-x/(x^2+1)$, (b) $\ln\sqrt{2}$, $-(1+2x+x^2+x^4+2x^5+x^6)$.
15 (i) $(9e-65e^{-1})/2$, (ii) $6-e-8e^{-1}$. **16** (i) $5\pi/32$, (ii) $16/15$, (iii) 0, (iv) $35\pi/256$. **17** 1·15, $28\pi/15$. **18** $(\operatorname{arsinh} 2)/4+\sqrt{5}/2$, $\pi(18\sqrt{5}-\operatorname{arsinh} 2)/32$.
19 (i) $8\pi(8-2\sqrt{2})/3$, (ii) (a) converges to 1/2, (b) diverges.
20 $120-44e$, $\pi a^2/16$. **21** $x^3(\ln x)^{n-1}(\ln x+n/4)$, $(5e^4-1)/32$, 0·467.
22 $(13/3, 0)$, $49\pi/3$. **23** (i) $\sqrt{(1+y^2)}\operatorname{arsinh} y - y$. **24** (i) $\pi/2-1$, (ii) $(e^x-e^{-x})/(e^x+e^x)$, $e-1+\pi/2-2\arctan e$, (iii) $3\pi/16$. $243/(4\ln 2)$, $3+\ln 2$.
26 $1/(x-1)^2+x/(2x^2+1)$, (a) $-1·9$, (b) $2/3+\{\ln(11/3)\}/4$, $|x|<1/\sqrt{2}$.
28 $(2-n)\operatorname{sech}^{n-2}x+(n-1)\operatorname{sech}^n x$, 3/4, 5/4. **29** $2-\pi/4$, $4\sqrt{2}\pi$.
30 6, $1-4/e$, 76/105. **31** πa^2. **32** $f(x)=(x^n e^{ax})/a$, $(5e^3-2)/27$. **33** 4·84.

Exercise 10.1 (p. 267)

1 (i) $5-i$ (ii) $8+7i$ (iii) $1+i$ (iv) $-4-17i$ (v) $7+i$ (vi) 13 (vii) $4+3i$ (viii) $(4+22i)/25$ (ix) $\{(ax-by)+i(ay+bx)\}/(x^2+y^2)$. **2** (i) $\{(2-3i)/13\}$ (ii) $\{(3-i)/5\}$ (iii) $\{(1-i\sqrt{3})/2, (1+i\sqrt{3})/2\}$ (iv) $\{6+2i, 6-2i\}$. **3** (i) 5, $-0·93$ (ii) $\sqrt{2}$, $\pi/4$ (iii) 13, 0·39 (iv) 1, $-\pi/6$ (v) 0·277, 2·16 (vi) 5, 0·64 (vii) 1, $-\pi/2$.
4 (i) $-5+12i$ (ii) a^2-b^2+2iab (iii) $4(\cos 2\pi/3+i\sin 2\pi/3)$ (iv) $r^2(\cos 2\theta+i\sin 2\theta)$ (v) $(9, 4\pi/3)$. **5** (i) $(8, \pi/2)$, $(2, -\pi/6)$ (ii) (r, θ), $(1/r, -\theta)$ (iii) $(9, 0)$, $(1, \pi)$ (iv) $(2\sqrt{2}, -\pi/12)$, $(1/\sqrt{2}, 7\pi/12)$.

Exercise 10.2A (p. 270)

1 $\{-3, (3+3i\sqrt{3})/2, (3-3i\sqrt{3})/2\}$, $(27, \pi)$, $(3, \pi)$, $(3, \pi/3)$, $(3, -\pi/3)$.
2 $\{1, -1, i, -i\}$, $(1, 0)$, $(1, \pi/2)$, $(1, \pi)$, $(1, 3\pi/2)$.
3 (i) $(2, 0)$, $(-2, 0)$ (ii) $(2, \pi)$, $(-2, \pi)$ (iii) $(2, \pi/4)$, $(2, 5\pi/4)$ (iv) $(2, -\pi/4)$, $(2, 3\pi/4)$ (v) $(5, \pi/4)$, $(5, 5\pi/4)$ (vi) $(5, \alpha/2)$, $(5, \alpha/2+\pi)$ (vii) $(5, -\alpha/2)$, $(5, \pi-\alpha/2)$ (viii) $(5, (2\alpha-\pi)/4)$, $(5, (2\alpha+3\pi)/4)$ (ix) $(5, -(\pi+2\alpha)/4)$, $(5, (3\pi-2\alpha)/4)$, all have modulus either 2 or 5 and so lie on one of the circles, centre O of radius 2 and of radius 5.

Exercise 10.2B (p. 273)

1 (i) $\sqrt{3}+i$ (ii) $(\sqrt[6]{2}, \pi/4)$ (iii) $(1, \pi/5)$ (iv) $(\sqrt[4]{8}, 3\pi/8)$ (v) $\cos\theta/4 - i\sin\theta/4$ (vi) $(4\sqrt[3]{2}, -\pi/6)$. **3** $(2, 2\pi/3)$ (i) $(2^{10}, 2\pi/3)$ (ii) $2^9(-1+i\sqrt{3})$.
4 (i) $(1, \pi/4)$, $(1, -\pi/4)$, $(1, 3\pi/4)$, $(1, -3\pi/4)$ (ii) $(1, \pi/8)$, $(1, 5\pi/8)$, $(1, -3\pi/8)$, $(1, -7\pi/8)$ (iii) $(1, -\pi/8)$, $(1, -5\pi/8)$, $(1, 3\pi/8)$, $(1, 7\pi/8)$ (iv) $(\sqrt{2}, \pi/24)$, $(\sqrt{2}, 13\pi/24)$, $(\sqrt{2}, -11\pi/24)$, $(\sqrt{2}, -23\pi/24)$. **5** (i) $(\sqrt[6]{2}, -\pi/12)$, $(\sqrt[6]{2}, 7\pi/12)$, $(\sqrt[6]{2}, -3\pi/4)$ (ii) $(2, \pi/4)$, $(2, 3\pi/4)$, $(2, -\pi/4)$, $(2, -3\pi/4)$ (iii) i, $-i$, $2i$, $-2i$.
6 $(z-1)(z+1)(z^2-z+1)(z^2+z+1)$. **7** G, $\{1, -1\}$, E, i, $-i$; G, E, every element except 1 is a generator. **8** $4\sin^3\theta = 3\sin\theta - \sin 3\theta$.
9 $8\cos^4\theta = \cos 4\theta + 4\cos 2\theta + 3$, $8\sin^4\theta = \cos 4\theta - 4\cos 2\theta + 3$.
11 All modulus 3, arguments $2k\pi/5$, $k\in\{0, 1, 2, 3, 4\}$.
13 (i) 64, π (ii) 1/2, $\pi/6$ (iii) 32, $7\pi/6$. **14** $\{-(1+i\cot k\pi/5)/2: k = 1, 2, 3, 4\}$.

Exercise 10.3 (p. 278)

2 (i) $z+z^*=-2$ (ii) $z=z^*+6i$ (iii) $(1+i)z+(1-i)z^*=0$ (iv) $(z+1)(z^*+1)=4$ (v) $(z-1-2i)(z^*-1+2i)=9$ (vi) $z=1+2i$.

3 (i) $(1-\mathrm{i}\sqrt{3})/2$ (ii) $(1, -\pi/3)$, 1. **5** (i) 6 (ii) $\pi/2$.
6 (i) $u=1$ (ii) $(u-1)^2+v^2=1$ (iii) $(u-4)^2+(v+1)^2=4$ (iv) $u+v+1=0$.
7 circle $|w|=1$. **8** $1/(2\sin\theta)$, $-\theta$. **9** (a) $|w-5|=8$ (b) $|w-5-2\mathrm{i}|=|w-5-6\mathrm{i}|$.
10 circle $|2w-1|=1$. **11** (a) $4\sqrt{10}/3$ (b) $8/3-2\mathrm{i}$, $4+2\mathrm{i}$.

Exercise 10.4 (p. 281)

2 (i) $\mathrm{e}^{\mathrm{i}\pi/2}$ (ii) $2\mathrm{e}^{3\mathrm{i}\pi/2}$ (iii) $2\mathrm{e}^{\mathrm{i}\pi/3}$ (iv) $2\sqrt{2}\mathrm{e}^{\mathrm{i}\pi/4}$ (v) $2\sqrt{2}\mathrm{e}^{3\mathrm{i}\pi/4}$
(vi) $5\mathrm{e}^{\mathrm{i}\alpha}$, $\alpha=\pi+\arccos(3/5)$. **5** $5\pi/32$. **6** $2\mathrm{e}^{-\mathrm{i}\pi/6}$.
7 (i) line $\mathrm{Re}(w)=1$, $\mathrm{Im}(w)\leqslant 0$ (ii) semi-circle centre 1/2, radius 1/2, imaginary part negative.

Miscellaneous Exercise 10 (p. 282)

2 $\mathrm{e}^{3\mathrm{i}\pi/5}$, $\mathrm{e}^{\mathrm{i}\pi/5}$, $\mathrm{e}^{7\mathrm{i}\pi/5}$, $\mathrm{e}^{9\mathrm{i}\pi/5}$, -1. **3** $(\sqrt{3}+\mathrm{i})/2$, $(-\sqrt{3}+\mathrm{i})/2$, $-\mathrm{i}$. **4** (i) $1-\mathrm{i}$
(ii) $\cos 4\theta=\cos^4\theta-6\cos^2\theta\sin^2\theta+\sin^4\theta$, $\sin 4\theta=4\sin\theta\cos\theta(\cos^2\theta-\sin^2\theta)$.
5 (ii) $32\cos^6\theta-48\cos^4\theta+18\cos^2\theta-1$. **6** $2\mathrm{e}^{\mathrm{i}\pi/3}$, $2\mathrm{e}^{-\mathrm{i}\pi/3}$, $-(1+\mathrm{i}\sqrt{3})/2$.
7 0, e^{-k}. **8** $22/3+5\pi/2$. **9** $\mathrm{e}^{\mathrm{i}\pi/3}$. **10** (i) 1, $\cos\pi/3+\mathrm{i}\sin\pi/3$, -1,
$2\cos 2\pi/3+2\mathrm{i}\sin 2\pi/3$ (ii) $\alpha=\cos\theta+\mathrm{i}\sin\theta$, $\beta=\cos\theta-\mathrm{i}\sin\theta$.
11 (ii) 5. **12** (i) $\{\mathrm{e}^{2k\pi\mathrm{i}/5}: k=0, 1, 2, 3, 4\}$. **13** $(z^n+z^{-n})/2$, $(z^n-z^{-n})/2\mathrm{i}$, $\pi/32$.
14 (i) $\sqrt{2}$, $-\pi/4$, $(1+\mathrm{i})/2$, $-4\sqrt{2}(1+\mathrm{i})$. **15** $v^2=16-4u$.
16 $\{\mathrm{e}^{2k\pi\mathrm{i}/5}: k=0, 1, 2, 3, 4\}$, $\{2\mathrm{e}^{2k\pi\mathrm{i}/5}: k=0, 1, 2, 3, 4\}$, not closed, C_5.
17 $1+\mathrm{i}\sqrt{3}, (2, \pi/3), -512$. **18** $\cos 2k\pi/5+\mathrm{i}\sin 2k\pi/5, k=0,1,2,3,4$, $4x^2+2x-1=0$.
19 (i) $2\mathrm{e}^{(1+3k)\pi\mathrm{i}/6}$, $k=0, 1, 2, 3$, (ii) $cww^*=w+w^*$, $(c, 0)$, $(0, 0)$, $(2/c, 0)$.
20 0, $\pi/9$, $3\pi/9$, $5\pi/9$, $7\pi/9$, $16c^4-12c^2+1$.
21 α^k, $k=0, 1, 2, 3, 4, 5, 6$, -1, $-1/2$, $\sqrt{7}/2$. **22** $b=1$, $a=\pm r^2$.
23 $a=4$, $b=9$, $c=-4$ (i) centre 4, radius 5.
24 let $\alpha=\mathrm{e}^{\mathrm{i}\pi/4}$, Subgroups are $\{1, \alpha^4\}$, $\{1, \alpha^2, \alpha^4, \alpha^6\}$,

$\left\{\begin{pmatrix}\cos k\pi/4 & -\sin k\pi/4\\ \sin k\pi/4 & \cos k\pi/4\end{pmatrix}: k=0, 1, 2, 3, 4, 5, 6, 7\right\}$. **25** $(3+\mathrm{i}\sqrt{3})/2$, $3(1+\mathrm{i}\sqrt{3})/2$, 0.

Exercise 11.1 (p. 289)

2 (i) $y=x^3/3+A$, $3y=x^3+3$ (ii) $y^2+x^2=A$, $y^2+x^2=25$
(iii) $y=Ax$, $y=2x$ (iv) $y=A\mathrm{e}^{4x}$, $y=2\mathrm{e}^{4x}$. **3** (i) $y^2=x^2+A$
(ii) $(y-1)^2=A\mathrm{e}^{-x^2}$ (iii) $x(\operatorname{cosec}y+\cot y)=A$ (iv) $3y^2=\ln(x^2)-x^2+A$
(v) $y-y^3/3=\sin x+A$ (vi) $y=A\mathrm{e}^{x-x^3/3}$ (vii) $x=1+A(x+1)\mathrm{e}^{t^2}$
(viii) $x^2=4t-t^2+A$. **4** (i) $(y+x)^3=A(y-x)$ (ii) $x\mathrm{e}^{x/y}=A$ (iii) $y\mathrm{e}^{x^2/2y^2}=A$
(iv) $y+\sqrt{(x^2+y^2)}=Ax^2$. **5** 2. **6** $P=A\mathrm{e}^{kt}$, 57 million. **7** $x=A(20-t)^3$.
8 Separable (i), (vi), Exact (iii), (v), Neither (ii), (iv). **9** $y=3\mathrm{e}^{x^2+x}-1$.
10 $\tan x$, $y=\sec x\,\mathrm{e}^{\tan x}$. **11** $\dfrac{\mathrm{d}x}{\mathrm{d}t}=-k\sqrt{x}$, $1/(\sqrt{2}-1)$ hours.
12 $x=100-99\mathrm{e}^{-t^2/50}$.

Exercise 11.2 (p. 293)

1 (i) e^{-6x} (ii) e^{x^3-2x} (iii) x (iv) $\sec x$ (v) $1/x^2$ (vi) $\ln x$.
2 (i) $y=A\mathrm{e}^{6x}$ (ii) $y\mathrm{e}^{x^3-2x}=A$ (iii) $xy=A$ (iv) $y=A\cos x$ (v) $y=Ax^2$
(vi) $y\ln x=A$. **3** (i) $y=A\mathrm{e}^{-2x}+3(2x-1)/4$ (ii) $y=Ax^2+x^3-1/2$
(iii) $y=2\sec x\tan x+A\sec x$ (iv) $y=\ln x\{2\ln(\ln x)+A\}$ (v) $xy=A\cos x+2\sin x$
(vi) $y=A(x-1)^3+(x-1)^5/2$. **4** (i) $2y=3x+1$ (ii) $y=6(x^2-x)\mathrm{e}^x+x$
(iii) $(1+x^2)^2y=3x+x^3$ (iv) $y=\mathrm{e}^{5(1-x)/2}-4\mathrm{e}^{-3x}+4\mathrm{e}^{-(1+5x)/2}$.
5 $y=-x^2-1$. **6** $y^2(x^3+2)=3x$. **7** $9t^3x=(3t^2+4)^{3/2}+A$.
8 $y\sin x=(4\cos 2x-2\cos 4x-1)/16$. **9** $y=2x^2-5x+3$. **10** $y=\mathrm{e}^{(1-x)/x}$.

Exercise 11.3 (p. 298)

1 (i) $y = Ae^{4x/3}$ (ii) $x = Ae^{-2t}$ (iii) $y = Ae^{x} + Be^{-2x}$ (iv) $x = (At + B)e^{t}$
(v) $y = e^{3x/2}\{A\cos(\sqrt{3}x/2) + B\sin(\sqrt{3}x/2)\}$ (vi) $x = A + Be^{-9t}$
(vii) $y = Ae^{4x} + Be^{-4x}$ (viii) $y = A\cos 5x + B\sin 5x$.
2 (i) $y = 4e^{8x}$ (ii) $y = e^{6x} + 2e^{-x}$ (iii) $y = 2xe^{-3x}$.
3 $y = e^{-x/2}\{A\cos(\sqrt{3}x/2) + B\sin(\sqrt{3}x/2)\}$. **4** $y = \cos kx + (1/k)\sin kx$.
5 $x = e^{kt}(X - Y)/2 + e^{-kt}(X + Y)/2$, $y = e^{kt}(Y - X)/2 + e^{-kt}(X + Y)/2$.
6 $x = (-e^{kt} + e^{-kt})/2$, $y = (e^{kt} + e^{-kt})/2$.

Exercise 11.4A (p. 301)

1 (i) $y = Ae^{-2x} + e^{x}$ (ii) $y = Ae^{4x} + (3\sin 3x - 4\cos 3x)/25$
(iii) $y = Ae^{5x} - 7/5$ (iv) $y = Ae^{-x} - x + 5$ (v) $y = Ae^{x/2} - x^2 - 4x - 8$
(vi) $y = (A + 3x)e^{-7x}$. **2** (i) $y = Ae^{-x/3} + x - 3$
(ii) $y = Ae^{-2x} + 7(\sin 2x - \cos 2x)/4$ (iii) $y = Ae^{-x} + (e^{2x})/3$ (iv) $y = Ae^{x^3/6}$.
3 $x = (e^{-2t} + 2t - 1)/4$. **4** $y = (3e^{x} - e^{-x})/2$.
5 $x = ae^{-2t}$, $y = 4a(e^{-t/2} - e^{-2t})/3$, $t = 2(\ln 4)/3$.

Exercise 11.4B (p. 304)

1 (i) $C = -1/2$ (ii) $C = 1/2$, $D = -7/4$ (iii) $C = 0$, $D = 1/2$
(iv) $C = -1/4$, $D = -3/8$, $E = 3/4$. **2** $y = (Ax + B)e^{-2x} + e^{x}$.
3 $p = 2$, $y = e^{3x}(\sin 2x - 2\cos 2x) + 2e^{2x}$. **4** $x = \{(3 + 4t)e^{-t} - 3e^{-5t}\}/16$.
5 $a = 10$, $b = 3$, $y = 10x + 3 - 10(e^{-x}\sin 3x)/3$.
6 $x = Ae^{t} + Be^{-3t}$, $x = e^{-t}(A\cos t + B\sin t)$, $x = (e^{-t}\cos t + t - 1)/2$.
7 $y = e^{-x}(A\cos x + B\sin x) + (\cos 2x + 2\sin 2x)/6$.
8 (i) $y = Ae^{3x} + Be^{-3x} - 4/9$ (ii) $a = 1/10$, $b = -1/5$.
9 $x = 6(e^{-2t} - 1)\cos 3t + 2\sin 3t$. **10** $y = (3 + 2x)e^{-x} - 2e^{-2x}$.
11 (i) $y = A\cos kx + B\sin kx + (5\sin 2x)/(k^2 - 4)$ (ii) $p = -5/4$, $y = (2 - 5x/4)\cos 2x$.

Exercise 11.5 (p. 308)

1 $y = x\sin^{-1}(x + A)$. **2** $y = A\cos 3x^2 + B\sin 3x^2 + 2/3$. **3** $y = x\cos x$.

Exercise 11.6 (p. 314)

1 (a) 1·1, 1·231, 1·403 (b) −2, −2·08, −2·246 (c) 0·25, 0·438, 0·568
(d) 1·558, 0·811, 0·114. **2** (a) 1·1, 1·262, 1·459 (b) −2, −2·16, −2·346
(c) 0·25, 0·375, 0·578 (d) 1·557, 1·508, −1·194. **3** 0·814. **4** (and **5**) 1·2, 1·403, 1·611.
6 (i) $2 + x^2 + x^4/12$, 3·083, 5·333 (ii) $x - x^3/6 + x^4/12$, 0·917, 2
(iii) $1 - 4x - 2x^3 + x^4/12$, −4·917, −21·67.
7 0·825. **8** 0·880, 0·842, 0·869. **9** 0·226, 0·270.

Miscellaneous Exercise 11 (p. 316)

1 $y = (\sin^3 x + 5)/3\sin^2 x$. **2** (a) $x = t$ (b) $y = \tan^{-1}x + 1 + Ae^{-\tan^{-1}x}$.
3 (a) $y = (1 - 2x)e^{-2x}$ (b) x^2e^{-2x}.
4 $l = -1$, $m = 0$, $y = 2\cos 2x - \cos 4x$, $x \in \{n\pi/2, n\pi \pm \pi/6 : n \in \mathbb{Z}\}$.
5 (a) $y = \{e^{3x}(2\cos 2x - 3\sin 2x)/2 + 1\}/13$ (b) $z = (e^{-2t} - e^{-t})/3$.
6 $y = x\{e^{2x} + 5e^{(3-x)}\}/3$. **7** 1·020, 1·065, 1·127. **8** $y = x^2 + Ae^{-x^2}$.
9 (i) $y = (x^4 - 40/x^2)/16$ (ii) $p = 1$, $q = 2$, $y = Ae^{-x} + Be^{-3x} + \sin 3x + 2\cos 3x$.
10 (i) $x = 4(\sin 3t)/3 + 2t$ (ii) $x = 2(1 - e^{-9t})/3$.
11 (a) $y = (x^2 + A)\sec x$ (b) $\sqrt{5}\,\mathrm{m\,s^{-1}}$. **12** (i) $y = 1/x$
(ii) $y = \sqrt[3]{\{2/(x(2 - (\ln x)^2)\}}$.

13 1·1, 1·19, 1·2072, 1·1%. **14** (a) $y = 2\sin x$ (b) $x = (1-t)e^{-t}$.
15 $a = 0$, $b = -1/2n$, $y = (\sin nx - xn\cos nx)/2n^2$. **16** (b) $y = x^3(\ln x - 1) + 3x^2$.
17 $x^2 = 2y^2(\ln y + A)$. **18** $y = e^{(x^2-x+2)}$. **19** $y = (e^{5kt/3} - 1)/(4e^{5kt/3} + 1)$.
20 0·22, 0·2442, 0·0516, 0·1614. **21** 1·11, 1·38.
22 (i) $y = 5e^{x^2/2}$ (ii) 0·015, 0·045. **23** $y = e^{-2x}(A\cos 2x + B\sin 2x)$.
24 $\dfrac{dy}{dx} = 2u\dfrac{du}{dx}, \dfrac{d^2y}{dx^2} = 2\left(\dfrac{du}{dx}\right)^2 + 2u\dfrac{d^2u}{dx^2}, \dfrac{d^2y}{dx^2} + \dfrac{dy}{dx} = 0, u = 2e^{-x/2}$.
25 (i) $a = 4$, $b = 3$, $y = e^{-x}(A\cos 2x + B\sin 2x) + 3x + 4$ (ii) $y = x\tan(\ln x)$.
26 $x = -\omega k e^{-t}(\omega^2+1) + x_0 + k/\omega$,
$-k(\omega\sin\omega t + \cos\omega t)/\omega(\omega^2+1)$, $x_0 + k\{1 \pm \sqrt{(\omega^2+1)}\}/\omega$.
27 1·190, 1·359. **28** $x = 1$, $x = -1$, (2, 0), (1/2, −3).
29 $y = e^{-x}(x^2-1)/x$.

Exercise 12.1 (p. 325)

2 (i) $\begin{pmatrix}0\\0\\5\end{pmatrix}$ (ii) $\begin{pmatrix}0\\2\\0\end{pmatrix}$ (iii) $\begin{pmatrix}-2\\5\\4\end{pmatrix}$ (iv) $\begin{pmatrix}3\\0\\-3\end{pmatrix}$ (v) $\begin{pmatrix}-1\\17\\5\end{pmatrix}$ (vi) $\begin{pmatrix}vz-wy\\wx-uz\\uy-wx\end{pmatrix}$.

3 (i) $2\mathbf{k}$ (ii) $\mathbf{i}+\mathbf{j}+\mathbf{k}$ (iii) $-\mathbf{i}+2\mathbf{j}-\mathbf{k}$ (iv) $-6\mathbf{i}+3\mathbf{j}-2\mathbf{k}$ (v) $\mathbf{i}-3\mathbf{j}-5\mathbf{k}$ (vi) $\mathbf{0}$.
6 $2\sqrt{2}$. **7** (i) −14 (ii) $-8\mathbf{i}-24\mathbf{j}-8\mathbf{k}$, $-(\mathbf{i}+3\mathbf{j}+\mathbf{k})/11$. **8** 1, −1, 1, $2\sqrt{2}/3$.
9 (ii) $2\mathbf{a}\times\mathbf{b}$. **11** $\sqrt{(u^2+9v^2)}/u$.

Exercise 12.2 (p. 330)

1 $\mathbf{x} = \lambda\mathbf{a}$. **2** (i) plane through A parallel to OB (ii) line through A parallel to OB.
3 (i) $\mathbf{w} = \mathbf{j}$ (ii) $\mathbf{w} = 2\mathbf{k}-\mathbf{j}$. **4** (i) $\mathbf{r} = \mathbf{k}+t\mathbf{i}$, line through (0, 0, 1) parallel to Ox
(ii) $\mathbf{r} = -\mathbf{i}+\mathbf{j}+t(\mathbf{j}+\mathbf{k})$, line through (−1, 1, 0) parallel to $(\mathbf{j}+\mathbf{k})$

(iii) $\mathbf{r} = t\begin{pmatrix}1\\2\\3\end{pmatrix} - \begin{pmatrix}0\\1\\2\end{pmatrix}$, line through (0, −1, −2) parallel to $(\mathbf{i}+2\mathbf{j}+3\mathbf{k})$.

5 (i) line AB (ii) sphere, diameter AB, $\mathbf{r} = \mathbf{a}$, $\mathbf{r} = \mathbf{b}$.

6 (i) $(\mathbf{r}-\mathbf{a}).\mathbf{b} = 0$ (ii) $(\mathbf{r}-\mathbf{c}).(\mathbf{a}\times\mathbf{b}+\mathbf{b}\times\mathbf{c}) = 0$.

7 $\begin{pmatrix}-8\\-10\\-4\end{pmatrix}$ (i) $(\mathbf{r}-\mathbf{a}).\begin{pmatrix}4\\5\\-2\end{pmatrix} = 0$ (ii) $3\sqrt{5}$, $x = 2z+5$. **8** (a) (i) $\begin{pmatrix}4\\-4\\2\end{pmatrix}$

(ii) $\begin{pmatrix}0\\-3\\3\end{pmatrix}$ (iii) $\begin{pmatrix}1\\-8\\4\end{pmatrix}$ (iv) $\begin{pmatrix}-2\\1\\1\end{pmatrix}$ (b) (i) 3 (ii) $3/\sqrt{2}$

(iii) 9/2 (iv) $\sqrt{(3/2)}$ (c) (i) $\mathbf{r}.\begin{pmatrix}2\\-2\\1\end{pmatrix} = 1$ (ii) $\mathbf{r}.\begin{pmatrix}0\\-1\\1\end{pmatrix} = 2$ (iii) $\mathbf{r}.\begin{pmatrix}1\\-8\\4\end{pmatrix} = 0$

(iv) $\mathbf{r}.\begin{pmatrix}-2\\1\\1\end{pmatrix} = 4$ (d) $2x-2y+z = 1$ (ii) $z-y = 2$ (iii) $x+4z = 8y$, $y+z = 2x+4$
(e) (i) 1/3 (ii) 1/2 (iii) 1/3 (iv) 1/6.
9 (i) $1/\sqrt{5}$ (ii) $1/\sqrt{2}$ (iii) $45/\sqrt{(293)}$. **10** (i) $2x-y+4z = 7$
(ii) (2/3, −1/3, 4/3) (iii) $7\sqrt{(21)}/6$. **12** (b) $bc/\sqrt{(b^2+c^2)}$, $ac/\sqrt{(a^2+c^2)}$.

Exercise 12.3 (p. 334)

2 $\mathbf{r}.\ddot{\mathbf{r}}+\dot{\mathbf{r}}^2$, $\mathbf{r}\times\ddot{\mathbf{r}}$, $2\mathbf{r}.\dot{\mathbf{r}}$. **5** $4\mathbf{i}-\mathbf{j}+10\mathbf{k}$. **7** $(\mathbf{r}-\mathbf{a})\times\mathbf{F}$, $\pm(1/\sqrt{3}, 1/\sqrt{3}, -1/\sqrt{3})$.

Miscellaneous Exercise 12 (p. 335)

1 -8, $-12\mathbf{i}-4\mathbf{j}-2\mathbf{k}$. **2** (b) (i) T, $2\mathbf{a}\times\mathbf{b}=\mathbf{0}$ (ii) F, $\mathbf{a}=\mathbf{i}=\mathbf{b}$, $\mathbf{c}=\mathbf{j}$ (iii) T, $(\mathbf{a}\times\mathbf{b})\times\mathbf{a}=-\mathbf{a}\times(\mathbf{a}\times\mathbf{b})=\mathbf{a}\times(\mathbf{b}\times\mathbf{a})$.
4 $\mathbf{i}-5\mathbf{j}+2\mathbf{k}$, parallel to the join of points.
5 (i) T (ii) F, $\mathbf{a}=\mathbf{i}=\mathbf{c}$, $\mathbf{b}=\mathbf{k}$ (iii) F, $\{\mathbf{b}-\mathbf{a}, \mathbf{c}-\mathbf{a}, \mathbf{b}+\mathbf{c}\}$ is L. I. (iv) F, line is parallel to plane. **7** $(1/\sqrt{2}, -1/\sqrt{2}, 0)$. **8** 1. **9** (ii) $3\mathbf{i}-2\mathbf{j}+\mathbf{k}$.
10 $-6\mathbf{i}+8\mathbf{j}$ (i) $4y=3x+6$ (ii) 5 (iii) $25/3$, $3x=4y+19$, $3x=4y-31$, $(-5, 4, 1)$.

12 (i) $-36\mathbf{i}+27\mathbf{k}$ (ii) $4x-3z=16$ (iii) $45/2$. **13** (a) (i) $\left\{\begin{pmatrix}1+2t\\ t\\ -3-3t\end{pmatrix}: t\in\mathbb{R}\right\}$

(ii) ϕ (b) $\mathbf{x}=\mathbf{c}/2-(\mathbf{c}.\mathbf{b})\mathbf{a}/(4+2\mathbf{a}.\mathbf{b})$. **14** $\mathbf{r}.(2\mathbf{i}-6\mathbf{j}+3\mathbf{k})=4$, $5/3\sqrt{(21)}$.
15 (a) line AC (b) sphere centre A and radius k (c) plane perpendicularly bisecting AB (d) line through O perpendicular to plane ABC (e) plane ABC, $|\mathbf{r}-\mathbf{a}|=|\mathbf{r}-\mathbf{b}|$ with (c) and (e). **16** (i) 23 (ii) $2\mathbf{i}+5\mathbf{j}-6\mathbf{k}$ (iii) 19° (iv) $\sqrt{(65)}/2$ (v) $2x+5y-6z=6$. **17** (a) $2\pi/3$ (b) $(\mathbf{i}+\mathbf{j}+\mathbf{k})/\sqrt{3}$ (c) $1/\sqrt{3}$.
18 (i) $(-x+6)/2=y-4=(z-5)/2$ (ii) $52\mathbf{j}+26\mathbf{k}$, $13\sqrt{5/3}$.
19 (b) $\mathbf{r}.(2\mathbf{i}-3\mathbf{j}+6\mathbf{k})=24$, 12. **20** $\mathbf{B}=\mathbf{j}$, $\mathbf{E}=3\mathbf{i}$.

Exercise 13.1 (p. 342)

1 (i) C, not D (ii) C, D (iii) not C, not D (iv) C, D (v) C, D (vi) C, D.
2 (i) $a=1, b=0$ (ii) $a=2, b=0$ (iii) $a=-2, b=1$ (iv) $a=-3, b=0$
3 $f(0+)=f(1-)$, $f'(0+)=f'(1-)$. **4** (i) $3/2$ (ii) $1/3$ (iii) 2 (iv) $-11/6$.
5 (i) $1/3$ (ii) 0 (iii) 12 (iv) 0 (v) 0.

Exercise 13.2 (p. 345)

1 (i), (iii), (v) diverge to $+\infty$, (ii) diverges oscillating infinitely, (iv), (vi), (viii) converge to 0, (vii) diverges oscillating finitely, (ix) converges to 1. **2** (a) (i), (iii), (v) are monotonic increasing, (b) (iv), (vi), (viii), (ix) are monotonic decreasing, (c) (ii), (vii) are not monotonic. **3** (a) (iv), (vi), (vii), (viii), (ix) are bounded, (b) none are bounded above and not below, (c) (i), (iii), (v) are bounded below and not above, (d) (ii) is unbounded above and below. **4** (i) 0 (ii) 0 (iii) 1 (iv) -3 (v) -2.
5 1·732051. **6** $\sqrt{5}$. **7** 1·709976.

Exercise 13.3 (p. 350)

1 Convergent (i), (ii), (iii), (iv), (v), (vi), (viii), (ix) Divergent (vii).
2 (i) $x<0$ (ii) $|x|<1$ (iii) $|x|>2$ (iv) $\{x: x>2\}\cup\{x: x<0\}$.
3 Convergent (i), (iii) Divergent (ii). **4** (i) $|x|<1$ (ii) $|x|<\ln 2$ (iii) $|x|>1$ (iv) all x (v) all x (vi) $|x|<1/3$.

Exercise 13.4 (p. 352)

1 (i) $(e^x-1)/2$, all x (ii) $-\ln(1-2x)$, $-1/2\leqslant x<1/2$ (iii) $e^{x-1}-1$, all x (iv) $-\ln(2-x)$, $0\leqslant x<2$. **2** (i) $(\sinh\sqrt{x})/\sqrt{x}-1$ for $x\neq 0$, 0 for $x=0$

(ii) $(\sin\sqrt{x})/\sqrt{x}-1$ for $x \neq 0$, 0 for $x=0$ (iii) $x(\cosh x+\sinh x-1)$
(iv) $(x-1)(\cosh x+\sinh x)+1$.

Miscellaneous Exercise 13 (p. 352)

1 (i) 2 (ii) 1. **2** 1/3. **3** Both continuous. **4** 2.
5 (i) D (ii) C (iii) C (iv) C. **6** -1, not differentiable. **7** $\sqrt{7}$. **8** 11.
9 $a=0, b=9$. **10** (i) $|x| \leqslant 1$ (ii) $|x| \leqslant 1$ (iii) $-1 \leqslant x < 1$.
11 Discontinuous at $x=3$, $f(3-)=1$, $f(3+)=-1$, 0. **12** 2.
13 (i) (a) C (b) D (c) C (ii) $1+2x-x^3/2$. **14** -1, 0.
15 (i) $\cos\sqrt{x}-1$, $0 \leqslant x$ (ii) $-\ln(-x)$, $-2 \leqslant x < 0$
(iii) $2/(1-2x)^2$, $|2x|<1$ (iv) $(xe^x-e^x+1)/x$ for $x \neq 0$, 0 for $x=0$, valid for all x.
16 $x+x^3/3+2x^5/15$, 1/2.

Exercise 14.1 (p. 358)

1 (i), (ii), (iv), (v) are closed under linear combinations but (iii) is not.
3 (i) $\mathbb{R}^2$ (ii) line $y=x$ (iii) $\{0\}$.

6 (i) $\mathbb{R}^2, \{0\}$ (ii) $\left\{\begin{pmatrix} t \\ -2t \end{pmatrix} : t \in \mathbb{R}\right\}, \left\{\begin{pmatrix} 3t \\ -2t \end{pmatrix} : t \in \mathbb{R}\right\}$ (iii) $\mathbb{R}^2, \left\{\begin{pmatrix} -3t \\ 2t \\ t \end{pmatrix} : t \in \mathbb{R}\right\}$

(iv) $\mathbb{R}^3, \{0\}$ (v) $\left\{\begin{pmatrix} s \\ s+t \\ t \end{pmatrix} : s, t \in \mathbb{R}\right\}, \left\{\begin{pmatrix} -4s \\ 5s \\ 2s \end{pmatrix} : s \in \mathbb{R}\right\}$

(vi) $\left\{\begin{pmatrix} s \\ 3s+t \\ 2s+t \end{pmatrix} : s, t \in \mathbb{R}\right\}, \left\{\begin{pmatrix} s \\ s+2t \\ -s-t \\ -t \end{pmatrix} : s, t \in \mathbb{R}\right\}$

Exercise 14.2 (p. 365)

6 $\{(3,0,1), (0,3,2)\}$, dimension 2. **7** $\{(1,1,-1)\}$, dimension 1. **8** (i) $\{x, x^2, x^3\}$
(ii) $\{x-2, x^2-2x, x^3-2x^2\}$ (iii) $\{x^2-1, x^3-x\}$.

10 $\mathbf{M} = \begin{pmatrix} 2 & 1 & 4 \\ 1 & 2 & -1 \\ 0 & 1 & -2 \end{pmatrix}$, (i) $\left\{\begin{pmatrix} 2 \\ 1 \\ 0 \end{pmatrix}, \begin{pmatrix} 1 \\ 2 \\ 1 \end{pmatrix}\right\}$ (ii) $\left\{\begin{pmatrix} -3 \\ 2 \\ 1 \end{pmatrix}\right\}$.

11 (i) $\{\mathbf{i}, \mathbf{i}+\mathbf{j}, \mathbf{k}\}$ (ii) $\{\mathbf{i}+\mathbf{j}, \mathbf{i}-\mathbf{k}, \mathbf{k}\}$ (iii) L. D. (iv) $\{\mathbf{i}+\mathbf{j}+\mathbf{k}, \mathbf{i}-\mathbf{j}+\mathbf{k}, \mathbf{k}\}$.
12 Dimension 4, $\{1, x, x^2, x^3\}$ (i) $\{(x-2), (x-2)^2, (x-2)^3\}$
(ii) $\{1, (x-1)^2, (x-1)^3\}$ (iii) $\{x^2-2x, x^3-3x^2+3x\}$, $T=V$, basis $\{1, x, x^2, x^3\}$.

13 No. **14** $2\mathbf{a}+2\mathbf{b}+\mathbf{c}$. **15** $\left\{\begin{pmatrix} -1 \\ 4 \\ -1 \\ 1 \end{pmatrix}\right\}, \left\{\begin{pmatrix} 4-z \\ 3+4z \\ 3-z \\ z \end{pmatrix} : z \in \mathbb{R}\right\}$.

16 $\left\{\begin{pmatrix} 1 \\ 1 \\ -1 \end{pmatrix}\right\}, \left\{\begin{pmatrix} 1 \\ 0 \\ 1 \end{pmatrix}, \begin{pmatrix} 0 \\ 1 \\ 1 \end{pmatrix}\right\}$ (i) line $\mathbf{r} = \begin{pmatrix} 3-t \\ 2t \\ 3+t \end{pmatrix}$ (ii) point (3, 0, 3)

(iii) line $\mathbf{r} = t\begin{pmatrix} 3 \\ 2 \\ 5 \end{pmatrix}$. **17** (a) 3 (b) $5x=6y+7z$, 2 (c) 7/9. **18** (ii) $\begin{pmatrix} 3 \\ 3 \\ 1 \end{pmatrix}$

(iii) $\left\{\begin{pmatrix} 3-2t \\ t-1 \\ t \end{pmatrix} : t \in \mathbb{R}\right\}$.

Exercise 14.3A (p. 371)

2 (i) (iv) (v) are orthogonal. **4** (ii) (iii) (vii). **5** (i) $\begin{pmatrix} -1/\sqrt{2} & -1/\sqrt{2} \\ 1/\sqrt{2} & -1/\sqrt{2} \end{pmatrix}$

(iv) $\begin{pmatrix} -2/3 & 1/3 & 2/3 \\ -1/3 & 2/3 & -2/3 \\ 2/3 & 2/3 & 1/3 \end{pmatrix}$ (vi) see (vii) (viii) $\begin{pmatrix} 3/7 & -6/7 & 2/7 \\ 2/7 & 3/7 & 6/7 \\ 6/7 & 2/7 & -3/7 \end{pmatrix}$.

Exercise 14.3B (p. 377)

1 (a) (i) $\left\{\begin{pmatrix}1\\2\end{pmatrix}, \begin{pmatrix}3\\4\end{pmatrix}\right\}$ (ii) $\left\{\begin{pmatrix}1\\3\end{pmatrix}, \begin{pmatrix}2\\4\end{pmatrix}\right\}$ (b) (i) $\left\{\begin{pmatrix}1\\2\end{pmatrix}\right\}$ (ii) $\left\{\begin{pmatrix}1\\2\end{pmatrix}\right\}$

(c) (i) $\left\{\begin{pmatrix}1\\-1\\0\end{pmatrix}, \begin{pmatrix}0\\1\\1\end{pmatrix}\right\}$ (ii) $\left\{\begin{pmatrix}2\\-1\\1\end{pmatrix}, \begin{pmatrix}3\\2\\0\end{pmatrix}\right\}$ (d) (i) $\left\{\begin{pmatrix}1\\2\\3\end{pmatrix}\right\}$ (ii) $\left\{\begin{pmatrix}1\\2\\-1\end{pmatrix}\right\}$

(e) (i) $\left\{\begin{pmatrix}1\\-1\\0\end{pmatrix}, \begin{pmatrix}2\\1\\3\end{pmatrix}\right\}$ (ii) $\left\{\begin{pmatrix}1\\2\end{pmatrix}, \begin{pmatrix}0\\3\end{pmatrix}\right\}$ (f) (i) $\left\{\begin{pmatrix}2\\-3\end{pmatrix}, \begin{pmatrix}1\\4\end{pmatrix}\right\}$

(ii) $\left\{\begin{pmatrix}2\\1\\3\\5\end{pmatrix}, \begin{pmatrix}-3\\4\\1\\-2\end{pmatrix}\right\}$ (g) (i) $\left\{\begin{pmatrix}1\\2\\5\\3\end{pmatrix}, \begin{pmatrix}2\\1\\4\\2\end{pmatrix}, \begin{pmatrix}4\\-1\\2\\1\end{pmatrix}\right\}$

(ii) $\left\{\begin{pmatrix}1\\2\\4\end{pmatrix}, \begin{pmatrix}2\\1\\-1\end{pmatrix}, \begin{pmatrix}3\\2\\1\end{pmatrix}\right\}$.

2 (i) (a) $\left\{\begin{pmatrix}0\\0\end{pmatrix}\right\}$ (b) $\left\{\begin{pmatrix}2\\-1\end{pmatrix}\right\}$ (c) $\left\{\begin{pmatrix}1\\1\\-1\end{pmatrix}\right\}$

(d) $\left\{\begin{pmatrix}2\\-1\\0\end{pmatrix}, \begin{pmatrix}3\\0\\-1\end{pmatrix}\right\}$ (e) $\left\{\begin{pmatrix}1\\1\\-1\end{pmatrix}\right\}$ (f) $\left\{\begin{pmatrix}0\\0\end{pmatrix}\right\}$ (g) $\left\{\begin{pmatrix}1\\2\\-1\\0\end{pmatrix}\right\}$

(ii) column spaces as in Qu 1. **3** (i) span $\begin{pmatrix}1\\2\\-1\end{pmatrix}$ (ii) $\mathbb{R}^2$ (iii) 2. **4** 2, 2 and 3, 3.

5 $2\mathbf{r}_1 + \mathbf{r}_2 + 5\mathbf{r}_3 = \mathbf{0}$, 2, 2, $\left\{\begin{pmatrix}1\\-2\\0\end{pmatrix}, \begin{pmatrix}0\\-5\\1\end{pmatrix}\right\}$. **6** (i) $(\mathbf{e}_1 + \mathbf{e}_2)/2$

(ii) $\{\mathbf{e}_3 + \mathbf{e}_2 - \mathbf{e}_1\}$ (iii) $\{s(\mathbf{e}_1 - \mathbf{e}_2 - \mathbf{e}_3) + \mathbf{e}_3/2 : s \in \mathbb{R}\}$.

7 $\left\{\begin{pmatrix}-5\\1\\3\\0\end{pmatrix}, \begin{pmatrix}-9\\0\\4\\2\end{pmatrix}\right\}$, $\left\{\begin{pmatrix}1\\0\\2\end{pmatrix}, \begin{pmatrix}1\\-3\\5\end{pmatrix}\right\}$, $\left\{\begin{pmatrix}-5s-9t-2\\s\\3s+4t+3\\2t\end{pmatrix} : s, t \in \mathbb{R}\right\}$ no, it does not contain $\mathbf{0}$. **8** 2 (i) $\left\{\begin{pmatrix}0\\2t\\t\end{pmatrix} : t \in \mathbb{R}\right\}$ (ii) ϕ

(iii) $\left\{\begin{pmatrix}1\\2+2t\\t\end{pmatrix} : t \in \mathbb{R}\right\}$. **9** (b) $c = 1$, $\left\{\begin{pmatrix}3-5t\\t\\4t-2\end{pmatrix} : t \in \mathbb{R}\right\}$.

10 $\left\{\begin{pmatrix}1\\2\\0\\-2\end{pmatrix}, \begin{pmatrix}2\\0\\1\\-3\end{pmatrix}\right\}$ (i) $\left\{\begin{pmatrix}s+2t\\2s\\t\\-2s-3t\end{pmatrix} : s, t \in \mathbb{R}\right\}$

(ii) $\left\{\begin{pmatrix}s+2t+1\\2s+1\\t\\-2s-3t\end{pmatrix} : s, t \in \mathbb{R}\right\}$ (iii) ϕ.

Exercise 14.4 (p. 386)

1 (a) $1, \begin{pmatrix}1\\-1\end{pmatrix}, 5, \begin{pmatrix}3\\1\end{pmatrix}$ (b) $2, \begin{pmatrix}1\\1\end{pmatrix}, -1, \begin{pmatrix}0\\1\end{pmatrix}$ (c) $-3, \begin{pmatrix}1\\-1\end{pmatrix}, -4, \begin{pmatrix}1\\-2\end{pmatrix}$

(d) $2, \begin{pmatrix}1\\0\\0\end{pmatrix}, -1, \begin{pmatrix}0\\1\\1\end{pmatrix}$ (e) $2, \begin{pmatrix}1\\-5\\-3\end{pmatrix}, 4, \begin{pmatrix}1\\1\\1\end{pmatrix}, -1, \begin{pmatrix}2\\2\\-3\end{pmatrix}$.

2 $1, \begin{pmatrix}3\\2\\1\end{pmatrix}, 2, \begin{pmatrix}1\\3\\1\end{pmatrix}, -1, \begin{pmatrix}1\\1\\0\end{pmatrix}$. **3** $-1/4, \begin{pmatrix}1\\-1\end{pmatrix}, 1, \begin{pmatrix}2\\3\end{pmatrix}, \begin{pmatrix}1 & 2\\-1 & 3\end{pmatrix}$.

4 $\begin{pmatrix}-1/2 & \sqrt{3}/2\\ \sqrt{3}/2 & 1/2\end{pmatrix}, 1, \begin{pmatrix}1\\ \sqrt{3}\end{pmatrix}, -1, \begin{pmatrix}\sqrt{3}\\-1\end{pmatrix}$, reflection in $y = x\sqrt{3}$.

5 $k = 0$, $1:0:1$, $k = -4$, $3:4:-9$. **7** $3x = 6y = 2z$.

8 $\begin{pmatrix}1 & 2 & 2\\2 & -2 & 1\\2 & 1 & -2\end{pmatrix}, \begin{pmatrix}8 & 1 & 4\\4 & -4 & -7\\1 & 8 & -4\end{pmatrix}$, $1, w, w^2$, where $w = e^{2\pi i/3}$, $\begin{pmatrix}1\\1\\1\end{pmatrix}, \begin{pmatrix}1\\w^2\\w\end{pmatrix}, \begin{pmatrix}1\\w\\w^2\end{pmatrix}$,

$\begin{pmatrix}5\\1\\1\end{pmatrix}, \begin{pmatrix}2+2w+w^2\\1-2w+2w^2\\-2+w+2w^2\end{pmatrix}, \begin{pmatrix}2+w+2w^2\\1+2w-2w^2\\-2+2w+w^2\end{pmatrix}$.

9 $a+d$, $ad-bc$. **10** Reflection in line $\mathbf{u}$, $(-2, 14)$.
11 $\mathbf{Av} = \lambda\mathbf{v}$, $\mathbf{A}$ is diagonal, any vector is an eigenvector, an enlargement, $(\alpha\lambda + \beta\kappa)\mathbf{v}_1 + \beta\mu\mathbf{v}_2$, $\beta(\alpha\lambda_1 + \beta\kappa - \alpha\mu) = 0$. **12** $z = 4x + 4y$.

Miscellaneous Exercise 14 (p. 388)

1 $\{(2, 1, 0), (2, 0, 1)\}$, 2. **2** subspace (i) $\{\mathbf{a}_2, \mathbf{a}_3, \mathbf{a}_4\}$ (ii) not closed under addition (ii) not closed under multiplication by scalars. **3** (v) $\{(1, 1), (i, 0)\}$
4 $\mathbf{X}_1 - \mathbf{X}_2 + 2\mathbf{X}_3$, 6.

5 $\begin{pmatrix}0 & 1 & -1\\-1 & 1 & 0\\1 & -1 & 1\end{pmatrix}$, $-4\mathbf{p} - 4\mathbf{q} + 5\mathbf{r}$, $-7\mathbf{p} - 6\mathbf{q} + 10\mathbf{r}$ $\mathbf{r} - \mathbf{p}$. **6** $(s, s+t, t+u, u)$.

7 (i) span $\begin{pmatrix}1\\2\\-1\end{pmatrix}$ (ii) $\begin{pmatrix}1\\0\\1\end{pmatrix} + \frac{1}{2}\begin{pmatrix}0\\-1\\1\end{pmatrix} + \frac{1}{2}\begin{pmatrix}0\\1\\2\end{pmatrix}$

(iii) $-\begin{pmatrix}1\\0\\1\end{pmatrix} - \frac{7}{2}\begin{pmatrix}0\\-1\\0\end{pmatrix} - \frac{3}{2}\begin{pmatrix}0\\1\\2\end{pmatrix}$. **8** $\begin{pmatrix}1 & 2\\-2 & 1\end{pmatrix}, \begin{pmatrix}-3a+4b\\-4a-3b\end{pmatrix}, \begin{pmatrix}(a-2b)/5\\(2a+b)/5\end{pmatrix}$.

9 (i) $\{(0, 1, 0, 2)\}$ (ii) $\{(-14, 7, 0, 3), (0, 0, 1, 0)\}$ (iii) $\left\{\begin{pmatrix}5\\-3\\1\end{pmatrix}\right\}$.

10 5, if $\alpha = 5$ $\left\{\begin{pmatrix} 9+9t \\ 1-11t \\ 5t \end{pmatrix} : t \in \mathbb{R}\right\}$, if $\alpha \neq 5$ $\left\{\begin{pmatrix} 1/5 \\ -6/5 \\ 1 \end{pmatrix}\right\}$.

12 $\begin{pmatrix} 3 & 2 & 3 \\ 1 & 3 & 2 \\ 2 & 1 & 1 \end{pmatrix}$, $\begin{pmatrix} 8 \\ 6 \\ 4 \end{pmatrix}$, $\begin{pmatrix} 15 \\ 11 \\ 10 \end{pmatrix}$, $\begin{pmatrix} 23 \\ 17 \\ 14 \end{pmatrix}$. **13** (i) Yes, $\{e^x, e^{2x}\}$
(ii) No, $x \in S_2$, $2x \notin S_2$ (iii) No, $(0, 1) \in S_3$, $2(0, 1) \notin S_3$ (iv) Yes, $\{(1, -1, 0), (1, 0, -1)\}$
(v) Yes, $\{1, i\}$. **14** $\{(1\ 0\ 0\ 0), (0\ 0\ 1\ 1)\}$, $\{(1\ -1\ 0\ 1), (0\ 1\ 1\ 0)\}$, $\{(1\ 0\ 1\ 1)\}$,
$\{(1\ 0\ 0\ 0), (0\ 0\ 1\ 1), (0\ 1\ 1\ 0)\}$, $\{(0\ 1\ 1\ 0), (0\ 0\ 0\ 1)\}$.

15 (i) T (ii) F, $\mathbf{d} = -\mathbf{a}$ (iii) T. **16** $\mathbf{p} = 2\begin{pmatrix} -1 \\ 2 \\ -1 \end{pmatrix} + 3\begin{pmatrix} 1 \\ -1 \\ 1 \end{pmatrix}$, $\begin{pmatrix} 0 \\ 2 \\ 3 \end{pmatrix}$, $\begin{pmatrix} 0 \\ 2 \\ 3 \end{pmatrix}$.

17 $3\mathbf{c}_1 = \mathbf{c}_2 + 2\mathbf{c}_3$, $5\mathbf{c}_2 + 4\mathbf{c}_3 = 3\mathbf{c}_4$, span$\left\{\begin{pmatrix} 3 \\ -1 \\ -2 \\ 0 \end{pmatrix}, \begin{pmatrix} 0 \\ 5 \\ 4 \\ -3 \end{pmatrix}\right\}$ $x = 3 + 3p$,
$y = 2 - p + 5q$, $z = -2p + 4q$, $t = -3q$, $p, q \in \mathbb{R}$.

18 $\left\{\begin{pmatrix} -11 \\ 1 \\ 3 \\ 0 \end{pmatrix}, \begin{pmatrix} -3 \\ 0 \\ 1 \\ 1 \end{pmatrix}\right\}$, $\left\{\begin{pmatrix} 1 \\ 0 \\ 4 \end{pmatrix}, \begin{pmatrix} -4 \\ 9 \\ 5 \end{pmatrix}\right\}$, $\left\{\begin{pmatrix} 4-11s-3t \\ 1+s \\ 3s+t \\ t \end{pmatrix} : s, t \in \mathbb{R}\right\}$.

19 (i) 3, -1 (ii) squash on to line $2x = y = z$, null map.

20 $\begin{pmatrix} 3/8 & -9/8 & 1 \\ -1/8 & 3/8 & 0 \\ -1/8 & 11/8 & -1 \end{pmatrix}$, -1, $\begin{pmatrix} 11 \\ 1 \\ -14 \end{pmatrix}$. **21** 2, (ii) $x + y = 0 = z$.

22 $\alpha = 7x - 3y$, $\beta = -2x + y$, $\begin{pmatrix} -1 \\ 1 \end{pmatrix}$, $\begin{pmatrix} 1 \\ 0 \end{pmatrix}$, $\begin{pmatrix} 2 \\ 2 \end{pmatrix}$, $\begin{pmatrix} -10 & 7 & 8 \\ 3 & -2 & -2 \end{pmatrix}$.

Exercise 15 (p. 397)

1 (i) $y^2 = -4x$, $(-1, 0)$, $x = 1$ (ii) $x^2 = y$, $(0, 1/4)$, $4y = -1$
(iii) $x^2 = -9y$, $(0, -9/4)$, $4y = 9$ (iv) $(y-2)^2 = 4(x-1)$, $(2, 2)$, $x = 0$
(v) $(y+1)^2 = 1 - x$, $(3/4, -1)$, $4x = 5$ (vi) $(2x+1)^2 = 5 - 4y$, $(-1/2, 1)$, $2y = 3$.
2 (i) $y^2 = -4ax$ (ii) $y^2 + 2x + 1 = 0$ (iii) $x^2 + 4y = 8$ (iv) $6y = x^2 - 2x + 4$
(v) $x^2 - 2xy + y^2 = 8(x + y - 2)$. **3** (i) $\sqrt{5}/3$, $(\sqrt{5}, 0)$, $(-\sqrt{5}, 0)$, $x = 9/\sqrt{5}$,
$x = -9/\sqrt{5}$ (ii) $1/\sqrt{2}$, $(\sqrt{2}, 0)$, $(-\sqrt{2}, 0)$, $x = 2\sqrt{2}$, $x = -2\sqrt{2}$
(iii) $3/5$, $(0, 3)$, $(0, -3)$, $3y = 25$, $3y = -25$. **4** (i) $6x^2 + 16y^2 = 24$
(ii) as (i) (iii) $9x^2 + 8y^2 = 288$ (iv) $x^2 + y^2(1 - e^2) = k^2(1 - e^2)$.
5 $x^2/25 + y^2/16 = 1$. **6** $(3/2, 0)$, $(-3/2, 0)$.
7 (i) $3x^2 = y^2 + 3$ (ii) $y^2 + 9 = 15x^2 + 22x$ (iii) $8x^2 = y^2 + 8$
(iv) $225x^2 - 400y^2 = 2304$. **8** $|h| > \sqrt{(15)}$. **9** $y = 5x$, $\pi/2$.
10 $(0, \sqrt{(0{\cdot}39)})$, $(0, -\sqrt{(0{\cdot}39)})$. **11** $(x-4)^2 = 3 - 2y$.
12 $ty = x + at^2$, $x = a + ty$.
13 $y^2 = 4(x-1)$, $my = 1 + m^2(x-1)$, $my + m^2 + x = 1$, $(0, (1 - m^2)/m)$, $x = 0$.
14 $5x^2 + 9y^2 = 180$. **17** $3x^2 + 4y^2 = 180$. **18** $bx = ay$, $bx = -ay$.
19 $y^2 = m^2(x^2 - c^2)$. **20** $y = 2x$, hyperbola.

Index